Soil Mechanics and Foundation Engineering

Soil Mechanics and Foundation Engineering

Fundamentals and Applications

Nagaratnam Sivakugan, Ph.D.

McGraw Hill

New York Chicago San Francisco
Athens London Madrid
Mexico City Milan New Delhi
Singapore Sydney Toronto

Library of Congress Cataloging-in-Publication Data

Names: Sivakugan, N. (Nagaratnam), date, author.
Title: Soil mechanics and foundation engineering : fundamentals and applications / Nagaratnam Sivakugan, Ph.D.
Description: New York : McGraw Hill, [2021] | Includes bibliographical references and index.
Identifiers: LCCN 2021011140 | ISBN 9781260468489 (acid-free paper) | ISBN 9781260468496 (ebook)
Subjects: LCSH: Foundations. | Soil mechanics.
Classification: LCC TA775 .S553 2021 | DDC 624.1/5—dc23
LC record available at https://lccn.loc.gov/2021011140

McGraw Hill books are available at special quantity discounts to use as premiums and sales promotions or for use in corporate training programs. To contact a representative, please visit the Contact Us page at www.mhprofessional.com.

Soil Mechanics and Foundation Engineering: Fundamentals and Applications

1 2 3 4 5 6 7 8 9 CCD 26 25 24 23 22 21

ISBN 978-1-260-46848-9
MHID 1-260-46848-8

Sponsoring Editor
Ania Levinson

Editing Supervisor
Stephen M. Smith

Production Supervisor
Pamela A. Pelton

Acquisitions Coordinator
Elizabeth M. Houde

Project Manager
Radhika Jolly, MPS Limited

Copy Editor
Nikhil Roshan, MPS Limited

Proofreader
Raama Thenarasu, MPS Limited

Indexer
Michael Ferreira

Art Director, Cover
Jeff Weeks

Composition
MPS Limited

To Sivakumaran,
Shirley, and Rohini

About the Author

Dr. Nagaratnam Sivakugan received his bachelor's degree in civil engineering from the University of Peradeniya, Sri Lanka, with First Class Honours. He earned his MSCE and Ph.D. from Purdue University, West Lafayette, Indiana. His writings include 10 books, 160 refereed journal papers, 100 refereed international conference papers, and more than 100 consulting reports. His h-index is 33, with more than 5000 citations. As a registered professional engineer in Queensland and a chartered engineer, Dr. Sivakugan works closely with the geotechnical and mining industries and carries out substantial consulting work in Australia and overseas, including for the World Bank. He has supervised 16 Ph.D.s to completion at James Cook University, Australia, where he was the head of civil engineering from 2003 to 2014. He was the associate editor for the *International Journal of Geotechnical Engineering* for many years, and currently serves on the editorial boards of the *Canadian Geotechnical Journal, International Journal of Geotechnical Engineering, International Journal of Geosynthetics and Ground Engineering*, and *Indian Geotechnical Journal*. Dr. Sivakugan is a Fellow of the American Society of Civil Engineers and of Engineers Australia.

Contents

Part 2 Applications

Preface

Soil Mechanics and Foundation Engineering: Fundamentals and Applications is divided into two parts, Fundamentals (Chaps. 2–10) and Applications (Chaps. 11–16). Having authored a few bestselling textbooks, and some research monographs, I present here soil mechanics and foundation engineering in their simplest form without sacrificing the nuances, intricacies, and richness of the subject. I have ensured the writing is succinct while flowing seamlessly within and between chapters. You will not want to merely scan the text. You miss a sentence and you miss some useful information!

The content is covered in a hands-on way, with a strong practical bias, relating to real-world problems and applications. Each chapter contains several worked examples, spread evenly within the chapter. In addition, there are many review exercises at the end of each chapter, with the answers provided. A complete, 180+ page solution manual is available online (https://www.mhprofessional.com/SoilMechanics) to qualified adopters of the text. The problems are at different difficulty levels reflecting the nature of the geotechnical engineering profession.

The book includes everything you want to know about soil mechanics and foundation engineering as a graduate geotechnical engineer. It can be used as the text for one or two courses on this subject, taught as Soil Mechanics, Geotechnical Engineering, Geomechanics, Geo-Engineering, or Foundation Engineering and covering the chapters relevant to the course content. Appropriate references are included in every chapter, nurturing the habit of lifelong learning expected in the geotechnical engineering profession. Some references are as recent as the year 2020.

The chapters are ordered in a logical sequence, *fundamentals* from Chap. 2 to Chap. 10, followed by *applications* from Chap. 11 to Chap. 16. For example, when site investigation is discussed in Chap. 11, it is necessary for the reader to have a good understanding of soil classification, compaction, effective stresses, permeability, consolidation, and shear strength, which are all covered in the preceding chapters. Similarly, for shallow foundations in Chap. 12, it is necessary to have adequate knowledge of the previous chapters, including site investigation. While Chaps. 2–15 cover the traditional soil mechanics and foundation engineering topics, Chap. 16 introduces the reader to load and resistance factor design (LRFD) and reliability-based designs. A brief introduction to unsaturated soils is given in App. A. The traditional summary at the end of a chapter is replaced by "Main Points," which lists the main points the reader should understand from the chapter.

Geotechnical engineering theory and literature comprise much empiricism, indeed more than most other professions. There are hundreds of empirical equations and graphs that were developed based on large databases, some rationale, and intuitive reasoning. Most empirical equations and design charts that are widely used in the profession are included in this book. They are very valuable for preliminary designs and estimates, especially in the absence of soil samples or test results.

A picture is worth a thousand words. This is very true in geotechnical engineering, especially when it comes to applications. The book includes more than 70 photographs showing some geotechnical applications, laboratory tests, and *in situ* test equipment. A suite of PowerPoint slideshows can be downloaded from https://www.mhprofessional.com/SoilMechanics. There are animated slideshows, for each of the 16 chapters, and some pictorials that are simply a collection of photographs. Traditionally, PowerPoint slides are used by instructors as a teaching tool. The slideshows here are meant for the reader, as a quick summary of the chapter and an effective revision tool.

There are about 130 solved examples and 190 review exercises with answers. I thank my former Ph.D. student Dr. Thushara Madanayaka (civil engineering program lead and research and innovation head at SLTC, Sri Lanka) for so painstakingly going through them and the entire manuscript with his eagle eyes. I am grateful for his thorough review. I take this opportunity to thank Dr. Prapa Haran (Branch Chief, Federal Energy Regulatory Commission, Division of Dam Safety and Inspections, New York) for reviewing some chapters. Prapa was my partner in crime at Purdue University during our graduate studies. In most of the chapters, there are questions that require browsing the Internet for a quick look-up of some special topics and case studies.

I have made this book a good reference and a one-stop shop so that you will not want to sell it to a second-hand bookshop at the end of the semester. This is a geotechnical keepsake. Rich content with clear explanations, numerous design aids with empirical correlations, and many worked examples make it a must-have reference in your professional library.

The key to success in studying geotechnical engineering is to imagine the structure and the process, whether it is seepage beneath a concrete dam or bearing capacity failure of a foundation. Simply plugging numbers into equations will not take you too far. You have to understand the process and mechanism, and decide on the right equations and parameters to use. Always try to go from first principles. A good insight into the problem is a key attribute for a good geotechnical engineer.

I would like to thank Ms. Ania Levinson, Ms. Summers Scholl, and Ms. Elizabeth Houde of McGraw Hill for promptly answering my questions and their assistance during the many stages of publishing this book. I look forward to receiving readers' feedback that can help in the next edition.

Nagaratnam Sivakugan, Ph.D.
Siva.Sivakugan@gmail.com

Symbols

α	Adhesion factor for piles in clays
β	Reliability index
γ	Unit weight
γ_d	Dry unit weight
$\gamma_{d,\max}$	Maximum dry unit weight of a granular soil
$\gamma_{d,\min}$	Minimum dry unit weight of a granular soil
γ_m	Bulk unit weight
γ_w	Unit weight of water
γ_s	Unit weight of soil grains
γ_{sat}	Saturated unit weight
γ'	Submerged unit weight
δ	Interface friction angle between soil and another material
ε	Normal strain
η	Efficiency
η_H	Hammer efficiency in standard penetration test
κ	Slope of recompression (swelling) line when stress is in natural log scale
λ	Slope of virgin consolidation line when stress is in natural log scale
μ_w	Dynamic viscosity of water
ν	Poisson's ratio
ν_d	Drained Poisson's ratio
ρ	Density (also electrical resistivity)
ρ_d	Dry density
$\rho_{d,\max}$	Maximum dry density of a granular soil

$\rho_{d,\min}$	Minimum dry density of a granular soil
ρ_m	Bulk density
ρ_s	Density of soil grains
ρ_{sat}	Saturated density
ρ'	Submerged density
ρ_w	Density of water
σ	Normal stress
σ'	Effective stress
σ'_p	Preconsolidation pressure
σ'_{vo}	Effective vertical overburden stress
τ	Shear stress
τ_f	Shear strength or shear stress at failure
τ_{mob}	Mobilized shear strength
ϕ	Friction angle
Φ	Resistance factor
Ψ	Load factor
Δe	Change in void ratio
ΔH	Change in height
Δu	Change in pore water pressure
$\Delta\sigma'$	Change in effective stress
ΔV	Change in volume
a	Air content
A	Activity
A_f	Skempton's A-parameter at failure
A_r	Area ratio
A_s	Specific surface of a soil grain
B	Foundation breadth (also Skempton's B-parameter)
c	Cohesion
c'	Cohesion in terms of effective stresses
c_a	Adhesion
c_u	Undrained cohesion or undrained shear strength

c_v	Coefficient of consolidation
C	Capacity
CBR	California bearing ratio
C_α	Coefficient of secondary compression
$C_{\alpha\varepsilon}$	Modified secondary compression index
C_c	Coefficient of curvature of a grain-size distribution curve
C_c	Compression index
C_N	Overburden correction factor in standard penetration test
CR	Compression ratio
$C_{c\varepsilon}$	Modified compression index or compression ratio
C_r	Recompression index (also inside clearance ratio)
$C_{r\varepsilon}$	Modified recompression index or recompression ratio
C_u	Coefficient of uniformity of a grain-size distribution curve
d	Pile diameter
D	Grain diameter (also demand, constrained modulus)
D_{10}	Grain size corresponding to 10% passing; the effective grain size
D_{15}	Grain size corresponding to 15% passing
D_{30}	Grain size corresponding to 30% passing
D_{50}	Grain size corresponding to 50% passing; median grain size
D_{60}	Grain size corresponding to 60% passing
D_{85}	Grain size corresponding to 85% passing
D_i	Internal diameter
D_e	External diameter
D_f	Foundation depth
$DCPI$	Dynamic cone penetration index
D_r	Relative density
e	Void ratio (also eccentricity)
e_0	Initial void ratio
e_{max}	Maximum void ratio of a granular soil
e_{min}	Minimum void ratio of a granular soil
E	Compaction effort

E	Young's modulus
E_u	Undrained Young's modulus
f_s	Sleeve resistance in cone penetration test (also skin friction along pile shaft)
F	Factor of safety
g	Gravitational acceleration
G	Shear modulus
G_s	Specific gravity of soil grains
H	Thickness
H_0	Initial thickness
H_{dr}	Average maximum length of drainage path during consolidation
I	Moment of inertia (also influence factor)
k	Hydraulic conductivity or permeability
K	Bulk modulus
K_0	Coefficient of earth pressure at rest
K_a	Active earth pressure coefficient
K_p	Passive earth pressure coefficient
K_E	Modulus number
l	Pile length
L	Foundation length
LL	Liquid limit
LI	Liquidity index
m_v	Coefficient of volume compressibility
M	Slope of the critical state line in p'-q plane
n	Porosity (also coefficient of restitution)
N	Blow count or N-value in standard penetration test
N_{60}	Blow count adjusted for hammer efficiency
N_1	Blow count corrected for overburden
N_{kt}	Cone factor
N_c, N_q, N_γ	Bearing capacity factors
OCR	Overconsolidation ratio
p_a	Atmospheric pressure

p_f	Probability of failure
PL	Plastic limit
PI	Plasticity index
P_a	Active thrust
P_p	Passive thrust
q	Foundation pressure
Q	Foundation load
q_c	Cone resistance
q_t	Cone resistance corrected for pore water effects
q_u	Unconfined compressive strength
R_f	Friction ratio in cone penetration test
RR	Recompression ratio
s	Settlement (also set of a pile when driving)
s_c	Final consolidation settlement
s_i	Immediate settlement
s_s	Secondary compression settlement
S	Degree of saturation
SL	Shrinkage limit
SM	Safety margin
t	Time
t_p	Time to complete primary consolidation
T	Time factor (also torque in vane shear test)
u	Pore water pressure
U	Degree of consolidation
U_{avg}	Average degree of consolidation
v	Velocity
V_s	Shear wave velocity
V_p	Compression wave velocity
w	Moisture content or water content
w_n	Natural moisture content
W	Weight

W_P	Weight of pile
W_R	Weight of the ram in a pile hammer
z	Depth
Z	Depth factor (also section modulus of a sheet pile section)

Soil Mechanics and Foundation Engineering

CHAPTER 1

Geotechnical Engineering

1.1 Introduction

Geotechnical engineering is a relatively young branch of the civil engineering discipline, which would not have been in the curriculum a century ago. It is also taught as *geomechanics* or *geo-engineering*, with slightly different interpretations on what it includes. The two key components here are *soil mechanics* and *foundation engineering*. It has some overlap with rock mechanics, geology, and pavement engineering. A good geotechnical engineer must have a thorough understanding of soil mechanics and foundation engineering, and at least a fair knowledge of rock mechanics, geology, and pavement engineering.

"Geo" is earth. Geology, geodesy, geomorphology, and geography are distinct disciplines that deal with specific aspects of earth. All humankind's activities occur on, in, or with the earth, and many of our most needed resources such as water, oil and gas, and mineral ores come from the earth (Mitchell 2020). In addition, earth itself provides the most abundant and low-cost construction materials in the form of soils, rocks, and aggregates. Geotechnical engineering deals with the engineering aspects of *geomaterials*, which include soils, rocks, and aggregates. Most of our discussions in this book will be limited to soils and aggregates. Rocks are discussed in rock mechanics, a specialized area covered separately, often as a civil engineering elective or a postgraduate course.

The aqueducts, roads, and bridges built by the Romans 2000 years ago, some ancient earth dams built in India, the Great Wall of China, and the pyramids are testimonials to the early contributions by our ancestors to geotechnical engineering. The very early works of geotechnical engineering were mostly empirical, based on past experiences. This includes the ancient civilization at the banks of Nile and Indus, pyramids (Fig. 1.1a), the Leaning Tower of Pisa (Fig. 1.1b), and more. Some pioneering work by Charles Augustine Coulomb (1736–1806), Henri Philibert Gaspard Darcy (1803–1858), William John Macquorn Rankine (1820–1872), Joseph Valentin Boussinesq (1842–1929), Albert Mauritz Atterberg (1846–1916), Wolmar Fellenius (1876–1957), Otto Mohr (1835–1918), and Karl Terzaghi (1883–1963) still remains the backbone of the soil mechanics theory and fundamentals, which have evolved nicely since then into the modern soil mechanics.

Mechanics is the physical science that deals with forces and equilibrium, covered in courses like Engineering Mechanics or Strength of Materials. In soil mechanics and rock mechanics, we apply these principles to soils and rocks. Karl Terzaghi, acknowledged as the father of modern soil mechanics, wrote the first book on soil mechanics *Erdbaumechanik* in 1925. In the twentieth century, along with Karl Terzaghi, there have

a b

FIGURE 1.1 (a) Pyramids and (b) Leaning Tower of Pisa. (Courtesy of Dr. N. Sivakugan, James Cook University, Australia.)

been several others who made significant contributions to geotechnical engineering. This includes Arthur Casagrande (1902–1981), Ralph Peck (1912–2008), Alec Skempton (1914–2001), Gregory Tschebotarioff (1899–1985), and Bolton Seed (1922–1989). The notable early contributions in soil mechanics were summarized by Skempton (1979).

1.2 Soils and Other Engineering Materials

Civil engineers work with many different materials including concrete, steel, and timber. While we have the luxury of specifying a special grade of concrete, steel, or timber, to suit a certain application, when it comes to soils, we often have to live with what we have. It may not be viable to replace the soil with a better quality soil and we have to design the structure to suit the soil conditions. Soils are different for most engineering materials.

Soils are an assemblage of particles, and hence known as *particulate media*. All other engineering materials are *continuous media*, known as *continuum*. Most of the theories developed in mechanics (e.g., Mohr's circles, bending of beams) are for a continuum and hence it is an approximation when we use them for soils.

The voids between the soil grains are filled with water and/or air, making soil a three-phase medium. Most engineering materials are single-phase media. The soils can be different few meters apart, and are known to be *heterogeneous*. Other engineering materials such as concrete, steel, or timber are relatively *homogeneous*.

The engineering properties of soils vary with the direction. This is known as *anisotropy*, meaning unequal properties in different directions. For example, water can flow through soils faster horizontally than vertically. A simplified version of anisotropy is *cross-anisotropy*, where the properties do not vary radially in the horizontal plane, but are different in the horizontal and vertical directions. Steel and concrete can be treated as isotropic. Timber also shows directional variation in properties and is known as *orthotropic* for its radial, tangential, and axial properties being different.

Most engineering materials have tensile strengths. In soils, the tensile strength is little to none. Granular soils have no tensile strength. Clays and cemented sands have very little tensile strength. The stress-strain plots for soils can vary significantly, depending on the soil type. There is little variability in the stress-strain plots for steel or concrete.

Concrete, steel, or timber can fail in tension, compression, or shear. When designing with these materials, we check for all three possibilities. Interestingly, soils can only fail in shear. Even when they are subjected to large compressive loads, failure occurs along a shear plane or surface, where the grains slide over each other. The grains rarely get crushed at these stress levels.

1.3 Geotechnical Applications

All buildings have to be placed on the ground, and it is necessary to ensure that the ground does not fail and the settlements are not excessive. When the ground conditions are good and/or the loads are not high, the foundations are in the form of *pad footings* (Fig. 1.2a) which spread the column load through a larger contact area with the soil, thus reducing the contact pressure applied to the underlying soil. When the ground conditions are poor and/or the column loads are high, *pile foundations* are used. Figure 1.2b shows a row of piles being driven into the ground to support an overhead bridge. Other geotechnical applications include the following:

- Retailing walls
- Dams
- Excavations
- Sheet piles
- Diaphragm walls
- Tunneling
- Geosynthetics
- Mining
- Hazardous waste disposal
- Ground improvement

Figures 1.2c and d show two different construction sites, which have been subjected to deep excavation where the walls are supported to limit the lateral deformations and possible failure.

When a building or an embankment is built on the ground, the entire weight is transferred to the underlying soils. It is necessary to ensure that the structure remains safe during its design life and the settlements are within tolerable limits. Therefore, prior to the design stages, it is necessary to carry out a proper *site investigation* to understand the soil profile at the site, and determine the soil properties of the different layers. This is usually carried out using a drill rig to drill boreholes and take samples for laboratory testing. In addition, *in situ* tests are carried out which involve pushing or driving

FIGURE 1.2 Geotechnical applications: (a) Pad footing, (b) pile driving, (c) construction site for a proposed building, and (d) construction site at a railway station. (Courtesy of Dr. N. Sivakugan, James Cook University, Australia.)

a probe into the ground where the resistance is correlated to the soil property. Figure 1.3 shows a drill rig on top of an abandoned bridge, drilling a borehole in the middle of a creek. The detailed *site investigation report* summarizes the soil properties determined from the in situ and laboratory tests. This report plays a vital role in the subsequent design and construction.

1.4 Standards, Measurements, and Significant Digits

The in situ and laboratory tests are carried out according to some standards that are developed by specialists in the area, which are updated regularly. Some of these specialized organizations include ASTM International (known as American Society for Testing Materials prior to 2001), American Association of State Highway and Transportation Officials (AASHTO), International Standardisation Office (ISO), British Standards (BS), German Institute for Standardisation (DIN), and Australian Standards (AS). Out of these, ASTM is the most popular and its standards are being used worldwide.

FIGURE 1.3 Site investigation.

ASTM (http://www.astm.org) standards cover a wide range of materials, including steel, concrete, timber, soils, rocks, plastics, textile, ceramics, paint, water, and more. The standards are published in 82 volumes. Luckily, the ones on soils and rocks are available in just two volumes—04.08 and 04.09. When carrying out in situ or laboratory tests, generally we are expected to follow the standards religiously, and report the results in the prescribed way.

In the in situ and laboratory tests as well as in the routine computations, we come across a lot of measurements and numbers. These should be reported to the correct *significant digits* (ASTM D 6026-13). Let us not let Microsoft Excel or the calculator decide this for us. The precision of a numerical value presented in a report must reflect the precision of the measurements. Reporting the mass of a specimen as 26.4 g is different to 26.400 g. In the latter, it is implied that the measurement is precise to three decimals. No one wants to report the strength of concrete as 32.2136 MPa, to six significant figures, based on a standard uniaxial compression test. The number of significant digits should reflect the specimen size, resolution of the measuring device, and the sensitivity of the instrument. The numerical values must be reported to appropriate significant figures. For example, when reporting the mass of a test specimen as 318.6 g, there are four significant digits. Reporting too many digits can give a false sense of accuracy and show your lack of appreciation of the quality of the measurements.

1.5 Physical and Numerical Modeling

It is not always possible to obtain closed form solutions to all problems. It may be because of the complex boundary conditions, or that the theoretical formulations are inadequate to solve the problem. Under such circumstances, physical and numerical modelings are viable alternatives. They are also powerful validation tools to compare one against the other, or to verify new theories.

In physical modeling, small-scale model of the prototype structure is built and tests are carried out simulating the prototype situation. The results have to be scaled back from the model to the prototype. Physical modeling is common for studying the behaviors of pad footings, pile foundations, reinforced soil structures, etc.

Numerical modeling is an effective way to study a range of geotechnical problems and is widely used in universities and in geotechnical design offices. Finite element (e.g., Plaxis, Geostudio, RocScience, Abaqus), finite difference (e.g., FLAC), and distinct element (e.g., UDEC) software packages are used to discretize the problem domain into hundreds or thousands of small elements and nodes. Applying appropriate *boundary conditions* and *constitutive models*, solutions are derived. The boundary conditions define how the boundaries of the problem domain behave (e.g., fixed or roller). The constitutive model (e.g., Hooke's law) tells the program how the material behaves.

Especially in situations where the boundary conditions are complex or poorly defined, numerical modeling can be the way to go. In addition, numerical models can be very effective tools for carrying out sensitivity analysis where different scenarios can be explored, by varying the design parameters. They can be also effective in validating new theories and approximations used in the analytical models.

1.6 Geotechnical Engineering Literature

The textbooks have a purpose, and cater to the needs of university students. They cover the course content that includes a range of topics. There are research monographs that are essentially books on specialized topics such as ground improvement, geosynthetics, etc.

Some of the early textbooks in soil mechanics and foundation engineering were written by Terzaghi (1943), Terzaghi and Peck (1948), Taylor (1948), Peck et al. (1974), and Lambe and Whitman (1979). These are well-reputed classics, rich in content and will always have their place. There has been lot of developments since then, but the fundamentals remain the same. The recent textbooks with more illustrations and worked examples are a lot more reader-friendly and may appeal to the present generation.

Geotechnical journals and proceedings of conferences are a good source of reference on some topics that are not covered in the textbooks or even research monographs. The peer-reviewed articles in a journal or a conference proceeding discuss the recent developments and cutting-edge research in many areas of geotechnical engineering. Journals and conference proceedings can be accessed through the libraries.

There have been major failures in geotechnical engineering—it is not always smooth sailing! We learn from our mistakes by investigating these failures, so that the mistakes are not repeated. They are also valuable in calibrating our understanding of the problem. Practicing engineers often discuss their geotechnical projects, known as case histories, giving the readers insight into some real-world engineering problems. Investigations of failures and case histories are widely discussed in geotechnical literature.

1.7 Workplace Health and Safety and Risk Assessment

Workplace health and safety plays a major role in all construction sites and material testing laboratories. The importance and the emphasis on safety are increasing every year, with little or no tolerance for any workplace-related accidents or injuries. Laboratories

Likelihood		Highly unlikely	Unlikely	Likely	Almost certain
Consequences (Hazard severity)	Fatality	MODERATE	HIGH	CRITICAL	CRITICAL
	Major injuries	LOW	MODERATE	HIGH	CRITICAL
	Minor injuries	LOW	LOW	MODERATE	HIGH
	Negligible injuries	LOW	LOW	LOW	MODERATE

CRITICAL	Can cause permanent disability, fatality, or serious harm. Stop the activity and improve the control measures to reduce the risk and critically assess.
HIGH	Moderate degree of injuries or harm likely. Reconsider the need to proceed. Implement appropriate control measures, and high level of competence required.
MODERATE	Minor injuries likely but major harm unlikely. Implement control measures and proceed with caution and continued risk assessments.
LOW	Minor harm/injury is still possible. Take necessary precautions and controls in carrying out the activity.

Table 1.1 A Risk Assessment Matrix

with machinery, equipment, and hazardous material are some of the high-risk areas that are prone to accidents. MSDS (Material Safety Data Sheets) must be used to assess whether special protective clothing and/or eye protection are required. Hazardous chemicals should be disposed of in an appropriate manner. Every possible step must be taken to minimize the risk. There are designated workplace health and safety officers to give induction for the beginners and ensure that those working in the laboratories follow the best practices as far as the workplace health and safety is concerned. A risk assessment has to be carried out for any activity that may involve risk. A simplified risk assessment matrix is shown in Table 1.1. Depending on the likelihood of the hazard and its severity, a risk rating is assigned.

1.8 Factor of Safety

When the geotechnical engineers design and construct foundations, embankments, or retaining walls, it is necessary to ensure there is no failure during the *design life*. The stresses under the working loads have to be well below the failure stresses. Here comes the concept of *factor of safety*. It is the ratio of the failure stress to the working stress.

When we know with *certainty* that a rope can carry 10 kN, we will not worry about using the rope to carry 9 kN. However, if we are not very confident about the load-carrying capacity of the rope and that 10 kN is only an estimate, it is prudent to apply a factor of safety and consider using the rope only for much lower loads (say up to 5 kN). This applies to soil mechanics and foundation engineering too. There is a lot of uncertainty associated with the loadings and the soil properties, making it necessary to apply factors of safety in the designs. The higher the uncertainty, the larger the factor of safety. In the design of pad footings, factor of safety of 3 is generally used. Increasing the factor of safety makes the design more conservative and increases the cost significantly. The magnitude of the factor of safety selected depends on the probability of failure and its consequences.

The factor of safety is strongly related to the probability of failure. Lower the probability of failure, higher is the factor of safety.

1.9 Professional Registration and Continuing Professional Development

As in most professions, civil or geotechnical engineering practice is also regulated by professional bodies. With some years of experience and formal assessments, the engineer earns the "professional engineer (PE)" or "chartered engineer (CEng)" status, which is often the license to practice unsupervised. The engineering designs and any professional engineering services are required to be signed by a PE or a CEng, who takes the responsibility. Becoming a PE or CEng is one of the milestones in your career.

With so much happening in all civil engineering branches, it is necessary that we remain up to date with the current developments and state-of-the-art. This requires a lifelong commitment to update our knowledge, which is a key requirement for a PE or CEng. All professional or chartered engineers are expected to maintain a diary and record the *continuing professional development* they have undertaken in a year, which is expected to exceed certain hours specified by the governing body. These are audited regularly.

References

ASTM D 6026-13. Standard practice for using significant digits in geotechnical data.

Lambe, T. W., and R. V. Whitman. *Soil Mechanics*, John Wiley & Sons, New York, 1979.

Mitchell, J. K. "Geotechnics in Addressing Engineering Grand Challenges," *Geostrata*, ASCE, March/April, 12–14, 2020.

Peck, R. B., W. E. Hanson, and T. H. Thornburn. *Foundation Design*, John Wiley & sons, New York, 1974.

Skempton, A. W. "Landmarks in Early Soil Mechanics," *Proceedings of 7th European Conference on Soil Mechanics and Foundation Engineering*, Brighton, UK, 1979, pp. 1–26.

Taylor, D. W. *Fundamentals of Soil Mechanics*, John Wiley & Sons, New York, 1948.

Terzaghi, K. *Erdbaumechanik auf Bodenphysikalischer Grundlage*, Franz Deuticke, Leipzig und Wein, 1925, p. 399.

Terzaghi, K. *Theoretical Soil Mechanics*, John Wiley, New York, 1943.

Terzaghi, K., and R. B. Peck. *Soil Mechanics in Engineering Practice*, John Wiley & Sons, New York, 1948.

PART 1

Fundamentals

CHAPTER 2
Phase Relations

2.1 Introduction

Let us consider a soil shown in Fig. 2.1a, which would generally consist of soil grains and voids that are filled with water and/or air. The soil grains (solids), water (liquid), and the air are the three phases that can be present in any soil. When all the voids are filled with air only (no water), it is a *dry soil*. When all the voids are filled with water only (no air), it is called a *saturated soil*. When the soil contains all three phases, it is a *partially saturated* or *unsaturated soil*. This chapter deals with the relationships among the masses and volumes of the different phases.

2.2 Phase Diagram and Definitions

Let us consider the quantity of the soil shown in Fig. 2.1a within the circle, where the *total* volume and mass including all three phases present are V_t and M_t, respectively. For simplicity, let us separate the three phases as rectangular boxes shown in Fig. 2.1b. This is known as the *phase diagram*, where the volumes (denoted by V) and masses (denoted by M) are shown on the left and right, respectively. The subscripts s, w, and a refer to soil grains, water, and air, respectively. Subscripts v and t refer to voids and total, respectively. It can be seen that

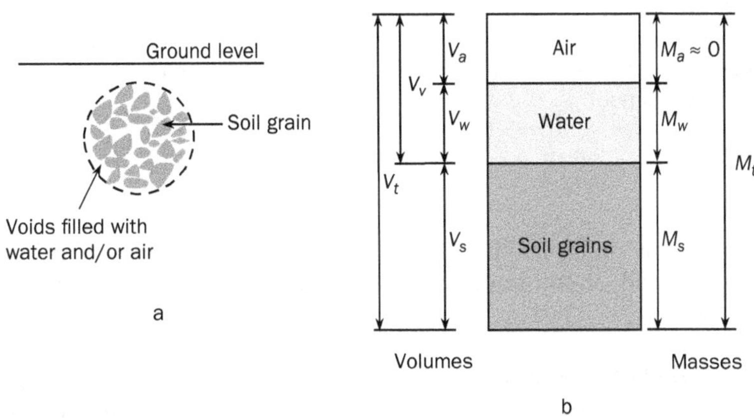

FIGURE 2.1 The three phases: (a) soil grains and voids filled with water and/or air and (b) phase diagram.

$$V_t = V_s + V_v = V_s + V_w + V_a \tag{2.1}$$

Neglecting the mass of air,

$$M_t = M_s + M_w \tag{2.2}$$

2.2.1 Void Ratio, Porosity, and Degree of Saturation

The relative proportion of the voids within the soil is quantified in terms of *void ratio* (e) or *porosity* (n). Void ratio is defined as the ratio of the void volume to the soil grain volume and hence

$$e = \frac{V_v}{V_s} \tag{2.3}$$

Porosity is defined as

$$n = \frac{V_v}{V_t} \tag{2.4}$$

and is generally multiplied by 100 and expressed as a percentage. A soil with 25% porosity means that 25% of the total volume consists of voids. Based on the definition, it can be seen that while the porosity must be less than 100%, the void ratio can exceed 1.

The relative void fraction filled by the water is known as the *degree of saturation* (S), defined as the ratio of the volume of water to volume of voids. Therefore,

$$S = \frac{V_w}{V_v} \tag{2.5}$$

S is expressed as a percentage in the range of 0–100%, with 0 for the dry soils and 100% for the saturated soils. A soil with 35% degree of saturation means that 35% of the voids are filled with water.

2.2.2 Moisture Content

The relative quantity of water within the soil is quantified by *moisture content* (or *water content*), denoted by w. It is defined as the ratio of the mass of water to the mass of soil grains and hence

$$w = \frac{M_w}{M_s} \tag{2.6}$$

It is generally multiplied by 100 and expressed as a percentage. It can be larger than 100%, especially in soft saturated clays which can have large void ratios as high as 2 or even more.

2.2.3 Densities and Unit Weights

Density (ρ) is the ratio of mass to volume. Depending on whether the soil is dry, moist, or saturated, the density can be defined slightly differently. For example, *bulk density*

(ρ_m) is defined considering the soil with all phases present and is defined as the ratio of the total mass to the total volume. Therefore,

$$\rho_m = \frac{M_t}{V_t} \tag{2.7}$$

It is also known as *moist*, *total*, or *wet density*.

Dry density (ρ_d) is defined as the ratio of the mass of the soil grains to the total volume, assuming there is no water in the voids, and the void volume remains the same. Therefore,

$$\rho_d = \frac{M_s}{V_t} \tag{2.8}$$

Saturated density (ρ_{sat}) assumes that the voids are filled with water. Therefore,

$$\rho_{sat} = \frac{M_{t,\,S=1}}{V_t} \tag{2.9}$$

Therefore, $\rho_{sat} \geq \rho_m \geq \rho_d$. In submerged (and hence saturated) soils, the effective density, considering the buoyancy effects, is known as the *submerged density* or *buoyant density* (ρ'). It is defined as

$$\rho' = \rho_{sat} - \rho_w \tag{2.10}$$

The common units of density are g/cm^3, Mg/m^3, t/m^3, and kg/m^3. Density of water is 1.0 g/cm^3, 1.0 Mg/m^3, 1.0 t/m^3, or 1000 kg/m^3. Here, metric ton or tonne is denoted by t, which is 1000 kg.

Unit weight (γ) is similar to density, where the mass (e.g., kg, g, t) is replaced by the weight (e.g., kN). It is the weight per unit volume. The *bulk unit weight* (γ_m), *dry unit weight* (γ_d), *saturated unit weight* (γ_{sat}), and *submerged* or *buoyant unit weight* (γ') are defined similar to the way we defined densities in Eqs. (2.7)–(2.10). The unit weight is generally expressed in kN/m^3. Unit weight of water γ_w is 9.81 kN/m^3. From Newton's law, force = mass × acceleration. Therefore, weight = mass × gravitational acceleration (g), suggesting $\gamma = \rho g$.

2.2.4 Specific Gravity of Soil Grains

Specific gravity (G_s) of the soil grains says how much heavier the soil grains are compared to water. It is the ratio of the density of a soil grain (ρ_s) to density of water (ρ_w).

Remember that the specific gravity of water is 1.0. Soil grains in general have G_s lying in a narrow range of 2.6–2.9. Some organic soils and fly ash can have G_s significantly less. Some mine tailings rich in heavy minerals can have G_s as high as 4.0. Light-colored quartz sand, commonly encountered in practice, has G_s of about 2.65. The specific gravity of soil grains must be reported to two decimals. Typical values of some soils and rock minerals are summarized in Table 2.1.

Mineral	G_s	Mineral	G_s
Aragonite	2.94	Illite	2.65–2.72
Attapulgite	2.61	Kaolinite	2.60–2.65
Augite	3.2–3.4	Limonite	3.8
Biotite	3.0–3.1	Magnetite	5.17
Calcite	2.70–2.72	Mica	2.80–3.20
Chlorite	2.6–3.0	Montmorillonite	2.5–2.8
Dolomite	2.85–2.90	Muscovite	2.80–2.90
Galena	7.5	Orthoclase	2.56
Gypsum	2.32	Peat	1.2–1.8
Halite	2.1–2.3	Pyrite	5.0
Hematite	4.9–5.3	Quartz	2.65–2.67
Hornblende	3.2–3.5	Smectite	2.2–2.7
Humus	1.4	Talc	2.7

TABLE 2.1 Typical Values of Specific Gravity

2.3 Phase Diagram for $V_s = 1$

In Fig. 2.1a, let us consider a small fraction of the soil where $V_s = 1$ (in any unit). From the definition of void ratio [Eq. (2.3)], $V_v = e$. From the definition of degree of saturation, $V_w = Se$. These are shown on the left of Fig. 2.2.

Now, let us determine the masses of the different phases in Fig. 2.2. The densities of the soil grains and water are $G_s\rho_w$ and ρ_w, respectively. Noting that the mass is the product of volume and density, the masses of the soil grains and water are $G_s\rho_w$ and $Se\rho_w$, respectively.

From Fig. 2.2, porosity can be written as

$$n = \frac{V_v}{V_t} = \frac{e}{1+e}$$

$$n = \frac{e}{1+e} \tag{2.11}$$

The moisture content can be written as

$$w = \frac{M_w}{M_s} = \frac{Se\rho_w}{G_s\rho_w} = \frac{Se}{G_s}$$

$$w = \frac{Se}{G_s} \tag{2.12}$$

The bulk density can be written as (see Fig. 2.2)

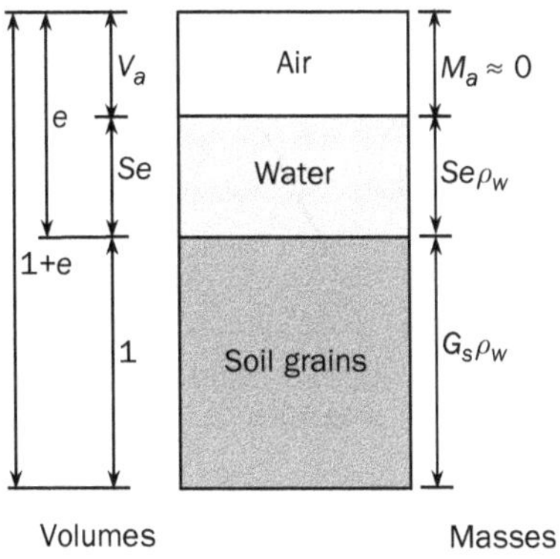

Figure 2.2 Phase diagram for $V_s = 1$.

$$\rho_m = \frac{M_t}{V_t} = \frac{G_s\rho_w + Se\rho_w}{1+e} = \left(\frac{G_s + Se}{1+e}\right)\rho_w$$

$$\rho_m = \left(\frac{G_s + Se}{1+e}\right)\rho_w \tag{2.13}$$

The dry density can be written as

$$\rho_d = \frac{M_s}{V_t} = \frac{G_s\rho_w}{1+e} = \left(\frac{G_s}{1+e}\right)\rho_w$$

$$\rho_d = \left(\frac{G_s}{1+e}\right)\rho_w \tag{2.14}$$

Equation (2.14) can be also deduced from Eq. (2.13), substituting $S = 0$. The expression for the saturated density can be deduced from Eq. (2.13), assuming $S = 1$ and hence

$$\rho_{sat} = \frac{M_{t,\,S=1}}{V_t} = \frac{G_s\rho_w + e\rho_w}{1+e} = \left(\frac{G_s + e}{1+e}\right)\rho_w$$

$$\rho_{sat} = \left(\frac{G_s + e}{1+e}\right)\rho_w \tag{2.15}$$

Identical equations can be developed for the different unit weights, replacing ρ by γ, and mass by weight.

2.4 Laboratory Measurements

With so many different variables connecting the three phases, only moisture content, specific gravity, and bulk density are generally measured. The others, such as void ratio, porosity, and degree of saturation, are calculated from the phase relations.

Moisture content measurement (ASTM D 2216-10) is fairly straightforward. The mass of a soil specimen is determined (m_1) and placed in an oven at 105–110°C for 24 h

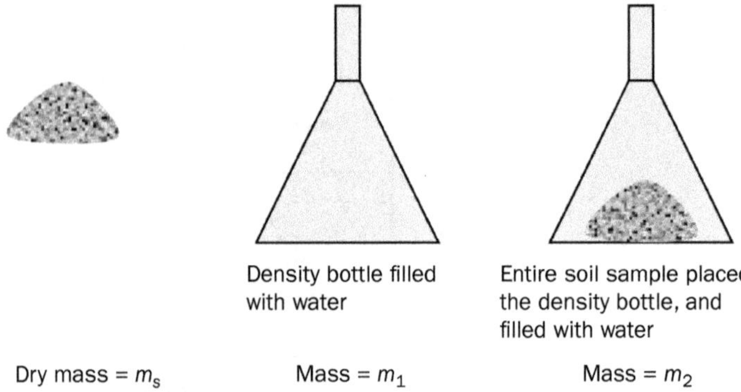

FIGURE 2.3 Specific gravity measurement.

to drive out the entire water present within the voids. Further drying beyond 24 h will make no difference. From the dry mass (m_2), the mass of the water ($m_1 - m_2$) is determined. The moisture content is computed as

$$w = \frac{m_1 - m_2}{m_2} \tag{2.16}$$

The specific gravity G_s is determined using a volumetric flask, in the form of a *density bottle* or *pycnometer* (ASTM D 854-14), and reported to two decimals (i.e., nearest 0.01) at a reference temperature of 20°C.

Let us take some dry soil grains with mass of m_s. If the soil is moist, knowing the moisture content, the dry mass can determined. The density bottle filled with water has a mass of m_1. Empty the density bottle and place the entire soil sample in the bottle and fill with water and the mass is determined as m_2 (see Fig. 2.3). The specific gravity of the soil grains is the ratio of the dry mass of the grains to the mass of the water displaced by the grains. Therefore,

$$G_s = \frac{m_s}{m_s + m_1 - m_2} \tag{2.17}$$

The bulk density is measured in different ways, depending on the soil conditions. For example, in clays thin-walled tubes can be pushed in to obtain an undisturbed cylindrical specimen, and the bulk density is determined from the mass and volume. When it is difficult to obtain undisturbed samples for measurements, field density measurements can be carried out as discussed in Chap. 4.

Example 2.1

A 50-mm-diameter and 150-mm-long undisturbed soil sample was collected from a site. The sample has a mass of 480 g, which becomes 350 g when dried in the oven at 105°C for 24 h. The specific gravity of the soil grains is 2.71. Determine the moisture content, void ratio, degree of saturation, and the bulk density.

Part a At the runway, $V_t = 100 \times 2000 \times 0.5 = 100{,}000\ \text{m}^3$
$M_s = 1.89 \times 100{,}000 = 189{,}000$ tonnes
$M_w = 0.142 \times 189{,}000 = 26{,}838$ tonnes
In the truck, $V_t = 8.0\ \text{m}^3$
$M_t = 13.2$ tonnes
Assuming the same 6% moisture content for the soil in the truck as in the borrow,

$$0.06 = \frac{13.2 - M_s}{M_s}$$

$$M_s = 12.45 \text{ tonnes}$$

Therefore, the number of truckloads required = 189,000/12.45 = **15,181.**

Part b For each truckload, the moisture content will be increased from 6% to 14.2%. Therefore, the mass of water to add per truckload = 12.45(0.142 – 0.06) = 1.0209 tonnes or **1020.9 L.**

Part c

$$\rho_d = \frac{G_s \rho_w}{1 + e}$$

$$e = \frac{2.721.0}{1.89} - 1 = 0.439$$

$$w = \frac{Se}{G_s}$$

If the subbase becomes saturated, $S = 1$ and hence

$$w = \frac{1 \times 0.439}{2.72} = 0.161 \text{ or } 16.1\%$$

There will be a slight increase in the moisture content.

Example 2.4

It is required to carry out a compaction test on a soil, which will be placed into a cylindrical mold and compacted at moisture content of 10.0%. The current moisture content of the soil is 2.5%. How much water would you add to 5.0 kg of the above soil to bring the moisture content to 10.0%?

The specific gravity of the soil grains is 2.68. If the soil at moisture content of 10% is compacted to a bulk unit weight of 17.5 kN/m³, determine the void ratio and degree of saturation.

Solution Let the mass of soil grains in the 5 kg of soil be x.

$$0.025 = \frac{5 - x}{x}$$

Therefore, $x = M_s = 4.878$ kg and hence $M_w = 5 - 4.878 = 0.122$ kg

At $w = 10\%$, $M_w = 0.10 \times 4.878 = 0.488$ kg
Therefore, the water to be added $= 488 - 122 =$ **366 g**
When compacted to $w = 10\%$,

$$w = \frac{Se}{G_s}$$

$$Se = 0.10 \times 2.68 = 0.268$$

$$\gamma_m = \left(\frac{G_s + Se}{1+e}\right)\gamma_w$$

$$17.5 = \left(\frac{2.68 + 0.268}{1+e}\right)9.81$$

Therefore, $e = 0.653$

$$S = \frac{wG_s}{e} = \frac{0.10 \times 2.68}{0.653} = 0.410 \text{ or } 41.0\%$$

Example 2.5

Some perfectly spherical grains of the same diameter d were stacked on top of each other as shown in Fig. 2.6. This gives the loosest possible packing. What is the void ratio? Volume of a sphere $= \pi d^3/6$.

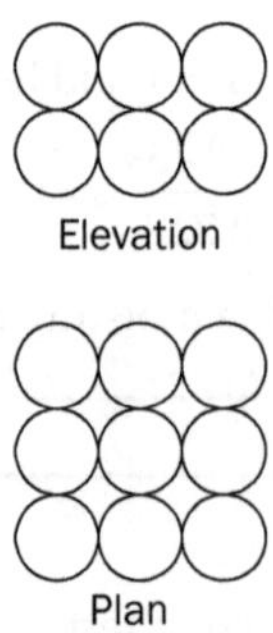

FIGURE 2.6 Elevation and plan views of the grain packing.

Solution Considering a cube of width d,

$$V_t = d^3$$
$$V_s = \pi d^3/6 = 0.5236\, d^3$$

Therefore, $V_v = d^3 - \pi d^3/6 = 0.4764\, d^3$

$$e = \frac{V_v}{V_s} = \frac{0.4764}{0.5236} = 0.91$$

Example 2.6

A saturated clay specimen with volume of 162.1 cm^3 has a mass of 295.3 g. When dried in the oven for 24 h at 105°C, the mass is 209.6 g. Determine the specific gravity of the soil grains.

Solution

The mass of water in the sample = 295.3 – 209.6 = 85.7 g
Volume of water (i.e., voids too) in the sample = 85.7 cm^3
Volume of the soil grains = 162.1 – 85.7 = 76.4 cm^3
Mass of soil grains = 209.6 g

Therefore, G_s = 209.6/76.4 = **2.74**

Example 2.7

A 3-m-thick saturated soft clay deposit is currently at natural moisture content of 115.0%. The site is surcharged with some fill placed at the ground level, and the clay layer settled by 500 mm while remaining saturated. What would be the new moisture content? Assume specific gravity of the soil grains as 2.75.

Solution

$S = 1, w = 1.15, G_s = 2.75$

Therefore, the initial void ratio is given by

$e = 1.15 \times 2.75 = 3.16$

Let us consider 1.0 m^2 plan area, where $V_t = 1 \times 3 = 3.0$ m^3

$$e = 3.16 = \frac{V_v}{V_s} = \frac{3.0 - V_s}{V_s}$$

$V_s = 0.721$ m^3

After settling by 0.5 m to 2.50 m, from 3.00 m to 2.50 m,

$V_t = 2.5 \text{ m} \times 1 \text{ m}^2 = 2.5 \text{ m}^3$

Therefore, the new volume of voids is given by $V_v = 2.5 - 0.721 = 1.779$ m^3

The new void ratio is

$$e = \frac{1.779}{0.721} = 2.467$$

The new moisture content is 2.467/2.75 = **0.897 or 89.7%.**

Example 2.8

The soil at a borrow area is at moisture content of 10.5% and bulk unit weight of 18.5 kN/m^3. The specific gravity of the soil grains is 2.72. This soil is taken in a dump truck to the site for making a compacted fill, carrying 90 kN per load. The fill at the site is compacted to dry unit weight of 18.5 kN/m^3 at moisture content of 15%. What would be the volume of the compacted fill from 100 truckloads? What would be the volume of the hole in the borrow area for the 100 truckloads? Assume that the moisture content of the soil in the truck is the same as that in the borrow area. What is the degree of saturation of the compacted fill?

Solution

The weight of soil grains carried per truckload = 90.0/(1 + 0.105) = 81.45 kN
Weight of the soil grains at the compacted fill = 100 × 81.45 = 8145 kN
Volume of the compacted fill = 8145/18.5 = **440.27 m³**
Dry unit weight of the soil at the borrow = 18.5/(1 + 0.105) = 16.74 kN/m³

Therefore, the volume of the hole = 8145/16.74 = **486.56 m³**

At the compacted fill,

$$\gamma_d = \frac{G_s \gamma_w}{1+e}$$

Therefore,

$$e = \frac{G_s \gamma_w}{\gamma_d} - 1 = \frac{2.72 \times 9.81}{18.5} - 1 = 0.442$$

Since

$$w = \frac{Se}{G_s}$$

$$S = \frac{0.15 \times 2.72}{0.442} = \mathbf{0.923 \text{ or } 92.3\%}$$

2.5 Main Points

1. The computations of masses and volumes of the different phases and the parameters such as moisture content, void ratio, and densities are an integral part of most laboratory tests. These computations can be carried out in terms of (a) masses and densities or (b) weights and unit weights—never mix them.
2. The definitions of void ratio, moisture content, degree of saturation, specific gravity, and the different densities are logical and make sense.
3. G_s varies in a narrow range of 2.6–2.9 for most soils. In the absence of laboratory-determined values, it is possible to assume a value for G_s.
4. With the different variables (w, e, n, S, ρ_m, ρ_d, ρ_{sat}, ρ') discussed in this chapter, it is possible to develop more than a dozen equations relating each of them in terms of few other variables. The equations presented in this chapter are adequate for all computations.
5. You do not have to memorize all the equations here. By recollecting the phase diagram for $V_s = 1$ (Fig. 2.2), all phase relations can be derived from the first principles. If you really want to remember two equations from this chapter, they are Eqs. (2.12) and (2.13). Equations (2.14) and (2.15) can be deduced from Eq. (2.13).
6. The density of water is $\rho_w = 1\ \text{g/cm}^3 = 1\ \text{Mg/m}^3 = 1\ \text{t/m}^3 = 1000\ \text{kg/m}^3$. The unit weight of water is $\gamma_w = 9.81\ \text{kN/m}^3$.
7. The equations for the density and unit weights are literally the same; just replace the ρ by γ on both sides of Eqs. (2.13)–(2.15).
8. When going between mass and weight, remember that 1 kg is equivalent to 9.81 N.

Review Exercises

2.1 Circle the correct answer.

a. Which of the following is likely to be the weight or mass of 1.0 m^3 of rock?

28 kN   28 metric tonnes   28 kg

b. Which of the following is the smallest?

Dry density   Submerged density   Saturated density

c. Which of the following is the largest?

Dry density   Submerged density   Saturated density

d. Which of the following can exceed 100%?

Porosity   Degree of saturation   Moisture content

Answer: 28 kN, submerged density, saturated density, moisture content

2.2 Determine the moisture content and the bulk unit weight of a saturated sand ($G_s = 2.65$) at void ratio of 0.75.

Answer: 28.3%, 19.06 kN/m³

2.3 A 500 g of soil sample ($G_s = 2.70$) is at 5% moisture content. If 50 ml of water is added to this soil and mixed homogeneously, what would be the new moisture content?

Answer: 15.5%

2.4 2.0 kg of soil *A* at a moisture content of 5.0% is mixed with 2.0 kg of soil *B* which is at a moisture content of 20.0%. The specific gravity of the soil grains of both the soils is 2.70. If the mix is compacted to a dry unit weight of 18.0 kN/m^3, what would be the degree of saturation?

Answer: 68.6%

2.5 Specific gravity of the soil grains is measured as per ASTM D 854 using a 250 ml density bottle. The mass of the soil used in the test is 82.4 g which is at a moisture content of 8.3%. The density bottle filled with water to the 250 ml mark, has a mass of 329.5 g. When the entire soil sample is placed in the dry density bottle, and the bottle is filled to the 250 ml mark, the mass is 377.3 g. Find the specific gravity of the soil grains.

Answer: 2.69

2.6 A large piece of dry rock has a mass of 2450 kg and volume of 0.925 m^3. The specific gravity of the rock mineral is 2.80. Determine the porosity of the rock.

Answer: 5.4%

2.7 A 1000-cm^3 soil suspension is made for a hydrometer test, with the dry soil sample of 50 g. The specific gravity of the soil grains is 2.72. What is the density of the suspension?

Answer: 1.0316 g/cm³

2.8 Prove the following:

a. $$\rho_d = \frac{\rho_m}{1 + w}$$

b. $$\rho' = \left(\frac{G_S - 1}{1 + e}\right)\rho_w$$

2.9 The water table at a site is at a depth of 1.0 m below the ground level. The void ratio of the soil is 0.625 and the specific gravity of the soil grains is 2.68. It can be assumed that the soil below the water table is saturated and the soil above is dry. It is proposed to cut a 2-m-wide and 1.5-m-deep trench in the ground. The excavated soil will be mixed and stockpiled for other uses.

a. What would be the moisture content of the stockpile?
b. If a 100 m long trench is cut, what would be the weight of the stockpile?

Answer: 7.77%, 5231 kN

2.10 The water table at a site lies 2 m below the ground level. The void ratio above and below the water table are 0.82 and 0.98, respectively. The specific gravity of the soil grains is 2.70. The soil above the water table is at moisture content of 10.0%. A 3.0-m-deep and 4-m-wide canal is cut through the ground and the excavated soil is mixed together and stockpiled. What would be the average moisture content of the stockpile?

Answer: 18.3%

2.11 A contractor requires 300 m^3 of aggregate base for roadwork. The aggregates will be compacted to a dry unit weight of 19.8 kN/m^3. The aggregates are available at a local quarry, at a moisture content of 7.0%, and are sold by metric tonne and not by cubic meters. Note: 1 metric tonne = 1000 kg.

a. How many tonnes of aggregates should the contractor order?
b. A few weeks later, an intense rainstorm increased the moisture content of the stockpile to 15.0%. If the contractor orders the same quantity of aggregates, how many cubic meters of compacted aggregate base will they produce?

Answer: 648 tonnes, 279.2 m^3

2.12 The moisture content of a saturated soil is 18.5%, and the specific gravity of the soil grains is 2.72. At the same void ratio, what is the dry unit weight?

Answer: 17.75 kN/m^3

2.13 The depth of a lake is increased by 1.0 m, over an area covering 300 m $\times$ 600 m. The in-place unit weight of the saturated soil at the bottom of the lake is 19.0 kN/m^3, and the specific gravity of the soil grains is 2.72. The soil removed from the bottom of the lake is placed under the sun to reduce the moisture content and is later used in a compacted fill placed at dry unit weight of 18.0 kN/m^3. What would be the volume of the compacted fill that can be made from the soil dredged from the bottom of the lake?

Answer: 145,371 m^3

2.14 A 250-mm soil layer, with initial void ratio of 0.88 is compacted into 220 mm. What would be the new void ratio?

Answer: 0.654

2.15 A 50-mm-diameter and 100-mm-long saturated soil specimen was recovered from a submerged soil. The specimen has a mass of 380.1 g. When the entire specimen was oven dried, it has a mass of 290.5 g. What is the specific gravity of the soil grains?

Answer: 2.72

2.16 The bulk unit weight and the moisture content at the borrow area are 19.0 kN/m^3 and 13.0%, respectively. The soil from this site will be compacted (not necessarily at the same moisture content) to build an embankment at dry unit weight of 18.0 kN/m^3. With 100,000 m^3 soil removed from the borrow area, how many cubic meters of embankment can be built?

Answer: 93,412 m^3

2.17 A soil with $G_s = 2.60$ and some mine tailings (crushed waste rocks) with $G_s = 3.20$ are mixed in equal proportions (by weight), when they are both dry. What is the specific gravity of the grains in the mix.

Answer: 2.87

2.18 In underground mining, large ore bodies are removed from the ground, leaving large voids in the ground. Valuable minerals are extracted from the ore, and the waste rock is crushed into tailings, consisting of gravel, sand, silt, and clay fractions. The tailings are used to backfill the underground voids. The in situ ground at a mine site is entirely rock that can be assumed dry, with porosity of 5%. From the site, 100,000 tonnes of ore is removed from the ground, producing 3000 tonnes of valuable minerals, thus leaving 97,000 tonnes of tailings that have to be disposed. The average specific gravity of the rock mineral is 2.85.

a. Determine the volume of the underground void created by the ore removal.
b. It is proposed to backfill the above void using the crushed waste rock (i.e., tailings), placed at porosity of 40%. What percentage of the tailings can be placed in the underground voids?

Answer: 36,934 m^3*, 65.1%*

2.19 An irregularly shaped sample of a saturated clay specimen, with a mass of 605.2 g, was coated with wax. The total mass of the coated sample was 614.2 g. The volume of the coated sample was determined (by water displacement method) as 311 cm^3. The wax was removed carefully and the soil sample was dried to obtain the dry mass as 479.2 g. The specific gravity of the wax is 0.90. Determine the moisture content and dry density of the sample, and the specific gravity of the soil grains.

Answer: 26.3%, 1.592 g/cm^3*, 2.74*

2.20 The soil from a borrow area ($e = 0.95$, $w = 8.5\%$, $G_s = 2.70$) is excavated and transported in trucks to the site, where it is compacted to a dry unit weight of 16.2 kN/m^3 and moisture content of 12.0% in the construction of an embankment. Each truck is loaded to its capacity of 80 kN. How many truckloads would be required to build 100,000 m^3 of embankment? How much water should be added at the site per truckload?

Answer: 21972, 263.1 L

2.21 An irregular lump of moist clay specimen has a mass of 564.2 g. To determine its density, it was coated in paraffin wax and the total mass of the coated specimen was 572.6 g. The volume of the coated lump was found to be 263.6 mL by displacement in water. The specific gravity of the hardened wax and the soil grains are 0.91 and 2.72, respectively. After the paraffin wax was carefully removed, the clay specimen was oven dried and the mass was 503.1 g. Determine the bulk density of the clay specimen and its degree of saturation.

Answer: 2.22 g/cm^3*, 88.0%*

References

ASTM D 854-14. *Standard Test Method for Specific Gravity of Soil Solids by Water Pycnometer.*

ASTM D 2216-10. *Standard Test Methods for Laboratory Determination of Water (Moisture) Content of Soil and Rock by Mass.*

CHAPTER 3

Soil Classification

3.1 Introduction

In this chapter, we will study how the soils are formed, which requires some understanding of geology. There are so many different types of soils and we will form groups where they exhibit similar behavior. Soil grains larger than 0.075 mm (or 75 μm) are called *coarse-grained soils*, where the grains are visible to a naked eye. This includes *sands, gravels, cobbles,* and *boulders*. *Fine-grained soils* include clays and silts, where it is difficult to distinguish individual grains by an unaided eye.

Fine-grained soils and coarse-grained soils are different in their characteristics, and can be found mixed in different proportions. Depending on the usage, there are different ways of classifying soils. They are classified differently for agriculture, roadwork, and general geotechnical applications. For most geotechnical applications, soils are classified using *Unified Soil Classification System* (*USCS*). For roadwork, they are classified using the guidelines developed by the *American Association of State Highway and Transportation Officials* (*AASHTO*). For agricultural work on soils, *U.S. Department of Agriculture textural classification system* is appropriate. There are also country specific standards that are slight variations of the USCS. Depending on the grain sizes, soils can be grouped as shown in Fig. 3.1 based on USCS, British, or Australian standards. The cut-off limits separating the groups vary slightly between the different standards. We will follow USCS throughout this book.

Gravels, sands, and silts are also called *granular soils* or *cohesionless soils*. Clays are called *nongranular soils* or *cohesive soils*.

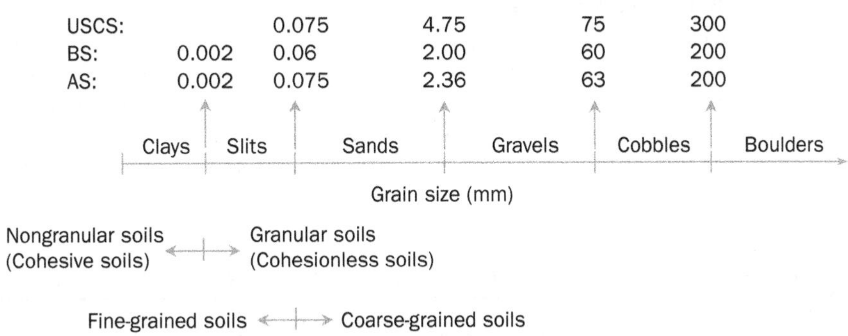

FIGURE 3.1 Major soil groups based on grain sizes.

3.2 Origin of Soils

Earth is approximately 12,500 km in diameter. The core of the earth is surrounded by a thick mantle, which is enclosed by a relatively thin crust. The core is further divided into inner core and outer core. The mantle can be divided as lower mantle with molten rock and upper mantle with solid rock. While the average soil grain density is 2.7 g/cm^3, the average density of the entire earth is about 5.5 g/cm^3, due to the presence of heavy metals, including high iron content, in the core. The temperature at the crust-mantle interface can be about 1000°C and increases to about 3500°C at the mantle-core interface. For most engineering applications, our activities are limited to the top few hundred meters, which form the *crust*. The highest mountain Everest is 8.85 km above the mean sea level. The deepest ocean is 11.0 km at Mariana trench in the Western Pacific ocean. The *oceanic crust* (5–10 km thick) lies under the oceans and is thinner than the *continental crust* (30–50 km thick) lying under the continents. Oceanic crust is slightly denser than the continental crust. The eight common elements, making up 98% of the earth's crust are (in the order of abundance) O, Si, Al, Fe, Ca, Na, K, and Mg (Table 3.1).

While there are thousands of minerals present on the earth, the eight most common rock forming minerals are feldspar, quartz, amphibole, mica, olivine, garnet, calcite, and pyroxene. Depending on which of the three different processes by which the rocks are formed, they can be grouped under *igneous rocks*, *sedimentary rocks*, and *metamorphic rocks*.

3.2.1 Igneous Rocks

Igneous rocks are formed by cooling of molten rocks known as *magma* when beneath the surface and *lava* when above the surface. Magma and lava are of the same mineral composition, which includes Si, Fe, Na, K, and others. When magma cools below the surface, the process is very slow, taking thousands of years and the grains are coarse and can be seen by naked eye and easy to identify. Such rocks are called *intrusive* or *plutonic*. Some examples of intrusive rocks are granite, diorite, gabbro, and peridotite. When the lava cools on the surface, the process is faster, taking few days to weeks, not giving enough time for the grains to grow. The grains are finer and are difficult to identify by naked eye. These rocks are called *extrusive* or *volcanic*. Some examples of volcanic rocks are rhyolite, andesite, basalt, and komatite. Basalt and granite are the two most common igneous rocks—one volcanic and the other plutonic.

3.2.2 Sedimentary Rocks

Sedimentary rocks are made of fine sediments of clays, silts, sands, gravels and other fragments, deposited in layers over a long time. They are formed by deposition (i.e., sedimentation) of geomaterials on the earth's surface and in water bodies. The sedimentary rocks can be broadly categorized as *detrital* (*clastic*) or *chemical sedimentary rocks*. Detrital sedimentary rocks are made of broken fragments of pre-existing parent rocks

Element	O	Si	Al	Fe	Ca	Na	K	Mg	Others
%	46.1	27.7	8.1	5.0	3.6	2.8	2.6	2.0	2.1

Table 3.1 Elements in the Crust

and hence retain the properties of the parent rocks. They are cemented together by minerals such as carbonates and sulfates. For example, clay becomes *shale*; silt becomes *siltstone*; sand becomes *sandstone*; and gravels and boulders become *conglomerate* (rounded grains) or *breccia* (angular grains). These are the five major clastic sedimentary rocks. The chemical sedimentary rocks are formed by a chemical process, where the minerals are dissolved in water and form new minerals. These include *limestone* [$CaCO_3$], *dolomite* [$CaMg(CO_3)_2$], *chert* [SiO_2], *rock salt* (or *halite*) [NaCl], and *gypsum* [$CaSO_4.2H_2O$]. These are the five major chemical sedimentary rocks. Limestone comes in several forms, all made of the mineral calcite [$CaCO_3$] and form bubbles when in contact with hydrochloric acid. From ancient times, it has been used to make lime for mortar. Dolomite looks very similar to limestone. It is limestone that has been chemically altered by replacing some calcium with magnesium. *Organic sedimentary rocks* are formed by decomposed plants, bones, and shells (e.g., Coal). *Coal* is formed by decomposed vegetation. It is used as a cheap fuel because of its ability to burn. However, it pollutes the environment.

Shale is the most abundant type of sedimentary rock. It is a soft rock that has a finely layered bedding structure, which shows a clear plane of weakness. The layers were developed from soil deposits consolidated under very high overburden pressures, over long geologic times. The layering is generally present in claystone or siltstone. Sandstone is made by sand grains that are bonded together by cementing agents such as clays, carbonates, or iron oxides.

3.2.3 Metamorphic Rocks

When pre-existing igneous and sedimentary rocks (or even metamorphic rocks) are subjected to high temperature, pressure, and shear, they undergo major structural changes and form metamorphic rocks. This generally occurs deep inside the earth crust. Since there is no melting, the original chemical composition is often preserved. Low grades of metamorphic rocks (e.g., *slate*) retain the characteristics of the parent rocks. High grades metamorphic rocks (e.g., *schist*, *gneiss*), which are subjected to high temperatures and pressures can be very different to their parent rocks. At low temperatures and pressures, shale transforms into slate; at intermediate pressures and temperatures, shales transform into *phyllites*; and at very high temperatures and pressures it transforms into schist or gneiss. Limestone ($CaCO_3$) and dolomite [$CaMg(CO_3)_2$] transform into *marble*. Sandstone transforms into *quartzite*. Due to the high temperatures and pressures, some of the minerals get elongated and form sheets and occur in thin layers. This is called *foliation*. Slate, schist, phyllite, and gneiss are examples of foliated (i.e., layered) metamorphic rocks. *Marble* and *quartzite* are nonfoliated metamorphic rocks. Gneiss can also be a highly metamorphosed granite.

Earth crust, making only 1% of earth's volume, consists of approximately 65% igneous rocks, 27% metamorphic rocks, and 8% sedimentary rocks. The igneous rocks are mainly granite and basalt. While sedimentary rocks make up only about 8% of the earth crust, they cover 75% of the earth's surface.

The parent rock breaks down into fine sediments by mechanical or chemical weathering. When the weathered sediments remain at the same location, they are called *residual soils*. They retain the same characteristics as the parent rock. When they are transported by geological agents such as wind (Aeolian soils—e.g., loess), water (alluvial, marine, or lacustrine soils), ice (glacial soils), or gravity (colluvial soils—e.g., down

Soil name	Description
Bentonite	Decomposed volcanic ash with high content of montmorillonite clay mineral.
Black cotton soil	A highly expansive clay with high content of montmorillonite minerals. Cotton grows well in this soil and hence the name.
Boulder clay	A mixture of all sizes of rock fragments from boulder to clay. Also called glacial till.
Calcareous soil	A soil rich in calcium (or magnesium) carbonate. Alkaline with high pH.
Caliche	Gravel, sand, and clay mixture cemented by calcium carbonate. Generally formed in arid or semiarid climate.
Gumbo	A clayey or loamy soil which is sticky when wet.
Hardpan	Densely cemented soil, often well graded and cohesive, that is difficult to penetrate or excavate. Often extremely dense. Some boulder clays and glacial tills are hardpans.
Laterite	Red-colored soils rich in iron and aluminum, formed in tropics.
Loam	Approximately equal proportions of sand, silt, and clay mixed, sometimes with organic matter.
Loess	Uniform wind-blown silt. Shows cohesion when dry which is lost on wetting. Near vertical cuts are possible in dry condition. Due to the presence of root holes, permeability is greater in the vertical direction than horizontally. Grain size 0.01–0.05 mm.
Marl	Mix of calcareous sand or clay or loam. Stiff or very stiff. Clay content not more than 75% and lime content not less than 15%.
Till	Unstratified glacial deposit of soils mixed with boulders.
Tuff	Water/wind laid aggregate of very small rock fragments ejected from volcanoes.
Varved clay	Clay or silt of glacial origin, of lacustrine deposit. Varve is a Swedish term for layer. Thicker silt varves from summer and thinner clay varves of winder alternate. The thickness of the layers is less than 10 mm generally.

Table 3.2 Special Names for Soils

a slope in landslides), they are called *transported soils*. During the transportation by the geological agents, there can be further breakdown and some sorting taking place. Larger grains are deposited early and smaller grains are carried further away. Gravity transports the weathered sediments over smaller distances, often down a slope, with no changes in the material characteristics. The transported soils are generally loose or soft to several meters of depth. Table 3.2 lists some special names for specific soils.

3.3 Grain Size Distribution

The soils are generally found as a mixture of grains in different sizes and shapes. Especially in coarse-grained soils, where the engineering properties depend on the distribution of grain sizes, it is necessary to know what sizes are present in what proportions. In the case of fine-grained soil such as clays, grain size distribution is not important. Their properties are controlled by other characteristics discussed in Sec. 3.4.

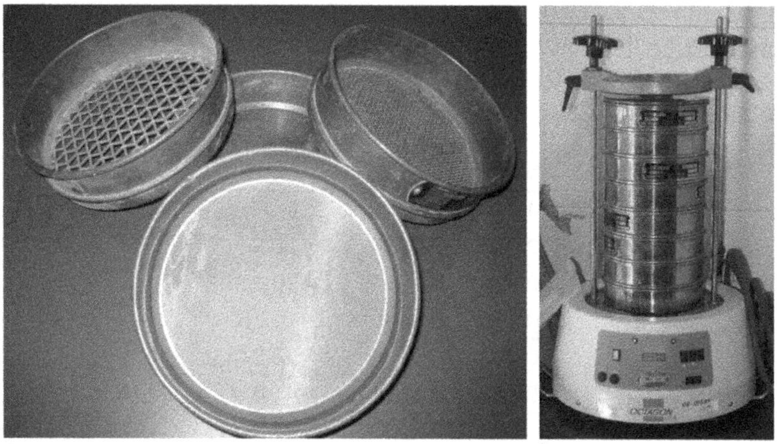

a b

Figure 3.2 Sieve analysis: (a) set of sieves, (b) sieve shaker.

75 mm (3 in)	50 mm (2 in)	38 mm (1 ½ in)	25 mm (1 in)
19 mm (3/4 in)	9.5 mm (3/8 in)	4.75 mm (No. 4)	2.00 mm (No. 10)
0.85 mm (No. 20)	0.425 mm (No. 40)	0.250 mm (No. 60)	0.150 mm (No. 100)
0.106 mm (No. 140)	0.075 mm (No. 200)		

Table 3.3 Common ASTM Sieve Sizes

Grain size distribution of a soil is determined using a set of *sieves* (ASTM D 6913-17) and a *hydrometer* (ASTM D 422-07). Sieves are used for coarse-grained soils and hydrometer is used for the fine-grained soils.

3.3.1 Sieve Analysis

A sieve consists of a circular tray where the base is made of a mesh with uniform openings (Fig. 3.2a). The diameter of the tray can be 200 mm, 300 mm, or 450 mm. Sieves with opening smaller than 4.75 mm are also assigned a sieve number, which is the number of openings per inch (25.4 mm) each way. A No. 10 sieve has 100 openings per square inch. The sieves are available with mesh openings in different sizes. The commonly used sieve sizes are listed in Table 3.3.

A set of sieves are stacked with the opening sizes progressively increasing upward. The bottom of the stack consists of a pan without any holes. The soil is passed through the sieves, placed in a mechanical sieve shaker (Fig. 3.2b), and agitated for about 10 min. The mass of grains retained on each sieve is recorded and the *cumulative* percentage of grains passing each sieve is determined. The *percentage passing*, also called *percentage finer* (always cumulative), is plotted against the grain size, generating the grain size distribution curve. When the fines are included (from hydrometer test discussed in Sec. 3.3.2), the grain size varies in orders of magnitude, and hence the grain size is always presented in logarithmic scale for clarity (see Fig. 3.3 in Example 3.1).

Example 3.1

A sieve analysis is carried out on a sandy soil and the test data are summarized below. Plot the grain size distribution curve.

Size (mm)		Mass retained (g)
4.75	(No. 4)	0
2.00	(No. 10)	56.8
0.85	(No. 20)	90.3
0.425	(No. 40)	112.5
0.25	(No. 60)	78.3
0.15	(No. 100)	41.6
0.075	(No. 200)	33.2
Pan		40.1
Total		452.8

What are the percentages of gravels, sands, and fines based on USCS (see Fig. 3.1)?

Solution The cumulative % passing 0.25 mm sieve is calculated as

$$\frac{41.6 + 33.2 + 40.1}{452.8} \times 100 = 25.4\%$$

These calculations were carried out for every sieve size and they are summarized in the table below.

Size (mm)		Mass retained (g)	Mass passing (g)	% passing
4.75	(No. 4)	0	452.8	100.0
2.00	(No.10)	56.8	396.0	87.5
0.85	(No. 20)	90.3	305.7	67.5
0.425	(No. 40)	112.5	193.2	42.7
0.25	(No. 60)	78.3	114.9	25.4
0.15	(No. 100)	41.6	73.3	16.2
0.075	(No. 200)	33.2	40.1	8.9
Pan		40.1		
Total		452.8		

The grain size distribution curve is shown in Fig. 3.3.

Percentage passing 0.075 mm (No. 200) = 8.9, which is the percentage of fines.

Percentage passing 4.75 mm (No. 4) = 100, which includes sands and fines.

Therefore,

Gravels = 0%, sands = 91.1%, and fines = 8.9%.

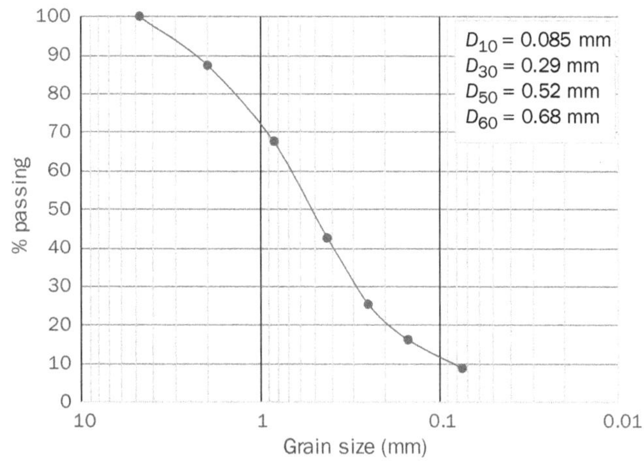

FIGURE 3.3 Grain size distribution curve for Example 3.1.

The North American practice is to have the grain size increasing from right to left in the plot, as shown in Fig. 3.3. In many other countries, it is shown increasing from left to right.

3.3.2 Hydrometer Analysis

Hydrometer (Fig. 3.4a) is a floating device that measures the density of the liquid in which it floats. The steel balls at the bottom bulb keep the device floating, with the graduated stem in the vertical position. The hydrometer reading R_h corresponds to the mass of soil grains in the suspension, which is also related to the specific gravity of the suspension (G_{susp} in Fig. 3.4a). The fines passing No. 200 (0.075 mm) sieve in the sieve analysis are subjected to a hydrometer test. About 50 g of fines are mixed with water, making 1000 mL suspension in a cylinder, and allowed to settle over at least 2 days. Larger grains settle first and the density of the suspension decreases with time. The density versus time variation is translated into grain size versus % passing using the principle of Stokes' law which suggests that the terminal velocity v of a spherical grain falling through water can be related to the grain diameter D by

$$D = \sqrt{\frac{18\mu_w v}{\gamma_s - \gamma_w}} \tag{3.1}$$

where μ_w = dynamic viscosity of water, γ_s = unit weight of the soil grain, and γ_w = unit weight of water. Equation (3.1) is valid for D in the range of 0.0002–0.2 mm, which covers silts and clays. While the hydrometer floats in the suspension, it measures the density of the suspension near the center of gravity of the bulb, at depth H_R (known as *effective depth*) below the water level. At time t, all soils grains larger than a certain diameter would have fallen below the center of the bulb. Since H_R is the maximum distance through which the grains have fallen, the velocity v can be written as

$$v = \frac{H_R}{t} \tag{3.2}$$

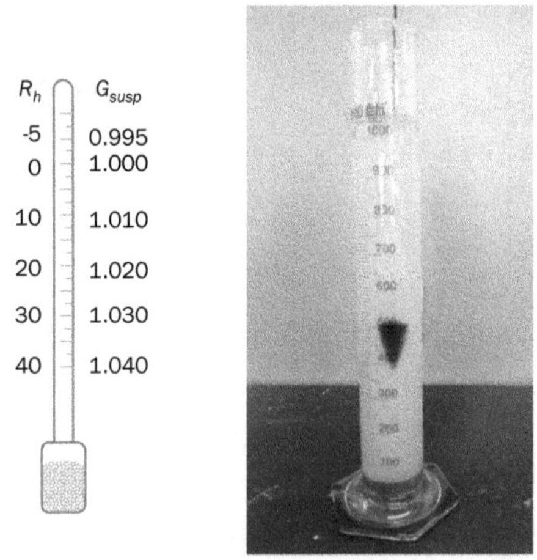

a b

Figure 3.4 Hydrometer: (a) diagram of a hydrometer showing the bulb and the graduated stem, (b) hydrometer floating in a soil-water suspension.

Substituting Eq. (3.2) in Eq. (3.1) and adjusting the units,

$$D(\text{mm}) = \sqrt{\frac{30\mu_w(\text{poise})}{981(\rho_s - \rho_w)}}\sqrt{\frac{H_R(\text{cm})}{t(\text{min})}} = K\sqrt{\frac{H_R(\text{cm})}{t(\text{min})}} \tag{3.3}$$

where

$$K = \sqrt{\frac{30\mu_w(\text{poise})}{981(\rho_s - \rho_w)}} \tag{3.4}$$

with ρ_s and ρ_w in g/cm^3. At 20°C (68°F), $\mu_w \approx 0.01$ poise (1 poise = 0.1 Pa·s). At time t, only grains smaller than D would be present in the suspension near the center of gravity of the hydrometer.

Figure 3.4b shows a hydrometer placed in a soil-water suspension just before a reading is taken. During the hydrometer test, there is a tendency for the clay particles to form flocs or clusters, which settle faster and distort the measurements. Generally, a dispersing agent such as sodium hexametaphosphate is added to disperse the clay particles and prevent them from flocculating.

There are two types of ASTM recommended hydrometers, 151H and 152H, which are used with soil suspensions. Their dimensions are the same. 151H has the stem graduated with specific gravity values of the suspension (near the centroid of the hydrometer) ranging from 0.995 to 1.038. 152H has its stem graduated with gram of soil grains per liter of suspension (near the centroid of the hydrometer), ranging from –5 to 60 g/L. Both scales are shown in Fig. 3.4a only for illustration. Note that specific gravity of 1.03 does not exactly correspond to 30 g/L of solids in the suspension—the effective depth H_R is different (see Table 3.4). The hydrometer readings and their corresponding effective depths H_R for the two types of hydrometers are given in Table 3.4.

Type 151H		Type 152H			
Reading	H_R (cm)	Reading (g/L)	H_R (cm)	Reading (g/L)	H_R (cm)
1.000	16.3	0	16.3	31	11.2
1.001	16.0	1	16.1	32	11.1
1.002	15.8	2	16.0	33	10.9
1.003	15.5	3	15.8	34	10.7
1.004	15.2	4	15.6	35	10.6
1.005	15.0	5	15.5	36	10.4
1.006	14.7	6	15.3	37	10.2
1.007	14.4	7	15.2	38	10.1
1.008	14.2	8	15.0	39	9.9
1.009	13.9	9	14.8	40	9.7
1.010	13.7	10	14.7	41	9.6
1.011	13.4	11	14.5	42	9.4
1.012	13.1	12	14.3	43	9.2
1.013	12.9	13	14.2	44	9.1
1.014	12.6	14	14.0	45	8.9
1.015	12.3	15	13.8	46	8.8
1.016	12.1	16	13.7	47	8.6
1.017	11.8	17	13.5	48	8.4
1.018	11.5	18	13.3	49	8.3
1.019	11.3	19	13.2	50	8.1
1.020	11.0	20	13.0	51	7.9
1.021	10.7	21	12.9	52	7.8
1.022	10.5	22	12.7	53	7.6
1.023	10.2	23	12.5	54	7.4
1.024	10.0	24	12.4	55	7.3
1.025	9.7	25	12.2	56	7.1
1.026	9.4	26	12.0	57	7.0
1.027	9.2	27	11.9	58	6.8
1.028	8.9	28	11.7	59	6.6
1.029	8.6	29	11.5	60	6.5
1.030	8.4	30	11.4		
1.031	8.1				
1.032	7.8				
1.033	7.6				
1.034	7.3				
1.035	7.0				
1.036	6.8				
1.037	6.5				
1.038	6.2				

Table 3.4 Hydrometer Reading and Corresponding Effective Depth H_R (Adapted from ASTM D 422–07)

Example 3.2

A soil suspension of 1000 mL is prepared with dry soil of 50 g. The hydrometer test is carried out at 20°C using a 152H hydrometer. After 30 min, the hydrometer reads 29 g/L. Find the grain diameter and percentage passing corresponding to this reading. Specific gravity of the soil grains is 2.67.

Solution From Table 3.4, the effective depth H_R at 30 min = 11.5 cm

$$K = \sqrt{\frac{30\mu_w(\text{poise})}{981(\rho_s - \rho_w)}} = \sqrt{\frac{30 \times 0.01}{981(2.67 - 1.0)}} = 0.0135$$

$$D(\text{mm}) = K\sqrt{\frac{H_R(\text{cm})}{t(\text{min})}} = 0.0135 \times \sqrt{\frac{11.5}{30}} = 0.0084$$

% finer than this diameter = (29/50)(100) = 58%

About 58% of the grains are smaller than 8.4 μm.

The grain size distribution data obtained from the hydrometer test are combined with the sieve analysis data in developing the complete grain size distribution curve, which includes the coarse and fine grains within the soil. When it is difficult to separate the fines from the coarse grains, it is recommended to carry out *wet sieving*, where the soil is washed through the sieves. Nowadays, determining the grain size distributions of fines by laser diffraction is increasingly becoming popular. It is quick and gives reasonable results.

3.3.3 Well or Poorly Graded Soils

The grain size corresponding to 10% passing is denoted by D_{10}. This means 10% of the grains are smaller than D_{10}. Similarly, it is possible to define D_{30}, D_{50}, and D_{60} from the grain size distribution curve. D_{10} is known as the *effective grain size*, which governs the flow of water through soils. D_{50} is the *median grain size*, meaning 50% of the grains are smaller than this size and 50% larger. Figure 3.3 shows these values determined for the sandy soil in Example 3.1. In the design of filters discussed in Chap. 6, D_{15} and D_{85} are also widely used.

Coefficient of uniformity C_u is defined as

$$C_u = \frac{D_{60}}{D_{10}} \tag{3.5}$$

Coefficient of curvature C_c is defined as

$$C_c = \frac{D_{30}^2}{D_{10}D_{60}} \tag{3.6}$$

When the grain size distribution stretches over a wide range of grain sizes, C_u is large and it implies that the grains are present in a wide range of sizes. Here, the large voids between the large grains are filled by smaller grains and overall there will be good packing with a dense matrix and small void ratio. A sand is said to be *well graded* if $C_u > 6$ and C_c is in the range of 1–3. Otherwise, it is *poorly graded*. A gravel is said to be well graded

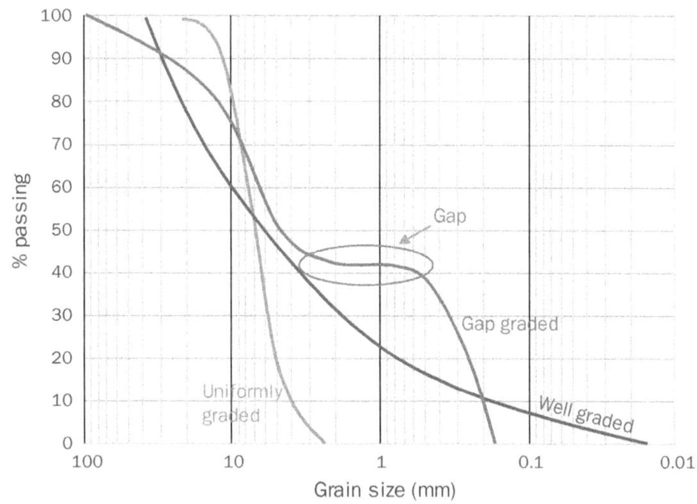

FIGURE 3.5 Well and poorly graded soils.

if $C_u > 4$ and C_c is in the range of 1–3. Otherwise, it is poorly graded. When the grains are all about the same size, it is a *uniformly graded soil*, a special case of poorly graded soil. A *gap-graded soil* is the one where there is a gap in grain sizes, showing a horizontal step in the grain size distribution curve. It is also a special case of poorly graded soils. Figure 3.5 shows three different types of grain size distribution curves. It can be seen from the figure that the gap graded soil has no grains in the size range of 0.7–2.5 mm. The uniformly graded soil has most grains in a very narrow size range of about 4–10 mm. The well-graded soil has grains in a wide range of sizes from 0.02 to 38 mm.

3.3.4 Relative Density

A coarse-grained soil can show different behavior depending on how densely the grains are packed. When the same soil grains are packed in their loosest possible state (ADTM D 4254-16), achieved by raining the grains from a small height, the void ratio is the maximum possible, known as the *maximum void ratio* (e_{max}) and the dry density is the minimum possible, known as the *minimum dry density* ($\rho_{d,min}$). When the same soil grains are packed in their densest possible state (ASTM D 4253-16), achieved through vibratory compaction, the void ratio is the minimum possible, known as the *minimum void ratio* (e_{min}) and the dry density is the maximum possible, known as the *maximum dry density* ($\rho_{d,max}$). Between these two extreme conditions, the granular soil can exist at any void ratio (e) and dry density (ρ_d), where the *relative density* D_r is defined as

$$D_r(\%) = \frac{e_{\max} - e}{e_{\max} - e_{\min}} \times 100 \tag{3.7}$$

From Eq. (2.14), in terms of dry densities, D_r can be written as

$$D_r(\%) = \frac{\rho_d - \rho_{d,\min}}{\rho_{d,\max} - \rho_{d,\min}} \times \frac{\rho_{d,\max}}{\rho_d} \times 100 \tag{3.8}$$

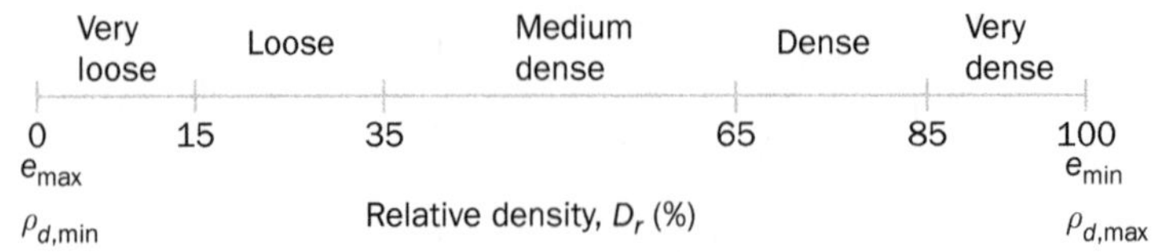

FIGURE 3.6 Coarse-grained soil descriptions based on D_r.

Soil	e_{max}	e_{min}	Reference
Uniform subangular sand	0.85	0.50	Sowers and Sowers 1961
Well-graded subangular sand	0.70	0.35	
Very well-graded silty sandy gravel	0.65	0.25	
Micaceous sand and silt	1.25	0.80	
Well-graded fine to coarse sand	0.70	0.35	McCarthy 1977
Uniform fine to medium sand	0.85	0.50	
Silty sand and gravel	0.80	0.25	
Uniform spheres	0.92	0.35	Lambe and Whitman 1979
Standard Ottawa sand	0.80	0.50	
Clean uniform sand	1.00	0.40	
Uniform inorganic silt	1.10	0.40	
Silty sand	0.90	0.30	
Fine to coarse sand	0.95	0.20	
Micaceous sand	1.20	0.40	
Silty sand and gravel	0.85	0.14	

TABLE 3.5 Typical Values for e_{max} and e_{min}

The relative density is also called the *density index* I_D. It is determined only when the fine content is less than 15%. It varies between 0% and 100%; 0% for the loosest state, and 100% for the densest state. Figure 3.6 shows the terms used to describe a coarse-grained soil based on relative density. Some typical values of e_{max} and e_{min} for granular soils are summarized in Table 3.5.

The test methods for determining the maximum and minimum densities vary between countries and the standards followed. In determining the maximum densities, the energy is applied in the form of vibration or tamping which may crush the sand grains and alter the grain size distribution curve. Lunne et al. (2019) tested six different sands by seven different methods and noted significant variations in the results.

The void ratio of a coarse-grained soil in relation to the maximum and minimum void ratios defines how densely the grains are packed. For example, two granular soils A ($e_{min} = 0.5$ and $e_{max} = 0.9$) and B ($e_{min} = 0.7$ and $e_{max} = 1.1$) at the same void ratio of 0.80 can be classified very differently. Here A is a loose sand with relative density of 25% and B is a dense sand with relative density of 75%.

Example 3.3

The maximum and minimum possible void ratios of a sand determined in the laboratory are 0.89 and 0.42, respectively. Determine the moist unit weight of this sand when compacted to relative density of 70% and moisture content of 12%, assuming $G_s = 2.65$. Also, determine the maximum and minimum dry unit weights of this sand.

Solution

$e_{max} = 0.89$, $e_{min} = 0.42$

$$D_r(\%) = \frac{e_{max} - e}{e_{max} - e_{min}} \times 100$$

$$70(\%) = \frac{0.89 - e}{0.89 - 0.42} \times 100$$

Therefore, at $D_r = 70\%$, $e = 0.56$

$$w = \frac{Se}{G_s}$$

$$S = \frac{wG_s}{e} = \frac{0.12 \times 2.65}{0.56} = 0.568$$

$$\gamma_m = \left(\frac{G_s + Se}{1+e}\right)\gamma_w = \left(\frac{2.65 + 0.568 \times 0.56}{1 + 0.56}\right) \times 9.81 = \mathbf{18.66\ kN/m^3}$$

$$\gamma_d = \frac{G_{sw}}{1+e}$$

At loosest state, with $e_{max} = 0.89$,

$$\gamma_{d,\min} = \frac{2.65 \times 9.81}{1 + 0.89} = \mathbf{13.75\ kN/m^3}$$

At densest state, with $e_{min} = 0.42$,

$$\gamma_{d,\max} = \frac{2.65 \times 9.81}{1 + 0.42} = \mathbf{18.31\ kN/m^3}$$

For spherical grains of the same size, the loosest possible packing when the grains are stacked in a rectangular pattern horizontally, with layers stacked directly above. This gives the maximum possible void ratio as 0.91 (see Example 2.5). The densest packing gives the minimum possible void ratio of 0.35.

From an extensive database of more than 300 natural sandy soils including clean sands, and sands with fines, Cubrinovski and Ishihara (2002) suggested the following empirical correlations with high coefficients of correlation.

For clean sands with less than 5% fines:

$$e_{max} = 1.53e_{min} + 0.072 \tag{3.9}$$

For sands with 5–15% fines:

$$e_{\max} = 1.37e_{\min} + 0.25 \tag{3.10}$$

For sands with 15–30% fines which includes 5–20% clays:

$$e_{\max} = 1.21e_{\min} + 0.44 \tag{3.11}$$

For silty soil containing 30–70% fines which includes 5–20% clays:

$$e_{\max} = 1.32e_{\min} + 0.44 \tag{3.12}$$

These equations can be used to determine e_{max} from e_{min} or vice versa. Cubrinovski and Ishihara (2002) also noted that the void ratio range $e_{max} - e_{min}$ increases with increasing fine content. They suggested that

$$e_{\max} - e_{\min} = 0.23 + \frac{0.06}{D_{50}(\text{mm})} \tag{3.13}$$

which is valid for gravels as well. Here, D_{50} is the median grain size.

When there is a very small percentage of nonplastic fines present in a sand, the entire fines occupy the voids within the sand matrix. Increase in the percentage of nonplastic fines such as silts, changes the classification from sands to silty sands or sandy silts. Sands are less compressible than silts and are less susceptible to liquefaction. At silt content higher than about 21–27%, silt-like behavior can be expected from a soil that is predominantly sand (Polito and Sibley 2020).

3.4 Atterberg Limits

Atterberg limits were originally developed by a Swedish soil scientist Albert Atterberg, working in the ceramics industry in the early 1900s. Casagrande (1932) modified them to suit geotechnical applications. When the moisture content of a fine grained soil is increased from zero, its consistency ("how it feels between your fingers") changes gradually, going through different states—solid, semisolid, plastic, and liquid. The three borderline moisture contents separating the four states are known as the Atterberg limits (Fig. 3.7). They are, *shrinkage limit* (*SL* or w_S), *plastic limit* (*PL* or w_P), and *liquid limit* (*LL* or w_L).

Shrinkage limit (*SL*) is the highest moisture content below which there will be no reduction in volume due to loss of moisture. It is also the lowest moisture content a saturated clay can have. Above *SL*, soil remains saturated while the moisture content decreases; the reduction in volume is due to the reduction in voids. Below *SL*, there is no reduction in voids when moisture is lost; there is only reduction in water within the voids, and the degree of saturation decreases.

When water is removed from a saturated soil slowly from its liquid limit or nearby moisture content (from Point *I* in Fig. 3.8), because of the loss of moisture there is reduction in volume and moisture content. The moisture content reaches the plastic limit and gets even lower. Until it reaches the shrinkage limit, the soil is still saturated and the volume continues to decrease due to the loss of moisture and reduction in voids. When the moisture content is reduced below the shrinkage limit, there is no further reduction in the void volume and the soil becomes unsaturated, with no further reduction in void ratio below the shrinkage limit.

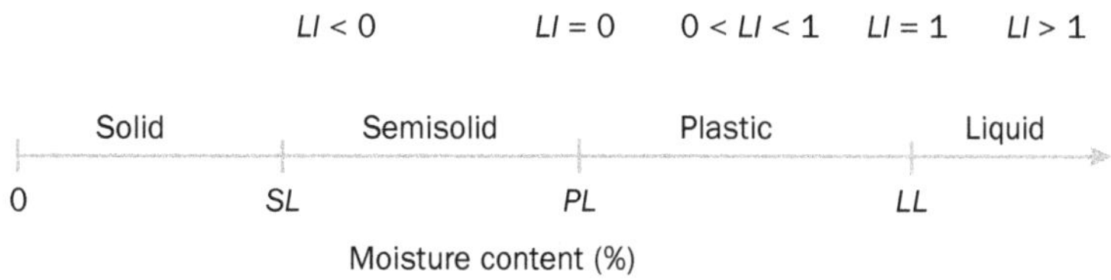

FIGURE 3.7 Atterberg limits—the border-line moisture contents.

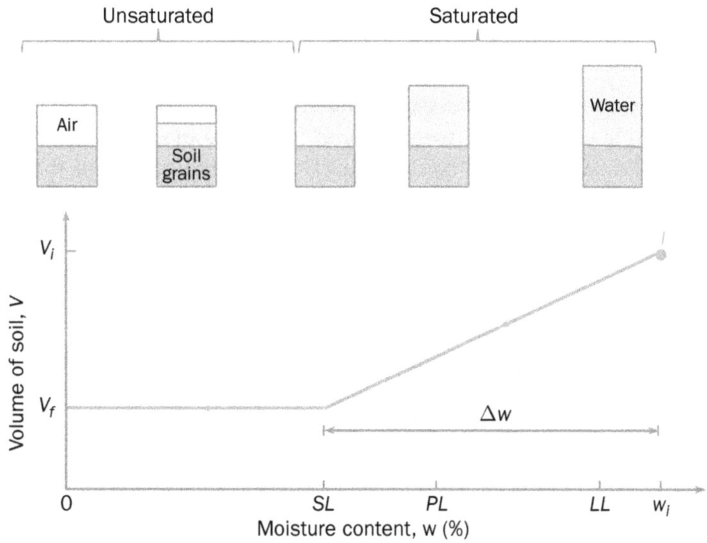

FIGURE 3.8 Definition of shrinkage limit.

From the initial moisture content of a sample w_i (%), it requires a reduction of Δw to reach the shrinkage limit. The shrinkage limit test is all about determining these two values and hence the shrinkage limit.

Shrinkage limit is the limiting water content between solid and semisolid states, where the soil volume is the minimum and it is saturated. This minimum void ratio occurring at the shrinkage limit, assuming $S = 100\%$, is given by

$$e_{SL} = SL \times G_s \tag{3.14}$$

Example 3.4

A clay sample has dry density of 1.85 g/cm³ and $G_s = 2.71$. Estimate the shrinkage limit of the clay. State your assumptions.

Solution

$$\rho_d = \frac{G_s \rho_w}{1 + e}$$

Therefore,

$$e = \frac{G_s \rho_w}{\rho_d} - 1 = \frac{2.71 \times 1.0}{1.85} - 1 = 0.460$$

Assuming that the soil has dried to the void ratio at shrinkage limit, with no further reduction in void ratio during drying, $e_{SL} = 0.460$.

Assuming the clay to be saturated at shrinkage limit,

$$w = \frac{Se}{G_s} = \frac{0.460}{2.71} = \mathbf{0.17 \text{ or } 17\%}$$

At moisture content greater than the liquid limit (*LL*), the clay behaves like a liquid with negligible strength. Between *PL* and *LL*, in the plastic range, the soil behaves like a plastic solid. At moisture content less than *PL*, the soil is no more plastic. It is in a semisolid state.

The range of moisture content, over which the fine-grained soil remains plastic, is called *plasticity index* (*PI* or I_p), and is defined as

$$PI = LL - PL \tag{3.15}$$

Plasticity index typically lies in the range of 0–50, with larger values for some special clays such as *montmorillonites* (see Sec. 3.8). Burmister (1949) classified fine-grained soils based on plasticity index as shown in Table 3.6. Pure silts are nonplastic, with $PI \approx 0$.

The in-place *natural moisture content* (w_n) of a fine-grained soil can be compared to the moisture content scale shown in Fig. 3.7, to know how the soil will behave. This is quantified through *liquidity index* (*LI* or I_L) defined as

$$LI = \frac{w_n - PL}{LL - PL} \tag{3.16}$$

A fine-grained soil with $LI > 1$ has moisture content larger than the liquid limit and flows like a liquid. The position of current moisture content in relation to the liquid and plastic limits better defines the state of the fine-grained soil than simply the moisture content alone. For example, clay *A* at moisture content of 40% ($PL = 35$ and $LL = 80$) will

PI	Classification
0	Nonplastic
1–5	Slightly plastic
5–10	Low plastic
10–20	Medium plastic
20–40	High plastic
>40	Very high plastic

Table 3.6 Classification of Clays Based on *PI*

be stiffer and display higher strength than clay B at the same moisture content of 40% ($PL = 17$ and $LL = 42$). Here, A is very close to the plastic limit and B is very close to the liquid limit and would be at the verge of flowing like a liquid. Liquidity index correlates well with strength, compressibility, and sensitivity (discussed in Chap. 9) of clays.

LL, *PL*, and *PI* are widely used in classifying fine-grained soils and in estimating soil parameters using empirical correlations. *SL* is rarely used in common geotechnical engineering problems. The laboratory tests for *LL* and *PL*, are generally carried out on the soil fraction passing No. 40 sieve (0.425 mm), which contains some fine sands as well. This is due to the difficulty in sieving out sufficient fines (less than 0.075 mm) from the soils. ASTM D 4318-17 suggests that Atterberg limits and plasticity index are expressed as integers without the percentage sign. Liquid limits can range from 0 to 1000, with the high values coming from montmorillonitic (see Sec. 3.8) clays. However, for most soils, liquid limit is less than 100.

3.4.1 Liquid Limit Determination

The two different ways of determining the liquid limit are Casagrande's percussion cup method (Fig. 3.9a) used in North America, and the Swedish fall cone method (Fig. 3.9b) used in Europe.

In Casagrande's percussion cup method (ASTM 4318-17), a moist soil pat is placed into the percussion cup and a standard groove is cut using a grooving tool (see Fig. 3.9a). By rotating the handle, the cup is raised and dropped by 10 mm repeatedly, hitting a hard rubber or micarta base. The number of blows required to close the groove over 12.5 mm (½ in.) is recorded for mixes at different moisture contents. By plotting the moisture content against the number of blows, and drawing the *line of best fit*, the moisture content corresponding to 25 blows is determined as the liquid limit. ASTM D 4318-17 suggests plotting the number of blows as abscissa on a logarithmic scale and the moisture contents as ordinates on an arithmetic scale.

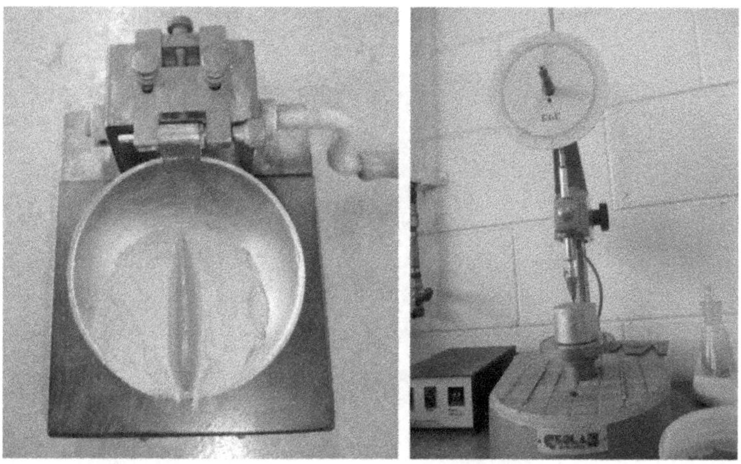

a b

Figure 3.9 Liquid limit test apparatus: (a) Casagrande's percussion cup, (b) Swedish fall cone.

In the Swedish fall cone test, an 80 g cone with 30° apex angle is allowed to fall freely for 5 s from the surface into the soil paste contained in a cup (see Fig. 3.9b), and the penetration is recorded for different moisture contents. By plotting the moisture content against penetration, and drawing the *line of best fit*, the moisture content corresponding to 20 mm penetration is determined as the liquid limit. Here, the penetrations are plotted on abscissa on a logarithmic scale, and the moisture contents are plotted as ordinates using arithmetic scale.

3.4.2 Plastic Limit Determination

Plastic limit is defined as the lowest moisture content at which the fine-grained soil can be rolled into a 3-mm-diameter (⅛-in.) thread (Fig. 3.10). A moist soil mix in the form of an ellipsoid is rolled on a glass sheet until it crumbles and the diameter of the crumbled thread is measured. If the diameter is less than 3 mm, the mix is too wet. If the diameter is larger than 3 mm, it is too dry. When the diameter is 3 mm, the crumbled threads are collected and the moisture content (i.e., plastic limit) determined. A 3-mm-diameter aluminum rod (Fig. 3.10) is used to compare the diameter of the crumbled threads.

3.4.3 Shrinkage Limit Estimation

Shrinkage limit is rarely determined in soil laboratories. Unlike *LL* and *PL*, it is not an independent parameter—it is related to *LL* and *PL*. It can be determined by the water submersion method (ASTM D 4943-18). Using the procedure suggested by Professor Casagrande at Harvard University, shrinkage limit can be estimated as

$$SL = 46.4\left(\frac{LL + 43.5}{PI + 46.4}\right) - 43.5 \tag{3.17}$$

Here, $LL = -43.5$ and $PI = -46.4$ define the point of intersection of A-line and U-line in Casagrande's *PI-LL* chart (Fig. 3.11) discussed in Sec. 3.5. In the *PI-LL* chart, the line joining the point defining the clay and the above intersection point intersects the *LL* axis at the shrinkage limit.

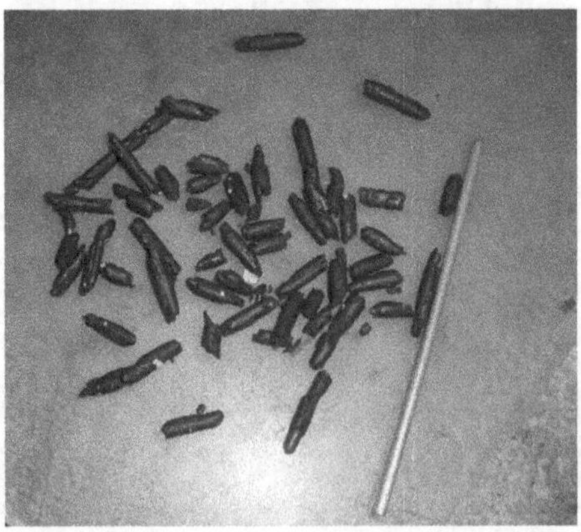

FIGURE 3.10 The crumbled threads in a plastic limit test and the 3 mm diameter aluminum rod for comparison.

Linear shrinkage is a measure of one-dimensional shrinkage, which is easy to measure. Soil paste at the liquid limit is placed in a semicylindrical trough and dried for 24 h. The reduction in the length, expressed as a percentage, is the linear shrinkage. It is an indirect measure of plasticity, but has nothing to do with the shrinkage limit.

3.5 Unified Soil Classification System

The unified soil classification system, fondly known as USCS (ASTM D 2487-17), is the most popular soil classification system, used worldwide, including North America. The coarse-grained soils are classified mainly based on the grain size distribution. The fine-grained soils are classified based on the Atterberg limits. The rules followed in USCS are summarized below.

1. The cut-off between *coarse-grained soils* and *fine-grained soils* is 0.075 mm (75 μm).
2. Soils where more than 50% of the grains are coarse grains are classified as coarse-grained soils. Otherwise, they are fine-grained soils.
3. The major soil groups are: *Gravels* (4.75–75 mm), *Sands* (0.075–4.75 mm), and fines (<0.075 mm) which are separated into *Silts* and *Clays* depending on Atterberg limits (not grain sizes). These four major soil groups have symbols of G, S, M, and C, respectively.
4. A coarse-grained soil having more gravels than sands is classified as a gravel and vice versa.
5. The fine-grained soil is classified as clay or silt depending on whether it plots above or below the A-line in Casagrande's *PI-LL* chart shown in Fig. 3.11.

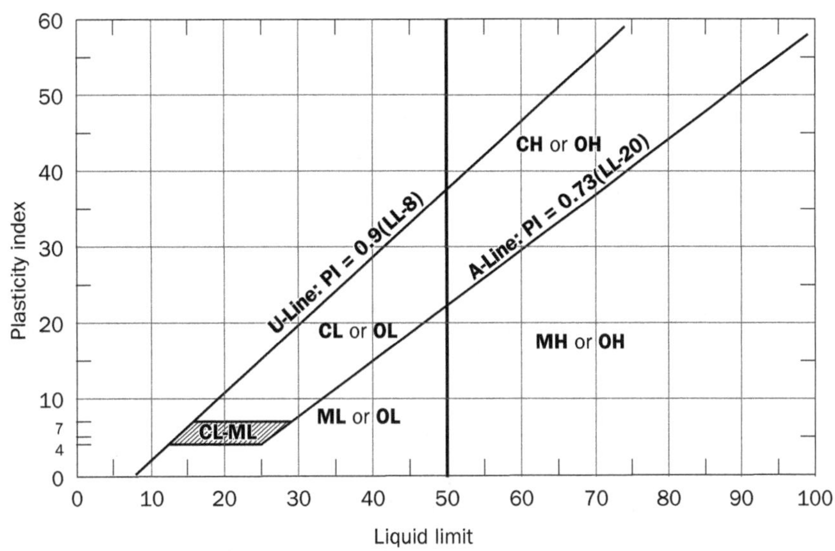

FIGURE 3.11 Casagrande's *PI-LL* chart.

6. The soils are assigned a two-letter symbol (XY), where the first letter (X) is the symbol for the major soil group and the second (Y) is the descriptor. The coarse-grained soils (G and S) are described as well graded (W), poorly graded (P), silty (M), or clayey (C), with possible symbols of GW, GP, GM, GC, SW, SP, SM, and SC. The fine-grained soils (M and C) are described as low plastic (L) or high plastic (H), depending on whether the liquid limit is less or greater than 50. They can have a symbol of ML, MH, CL, or CH. Fine-grained soils with high organic content are given the group symbol O (for organic) and described as low or high in plasticity (L or H). Table 3.7 summarizes the two-letter symbols for the different soil groups.

   When there is 5–12% fines within the soil, it is given a dual symbol in the form XY-XZ, where X is the major soil type (G or S), Y denotes the gradation (W or P), and Z defines the fines as silty or clayey (M or C).

7. In describing a gravel containing ≥15% sands, include "with sand" in the group name. In describing a sand containing ≥ 15% gravels, include "with gravel" in the group name.
8. CH and CL are described as *fat* and *lean* clays, respectively, in the group names. MH and ML are described as elastic silt and silt, respectively. For the precise group names of the different fine-grained soils, the reader is suggested to refer the flow chart in ASTM D 2487-17. The CL-ML soil (see Fig. 3.11) is called a silty clay.
9. In a fine-grained soil containing more than 30% coarse grains, depending on whether sands or gravels are predominant, add "sandy" or "gravelly" to the group name. When both sands and gravels are more than 15%, add "with sand" or "with gravel" to the group name.
10. In general, the symbols are assigned based on the percentage of fines contained within the soil.

    - <5% fines: GW, GP, SW, SP
    - 5–12% fines: GW-GM, GW-GC, GP-GM, GP-GC, SW-SM, SW-SC, SP-SM, SP-SC
    - 12–50% fines: GM, GC, SM, SC (When the fines classify as CL-ML, use dual symbol GC-GM or SC-SM, and include "silty clayey" to the group name)
    - >50% fines: ML, MH, CL, CH

Major soil group (X)	Descriptor (Y)	Group symbols (XY)
Gravel (G) Sand (S)	Well graded (W) Poorly graded (P) Silty (M) Clayey (C)	GW, GP, GM, GC SW, SP, SM, SC
Silt (M) Clay (C) Organic (O)	Low plasticity (L) High plasticity (H)	ML, MH CL, CH OL, OH

TABLE 3.7 Two-letter Symbols and the Soil Groups

11. Organic soils show a reduction in liquid limit when oven dried. If the reduction in *LL* is more than 25%, it is assigned a group symbol OH or OL, depending on whether the *LL* is greater or less than 50. It is called organic clay if it plots above the A-line and organic silt if it plots below the A-line.

Organic soils have dark grey, black, or blue color and an unpleasant odor. Peats have very high organic content, with a symbol Pt.

3.6 AASHTO Soil Classification System

AASHTO (American Association of State Highway and Transportation Officials) soil classification system (ASTM D 3282-15) is used mainly for roadwork. The size-limit separating gravels and sands is 2.0 mm (No. 10 sieve). The soils are divided into eight groups A-1 to A-8. A-1 to A-3 are coarse-grained soils (defined as soils where % fines ≤ 35), and A-4 to A-7 are fine-grained soils (defined as soils where % fines > 35), while A-8 includes highly organic soils (e.g., peats). As with other classification systems, sieve analysis, and Atterberg limits are used in assigning the above symbols. A-1 soils are well-graded gravels or sands with fines ($\leq 25\%$) of little plasticity ($PI \leq 6$), and are further subdivided into A-1-a (% finer than 2 mm ≤ 50; % finer than 0.425 mm ≤ 30; and % fines ≤ 15) and A-1-b (% finer than 0.425 mm ≤ 50; and % fines ≤ 25). A-3 soils are clean poorly graded fine sands, with less than 10% nonplastic fines and $\geq 51\%$ finer than 0.425 mm. A-2 soils are coarse-grained soil with $\leq 35\%$ fines and are differentiated on the basis of *PI* and *LL* using Fig. 3.12. Depending on the quadrant they fall into, they are assigned symbols of A-2-4, A-2-5, A-2-6, and A-2-7. When the fine fraction is greater than 35%, the soil is grouped as A-4, A-5, A-6, or A-7 on the basis of *PI* and *LL* values, as shown in Fig. 3.12. Here, the horizontal line of $PI = 10$ separates clays from silts. When the *PI* and *LL* are both high, the soil is subdivided into A-7-5 and A-7-6 by a 45° line. Casagrande's A-line and U-line are also shown in the figure for comparison.

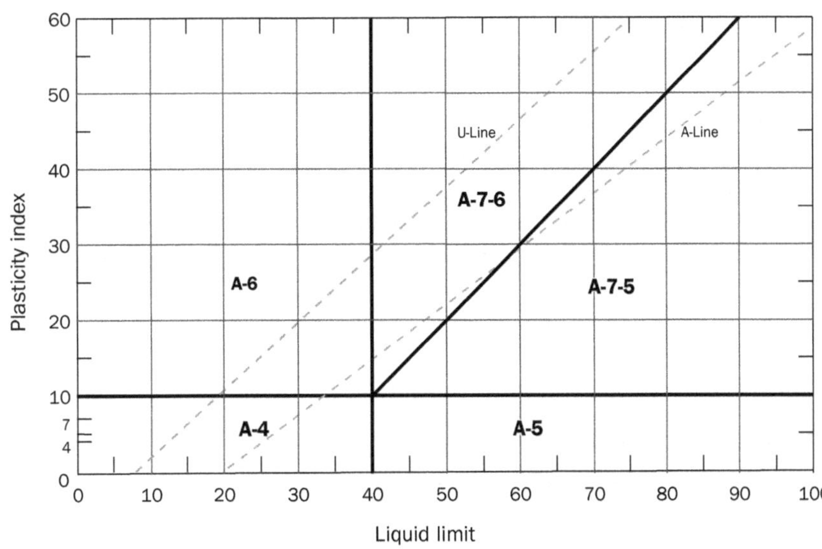

FIGURE 3.12 AASHTO classification of the fine-grained soils.

Within the groups A-1 to A-7, a *group index* (*GI*) is used to further categorize the soils. It is defined as

$$GI = (F - 35)[0.2 + 0.005(LL - 40)] + 0.01(F - 15)(PI - 10) \tag{3.18}$$

consisting of two partial group indices, one based on *LL* and the other based on *PI*. *F* is the percentage of fines. With no upper limit, *GI* is rounded off to the nearest integer, and negative values are taken as 0. For A-1-a, A-1-b, A-3, A-2-4, and A-2-5, $GI = 0$. For A-2-6 and A-2-7, *GI* is calculated from the second partial group index as

$$GI = 0.01(F - 15)(PI - 10) \tag{3.19}$$

The group classifications and the limiting values are summarized in Table 3.8. The suitability of the soil for roadwork decreases in the following order:

A-1-a A-1-b A-3 A-2-4 A-2-5 A-2-6 A-2-7 A-4 A-5 A-6 A-7-5 A-7-6

A process of elimination (see Table 3.8), coming from *left to right*, assigns the AASHTO group symbol. The first group that fits the data gives the correct classification.

Example 3.5

The following data are obtained for a soil from sieve analysis.

% passing 2.0 mm (No. 10) sieve = 48
% passing 0.425 mm (No. 40) sieve = 38
% passing 0.075 mm (No. 200) sieve = 22
Liquid limit = 54
Plastic limit = 32

Classify the soil using AASHTO.

Solution

% gravel = 52, % sands = 26, % fines = 22

$PI = 22$

With $PI = 22$, the soil cannot be A-1 or A-3. With fines less than 35%, it is an A-2.

With $LL = 54$ and $PI = 22$, falling within the top right quadrant, it is an A-2-7.

$$GI = 0.01(F - 15)(PI - 10)$$

$$GI = 0.01(22 - 15)(22 - 10) = 0.84 \approx 1$$

The AASHTO symbol is **A-2-7(1).**

General classification	Granular soils (<35% fines)							Silt-clay soils (>35% fines)			
	A-1			A-2							A-7
Group classification	A-1-a	A-1-b	A-3	A-2-4	A-2-5	A-2-6	A-2-7	A-4	A-5	A-6	A-7-5 A-7-6
% passing:											
2.00 mm (No. 10)	50 max										
0.425 mm (No. 40)	30 max	50 max	51 min								
0.075 mm (No. 200)	15 max	25 max	10 max	35 max	35 max	35 max	35 max	36 min	36 min	36 min	36 min
Liquid limit				40 max	41 min	40 max	41 min	40 max	41 min	40 max	41 min
Plasticity index	6 max		NP	10 max	10 max	11 min	11 min	10 max	10 max	11 min	11 min
Soil types	Stone fragments, gravel, and sand		Fine sand	Silty or clayey gravel and sand				Silty soils		Clayey soils	
Subgrade rating	Excellent to good							Fair to poor			

Table 3.8 Summary of AASHTO Limiting Values

Example 3.6

Classify the following soils using USCS, giving them the symbol and description.

	Soil *A*	Soil *B*	Soil *C*
% retained on 75 mm sieve	6	0	0
% passing 4.75 mm sieve	68	100	56
% passing 75 μm sieve	20	82	3
Coefficient of curvature	2.4	1.2	2.7
Coefficient of uniformity	15.2	9.3	35.8
Liquid limit	40	60	-
Plastic limit	18	28	-

Solution

Soil A:

Percentage of fines = 20, % sands = 68 – 20 = 48, % gravels = 94 – 68 = 26

There are about 6% cobbles.

The fines with $LL = 40$ and $PI = 22$ plot above the A-line, and hence are classified as clays.

The soil is a clayey sand with gravel, with the symbol SC.

Soil B:

Percentage of fines = 82, % sands = 100 – 82 = 18, % gravels = 0

The fines with $LL = 60$ and $PI = 32$ plot above the A-line, and hence are classified as clays.

The soil is a fat clay with sand, with symbol CH.

Soil C:

Percentage of fines = 3, % sands = 56 – 3 = 53, % gravels = 44

The soil is a well-graded sand with gravels, with symbol of SW.

AASHTO and USCS are two different classification systems and were developed for different purposes. Table 3.9 compares the soil classification symbols derived by the two systems, as suggested by Liu (1970).

3.7 Visual Classification and Description

When the soil samples are encountered in the laboratory or field, it is desirable to classify and describe the soil right away, by the look and feel, without waiting for the grain size distribution or Atterberg limit tests in the laboratory. This is a skill one develops with practice or experience. Coarse-grained soils are relatively easier to identify and the visual description can include the following information qualitatively.

a. Grain size distribution (well or poorly graded)

b. Grain sizes (e.g., fine, medium, or coarse sands)

AASHTO Symbol	USCS Symbol		
	Most probable	**Possible**	**Possible but improbable**
A-1-a	GW, GP	SW, SP	GM, SM
A-1-b	SW, SP, GM, SM	GP	-
A-3	SP	-	SW, GP
A-2-4	GM, SM	GC, SC	GW, GP, SW, SP
A-2-5	GM, SM	-	GW, GP, SW, SP
A-2-6	GC, SC	GM, SM	GW, GP, SW, SP
A-2-7	GM, GC, SM, SC	-	GW, GP, SW, SP
A-4	ML, OL	CL, SM, SC	GM, GC
A-5	OH, MH, ML, OL	-	SM, GM
A-6	CL	ML, OL, SC	GC, GM, SM
A-7-5	OH, MH	ML, OL, CH	GM, SM, GC, SC
A-7-6	CH, CL	ML, OL, SC	OH, MH, GC, GM, SM

TABLE 3.9 Comparison of AASHTO and USCS Symbols for the Same Soils

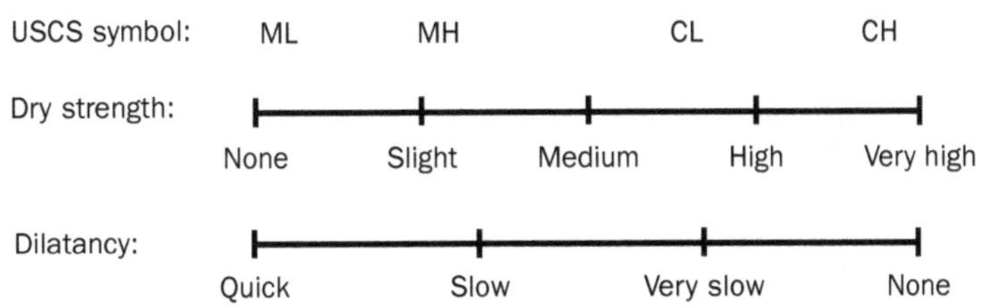

FIGURE 3.13 Dry strength and dilatancy in classifying fines.

c. Grain shape (e.g., angular, subangular, subrounded, or rounded)

d. Color

e. Presence of fines (e.g., trace, little, or significant)—Are the fines clayey or silty?

f. Homogeneity

g. State of compaction or cementation

In USCS, fine-grained soils are those where fines are more than 50%. In some countries such as the United Kingdom, Australia, and New Zealand, the soil is considered fine grained, if fines exceed 35% (similar to AASHTO).

Fine-grained soils are identified by dry strength, by crushing a dry lump of fines. Clays have high dry strength and silts have low *dry strength*. In addition, clays feel sticky between fingers and silts feel gritty. Fines can also be identified as clays or silts based on their *dilatancy*. The test for dilatancy involves placing a moist pat of fines in the palm and shaking it vigorously to see how quickly the water rises to the surface of the soil pat. Dilatancy is quick in silts and very slow in clays. Figure 3.13 shows a rough guide for visually identifying the fines on the basis of dry strength and dilatancy. Note the standard terms used to describe dry strength and dilatancy.

Example 3.7

The grain size distribution data for three soils *A*, *B*, and *C* are given in the table below. The liquid limit and plastic limit for soil C are also given. The grain size distribution curves are shown in Fig. 3.14. Classify the three soils according to USCS, giving their descriptions and symbols.

	% Passing		
Grain size (mm)	**Soil *A***	**Soil *B***	**Soil *C***
9.50	100		
4.75	92.5	100	
2.00	73.4	92.6	100
0.85	46.7	28.9	92.3
0.60	33.9	14.5	83.5
0.425	24.8	7.2	70.5
0.25	14.7	0	52.7
0.106	4.5		42.1
0.075	2.3		40.1
Liquid limit			65
Plastic limit			26

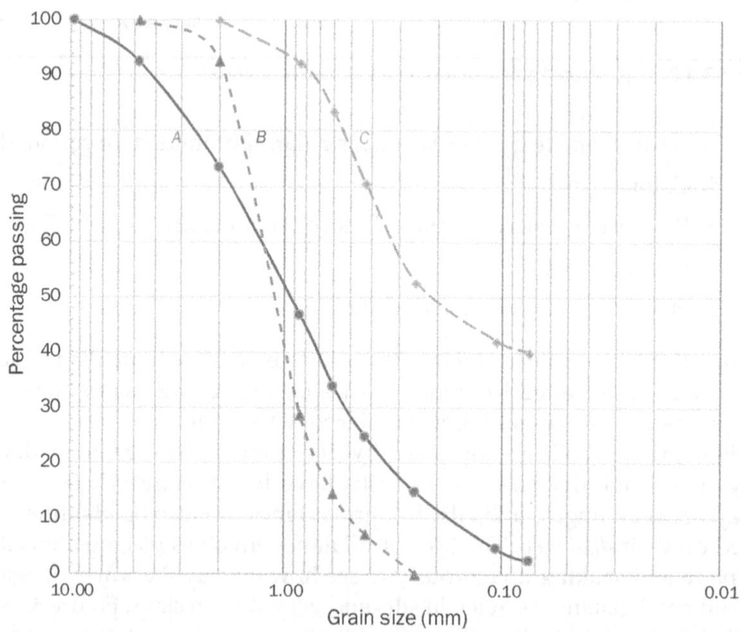

FIGURE 3.14 Grain size distribution curves for the three soils.

Solution The percentages of the fines, sands, and gravels are summarized below along with the D_{10}, D_{30}, and D_{60} values for the soils *A* and *B*, which are coarse-grained soils.

Soil	% fines	% sands	% gravels	D_{10} (mm)	D_{30} (mm)	D_{60} (mm)	C_u	C_c
A	2.3	90.2	7.5	0.17	0.52	1.25	7.35	1.27
B	0	100	0	0.50	0.87	1.20	2.40	1.26
C	40.1	59.9	0	-	-	0.32	-	-

Soil *A* is a well-graded sand (SW).

Soil *B* is a poorly graded sand (SP).

Soil *C* is a clayey sand (SC).

3.8 Clay Mineralogy

Clay minerals are phyllosilicates (in Greek, "phyllo" means layered). They are made using two distinct structural units, *tetrahedrons* and *octahedrons*, as the building blocks. The four-sided tetrahedron (Fig. 3.15a) has an oxygen atom at each of the four corners and a silicon atom at the center. The distance between the oxygen atoms is 2.6 Å (ångström) (1 Å = 10^{-10} m). Hundreds of these units joined at the base along a plane form a tetrahedral sheet, represented by a trapezoid (Fig. 3.15b). The eight-sided octahedrons consist of OH atoms at each of the six corners and an aluminum atom at the center (Fig. 3.15c). The OH atoms are 2.9 Å apart. Hundreds of these units connected together form an octahedral sheet, represented by a rectangle (Fig. 3.15d). The octahedral sheet with aluminum atoms is called *gibbsite sheet*. Sometimes, the aluminum atoms are replaced by magnesium, and the sheet is called *brucite sheet*.

3.8.1 Types of Clay Minerals

By stacking the tetrahedral and octahedral sheets in different ways, clay minerals are formed. Depending on the ratio of silica tetrahedral sheets to alumina octahedral sheets, the mineral is called 1:1 or 2:1. In 1:1 mineral tetrahedral and octahedral sheets alternate. In 2:1 minerals, an octahedral sheet is sandwiched between two tetrahedral sheets.

Alternate layers of the two sheets as in Fig. 3.16a forms a 1:1 clay mineral called *kaolinite*. The successive layers, each unit with a thickness of 7.2 Å (1 Å = 10^{-10} m), are held together by a strong hydrogen bond, which makes it difficult to separate the layers.

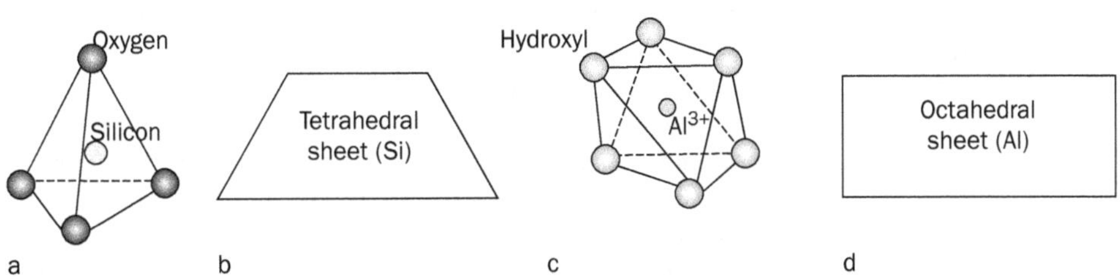

FIGURE 3.15 Atomic structural units of clay minerals: (a) tetrahedron, (b) tetrahedral sheet, (c) octahedron, (d) octahedral sheet.

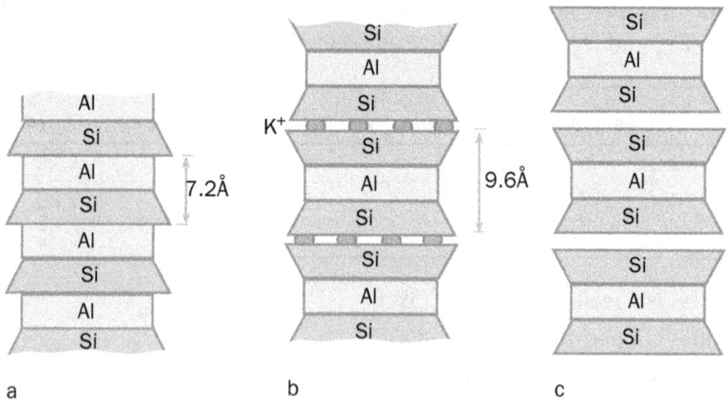

FIGURE 3.16 Three main clay minerals: (a) kaolinite, (b) illite, (c) montmorillonite.

The atomic structure of kaolinite is given by $Al_2Si_2O_5(OH)_4$. A kaolinite particle can be 0.1–4.0 μm in width and 0.05–2.0 μm in thickness. Kaolinite is used in manufacturing paper, paint, and ceramics. *Illite* (Fig. 3.16b) and *montmorillonite* (Fig. 3.16c) are made of units where an octahedral sheet is sandwiched between a tetrahedral and an inverted tetrahedral sheet. They are both 2:1 minerals. The main difference is the way the units are held together. The atomic structure of illite is such, potassium ions comfortably fit into the hexagonal holes in the tetrahedral sheets, holding the successive layers together. In montmorillonites, there are only very weak van der Waal's forces holding the units together, which can be easily separated by water entering between the layers, causing swelling. The other clay minerals include chlorite, halloysite, attapulgite, vermiculite, and more. Chlorite is classified as a 2:1:1 mineral. It has a tetrahedral, Al-octahedral, and tetrahedral sheets followed by an Mg-octahedral sheet. The basal spacing is 14 Å.

Not all clay minerals have a platey structure. Halloysite is a hydrated form of kaolinite, which occurs as tubular structure, with a 2.9 Å thick layer of water separating the 7.2 Å units, making the unit thickness 10.1 Å. Attapulgite has needle-like structure. Some amorphous clay minerals such as hematite or bauxite have granular structure.

3.8.2 Isomorphous Substitution, Specific Surface, and Cation Exchange Capacity

When high valence cations (e.g., Al^{3+}) in the atomic structure are replaced by low valence cations (e.g., Mg^{2+}) of similar size, there will be charge imbalance, resulting in net negative charge to the clay particle, mainly on the surface. For example, in the octahedral sheet Mg^{2+} ions can replace Al^{3+} ions, and in the tetrahedral sheets Al^{3+} can replace Si^{4+}. This is called *isomorphous substitution*. As a result of the net negative charge, the clay particle attains an ability to attract cations, known as *cation exchange capacity*, measured in milliequivalents (meq) per 100 g of dry clay. Kaolinites, illites, and montmorillonites have typical cation exchange capacities of about 4, 20, and 100 meq per 100 g, respectively.

Specific surface area is a term used to quantify the fineness of a clay. It is defined as the surface area per unit mass, generally given in m^2/g. Kaolinite, illite, and montmorillonite have specific surfaces of 15, 80, and 800 m^2/g, respectively. These compare with the specific surface of 2.22×10^{-3} m²/g for a 1 mm diameter sand grain. In general, the specific surface has to be more than 25 m²/g for the clay to exhibit colloidal behavior. With larger specific surface, it requires more water to wet the surfaces and hence these minerals have larger liquid limit and plastic limit. Based on data from 19 British clays, Farrar and Coleman (1967) suggested that, within ± 20%,

$$LL = 19 + 0.56A_s \tag{3.20}$$

where A_s is the specific surface in m²/g. The cation exchange capacity increases with the specific surface. Mehta and Sachan (2017) studied 10 expansive clays from Gujarat, India, and observed that the higher percentage of montmorillonite within the clay was strongly correlated to higher *LL*, *PI*, specific surface area, cation exchange capacity, and swelling.

The cations in the pore water are attracted by the negatively charged clay particles, which makes the cation concentration decrease with the distance from the clay particle. The cations (e.g., Ca^{2+}, Mg^{2+}, Na^+, K^+) attracted to the clay particle surface are weakly held and are called *exchangeable cations*. They can be easily replaced by other cations. The replacing power depends on the valence and the size of the cation. Higher valence cations easily replace lower valence cations, and larger cations can easily replace smaller cations. In the order of increasing replacement power (Holtz and Kovacs 1981),

$$Li^+ < Na^+ < H^+ < K^+ < NH_4^+ << Mg^{2+} < Ca^{2+} << Al^{3+}$$

In addition to the cations, the dipolar molecules of water are also attracted by the clay particles. There is always a "skin" of water surrounding the clay particle, approximately 1–10 nm thick, called *adsorbed water*. The adsorbed water is more viscous than the free water, and is strongly held to the clay particle. Some of the adsorbed water can only be removed by heating the clay to a temperature much greater than 100°C. The adsorbed water layer and the surrounding water with cations trying to diffuse away from the particle form the *diffuse double-layer*.

3.8.3 Identifying Clay Minerals

Clay minerals can be identified by differential thermal analysis, X-ray diffraction, and electron microscopy. Figure 3.17 shows the scanning electron micrograph of kaolinite.

Clays behave differently compared to silts or sands. Clay particles generally have plate-like structure like flakes (Fig. 3.17). A kaolinite platelet can have lateral dimension of 1000–20,000 Å with thickness of 100–1000 Å. Halloysites have a tubular structure. The electrically charged particles with larger surface areas make them plastic and cohesive. When the clay particles are oriented in the same direction, the clay fabric is called *oriented* or *dispersed*. When the clay particles are randomly oriented in all directions, the fabric is called *flocculated*.

Plasticity of a fine-grained soil comes from its clay content. For a particular fine-grained soil group, the plasticity index is proportional to the percentage of the clay fraction. Activity is defined as

$$\text{Activity}, A = \frac{\text{Plasticity index}}{\text{\% of clay fraction finer than 2 } \mu\text{m}} \tag{3.21}$$

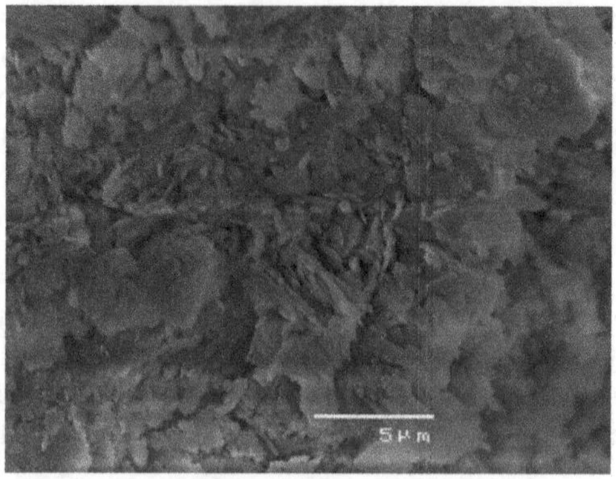

FIGURE 3.17 Scanning electron micrograph of kaolinte.

Activity depends on the type of clay mineral. It is typically in the range of 0–7, with values higher than 1.0 (e.g., montmorillonites) suggesting potential swell-shrink problems which are characteristics of expansive clays.

There is significant difference between clays and silts. Silts, sands, and gravels are equidimensional, where they have the same order of dimensions in all three directions. Clays are mostly two-dimensional, in the form of flakes, or one-dimensional, in the form of needles. Silts are nonplastic and noncohesive. Clays are plastic and cohesive. Silts, sands, and gravels are inert. Clays have net negative charge. Clays have very large specific surface compared to silts and other granular soils.

Where the *PI* and *LL* values plot on Casagrande's *PI* versus *LL* chart may give some indication about the clay mineral present. For example, montmorillonites plot very close to the U-line, and kaolinites plot just below the A-line (Holtz and Kovacs 1981). Most fine-grained soils plot near the A-line.

3.9 Main Points

1. Coarse-grained soils (gravels and sands) are larger than 0.075 mm; fine-grained soils (clays and silts) are smaller than 0.075 mm.
2. Coarse-grained soils are classified based on the grain size distribution, and fine-grained soils are classified based on Atterberg limits.
3. Grain size distribution is determined by sieve analysis for coarse-grained soils and hydrometer for fine-grained soils.
4. In coarse-grained soils, relative density quantifies how densely the grains are packed.
5. Plasticity index is a measure of the plasticity of a fine-grained soil. Pure silts are nonplastic with $PI \approx 0$.
6. Soils are classified mainly by USCS in geotechnical work. AASHTO is used only for road work.
7. Kaolinite, illite, and montmorillonite are the three major clay minerals.
8. Silts and clays have significant differences.

Review Exercises

3.1 Surf the Web and give five examples of igneous, sedimentary, and metamorphic rocks.

3.2 Surf the Web and discuss the differences between alluvial, lacustrine, and marine soils.

3.3 State whether the following are true or false.

a. Alluvial soils are transported by wind.
b. The larger the sieve number, the larger the sieve opening.
c. In a hydrometer test, the hydrometer reading increases with time.
d. Plastic limit is always less than the plasticity index.
e. D_{10} is always less than D_{15}.
f. Glacial soils are transported soils.
g. In stacking the sieves in sieve analysis, the opening sizes increase from top to bottom.
h. The ion at the center of a tetrahedron is silicon.
i. The larger the specific surface, the larger the liquid limit.
j. Liquidity index can be negative.

Answer: False, False, False, False, True, True, False, True, True, True.

3.4 A soil suspension of 1000 mL is made with 47 g of dry soil and a hydrometer test is carried out at 20°C using a Type 152H hydrometer. The specific gravity of the soil grains is 2.70. The hydrometer readings with time are given below. Plot the grain size distribution curve.

Time (min)	Hyd. Reading (g/L)
0.5	45
1	44.5
2	43
4	41.5
8	38.5
15	36
30	33
60	30
120	26.5
240	23.5
480	21
1440	17.5
1920	16.5
2880	15.5

3.5 Figure RE3.5 shows the grain size distribution curves for three soils *A*, *B*, and *C*. Answer the following without any calculations.

a. Which of the three soils has the largest fine content?
b. Which of the three soils has the largest D_{10}?
c. Which of the three is likely to be a well-graded soil?
d. Which of the three is likely to be a uniformly graded soil?
e. Which of the three soils is entirely made up of sands?

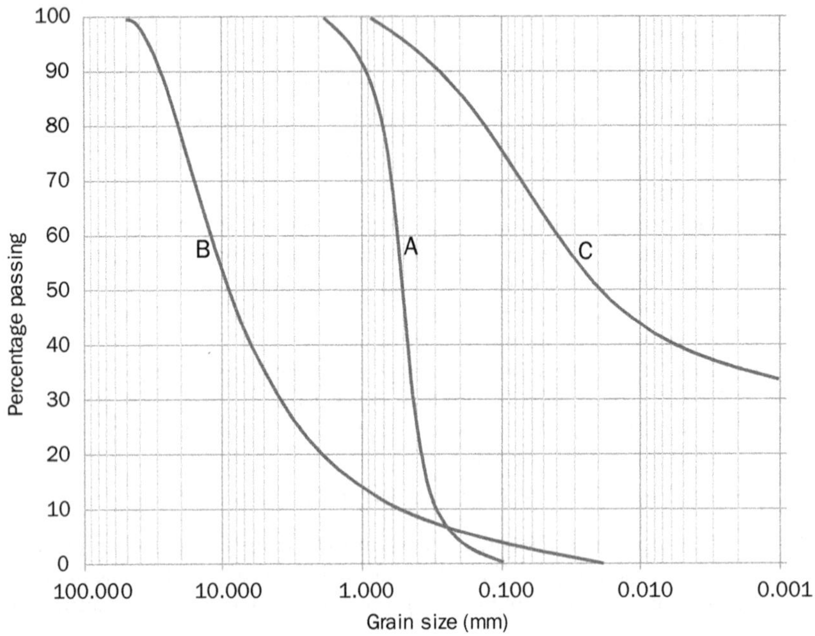

FIGURE RE3.5

Answer: Soil C, Soil B, Soil B, Soil A, Soil A.

3.6 Figure 3.5 shows the grain size distribution of a well-graded soil, uniformly graded soil, and a gap-graded soil. Find the percentages of gravels, sands, and fines in the three soils and check whether they are well graded or poorly graded. Use USCS.

Answer: 53.5%, 40.0%, 6.5%, well graded; 84.0%, 16.0%, 0, poorly graded; 50.0%, 50.0%, 0, poorly graded.

3.7 A sandy soil has $G_s = 2.65$, $e_{max} = 0.89$, and $e_{min} = 0.64$. If the sand is placed at dry unit weight of 15.3 kN/m^3, find the relative density and say how you would describe the density (see Fig. 3.6).

Answer: 76%, dense sand.

3.8 Classify the following soils, giving their symbols and descriptions.

a. % passing No. 4 sieve (4.75 mm) = 84, % passing No. 200 sieve (0.075 mm) = 25; $LL = 45$, $PL = 23$

b. % passing No. 4 sieve (4.75 mm) = 90, % passing No. 200 sieve (0.075 mm) = 72; $LL = 65$, $PL = 36$

c. % passing No. 4 sieve (4.75 mm) = 100, % passing No. 200 sieve (0.075 mm) = 4; C_u = 65.3, $C_c = 2.1$

d. % passing No. 4 sieve (4.75 mm) = 98, % passing No. 200 sieve (0.075 mm) = 11.5, C_u = 16.1, $C_c = 3.6$; The fines has very low dry strength.

Answer: clayey sand with gravel (SC), elastic silt with sand (MH), well-graded sand (SW), poorly graded sand with silt (SP-SM)

3.9 Grain size distribution test data for four soils are given below. Plot their grain size distribution curves.

a. Determine the percentages of gravels, sands, and fines.
b. For soils *A*, *B*, and *D*, determine the D_{10}, D_{30}, D_{60}, C_u, and C_c.
c. Check whether the soils *A*, *B*, and *D* are well or poorly graded.

	Size (mm)	Soil *A*	Soil *B*	Soil *C*	Soil *D*
Sieve	19.00	100			99.4
	19.50	96.7			89.1
	4.75	83.1			66.2
	2.00	62.5	100		43.3
	0.85	40	97.2	100	28.4
	0.425	24.1	82.5	97.1	25.3
	0.25	12.5	40	93.4	24.1
	0.15	6.5	6	87.5	21.4
	0.075	2.5	0.5	73.9	9.5
Hydrometer	0.036			58.3	
	0.015			40.9	
	0.008			32.8	
	0.003			25.5	
	0.001			19.8	

Answer: 2.5%, 80.6%, 16.9%; 0.5%, 99.5%, 0%; 73.9%, 26.1%, 0%; 9.5%, 56.7%, 33.8%; 0.21 mm, 0.54 mm, 1.8 mm, 8.6, 0.77; 0.16 mm, 0.21 mm, 0.31 mm, 1.9, 0.89; 0.078 mm, 0.96 mm, 3.8 mm, 48.7, 3.11; all three are poorly graded.

3.10 Find the specific surface of a sand with 1 mm diameter spherical grains. Assume $G_s = 2.65$.

Answer: 0.0023 m²/g

3.11 The grain size distribution curve of a coarse-grained soil is described by

$$p(\%) = \sqrt{\frac{D}{D_{max}}} \times 100$$

where p = percentage passing, D = grain size, and $D_{\max}$ = size of the largest grain.

(a) Check whether the soil is well graded or poorly graded.
(b) If $D_{\max} = 40$ mm, give the USCS symbol and describe the soil.

Answer: Well graded; GW, Well-graded gravel with sand.

3.12 The following data are obtained for a soil from sieve analysis and Atterberg limit tests.

% passing 2.0 mm (No. 10) sieve = 74
% passing 0.425 mm (No. 40) sieve = 61
% passing 0.075 mm (No. 200) sieve = 45
Liquid limit = 53, Plastic limit = 34

Classify the soil using AASHTO.

Answer: A-7-7(5)

3.13 Extract the Atterberg limits (LL, PL, and SL) data from the following:

- Lambe, T.W., and R.V. Whitman. *Soil Mechanics*, John Wiley & Sons, 1979.
- Sridharan, A., and K. Prakash. "Mechanisms Controlling Shrinkage Limits of Soils," *Geotechnical Testing Journal*, ASTM, 21(3):240–250, 1998.

Compute the shrinkage limit from the following equation [Eq. (3.17)], developed from Casagrande's empirical procedure, and check whether the predictions are comparable to the measured values.

$$SL = 46.4\left(\frac{LL + 43.5}{PI + 46.4}\right) - 43.5$$

3.14 A clay has liquid limit of 71 and shrinkage limit of 19. A sample has volume of 40.5 cm^3 at the liquid limit, which reduced to 20.9 cm^3 at the shrinkage limit. What is the specific gravity of the soil grains?

Answer: 2.74

3.15 Prepare a table summarizing the possible USCS symbols for all the AASHTO soil groups. This table would be the reverse of Table 3.9.

References

ASTM D 422-07 *Standard Test Method for Particle-size Analysis of Soils.*

ASTM D 2487-17 *Standard Practice for Classification of Soils for Engineering Purposed (Unified Soil Classification System).*

ASTM D 3282-15 *Standard Practice for Classification of Soils and Soil-aggregate Mixtures for Highway Construction Purposes.*

ASTM D 4253-16 *Standard Test Method for Maximum Index Density and Unit Weight of Soils Using a Vibratory Table.*

ASTM D 4254-16 *Standard Test Method for Minimum Index Density and Unit Weight of Soils and Calculation of Relative Density.*

ASTM 4318-17 *Standard Test Method for Liquid Limit, Plastic Limit and Plasticity Index of Soils.*

ASTM D 4943-18 *Standard Method for Shrinkage Factors of Soils by Wax Method.*

ASTM D 6913-17 *Standard Test Method for Particle-size Distribution (Gradation) of Soils Using Sieve Analysis.*

Burmister, D. M. "Principles and Techniques of Soil Identification," *Proceedings of Annual Highway Research Board meeting*, National Research Council, Washington, DC, 29, 1949, pp. 402–433.

Casagrande, A. "Research on Atterberg Limits of Soils," *Public Roads*, 13(8):121–136, 1932.

Cubrinovski, M., and K. Ishihara. "Maximum and Minimum Void Ratio Characteristics of Sands," *Soils and Foundations*, 42(6):65–78, 2002.

Farrar, D. M., and J. D. Coleman. "The correlation of surface area with other properties of nineteen British clay soils," *Journal of Soil Science*, 18(1):118–124, 1967.

Holtz, R. D., and W. D. Kovacs. *An Introduction to Geotechnical Engineering*, Prentice Hall, Reston, 1981.

Lambe, T. W., and R. V. Whitman (1979). *Soil Mechanics SI Version*, John Wiley & Sons, New York, p. 553.

Liu, T. K. "A Review of Engineering Soil Classification Systems," *Special Procedures for Testing Soil and Rock for Engineering Purposes*, 5th Edition, ASTM, Special Technical Publication 479, 1970, pp. 361–382.

Lunne, T., S. Knudsen, O. Blaker, T. Vestgarden, J. J. M. Powell, C. F. Wallace, L. Krogh, N.V. Thomsen, G. Yetginer, and R. K. Ghanekar. "Methods Used to Determine Maximum and Minimum Dry Unit Weights of Sand: Is There a Need for New Standard?" *Canadian Geotechnical Journal*, 56(4):536–553, 2019.

McCarthy, D. F. *Essentials of Soil Mechanics and Foundations*, Prentice Hall, Reston, 1977, p. 505.

Mehta, B., and A. Sachan. "Effect of Mineralogical Properties of Expansive Soil on Its Mechanical Behaviour," *Geotechnical and Geological Engineering*, 35:2923–2934, 2017.

Polito, C. P., and E. L. D. Sibley. "Threshold Fines Content and Behavior of Sands with Nonplastic Silts," *Canadian Geotechnical Journal*, 57(3):462–465, 2020.

Sowers, G. B., and G. F. Sowers. *Introductory Soil Mechanics and Foundations*, 2nd Edition, The Macmillan Co., New York, 1961, p. 386.

CHAPTER 4
Compaction

4.1 Introduction

In designing a steel or concrete structure, the engineer specifies a suitable grade of the material to meet the requirements. When it comes to soils, we do not have that luxury and the structure is designed to suit the soil conditions. In some circumstances, it is necessary to improve the ground conditions so that better economies can be achieved in the design of the structure. The poor soil conditions include loose granular deposits, soft clays, collapsible soils, expansive soils, organic soils, and contaminated soils. The *ground improvement* is generally well worth the expenditure and reduces the overall cost of the project.

When the soil conditions are not favorable to the proposed project, the options at hand include the following:

a. Bypass the site and relocate to a site with better soil conditions.
b. Remove the undesirable soils (if at shallow depth) and replace them with better soils.
c. Design the proposed structure to suit the poor soil conditions.
d. Transfer the loads to a more competent layer below the poor ground.
e. Abandon the project as it may not be economically viable.
f. Adopt appropriate ground improvement.

4.2 Ground Improvement Techniques

When construction is carried out in, on, or with soils, the natural soil needs to be modified by densification, reinforcement, drainage, cementation, or by drying. Some of these techniques were developed several centuries ago and have evolved with time. There are several methods available to improve the soil conditions. The appropriate method is selected based on the soil type, project constraints, and the budget. *Mechanical compaction* discussed in this chapter is one of the oldest, simplest, and economical methods widely used for ground improvement. Some of the other ground improvement methods are:

a. Soil replacement;
b. Preloading with or without vertical drains;
c. Mass stabilization using lime, cement or fly-ash;

d. Vibroflotation;
e. Stone columns;
f. Semi-rigid inclusions;
g. Electro-osmotic consolidation/dewatering;
h. Jet grouting;
i. Dynamic compaction; and
j. Deep soil mixing.

These ground improvement methods can be implemented during the early works or at the design stage, during construction (e.g., unexpected soft soils), or postconstruction (e.g., as remedial measures).

4.3 Compaction Curve

When a site is cleared prior to the construction, there is generally a need for cut and fill and earth moving work in preparing the site for the proposed structure (Fig. 4.1). During the cut and fill operation, the soil is loosened (Fig. 4.1a), and therefore requires densification through mechanical compaction (Fig. 4.1b).

When a weak soil is compacted, the void ratio is reduced and the dry density is increased. This generally makes the soil stronger, less permeable, and prone to lesser deformations while in service. When the soil is compacted, water is added to act as a lubricant between the soil grains, facilitating the compaction process. The properties of the compacted soil are mainly influenced by (a) the moisture content and (b) the effort used in compaction, known as *compaction effort* or *compaction energy*, quantified in N·m (same as Joules) per cubic meter of compacted soil.

Figure 4.2 shows the phase diagrams for five mixes (1–5) of the same soil compacted using the same energy but at different moisture contents, increasing from mix 1 to mix 5. In the phase diagrams, the volume of the soil grains is taken as the same for all five mixes, so that the relative changes in the other two phases can be compared.

a b

FIGURE 4.1 Site preparation: (a) before, (b) after.

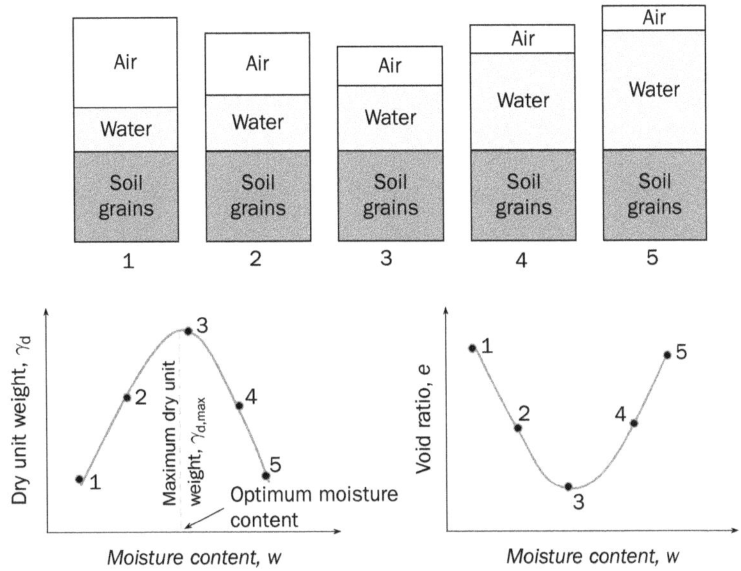

FIGURE 4.2 Effects of moisture content on dry unit weight and void ratio.

The variations of the dry unit weight (γ_d) and the void ratio (e) with the moisture content (w) are also shown in the figure. While the dry unit weight increases and decreases with the moisture content, the trend is opposite for the void ratio. The moisture content at which the dry unit weight is the maximum is known as the *optimum moisture content*. The plot of dry unit weight (or dry density) against the moisture content is the *compaction curve*.

The dry unit weight can be written as

$$\gamma_d = \frac{G_s \gamma_w}{1+e} \tag{4.1}$$

It can be deduced from Eq. (4.1) that when void ratio decreases, dry unit weight increases and vice versa. When the dry unit weight is the maximum, the void ratio is the minimum. The soils are generally compacted at moisture content close to the optimum moisture content, where they have the best engineering properties and a dense matrix. Therefore, there is good reason for us to determine the optimum moisture content by plotting the compaction curve.

Compacting a soil at moisture content less than the optimum moisture content is called "compacting dry of optimum" and compacting at moisture content greater than the optimum moisture content is known as "compacting wet of optimum."

The compaction curve, the optimum moisture content, and the *maximum dry unit weight* depend on the type of soil and the compaction energy. Figure 4.3 shows the compaction curves for the same soil under three different compaction energies E_1, E_2, and E_3 ($E_3 > E_2 > E_1$). When the compaction energy (E) is increased the compaction curve

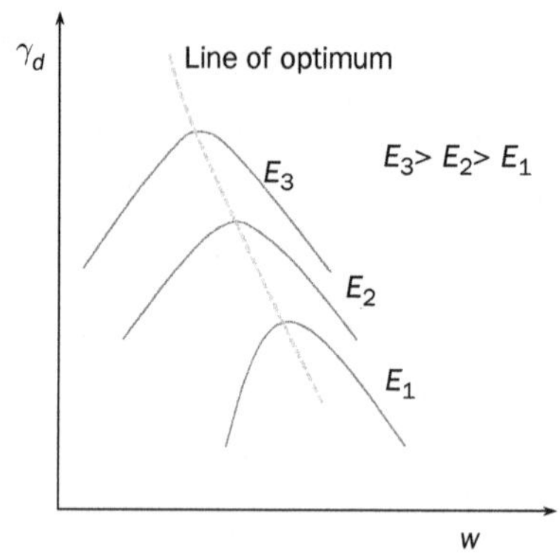

FIGURE 4.3 Effects of compaction energy on dry unit weight and optimum moisture content.

moves upwards and to the left, giving higher maximum dry unit weight and lower optimum moisture content. The line connecting the points of optimum is called the *line of optimum.*

Void ratio or porosity of the compacted soil is a good parameter to quantify the efficiency of compaction. They are not directly measured, and are calculated from the densities or unit weights. Therefore, in compaction dry unit weight (or dry density) is seen as a better parameter to measure and quantify the efficiency of compaction. Dry unit weight and void ratio are directly related by Eq. (4.1).

The bulk unit weight is given by

$$\gamma_m = \left(\frac{G_s + Se}{1+e}\right)\gamma_w \tag{4.2}$$

Dividing Eq. (4.2) by Eq. (4.1),

$$\frac{\gamma_m}{\gamma_d} = \frac{G_s + Se}{G_s} = 1 + \frac{Se}{G_s}$$

Substituting Eq. (2.12) in the above expression,

$$\frac{\gamma_m}{\gamma_d} = \frac{G_s + Se}{G_s} = 1 + \frac{Se}{G_s} = 1 + w$$

$$\gamma_d = \frac{\gamma_m}{1+w} \tag{4.3}$$

Equation (4.3) is a very useful equation in going between γ_d and γ_m, for a certain moisture content.

Example 4.1

A soil sample was compacted in a cylindrical mold of 1000 cm^3 volume at six different moisture contents, using the same compaction energy. The data are summarized below.

Moisture content (%)	Total mass (g)
15.6	1838.3
16.9	1966.2
18.2	2036.3
19.9	2038.5
21.2	1995.2
22.4	1946.4

a. Develop the compaction curve by plotting the dry density against the moisture content.

b. Find the optimum moisture content and the maximum dry density.

Solution

Part a

$V_t = 1000$ cm^3

The total mass M_t of the wet soil in the mold is given in the table for each moisture content.

The bulk density ρ_m (g/cm^3) can be computed as M_t/V_t.

The dry density ρ_d (g/cm^3) can be determined as

$$\rho_d = \frac{\rho_m}{1+w}$$

The computed values and the plot (Fig. 4.4) are shown below.

Moisture content (%)	Total mass (g)	Bulk density (g/cm^3)	Dry density (g/cm^3)
15.6	1838.3	1.838	1.590
16.9	1966.2	1.966	1.682
18.2	2036.3	2.036	1.723
19.9	2038.5	2.039	1.700
21.2	1995.2	1.995	1.646
22.4	1946.4	1.946	1.590

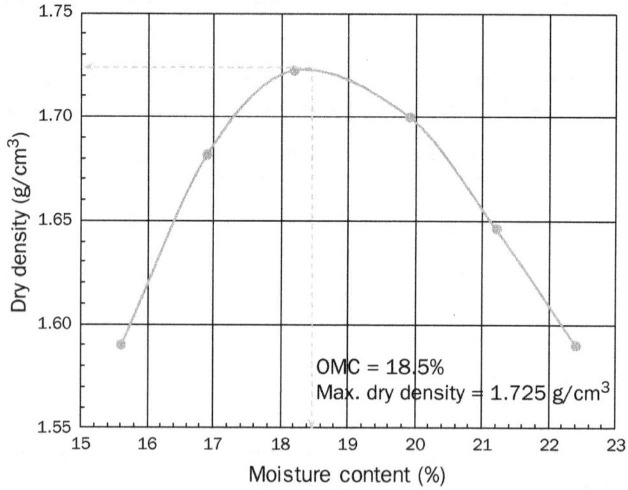

FIGURE 4.4 Compaction curve.

Part b From the figure, the optimum moisture content is 18.5% and the maximum dry density is 1.725 g/cm^3.

4.4 Laboratory Compaction

The soil is compacted in the field in layers, known as *lifts*, having *lift thicknesses* typically in the range of 150–1000 mm, with clays at the lower end of the range and granular soils at the higher end. Therefore, the laboratory compaction tests are developed to replicate the field situation by compacting the soil in layers in a metallic cylindrical mold. ASTM suggests a cylindrical mold of 943.3 cm^3 (1/30 ft^3) volume for the compaction tests on most soils. R. R. Proctor developed compaction tests in 1930s. Two different compaction energies are generally used in laboratory tests. The tests are called (a) standard Proctor and (b) modified Proctor compaction tests. A hammer, consisting of a steel mass falling over a certain height is used to give repeated blows to the soil in the mold. The test details are summarized in Table 4.1. A photograph of the compaction mold and the two hammers is shown in Fig. 4.5.

When the sample contains substantial gravels, a larger mold is used in the compaction test, with increased number of blows per layer to maintain the same energy per unit volume expected in a smaller mold. The larger mold has diameter of 152.4 mm (6.0 in), height of 116.4 mm (4.584 in), with volume of 2124 cm^3 (1/13.33 ft^3). The mold dimensions and the hammer weights vary slightly for countries using SI units. ASTM suggests removing the very coarse fraction for compaction tests as summarized in Table 4.2. The methods A and B are almost the same, except for the soil fraction used in the compaction test. When oversized fraction is in the range of 5–25%, it is removed from the sample for testing, and the moisture content and the dry density have to be adjusted as per ASTM D 4718-15. The method C requires the larger mold, using the soil fraction passing 19 mm sieve.

	Standard Proctor	Modified Proctor
ASTM Standard	D 698	D 1557
Hammer mass	2.495 kg (5.50 lbs)	4.536 kg (10.00 lbs)
Hammer drop	304.8 mm (12 in.)	457.2 mm (18.0 in.)
No. of soil layers	3	5
No. of blows per layer	25	25
Diameter of the mold	101.6 mm (4.0 in.)	101.6 mm (4.0 in.)
Height of the mold	116.4 mm (5.584 in.)	116.4 mm (5.584 in.)
Volume of the mold	943.3 cm^3 (1/30 ft^3)	943.3 cm^3 (1/30 ft^3)
Compaction energy	593 kJ/m^3	2696 kJ/m^3

TABLE 4.1 Standard and Modified Proctor Compaction Tests

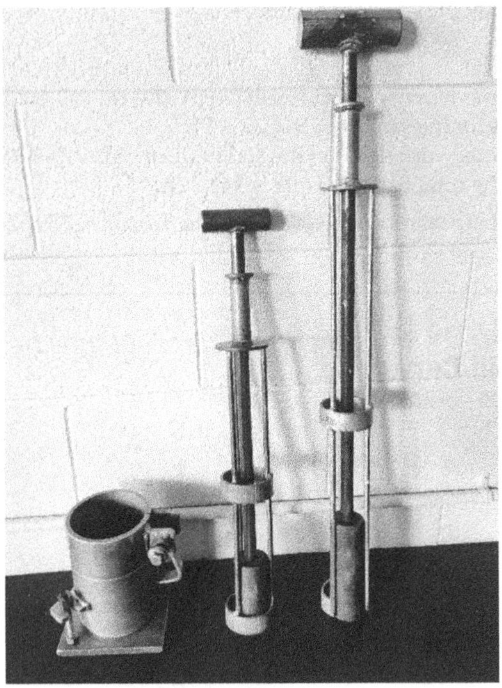

FIGURE 4.5 Compaction mold and the two hammers.

Method	Mold	No. of blows	When to use?	What fraction to use?
Method A	Standard	25	<25% retained on 4.75 mm (No. 4) sieve	Fraction passing 4.75 mm (No. 4) sieve
Method B	Standard	25	<25% retained on 9.5 mm (3/8 in) sieve	Fraction passing 9.5 mm (3/8 in) sieve
Method C	Large	56*	<30% retained on 19 mm (3/4 in) sieve	Fraction passing19 mm (3/4 in) sieve

*Ensures same compaction energy per unit volume.

TABLE 4.2 Soil Fractions and the Molds to Be Used when Gravels Are Present

Example 4.2

In a standard Proctor compaction test, the soil is compacted in a cylindrical mold of 943.3 cm^3 volume in three equal layers, with 25 blows per layer. For each blow, the 2.495 kg hammer is dropped over 304.8 mm in delivering the blows. Determine the compaction energy per unit volume in kJ/m^3.

Solution

Work done per blow = (2.495 × 9.81 N) (0.3048 m) = 7.4603 N·m
No of blows for the sample = 3 × 25 = 75
Therefore, total work done on the soil in the mold = 7.4603 × 75 = 559.52 N·m
Volume of the compacted sample = 943.3 cm^3

Therefore, the compaction energy delivered to the soil = 559.52 N·m/(943.3 × 10^{-6} m^3)
= **593 kN·m/m³ or 593 kJ/m³**

4.5 Zero Air Void Curve

$$\gamma_d = \frac{G_s \gamma_w}{1+e} \tag{4.4}$$

$$w = \frac{Se}{G_s} \tag{4.5}$$

From Eqs. (4.4) and (4.5)

$$\gamma_d = \frac{G_s \gamma_w}{1+\dfrac{wG_s}{S}} \tag{4.6}$$

From Eq. (4.6) it can be seen that in a specific soil (i.e., for a known value of G_s), every point on the γ_d-w plane will have a specific value for the degree of saturation S. Therefore, Eq. (4.6) can be used to define the curves representing different values of degree of saturation. The curve representing S = 100% is called the *zero air void curve* (see Fig. 4.6a).

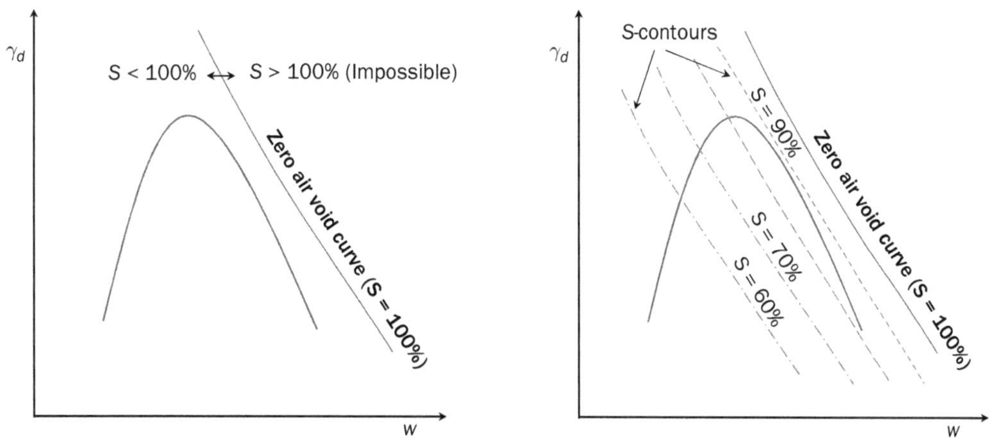

a b

FIGURE 4.6 (a) Zero air void curve, (b) degree of saturation contours.

The zero air void curve separates the γ_d-w space into two regions where $S < 100\%$ and $S > 100\%$ (which is not possible). The regions to the left and right of the zero air void curve represent the points where $S < 100\%$ and $S > 100\%$, respectively. Therefore, all compaction test data and the entire compaction curve must lie on the left of the zero air void curve, which serves as a good check. It is a good practice to always include the zero air void curve with the compaction curves. Figure 4.6b shows the degree of saturation contours along with the compaction curve. When a soil is compacted wet of optimum, the degree of saturation is very high, in the range of 85–95%.

Example 4.3

In Example 4.1, if $G_s = 2.68$.

(a) Plot the zero air void curve.

(b) Plot the $S = 80\%$ curve.

(c) At what moisture content should the soil be compacted to achieve degree of saturation of 80%?

Solution

Part a Zero air void curve ($S = 100\%$)

$$\rho_d = \frac{2.68 \times 1.0}{1 + \dfrac{2.68w}{1}}$$

Let us select some values for w and compute corresponding ρ_d from the above equation (see the following table).

Part b $S = 80\%$ curve

$$\rho_d = \frac{2.68 \times 1.0}{1 + \dfrac{2.68w}{0.8}}$$

Let us select some values for w and compute corresponding ρ_d from the above equation.

The computed values for $S = 100\%$ and 80% are summarized below. The two curves are plotted along with the compaction curve in Fig. 4.7.

	ρ_d (g/cm³)	
Moisture content (%)	**S = 100%**	**S = 80%**
16	1.876	1.745
17	1.841	1.708
18	1.808	1.672
19	1.776	1.638
20	1.745	1.605
21	1.715	1.573
22	1.686	1.543

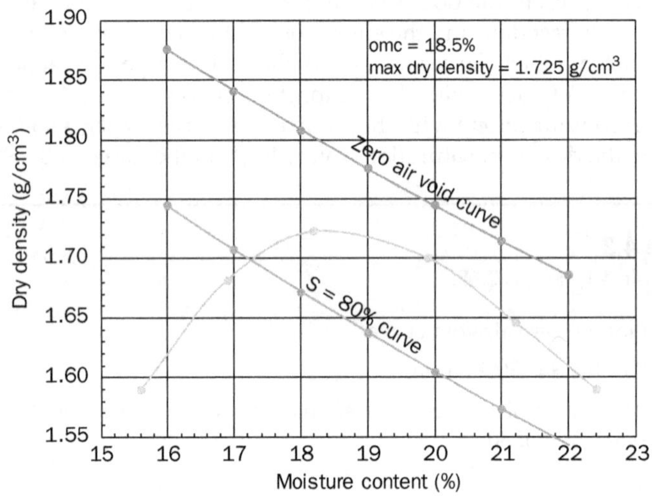

FIGURE 4.7 Compaction curve.

Part c From the plot, it can be seen that the $S = 80\%$ curve intersects the compaction curve at $w = 17.3\%$. **Therefore, the soil must be compacted at moisture content of 17.3% to achieve 80% saturation.**

4.5.1 Air Content

Air content (a) of a compacted soil is defined as

$$a = \frac{V_a}{V_t} \tag{4.7}$$

From the phase diagram for $V_s = 1$ (Fig. 2.2), it can be deduced that

$$a = \frac{V_a}{V_t} = \frac{(1 - S)e}{1 + e} \tag{4.8}$$

Air content in Eq. (4.7) or (4.8) is generally multiplied by 100, and expressed as a percentage. It is also possible to draw curves representing specific values of air content, similar to the curves for different S values. Zero air void corresponds to $a = 0\%$ curve. Note that $S = 90\%$ curve and $a = 10\%$ curve are different.

Example 4.4

In Example 4.3, when the soil is compacted to achieve 80% saturation, determine the air content.

Solution The soil is compacted at $w = 17.3\%$ and $\rho_d = 1.70 \text{ g/cm}^3$

$$\rho_d = \frac{G_s \rho_w}{1 + e}$$

Therefore,

$$e = \frac{2.68 \times 1.0}{1.70} - 1 = 0.5765$$

$$a = \frac{(1 - S)e}{(1 + e)} = \frac{(1 - 0.8)(0.5765)}{(1 + 0.5765)} = \mathbf{0.0731 \text{ or } 7.31\%}$$

4.6 Field Compaction

There are different types of compaction rollers available to suit the soil conditions, applying the compaction energy in different forms. *Smooth-wheeled rollers* (Fig. 4.8a left) apply static pressure of 300–400 kN/m^2, with 100% coverage under the wheels. The *rubber tyred* or *pneumatic rollers* have about 80% coverage but have higher contact pressure of about 700–1000 kN/m^2. They both are suitable for most soils.

Clays are better compacted with some kneading action, similar to kneading dough in the kitchen, which can be provided by *sheepsfoot rollers* (Fig. 4.8a right). The 150–250 mm long protruding feet on the surface of the drum get into the soil, compact with some kneading action and contact pressure of 1500–7000 kN/m^2, and walk out, allowing the next set of feet to do the same. *Harvard miniature compaction* device can be used

a

b

Figure 4.8 (a) A smooth wheeled roller and a sheepsfoot roller, (b) an impact roller.

to simulate the kneading action in compacting cohesive soils in the laboratory, producing small clay specimens.

Impact rollers (Fig. 4.8b) have three-, four-, or five-sided rollers with rounded corners. Every time a side comes into contact with the ground, there is an impact which provides impact energy of up to 100 kJ to the underlying ground.

Granular soils are more effectively compacted with vibratory rollers. Vibrating plates or other types of roller with vibrating loads can provide the necessary vibratory action. Frequencies in the range of 25–30 Hz are the most effective for vibratory compaction (Xanthakos et al., 1994). Most of the rollers discussed above can have vibrating capability.

In compacting small volumes of soils, especially when there is limited access for big machinery, a wacker packer or plate compactor can be used.

4.7 Compaction Specifications and Control

With the maximum dry density and minimum void ratio occurring at the optimum moisture content, it is expected that the better engineering properties are achieved at moisture contents in the vicinity of optimum. Especially in clays, the engineering behavior of the compacted earthwork is highly sensitive to the moisture content. Compacting to the dry or wet of optimum can make a big difference in the engineering behavior of the compacted clay. Therefore, it is necessary to understand the effects of moisture content on the clay behavior, and apply stringent controls in the field compaction through proper specifications.

4.7.1 Effects of Moisture Content on Clay Behavior

When a clay is compacted *dry of optimum* (i.e., at moisture content less than the optimum moisture content), the clay fabric would tend to be more flocculated than when it is compacted *wet of optimum* (i.e., at moisture content greater than the optimum moisture content). Increasing the moisture content makes the fabric more dispersed. Compacting dry of optimum gives higher strength and higher stiffness, but makes the clay more brittle and prone to cracking. Increasing the moisture content toward wet of optimum and beyond, makes the clay softer and more ductile. Permeability is the lowest near the optimum moisture content, but is larger at dry of optimum than at wet of optimum. Clays compacted dry of optimum have greater affinity to water and hence swell more. Clays compacted wet of optimum have more tendency to shrink.

Compacting dry or wet of optimum has its own advantages and disadvantages. Depending on the structure being built, it is necessary to select the moisture content prudently, considering the above. For compacting the clay core of an earth dam or a landfill liner, it is better to compact wet of optimum to minimize the seepage and any potential cracking. When compacting soils beneath a foundation, where high strength and stiffness are preferred, compacting dry of optimum is desirable. Lambe (1958a, b) discussed the effects of moisture content on the clay fabric and other engineering properties in detail.

Due to the sensitivity of the properties of the compacted clay to the moisture content, it is necessary to have stringent specification and control measures to ensure that the field compaction is carried out at the correct moisture content.

4.7.2 Method Specifications and End-Product Specifications

There are two ways of specifying the compaction requirements: (a) *method specification*, and (b) *end-product specification*. In the method specification, the engineer representing the client specifies to the contractor every detail (e.g., roller type, number of passes, moisture content, lift thickness) and takes full responsibility of the final product. The contractor is not taking any risk and is simply following the engineer's instructions. In the end-product specification, the client specifies what is required and it is up to the contractor to select the variables and take the risk and achieve the required outcome. In large projects, method specification may be economical, as the contractor will otherwise price in the risk associated with the end-product specification. For the method specifications, the engineer carries out trial test pads to decide the variables, and takes the responsibility for the end product. Figure 4.9 shows a test pad being compacted by a sheepsfoot roller for a clay liner of a landfill.

FIGURE 4.9 Compaction of a test pad.

4.7.3 Field Measurements

Since the properties of the compacted earthwork are sensitive to the moisture content at which the soil is compacted, the required field moisture content (w_{field}) is specified in a narrow range within ± 3% from the optimum moisture content. In addition, the minimum required dry density is also specified. The target moisture content (w_{field}) and dry density ($\rho_{d,field}$) in the field are specified based on the laboratory compaction test data, which define the optimum moisture content and the maximum dry density ($\rho_{d,max\text{-}lab}$) of the soil being compacted. *Relative compaction* is a term defined to compare the dry density in the field ($\rho_{d,field}$) with the maximum dry density achievable in the laboratory ($\rho_{d,max\text{-}lab}$). It is defined as

$$\textit{Relative compaction}\,(\%) = \frac{\rho_{\mathrm{d,field}}}{\rho_{\mathrm{d,max-lab}}} \times 100 \tag{4.9}$$

Relative compaction can exceed 100%, suggesting that the field dry density is larger than the maximum dry density achieved in the laboratory compaction test. When specifying relative compaction, one must clearly state whether it is with respect to standard or modified Proctor compaction energy. Typically, compaction specifications require relative compaction of 90–105% with respect to modified Proctor compaction energy. In granular soils, relative density can be specified instead of relative compaction. Typical compaction requirements for some geotechnical applications are given in Table 4.3.

The contractor carries out compaction with the objective of meeting the specifications at minimum cost. The client or client's engineer carries out *control tests* at random locations on the compacted fill, measuring the moisture content and density to ensure

Fill used for	% of $\rho_{d,\max}$ from modified Proctor	Moisture content range about optimum (%)
Roads:	90–105*	−2 to +2
Depth of 0–0.5 m	90–95*	−2 to +2
Depth > 0.5 m		
Small earth dam	90–95	−1 to +3
Large earth dam	95	−1 to +2
Railway embankment	95	−2 to +2
Foundation for structure	95	−2 to +2
Backfill behind walls/trenches	90	−2 to +2
Canal linings of clays	90	−2 to +2
Drainage blanket or filter	90	Thoroughly wet

*Depending on soil type, traffic, and function of the layer.

Table 4.3 Typical Compaction Requirements (After US Navy 1982 and Hausmann 1990)

a b

Figure 4.10 (a) Nuclear density meter in use, (b) adding water during compaction.

the compaction is carried out properly, meeting the specifications. Typically, one control test is carried out for every 500–1000 m^3.

The control tests can be carried out using a *nuclear density meter* (Fig. 4.10a), measuring both moisture content and density. Alternatively, *sand cone* (or *sand replacement*) test can be carried out where a hole is dug into the compacted ground and the moisture content and the density are determined. Here the soil removed from the hole is weighed. The volume is determined by filling the hole with *uniform sand* of known density. From the two, the density of the compacted soil can be determined. The uniform sand (e.g., Ottawa sand) ensures minimum variability in its density. Figure 4.10b shows a truck spraying water during compaction.

In a fine-grained soil compacted with standard Proctor compaction energy, the optimum moisture content is typically 2–4% less than the plastic limit. The optimum moisture content with modified Proctor compaction energy can be a further 2–5% less. The maximum dry density from a standard Proctor test can be 85–97% of that from a modified Proctor compaction test. The difference is more pronounced for clays than sands. In sands, the lowest moisture content at which a drop of water can be squeezed out is the optimum moisture content. Soils compacted wet of optimum have degree of saturation in the range of 90–95%.

The maximum dry density of a granular soil determined as per ASTM D 4253 is used in defining the relative density D_r. The maximum dry density determined from a compaction test (ASTM D 698 and ASTM D 1557) is used in defining the relative compaction. Having obtained in the maximum dry density in different ways, they are not necessarily the same in granular soils. Assuming they are the same, and using Eqs. (3.8) and (4.9), it can be shown that

$$\textit{Relative compaction}\,(\%) = \frac{R_0}{1 - D_r(1 - R_0)} \times 100 \tag{4.10}$$

where

$$R_0 = \frac{\rho_{d,\max}}{\rho_{d,\min}} \tag{4.11}$$

While the relative density can take any value in the range of 0–100%, relative compaction is always confined to a narrow range of about 80–100%, with some exceptions outside this range. In a granular soil, relative density and relative compaction can be approximately correlated as

$$\textit{Relative compaction}\,(\%) = 80 + 0.2D_r \tag{4.12}$$

Example 4.5

The modified Proctor compaction curve for the soil sample from a construction site is shown in Fig. 4.11, along with the zero air void curve. The specific gravity of the soil grains is 2.70. The compaction specifications require that the soil at the site to be compacted to at least 95% relative compaction, with the moisture content to be in the range of +1% to +3% from the optimum moisture content. A field-density test was carried out where a hole is dug in the compacted ground and the volume was determined to be 1250 cm^3, and the soil removed from the hole has a mass of 2642 g. The entire soil from the hole is placed in the ground and the dry mass was 2287 g. Does the compaction meet the specifications?

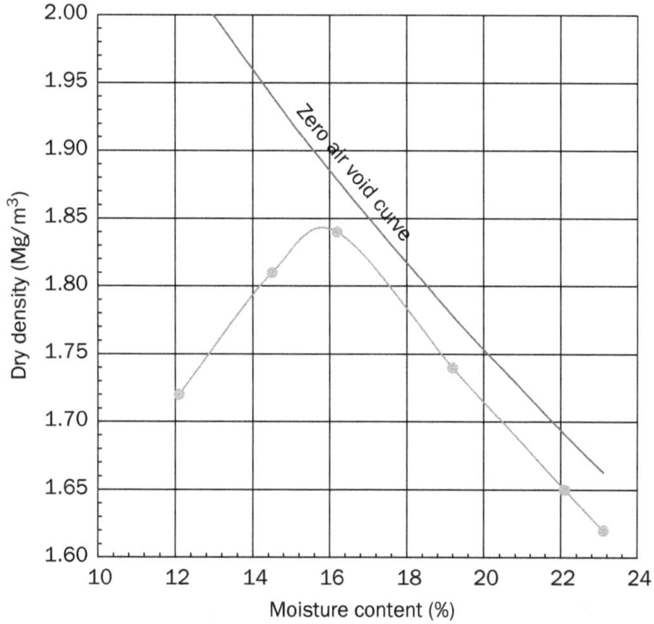

FIGURE 4.11 Compaction curve.

Solution From Fig. 4.11,

Optimum moisture content = 16.0%

Maximum dry density = 1.84 Mg/m^3

The moisture content of the compacted soil is

$$w_{\text{field}} = \frac{2642 - 2287}{2287} = 15.5\%$$

The dry density of the compacted soil is

$$\rho_{d,\text{field}} = \frac{2287}{1250} = 1.83\,\text{Mg/m}^3$$

Specifications require that

$\rho_{d,\text{field}} \geq 0.95 \times 1.84\ (= 1.75\ \text{Mg/m}^3)$

and

$17.0\%\ w_{\text{field}} \leq 19.0\%$

While the field dry density is satisfactory, the field moisture content lies outside the required range. **Therefore, the compaction does not meet the specifications.**

4.8 California Bearing Ratio

The *California Bearing Ratio* (*CBR*) test (ASTM D 1883) is an empirical test developed in 1930s in California, USA, for estimating the bearing values of materials for subgrades, subbases, and bases that were used in highways and airfields. The test is performed in the laboratory by pushing a 49.63-mm-diameter standard plunger into the soil contained in a 152-mm-diameter cylindrical mold at a fixed rate of penetration (1 mm/min), and continuously measuring the force required to maintain that rate. From the resulting load versus penetration relationship, the California Bearing Ratio (*CBR*) value can be derived. The load versus penetration curve is compared against that of a *compacted crushed Californian limestone rock*, which is defined to have *CBR* of 100. *CBR* is defined as

$$CBR = \frac{\text{Plunger load on the soil specimen}}{\text{Plunger load on compacted crushed rock}} \times 100 \tag{4.13}$$

The comparison is made of the loads at plunger penetration of 2.5 and 5.0 mm, from which the higher value is taken as the *CBR*. The plunger loads at these penetrations for the reference material, the compacted crushed rock, are 13.24 and 19.96 kN, respectively (see Table 4.4). Typical *CBR* values for different soils are given in Table 4.5.

Penetration (mm)	2	2.5	4	5	6	8	10	12
Load (kN)	11.5	13.24	17.6	19.96	22.2	26.3	30.3	33.5

Table 4.4 Penetration-load Data for the Reference Compacted Crushed Rock with *CBR* of 100

USCS symbol	Density (Mg/m³)	*CBR*
GW	2.00–2.25	40–80
GP	1.75–2.25	30–60
GM	1.85–2.30	20–60
GC	2.05–2.35	20–40
SW	1.75–2.10	20–40
SP	1.70–2.15	10–40
SM	1.60–2.15	10–40
SC	1.60–2.15	5–20
ML	1.45–2.10	<15
CL	1.45–2.10	<15
OL	1.45–1.70	<5
MH	1.30–1.70	<10
CH	1.45–1.85	<15
OH	1.30–1.75	<5

Table 4.5 Typical Values of *CBR*

Example 4.6

In a *CBR* test carried out on a soil sample, the loads corresponding to 2.5 and 5.0 mm are 2.3 and 3.8 kN, respectively. What is the *CBR*?

Solution

Load at 2.5 mm penetration = 2.3 kN
Load at 5.0 mm penetration = 3.8 kN

Therefore is the larger of $\dfrac{2.3}{13.24} \times 100 = 17.4$ and $\dfrac{3.8}{19.96} \times 100 = 19.0$

$CBR = 19$

4.8.1 Dynamic Cone Penetration Test

The lightweight dynamic cone penetrometer (ASTM D 6951) shown in Fig. 4.12a is used for profiling shallow depths, especially for roadworks. An 8.0 kg hammer falling through 575 mm (45.1 Joules/blow) or a 9.0 kg hammer falling through 510 mm (45.0 joules/blow) is used for driving a 60° cone into the ground. The penetration resistance is correlated with the California bearing ratio *CBR*. From Fig. 4.12b, for a wide range of soils, the penetration per blow is related to the *CBR* by (Livneh et al., 1994)

$$CBR = \frac{292}{DCPI^{1.12}} \tag{4.14a}$$

or

$$\log CBR = 2.46 - 1.12 \log DCPI \tag{4.14b}$$

where *DCPI* is the *dynamic cone penetration index*, which is the penetration in mm per blow.

4.9 Other Ground Improvement Techniques

4.9.1 Soil Replacement

Soil replacement is a common-sense technique, where the weaker soil is replaced by a more competent soil. There are two ways of going about this: (a) *Excavation and replacement*, and (b) *Displacement and replacement*. Excavation and replacement is suitable when the volume to be replaced is small, typically up to 4 m in depth. The advantage of this technique is that one can clearly see the soil being removed and the soil that remains at the excavation level.

For larger volumes to replace, displacing the weaker soil near the surface by more competent soil is recommended. It may not be possible to displace the entire weak soil and hence there is only partial replacement. For example, by placing a competent granular fill at the ground level, on top of a bed of soft clay, and rolling it down using heavy machinery, the soft clay can be displaced and replaced by the granular fill. The disadvantage of this method is that the extent of ground replacement is unknown.

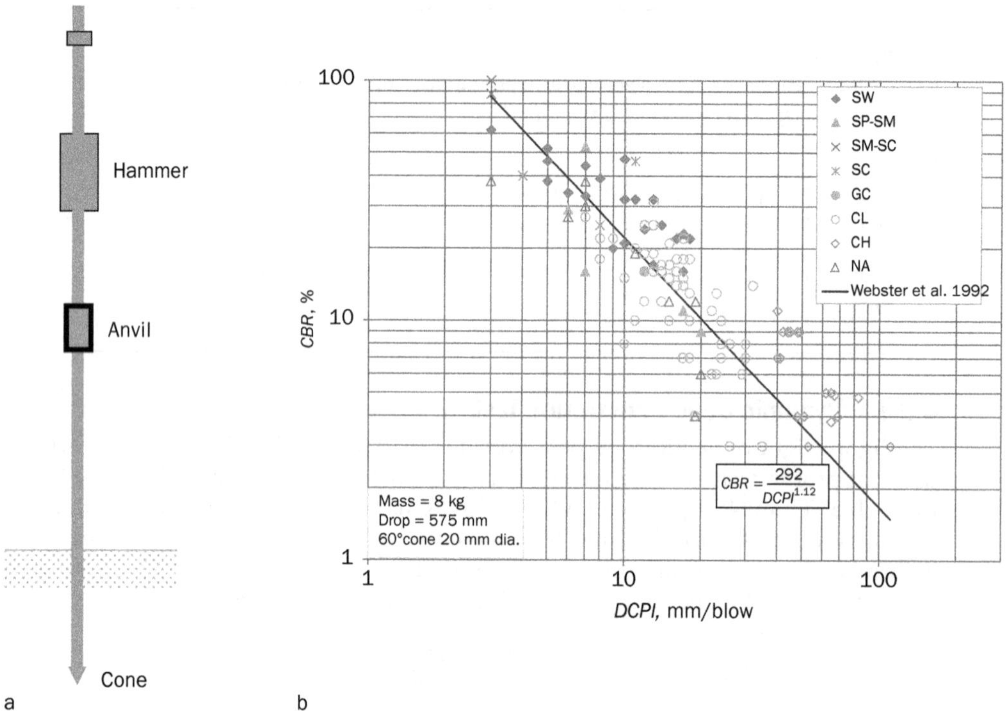

FIGURE 4.12 Lightweight DCPT: (a) schematic diagram, (b) correlation with *CBR*. (Data from Webster et al. 1992.)

4.9.2 Preloading with or without Vertical Drains

Preloading is a popular ground improvement technique carried out in clays, which otherwise take a long time to settle. To minimize the postconstruction settlement of a clay soil, it is subjected to a large and sustained load at the ground level over a certain period using a soil fill. The removal of soil makes the clay stronger than it was originally, and ensures lower postconstruction settlement. The theory behind preloading is discussed briefly in Sec. 8.7. During surcharging, prefabricated vertical drains, wick drains, or sand drains can be installed at intervals in a rectangular or triangular grid to accelerate the water flow, and make the consolidation process faster.

4.9.3 Mass Stabilization

Mass stabilization is carried out by mixing a binder such as cement, fly ash, slag, lime, or gypsum into the weak soil, up to a depth of about 5 m. The stabilization can be mixing the binder *in situ* where the soil is mixed in place or *ex situ*, which requires removal, mixing, and replacing the weak soil. *In situ* mass stabilization is carried out on very weak soils with high water content. *Ex situ* mass stabilization is recommended in situations where the removed material can be used as a good-quality engineered fill.

Mass stabilization is used in road or rail embankments, ports, land reclamation, and landfills. The mass stabilization reduces the plasticity and potential expansiveness of the clay soils, and increases the strength and stiffness of all soils. Mass stabilization can be carried out along with other ground improvement techniques such as controlled modulus columns (Sec. 4.9.6) or deep soil mixing (Sec. 4.9.10). The top few meters can be mass stabilized to provide a working platform for the machinery used for the controlled modulus columns or deep soil mixing carried out to a larger depth.

4.9.4 Vibroflotation

Vibroflotation is a technique developed in Germany in 1930s to densify thick layers of loose granular soil deposits. About 2-m-long vibrating cylinder called *vibroflot*, with water jet at the tip to facilitate the soil removal while the probe advances into the ground. The vertical cavity produced in the ground is backfilled from the ground level.

4.9.5 Stone Columns

These are columns of stones placed in a weak soil, and hence serve as reinforcement and create a stiffer composite soil mass. A vibroflot is used to produce the hole, which is backfilled with stones. In clays, the closely spaced stone columns shorten the drainage path and increase the horizontal drainage, making the clays settle fast. Stone columns are often used in combination with preloading. Stone columns cannot carry tensile stresses and have very low capacity to carry bending moments. Since they are not structural elements, they are often seen as semirigid inclusions. They are generally 800–1200 mm in diameter, 20 m to 25 m long, with load carrying capacity up to 300 kN. They are constructed in square or triangular grids at 1.5–3.5 m spacing.

4.9.6 Semirigid Inclusions

Semirigid inclusions (SRI), also known as Controlled modulus columns (CMC™) are semirigid columns, often made of concrete or other materials (e.g., stones), to improve the overall strength of a weak soil layer. The soil-SRI continuum acts as a composite system in reducing the expected settlements and lateral deformations, and is widely used to support embankments on weak ground conditions. The structure (e.g., embankment) is generally placed on a load transfer platform made of compacted granular soil (e.g., gravel) reinforced with geotextiles, which transfers the load to the soil-SRI system. The load transfer platform between the SRIs and the structure makes the load transfer mechanism quite different to a pile group or a pile raft. During the early stages, the weak soil consolidates and there is downdrag (negative skin friction) on the upper regions of the SRIs. With time, the SRIs begin to carry loads and develop upward skin friction at the lower regions along with base resistance. Finally, equilibrium is reached between negative downdrag at the upper regions of the SRIs and the upward skin friction and base resistance at the lower regions.

SRIs are generally placed in a rectangular or triangular grid pattern, with center-to-center spacing of three to four times the diameter. The concrete columns are commonly 0.3–0.6 m in diameter. Sometimes, timber columns or stone columns are also used as the support columns.

4.9.7 Electroosmotic Consolidation

Electroosmotic consolidation involves using electrodes embedded in a soft clay and applying an electric potential across them, which produces an electric current and water movement from anode to cathode. The water can be pumped out from the cathode. While there have been several bench-scale trials to support this method, this has not been used much in large scale. Some of the problems include high power requirements and deteriorating anodes during the process. Recently developed electrokinetic geosynthetics, known as EKGs, eliminate the problem with deteriorating electrodes. The method works better with high water content materials like soft clays and soil slurries.

4.9.8 Jet Grouting

In jet grouting, developed by Yahiro and Yoshida in 1973, cement grout is injected into the ground at high velocity, forming a soilcrete column. The required depth is reached by hydraulic rotary drilling, from where the grout is injected. The high-pressure jet breaks the soil structure and the disturbed soil is mixed with cement grout to form the soilcrete. The method can be applied on a wide range of soils from clays to gravels, typically to a depth of up to 30 m. It can be used under settling embankments or in underpinning building foundations.

Jet grouting is done by a single fluid system (grout only), double fluid system (grout and air), or triple fluid system (grout, air, and water). In the single fluid system, the grout is injected at high pressure to erode the soil, mix, and form the soilcrete. In the double fluid system, the grout it sheathed in a cone of air, at high velocity to break the soil, mix and form the soilcrete. In the triple fluid system, the soil is broken down by a water jet sheathed in a cone of air. The grout is injected through a separate low-pressure grout line. The double and triple fluid systems are more popular.

4.9.9 Dynamic Compaction

Dynamic compaction is used in granular soils, landfills, and in karst topography with sinkholes. A heavy weight in the order of 10–30 tonnes is raised by a height of 10–30 m above the ground and dropped repeatedly (Fig. 4.13). The impact formed by each blow densifies the ground. Leonards et al. (1980) suggested that the effective depth of ground improvement Z_e can be estimated as

$$Z_e(\text{m}) = 0.5\sqrt{W(\text{tonne})\,H(\text{m})} \tag{4.15}$$

4.9.10 Deep Soil Mixing

Deep soil mixing (DSM) or deep mixing method (DMM) is similar to jet grouting, where the wet or dry binder (e.g., cement, fly ash, lime, or bentonite) is mechanically mixed with the weak *in situ* soil. Dry method is preferred in very soft cohesive soils with high moisture contents (e.g., > 60%), and wet method is preferred in cohesive and fine granular soils with lower moisture contents (e.g., < 60%). Deep soil mixing can be carried out as isolated or overlapping columns, or for the entire area as mass mixing. Typical diameters of the columns are in the range of 0.6–1.5 m, and the depths generally up to 25 m. The unconfined compressive strength of the column can vary between 200 and 2000 kN/m^2. The technique originated in the United States in the 1950s, and was refined in Japan and Scandinavia during the past four decades. *Cutter soil mixing* (CSM) is a form of DSM in the wet method. A set of cutter wheels (Fig. 4.14) dig a narrow trench, 2–3 m wide and 0.5–1.0 m thick, where the binder is mixed.

4.9.11 Lightweight Fills

Very often, it is required to raise the ground level without imposing high loads to the underlying ground. When compacted soils are used for this purpose, with unit weights

Figure 4.13 Dynamic compaction.

Figure 4.14 Cutter wheels used in CSM (from Ameratunga et al. 2009).

of 18.0–20.0 kN/m^3, they can impose substantial load to the underlying ground. There are lower density alternatives such as pumice, shredded tire, or fly ash which have been successfully used to replace fills. Lately polystyrene blocks, known as EPS (for expanded polystyrene), are increasingly becoming popular as very lightweight fills for this purpose. They have unit weight of about 0.1–0.3 kN/m^3, relatively high strength and durability, very little material variability and are easy to use in construction.

4.9.12 Basal High Strength Geotextiles

For construction of an embankment or placement of fill on a soft ground which can experience instability and/or high settlements, basal high strength geotextiles can be used. They are high strength geotextiles or geogrids, laid at the interface of the embankment/fill and the soft ground, to function as reinforcement. They can be used along with preloading, wick drains, or stone columns. The basal high strength geotextile reinforcements enable steeper embankment slopes, reduce the need for berms, and make the footprint of the embankment smaller.

4.10 Main Points

1. Compaction densifies the soil, improving its engineering properties.
2. The void ratio is the minimum at the optimum moisture content.
3. The best soil properties are achieved near the optimum moisture content. Therefore, field compaction is carried out near optimum moisture content.
4. Increasing the compaction energy reduces the optimum moisture content, and increases the maximum dry density.
5. All compaction or control test points should lie to the left of zero air void curve.
6. Granular soils are better compacted with vibratory action; clays are better compacted through kneading.
7. Clay behavior is very sensitive to the moisture content. Therefore, it is necessary to specify the moisture content clearly, especially whether dry or wet of optimum.
8. Compaction is only one of the many ways the weak ground can be improved.

Review Exercises

4.1 State whether the following are true or false.

a. Increasing the compaction energy, increases the optimum moisture content.
b. The maximum dry density is larger for modified Proctor compaction than standard Proctor compaction.
c. The zero air void curve shifts upward with increasing G_s.
d. Sheepsfoot rollers are suitable for granular soils.
e. Relative compaction cannot be greater than 100%.

Answers: False, True, True, False, False

4.2 In a modified Proctor compaction test, the soil is compacted in a cylindrical mold of 943.3 cm^3 volume in five equal layers, with 25 blows per layer. For each blow, the 4.536 kg hammer is dropped over 457.2 mm in delivering the blow. Determine the compaction energy.

Answer: 2696 kJ/m^3

4.3 It is required to carry out a standard Proctor compaction test, with compaction energy of 593 kJ/m^3, on a gravelly soil in the large compaction mold having a volume of 2124 cm^3. The soil is to be compacted in three layers. How many blows would be required using the standard Proctor hammer?

Answer: 56 blows

4.4 A standard and modified compaction tests were carried out on the same soil using soils mixed at six different moisture content and the test data are as follows.

Moisture content (%)	Mass of the soil specimen (g)	
	Standard Proctor	Modified Proctor
14.5	1743	1835
15.9	1802	1928
17.1	1848	1956
18.3	1878	1943
19.5	1856	1923
20.7	1824	1897

The volume of the compaction mold is 943 cm^3. The specific gravity of the soil grains is 2.64. Plot the compaction curves and determine the optimum moisture contents and the maximum dry densities, for standard and modified compaction energies.

Answer: 18.0%, 1685 kg/m^3; 16.5%, 1775 kg/m^3

4.5 A standard Proctor compaction test was carried out on a soil ($G_s = 2.67$) and the compaction curve is shown in Fig. RE4.5. At what moisture content should the soil be compacted with standard Proctor energy to achieve degree of saturation of 70%?

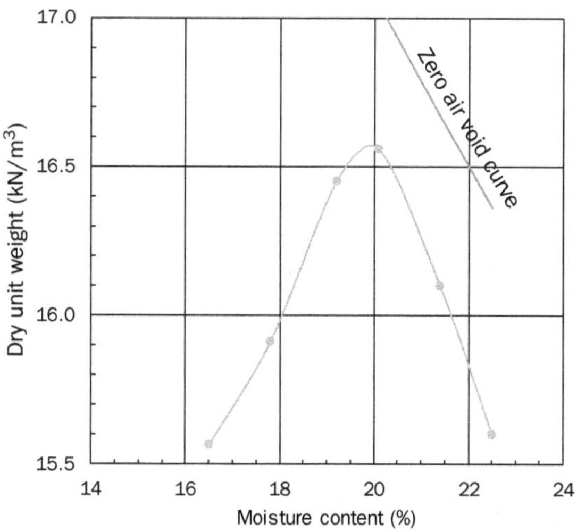

FIGURE RE4.5 Compaction curve.

Answer: 17.2%

4.6 The standard Proctor compaction curve for a soil is shown in Fig. RE4.5. At the site, the specifications require the soil to be compacted to a moisture content of 18–20%, with a relative compaction (with respect to standard Proctor) exceeding 98%. A field density test is carried out using a sand cone apparatus, where a hole is dug into the compacted soil. The mass of the soil removed is 2063 g, and the volume of the hole is determined as 1014 cm³. A small specimen of 30.1 g of this soil is dried in the oven for 24 h at 105°C and the dry mass is determined as 25.2 g. Has the contractor met the specifications?

Answer: Yes

4.7 (a) Modified Proctor compaction curve for a soil ($G_s = 2.74$) is given in Fig. RE4.7. Find the optimum moisture content, maximum dry unit weight, and the degree of saturation at the optimum.

(b) The compaction specifications require this soil be compacted at the site to

- Moisture content in the range of 17–19%, and
- Relative compaction of at least 98% (Modified Proctor)

The engineer carries out control tests at five locations, measuring moisture content and dry unit weight, and the following results are obtained. For each case, check whether the compaction meets the specifications and give reasons for rejection when they do not meet the specifications.

Control test no.	Moisture content (%)	Dry unit weight (kN/m³)
1	16.5	17.5
2	19.5	17.4
3	17.8	17.6
4	18.5	17.1
5	18.6	18.1

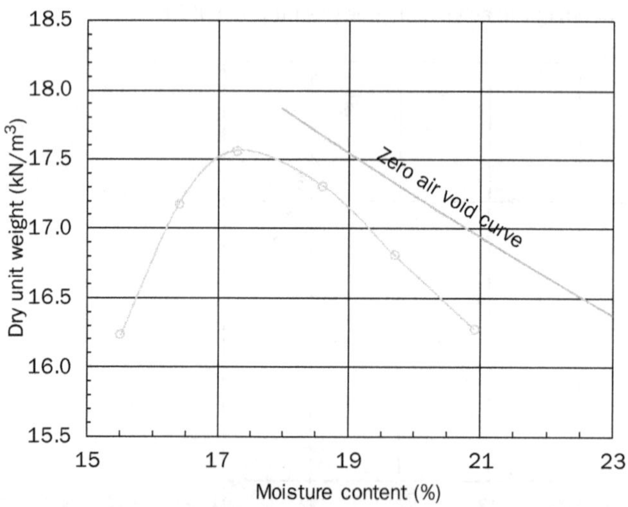

FIGURE RE4.7 Compaction curve.

Answer: 17.5%, 17.6 kN/m³, 91.0%; Reject, reject, accept, reject, invalid data

4.8 A soil at moisture content of 12% is compacted to an air content of 10%. The specific gravity of the soil grains is 2.72. Determine its saturated unit weight.

Answer: 21.3 kN/m³

References

Ameratunga, J., D. Brown, M. Ramachandran, and R. Denny. "Ground improvement for a large above ground storage tank using cutter soil mix columns." *Proceedings of the 17th International Conference on Soil Mechanics and Geotechnical Engineering*, Alexandria, 2009.

ASTM D 698-12. Standard Test Method for Laboratory Compaction Characteristics of Soil Using Standard Effort (12,400 ft-lbf/ft^3 (600 kN-m/m^3)).

ASTM D 1557-12. Standard Test Method for Laboratory Compaction Characteristics of Soil Using Modified Effort (56,000 ft-lbf/ft^3 (2700 kN-m/m^3)).

ASTM D 1883-16. Standard Test Method for California Bearing Ratio (CBR) of Laboratory Compacted Soils.

ASTM D 4718-15. Standard Practice for Correction of Unit Weight and Water Content for Soils Containing Oversize Particles.

Hausmann, M. *Engineering Principles of Ground Modification*, McGraw Hill, Singapore, 1990, p. 632.

Lambe, T. W. "The Structure of Compacted Clay," *Journal of Soil Mechanics and Foundations Division*, ASCE, 84(SM2):1654-1–1654-34, 1958a.

Lambe, T. W. "The Engineering Behaviour of Compacted Clay," *Journal of Soil Mechanics and Foundations Division*, ASCE, 84(SM2):1655-1–1655-35, 1958b.

Leonards, G. A., W. A. Cutter, and R. D. Holtz. "Dynamic Compaction in Granular Soils," *Journal of Geotechnical Engineering Division*, ASCE, 106(1):35–44, 1980.

Livneh, M., I. Ishai, and N. A. Livneh. "Effect of Vertical Confinement on Dynamic Cone Penetrometer Strength Values in Pavement and Subgrade Evaluations," *Transportation Research Record*, 1473:1–8, 1994.

Xanthakos, P. P., L. W. Abramson, and D. A. Bruce. *Ground Control and Improvement*, John Wiley & Sons, New York, 1994, p. 910.

U.S. Navy. *Soil Mechanics—Design Manual 7.1*, Department of the Navy, Naval Facilities Engineering Command, U.S. Government Printing Office, Washington, DC, 1982.

Webster, S. L., R. H. Grau, and T. P. Williams. *Description and Application of Dual Mass Dynamic Cone Penetrometer*, US Army Engineer Waterways Experiment Station, Instruction Report, No. GL-92-3, 1992.

Yahiro, T., and H. Yoshida. "Induction Grouting Method Utilizing High-speed Water Jet," *Proceedings of the Eighth International Conference on Soil Mechanics and Foundation Engineering*, Moscow, Russia, 1973, pp. 402–4.

References

CHAPTER 5

Effective Stresses

5.1 Introduction

There are stresses within the soil mass, acting in all directions. When the ground level is horizontal, with no other load acting at the ground level, it is reasonable to assume that the *vertical overburden pressure* (σ_{vo}) and the horizontal earth pressure within the soil mass are *principal stresses,* with no shear stresses acting horizontally or vertically. With construction activities at the ground level, the stresses within the soil mass are altered. To analyze the stability of the structure, it is necessary to determine the initial *in situ* overburden pressure σ_{vo} *and* the change in the vertical normal stress ($\Delta\sigma$) caused by the structure. For now, let us not worry about the changes in the vertical stresses caused by loads at the ground level, which are discussed in Chap. 7, or the horizontal stresses that are discussed later in Chap. 10. The main focus of this chapter is on computing the vertical overburden pressure σ_{vo}. The techniques to determine the vertical normal stress increases $\Delta\sigma$ under different types of surface loads are discussed in Chap. 7.

5.2 Vertical Overburden Stresses

Let us consider a homogeneous soil shown in Fig. 5.1a with bulk unit weight γ_m. At a depth z below the ground level, the in situ vertical overburden pressure is given by

$$\sigma_{vo} = \gamma_m z \tag{5.1}$$

If a uniformly distributed surcharge q kN/m^2 is placed at the ground level as shown in Fig. 5.1b, the new vertical normal stress σ_v is given by

$$\sigma_v = \gamma_m z + q \tag{5.2}$$

Figure 5.1c shows a soil profile with three soil layers of different unit weights. The vertical overburden stress at point A is given by

$$\sigma_{vo} = \gamma_{m1} h_1 + \gamma_{m2} h_2 + \gamma_{m3} h_3 \tag{5.3}$$

Groundwater table is the level to which the water will be present within an open borehole. This water table is at atmospheric pressure. Soils below the water table are saturated, where the bulk unit weight has to be replaced by saturated unit weight γ_{sat} in computing the overburden pressure.

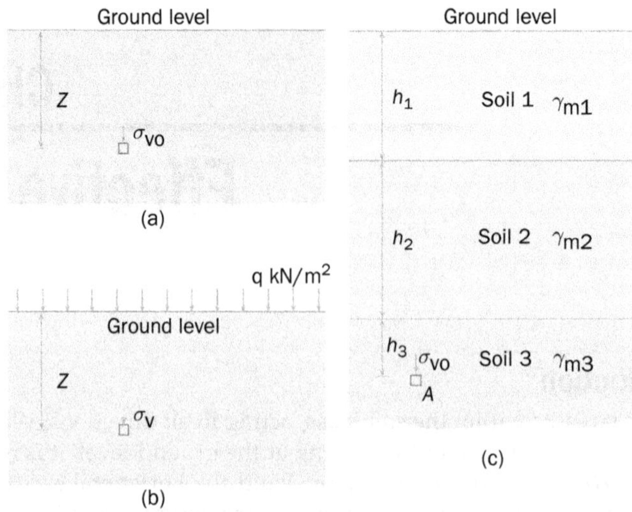

FIGURE 5.1 Overburden pressures within: (a) homogeneous soil (b) with surcharge q, (c) layered soils.

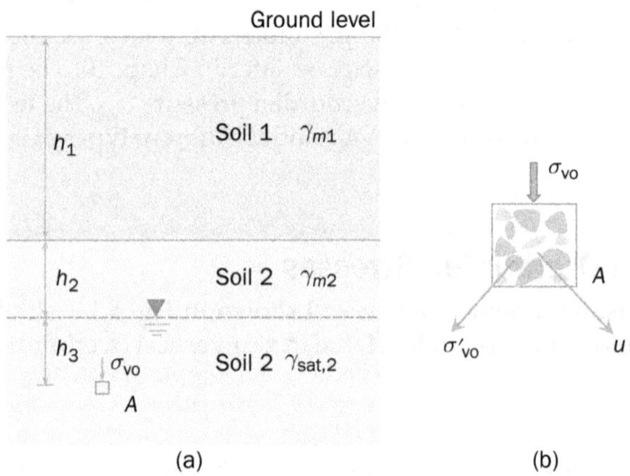

FIGURE 5.2 (a) Overburden pressure in soils below the water table, (b) stress at point A.

Figure 5.2a shows a soil profile with two different soils, with water table as shown. The vertical overburden stress at point A is given by

$$\sigma_{vo} = \gamma_{m1}h_1 + \gamma_{m2}h_2 + \gamma_{sat,2}h_3 \tag{5.4}$$

Note the use of bulk (or dry if appropriate) unit weights above the water table and saturated unit weights below the water table in computing the vertical overburden stress σ_{vo}.

5.3 Terzaghi's Effective Stress Principle

In a *saturated soil*, the normal stress acting in any direction is shared by the soil grains and the pore water. As shown in Fig. 5.2b, it is hypothesized that the point A consists of soil grains and water within the pores. The *total vertical overburden stress* σ_{vo}, computed in Eq. (5.4) is shared by the soil grains and the pore water. The vertical overburden stress acting on the soil grains (or soil skeleton) is called the *effective vertical overburden stress*, denoted by σ'_{vo}. The water within the pores carries the remainder in the form of *pore water pressure*, denoted by u. Therefore,

$$\sigma_{vo} = \sigma'_{vo} + u \tag{5.5}$$

This idea is not limited to only the overburden stresses. It is true for any normal stress in any direction at any point within the soil mass. In general terms,

$$\sigma = \sigma' + u \tag{5.6}$$

where σ = total stress, σ' = effective stress, and u = pore water pressure. This is known as Terzaghi's principle of effective stress (Terzaghi, 1925). Effective stress is also called *inter-granular stress*, and it varies with direction. The pore water pressure is also known as *neutral stress*. It is hydrostatic and hence is the same in all directions.

In Fig. 5.2a, the pore water pressure at A is given by

$$u = \gamma_w h_3 \tag{5.7}$$

Therefore, at A, the effective overburden stress σ'_{vo} becomes

$$\sigma'_{vo} = \sigma_{vo} - u$$

From Eqs. (5.4) and (5.7)

$$\begin{aligned} \sigma'_{vo} &= \gamma_{m1}h_1 + \gamma_{m2}h_2 + \gamma_{sat,2}h_3 - \gamma_w h_3 \\ &= \gamma_{m1}h_1 + \gamma_{m2}h_2 + (\gamma_{sat,2} - \gamma_w)h_3 \\ \sigma'_{vo} &= \gamma_{m1}h_1 + \gamma_{m2}h_2 + \gamma'_2\, h_3 \end{aligned} \tag{5.8}$$

where γ'_2 is the submerged unit weight of soil 2. Water cannot carry any shear stress, which is carried only by the soil grains.

It is often necessary to assume appropriate values for the unit weights of different soils for estimating the overburden pressures. Typical unit weights of different soil types, as suggested by AS 4678-02 are given in Table 5.1. It is worth noting that for clays, bulk, and saturated unit weights are the same, implying that they are near saturation in their natural states, due to capillary effects discussed in this chapter (see Sec. 5.4). Saturated granular soils are heavier than clays. There is a difference of 2–4 kN/m^3 between the bulk and saturated unit weights of granular soils.

		Bulk unit weight (kN/m³)		Saturated unit weight (kN/m³)	
		Loose	Dense	Loose	Dense
Granular soils	Gravel	16.0	18.0	20.0	21.0
	Well-graded sand and gravel	19.0	21.0	21.5	23.0
	Coarse or medium sand	16.5	18.5	20.0	21.5
	Well-graded sand	18.0	21.0	20.5	22.5
	Fine or silty sand	17.0	19.0	20.0	21.5
	Rock fill	15.0	17.5	19.5	21.0
	Brick hardcore	13.0	17.5	16.5	19.0
	Slag fill	12.0	15.0	18.0	20.0
	Ash fill	6.5	10.0	13.0	15.0
Cohesive soils	Peat (high variability)	12.0		12.0	
	Organic clay	15.0		15.0	
	Soft clay	17.0		17.0	
	Firm clay	18.0		18.0	
	Stiff clay	19.0		19.0	
	Hard clay	20.0		20.0	
	Stiff or hard glacial clay	21.0		21.0	

TABLE 5.1 Typical Values for Bulk and Saturated Unit Weights (After AS 4678-2002)

Example 5.1

Plot the variation of the total and effective vertical stresses and the pore water pressure with depth for the soil profile shown in Fig. 5.3.

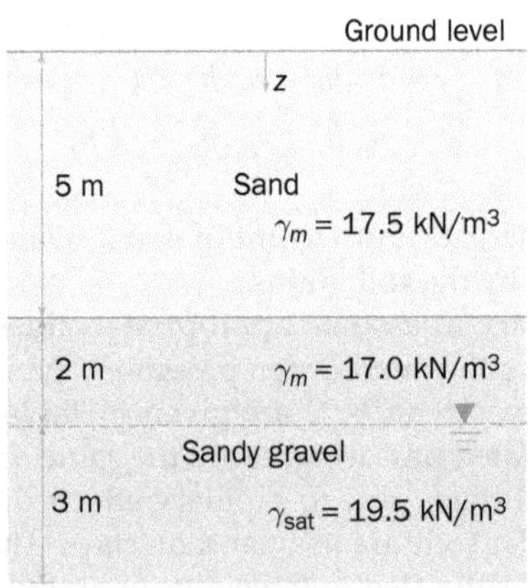

FIGURE 5.3 Soil profile.

Solution At the ground level, σ_v, σ'_v, and u are zero.

At $z = 5$ m (measuring the depth z from the ground level—see Fig. 5.3),

$\sigma_{vo} = 5 \times 17.5 = 87.5$ kN/m^2

$u = 0$ and hence $\sigma'_{vo} = 87.5$ kN/m^2

At $z = 7$ m,

$\sigma_{vo} = 5 \times 17.5 + 2 \times 17 = 121.5$ kN/m^2

$u = 0$ and hence $\sigma'_{vo} = 121.5$ kN/m^2

At $z = 10$ m,

$\sigma_{vo} = 5 \times 17.5 + 2 \times 17 + 3 \times 19.5 = 180.0$ kN/m^2

$u = 3 \times 9.81 = 29.4$ kN/m^2

$\sigma'_{vo} = 180.0 - 29.4 = 150.6$ kN/m^2

The computed values are summarized in the table below.

z(m)	σ_{vo}(kN/m^2)	u(kN/m^2)	σ'_{vo}(kN/m^2)
0	0	0	0
5	87.5	0	87.5
7	121.5	0	121.5
10	180	29.4	150.6

The variations of the total and effective overburden stresses and the pore water pressure with depth are shown in Fig. 5.4.

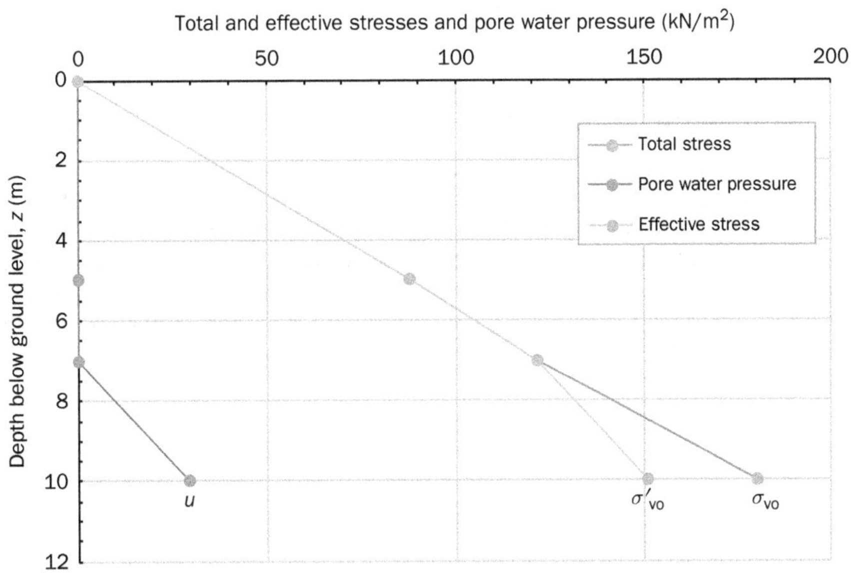

FIGURE 5.4 Variation of σ_{vo}, σ'_{vo}, and u with depth.

Example 5.2

During high tide, a site is fully submerged in water, with water level at 1 m above the ground level. The moisture content of the saturated soil is 25.5% and the specific gravity of the soil grains is 2.69. Determine the effective overburden stress and the pore water pressure at 3 m below the ground level.

If the water level drops to the ground level, what would be the new effective stress?

Solution

$$w = \frac{Se}{G_s}$$

$$e = 0.255 \times 2.69 = 0.686$$

$$\gamma_{sat} = \left(\frac{G_S + e}{1 + e}\right)\gamma_w = \left(\frac{2.69 + 0.686}{1 + 0.686}\right)9.81 = 19.6 \text{ kN/m}^3$$

At 3 m depth below the ground level,

$$\sigma_{vo} = 1 \times 9.81 + 3 \times 19.6 = 68.6 \text{ kN/m}^2$$

$$u = 4 \times 9.81 = \mathbf{39.2\ kN/m^2}$$

$$\sigma'_{vo} = 68.6 - 39.2 = \mathbf{29.4\ kN/m^2}$$

If the water level drops to the ground level,

$$\sigma_{vo} = 3 \times 19.6 = 58.8 \text{ kN/m}^2$$

$$u = 3 \times 9.81 = 29.4 \text{ kN/m}^2$$

$\sigma'_{vo} = 58.8 - 29.4 = \mathbf{29.4\ kN/m^2}$. There is no change in the effective stress.

5.4 Capillary Effects in Soils

Let us review some simple physics of *capillary* action. When a glass *capillary tube* of inner diameter d is placed in water, the water rises within the tube due to capillary, to a height h_c as shown in Fig. 5.5. This height is called *capillary rise*. The column of water is held by the surface tension T at the top of the water column at the interfaces of the glass, water, and air. The water column is in equilibrium under two forces: (a) surface tension T acting along the perimeter of the water column at the top at an angle α to the vertical, and (b) self-weight of the water column. Equating the two,

$$(T \cos \alpha)(\pi d) = \left(\frac{\pi d^2}{4}\right) h_c \gamma_w$$

Therefore,

$$h_c = \frac{4T \cos \alpha}{\gamma_w d} \tag{5.4}$$

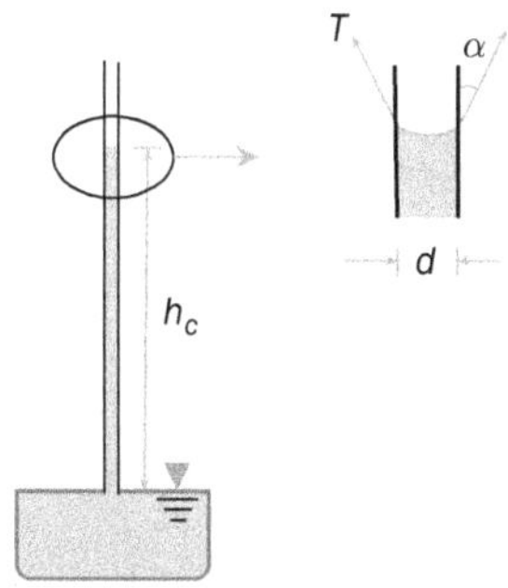

FIGURE 5.5 Capillary rise in a glass tube.

For glass tubes in pure water, $\alpha \approx 0$ and $T = 0.073$ N/m. With $\gamma_w = 9810$ N/m^3, h_c becomes

$$h_c(\text{m}) = \frac{4(0.073)(1)}{(9810)\left(\frac{d}{1000}\right)} = \frac{0.0298}{d(\text{mm})} \approx \frac{0.03}{d(\text{mm})} \tag{5.10}$$

It is clear from Eq. (5.10) that the smaller the capillary diameter the larger the capillary rise. What does this have to do with soils?

The soil skeleton consists of an assembly of grains and voids. The interconnected voids above the ground water table act as capillary tubes in conducting water, thus enabling capillary rise in soils. This makes the water rise to significant heights above the water table. The effective pore diameter is approximately 1/5 of D_{10} of the soil. Therefore, capillary rise in soils can be estimated as

$$h_c(\text{m}) \approx \frac{0.15}{D_{10}(\text{mm})} \tag{5.11}$$

Due to capillary rise, water rises to fill the voids above the water table up to a height h_c. Within this capillary zone, near the water table, the soil can be assumed to be almost saturated (hence use γ_{sat} in computing the total stress) *but not submerged*. There is no buoyancy effect. In reality, at the upper region of the capillary zone, degree of saturation is less than 100%, with larger voids still containing some air and smaller voids filled with water. The pore water pressure increases linearly with depth with negative values in the capillary zone, zero at the water table level, and positive values below the water table as shown in Fig. 5.6.

At height h above the water table, within the capillary zone, when the soil is not saturated ($S < 1$), the pore water pressure is approximated as

$$u = -Sh\gamma_w \tag{5.12}$$

where S is the degree of saturation.

Capillary rise is the highest in clays, where it can exceed 10 m. The negative capillary pressure has the effect of increasing the effective stress significantly, for the same

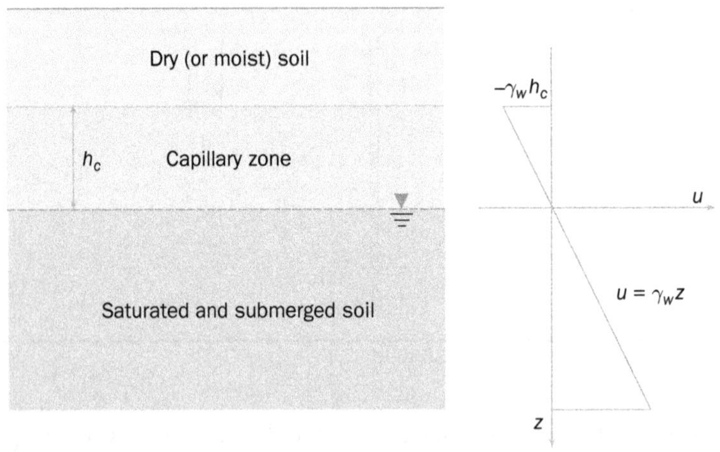

Figure 5.6 Pore water pressure variation with depth assuming the capillary zone is saturated.

total stress level. Capillary effects cause suction within the soil, leading to negative pore water pressures and increased effective stresses.

5.5 Main Points

1. The normal stress at a point within a saturated soil mass is shared by the soil skeleton (grains) and the pore water. The first is the effective stress and the second is the pore water pressure.
2. Terzaghi's effective stress principle states that $\sigma = \sigma' + u$, which is true at all points within the soil mass and in all directions.
3. Water cannot carry any *shear* stress, which is carried only by the soil grains.
4. Capillary rise is inversely proportional to the pore channel diameter.
5. Capillary pressure is negative, similar to suction.

Review Exercises

5.1 State whether the following are true or false.

a. Total stresses are always larger than the effective stresses.

b. Pore water pressure at a point is the same in magnitude in all directions.

c. Effective stress at a point is the same in magnitude in all directions.

d. Capillary pressure is negative.

e. Capillary rise is larger in sands than in silts.

Answer: False, True, False, True, False

5.2 Plot the variation of σ_{vo}, σ'_{vo}, and u with depth for the following soil profile (Fig. RE5.2).

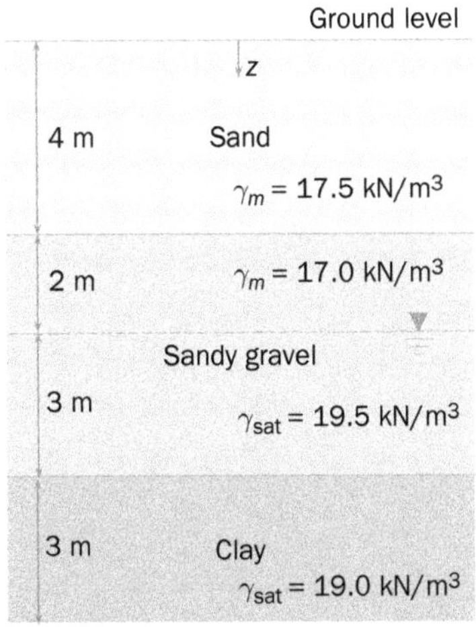

FIGURE RE5.2 Soil profile.

5.3 Find the total and effective overburden stresses and the pore water pressure at 5 m depth within the seabed, when the depth of the sea is 100 m and the saturated unit weight of the seabed soil is 19.0 kN/m^3.

Answer: 1076.0 kN/m^2, 45.9 kN/m^2, 1030.1 kN/m^2

5.4 The water table at a site is at 2.0 m below the ground level. The bulk and saturated unit weights of the soil are 17.0 and 19.5 kN/m^3, respectively. Determine σ_{vo}, σ'_{vo}, and u at 6.0 m below the ground level.

Answer: 112.0 kN/m^2, 72.8 kN/m^2, 39.2 kN/m^2

5.5 In Review Exercise 5.4, if the water table is permanently lowered by 2.0 m, what would be the values of σ_{vo}, σ'_{vo}, and u at 6.0 m below the ground level?

Answer: 107.0 kN/m^2, 87.4 kN/m^2, 19.6 kN/m^2

References

AS 4678-02. Earth Retaining Structures, Australian Standard.

Terzaghi, K. *Erdbaumechanik auf Bodenphysikalischer Grundlage*, Deuticke, Vienna, 1925.

References

CHAPTER 6
Permeability and Seepage

6.1 Introduction

Soils are particulate media and therefore they are porous. The interconnected voids between the soil grains can become pore channels and conduct water. *Permeability*, as the name implies (the ability to permeate), is a measure of how easily a fluid can flow through a porous medium. In the context of geotechnical engineering, the porous medium is soil and the fluid is water. A petroleum engineer may be interested in the flow of oil through rocks. An environmental engineer may be looking at the flow of leachate through the compacted clay liner at the bottom of a landfill.

Soils with high void ratios are more porous and have high permeability. In addition, coarse-grained soils are more permeable than fine-grained soils. The flow of water through soils is called *seepage*, which takes place when there is a difference between the *upstream* and *downstream* water levels as shown in Fig. 6.1. In Fig. 6.1a, the concrete is assumed to be impervious, similar to the impervious stratum lying beneath the soil. In Fig. 6.1b, the *cofferdam* enclosure, made of *sheet piles* driven into the ground, enables lowering of the water level within the enclosure by pumping the water out. Sheet piles are interlocking sections of wall driven into the ground, making a continuous wall. They are commonly made of steel. Figure 6.2 shows two sheet pile enclosures at locations with high water table, where excavation and/or dewatering will be required prior to construction. Figure 6.2a is a close-up view of a small enclosure, showing the interlocking sections. Figure 6.2b shows the sheet pile enclosure for an 8 m diameter excavation.

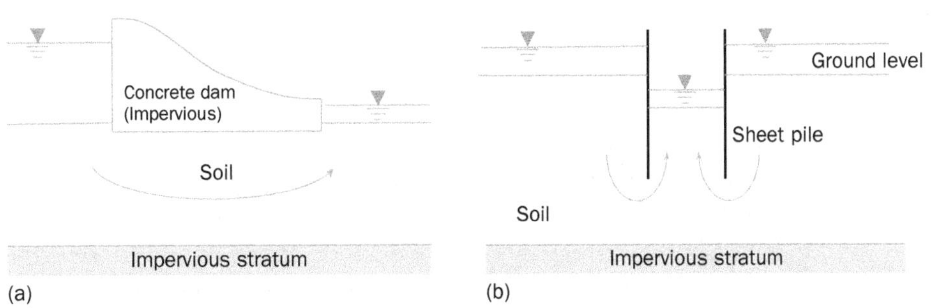

FIGURE 6.1 Seepage through soils: (a) beneath a concrete dam, (b) into a sheet pile cofferdam.

(a) (b)

Figure 6.2 Sheet pile enclosures: (a) a small enclosure, (b) a circular enclosure.

6.2 Bernoulli's Equation

Bernoulli's theorem in fluid mechanics suggests that the energy of a water molecule at point P in Fig. 6.3a consists of three components: *potential energy*, *strain energy*, and *kinetic energy*. Let the mass of the water molecule be m, velocity v, and it is under pressure p given by $\gamma_w h$, where h is the depth of the point P below the water level. We will select a *datum* (an arbitrary horizontal line) from which the height z to the point P is measured. The total energy of the water molecule can be written as

$$\text{Total energy} = \text{Strain energy} + \text{Kinetic energy} + \text{Potential energy}$$

$$\text{Total energy} = \frac{mp}{\rho_w} + \frac{1}{2}mv^2 + mgz$$

Dividing the above expression by the weight of the molecule mg on both sides,

$$\text{Total energy per unit weight} = \frac{p}{\rho_w g} + \frac{1}{2}\frac{v^2}{g} + z$$

$$\text{Total energy per unit weight} = \frac{p}{\gamma_w} + \frac{v^2}{2g} + z \tag{6.1}$$

In Eq.(6.1), the three terms on the right-hand side have unit of length. They are called *pressure head*, *velocity head*, and *elevation head*, respectively. The one on the left-hand side is the *total head*, which includes the three components. Therefore, Eq. (6.1) can be written as

$$\text{Total head} = \text{Pressure head} + \text{Velocity head} + \text{Elevation head} \tag{6.2}$$

The head is a measure of energy, expressed with the unit of length for simplicity.

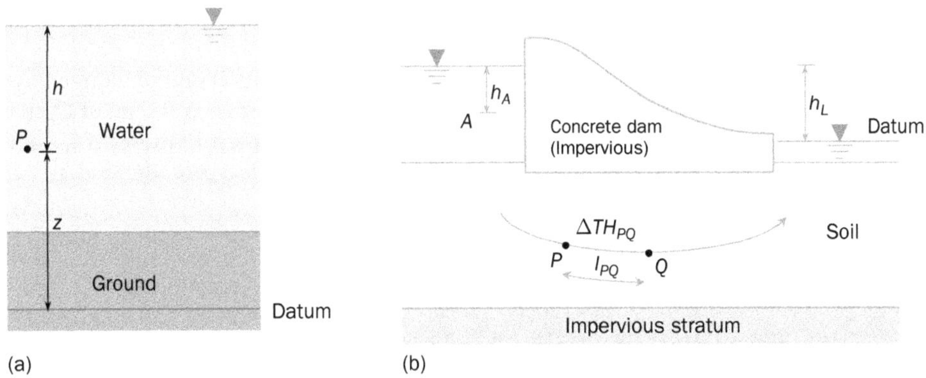

FIGURE 6.3 Energy of a water molecule at point *P*: (a) Bernoulli's theorem, (b) seepage beneath a concrete dam.

Figure 6.3b shows a water molecule at point P in the soil beneath a concrete dam where seepage takes place. The downstream water level has been selected as the datum. The point P is 6 m below the datum, where the velocity is 1 cm/s, and the pore water pressure is 45 kN/m^2. Therefore, the heads can be computed as follows:

$$\text{Pressure head} = p/\gamma_w = 45/9.81 = 4.6 \text{ m}$$

$$\text{Velocity head} = v^2/2g = \left(\frac{0.01^2}{2 \times 9.81}\right) = 5.1 \times 10^{-6} \text{ m}$$

$$\text{Elevation head} = -6.0 \text{ m}$$

In flow through soils, the velocity of flow is small and hence the velocity head becomes negligible. In this example, total head is 4.6 – 6.0 = –1.4 m. From now on, for flow through soils, neglecting the velocity head, Eq. (6.2) can be written as

$$\text{Total head} = \text{Pressure head} + \text{Elevation head} \tag{6.3}$$

In Fig. 6.3b, let us consider a point A within the upstream reservoir, which is at a depth of h_A below the water surface. The difference in the upstream and downstream water levels is h_L. Therefore at point A,

$$\text{Elevation head} = h_L - h_A$$

$$\text{Pressure head} = h_A$$

$$\text{Therefore, the total head} = h_L$$

Since the point A can be anywhere within the upstream reservoir, the total head at any point within the upstream reservoir is h_L. Similarly, the total head at any point in the downstream reservoir is zero. The total head loss from upstream to downstream is h_L. When water flows from upstream to downstream, there is no head loss within the reservoirs on either side. Head loss only occurs along the flow path while flowing through the soil, as the water has to shed some energy in overcoming the frictional resistance provided by the soil.

In the flow path in Fig. 6.3b, let us consider two points P and Q. Water flows from higher total head to lower total head. The total head is larger at P than at Q, the difference being ΔTH_{PQ}. *Along the flow path,* the distance between P and Q is l_{PQ}. *Hydraulic gradient* (i) is the total head loss per unit length *along the flow path,* defined as

$$i = \frac{\Delta TH_{PQ}}{l_{PQ}} \tag{6.4}$$

It is a dimensionless quantity. It can vary significantly along the flow path even in a homogeneous soil.

Example 6.1

Figure 6.4 shows a laboratory setup where water flows through a tube containing a cylindrical soil specimen, 400 mm in length. The water level at both sides (i.e., head water and tail water) are maintained at the levels shown in the figure. What is the hydraulic gradient in the soil specimen?

Solution Let us take the tail water level as the datum.

The total head at the top of the water on the left side = Elevation head + Pressure head
= 200 mm + 0 = 200 mm

This 200 mm total head remains the same within the entire water on the left.

Similarly, on the right side, the total head within the water is zero.

Therefore, the total head loss across the soil specimen is 200 mm. This head loss occurs over a length of 400 mm of soil.

Therefore, the hydraulic gradient $i = 200/400 = 0.5$.

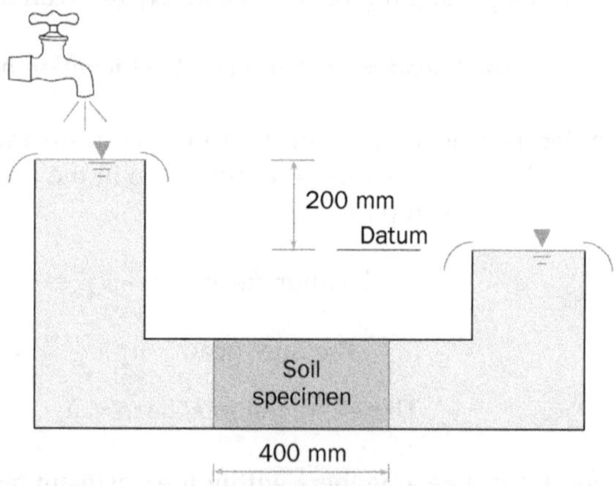

FIGURE 6.4 Diagram for Example 6.1.

Example 6.2

Figure 6.5 shows a laboratory setup where water flows through two cylindrical specimens from different soils. The water levels remain steady during the test. Find the hydraulic gradients in the two soil specimens.

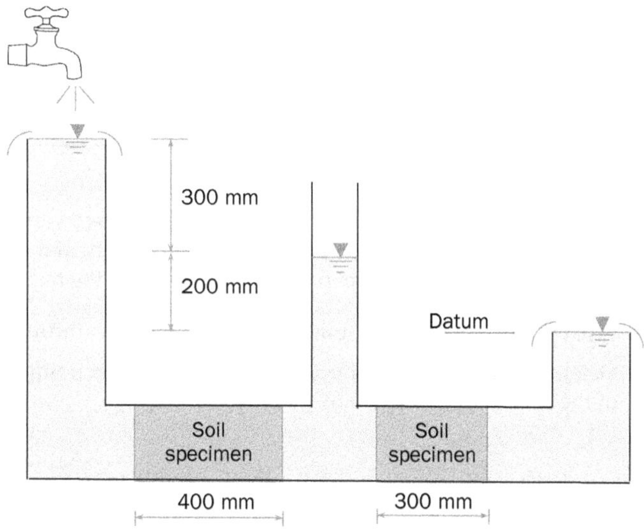

FIGURE 6.5 Diagram for Example 6.2.

Solution Let us take the tail water level as the datum (see figure).

The total head at the top of the water on the left side = Elevation head + Pressure head
= 500 mm + 0 = 500 mm

This 500 mm total head remains the same within the entire water on the left.

Similarly, the total head in the water in the middle column = 200 mm

On the right side, the total head within the water is zero.

For the soil specimen on the left,

Total head loss = 500 – 200 = 300 mm

Therefore, $i = 300/400 = 0.75$

For the soil specimen on the right,

Total head loss = 200 – 0 = 200 mm

Therefore, $i = 200/300 = 0.67$

Example 6.3

The average hydraulic gradient along a flow path is 0.25, and the total head at a point P on the flow path is 6.0 m. What would be the total head at (a) 4 m downstream and (b) 2 m upstream?

Solution

Part a At 4 m downstream, the total head is $6.0 - 4 \times 0.25 = 5.0$ m

Part b At 2 m downstream, the total head is $6.0 + 2 \times 0.25 = 6.5$ m

Example 6.4

Figure 6.6 shows a laboratory setup where water flows downward through a 400-mm-long soil specimen, while the water levels are maintained constant at the entry and exit. The saturated unit weight of the soil specimen is 18.5 kN/m³. Take the tail water level as the datum.

a. What is the average hydraulic gradient for the flow through the soil specimen?
b. Determine the total and effective vertical stresses at point A, 100 mm below the top of the specimen (see Fig. 6.6).

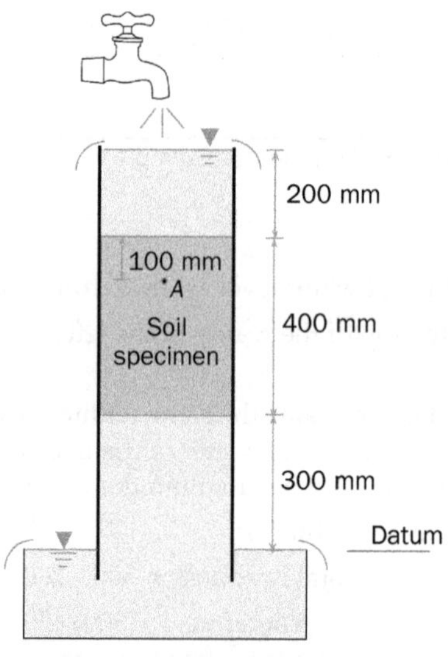

FIGURE 6.6 Diagram for Example 6.4.

Solution

Part a

The total head within the water at the top of the specimen = 900 mm

The total head within the entire water below the specimen = 0

Therefore, the total head loss across the specimen is 900 mm, which occurs over the length of 400 mm.

Thus, the average hydraulic gradient across the specimen = 900/400 = 2.25

Part b

Total head at $A = 900 - 2.25 \times 100 = 675$ mm

Elevation head at $A = 600$ mm

Therefore, the pressure head at $A = 675 - 600 = 75$ mm

Pore water pressure at $A = 9.81 \times 0.075 = 0.736$ kN/m^2

Total vertical stress at A, $\sigma_v = 9.81 \times 0.2 + 18.5 \times 0.1 = 3.812$ kN/m^2

Effective vertical stress at A, $\sigma'_v = 3.812 - 0.736 = 3.076$ kN/m^2

6.3 Darcy's Law

In 1856, a French engineer Henry Darcy showed that when there is laminar flow through soils, the *discharge velocity* v is proportional to the hydraulic gradient i. Thus,

$$v \propto i$$

$$v = ki \tag{6.5}$$

where k is the proportionality constant, known as the *hydraulic conductivity* or *permeability* of the soil. Since i is dimensionless, k has the unit of velocity, and commonly expressed in cm/s or m/s.

Typical variation of the discharge velocity against the hydraulic gradient is shown in Fig. 6.7. It can be seen that when the velocity is low, the variation is linear and the flow is laminar where Darcy's law [Eq. (6.5)] can be applied. Forchheimer (1901) and several others noted the slight deviation from Darcy's law at higher velocities and suggested that the trend can be explained better by

$$i = av + bv^2 \tag{6.6}$$

where a and b are constants. The slope of the line at $v = 0$ is a, and hence $a = 1/k$.

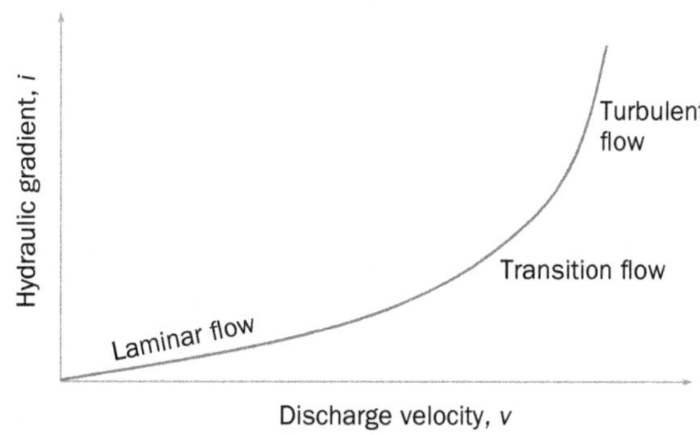

Figure 6.7 Hydraulic gradient—discharge velocity variation.

Reynolds (1883) number *Re*, used more in fluid mechanics, can be defined for flow through soils as

$$Re = \frac{vD_{50}\rho_w}{\mu_w} \tag{6.7}$$

where D_{50} is generally taken as the median grain size, ρ_w = density of water (1000 kg/m^3) and μ_w = dynamic viscosity of water (1.002×10^{-3} Pa·s or kg/m/s). Provided the Reynolds number as determined from Eq. (6.7) is less than approximately 1, it is reasonable to assume that the flow is laminar. Flow through coarse sands and gravels is generally turbulent, and hence Darcy's law is not valid.

When water flows through the soil specimen, it only flows through the voids and not the entire cross section. Therefore, the actual velocity of flow, known as *seepage velocity* v_s is larger than the discharge velocity v commonly used in the computations including Darcy's law. Seepage velocity and discharge velocity are related by

$$v_s = \frac{v}{n} \tag{6.8}$$

where n is the porosity.

Example 6.5

Check whether Darcy's law is valid for

a. A silt where $D_{50} = 0.04$ mm and $v = 0.05$ cm/s
b. A gravel where $D_{50} = 10$ mm and $v = 1$ cm/s

Solution

Part a

$$Re = \frac{(0.05 \times 10^{-2})(0.04 \times 10^{-3})(1000)}{10^{-3}} = 0.02 < 1$$

Therefore, the flow is laminar.

Part b

$$Re = \frac{(1 \times 10^{-2})(10 \times 10^{-3})(1000)}{10^{-3}} = 100 > 1$$

Therefore, the flow is not laminar.

6.3.1 Typical Values and Empirical Relations for Hydraulic Conductivity

Typical values of the hydraulic conductivity for different soil groups are shown in Fig. 6.8. In coarse-grained soils such as gravels, the flow is turbulent and Darcy's law is no longer valid. Soils such as stiff clays with hydraulic conductivity less than 10^{-7} cm/s are literally impervious. Clay liners for waste containment areas are expected to have $k < 10^{-7}$ cm/s to minimize contamination of the underlying soils and groundwater.

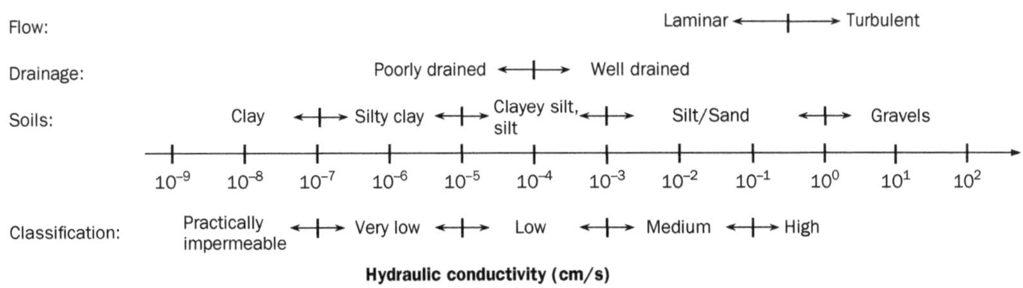

FIGURE 6.8 Typical values of hydraulic conductivity for different soil groups.

Hazen (1930) suggested that for uniform sands,

$$k(\text{cm/s}) = CD_{10}^2 \tag{6.9}$$

where D_{10} is in mm and is known as the *effective grain size*, which is a measure of the pore channel diameter. C is a constant in the range of 1.0–1.5. Hydraulic conductivity (permeability) increases with the void ratio e, and can be proportional to e^2, $e^2/(1+e)$, or $e^3/(1+e)$. In most clays and some sands, e is proportional to log k. Therefore,

$$\log k = \log k_0 - \frac{e_0 - e}{C_k} \tag{6.10}$$

where k is the permeability at void ratio e and k_0 is the permeability at void ratio e_0. The slope of the log k versus e line is $1/C_k$. Here, C_k is the dimensionless *permeability change index*, which can be taken as approximately $0.5e_0$. Equation (6.10) works well for $e_0 < 2.5$.

Permeability has slight dependence on the temperature. Higher the temperature lower the viscosity and higher the permeability. The standard practice is to report the value at 20°C (ASTM D2434-19). It can be obtained from:

$$k_{20^\circ\text{C}} = k_{T^\circ\text{C}} \frac{\mu_{T^\circ\text{C}}}{\mu_{20^\circ\text{C}}} \tag{6.11}$$

where $\mu_{20^\circ\text{C}}$ = dynamic viscosity at 20°C, and $\mu_{T^\circ\text{C}}$ = dynamic viscosity at T°C. ASTM D5084-03 suggests that $\mu_{T^\circ\text{C}}/\mu_{20^\circ\text{C}}$ can be approximated as:

$$\frac{\mu_{T^\circ\text{C}}}{\mu_{20^\circ\text{C}}} = \frac{2.2902 \times 0.9842^T}{T^{0.1702}} \tag{6.12}$$

6.3.2 Intrinsic Permeability

The hydraulic conductivity or permeability defined in Eq. (6.5) from Darcy's law depends on the viscosity and density of the permeant. In compacted clay landfill liners, the permeant is the leachate; in petroleum industry, the permeant is the oil flowing through the rocks. For geotechnical engineers, working with soils, the permeant is generally water. To eliminate the dependence of the density and viscosity of the permeant, *intrinsic permeability* or *absolute permeability* K is defined as

$$K = \frac{\mu_w}{\gamma_w} k \tag{6.13}$$

where μ_w and γ_w are the dynamic viscosity and the unit weight of water, respectively. The dimension of K is L^2 with unit of cm^2, m^2, etc. This intrinsic permeability of the soil matrix K is a measure of the geometry and size of the void network, and is independent of the permeant characteristics. In oil and gas industry, Darcy is a common unit for K (1 Darcy = 0.987 μm^2). For flow of water through soils, assuming $\gamma_w = 9810$ N/m^3 and $\mu_w = 1.002 \times 10^{-3}$ N·s/m^2 (or kg/ms) at 20°C,

$$K(\text{cm}^2) = k(\text{cm/s}) \times 1.02 \times 10^{-5} \tag{6.14}$$

$$K(\text{Darcy}) = k(\text{cm/s}) \times 1.035 \times 10^3 \tag{6.15}$$

6.3.3 Effect of Surcharge on Hydraulic Conductivity

Most of the time, the constant head or falling head test in the laboratories are carried out on soil specimens under little or no surcharge pressure. In reality, there is always an overburden pressure, which can reduce the hydraulic conductivity. The extent to which the hydraulic conductivity is reduced by the applied pressure depends on the soil type, as shown in Fig. 6.9. It can be seen that in granular soils, there is only a slight drop in the hydraulic conductivity due to surcharge.

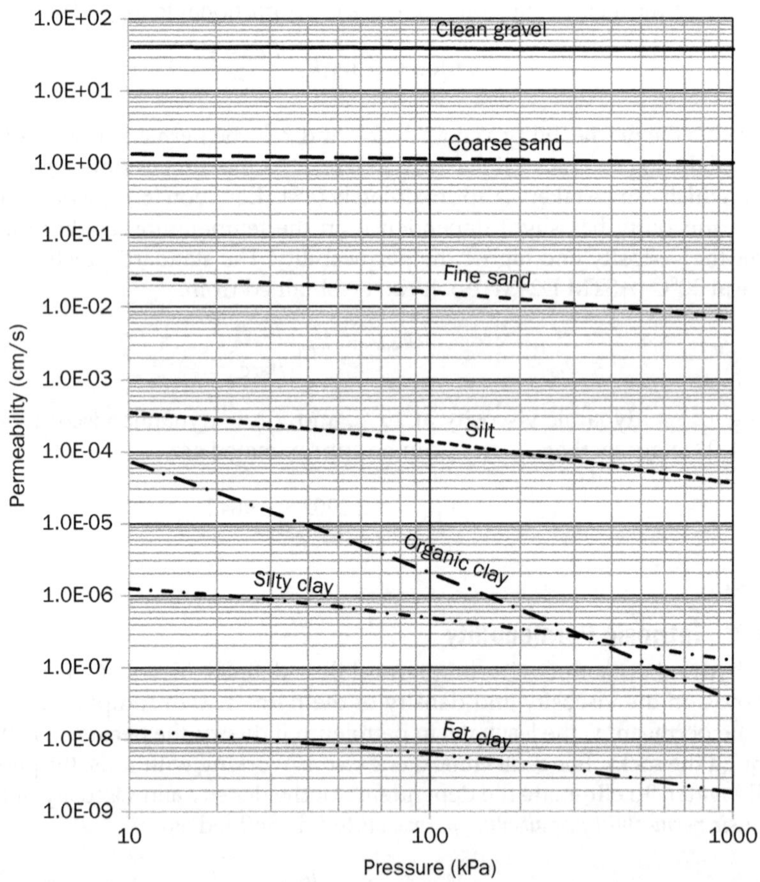

FIGURE 6.9 Effect of surcharge pressure on hydraulic conductivity (after Cedegren, 1967).

6.4 Laboratory Determination of Hydraulic Conductivity

The hydraulic conductivity k can be measured in the laboratory or in the field. It can be carried out on *undisturbed* or *reconstituted* soil samples in the laboratory. Undisturbed or intact samples are collected from the field. Reconstituted samples, especially in granular soils, are prepared in the laboratory to specific void ratios replicating desired state of compaction or density. Generally, a *constant head permeability test* (ASTM D 2434-19) is carried out on a coarse-grained soil sample and a *falling head permeability test* (ASTM D 5856-15) is carried out on a fine-grained soil sample.

6.4.1 Constant Head Test

Figure 6.10 shows the laboratory setup for a constant head test. Here, water flows through a soil specimen of length L and cross-sectional area A. The water levels on both ends, head water and tail water, are maintained at constant levels throughout the test, by allowing any overflow through the spouts as shown. The height difference between the water levels, and hence the total head loss across the soil specimen is h (see Fig. 6.10) throughout the test. Therefore, the hydraulic gradient across the soil specimen at all times is

$$i = \frac{h}{L}$$

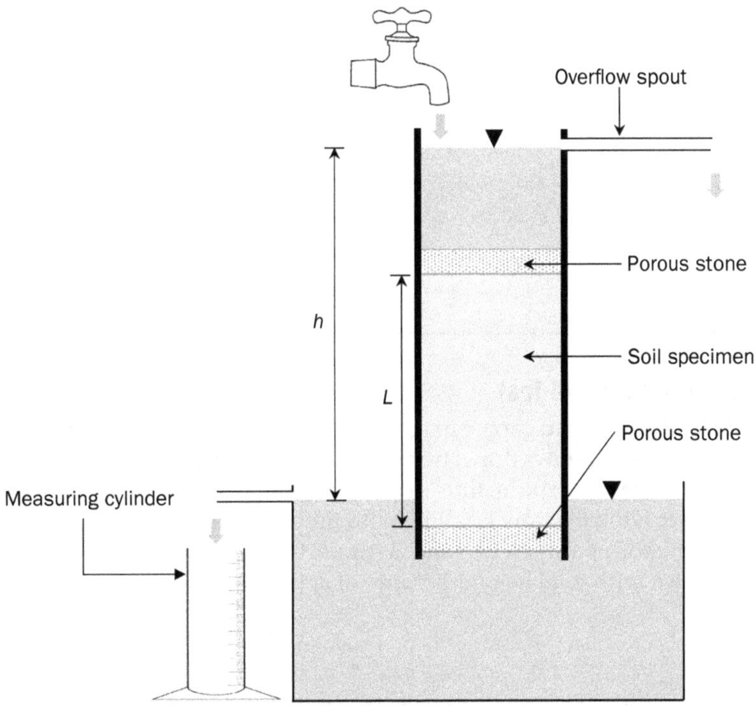

FIGURE 6.10 Constant head hydraulic conductivity test in the laboratory. (Courtesy of Dr. Thushara Madanayaka, SLTC, Sri Lanka.)

The water flowing through the specimen Q (cm^3) in time t (s) is collected in a measuring cylinder. Therefore, the discharge velocity v (cm/s) can be determined as

$$v = \frac{Q}{At}$$

Assuming laminar flow and applying Darcy's law,

$$v = ki$$

$$\frac{Q}{At} = k\frac{h}{L}$$

Therefore,

$$k = \frac{QL}{hAt} \tag{6.16}$$

The test is not suitable for fine-grained soils where it may take too long to collect measurable quantity of water.

Example 6.6

A constant head test was carried out on a 100 mm diameter and 200 mm long reconstituted silty sand specimen in a set up similar to that in Fig. 6.10. The height difference in the water levels h is 300 mm. If 2435 g water was collected in a bucket in 5 min., determine the hydraulic conductivity of the sand specimen.

Solution

$L = 200$ mm and $h = 300$ mm.

$$A = \frac{\pi}{4} \times 10^2 = 78.54 \text{ cm}^2$$

$Q = 2435$ cm^3 and $t = 300$ s.

$$k = \frac{QL}{hAt} = \frac{(2435)(20)}{(30)(78.54)(300)} = \mathbf{6.9 \times 10^{-2}\,cm/s}$$

6.4.2 Falling Head Test

The laboratory setup to carry out a falling head hydraulic conductivity test is shown in Fig. 6.11. The water flows down from the stand pipe through the soil specimen, and then through the overflow spout, thus maintaining the water level at exit. The total head difference (h) driving the flow is falling with time. Let the cross-sectional area of the standpipe and the soil specimen be a and A, respectively. The length of the soil specimen is L.

At time t, with total head difference of h, the flow rate Q through the standpipe is

$$Q = -a\frac{dh}{dt}$$

Negative sign is used to account for the reduction in h with time t.

The hydraulic gradient at time t is given by h/L. Thus, the discharge velocity is given by

$$v = k\frac{h}{L}$$

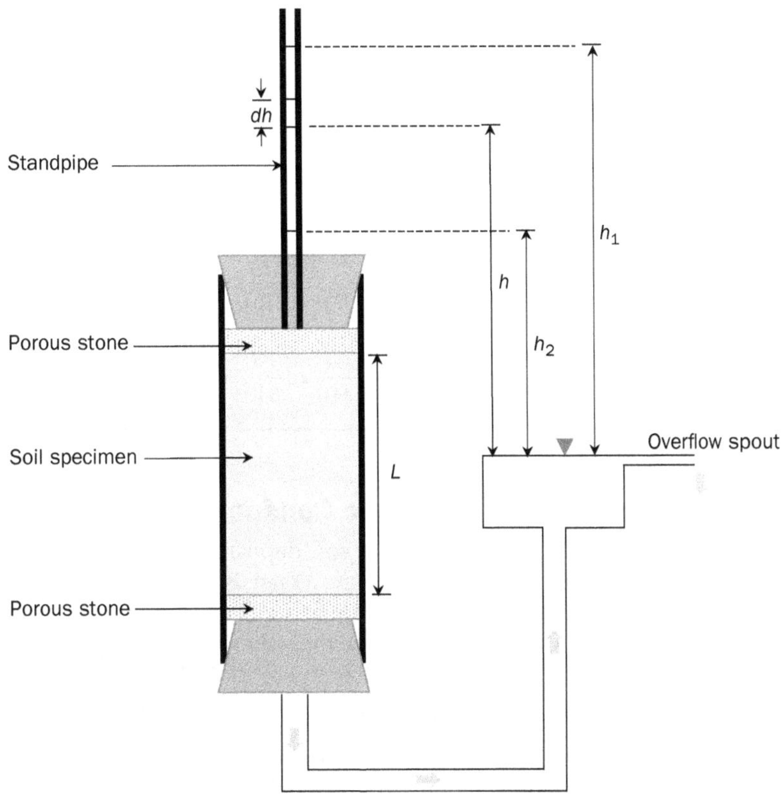

Figure 6.11 Falling head hydraulic conductivity test in the laboratory. (Courtesy of Dr. Thushara Madanayaka, SLTC, Sri Lanka.)

Therefore, the flow rate through the soil specimen at time t is

$$Q = vA = k\frac{h}{L}A$$

Equating the two flow rates at time t,

$$-a\frac{dh}{dt} = k\frac{h}{L}A$$

$$\frac{dh}{h} = -\frac{kA}{aL}dt$$

If the water level in the standpipe was at h_1 initially and has fallen to h_2 after time t,

$$\int_{h_1}^{h_2}\frac{dh}{h} = -\frac{kA}{aL}\int_0^t dt$$

$$k = \frac{aL}{At}\ln\left(\frac{h_1}{h_2}\right) \tag{6.17}$$

The test is not suitable for coarse-grained soils where the fall between h_1 and h_2 can be too quick to monitor.

Example 6.7

A 101.6-mm-diameter and 116.4-mm-high silty clay specimen in a compaction mold is subjected to a falling head permeability test where the inner diameter of the standpipe was 3 mm. The head falls from 990 to 340 mm in 5 min. Find the hydraulic conductivity of the silty clay.

Solution

$$a = \frac{\pi}{4} \times 0.3^2 = 0.0707 \text{ cm}^2$$

$$A = \frac{\pi}{4} \times 10.16^2 = 81.07 \text{ cm}^2$$

$$k = \frac{aL}{At}\ln\left(\frac{h_1}{h_2}\right) = \frac{(0.0707)(11.64)}{(81.07)(5 \times 60)}\ln\left(\frac{990}{340}\right) = \mathbf{3.62 \times 10^{-5}\,cm/s}$$

6.5 Field Determination of Hydraulic Conductivity

The average hydraulic conductivity of a soil deposit in the field can be determined through a pumping-out test described here. Water is pumped out from the test well until a steady state is reached with flow rate being Q m^3/s. Some observation wells are made at various radial distances to observe the water level during the steady state.

From the drawdown curve during the steady state (see Fig. 6.12), the hydraulic gradient at radial distance r can be taken as $\frac{dh}{dr}$, where h is the height of water above the impervious stratum. For the water flowing in through the cylindrical surface at the radial distance r, the steady state flow rate Q can be written as

$$Q = k\frac{dh}{dr}2\pi rh$$

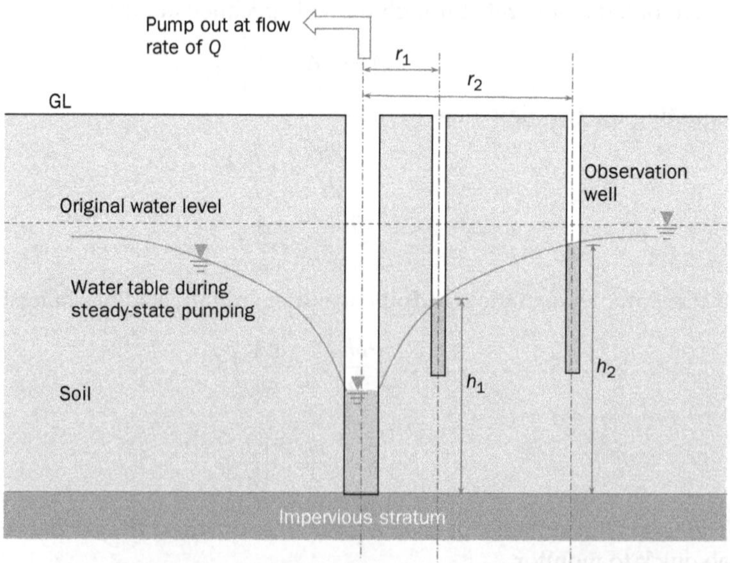

FIGURE 6.12 Pumping out test to determine the permeability in the field.

$$\int_{r_1}^{r_2} \frac{dr}{r} = \frac{2\pi k}{Q} \int_{h_1}^{h_2} h\,dh$$

$$k = \frac{Q}{\pi\left(h_2^2 - h_1^2\right)} \ln\left(\frac{r_2}{r_1}\right) \tag{6.18}$$

6.6 Stresses in Soils due to Flow

Let us discuss the stresses due to flow by considering the three scenarios shown in Fig. 6.13. In all three situations, we will study the total and effective vertical stresses and the pore water pressure at the same point X, at a depth of z within the soil specimen. While there is no flow in Fig. 6.13a, there is downward and upward flow through the soil specimen in Figs. 6.13b and 6.13c, respectively, under a constant head of h_L. In all three scenarios, the *total* vertical stress at X is given by

$$\sigma_v = \gamma_w h_w + \gamma_{sat} z \tag{6.19}$$

For static case (Fig. 6.13a),

$$u = \gamma_w (h_w + z) \tag{6.20}$$

Therefore,

$$\sigma_v' = \sigma_v - u = \gamma' z \tag{6.21}$$

When the flow is downwards (Fig. 6.13b), the average hydraulic gradient is h_L/L. Based on the datum shown on the right, the total head at X is $-iz$. The elevation head at X is $-(h_w + z)$. Therefore, the pressure head at X is $h_w + z - iz$. Therefore,

$$u = \gamma_w (h_w + z - iz) \tag{6.22}$$

$$\sigma_v' = \sigma_v - u = \gamma' z - iz\gamma_w \tag{6.23}$$

Comparing Eqs. (6.22) and (6.23) with (6.20) and (6.21), respectively, it can be seen that due to downward flow through the soil specimen, pore water pressure is decreased and the effective vertical stress is increased by the same quantity $iz\gamma_w$.

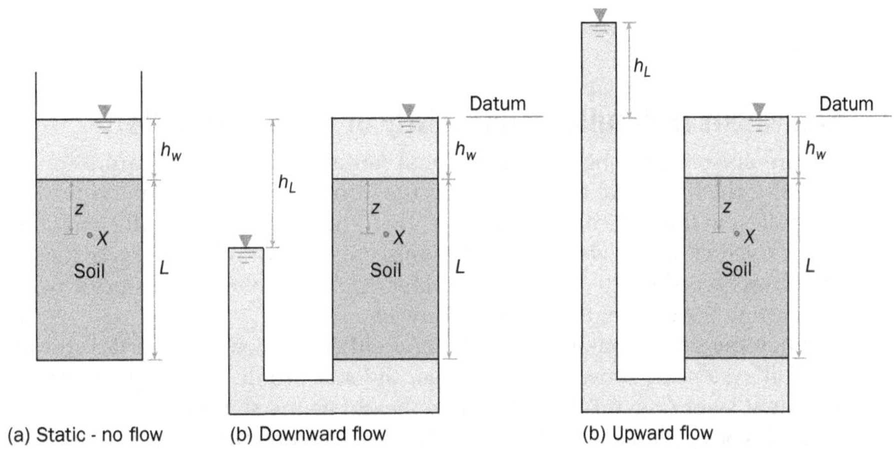

FIGURE 6.13 Three different scenarios: (a) static—no flow, (b) downward flow, (c) upward flow.

When the flow is upward (Fig. 6.13c), the total head at X is iz. The elevation head at X is $-(h_w + z)$. Therefore, the pressure head at X is $h_w + z + iz$. Therefore,

$$u = \gamma_w(h_w + z + iz) \tag{6.24}$$

$$\sigma'_v = \sigma_v - u = \gamma' z - iz\gamma_w \tag{6.25}$$

Comparing Eqs. (6.24) and (6.25) with (6.20) and (6.21), respectively, it can be seen that due to upward flow through the soil specimen, pore water pressure is increased and the effective vertical stress is decreased by the same quantity $iz\gamma_w$. In both scenarios (i.e., downward and upward flow), larger hydraulic gradients give larger changes in u and σ'_v.

6.6.1 Quick Conditions in Granular Soils during Upward Flow

Let us have a closer look at the upward flow scenario (Fig. 6.13c), assuming the soil is granular. The effective vertical stress in Eq. (6.25) can be written as

$$\sigma'_v = \gamma_w z\left(\frac{\gamma'}{\gamma_w} - i\right) \tag{6.26}$$

When the hydraulic gradient i reaches the critical value of $\dfrac{\gamma'}{\gamma_w}$, becomes zero and the grains are barely in contact. This is known as the *quick condition*. The critical value of i at this instant is called the *critical hydraulic gradient*, denoted by i_c. When the hydraulic gradient exceeds i_c, the granular soil fails. It can be shown that

$$i_c = \frac{\gamma'}{\gamma_w} = \frac{G_s - 1}{1 + e} \tag{6.27}$$

You will learn in Chap. 9 that the shear strength of granular soils is proportional to the effective normal stress. When the effective normal stress reduces to zero, the granular soil literally has no shear strength. This is what causes *quick sand* you may have seen in movies. When saturated granular soils are subjected to vibratory loads due to earthquakes, blasting or pile driving, there is a sudden increase of pore water pressures without any changes in the total stresses. This reduces the effective stresses to zero, and causes *liquefaction*.

6.7 Equivalent Hydraulic Conductivity of Stratified Soils

Soils are anisotropic, showing directional variations in the hydraulic conductivities. Generally, the hydraulic conductivity is the largest in the horizontal direction (k_h) and the smallest in the vertical direction (k_v). The ratio k_h/k_v can generally be in the range of 1–10. Varved clays are well known for anisotropic hydraulic conductivities with k_h/k_v sometimes exceeding 10. They are thin-layered sedimentary deposits formed in glacial lakes during the different seasons of the year.

When the flow is one-dimensional and either horizontal or vertical through a layered soil system as shown in Fig. 6.14a, the soil profile can be transformed into an equivalent layer (Fig. 6.14b) of the same height as the soil profile with hydraulic conductivities of $k_{v,eq}$ and $k_{h,eq}$ in the vertical and horizontal directions, respectively.

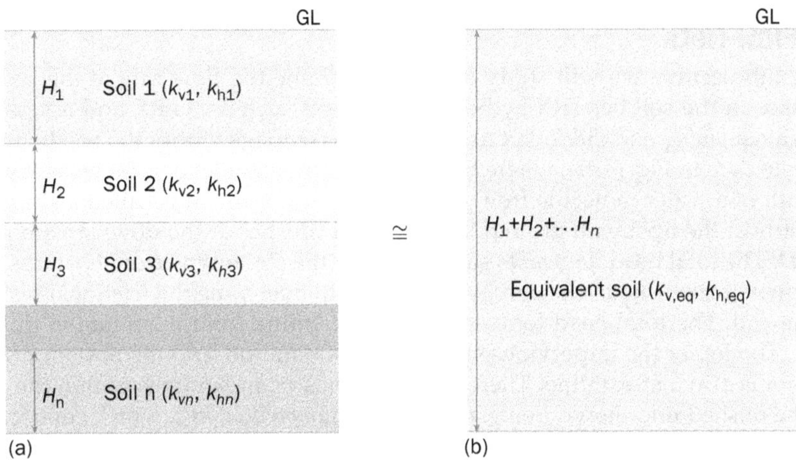

FIGURE 6.14 (a) Layered soil profile, (b) equivalent homogeneous soil.

6.7.1 Horizontal Flow

When water flows horizontally through the layered system under hydraulic gradient i, the flow rate through the layers can be written as $k_{h1}iH_1$, $k_{h2}iH_2$, $k_{h3}iH_3$, ..., $k_{hn}iH_n$ per unit width. In the equivalent homogeneous profile, the flow rate is given by $k_{h,eq}i(H_1 + H_2 \ldots + H_n)$ per unit width. Equating the two,

$$k_{h,eq}i(H_1 + H_2 + \cdots + H_n) = k_{h1}iH_1 + k_{h2}iH_2 + \cdots + k_{hn}iH_n$$

$$k_{h,eq} = \frac{k_{h1}H_1 + k_{h2}H_2 + \ldots + k_{hn}H_n}{H_1 + H_2 + \ldots + H_n} \tag{6.28}$$

6.7.2 Vertical Flow

When water flows vertically through the layered system, let the total head loss within the layers be h_1, h_2, h_3, ..., h_n. Therefore, the total head loss across the equivalent layer is $h_1 + h_2 + \ldots h_n$. Since the velocities have to be the same through the layers and the equivalent soil,

$$v = k_{v1}\frac{h_1}{H_1} = k_{v2}\frac{h_2}{H_2} = \ldots = k_{vn}\frac{h_n}{H_n} = k_{v,eq}\frac{(h_1 + h_2 + \ldots + h_n)}{(H_1 + H_2 + \ldots + H_n)}$$

$$v = k_{v,eq}\frac{\left(\dfrac{vH_1}{k_{v1}} + \dfrac{vH_2}{k_{v2}} + \ldots + \dfrac{vH_n}{k_{vn}}\right)}{(H_1 + H_2 + \ldots + H_n)}$$

$$k_{v,eq} = \frac{(H_1 + H_2 + \ldots + H_n)}{\left(\dfrac{H_1}{k_{v1}} + \dfrac{H_2}{k_{v2}} + \ldots + \dfrac{H_n}{k_{vn}}\right)} \tag{6.29}$$

6.8 Flow Nets

In this section, we will study how to determine the flow rate of the seepage taking place in the soil beneath hydraulic structures such as dams, and assess their safety against *piping* and *uplift*. Let us consider the seepage within the soil beneath the concrete dam in Fig. 6.15a, where the downstream water level is taken as the datum. The path of a water molecule from upstream to downstream is called a *streamline*, which starts at the upstream ground level *AB* and finishes at the downstream ground level *CD*. The total head decreases steadily along the streamline from h_L at upstream to 0 at downstream, due to the energy dissipated to overcome the frictional resistance from the soil. The total head loss along any streamline from upstream to downstream is h_L. The top of the impervious stratum is a streamline, and the bottom of the concrete dam is also a streamline. There are thousands of streamlines within the flow region. The dashed line shown in Fig. 6.15a is an *equipotential line*, which connects the points having the same total head in the flow region. *AB* and *CD* are both equipotential lines, with total heads of h_L and 0, respectively. There are thousands of equipotential lines in the flow region.

A *flow net* is a network of selected streamlines and equipotential lines. The streamlines are selected such that the flow rate between adjacent streamlines (i.e., through a *flow channel*) is the same. The equipotential lines are selected such that the total head difference between two adjacent equipotential lines is the same. Figure 6.15b shows a flow net with three flow channels (four streamlines including the ones along the flow boundaries) and 10 *equipotential drops* (11 equipotential lines).

6.8.1 Laplace's Equation

Figure 6.16 shows the same concrete dam discussed previously where the total head $h(x,y)$ and the discharge velocities at a point (shown as an element) are shown with respect to a two-dimensional Cartesian coordinate system.

At steady state, the net flow into the element must be zero. Therefore,

$$v_x dy + v_y dx = \left(v_x + \frac{\partial v_x}{\partial x} dx\right) dy + \left(v_y + \frac{\partial v_y}{\partial y} dy\right) dx$$

$$\frac{\partial v_x}{\partial x} + \frac{\partial v_y}{\partial y} = 0 \tag{6.30}$$

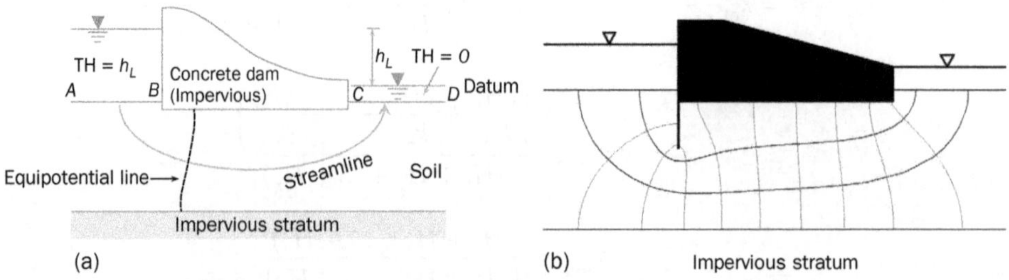

FIGURE 6.15 (a) Streamline and equipotential line, (b) flow net.

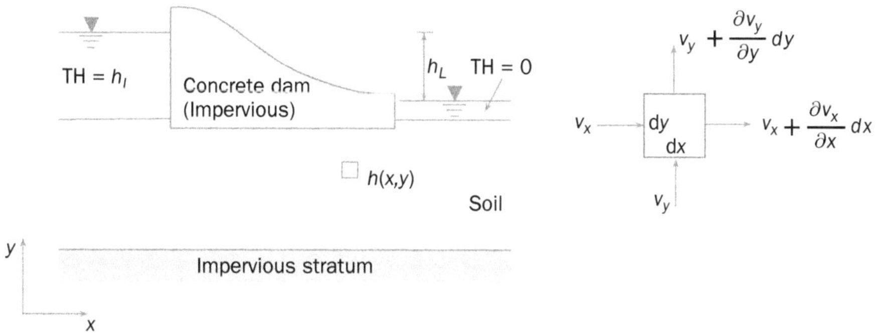

FIGURE 6.16 Total head and the discharge velocities at a point.

From Darcy's law,

$$v_x = -k_x \frac{\partial h}{\partial x} \tag{6.31}$$

$$v_y = -k_y \frac{\partial h}{\partial y} \tag{6.32}$$

where k_x and k_y are the hydraulic conductivities in x and y directions. Substituting Eqs. (6.31) and (6.32) in Eq. (6.30),

$$k_x \frac{\partial^2 h}{\partial x^2} + k_y \frac{\partial^2 h}{\partial y^2} = 0 \tag{6.33}$$

When the flow is in three dimensions, Eq. (6.33) becomes

$$k_x \frac{\partial^2 h}{\partial x^2} + k_y \frac{\partial^2 h}{\partial y^2} + k_z \frac{\partial^2 h}{\partial z^2} = 0 \tag{6.34}$$

In two-dimensional flow, it can be shown mathematically that the streamlines and equipotential lines intersect at 90° (see Fig. 6.15) if the hydraulic conductivity of the soil is isotropic (i.e., $k_x = k_y$). Equation (6.33) becomes

$$\frac{\partial^2 h}{\partial x^2} + \frac{\partial^2 h}{\partial y^2} = 0 \tag{6.35}$$

Flow net is the graphical solution to Laplace's equation [Eq. (6.35)] consisting of two families of curves—stream lines and equipotential lines, which are orthogonal.

6.8.2 Equation for Flow Rate

Let us consider a flow net similar to the one shown in Fig. 6.15b, where there are N_f flow channels carrying the same discharge of ΔQ, and N_d equipotential drops with the same total head loss Δh between any two adjacent equipotential lines. Thus,

$$Q = N_f \Delta Q \tag{6.36}$$

$$h_L = N_d \Delta h \tag{6.37}$$

A portion of this flow net is shown in Fig. 6.17. Let us consider the region *PQRS* where the streamlines and equipotential lines are perpendicular to each other. The width of the flow channel is a, and the distance between the two equipotential lines is b.

The hydraulic gradient between the equipotential lines *PS* and *QR* is $\Delta h/b$. By Darcy's law, and assuming isotropic permeability, the discharge velocity in the flow channel is

$$v = k\frac{\Delta h}{b}$$

Considering a unit width perpendicular to the plane,

$$\Delta Q = k\left(\frac{\Delta h}{b}\right)(a \times 1) \tag{6.38}$$

From Eqs. (6.36), (6.37), and (6.38),

$$Q = kh_L \frac{N_f}{N_d}\frac{a}{b}$$

If the flow net is drawn such that all the "boxes" are approximately *curvilinear squares* where it can be assumed that $a \approx b$,

$$Q = kh_L \frac{N_f}{N_d} \tag{6.39}$$

By counting N_f and N_d in a flow net, the flow rate *per unit width of the dam* can be computed using Eq. (6.39). A good curvilinear square must be able to accommodate a circle within it, touching all sides. Figure 6.18 shows a flow net around a vertical sheet pile, drawn using a software.

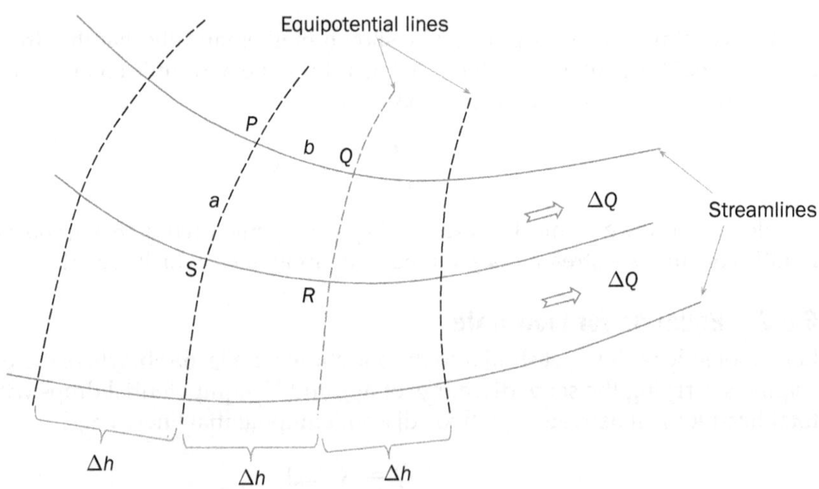

FIGURE 6.17 A portion of a flow net.

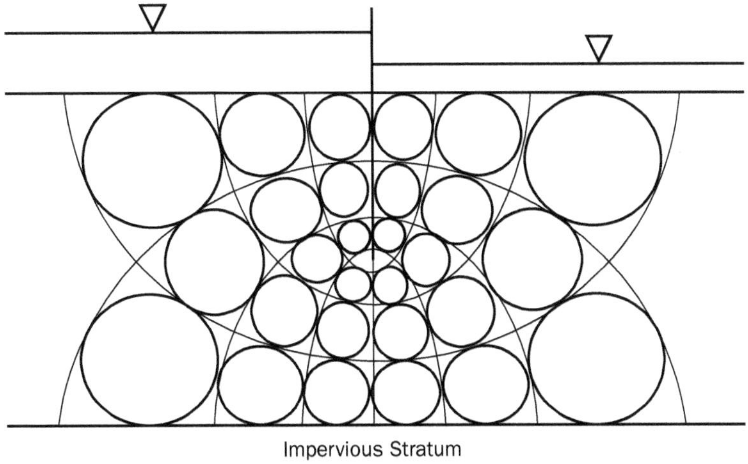

Figure 6.18 Flow net around a sheet pile. (Courtesy of Dr. Thushara Madanayaka, SLTC, Sri Lanka.)

Example 6.8

A concrete dam with a sheet pile cut-off wall at the upstream end is shown in Fig. 6.19, for which a flow net is shown in Fig. 6.15b. The total head loss from upstream to downstream is 1.0 m. Hydraulic conductivity of the underlying soil is 3.2×10^{-3} cm/s. Determine the flow rate per unit width of the dam per day.

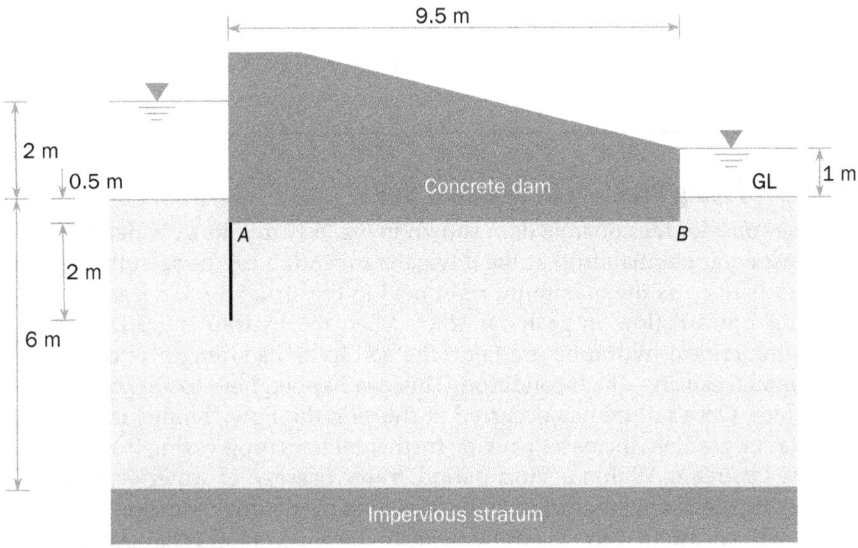

Figure 6.19 Concrete dam with upstream sheet pile.

Solution From Fig. 6.15b,

$$N_f = 3, N_d = 10$$

$$Q = kh_L \frac{N_f}{N_d} = (3.2 \times 10^{-5})(1.0)\left(\frac{3}{10}\right) = 0.96 \times 10^{-5} \text{ m}^3\text{/s per m width}$$

$$= (0.96 \times 10^{-5})(24 \times 3600) = \mathbf{0.83\ m^3\ per\ day\ per\ m\ width}$$

Example 6.9

In the concrete dam shown in Example 6.8 (Fig. 6.19), determine the pore water pressures at A and B and hence the uplift on the dam.

Solution Let us take the downstream water level as the datum.

The total head loss from upstream to downstream, $h_L = 1.0$ m

The total head loss between any two adjacent equipotential lines = 1.0/10 = 0.1 m

The total head at upstream and downstream ground levels are 1 and 0 m, respectively. There are about 3.5 equipotential drops from the upstream to point A, and hence the total head at A is 0.65 m. The elevation head at A is −1.5 m. Therefore, the pressure head at A is 2.15 m.

The pore water pressure at $A = 2.15 \times 9.81 =$ **21.1 kN/m²**

There is 0.9 equipotential drop from the downstream to point B (or 9.1 drops from upstream) where the total head is $0.9 \times 0.1 = 0.09$ m. The elevation head at B is −1.5 m. Therefore, the pressure head at B is 1.59 m.

The pore water pressure at $B = 1.59 \times 9.81 =$ **15.6 kN/m²**

Assuming linear variation of pore water pressure between the points A and B, the uplift force on the bottom of the concrete dam can be computed as

$$U = \frac{(21.1 + 15.6)(9.5)}{2} = \mathbf{174.3\,kN\ per\,m\,width}$$

By following the method discussed in Example 6.9, the pore water pressure at any point within the flow region can be determined.

6.8.3 Piping Problem in Granular Soils

Let us consider the concrete dam shown in Fig. 6.19 and its flow net in Fig. 6.15b. From the last equipotential drop at the downstream end, it can be seen that the exit hydraulic gradient i_{exit} is the maximum right next to the structure. We learned in Sec. 6.6 that during upward flow in granular soils, when the hydraulic gradient becomes larger than the critical hydraulic gradient, the soil loses its strength with the grains barely in contact, causing quick condition. This can happen here too at the exit location near the dam. Once failure has occurred at the exit, the flow channel is shortened and the hydraulic gradient increases causing further failures progressing from downstream end to the upstream. Within a short period, a free passage of water or a "pipe" is formed under the dam, causing catastrophic flooding at the downstream. Therefore, it is necessary to ensure that the maximum exit hydraulic gradient that occurs near the structure is very much less than the critical hydraulic gradient i_c. Many dam failures have occurred worldwide due to piping problems. The factor of safety with respect to piping is defined as

$$F_{\text{piping}} = \frac{i_c}{i_{\text{exit}}} \tag{6.40}$$

It is recommended that F_{piping} is at least 3. Larger values are sometimes suggested conservatively.

Figure 6.20 shows a concrete dam with a sheet pile, where h_L is the total head loss from upstream to downstream. The arrows along the structure show the shortest flow path from upstream to downstream, and let the length of this line be L. Bligh (1910) suggested the first *line of creep method*, where he proposed that for a structure to be safe from piping, L has to be greater than $C_w h_L$, where C_w is an empirical constant in the range of 4–18, depending on the soil type. This was later improved by Lane (1934), based on a database of 280 dams, where 24 have failed. He assumed anisotropic flow and used a weighting factor of 1/3 to the horizontal flow paths in computing the line of creep length L. It can be seen that the piping potential is (a) proportional to the head loss (h_L) from upstream to downstream, and (b) inversely proportional to the length of the flow path. Values of C_w (safe weighted creep ratio), for Lane's weighted creep method are given in Table 6.1 for different soils. Lane's C_w values are significantly less than the ones originally given by Bligh, due to the fact that the horizontal flow paths are multiplied by 1/3. Richards and Reddy (2007) provided a good review of piping phenomenon in earth dams and listed more than 250 case histories of piping-related failures.

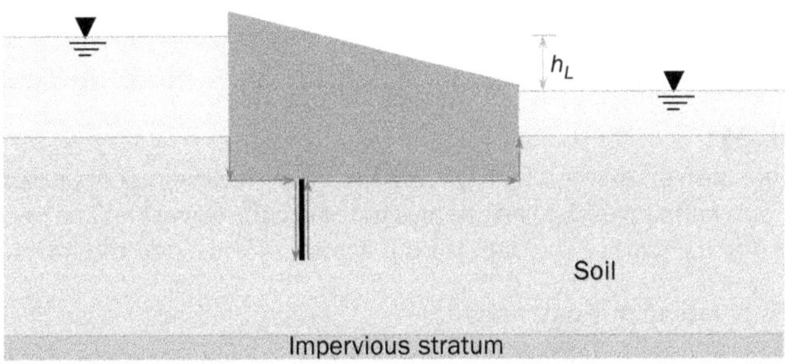

FIGURE 6.20 Line of creep method.

Soil type	C_w	Soil type	C_w
Very fine sand and silt	8.5	Coarse gravels and cobbles	3.0
Fine sand	7.0	Soft clay	3.0
Medium sand	6.0	Boulders with some cobbles and gravels	2.5
Coarse sand	5.0	Medium clay	2.0
Fine gravel	4.0	Hard clay	1.8
Medium gravel	3.5	Very hard clay	1.6

TABLE 6.1 Lane's C_w for the Weighted Creep Method

Example 6.10

In the concrete dam shown in Example 6.8 (Figs. 6.15b and 6.19), if the soil is granular and the saturated unit weight is 19.9 kN/m^3, find the factor of safety with respect to piping.

What is the weighted creep ratio for the dam?

Solution The critical hydraulic gradient is

$$i_c = \frac{\gamma'}{\gamma_w} = \frac{19.9 - 9.81}{9.81} = 1.03$$

Considering the curvilinear square at the downstream end of the dam, the maximum exit hydraulic gradient, which occurs along the vertical face of the dam, can be determined as

$$i_{\text{exit}} = \frac{0.09}{0.5} = 0.18$$

$$F_{\text{piping}} = \frac{i_c}{i_{\text{exit}}} = \frac{1.03}{0.18} = 5.7$$

The weighted creep ratio of the dam is given by

$$\frac{L}{h_L} = \frac{2.5 + 2.0 + \frac{1}{3} \times 9.5 + 0.5}{1.0} = 8.2$$

Example 6.11

The concrete dam shown in Fig. 6.21 consists of an impervious upstream blanket and a sheet pile at the downstream end, in an attempt to reduce the flow rate. The permeability of the granular soil is 3×10^{-3} cm/s. The dam is built across a 15-m-wide river. Determine the following.

a. Flow rate in m^3/day, and
b. Factor of safety with respect to piping, assuming saturated unit weight of 19.0 kN/m^3 for the soil.

Solution

Part a

$$Q = kh_L \frac{N_f}{N_d} = (3.0 \times 10^{-5})(3.0)\left(\frac{4}{12}\right) = 3 \times 10^{-5} \text{ m}^3/\text{s per m width}$$

$$= (3 \times 10^{-5})\,(24 \times 3600)\,(15) = 38.9 \text{ m}^3/\text{day}$$

Part b The critical hydraulic gradient is

$$i_c = \frac{\gamma'}{\gamma_w} = \frac{19.0 - 9.81}{9.81} = 0.94$$

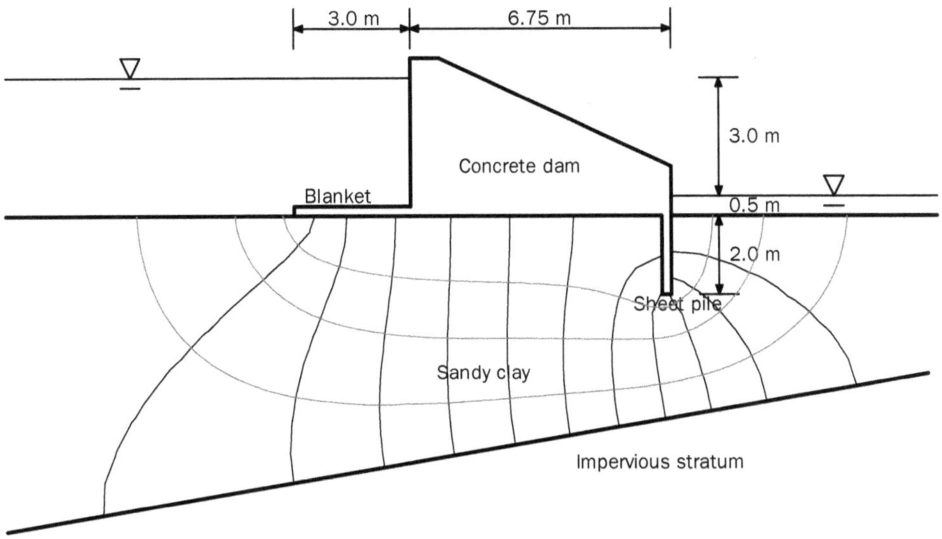

FIGURE 6.21 Diagram for Example 6.11.

Total head loss between two adjacent equipotential lines $\Delta h = 3/12 = 0.25$ m
From the curvilinear square at the downstream end next to the dam,

$$i_{\text{exit}} = \frac{0.25}{1.0} = 0.25$$

$$F_{\text{piping}} = \frac{i_c}{i_{\text{cexit}}} = \frac{0.94}{0.25} = \mathbf{3.8}$$

Example 6.12

To carry out an excavation in a submerged area, two sheet piles are driven in parallel, as shown in Fig. 6.22, forming an enclosure. The hydraulic conductivity of the sandy soil is 2.2×10^{-2} cm/s, and the saturated unit weight is 19.0 kN/m^3.

a. Draw a flow net.
b. At what rate should the water be pumped out from the excavation to maintain the water levels as shown?
c. What is the pore water pressure at the tip of the sheet pile?
d. What is the factor of safety with respect to piping?

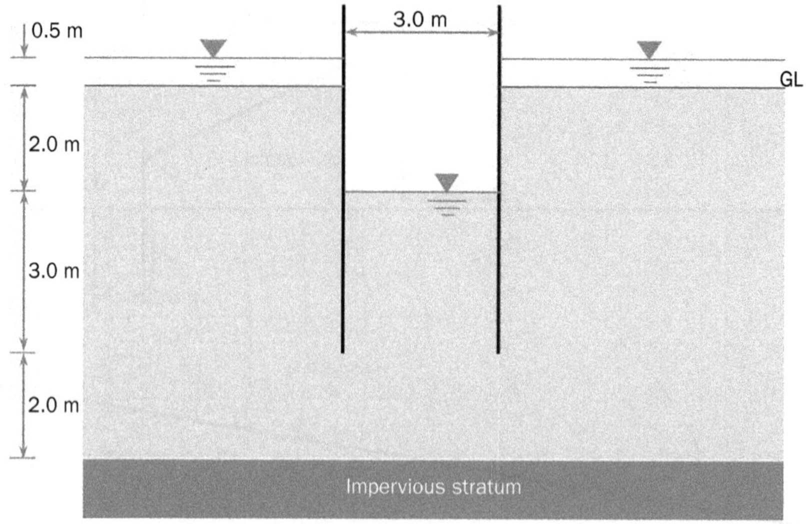

FIGURE 6.22 Excavation between sheet piles.

Solution

Part a See Fig. 6.23.

Part b Since the flow net is symmetrical about the centerline, let us consider the left half only for the computations.

$N_f = 2.5$ (note: only half a channel at the bottom); $N_d = 10$; $h_L = 2.5$ m

$$\frac{Q}{2} = kh_L \frac{N_f}{N_d} = (2.2 \times 10^{-4})(2.5)\left(\frac{2.5}{10}\right)(24 \times 3600) = 11.88 \text{ m}^3\text{/day per m width}$$

$$Q = 23.8 \text{ m}^3\text{/day per m width}$$

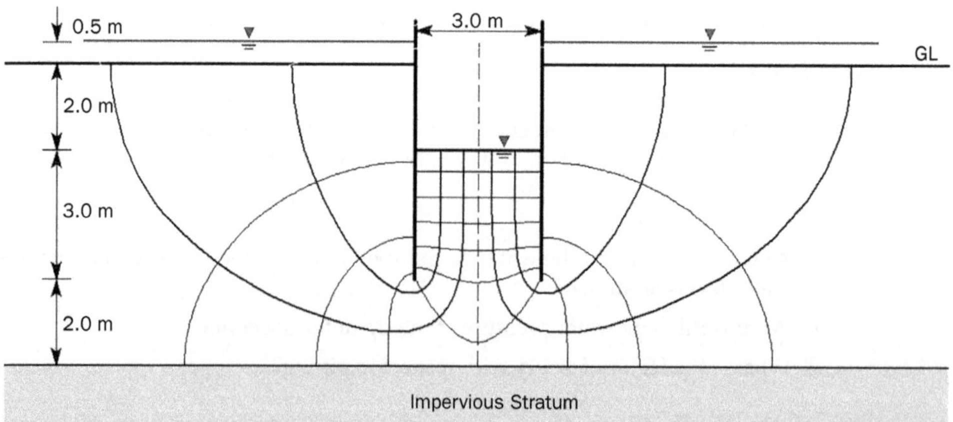

FIGURE 6.23 Flow net for Example 6.12. (Courtesy of Dr. Thushara Madanayaka, SLTC, Sri Lanka.)

Part c Let us take the excavation level within the sheet pile enclosure as the datum.

Total head loss between two adjacent equipotential lines, $\Delta h = 2.5/10 = 0.25$ m

Total heads at the upstream and downstream ends are 2.5 and 0 m, respectively. It takes about 6.2 (estimated from the flow net) equipotential drops from the downstream (excavation level) to reach the sheet pile tip.

Therefore, the total head at the tip of the sheet pile $= 6.2\ \Delta h = 1.55$ m

Elevation head at the sheet pile tip $= -3.0$ m

Therefore, the pressure head = Total head – elevation head $= 1.55 - (-3.0) = 4.55$ m

Pore water pressure at the sheet pile tip $= 4.55 \times 9.81 = 44.6\ \text{kN/m}^2$

Part d The critical hydraulic gradient is given by

$$i_c = \frac{\gamma'}{\gamma_w} = \frac{19 - 9.81}{9.81} = 0.94$$

Considering the curvilinear square near the sheet pile at the exit,

$$i_{\text{exit}} = \frac{0.25}{0.46\,(\text{measured from the figure})} = 0.54$$

$$F_{\text{piping}} = \frac{i_c}{i_{\text{cexit}}} = \frac{0.94}{0.54} = \mathbf{1.7}$$

6.8.4 Flow Nets in Anisotropic Soils

When the hydraulic conductivity is anisotropic, generally $k_x > k_y$. Therefore, Eq. (6.35) is not valid and the stream lines do not intersect equipotential lines at 90°. That makes it difficult to draw the flow net.

$$k_x \frac{\partial^2 h}{\partial x^2} + k_y \frac{\partial^2 h}{\partial y^2} = 0$$

$$\frac{1}{\left(\frac{k_y}{k_x}\right)} \frac{\partial^2 h}{\partial x^2} + \frac{\partial^2 h}{\partial y^2} = 0$$

Substituting $x' = \sqrt{\frac{k_y}{k_x}}x$,

$$\frac{\partial^2 h}{\partial x'^2} + \frac{\partial^2 h}{\partial y^2} = 0 \qquad (6.41)$$

Therefore, if the flow net is transformed from x-y coordinate system to x'-y coordinate system, Eq. (6.41) takes the same form as Eq. (6.35). Therefore, in x'-y coordinate system, the stream lines and equipotential lines are orthogonal. It just requires to rescale (i.e., shorten) the dimensions in x direction to $x' = \sqrt{\frac{k_y}{k_x}}x$. The system can be treated as having isotropic hydraulic conductivity of $\sqrt{k_x k_y}$. The computations of flow rate and pore water pressures are similar to the methods discussed in Sec. 6.8.2.

6.9 Design of Granular Filters

Protective granular filters are used in earth dams, trenches, and backfills behind retaining walls where seepage takes place. The purpose of the filter is to protect the upstream soils and ensure that the fines are not washed away. Here, the pore channels must be small enough to prevent the migration of the fines. This is known as the *retention criterion*. While doing this, pore channels must be large enough to allow free flow of water and prevent any build-up of pore water pressure. This is known as the *permeability criterion*. They are two contradicting criteria.

Terzaghi and Peck (1948) suggested the following condition to satisfy the retention criteria.

$$D_{15,\text{filter}} < 4 \text{ to } 5\, D_{85,\text{soil}} \tag{6.42}$$

To satisfy the permeability criterion, they suggested that

$$D_{15,\text{filter}} > 4 \text{ to } 5 D_{15,\text{soil}} \tag{6.43}$$

US Navy (1971) suggests two additional conditions to satisfy the retention criterion. They are:

$$D_{15,\text{filter}} < 20\, D_{15,\text{soil}} \tag{6.44}$$

$$D_{50,\text{filter}} < 25 D_{50,\text{soil}} \tag{6.45}$$

To avoid segregation, it is generally suggested that the filter material is not gap graded, has low uniformity coefficient C_u, and is free of grains larger than 75 mm.

The retention and permeability criteria define the upper and lower bounds for the grain size distribution curve of the filter material. From these, a possible band can be established within which the grain size distribution of the filter material should lie. Generally, the grain size distribution curves for the soil and the filter material are approximately of the same shape.

6.10 Seepage through an Embankment on an Impervious Base

Let us consider a homogeneous earth dam on an impervious base as shown in Fig. 6.24a. Here, the flow is between the impervious base at the bottom and the free surface *AB*. This is an *unconfined flow* problem, where one boundary of the flow domain is the free surface that has to be determined. In the previous sections, both upper and lower boundaries are known and they were *confined flow* problems where the flow is confined between the two boundaries.

It is assumed that the slope of the free surface is the hydraulic gradient, which does not vary with depth (Dupuit, 1863). The discharge through any vertical section of the dam is given by

$$q = -ky\frac{dy}{dx} \tag{6.46}$$

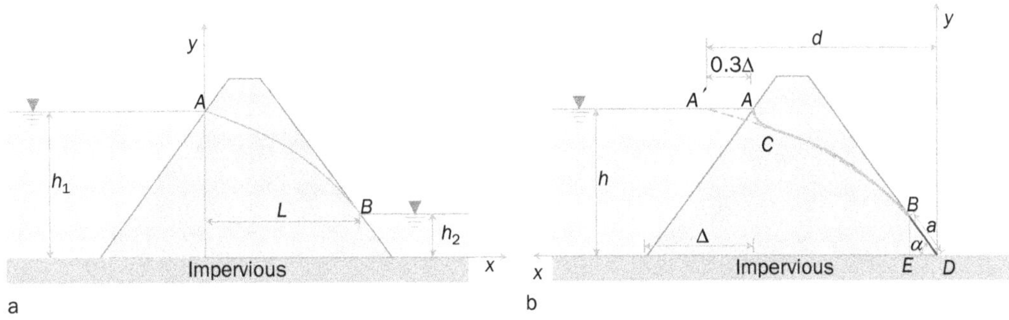

FIGURE 6.24 (a) Dupuit's solution, (b) Schaffernak's solution with Casagrande's correction.

Integrating Eq. (6.46) with boundary conditions $x = 0, y = h_1$ and $x = L, y = h_2$, gives

$$q = \frac{k(h_1^2 - h_2^2)}{2L} \tag{6.47}$$

Equation (6.47) is known as Dupuit's formula, based on a parabolic free surface.

Dupuit's solution was modified to reflect more realistic exit and entry conditions by Schaffernak (1917) and Casagrande (1937) as shown in Fig. 6.24b. Here, it is assumed that there is no tail water at the downstream. The free surface is *ACB*, with *A′CB* being parabolic. The length of the surface of seepage *BD* at the downstream side is *a*, which has to be determined. Let us take the *x* axis going from right to left, with the origin at *D*.

As before, it is assumed that the hydraulic gradient is given by the slope of the free surface, which does not vary with depth. Considering the flow across the vertical section *BE*,

$$q = -ky\frac{dy}{dx} = ka\sin\alpha\tan\alpha \tag{6.48}$$

Integrating Eq. (6.48),

$$\int_{a\sin\alpha}^{h} y\,dy = a\sin\alpha\tan\alpha \int_{a\cos\alpha}^{d} dx$$

$$a = \frac{d}{\cos\alpha} - \sqrt{\frac{d^2}{\cos^2\alpha} - \frac{h^2}{\sin^2\alpha}} \tag{6.49}$$

Once *a* is determined from Eq. (6.49), by substituting this value in Eq. (6.48), the flow rate can be computed.

6.11 Method of Fragments

Method of fragments is an approximate method for solving confined flow problems. It was originally developed by Pavlosky (1956) and was brought to the attention of the western world by Harr (1962). The method enables quick estimates of the flow rate and the exit hydraulic gradient, without drawing the flow net.

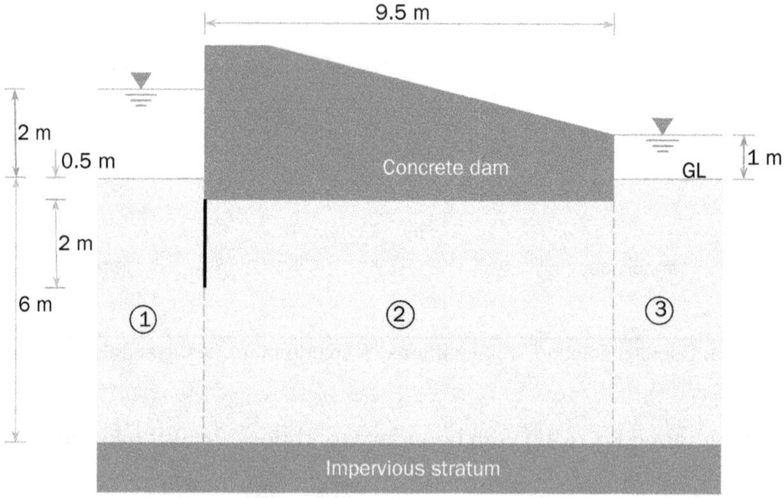

FIGURE 6.25 Flow domain divided into fragments.

It is assumed in this method that the equipotential lines at certain critical points (e.g., at the tip of a sheet pile) within the flow domain are vertical, which divide the flow domain into zones named as *fragments*. Figure 6.25 shows seepage beneath a concrete dam with a sheet pile at the upstream end. It is assumed that the two dashed vertical lines at the upstream and downstream ends (i.e., critical points) divide the flow domain into three fragments 1, 2, and 3. Hypothetically, if a flow net is drawn with N_f flow channels, there are N_d equipotential drops, with N_{d1}, N_{d2}, and N_{d3} drops within fragments 1, 2, and 3, respectively. The ratio of the number of equipotential drops to the number of flow channels in each fragment is defined as the *form factor* of the fragment, which is a dimensionless number that reflects the resistance to flow. In the ith fragment, the form factor is written as

$$\Phi_i = \frac{N_{di}}{N_f} \tag{6.50}$$

The flow rate Q beneath the dam per unit width is given by

$$Q = kh_L \frac{N_f}{N_d} \tag{6.39}$$

where h_L is the total head loss from upstream to downstream.

$$\frac{N_d}{N_f} = \frac{N_{d1} + N_{d2} + N_{d3}}{N_f} = \Phi_1 + \Phi_2 + \Phi_3 = \sum_{i=1}^{3} \Phi_i \tag{6.51}$$

When the flow domain is divided into n fragments,

$$Q = \frac{kh_L}{\sum_{i=1}^{n} \Phi_i} \tag{6.52}$$

If the total head loss within the fragments 1 to n are h_1 to h_n,

$$h_1 + h_2 + \ldots + h_n = h_L \tag{6.53}$$

Since the flow rate is the same through all the fragments,

$$Q = \frac{kh_1}{\Phi_1} = \frac{kh_2}{\Phi_2} = \ldots = \frac{kh_n}{\Phi_n} = \frac{kh_L}{\sum_{i=1}^{n} \Phi_i} \tag{6.54}$$

From Eq. (6.54), the head loss within a fragment can be determined from the form factors as

$$h_i = \frac{\Phi_i}{\sum_{i=1}^{n} \Phi_i} h_L \tag{6.55}$$

6.11.1 Form Factors

From the way the form factor is defined in Eq. (6.50), it is obvious that this factor depends only on the geometry of the fragment, including where the water enters and exits the fragment and the dimensions of the fragment. Harr (1962) defined six fragments that can be condensed into two, fragments A and B, as suggested by Griffiths (1984). For problems such as seepage beneath concrete dams with sheet piles, fragments A and B are adequate and these are discussed in some textbooks. Griffiths (1984) defined a new fragment C to cover some situations such as double-walled cofferdams, which cannot be handled by Harr's six fragments. The fragment geometries and the methods for deriving the form factors are shown in Fig. 6.26. Fragment A requires Fig. 6.27 for determining the form factor. Graphical solution for the form factor of type C fragment was presented by Griffiths (1984). This was approximated by Madanayaka and Sivakugan (2016) as

$$\Phi_C = \ln\left(1 + \frac{s}{T}\right) + 0.85\frac{s}{L} + 0.4\exp\left(-\frac{L}{2T}\right) \tag{6.56}$$

Equation (6.56) works well for $L \leq T$. For $L > T$, it is suggested to use Eq. (6.56) assuming $L = T$ and reducing Φ by 10%.

In the special case of type A fragments with $b = 0$, Madanayaka and Sivakugan (2016) showed that the form factor can be approximated as

$$\Phi_A = 0.43\exp\left(\frac{5}{3}\frac{s}{T}\right) \tag{6.57}$$

6.11.2 Exit Hydraulic Gradient

The streamlines exit in the last (i.e., nth) fragment, where the exit hydraulic gradient is the maximum right next to the structure. Within this fragment, the head loss is h_n. This fragment is generally of type A with $b = 0$. Harr (1962) proposed a graphical solution shown in Fig. 6.28 to determine the maximum exit gradient for type A fragment with $b = 0$. Griffiths (1984) provided graphical solutions for situations where the exit

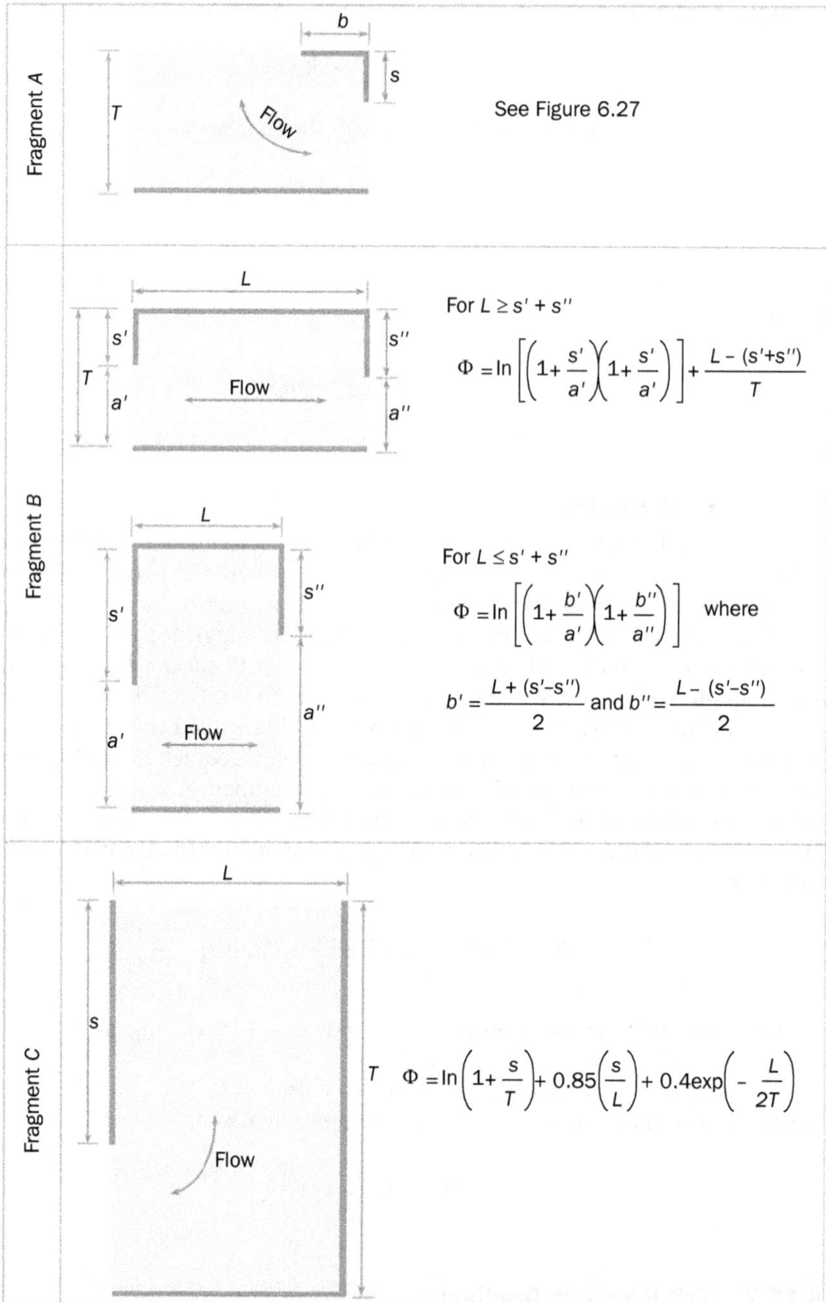

FIGURE 6.26 Form factors for the three fragments *A*, *B*, and *C*.

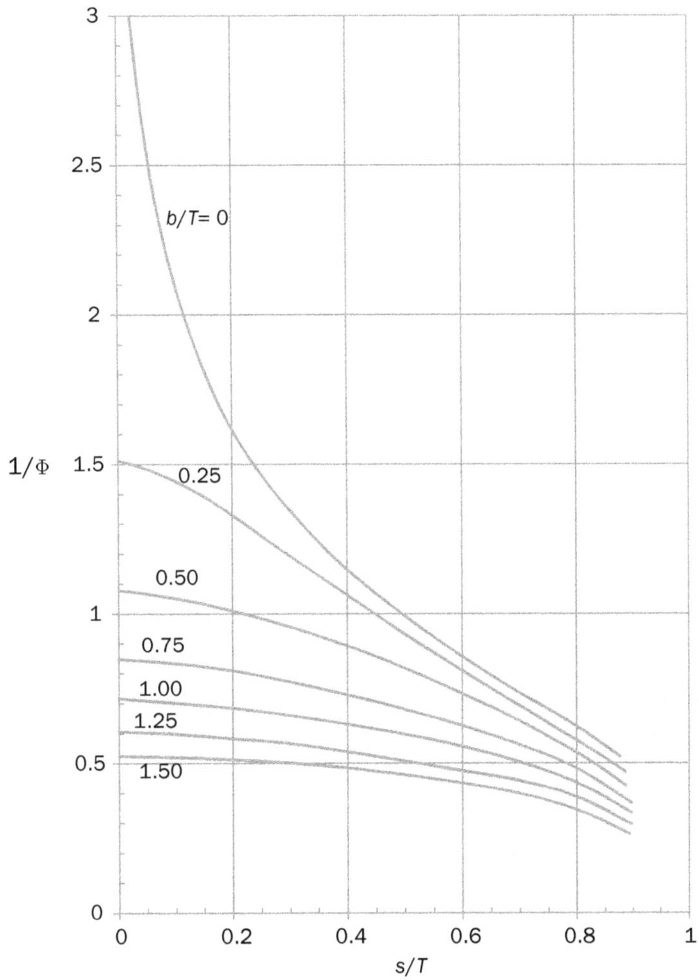

FIGURE 6.27 Form factor for fragment *A*.

fragment is of type *C*, which can happen in cofferdams. Madanayaka and Sivakugan (2016) approximated Griffiths' (1984) solution for type C fragment as

$$i_{exit} = 1.5\left[\frac{1-\exp\left(-\frac{s}{T}\right)}{\left(\frac{s}{T}\right)^{0.3\left(\frac{L}{T}\right)+0.65}}\right]\left[1-\frac{L}{2T}\right]\left(\frac{h_n}{s}\right) \tag{6.58}$$

Equation (6.58) works well for $L \le T$. For $L > T$, it is suggested that i_{exit} is estimated assuming $L = T$, which is generally conservative.

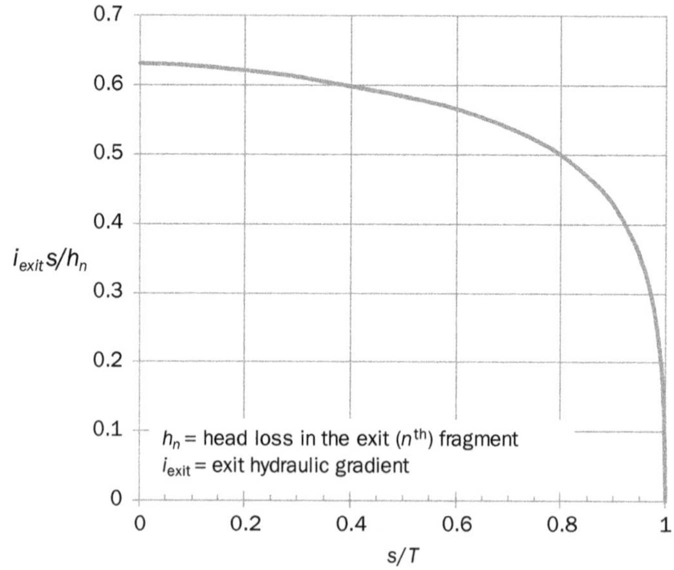

FIGURE 6.28 Exit hydraulic gradient for fragment *A* with $b = 0$.

Example 6.13

Flow takes place between two sheet piles as shown in Fig. 6.29a. Permeability of the soil is 2.2×10^{-2} cm/s. Determine

(a) the flow rate in m^3/day per m, and

(b) the maximum exit hydraulic gradient using the method of fragments.

This is the same problem as in Example 6.12. Compare the results from the two methods.

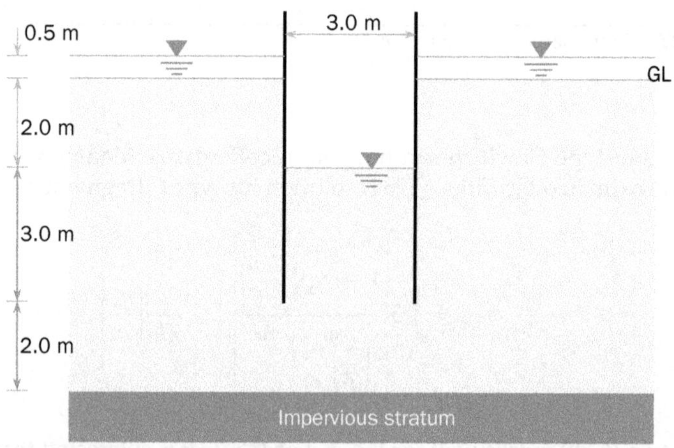

FIGURE 6.29a Seepage into a cofferdam.

Solution Let us analyze the left half of the configuration that is symmetrical about the vertical centerline, which can be treated as an impervious boundary. It is assumed that the dashed vertical line at the bottom of the sheet pile is an equipotential line that divides the left half of the flow domain into two fragments, 1 and 2. Fragment 1 is of type *A* and fragment 2 is of type *C*.

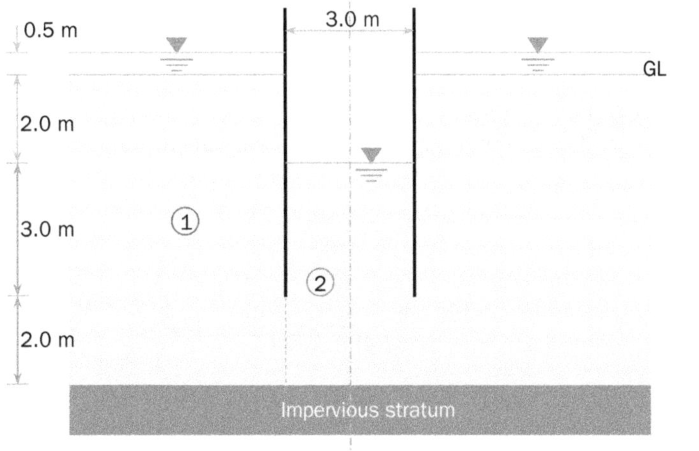

FIGURE 6.29b Flow domain divided into fragments.

Part a: Flow rate

The form factors can be calculated as follows.

For fragment 1, $s = 5$ m, $T = 7$ m.

Using Eq. (6.57),

$$\Phi_1 = 0.43\exp\left(\frac{5}{3}\frac{s}{T}\right) = 0.43\exp\left(\frac{5}{3}\times\frac{5}{7}\right) = 1.41$$

The same value can also be obtained from Fig. 6.27 with $s/T = 0.714$ and $b/T = 0$.

For fragment 2, $s = 3$ m, $T = 5$ m, and $L = 1.5$ m.

Using Eq. (6.56),

$$\Phi_2 = \ln\left(1 + \frac{s}{T}\right) + 0.85\frac{s}{L} + 0.4\exp\left(-\frac{L}{2T}\right)$$

$$\Phi_2 = \ln\left(1 + \frac{3}{5}\right) + 0.85\frac{3}{1.5} + 0.4\exp\left(-\frac{1.5}{2\times 5}\right) = 2.51$$

From Eq. (6.54),

$$Q = \frac{kh_L}{\sum_{i=1}^{n} \Phi_i} = \frac{(2.2 \times 10^{-4})(2.5)}{1.41 + 2.51} = 1.403 \times 10^{-4}\,\text{m}^3/\text{s per m width}$$

$$= 1.403 \times 10^{-4} \times 24 \times 3600 = 12.12\ \text{m}^3/\text{day per m width}$$

Considering the two halves,

$$Q = \mathbf{24.2\ m^3/day\ per\ m\ width.}$$

Part b: Maximum exit hydraulic gradient

The total head loss from fragment 2 can be determined from Eq. (6.55) as

$$h_2 = \frac{\Phi_2}{\sum_{i=1}^{2} \Phi_i} h_L$$

$$h_2 = \left(\frac{2.51}{1.41 + 2.51}\right) 2.5 = 1.60\ \text{m}$$

From Eq. (6.58),

$$i_{\text{exit}} = 1.5 \left[\frac{1 - \exp\left(-\frac{s}{T}\right)}{\left(\frac{s}{T}\right)^{0.3\left(\frac{L}{T}\right)+0.65}}\right]\left[1 - \frac{L}{2T}\right]\left(\frac{h_2}{s}\right)$$

$$i_{\text{exit}} = 1.5 \left[\frac{1 - \exp\left(-\frac{3}{5}\right)}{\left(\frac{3}{5}\right)^{0.3\left(\frac{1.5}{5}\right)+0.65}}\right]\left[1 - \frac{1.5}{2 \times 5}\right]\left(\frac{1.60}{3}\right) = \mathbf{0.45}$$

Note: A finite element analysis gives flow rate of 24.5 m³/day per m and maximum exit hydraulic gradient of 0.43, showing excellent agreement with the method of fragments solution above. The values determined from the flow net are given in Example 6.12. These are summarized as follows:

	Flow net	Method of fragment	Finite element analysis
Flow rate (m³/day per m)	23.8	24.2	24.5
i_{exit}	0.54	0.45	0.43

Example 6.14

Determine the pore water pressure at the tip of the sheet pile in Example 6.12 using the method of fragments.

Solution The form factors for the two fragments in the right half of the configuration were determined in Example 6.13 as

$\Phi_1 = 1.41$ and $\Phi_2 = 2.51$

$h_L = 2.5$ m

$$h_1 = \left(\frac{1.41}{1.41 + 2.51}\right) \times 2.5 = 0.90 \text{ m}$$

Taking the excavation level as the datum (see Fig. 6.29b), the total head at the upstream is 2.5 m.

Therefore, the total head at the tip of the sheet pile = 2.5 – 0.9 = 1.6 m

Elevation head at the tip = –3.0 m

Therefore, the pressure head at the tip of the sheet pile = 1.6 – (–3.0) = 4.6 m

Pore water pressure at the tip of the sheet pile = 4.6 × 9.81 = **45.1 kN/m²**, which compares well with the 44.6 kN/m² determined in Example 6.12.

Example 6.15

Let us consider the concrete dam shown in Fig. 6.25, with a sheet pile at the upstream end. The permeability of the underlying soil is 3.2×10^{-3} cm/s. Determine

(a) the flow rate, and

(b) the maximum exit gradient using the method of fragments.

This is the same problem as in Examples 6.8, 6.9, and 6.10. Compare the values determined in parts a and b.

Solution

Part a: Flow rate

The three fragments 1, 2, and 3 are of the types *A*, *B*, and *A*, respectively. The form factors can be computed as follows.

For fragment 1 (type *A*), $b = 0$, $s = 2.5$ m, and $T = 6.0$ m.

From Eq. (6.57),

$$\Phi_1 = 0.43\exp\left(\frac{5}{3}\frac{s}{T}\right) = 0.43\exp\left(\frac{5}{3} \times \frac{2.5}{6}\right) = 0.861$$

For fragment 2 (type *B*), $s' = 2.0$ m, $a' = 3.5$ m, $s'' = 0$, $a'' = 5.5$ m, $L = 9.5$ m, and $T = 5.5$ m (see Fig. 6.26).

From Fig. 6.26, since $L > s' + s''$,

$$\Phi_2 = \ln\left[\left(1+\frac{s'}{a'}\right)\left(1+\frac{s''}{a''}\right)\right] + \frac{L-(s'+s'')}{T}$$

$$\Phi_2 = \ln\left[\left(1+\frac{2.0}{3.5}\right)\left(1+\frac{0}{5.5}\right)\right] + \frac{9.5-(2.0+0)}{5.5} = 1.816$$

For fragment 3 (type A), $b = 0$, $s = 0.5$ m, and $T = 6.0$ m.
From Eq. (6.57),

$$\Phi_3 = 0.43\exp\left(\frac{5}{3}\frac{s}{T}\right) = 0.43\exp\left(\frac{5}{3}\times\frac{0.5}{6}\right) = 0.494$$

From Eq. (6.54),

$$Q = \frac{kh_L}{\sum_{i=1}^{n}\Phi_i} = \frac{(3.2\times10^{-5})(1.0)}{0.861+1.816+0.494} = 1.009\times10^{-5}\,\text{m}^3/\text{s per m width}$$

$$= 1.009\times10^{-5}\times24\times3600 = \mathbf{0.87\ m^3/day\ per\ m\ width}$$

Part b: Maximum exit hydraulic gradient
From Eq. (6.55),

$$h_3 = \frac{\Phi_3}{\sum_{i=1}^{3}\Phi_i}h_L$$

$$h_3 = \left(\frac{0.494}{0.861+1.816+0.494}\right)1.0 = 0.156\ \text{m}$$

From Fig. 6.28, for $s/T = 0.5/6.0 = 0.0833$

$$\frac{i_{exit}s}{h_3} = 0.625$$

Therefore,

$$i_{exit} = \frac{0.625\times0.156}{0.5} = \mathbf{0.195}$$

Note: A finite element analysis gives flow rate of 0.83 m³/day per m width and maximum exit gradient of 0.13. While the flow rate agrees well, there is difference in the i_{exit} value. A closer look at the flow net shows that the equipotential line at the downstream end of the dam is not at all vertical (as assumed in MoF), which is the reason for the discrepancy. MoF gives a conservative estimate of i_{exit}.

	Flow net	**Method of fragment**	**Finite element analysis**
Flow rate (m³/day per m)	0.83	0.87	0.83
i_{exit}	0.18	0.195	0.13

Cofferdam walls are not always parallel as shown in some examples herein. In plan view, they can be circular, square, rectangular, or any arbitrary shape to suit the excavation zone. Here, the flow is no more two-dimensional and it is three-dimensional. Some simple methods to determine the flow rate and the maximum exit hydraulic gradients for such situations were presented by Madanayaka and Sivakugan (2019).

Permeability and seepage discussed in this chapter are for saturated soils. It gets more complex in unsaturated soils, which are three phase media, containing soil grains, water, and air. These are discussed briefly in App. A.

6.12 Main Points

1. In flow through soils, the velocity head is negligible and hence
   Total head = Pressure head + Elevation head
2. Always define the datum clearly in seepage problems.
3. Hydraulic gradient is the total head loss per unit length along the flow path. It is a dimensionless quantity.
4. Pore water pressure = Pressure head × γ_w
5. In soils, the hydraulic conductivity k varies in a wide range from less than 10^{-7} cm/s to more than 10 cm/s.
6. Hydraulic conductivity can be determined in the laboratory by constant head tests for coarse-grained soils and falling head tests for fine-grained soils.
7. When water flows upward, effective stress decreases and pore water pressure increases; when the flow is downward, effective stress increases and pore water pressure decreases.
8. In upward flow through granular soils, when the hydraulic gradient becomes larger than the critical hydraulic gradient i_c, quick condition occurs and there is no intergranular contact and hence the effective stress is zero.
9. Flow net consists of selected streamlines and equipotential lines, forming curvilinear squares. Equipotential lines are perpendicular to the streamlines.
10. Factor of safety with respect to piping in granular soils must be greater than 3.

Review Exercises

6.1 Figure RE6.1 shows water flowing through a soil specimen contained in a pipe with two manometers (to measure the water levels) 450 mm apart. The height difference between the manometer water levels is 90 mm. The length of the soil specimen is 750 mm. What is the total head loss across the entire soil specimen?

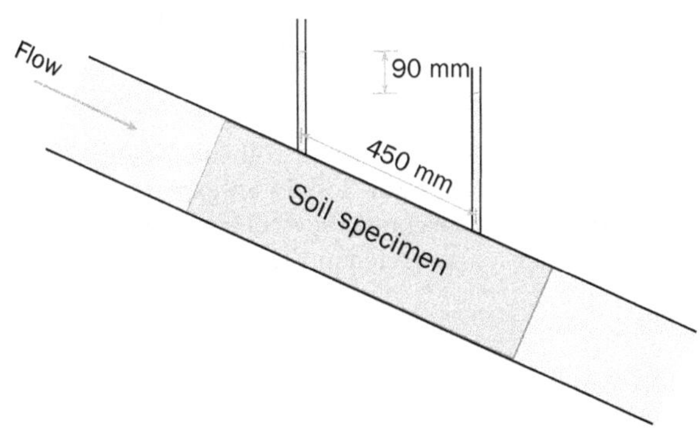

FIGURE RE6.1 Flow through the soil specimen.

Answer: 150 mm

6.2 Figure RE6.2 shows a laboratory setup where water flows upward through a 400-mm-long soil specimen, where the water levels at both sides are maintained at constant levels. The saturated unit weight of the soil is 18.0 kN/m^3. Determine the total and effective stresses at the point *A*. Take the tail water level as the datum.

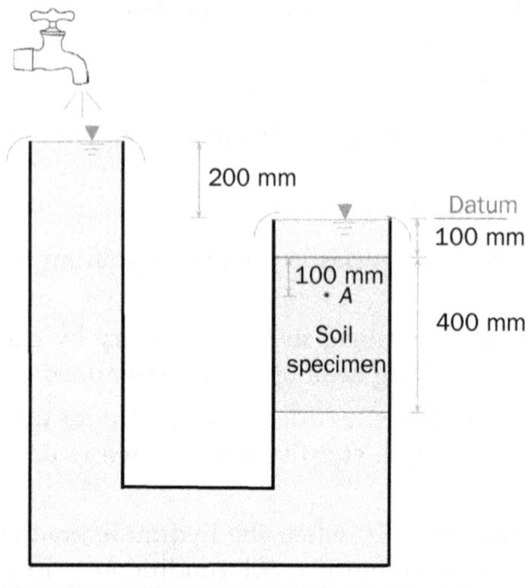

FIGURE RE6.2 Laboratory setup.

Answer: 2.78 kN/m², 0.33 kN/m²

6.3 A constant head permeability test was carried out on a 70-mm-diameter and 500-mm-long sand column using a setup similar to that shown in Fig. 6.10, under head difference *h* of 900 mm. If 3640 g of water flows through the column in 5 min, what is the hydraulic conductivity of the sand?

Answer: 0.175 cm/s

6.4 In the Review Exercise 6.1, if the diameter of the soil specimen is 50 mm, and the hydraulic conductivity is 8.0×10^{-2} cm/s, what would be the flow rate through the specimen? How much water will flow through the specimen in 15 min?

Answer: 0.324 cm³/s, 282.6 g

6.5 Two cylindrical specimens *A* and *B* of the same soil, of 50 and 75 mm diameter, respectively, are subjected to a constant head flow in an experimental setup shown in Fig. RE6.5. Water flowing through the specimens was collected in a bucket for measuring the flow rate, where 340 g of water was collected in 10 min. Find the hydraulic conductivity of the soil specimens.

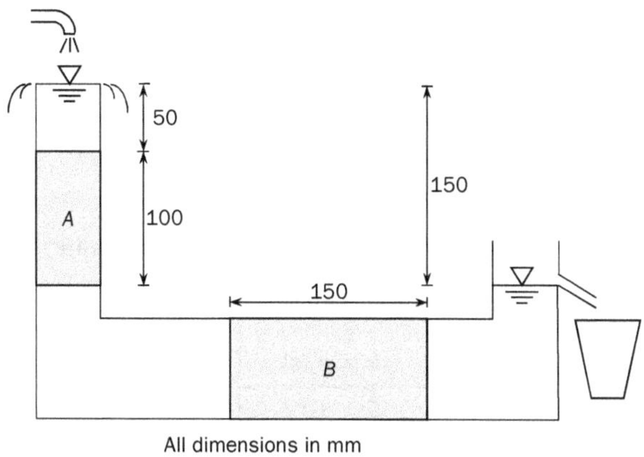

FIGURE RE6.5 Experimental setup.

Answer: 0.0321 cm/s

6.6 Three 60-mm-diameter soil specimens 1, 2, and 3, with hydraulic conductivity values of 2×10^{-2}, 3×10^{-3}, and 2×10^{-1} cm/s are subjected to a constant head permeability test as shown in Fig. RE6.6. The water levels in the manometer tubes are as shown in the figure, which are maintained constant throughout the test.

a. Determine the flow rate.

b. Which of the three specimens has the largest hydraulic gradient?

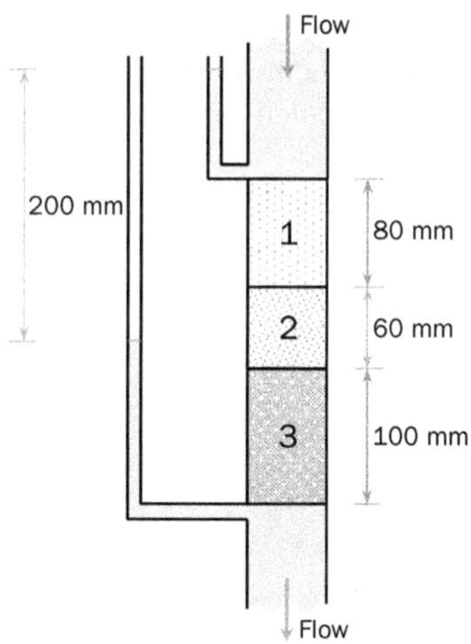

FIGURE RE6.6 Flow through three soil specimens.

Answer: 0.232 cm³/s; specimen 2

6.7 A 75-mm-diameter and 100-mm-long silty clay specimen in a tube is subjected to a falling head permeability test, where the standpipe had internal diameter of 4.5 mm. If the head falls from 870 to 385 mm in 5 min, what is the hydraulic conductivity of the specimen?

Answer: 9.78×10^{-5} cm/s

6.8 An 8-m-thick soil deposit overlies an impermeable stratum. The water table is at 2 m below the ground level. Water is pumped out from a test well and at steady state, the flow rate was 6200 liters/minute. In the two observation wells at radial distances of 3.5 and 6.2 m, respectively, from the centerline of the test well, the water table dropped by 2.1 and 1.4 m. Determine the average hydraulic conductivity of the soil deposit.

Answer: 0.316 cm/s

6.9 A soil profile consists of four layers where the hydraulic conductivities are isotropic and are summarized as follows.

Layer number (from the top)	Layer thickness *H* (m)	*k* (cm/s)
1	2.0	2×10^{-4}
2	3.0	4×10^{-6}
3	2.5	3×10^{-5}
4	4.0	6×10^{-4}

Find the horizontal and vertical hydraulic conductivity values of the homogeneous equivalent layer.

Answer: 2.51×10^{-4} *cm/s,* 1.35×10^{-5} *cm/s*

6.10 In the Review Exercise 6.6, determine the equivalent hydraulic conductivity and hence the flow rate.

Answer: 0.0098 cm/s, 0.231 cm^3*/s*

6.11 Two soil specimens 1 and 2, with diameter of 65 mm and lengths of 190 mm and 250 mm, respectively, are subjected to a constant head permeability test as shown in Fig. RE6.11. If 400 mL of water was collected in the bucket in 10 min, find the hydraulic conductivities of the two specimens.

What is the pore water pressure at the bottom of the specimen 2?

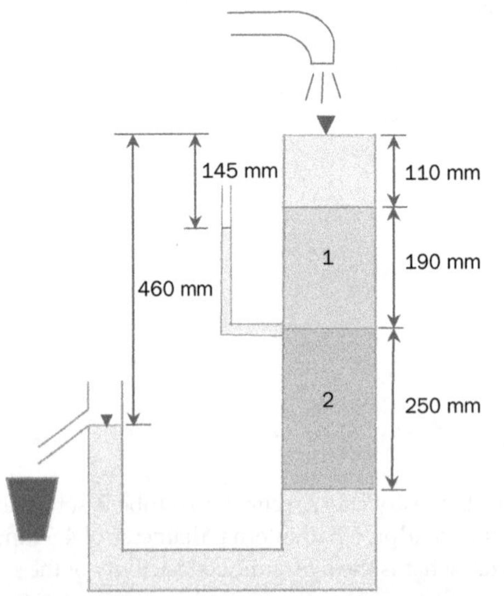

FIGURE RE6.11 Constant head permeability test setup.

Answer: 0.0263 cm/s, 0.0160 cm/s; 0.883 kN/m^2

6.12 Water flows through an 800-mm well-graded densely packed sand column as shown in Fig. RE6.12. The water levels on both sides are maintained at the levels shown. What would be the pore water pressure at *A*? If the saturated unit weight of the sand is 21.0 kN/m^3, is there a possibility that the column can fail by reaching quick condition?

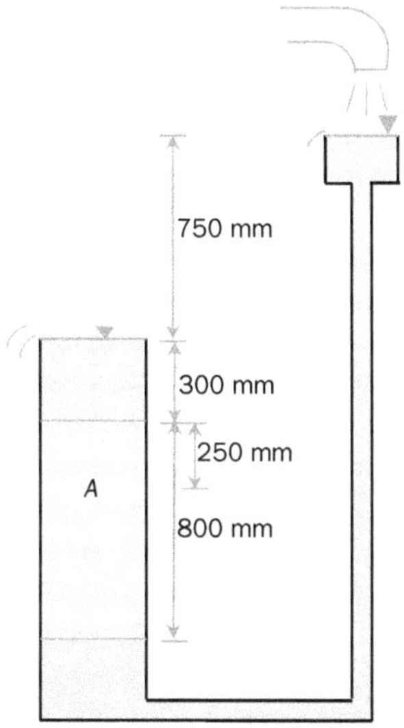

FIGURE RE6.12 Flow through a soil specimen.

Answer: 7.70 kN/m^2; current hydraulic gradient of 0.94 is less than the critical hydraulic gradient of 1.14, and hence the specimen will not fail.

6.13 In a remote location, bore wells are used to access the ground water. Houses 1 and 3 already have bore wells and the water table is located at 8.0 and 13.5 m depths, respectively (Fig. RE6.13). The elevation of the ground level is 36.3 m. The elevations of the underlying impervious stratum directly below the three houses are given in the figure. The owner of House 2 is considering a bore well at his property. What is the depth *h* of the water table there?

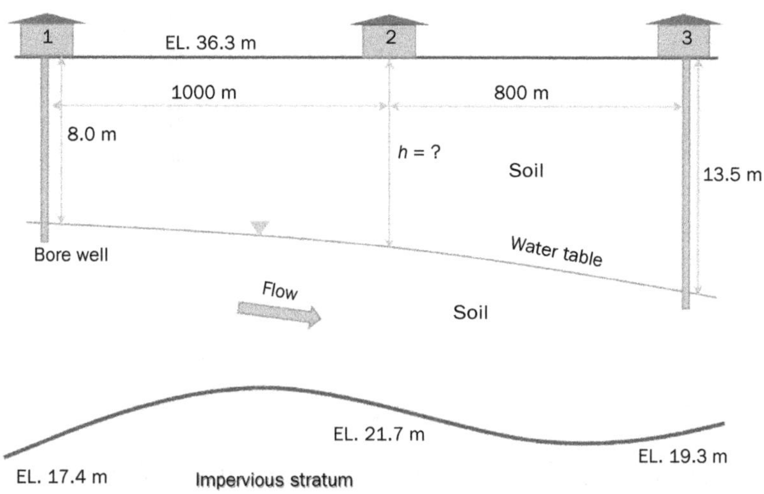

FIGURE RE6.13 Schematic diagram.

Answer: 10.16 m

6.14 In some waterfront construction, sheet piles are driven into the ground forming a watertight enclosure as shown in Fig. RE 6.14a. Within the sheet pile enclosure, excavation is carried out to 2 m depth. Within the enclosure, water level is maintained at the excavation level, and it is necessary to pump the water out continuously. Permeability of the soil is 2.5×10^{-3} cm/s.

The configuration shown in Fig. RE6.14a is symmetrical about the vertical centerline. Fig. RE6.14b shows the flow net for the left half. The specific gravity of the soil grains is 2.68 and the void ratio is 0.70.

a. Determine the rate at which water has to be pumped out, in m^3/day per m.

b. Determine the safety factor with respect to piping.

c. What is the pore water pressure at the tip of the sheet pile?

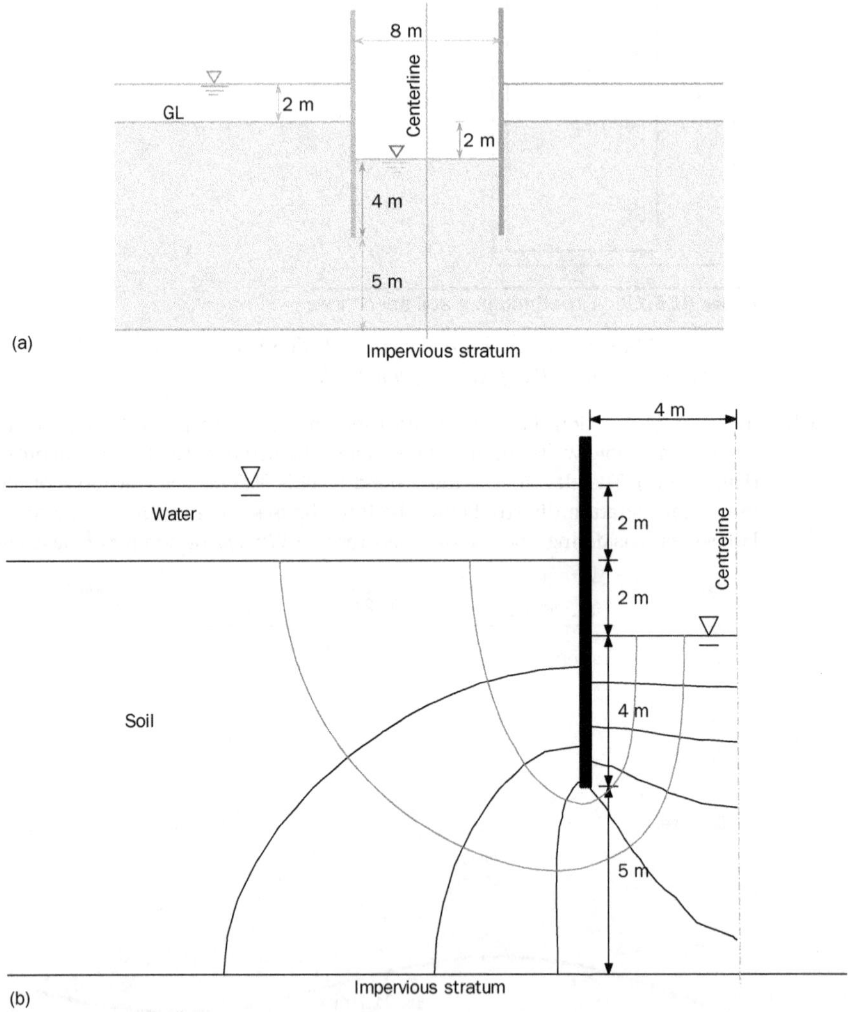

FIGURE RE6.14 (a) Excavation between the two parallel sheet piles, (b) flow net for the left half.

Answer: 6.48 m^3/day per m width, 2.48, 61.3 kN/m^2

6.15 Surface water flows through the soil into a horizontal drain as shown in Fig. RE6.15. The depth of the soil is 7.0 m that is underlain by an impervious stratum. The horizontal drain is at 3.5 m depth below the ground level. The hydraulic conductivity of the soil is 4.5×10^{-5} cm/s. The water within the drain is at atmospheric pressure.

a. Determine the flow rate into the drain per m length.
b. What is the pore water pressure at X, 1.8 m below the ground level?

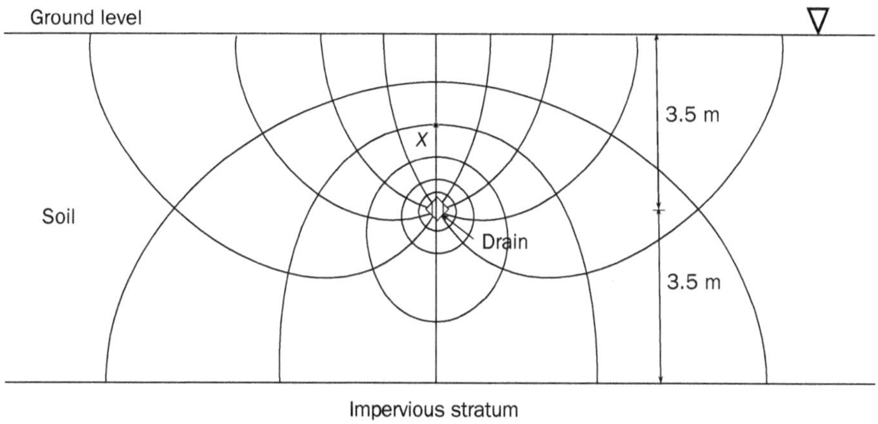

FIGURE RE6.15 Flow net. (Courtesy Dr. Thushara Madanayaka, SLTC, Sri Lanka.)

Answer: 0.226 m³/day per m, 40.5 kN/m²

6.16 Without drawing the flow net, determine the pore water pressure at the bottom tip of the sheet pile in Fig. RE6.16.

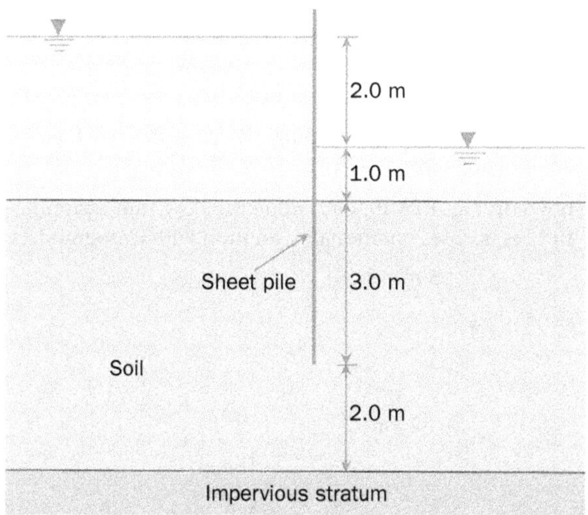

FIGURE RE6.16 Schematic diagram.

Answer: 49.1 kN/m²

6.17 Figure RE6.17 shows seepage taking place beneath a sheet pile in a silty sand, where the hydraulic conductivity is 2×10^{-3} cm/s and the saturated unit weight is 19.5 kN/m^3. Some equipotential lines have been drawn in the figure at total head intervals of 0.5 m.

a. Complete the flow net.
b. Determine the flow rate in m^3/day per m.
c. What is the safety factor with respect to piping?
d. Find the pore water pressure at the tip of the sheet pile.

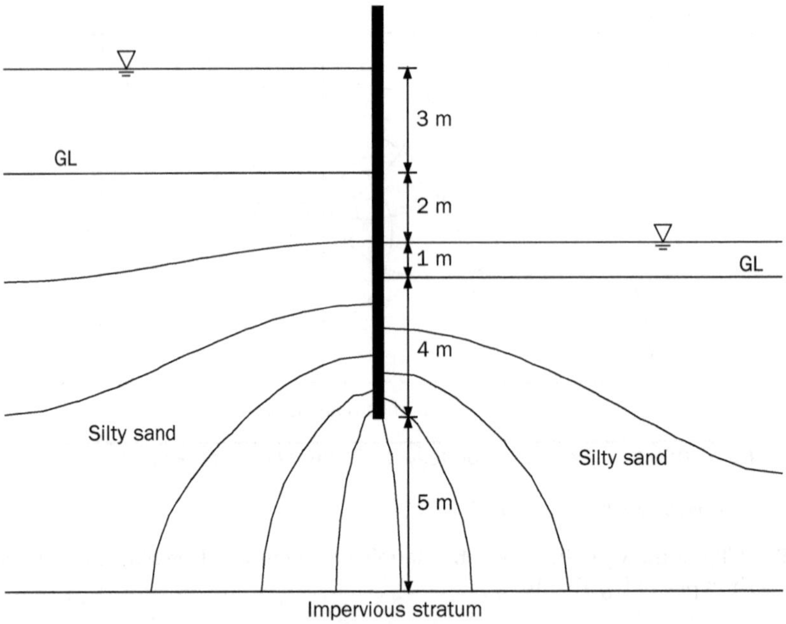

FIGURE RE6.17 Equipotential lines of a flow net.

Answer: 3.6 m³/day per m, 2.5, 73 kN/m²

6.18 List five major dam failures caused by piping. Discuss one of these in detail with photographs.

6.19 For the earth dam shown in Fig. RE6.19, determine the flow rate assuming the hydraulic conductivity of 2×10^{-5} cm/s. Use Schaffernak's solution with Casagrande's correction.

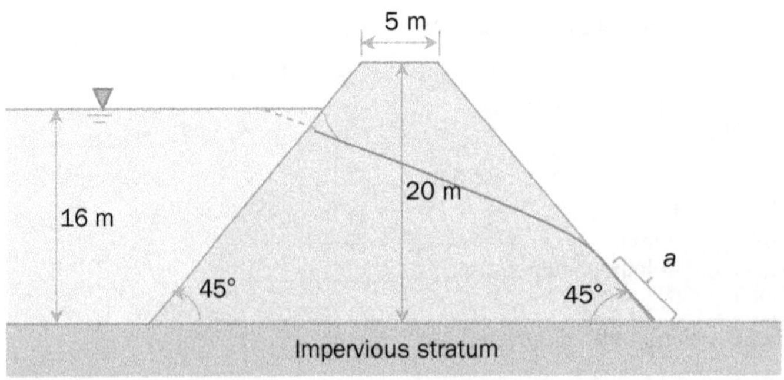

FIGURE RE6.19 Flow through an earth dam.

Answer: 0.0696 m³/day per m

6.20 Seepage takes place around a sheet pile as shown in Fig. RE6.20. The hydraulic conductivity of the soil is 3.8×10^{-3} cm/s. Using the method of fragments, determine the following:

a. Flow rate per unit width, and
b. Maximum exit hydraulic gradient.

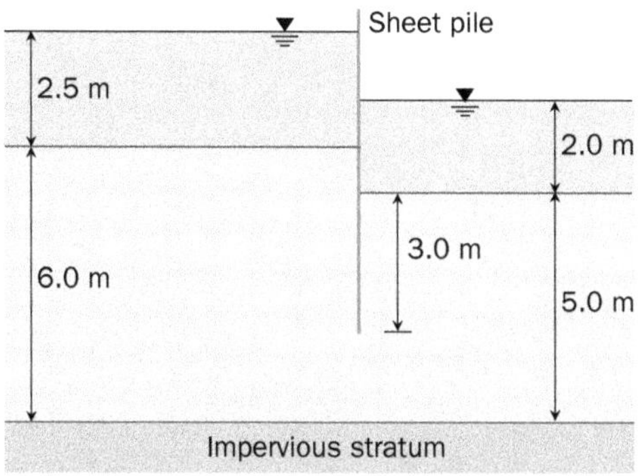

FIGURE RE6.20 Schematic diagram.

Answer: 1.99 m³/day per m width, 0.13

References

ASTM D 2434-19. Standard Test Method for Permeability of Granular Soils (Constant Head).

ASTM D 5856-15. Standard Test Method for Measuring Hydraulic Conductivity of Porous Material using a Rigid-wall, Compaction Mold Permeameter.

ASTM D 5084-03. Standard Test Methods for Measurement of Hydraulic Conductivity of Saturated Porous Materials using a Flexible Wall Permeameter.

Bligh, W. G. Dams, "Barrages and Weirs on Porous Foundations," *Engineering News*, 64(26):708–710, 1910.

Casagrande, A. "Seepage through Dams," *Contributions to Coil Mechanics 1925–1940*, Boston Society of Civil Engineers, Boston, 1937.

Cedegren, H. R. *Seepage, Drainage and Flow Nets*, John Wiley & Sons, New York, 1967.

Darcy, H. Les Fontaines Publiques de la Ville de Dijon, Dalmont, Paris, 1856.

Dupuit, J. Etudes Théoriqueset Pratiques sur le Movement des Eaux Dans les Canaux Découverts et à Travers les Terrains Perméables, Dunod, *Paris, 1863.*

Forchheimer, P. H. "Wasserbewegung Durch Boden." *Zeitschrift Verein Deutscher Ingenieure*, 45:1782–1788, 1901.

Griffiths, D. V. "Rationalized Charts for the Method of Fragments Applied to Confined Seepage," *Géotechnique*, 34(2):229–238, 1984.

Harr, M. E. *Groundwater and Seepage*, McGraw-Hill, New York, 1962.

Hazen, A. "Water Supply," *American Civil Engineers Handbook*, Wiley, New York, 1930.

Lane, E. W. "Security from Under-seepage Masonry Dams on Earth Foundations," *Transactions* ASCE, 60(4):929–966, 1934.

Madanayaka, A. T., and N. Sivakugan. "Approximate Equations for the Method of Fragments," *International Journal of Geotechnical Engineering*, 10(3):297–303, 2016.

Madanayaka, A. T. and N. Sivakugan. "Simple Solutions for Square and Rectangular Cofferdam Seepage Problems," *Canadian Geotechnical Journal*, 56:730745, 2019.

Pavlovsky, N. *Collected Works*, Leningrad, Doklady Akademii Nauk, USSR, 1956.

Reynolds, O. "An Experimental Investigation of the Circumstances which Determine Whether the Motion of Water shall He Direct or Sinuous, and of the Law of Resistance in Parallel Channels." *Philosophical Transactions of the Royal Society of London*, 174:935–982, 1883.

Richards, K. S., and K. R. Reddy. "Critical Appraisal of Piping Phenomena in Earth Dams," *Bulletin of Engineering Geology and Environment*, Springer, 66:381–402, 2007.

Schaffernak, F. Über die Standsicherheit Durchlaessiger Geschustteter Dämme, Allgem. Bauzeitung, 1917.

Terzaghi, K., and R. B. Peck. *Soil Mechanics in Engineering Practice*, 1st ed., John Wiley & Sons, New York, 1948.

U.S. Department of the Navy. *Naval Facilities Engineering Command Design Manual—Soil Mechanics, Foundations and Earth Structures, NAVFAC DM-7*, Washington, DC, 1971.

CHAPTER 7

Vertical Stresses under Loaded Areas

7.1 Introduction

In Chapter 5, we learned how to determine the total and effective vertical stresses, and pore water pressure at any depth within a soil profile. The total vertical stress σ_{vo}, known as the *total overburden stress*, is the sum of the product of the layer thicknesses and unit weights of all the soil layers above the point of interest. It can be written as

$$\sigma_{vo} = \sum_{i=1}^{n} \gamma_i H_i \tag{7.1}$$

What would be the increase in vertical stress $\Delta\sigma_v$ due to some surface loads? This depends on the magnitude of the loads, their shapes, and their locations.

When a uniform surcharge q of very large lateral extent is placed at the ground level,

$$\sigma_{vo} = \sum_{i=1}^{n} \gamma_i H_i + q \tag{7.2}$$

Here, the vertical stress at every point within the soil mass increases by q.

The methods to calculate $\Delta\sigma_v$ due to different types of surface loads are discussed in this chapter. In this chapter, we will not separate the effective stresses and pore water pressures, and deal mainly with the total stresses, treating the soil mass as a *linear elastic continuum* than a *particulate medium*. It is important to note that when the soil mass is assumed to be a linear elastic continuum, *principle of superposition* is valid.

7.2 Vertical Stress Increase due to a Point Load

Let's consider *an elastic half-space* (i.e., infinite extent laterally and with depth) representing the soil, where a point load Q is applied at the ground level (Fig. 7.1). Boussinesq (1883) showed that the vertical stress increase at point X is given by

$$\Delta\sigma_v = \frac{3Q}{2\pi z^2\left[1+\left(\frac{r}{z}\right)^2\right]^{5/2}} = \frac{Q}{z^2} I_B \tag{7.3}$$

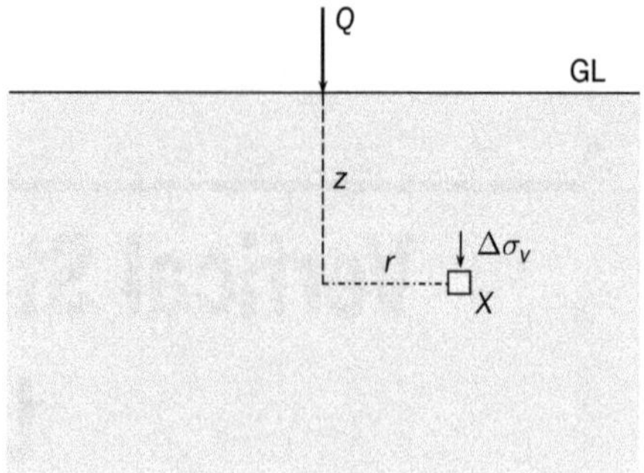

Figure 7.1 Vertical stress increase due to a point load.

where z = depth of the point below the ground level, r = horizontal distance to the point from the vertical centerline, and

$$I_B = \frac{3}{2\pi\left[1+\left(\frac{r}{z}\right)^2\right]^{5/2}} \tag{7.4}$$

Equation (7.3) suggests that the vertical stress increase does not depend on the soil properties such as Young's modulus or Poisson's ratio.

Westergaard (1935) treated the soil as an elastic material interspersed with a large number of thin and rigid reinforcements, which prevent any lateral strains. He showed that

$$\Delta\sigma_v = \frac{Q}{2\pi z^2}\frac{\sqrt{\frac{(1-2\nu)}{(2-2\nu)}}}{\left[\left(\frac{1-2\nu}{2-2\nu}\right)+\left(\frac{r}{z}\right)^2\right]^{3/2}} = \frac{Q}{z^2}I_W \tag{7.5}$$

where ν is the *Poisson's ratio*, which varies in the range 0–0.5 for any material, and

$$I_W = \frac{1}{2\pi}\frac{\sqrt{\frac{(1-2\nu)}{(2-2\nu)}}}{\left[\left(\frac{1-2\nu}{2-2\nu}\right)+\left(\frac{r}{z}\right)^2\right]^{3/2}} \tag{7.6}$$

Boussinesq and Westergaard coefficients I_B and I_W are compared in Fig. 7.2 for $\nu = 0$, 0.1, 0.2, and 0.3. It can be seen that the $\Delta\sigma_v$ determined by Boussinesq equation is larger than that from Westergaard's equation, for $\nu = 0$–0.2, especially when $r/z < 1.5$, when

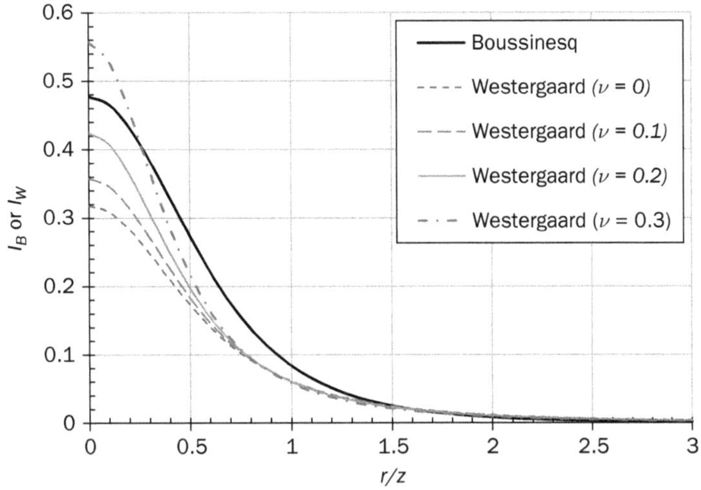

FIGURE 7.2 Comparison of Boussinesq and Westergaard values.

the stress increases are significant. For $r/z > 1.5$, they are the same. For simplicity and being larger and hence conservative, Boussinesq values are used more in practice. For larger values of ν and very small values of r/z, Westergaard equation gives larger $\Delta\sigma_v$.

7.3 Vertical Stress Increase due to a Line Load

The vertical stress increase $\Delta\sigma_v$ due to an infinitely long line load Q at depth z and horizontal distance x (Fig. 7.3) is given by

$$\Delta\sigma_v = \frac{2Q}{\pi z\left[1+\left(\frac{x}{z}\right)^2\right]^2} \tag{7.7}$$

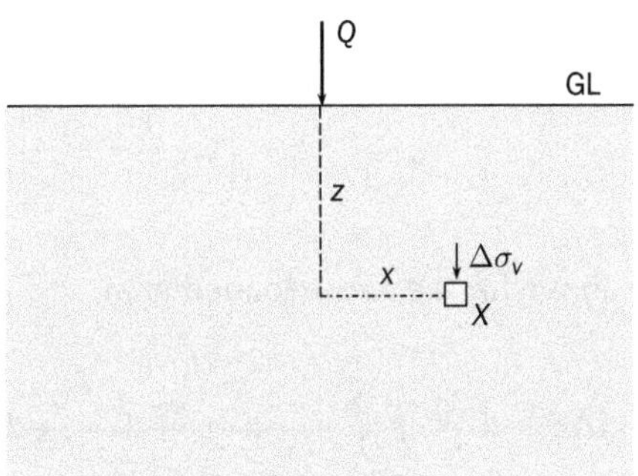

FIGURE 7.3 Vertical stress increase due to a line load.

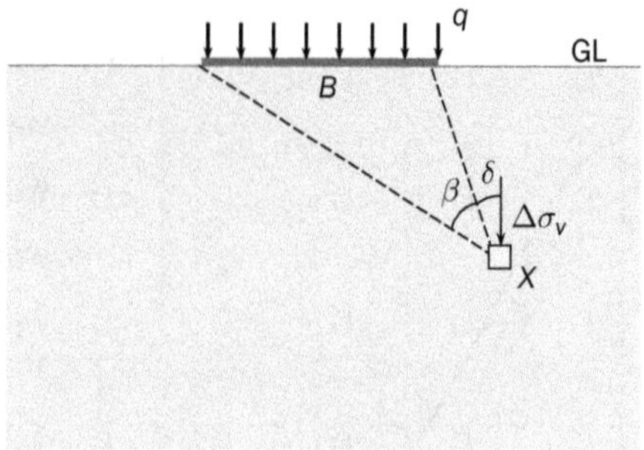

FIGURE 7.4 Vertical stress increase due to a strip load.

7.4 Vertical Stress Increase due to a Strip Load

The vertical stress increase $\Delta\sigma_v$ due to an infinitely long strip load of width B applying pressure of q at the ground level is given by (see Fig. 7.4)

$$\Delta\sigma_v = \frac{q}{\pi}[\beta + \sin\beta\ \cos(\beta + 2\delta)] \tag{7.8}$$

where the angles β and δ are in radians.

7.5 Vertical Stress Increase under the Corner of a Rectangular Load

Let us consider a flexible uniform rectangular load of width B and length L applying a pressure of q to the underlying soil as shown in Fig. 7.5. What is the vertical stress increase ($\Delta\sigma_v$) at X at depth z directly under a *corner*? This can be determined by discretizing the rectangular load into infinitesimal point loads and applying Boussinesq equation [Eq. (7.3)].

Due to the infinitesimal element shown shaded darker in Fig. 7.5, carrying a point load $q\ dx.dy$, the vertical stress increase at X is

$$d(\Delta\sigma_v) = \frac{3\,q\,dx\,dy}{2\pi z^2\left\{1 + \dfrac{x^2 + y^2}{z^2}\right\}^{5/2}} \tag{7.9}$$

Integrating the above for the entire loaded area,

$$\Delta\sigma_v = \frac{3q}{2\pi}\int_{y=0}^{B}\int_{x=0}^{L}\frac{1}{z^2\left\{1 + \dfrac{x^2 + y^2}{z^2}\right\}^{5/2}}dxdy = Iq \tag{7.10}$$

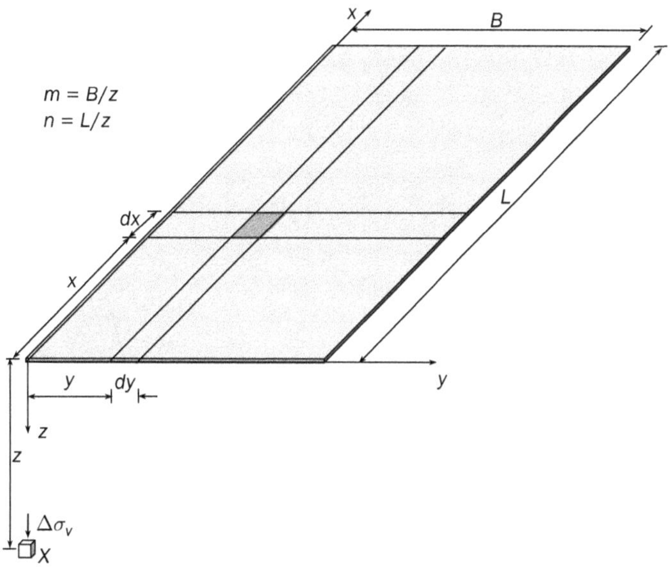

FIGURE 7.5 Vertical stress increase under the corner of a flexible uniform rectangular load.

where I is the *influence factor* defined as

$$I = \frac{1}{4\pi}\left\{\left(\frac{2mn\sqrt{m^2+n^2+1}}{m^2+n^2+m^2n^2+1}\right)\left(\frac{m^2+n^2+2}{m^2+n^2+1}\right) + tan^{-1}\left(\frac{2mn\sqrt{m^2+n^2+1}}{m^2+n^2-m^2n^2+1}\right)\right\} \tag{7.11}$$

where $m = B/z$ and $n = L/z$. Here, B and L are the breadth and length, respectively, of the loaded area, and z is the depth of the point of interest under a corner (see Fig. 7.5). The second term in the right hand side of Eq. (7.11) must be a positive angle in radians. When $m^2 + n^2 + 1 < m^2n^2$, the angle becomes negative and π must be added to make it positive. The values of the influence factor I as a function of m and n, determined from Eq. (7.11) are shown graphically in Fig. 7.6. Boussinesq's influence factors are also given in Table 7.1.

Using Westergaard's solution, the influence factor can be obtained as

$$I = \frac{1}{2\pi}\left\{\cot^{-1}\sqrt{\left(\frac{1-2\nu}{2-2\nu}\right)\left(\frac{1}{m^2}+\frac{1}{n^2}\right)+\left(\frac{1-2\nu}{2-2\nu}\right)^2\left(\frac{1}{m^2n^2}\right)}\right\} \tag{7.12}$$

Figure 7.7 shows the values of the Westergaard's influence factors in terms of m and n for Poisson's ratio of 0. These values are also given in Table 7.2. It can be clearly seen that the Westergaard values are significantly less than the Boussinesq values, especially

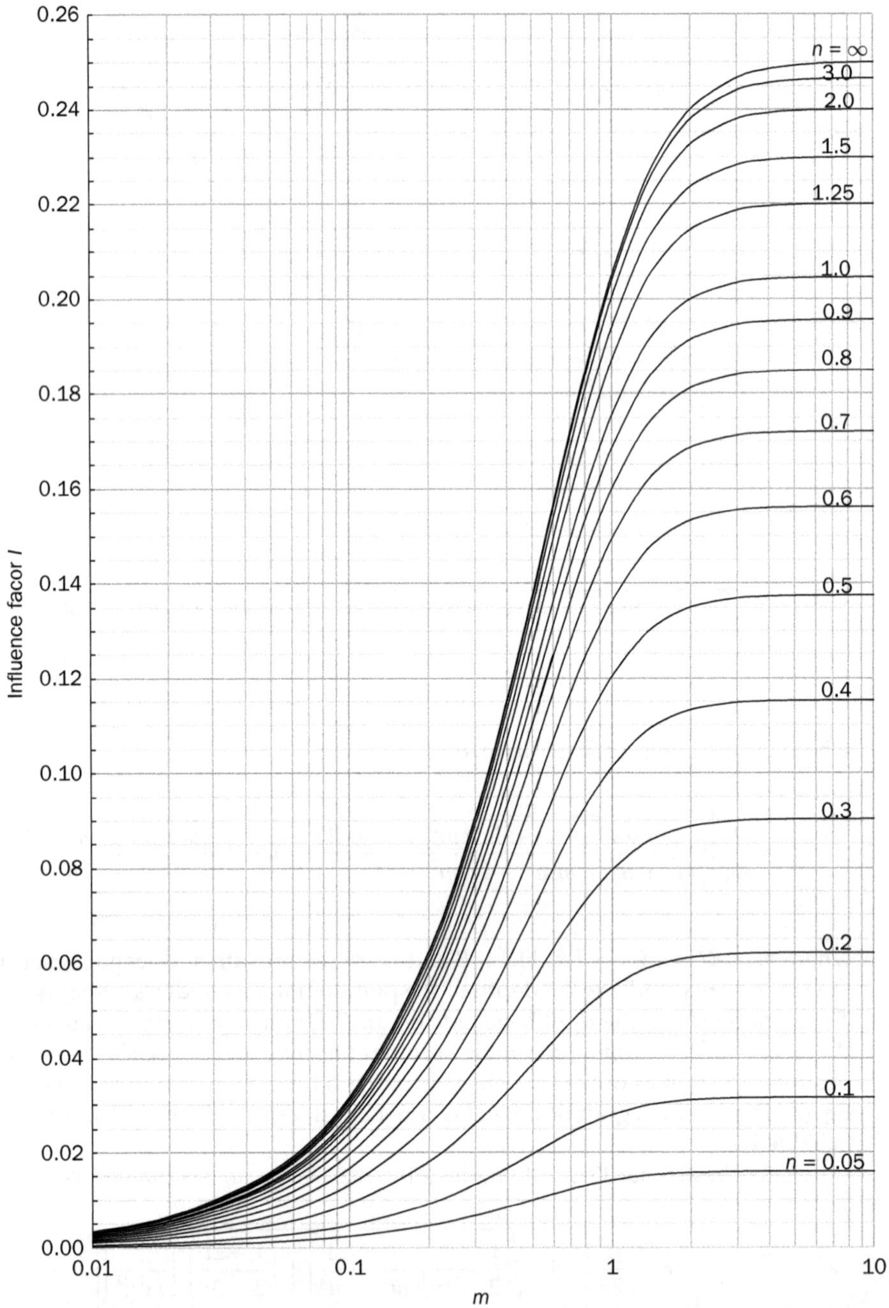

Figure 7.6 Influence factor I for $\Delta\sigma_v$ under the corner of a flexible uniform rectangular load from Boussinesq solution.

m or n	Influence factor I															
	0.05	**0.1**	**0.2**	**0.3**	**0.4**	**0.5**	**0.6**	**0.7**	**0.8**	**0.9**	**1**	**1.25**	**1.5**	**2**	**3**	**∞**
0.05	0.0012	0.0024	0.0046	0.0067	0.0084	0.0099	0.0112	0.0122	0.0129	0.0136	0.0140	0.0148	0.0153	0.0156	0.0158	0.0159
0.1	0.0024	0.0047	0.0092	0.0132	0.0168	0.0198	0.0222	0.0242	0.0258	0.0270	0.0279	0.0295	0.0304	0.0311	0.0315	0.0316
0.2	0.0046	0.0092	0.0179	0.0259	0.0328	0.0387	0.0435	0.0473	0.0504	0.0528	0.0547	0.0578	0.0595	0.0610	0.0618	0.0620
0.3	0.0067	0.0132	0.0259	0.0374	0.0474	0.0559	0.0629	0.0686	0.0731	0.0766	0.0794	0.0839	0.0864	0.0887	0.0898	0.0902
0.4	0.0084	0.0168	0.0328	0.0474	0.0602	0.0711	0.0801	0.0873	0.0931	0.0977	0.1013	0.1072	0.1105	0.1134	0.1150	0.1154
0.5	0.0099	0.0198	0.0387	0.0559	0.0711	0.0840	0.0947	0.1034	0.1103	0.1158	0.1202	0.1274	0.1314	0.1350	0.1368	0.1375
0.6	0.0112	0.0222	0.0435	0.0629	0.0801	0.0947	0.1069	0.1168	0.1247	0.1311	0.1360	0.1444	0.1490	0.1533	0.1555	0.1562
0.7	0.0122	0.0242	0.0473	0.0686	0.0873	0.1034	0.1168	0.1277	0.1103	0.1436	0.1491	0.1585	0.1637	0.1686	0.1711	0.1720
0.8	0.0129	0.0258	0.0504	0.0731	0.0931	0.1103	0.1247	0.1365	0.1461	0.1537	0.1598	0.1700	0.1758	0.1812	0.1841	0.1850
0.9	0.0136	0.0270	0.0528	0.0766	0.0977	0.1158	0.1311	0.1436	0.1537	0.1618	0.1684	0.1794	0.1857	0.1915	0.1947	0.1958
1	0.0140	0.0279	0.0547	0.0794	0.1013	0.1202	0.1360	0.1491	0.1598	0.1684	0.1752	0.1869	0.1936	0.1999	0.2034	0.2046
1.25	0.0148	0.0295	0.0578	0.0839	0.1072	0.1274	0.1444	0.1585	0.1700	0.1794	0.1869	0.1999	0.2075	0.2147	0.2188	0.2202
1.5	0.0153	0.0304	0.0595	0.0864	0.1105	0.1314	0.1490	0.1637	0.1758	0.1857	0.1936	0.2075	0.2157	0.2236	0.2282	0.2299
2	0.0156	0.0311	0.0610	0.0887	0.1134	0.1350	0.1533	0.1686	0.1812	0.1915	0.1999	0.2147	0.2236	0.2325	0.2378	0.2399
3	0.0158	0.0315	0.0618	0.0898	0.1150	0.1368	0.1555	0.1711	0.1841	0.1947	0.2034	0.2188	0.2282	0.2378	0.2439	0.2465
4	0.0159	0.0316	0.0619	0.0901	0.1153	0.1372	0.1560	0.1717	0.1847	0.1954	0.2042	0.2197	0.2293	0.2391	0.2455	0.2485
5	0.0159	0.0316	0.0620	0.0901	0.1154	0.1374	0.1561	0.1718	0.1849	0.1956	0.2044	0.2200	0.2296	0.2395	0.2461	0.2494
6	0.0159	0.0316	0.0620	0.0902	0.1154	0.1374	0.1562	0.1719	0.1850	0.1957	0.2045	0.2201	0.2297	0.2397	0.2463	0.2497
8	0.0159	0.0316	0.0620	0.0902	0.1154	0.1374	0.1562	0.1720	0.1850	0.1957	0.2045	0.2202	0.2298	0.2398	0.2465	0.2498
10	0.0159	0.0316	0.0620	0.0902	0.1154	0.1374	0.1562	0.1720	0.1850	0.1958	0.2046	0.2202	0.2299	0.2398	0.2465	0.2499
∞	0.0159	0.0316	0.0620	0.0902	0.1154	0.1375	0.1562	0.1720	0.1850	0.1958	0.2046	0.2202	0.2299	0.2399	0.2465	0.2500

TABLE 7.1 Influence Factors from Eq. (7.11) for a Corner of a Uniform Rectangular Load Based on Boussinesq's Equation

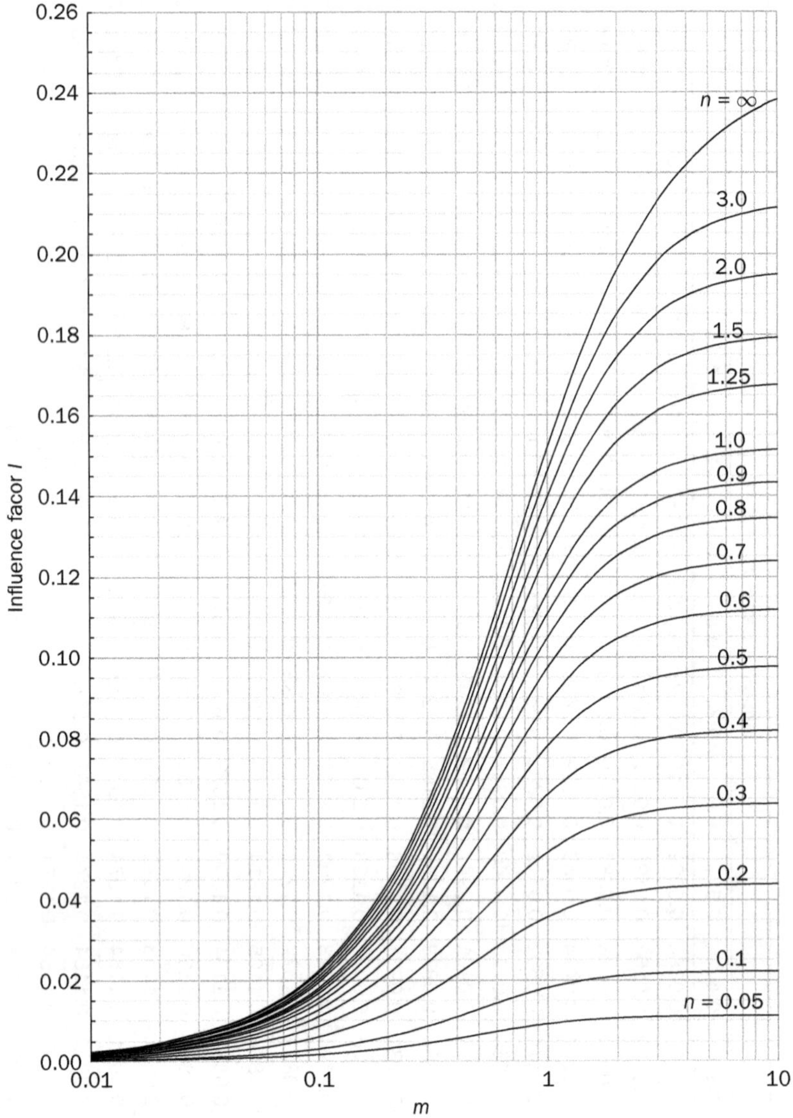

FIGURE 7.7 Influence factor I for $\Delta\sigma_v$ under the corner of a flexible uniform rectangular load from Westergaard solution, with Poisson's ratio $\nu = 0$.

in the vicinity of the load, for large m or n values. It is suggested that Fig. 7.6, developed from Boussinesq be used in the computations.

In Eqs. (7.11) and (7.12), and Figs. 7.6 and 7.7, the values of m and n are interchangeable. Using principle of superposition, $\Delta\sigma_v$ at any point within the soil mass under the uniform rectangular load can be computed. This is illustrated in Examples 7.1 and 7.2.

	Influence factor *I*															
m* or *n	**0.05**	**0.1**	**0.2**	**0.3**	**0.4**	**0.5**	**0.6**	**0.7**	**0.8**	**0.9**	**1**	**1.25**	**1.5**	**2**	**3**	**∞**
0.05	0.0008	0.0016	0.0031	0.0044	0.0055	0.0065	0.0073	0.0079	0.0084	0.0088	0.0092	0.0098	0.0102	0.0106	0.0109	0.0112
0.1	0.0016	0.0031	0.0061	0.0087	0.0110	0.0129	0.0144	0.0157	0.0167	0.0176	0.0182	0.0194	0.0202	0.0211	0.0218	0.0224
0.2	0.0031	0.0061	0.0118	0.0170	0.0214	0.0251	0.0282	0.0307	0.0327	0.0343	0.0357	0.0381	0.0396	0.0413	0.0427	0.0439
0.3	0.0044	0.0087	0.0170	0.0244	0.0308	0.0362	0.0407	0.0443	0.0473	0.0497	0.0517	0.0552	0.0575	0.0600	0.0621	0.0639
0.4	0.0055	0.0110	0.0214	0.0308	0.0390	0.0459	0.0516	0.0563	0.0601	0.0633	0.0658	0.0705	0.0735	0.0768	0.0795	0.0819
0.5	0.0065	0.0129	0.0251	0.0362	0.0459	0.0541	0.0609	0.0666	0.0712	0.0750	0.0781	0.0838	0.0875	0.0916	0.0950	0.0980
0.6	0.0073	0.0144	0.0282	0.0407	0.0516	0.0609	0.0687	0.0752	0.0805	0.0849	0.0886	0.0952	0.0995	0.1044	0.1084	0.1120
0.7	0.0079	0.0157	0.0307	0.0443	0.0563	0.0666	0.0752	0.0824	0.0712	0.0933	0.0974	0.1049	0.1098	0.1154	0.1200	0.1242
0.8	0.0084	0.0167	0.0327	0.0473	0.0601	0.0712	0.0805	0.0884	0.0949	0.1003	0.1048	0.1131	0.1185	0.1248	0.1301	0.1348
0.9	0.0088	0.0176	0.0343	0.0497	0.0633	0.0750	0.0849	0.0933	0.1003	0.1061	0.1110	0.1200	0.1259	0.1329	0.1387	0.1440
1	0.0092	0.0182	0.0357	0.0517	0.0658	0.0781	0.0886	0.0974	0.1048	0.1110	0.1161	0.1258	0.1322	0.1398	0.1462	0.1520
1.25	0.0098	0.0194	0.0381	0.0552	0.0705	0.0838	0.0952	0.1049	0.1131	0.1200	0.1258	0.1368	0.1443	0.1532	0.1608	0.1681
1.5	0.0102	0.0202	0.0396	0.0575	0.0735	0.0875	0.0995	0.1098	0.1185	0.1259	0.1322	0.1443	0.1525	0.1625	0.1714	0.1799
2	0.0106	0.0211	0.0413	0.0600	0.0768	0.0916	0.1044	0.1154	0.1248	0.1329	0.1398	0.1532	0.1625	0.1743	0.1850	0.1959
3	0.0109	0.0218	0.0427	0.0621	0.0795	0.0950	0.1084	0.1200	0.1301	0.1387	0.1462	0.1608	0.1714	0.1850	0.1981	0.2132
4	0.0111	0.0220	0.0432	0.0628	0.0806	0.0962	0.1099	0.1218	0.1321	0.1410	0.1487	0.1639	0.1749	0.1894	0.2040	0.2222
5	0.0111	0.0221	0.0434	0.0632	0.0810	0.0969	0.1107	0.1227	0.1330	0.1420	0.1498	0.1653	0.1766	0.1916	0.2070	0.2296
6	0.0112	0.0222	0.0436	0.0634	0.0813	0.0972	0.1111	0.1231	0.1336	0.1426	0.1505	0.1662	0.1776	0.1929	0.2088	0.2340
8	0.0112	0.0223	0.0437	0.0636	0.0816	0.0975	0.1115	0.1236	0.1341	0.1432	0.1512	0.1670	0.1786	0.1942	0.2106	0.2368
10	0.0112	0.0223	0.0438	0.0637	0.0817	0.0977	0.1117	0.1238	0.1343	0.1435	0.1515	0.1674	0.1791	0.1948	0.2115	0.2388
∞	0.0112	0.0224	0.0439	0.0639	0.0819	0.0980	0.1120	0.1242	0.1348	0.1440	0.1520	0.1681	0.1799	0.1959	0.2132	0.2500

TABLE 7.2 Influence Factors from Eq. (7.12) for a Corner of a Uniform Rectangular Load Based on Westergaard's Equation

Example 7.1

A 4 m × 3 m rectangular pad AJLH shown in Fig. 7.8 applies 120 kN/m^2 at the ground level. Find the vertical stress increase at 3 m below the points *A, B, C, D,* and *E*.

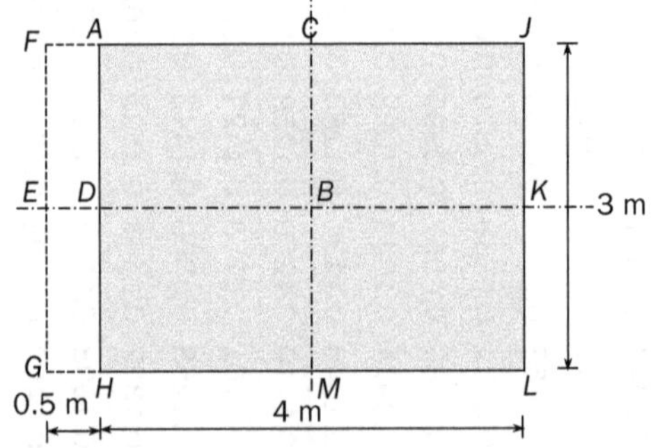

Figure 7.8 Rectangular pad.

Solution

Under *A*: $m = 3/3 = 1$ and $n = 4/3 = 1.33$

From Fig. 7.6, $I = 0.195$ and hence $\Delta\sigma_v = Iq = 0.195 \times 120 = \mathbf{23.4\ kN/m^2}$

Under *B*: The method can only be applied under a corner. Let us consider the top left quarter *ACBD* (where *B* is a corner), find *I* and multiply by 4.

$$m = 1.5/3 = 0.5 \text{ and } n = 2/3 = 0.67$$

From Fig. 7.6, $I = 0.100$ and hence $\Delta\sigma_v = Iq = (4 \times 0.100) \times 120 = \mathbf{48.0\ kN/m^2}$

Under C: Let us consider the left half *ACMH* (where *C* is a corner), find *I* and multiply by 2.

$$m = 2/3 = 0.67 \text{ and } n = 3/3 = 1.0$$

From Fig. 7.6, $I = 0.145$ and hence $\Delta\sigma_v = Iq = (2 \times 0.145) \times 120 = \mathbf{34.8\ kN/m^2}$

Under *D*: Let us consider the top half *AJKD* (where *D* is a corner), find *I* and multiply by 2.

$$m = 1.5/3 = 0.5 \text{ and } n = 4/3 = 1.33$$

From Fig. 7.6, $I = 0.130$ and hence $\Delta\sigma_v = Iq = (2 \times 0.130) \times 120 = \mathbf{31.2\ kN/m^2}$

Under *E*: Let us subtract the *I* value of *EFAD* from *EFJK*, and multiply by 2.

For *EFAD*, $m = 0.5/3 = 0.167$ and $n = 1.5/3 = 0.5$. From Fig. 7.6, $I = 0.03$

For *EFJK*, $m = 1.5/3 = 0.5$ and $n = 4.5/3 = 1.5$. From Fig. 7.6, $I = 0.132$

$$\Delta\sigma_v = Iq = 2(0.132 - 0.03) \times 120 = \mathbf{24.5\ kN/m^2}$$

Example 7.2

A uniform pressure of 100 kN/m^2 is applied to the ground through a 3 m × 3 m square foundation. Plot the lateral variation of $\Delta\sigma_v$ along the horizontal centerline at depths of 1 m and 3 m.

Solution Due to symmetry, let us study the right half of the foundation only, and select points A to H at 0.5 m intervals, as shown in Fig. 7.9.

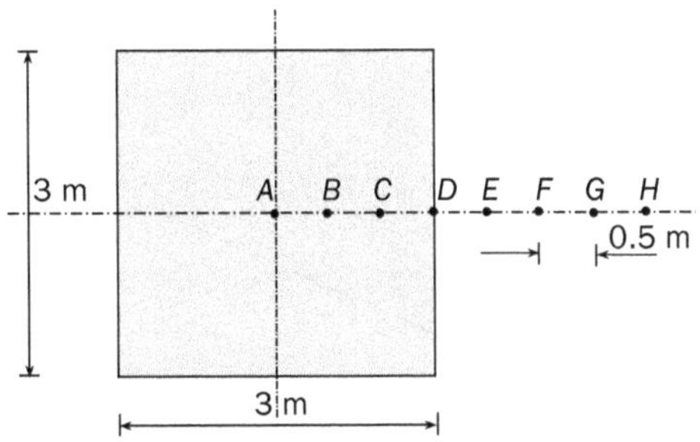

FIGURE 7.9 Selected points A to H.

The values of $\Delta\sigma_v$ computed at these points, following the steps in Example 7.1, are summarized in Table 7.3.

The plot is shown in Fig. 7.10. The bell-shaped variation is evident in the figure where the $\Delta\sigma_v$ decays with depth and horizontal distance. The maximum vertical stress increases occurs directly under the load. The magnitude of this maximum decays with depth.

Point	*A*	*B*	*C*	*D*	*E*	*F*	*G*	*H*
$\Delta\sigma_v$ at 1 m depth (kN/m^2)	86.4	83.6	71.6	45.6	19.6	7.0	2.6	1.2
$\Delta\sigma_v$ at 3 m depth (kN/m^2)	33.6	32.4	29.0	24.0	18.4	13.6	9.4	6.4

TABLE 7.3 $\Delta\sigma_v$ Computed under the Eight Points

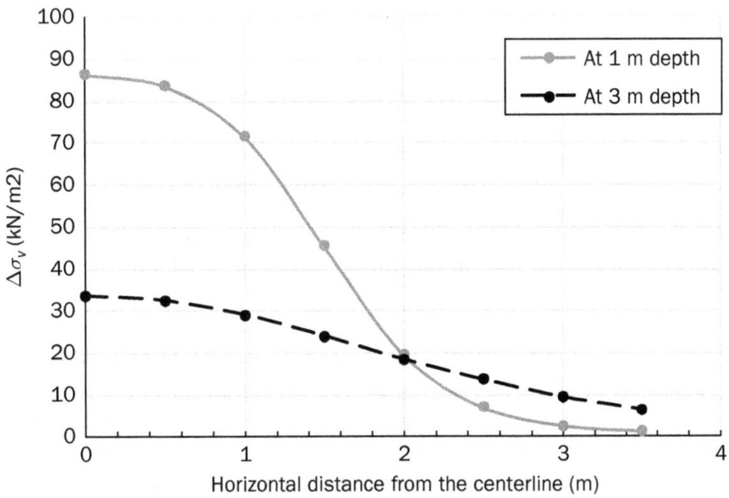

FIGURE 7.10 Variation of $\Delta\sigma_v$ with horizontal distance.

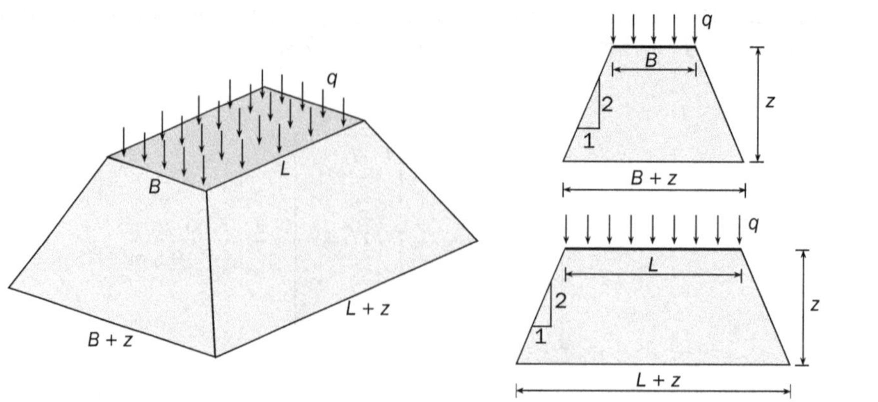

FIGURE 7.11 2:1 distribution.

7.6 2:1 Distribution for a Uniform Rectangular Load

It can be seen intuitively (and from Example 7.2) that the maximum vertical stress increase at any given depth occurs along the centerline of a uniform rectangular load. This $\Delta\sigma_v$ along the centerline can be estimated by a simple and crude method described herein.

A uniform pressure q is applied over a rectangular area $B \times L$. It is assumed that the load intensity decays with depth along a trapezoid in both perpendicular directions as shown in Fig. 7.11, where the slope of the trapezoid is 2 (vertical):1 (horizontal). The same load qBL at the ground level is transferred to larger areas, which increase with depth. At depth z, the base area is $(L + z)(B + z)$. Therefore, the vertical normal stress increase is given by

$$\Delta\sigma_v = \frac{qBL}{(B + z)(L + z)} \tag{7.13}$$

For strip foundations with $L >> B$, carrying a uniform pressure q, Eq. (7.13) becomes

$$\Delta\sigma_v = \frac{qB}{(B + z)} \tag{7.14}$$

Bowles (1988) noted that this method compares well with the theoretical methods for $z = B$ to $4B$, but not for very shallow depths of $z = 0$ to B.

7.7 Pressure Isobars under Square and Strip Flexible Uniform Loads

Let us consider a square or strip (infinitely long) foundation of width B, applying a uniform pressure q at the ground level. Using the influence factors discussed in Sec. 7.4, it is possible to determine the vertical stress increase $\Delta\sigma_v$ at any point in the underlying soil mass. In Example 7.2, we computed the values of $\Delta\sigma_v$ along the centerline of a square foundation at two depths. The process can be followed for any depth for a square or a

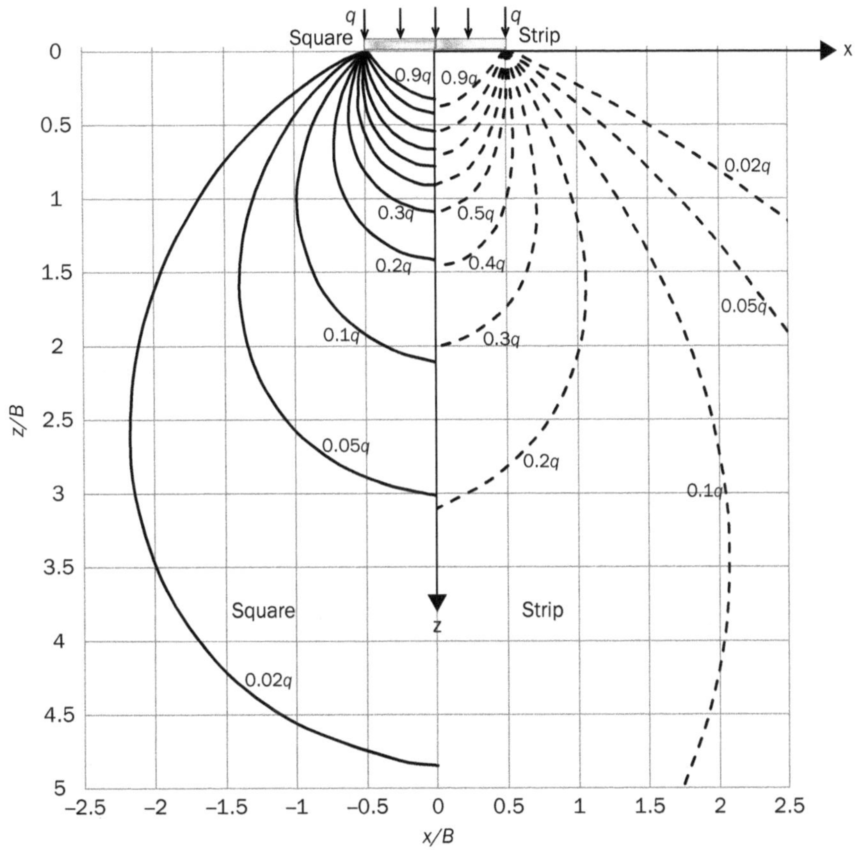

FIGURE 7.12 Pressure isobars.

strip load. Let us locate all the points where $\Delta\sigma_v = 0.1q$ and connect them. This gives the *stress contour* or *isobar* for $0.1q$, which is symmetrical about the vertical centerline. Such isobars can be drawn for any value of $\Delta\sigma_v$. The left half of Fig. 7.12 shows the pressure isobars for a square load and the right half shows the isobars for a strip load.

It can be seen from Fig. 7.12 that the pressure isobars extend to significantly higher depths for strip loads than the square loads. For example, the $0.1q$ isobar extends to about $2.1B$ for the square load and more than $5B$ for a strip load. At any depth, $\Delta\sigma_v$ is larger under a strip than a square load.

Example 7.3

A flexible square foundation of width B applies a uniform pressure q to the underlying ground. Determine the vertical normal stress increase $\Delta\sigma_v$ at a depth B below the center using

a. Figure 7.6,

b. 2:1 distribution, and

c. Figure 7.12.

Solution

Part a Considering a quarter of the foundation, $m = n = 0.5B/B = 0.5$
From Fig. 7.6, $I = 0.084$

$$\Delta\sigma_v = Iq = 4 \times 0.084q = 0.336\,q$$

Part b

$$\Delta\sigma_v = \frac{qBL}{(B+z)(L+z)} = q\frac{B \times B}{(B+B)(B+B)} = 0.25q$$

Part c From Fig. 7.12, $\Delta\sigma_v = 0.34q$

7.8 Vertical Stress Increase under an Embankment Load

The vertical normal stress at a depth z below a long embankment along the centerline (see Fig. 7.13) can be calculated as

$$\Delta\sigma_v = Iq \tag{7.15}$$

where $q = \gamma h$, h = height of the crest, γ = unit weight of the embankment soil, and I = influence factor from Fig. 7.13, proposed by Osterberg (1957). Considering the two halves of the embankment, $\Delta\sigma_v$ has to be multiplied by 2. As shown in Example 7.4, Fig. 7.13 can be used for points that are not necessarily along the vertical centerline.

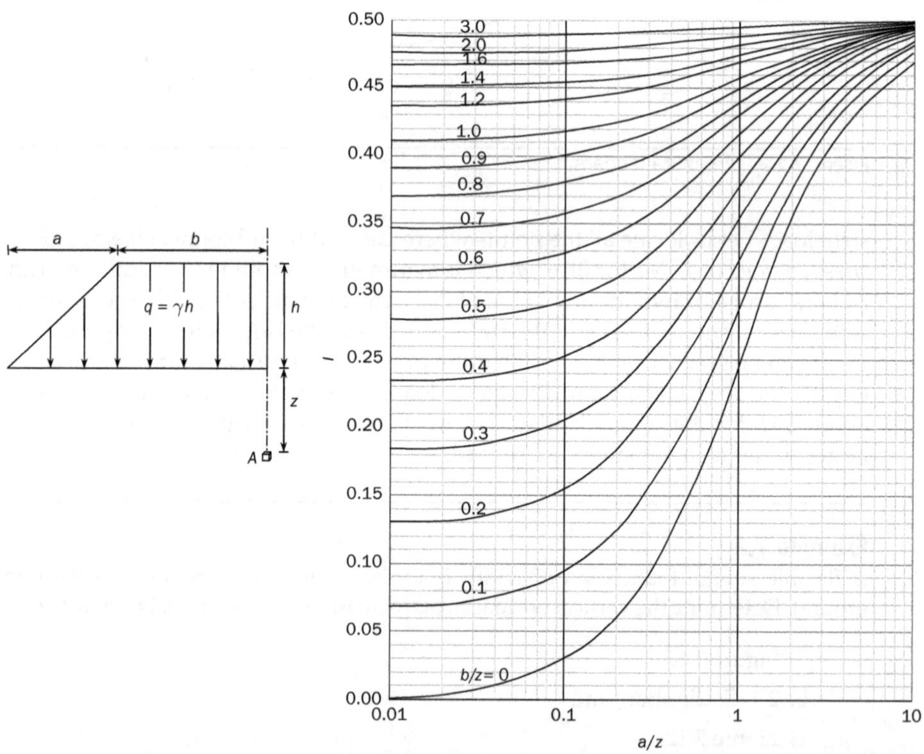

FIGURE 7.13 Influence factor I for embankment loads.

Example 7.4
An 8.0-m-high embankment shown in Fig. 7.14 is made of compacted clay with unit weight of 19.0 kN/m^3. Determine the vertical normal stress increase 10 m below A and B.

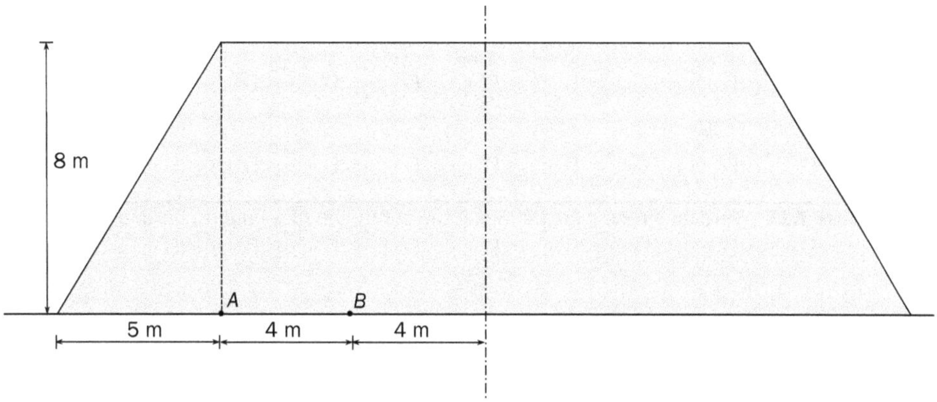

FIGURE 7.14 Embankment in Example 7.4.

Solution

Below point A:

Considering the region left of A, $a = 5$ m, $b = 0$, $h = 8$ m, and $z = 10$ m

$a/z = 0.5$ and $b/z = 0$. From Fig. 7.13, $I = 0.147$

Considering the region right of A, $a = 5$ m, $b = 16$ m, $h = 8$ m, and $z = 10$ m

$a/z = 0.5$ and $b/z = 1.6$. From Fig. 7.13, $I = 0.476$

$$\Delta\sigma_v = Iq = (0.147 + 0.476)(8 \times 19) = \mathbf{94.7\ kN/m^2}$$

Below point B:

Considering the region left of B, $a = 5$ m, $b = 4$ m, $h = 8$ m, and $z = 10$ m

$a/z = 0.5$ and $b/z = 0.4$. From Fig. 7.13, $I = 0.326$

Considering the region right of B, $a = 5$ m, $b = 12$ m, $h = 8$ m, and $z = 10$ m

$a/z = 0.5$ and $b/z = 1.2$. From Fig. 7.13, $I = 0.459$

$$\Delta\sigma_v = Iq = (0.326 + 0.459)(8 \times 19) = \mathbf{119.3\ kN/m^2}$$

7.9 Vertical Stress Increase beneath the Center of a Uniform Circular Load

When a flexible uniform circular load of radius a applies a pressure of q to the underlying soil, as shown in Fig. 7.15, the vertical stress increase at a depth z below the center is given by

$$\Delta\sigma_v = q\left\{1 - \frac{1}{\left[1 + \left(\frac{a}{z}\right)^2\right]^{3/2}}\right\} \tag{7.16}$$

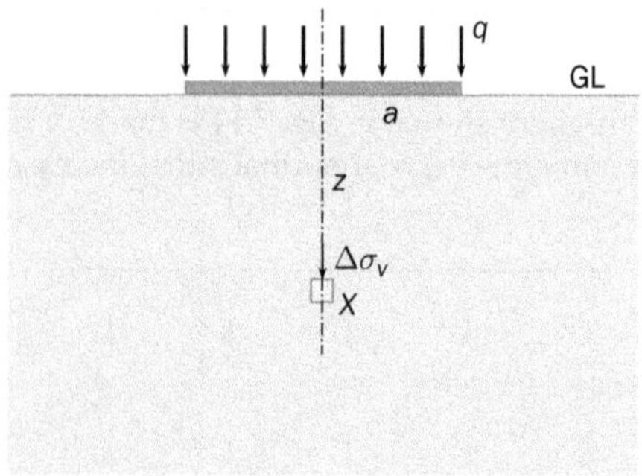

FIGURE 7.15 Vertical stress increase along the centerline of a uniform flexible circular load.

	r/a					
z/a	**0**	**0.2**	**0.4**	**0.6**	**0.8**	**1.0**
0	1.000	1.000	1.000	1.000	1.000	1.000
0.1	0.999	0.999	0.998	0.996	0.976	0.484
0.2	0.992	0.991	0.987	0.970	0.890	0.468
0.3	0.976	0.973	0.963	0.922	0.793	0.451
0.4	0.949	0.943	0.920	0.860	0.713	0.435
0.5	0.911	0.902	0.869	0.796	0.646	0.417
0.6	0.864	0.852	0.814	0.732	0.591	0.400
0.7	0.811	0.798	0.756	0.674	0.545	0.367
0.8	0.756	0.742	0.699	0.619	0.504	0.366
0.9	0.701	0.688	0.644	0.570	0.467	0.348
1.0	0.646	0.633	0.591	0.525	0.434	0.332
1.2	0.547	0.535	0.501	0.447	0.377	0.300
1.5	0.424	0.416	0.392	0.355	0.308	0.256
2.0	0.284	0.286	0.268	0.248	0.224	0.196
2.5	0.200	0.197	0.191	0.180	0.167	0.151
3.0	0.146	0.145	0.141	0.135	0.127	0.118
4.0	0.087	0.086	0.085	0.083	0.080	0.075

TABLE 7.4 Values of *I* for Eq. (7.17) (Data from Ahlvin and Ulery 1962)

The vertical stress increase at depth *z*, and radial distance *r* can be expressed as

$$\Delta\sigma_v = q.f\left(\frac{r}{a}, \frac{z}{a}\right) = qI \tag{7.17}$$

The influence factor *I* can be determined from Table 7.4 for $r/a < 1$ (Ahlvin and Ulery 1962; Jumikis 1971).

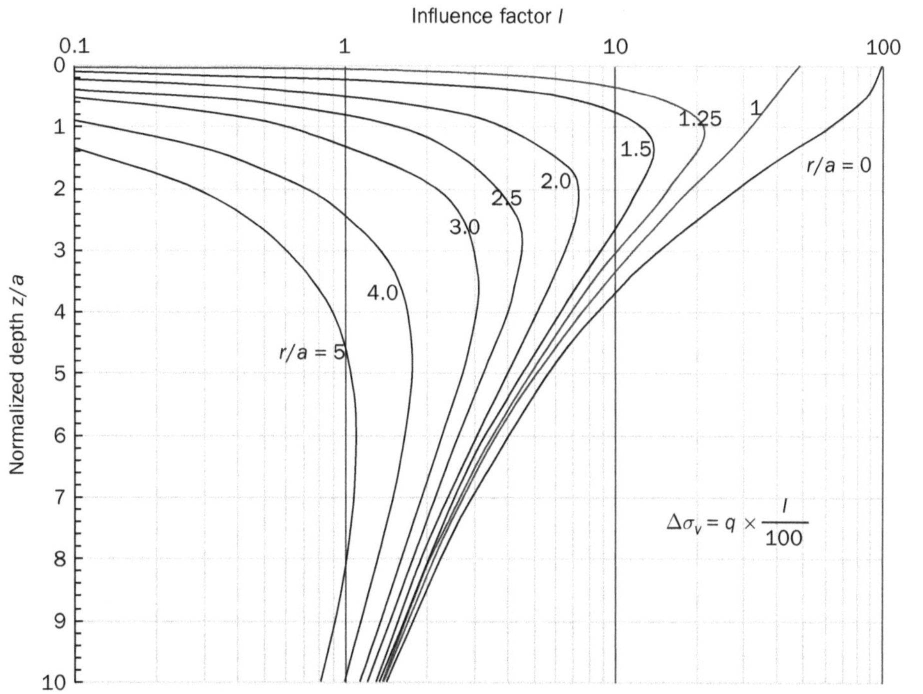

FIGURE 7.16 Influence factor I for uniformly loaded circular areas in Eq. (7.18). (Foster and Ahlvin, 1954.)

For situations where it is required to determine the vertical stress increase $\Delta\sigma_v$ for r/a values larger than 1.0, Fig. 7.16 can be used. Here, $\Delta\sigma_v$ is determined as

$$\Delta\sigma_v = q \times \frac{I}{100} \tag{7.18}$$

7.10 Newmark's Chart

Until now, we studied how to compute the vertical stress increase under a point load, line load, strip load, rectangular load, embankment load, or circular load. What if the loaded area is irregular in shape? Newmark (1942) developed a clever graphical method to compute $\Delta\sigma_v$ at any point below a uniformly loaded irregular area.

Equation (7.16) can be written as

$$\frac{a}{z} = \sqrt{\left(1 - \frac{\Delta\sigma_v}{q}\right)^{-\frac{2}{3}} - 1} \tag{7.19}$$

Values of a/z, corresponding to $\Delta\sigma_v/q = 0.1, 0.2 \ldots 1.0$ can be determined from Eq. (7.19) and are shown in Table 7.5.

$\Delta\sigma_v/q$	0	0.1	0.2	0.3	0.4	0.5
a/z	0	0.2698	0.4005	0.5181	0.6370	0.7664
$\Delta\sigma_v/q$	0.6	0.7	0.8	0.9	1.0	
a/z	0.9176	1.1097	1.3871	1.9084	∞	

Table 7.5 Values of a/z and Corresponding $\Delta\sigma_v/q$ from Eq. (7.19)

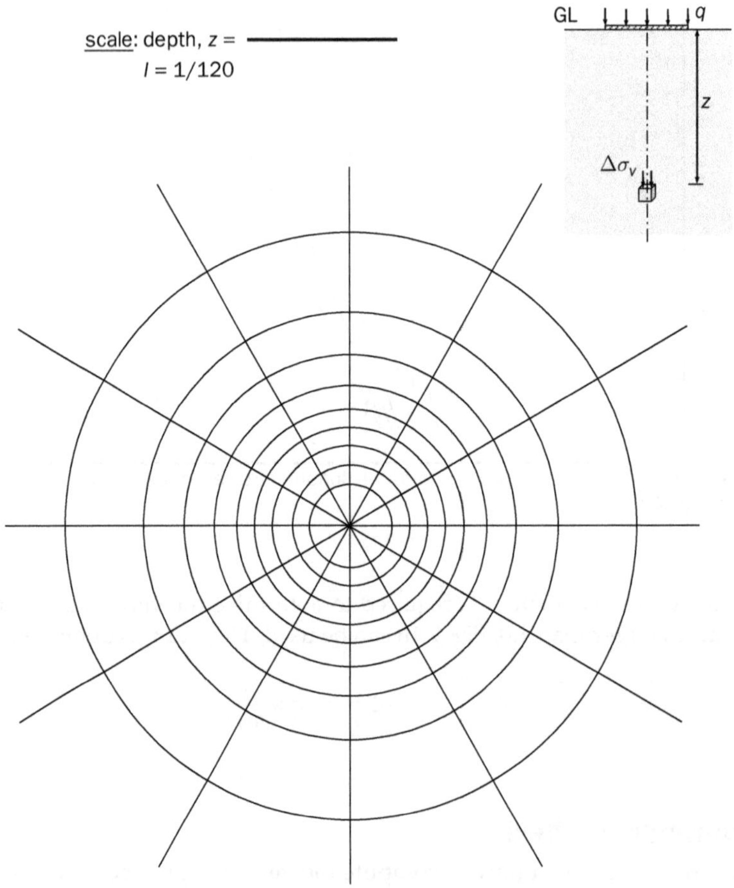

Figure 7.17 Newmark's chart.

Let us imagine a circular load of radius a applying a uniform pressure q at the ground level. Let us consider $\Delta\sigma_v$ at a point at depth z below the center from the ground level as shown in Fig. 7.17. It can be seen from Table 7.5 that, for $\Delta\sigma_v = 0.4q$, $a = 0.6370z$; and for $\Delta\sigma_v = 0.5q$, $a = 0.7664z$. Therefore, when a uniform pressure is applied over the annular region between $a = 0.6370z$ and $a = 0.7664z$, $\Delta\sigma_v = 0.5q - 0.4q = 0.1q$.

Figure 7.17 consists of a series of nine concentric circles, representing the different values for a given in Table 7.5, where the depth z is shown as the scale in the figure. Each annular region is divided into 12 equal blocks, making a total of 120 blocks. Thus, a uniform pressure of q in any one of these blocks will apply $\Delta\sigma_v$ of $1/120\ q$ at depth z below the center. If q is applied over n blocks, $\Delta\sigma_v = n\ I\ q$, where $I = 1/120$.

The use of Newmark's chart to calculate $\Delta\sigma_v$ at a certain depth z^* (e.g., 3 m) below a point X requires the following steps.

1. Redraw the loaded area to a new scale where the scale length "________" shown in the chart equals z^*.
2. Overlay the redrawn area on Newmark's chart such that the point of interest X lies at the center of the chart.
3. Count the number of blocks (including fractions) n covered by the loaded area.
4. The vertical normal stress increase $\Delta\sigma_v$ can be calculated as

$$\Delta\sigma_v = nIq \tag{7.20}$$

Example 7.5

Figure 7.18a shows an L-shaped area applying a uniform pressure of 50 kN/m² to the underlying ground. Find the vertical normal stress increase at 6 m below X.

Solution Let us rescale the area (Fig. 7.18a) to a scale of 6 m = scale length shown in the chart in Fig. 7.17b.

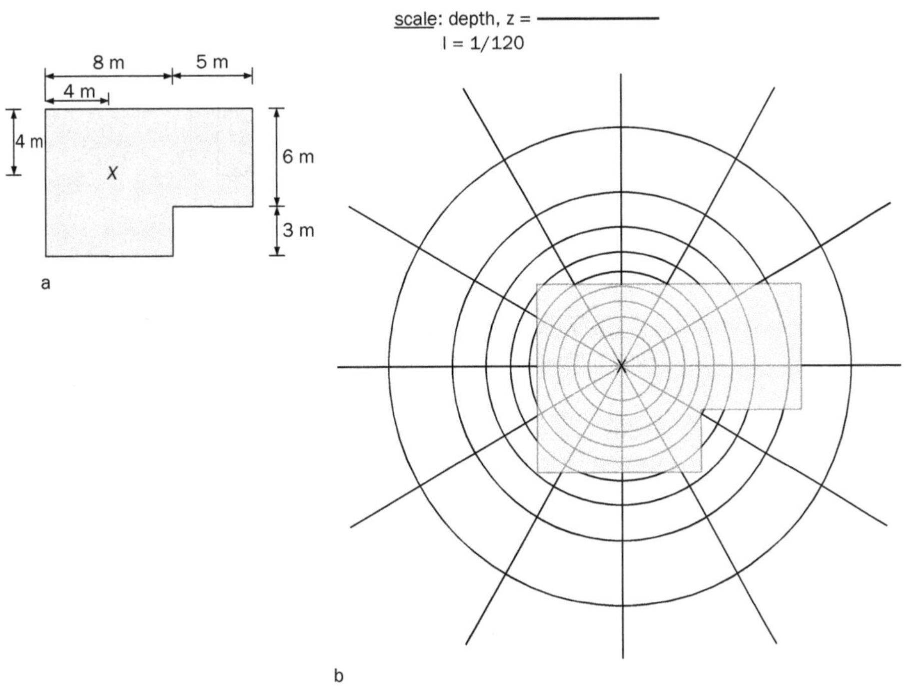

FIGURE 7.18 (a) Loaded area, (b) on top of Newmark's chart.

Counting the blocks covered by the loaded area (including fractions), $n = 68.7$

Therefore, at 6 m below X,

$$\Delta\sigma_v = 68.7 \times \frac{1}{120} \times 50 = \mathbf{28.63\ kN/m^2}$$

Example 7.6

Figure 7.19 shows a structure in elevation and plan, which consists of a circular tower surrounded by a podium. The circular tower applies 40 kN/m² to the under lying ground, and the surrounding podium applies 10 kN/m². Determine the vertical stress increase at 20 m below the center.

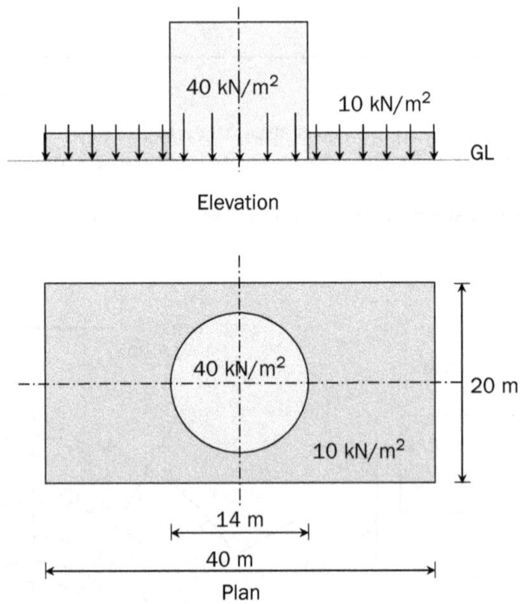

Figure 7.19 Elevation and plan views of the tower and podium.

Solution Let us treat this as a 20 m × 40 m rectangular area applying 10 kN/m², and a 14 m diameter circular area applying a 30 kN/m².

Due to the 20 m × 40 m rectangular area applying 10 kN/m²:

Let us divide the area into four quarters.

$$L = 20\ \text{m},\ B = 10\ \text{m},\ z = 20\ \text{m}.$$

$$m = B/z = 10/20 = 0.5$$

$$n = L/z = 20/20 = 1$$

From Fig. 7.6, $I = 0.120$

Therefore, $\Delta\sigma_v = 4 \times 0.120 \times 10 = 4.8\ \text{kN/m}^2$.

Due to the 14 m diameter circular area applying a 30 kN/m²:

From Eq. (7.16),

$$\Delta\sigma_v = q\left\{1 - \frac{1}{\left[1 + \left(\frac{a}{z}\right)^2\right]^{3/2}}\right\}$$

$$\Delta\sigma_v = 30\left\{1 - \frac{1}{\left[1 + \left(\frac{7}{20}\right)^2\right]^{3/2}}\right\} = 4.8\ \text{kN/m}^2$$

Therefore, the total increase in vertical stress at 200 m depth is $4.8 + 4.8 = 9.6\ \text{kN/m}^2$.

Example 7.7

A cylindrical silo is supported on a 1-m-wide ring foundation as shown in Fig. 7.20. The weight of the silo with the contents is 1600 kN.

a. Determine the vertical normal stress increase at 1 m below X.

b. Assuming the ring foundation as a 1-m-wide strip load, estimate the vertical normal stress increase at 1 m below X.

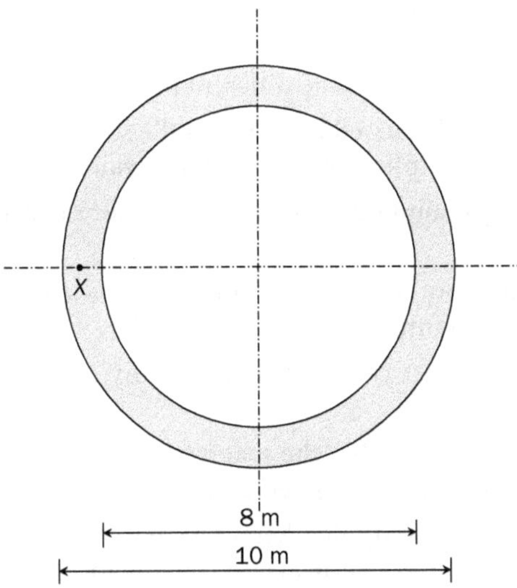

FIGURE 7.20 Plan view of the ring foundation carrying the silo.

Solution

Part a

Area of the foundation = $(\pi/4)(10^2 - 8^2) = 28.27\ \text{m}^2$

Therefore, the contact pressure at the soil foundation interface = $1600/28.27 = 56.6\ \text{kN/m}^2$

Let us subtract the $\Delta\sigma_v$ induced by the 8 m diameter circle from the 10 m diameter circle.

For the 10 m diameter circle,

$$z/a = 1/5 = 0.2$$
$$r/a = 4.5/5 = 0.9$$

From Table 7.4, $I = (0.890 + 0.468)/2 = 0.679$

Therefore, $\Delta\sigma_v = 0.679 \times 56.6 = 38.4\ \text{kN/m}^2$.

For the 8 m diameter circle,

$$z/a = 1/4 = 0.25$$
$$r/a = 4.5/4 = 1.125$$

From Figure 7.16, $I = 20$

Therefore, $\Delta\sigma_v = 56.6 \times 20/100 = 11.3\ \text{kN/m}^2$

Therefore, σ_v induced by the ring foundation at 1 m below X is 38.4 – 11.3 = **27.1 kN/m²**

Part b Approximating the ring foundation as a 1.0-m-wide strip load, and using Eq. (7.14),

$$\Delta\sigma_v = \frac{qB}{(B+z)} = \frac{56.6 \times 1}{(1+1)} = \mathbf{28.3\ kN/m^2}$$

7.11 Main Points

1. Very often, it is necessary to determine the vertical stress increase $\Delta\sigma_v$ at a point within the soil due to the various types of loads applied at the ground level. This chapter discusses the different methods to determine $\Delta\sigma_v$.
2. The methods discussed herein treat the soil mass as a linear elastic continuum, where the principle of superposition is valid.
3. Boussinesq's solution is simpler and more conservative than Westergaard's solution.
4. 2:1 distribution can be used for a quick estimate of $\Delta\sigma_v$ along the vertical centerline of a uniform rectangular load.
5. Newmark's chart can be applied to any irregular-shaped uniformly loaded area.
6. All methods imply that the magnitude of the vertical stress increase due to a surface load does not depend on the properties of the underlying soils.
7. The $\Delta\sigma_v$ values computed herein have to be added to the overburden stress to arrive at the current vertical stress at a point.

Review Exercises

7.1 A point load 100 kN is applied at the ground level. Plot the variation of $\Delta\sigma_v$ along a vertical line (a) directly under the load, (b) 1 m away from the load, and (c) 3 m away from the load.

7.2 The L-shaped area shown in Fig. RE7.2 applies a uniform pressure of 80 kN/m^2 to the underlying ground. Find the vertical stress increase at 5 m below *A*, *B*, and *C*.

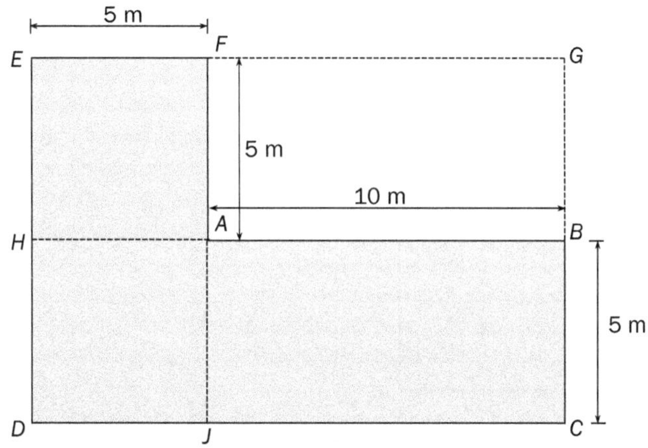

FIGURE RE7.2 Plan view of the building.

Answer: 44.0 kN/m², 16.6 kN/m², 16.4 kN/m²

7.3 A flexible square foundation applies a uniform pressure q to the underlying ground. At what depth below the center would the vertical normal stress increase $\Delta\sigma_v$ be $0.1q$? Determine this using

a. Figure 7.6,
b. 2:1 distribution, and
c. Pressure isobars (Fig. 7.12).

Answer: 2.0 B, 2.16 B, 2.1 B

7.4 A 6-m-high embankment with crest width of 8 m is made of a compacted clay with unit weight of 19.5 kN/m^3 and slope of 1(vertical):2(horizontal) on both sides (see Fig. RE7.4). Find the vertical normal stress increase at 10 m below the bottom of the embankment, along the centerline.

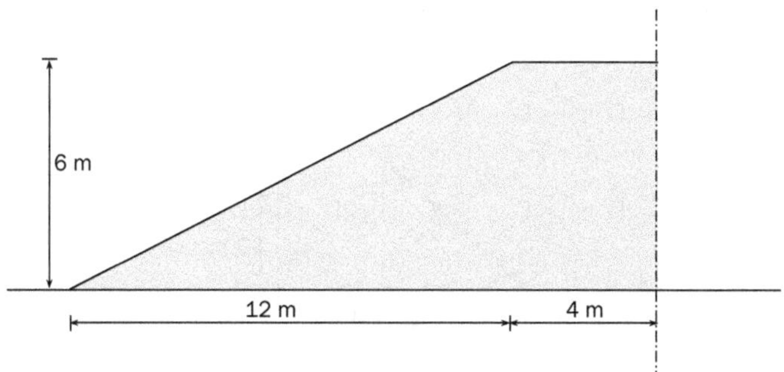

FIGURE RE7.4 Left half of the embankment.

Answer: 92.0 kN/m²

7.5 The area shown in Fig. RE7.5 is subjected to a uniform vertical stress of 40 kN/m^2. What would be the maximum vertical stress increase at a depth of 5 m below this area?

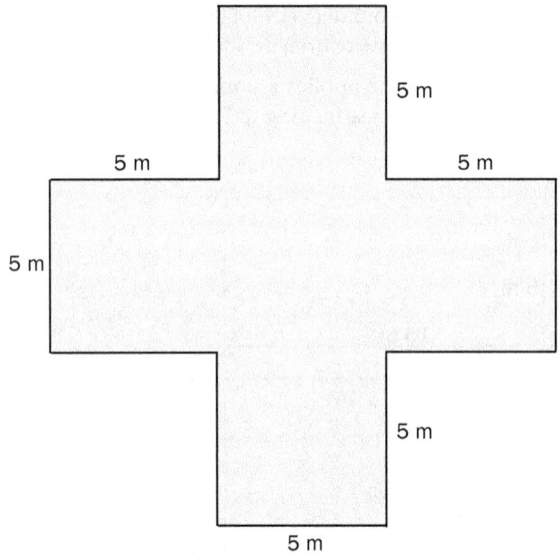

FIGURE RE7.5 Plan view of the loaded area.

Answer: 29.2 kN/m^2

7.6 A flexible square foundation applies a uniform pressure q to the underlying ground. What would the vertical normal stress increase $\Delta\sigma_v$ at a depth of B below the center? Determine this using

a. Figure 7.6,
b. 2:1 distribution, and
c. Pressure isobars (Fig. 7.12).

Answer: 0.34 q, 0.25 q, 0.34 q

7.7 Figure RE7.7 shows an 8 m × 16 m rectangular area which is under a uniform pressure of 60 kN/m^2 at the ground level. There is a pipe line at 5 m below the loaded area. Determine the vertical stress increase on the pipeline at X.

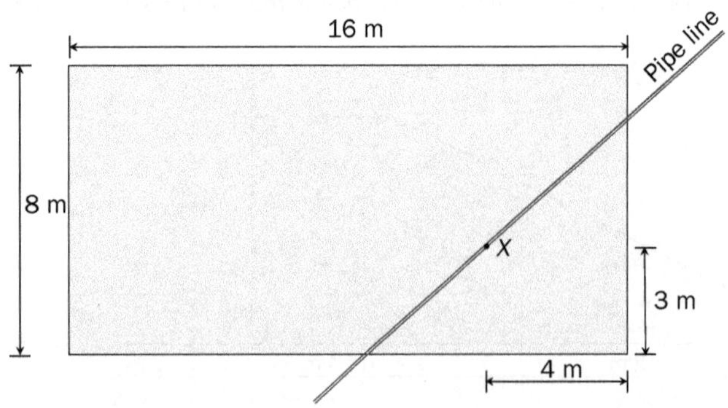

FIGURE RE7.7 Plan view.

Answer: 38.6 kN/m^2

References

Ahlvin, R. G., and H. H. Ulery. *Tabulated Values for Determining the Complete Pattern of Stresses, Strains and Deflections Beneath a Uniform Load on a Homogeneous Half Space*, Bulletin 342, Highway Research Record, 1962, pp. 1–13.

Boussinesq, J. *Application des Potentials a L'etude de L'equilibre et du Mouvement des Solides Elastiques*, Gauthier-Villars, Paris, 1885.

Bowles, J. E. *Foundation Analysis and Design*, 4th Edition, McGraw Hill, New York, 1988.

Foster, C. R., and R. G. Ahlvin. "Stresses and Deflections Induced by a Uniform Circular Load," *Proceedings of the Highway Research Board*, 33:467–470, 1954.

Jumikis, A. R. "Vertical stress tables for uniformly distributed loads on soil," Rutgers University Engineering Research Publication No. 52.

Newmark, N. M. *Influence Charts for Computation of Stresses in Elastic Soils*, University of Illinois Experiment Station, Bulletin No. 338, 1942.

Osterberg, J. O. "Influence Values for Vertical Stresses in Semi-infinite Mass due to Embankment Loading," *Proceedings of the 4th International Conference on Soil Mechanics and Foundation Engineering*, London, 1, 1957, pp. 393–396.

Westergaard, H. M. "A Problem of Elasticity Suggested by a Problem in Soil Mechanics: A Soft Material Reinforced by Numerous Strong Horizontal Sheets," *Contributions to the Mechanics of Solids, Stephen Timoshenko 60th Anniversary Volume*, Macmillan, New York, 1938, pp. 268–277.

CHAPTER 8
Consolidation

8.1 Introduction

Let us consider a homogeneous soil mass which is fully submerged to the ground level as shown in Fig. 8.1. A foundation is placed at the ground level, applying a pressure of q. How much will the foundation settle and how fast will it settle? As shown in the figure, the settlement in the granular soil is almost instantaneous. In the case of the saturated cohesive soil, the settlement is a very slow process, which can take several years to complete.

Consolidation is a time-dependent process, where some of the water is squeezed out of the voids from a saturated clay, by the loads applied at the ground level. With some water being squeezed out of a saturated clay, void ratio decreases, resulting in settlement. The low permeability of the clay makes the process very slow, taking several months or years. In granular soils, where the permeability is high, the consolidation process is almost instantaneous.

In this chapter, we will only discuss the consolidation of saturated clays. We will discuss this in two parts: (a) the magnitude of the *final* settlement in Secs. 8.2–8.7 and (b) the rate of settlement in Sec. 8.8. What happens after the consolidation is completed? This is discussed in Sec. 8.9 under *secondary compression*.

8.2 Fundamentals

Let us consider the same clay layer discussed before, and shown again in Fig. 8.2a. Initially, at a point X, at depth h below the ground level, the total vertical overburden stress σ_{vo}, pore water pressure u_o, and the effective vertical overburden stress σ'_{vo} are given by

$$\sigma_{vo} = \gamma_{sat} h \tag{8.1a}$$

$$u_o = \gamma_w h \tag{8.1b}$$

$$\sigma'_{vo} = \gamma' h \tag{8.1c}$$

When a uniform surcharge pressure q of infinite lateral extent is applied at the ground level (Fig. 8.2b), the total vertical stress, pore water pressure, and the effective vertical stress at X increase above the initial values given by Eq. (8.1), and the magnitude of the increase is time dependent. At time t,

$$\sigma_V(t) = \sigma_{vo} + \Delta\sigma(t) \tag{8.2a}$$

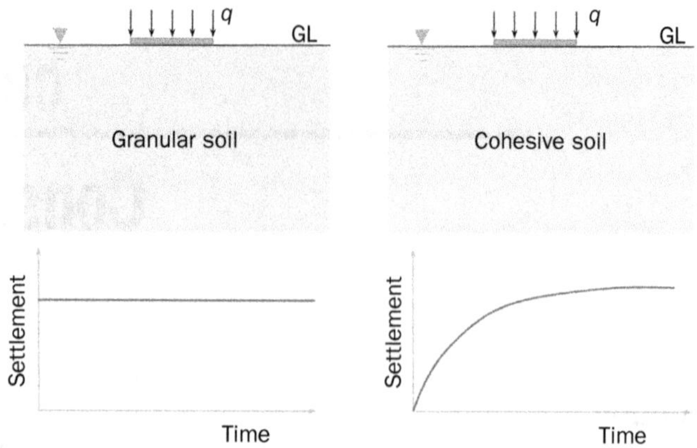

FIGURE 8.1 Settlements in granular and cohesive soils.

$$u(t) = u_o + \Delta u(t) \tag{8.2b}$$

$$\sigma'_v(t) = \sigma'_{vo} + \Delta\sigma'(t) \tag{8.2c}$$

Here, the total vertical stress increases by $\Delta\sigma(t)$, the pore water pressure increases by $\Delta u(t)$, and the effective vertical stress increases by $\Delta\sigma'(t)$. Figure 8.2c shows the variation of the consolidation settlement with time t. At time t,

$$\sigma_v(t) = \sigma'_v(t) + u(t) \tag{8.3}$$

Therefore, at any time during consolidation,

$$\Delta\sigma(t) = \Delta\sigma'(t) + \Delta u(t) \tag{8.4}$$

At time t, the total stress increase at any point within the clay remains the same at q and hence

$$\Delta\sigma(t) = q \tag{8.5}$$

This is shown by the horizontal line in Fig. 8.2d.

Immediately after the surcharge q is applied, at time $t = 0^+$, there is no drainage and the entire load is carried by the water in the form of excess pore water pressure $\Delta u(0^+)$. Thus, $\Delta u(0^+) = q$ and $\Delta\sigma'(0^+) = 0$, which satisfies Eqs. (8.4) and (8.5).

With time, drainage takes place and the excess pore water pressure slowly dissipates, with $\Delta u(t)$ decreasing and $\Delta\sigma'(t)$ increasing, thus transferring the load from the pore water to the soil skeleton.

From Eqs. (8.4) and (8.5),

$$\Delta\sigma'(t) + \Delta u(t) = q$$

$$d[\Delta\sigma'(t)] + d[\Delta u(t)] = dq = 0$$

$$d[\Delta\sigma'(t)] = -d[\Delta u(t)] \tag{8.6}$$

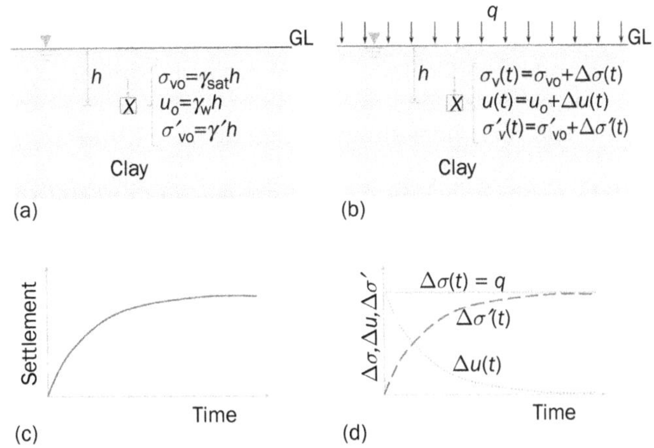

Figure 8.2 (a) Just before loading, (b) during the consolidation process, (c) settlement vs. time variation, (d) changes in σ, σ', and u during consolidation.

Thus, within any time interval during consolidation, the reduction in pore water pressure must be the same as the increase in the effective stress.

After a long time, at $t = \infty$, the entire excess pore water pressure is fully dissipated, bringing the pore water pressure back to the initial level prior to loading. At this time, the applied load is fully transferred to the soil skeleton with $\Delta u(\infty) = 0$ and $\Delta\sigma'(\infty) = q$, satisfying Eqs. (8.4) and (8.5). The variations of $\Delta\sigma(t)$, $\Delta u(t)$, and $\Delta\sigma'(t)$ with time are shown in Fig. 8.2d. It can be seen that the excess pore water pressure $\Delta u(t)$ decays during consolidation and $\Delta\sigma'(t)$ increases. During this process, the applied load is gradually transferred from the pore water to the soil skeleton.

A simple analogy for consolidation is squeezing a pack of potato chips. When the pack is squeezed, the load is taken by the air only and not the chips. When a hole is made on the pack, the air leaks out similar to water draining through clay, and the load is transferred to the chips.

8.3 One-Dimensional Consolidation

The clay layer shown in Fig. 8.3a is sandwiched between two sand layers. The initial thickness of the clay layer is H_0 and the initial *in situ* void ratio of the clay is e_0. When a surcharge q of large lateral extent is applied at the ground level, consolidation occurs. During consolidation, it is reasonable to assume that the *strains* and the *drainage* within the clay are both in the vertical direction. There will be no lateral deformations or strains within the clay at any location. While some water is drained out due to dissipation of the excess pore water pressure, the drainage path will be only vertical. There will be no drainage in the horizontal direction. Here, the clay is undergoing *one-dimensional consolidation*.

How do we simulate one-dimensional consolidation in the laboratory? Figure 8.3b shows a one-dimensional consolidation test setup (known as *oedometer*) in the laboratory. Here, an undisturbed clay specimen of thickness H_0 and initial void ratio e_0, is placed within a rigid metal ring (*oedometer* ring) to prevent lateral deformations and drainage. Note that we use the same notation H_0 in the figure for the laboratory and in situ clay thickness, mainly to signify the initial height. Two porous stones are placed at the top and bottom to facilitate vertical drainage, similar to the sand layers in the field. When the clay specimen is loaded, it undergoes one-dimensional consolidation, replicating the field conditions shown in Fig. 8.3a.

8.3.1 Relationship between Δe and ΔH during Consolidation

Figure 8.4a shows a saturated clay layer of initial height H_0 and void ratio e_0, with the phase diagram below. Figure 8.4b shows the same, during consolidation, after time t since the surcharge q was applied. Here, ΔH and Δe are the reduction in the height and the void ratio, respectively, due to consolidation.

The average vertical strain at time t can be written as

$$\varepsilon_{\text{avg}} = \frac{\Delta H}{H_0} \tag{8.7}$$

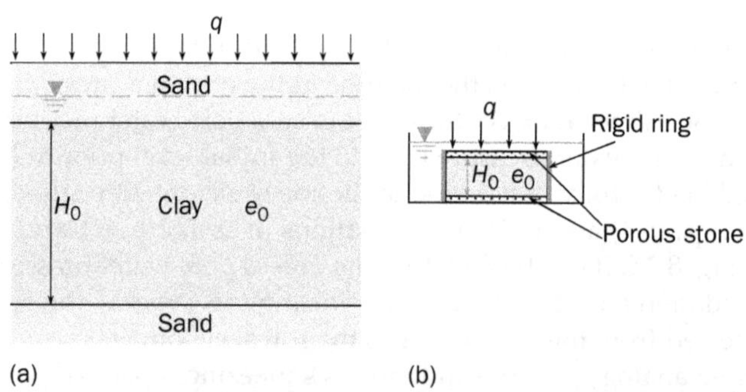

FIGURE 8.3 One-dimensional consolidation: (a) In the field, (b) in the laboratory.

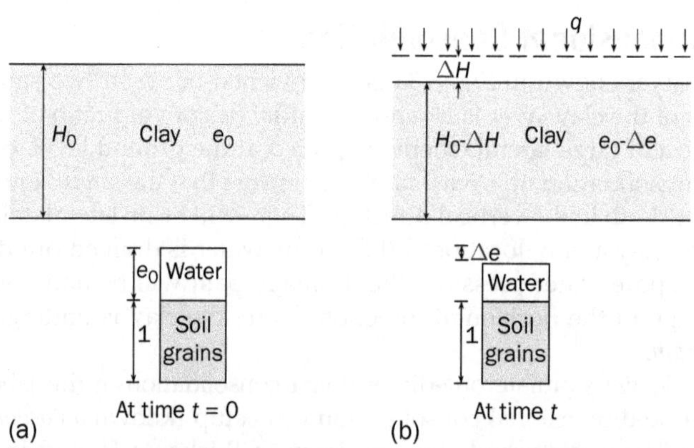

FIGURE 8.4 Clay layer and phase diagram: (a) At time $t = 0$, (b) at time t during consolidation.

From the phase diagram, the same average vertical strain can be written as

$$\varepsilon_{\text{avg}} = \frac{\Delta e}{1+e_0} \tag{8.8}$$

From Eqs. (8.7) and (8.8),

$$\frac{\Delta H}{H_0} = \frac{\Delta e}{1+e_0} \tag{8.9}$$

Equation (8.9) is valid at all times during consolidation, including at $t = \infty$. Therefore, if the reduction in void ratio Δe at $t = \infty$ can be determined, the *final consolidation settlement* s_c (i.e., ΔH at time $t = \infty$) can be calculated from Eq. (8.9).

8.3.2 Coefficient of Volume Compressibility

Coefficient of volume compressibility, denoted by m_v, is a measure of the compressibility of the clay during one-dimensional consolidation. It is defined as the volumetric strain per unit stress increase and hence

$$m_v = \frac{\varepsilon_{\text{vol}}}{\Delta\sigma'} = \frac{\left(\frac{\Delta V}{V_0}\right)}{\Delta\sigma'} = \frac{\left(\frac{\Delta V}{V_0}\right)}{q} \tag{8.10}$$

where ε_{vol} = volumetric strain during the entire consolidation process, V_0 = initial volume, ΔV = change in volume during the entire consolidation process, and $\Delta\sigma'$ = the effective stress increase applied on the clay, which is causing the consolidation (= q in one-dimensional consolidation). m_v is expressed in kPa^{-1}, MPa^{-1}, m^2/kN, or m^2/MN. Typically, m_v varies from 0 to 2 MPa^{-1}. In one-dimensional consolidation, where there is no change in the cross-sectional area,

$$\frac{\Delta V}{V_0} = \frac{\Delta H}{H_0} \tag{8.11}$$

From Eqs. (8.10) and (8.11), and noting that $\Delta H = s_c$ when the consolidation is complete, the final consolidation settlement is given by

$$s_c = m_v q H_0 \tag{8.12}$$

Equation (8.12) can be used to determine the final consolidation settlement from the values of m_v, q, and H_0. Table 8.1 shows a simple classification of clays based on m_v.

Type of soil	m_v (MPa^{-1})	Compressibility
Heavily overconsolidated clays	<0.05	Very low
Very stiff or hard clays, tills	0.05–0.10	Low
Varved and laminated clays, firm to stiff clays	0.10–0.30	Medium
Normally consolidated alluvial clays	0.3–1.5	High
Organic alluvial clays and peats	>1.5	Very high

TABLE 8.1 Classification of Clays Based on m_v (after Bell, 2000)

Example 8.1

A 4-m-thick clay layer is subjected to a surcharge of 45 kN/m^2 at the ground level. Assuming m_v = 0.3 MPa^{-1}, what would be the final consolidation settlement after very long time?

Solution The final consolidation settlement s_c is given by [Eq. (8.12)]

$$s_c = m_v q H_0 = (0.3)\left(\frac{45}{1000}\right)(4.0) = 0.054\,\text{m} = 54\,\text{mm}$$

The reciprocal of m_v is called the *constrained modulus* or *oedometer modulus D*. It is similar to Young's modulus E but here the specimen is constrained from any lateral expansion during the loading and hence D is larger than E. For a linear elastic material, it can be shown that

$$D = \frac{1}{m_v} = \frac{(1-\nu)}{(1+\nu)(1-2\nu)}E = K + \frac{4}{3}G \tag{8.13}$$

where v = Poisson's ratio which varies in the range of 0–0.5 for all materials, K = bulk modulus, and G = shear modulus (or modulus of rigidity). E in Eq. (8.13) is the *drained Young's modulus*, which is different to *undrained Young's modulus* E_u.

Equating the shear modulus G at drained and undrained conditions, it can be shown that

$$\frac{E}{E_u} = \frac{1+\nu}{1.5} \tag{8.14}$$

where E and E_u are the drained and undrained Young's moduli, respectively. Here, ν is the drained Poisson's ratio. Equation (8.14) is useful in determining the drained or undrained modulus from each other. For saturated clays during undrained loading, the Poisson's ratio is 0.5.

8.4 One-Dimensional Consolidation Test

A laboratory consolidation test (ASTM D 2435-20) is generally carried out on an undisturbed cylindrical clay specimen, typically of 50–75 mm diameter and 20–25 mm in thickness. The specimen is placed in a rigid metal ring as shown in Fig. 8.3b, submerged in water and subjected to a series of stress increments, typically at 24-h intervals, followed by unloading. A dial gauge is placed at the top of the oedometer to monitor the settlements during consolidation.

Figure 8.5a shows a series of three oedometer tests where the loads are applied using dead weights. Figure 8.5b shows a fully servo-controlled consolidation setup, which uses pneumatic loading. As shown in Fig. 8.3b, a porous stone at the top and bottom of the clay specimen facilitates drainage. The rigid metal ring does not allow lateral deformation or drainage, thus ensuring one-dimensional consolidation.

The initial void ratio e_0 and the height H_0 of the specimen must be known, and the dial gauge is initialized to zero. When the first stress increment Δq_1 is applied, consolidation is allowed to take place and the dial gauge readings are monitored. From the reduction in thickness ΔH_1 at the end of consolidation, the reduction in the void ratio Δe_1 can be determined using Eq. (8.9). Therefore, the new σ'_v and e at the end of

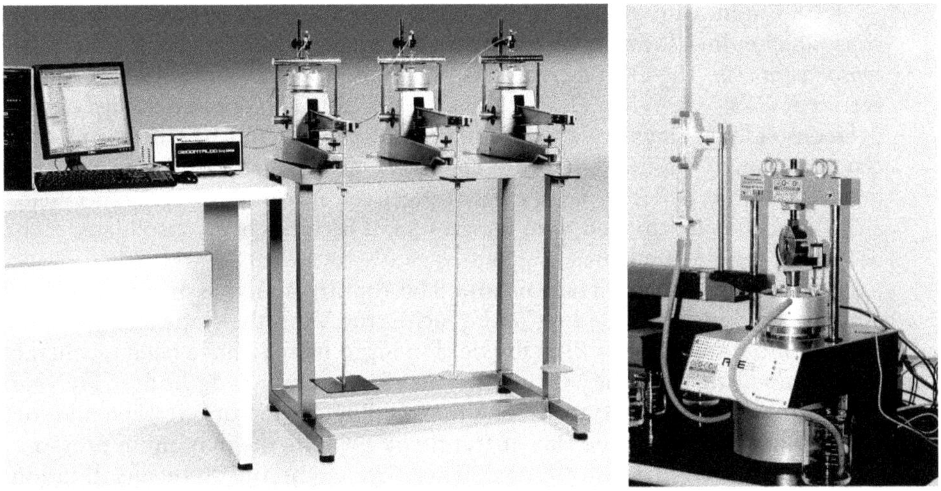

(a) (b)

FIGURE 8.5 Laboratory consolidation test: (a) A series of three oedometers using front-loaded dead weights (courtesy of Mr. Rodrigo Palma, Business Development and Technical Manager, Geo-Con Products Pty Ltd., Australia), (b) servo-controlled consolidation setup with pneumatic loading.

consolidation are determined. This can be repeated for a series of pressure increments, followed by unloading, and the pairs of σ'_v and e values are tabulated. From these test data, e is plotted against σ'_v (in logarithmic scale), as shown in Fig. 8.6a.

The void ratio versus effective stress plot in Fig 8.6a, starts at A and reaches B at the end of the final loading increment. Along this path AB, the void ratio e decreases with the increase in σ'_v. BC shows the unloading path, where the clay heaves with increasing void ratio with reduction in σ'_v. During loading, the plot takes a sharp turn at P where the effective stress σ'_p is called the *preconsolidation pressure*. It is the maximum effective overburden pressure ever experienced by the clay at that location (from where the specimen was taken from) since its deposition. The segment PB is approximately a straight line, where the slope C_c is known as the *compression index*. The unloading (or swelling) path BC is also a straight line, which has a slope of C_s or C_r, known as *swelling index* or *recompression index*. When the clay is reloaded from C, it follows the same path CB and hence is the term *recompression index*, which is used in this book. The paths AP, CB, and any further unload/reload paths have the same slope of C_r (or C_s).

There is a unique straight line called *virgin consolidation line* (VCL) for every clay. The straight line segment of the $e - \log \sigma'_v$ plot at stresses beyond the preconsolidation pressure coincides with this line. Therefore, VCL has the slope of C_c.

In Fig. 8.6b, the point E represents the initial values of e_0 and σ'_{vo} of the same clay in the field, for which the VCL is the same. When this clay element is loaded, it follows the path EFG, with the preconsolidation pressure σ'_p at F. The ratio of σ'_p to σ'_{vo} is called the *overconsolidation ratio* (OCR). Thus,

$$OCR = \frac{\sigma'_p}{\sigma'_{vo}} \tag{8.15}$$

OCR is a measure of how far the initial state (i.e., Point *E*) is away from the virgin consolidation line. When it lies on the VCL, $OCR = 1$, and the clay is called a *normally consolidated clay*. The *OCR* increases as the point moves away from the VCL. It can be seen that $OCR \geq 1$. When $OCR > 1$, the clay is called an *overconsolidated clay*. Slightly overconsolidated clays have *OCR* little more than 1.0 and heavily overconsolidated clays may have *OCR* in excess of 10. The term *OCR* is also applicable to clay specimens in the laboratory, where σ'_{vo} is the current effective vertical stress on the specimen. When a clay specimen is removed from the ground, it becomes overconsolidated right away because of the stress relief.

It can be seen from Figs. 8.6a and 8.6b that irrespective of the initial values e_0 and σ'_{vo}, as soon as the loading path reaches the VCL, the slope changes from C_r to C_c. The initial state (Point *E*) in the field in Fig. 8.6b may have been reached by previous unloading from the VCL, which may have happened hundreds of years ago, due to removal of overburden or other geological factors. Irrespective of when that has happened, for the clay currently at σ'_{vo}, σ'_p is the maximum pressure it has ever been subjected to in the past. Therefore, sometimes preconsolidation pressure is called the *maximum past pressure*. Preconsolidation pressure can be seen as the yield stress of the clay, showing elastic behavior when overconsolidated and elasto-plastic behaviour when normally consolidated. The magnitude of σ'_p is influenced by the locations of the initial state and the VCL on the $e - \log \sigma'_v$ plane.

The constrained modulus *D* and the preconsolidation pressure σ'_p are approximately related by (Canadian Geotechnical Society, 1992)

$$40\sigma'_p \leq D \leq 80\sigma'_p \tag{8.16}$$

Lower end of the range applies to soft clays and the upper end for stiff clays.

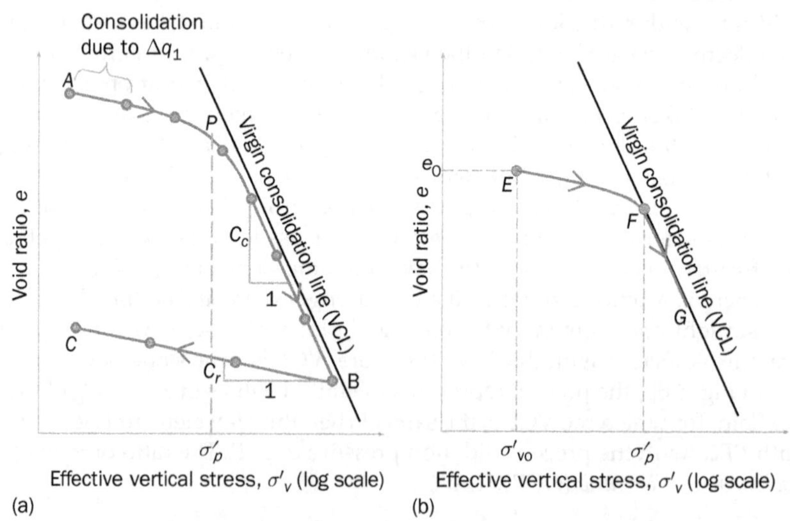

FIGURE 8.6 (a) e versus σ'_v plot from the laboratory, (b) e versus σ'_v plot in the field.

8.4.1 Compression Index C_c and Compression Ratio *CR*

Typically, compression index varies in the range of less than 0.1 for stiff clays to greater than 1.5 for very soft clays. It is generally correlated with *natural* (in situ) *moisture content* w_n, initial void ratio e_0, and liquid limit *LL* of the clay. Kulhawy and Mayne (1990) noted that, although hundreds of relationships are found in literature, correlations based on natural moisture content (w_n) are better than correlations based on liquid limit *LL* or initial void ratio e_0. Some of the popular empirical correlations for C_c are summarized in Table 8.2. Ameratunga et al. (2016) and others have listed more empirical correlations.

Instead of the void ratio, if the vertical strain is plotted against the effective stress (log scale), the slope of the virgin consolidation line is known as the *modified compression index* $C_{c\varepsilon}$, or *compression ratio CR*. Vertical strain is given by $\Delta e/(1 + e_0)$ and hence

$$C_{c\varepsilon} = CR = \frac{C_c}{1 + e_0} \tag{8.17}$$

CR typically varies in a narrow range of 0.2–0.4.

In sands, C_c is in the range of 0.001–0.1 at typical stress levels encountered in geotechnical work. At very high stress levels, when the grains start crushing, C_c increases significantly to values in the range of 0.1–1.0 (Zheng et al., 2017). Mesri and Vardhanabhuti (2009) noted that the grain crushing occurs at stresses in the order of 0.3 MN/m^2 for biogenic carbonated sands and 30 MN/m^2 for quartz sands.

8.4.2 Recompression Index C_r and Recompression Ratio *RR*

Recompression index C_r is typically 1/5 to 1/10 of C_c. *Modified recompression index* $C_{r\varepsilon}$, or *recompression ratio RR*, is defined as

$$C_{r\varepsilon} = RR = \frac{C_r}{1 + e_0} \tag{8.18}$$

Some practicing engineers prefer using *CR* and *RR* values instead of C_c and C_r, for computing consolidation settlements.

In critical state soil mechanics using models such as Cam Clay, natural logarithmic (ln) axis is used for the effective stress instead of the traditional base 10 logarithmic (log)

Equation	Comments	Reference
C_c = 1% of w_n	Chicago clays Fibrous peats Various clays	Holtz and Kovacs (1981) Mesri and Ajlouni (2007) Al Khafaji and Andersland (1992)
$C_c = 0.009(LL\text{-}10)$	Undisturbed clays	Skempton (1944)
$C_c = 0.007(LL\text{-}10)$	Remolded clays	Skempton (1944)
$C_c = G_s \times \frac{PI}{200}$	From critical state soil mechanics	Wroth and Wood (1978)

Table 8.2 Empirical Correlations for C_c

Description	$\frac{C_c}{1+e_0}$ for NC clays or $\frac{C_r}{1+e_0}$ for OC clays
Very slightly compressible	<0.05
Slightly compressible	0.05–0.10
Moderately compressible	0.10–0.20
Highly compressible	0.20–0.35
Very highly compressible	>0.35

Table 8.3 Classification Based on Soil Compressibility *CR* and *RR*

axis. This makes the slopes of the virgin consolidation line and swelling line λ and κ, respectively. They are related to C_c and C_r by

$$C_c = \lambda \ln 10 = 2.3026\lambda \tag{8.19}$$

$$C_r = \kappa \ln 10 = 2.3026\kappa \tag{8.20}$$

Based on the *CR* or *RR* values clays can be classified as shown in Table 8.3.

Example 8.2

A consolidation test was carried out on a 61.4-mm-diameter and 25.4-mm-thick saturated undisturbed specimen, obtained from a soft clay deposit, taken from a depth of 3.0 m. The water table is the ground level. The initial moisture content is 105.7% and the specific gravity of the soil grains is 2.70. The dial gauge readings, which measure the change in the specimen thickness, are summarized below.

Vertical stress (kN/m²)	0	5	10	20	40	80
Dial reading (mm)	0.000	0.348	0.406	0.569	1.476	3.647

Vertical stress (kN/m²)	160	320	640	260	40	5
Dial reading (mm)	6.035	8.096	10.152	9.953	9.167	8.350

a. Plot the $e - \log \sigma'_v$ curve and find the compression index, the recompression index, and the preconsolidation pressure.
b. Find the compression ratio and the recompression ratio.
c. Plot the coefficient of volume compressibility m_v against $\log \sigma'_v$.
d. Find the overconsolidation ratio of the clay.

Note: The secondary compression (Sec. 8.9) taking place between two successive increments is neglected here for simplicity.

Solution

Part a The initial void ratio can be computed as $e_0 = 1.057 \times 2.70 = 2.854$.

Initial height $H_0 = 25.4$ mm.

With these, let us calculate the values of e and m_v at the end of consolidation due to the *first* stress increment of 5 kPa, where the dial gauge reading has changed from 0.000 to 0.348 mm, suggesting 0.348 mm of settlement. From Eq. (8.9), the reduction in void ratio is given by

$$\Delta e = \frac{\Delta H}{H_0}(1 + e_0) = \frac{0.348}{25.4}(1 + 2.854) = 0.0528$$

From Eqs. (8.10) and (8.11),

$$m_v = \frac{\left(\frac{\Delta H}{H_0}\right)}{\Delta\sigma'} = \frac{\left(\frac{0.348}{25.4}\right)}{5} = 0.00274\ \text{kPa}^{-1} = 2.74\ \text{MPa}^{-1}$$

Void ratio at the end of consolidation = 2.854 – 0.0528 = 2.801

Specimen height at the end of consolidation = 25.4 – 0.348 = 25.052 mm

Using these values of e (= 2.801) and H (= 25.052 mm) at the end of consolidation as the *initial* values for the next increment, Table 8.4 can be prepared. The void ratio values are plotted against the effective stress values in Fig. 8.7. The compression index C_c can be computed between A and B as

$$C_c = \frac{2.631 - 1.626}{\log 320 - \log 40} = \frac{2.630 - 1.626}{\log\left(\frac{320}{40}\right)} = \mathbf{1.11}$$

σ'_v (kN/m²)	Dial guage reading (mm)	ΔH (mm)	H_0 (mm)	H_{100} (mm)	Δe	e	m_v (MPa⁻¹)	σ'_{avg} (kN/m²)
0	0					**2.854**		
5	0.348	0.348	**25.4**	25.052	0.0528	2.801	2.74	2.5
10	0.406	0.058	25.052	24.994	0.0088	2.792	0.46	7.5
20	0.569	0.163	24.994	24.831	0.0247	2.768	0.65	15
40	1.476	0.907	24.831	23.924	0.1376	2.630	1.83	30
80	3.647	2.171	23.924	21.753	0.3294	2.301	2.27	60
160	6.035	2.388	21.753	19.365	0.3623	1.938	1.37	120
320	8.096	2.061	19.365	17.304	0.3127	1.626	0.67	240
640	10.152	2.056	17.304	15.248	0.3120	1.314	0.37	480
260	9.953	–0.199	15.248	15.447	–0.0302	1.344		
40	9.167	–0.786	15.447	16.233	–0.1196	1.463		
5	8.35	–0.817	16.233	17.05	–0.1240	1.587		

H_0 = height of specimen before applying the stress increment
H_{100} = height of specimen at the end of consolidation due to the increment
ΔH = change in dspecimen height due to consolidation during the increment

TABLE 8.4 Consolidation Test Data Analysis

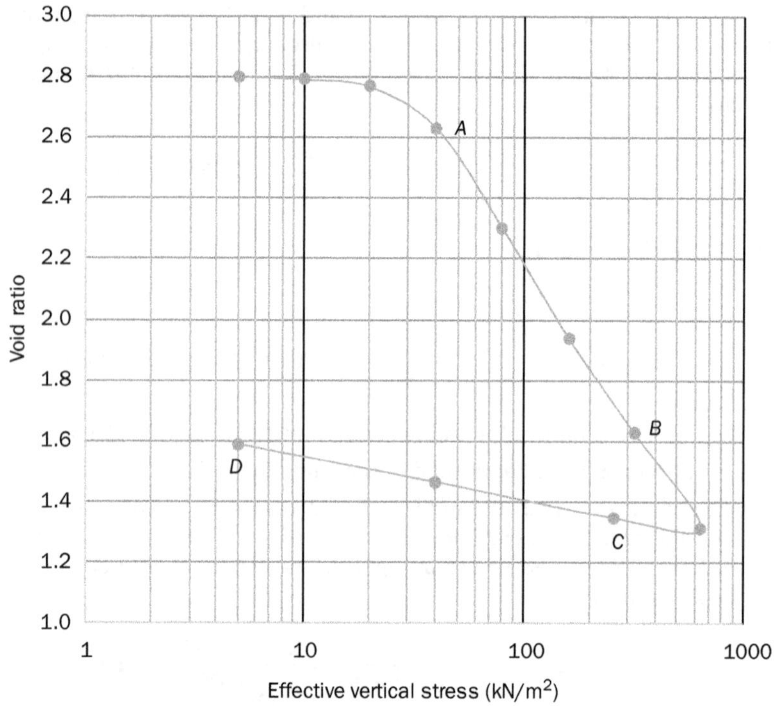

FIGURE 8.7 Void ratio vs. effective stress plot using the data in Table 8.4.

The recompression index C_r can be computed between C and D as

$$C_r = \frac{1.588 - 1.344}{\log 260 - \log 5} = \frac{1.587 - 1.344}{\log\left(\frac{260}{5}\right)} = \mathbf{0.14}$$

From the plot, the **preconsolidation pressure may be estimated as about 30.0 kN/m²**.

Part b

$$C_{c\varepsilon} = CR = \frac{C_c}{1 + e_0} = \frac{1.11}{1 + 2.854} = \mathbf{0.29}$$

$$C_{r\varepsilon} = RR = \frac{C_r}{1 + e_0} = \frac{0.14}{1 + 2.854} = \mathbf{0.036}$$

Part c The coefficient of volume compressibility is plotted against the effective stress in Fig. 8.8.

Part d

$$e_0 = 2.854$$

$$\gamma_{sat} = \left(\frac{G_s + e}{1 + e}\right)\gamma_w = \left(\frac{2.70 + 2.854}{1 + 2.854}\right)(9.81) = 14.14 \text{ kN/m}^3$$

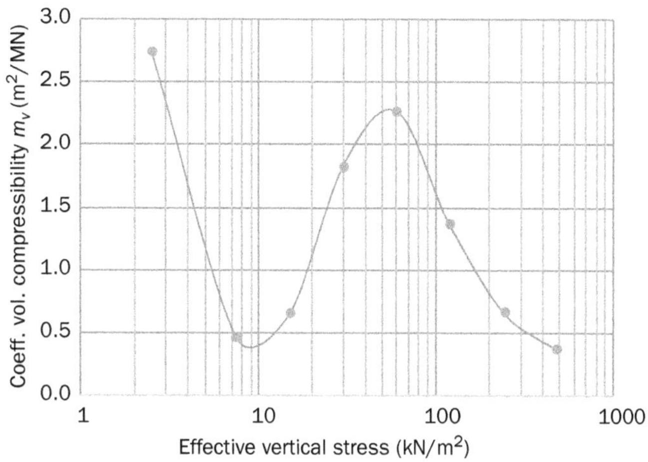

Figure 8.8 Coefficient of volume compressibility vs. effective stress.

At the sample location,

$$\sigma'_{vo} = 3(14.14 - 9.81) = 13.0\ \text{kN/m}^2$$

$$OCR = \frac{\sigma'_p}{\sigma'_{vo}} = \frac{30}{13} = \mathbf{2.3}$$

It can be seen from Fig. 8.8 that m_v varies significantly with the effective stress, which makes it difficult to assign a specific value for m_v in settlement calculations. A value appropriate to the stress level must be used in the settlement computations (see Example 8.1). It can be shown theoretically that m_v and C_c are related by

$$m_v = \frac{0.434C_c}{(1 + e_0)\sigma'_{\text{average}}} \tag{8.21a}$$

$$D = \frac{1}{m_v} = 2.30\frac{\sigma'_{\text{average}}}{CR} \tag{8.21b}$$

where e_0 = void ratio at the beginning of consolidation, and σ'_{average} = average effective vertical stress during consolidation. Equation (8.21) works well on the linear region of $e - \log \sigma'_v$ representing normally consolidated state. In the overconsolidated state, C_c or CR are replaced by C_r or RR, respectively.

Example 8.3

Use the consolidation test data in Example 8.2, and compute m_v for the stress increment from 160 to 320 kN/m² using Eq. (8.21).

Solution During consolidation,

$$\sigma'_{\text{average}} = \frac{160 + 320}{2} = 240\ \text{kN/m}^2$$

At the beginning of consolidation, $e = 1.939$ and $C_c = 1.11$ (see the solution for Example 8.2).

$$m_v = \frac{0.434C_c}{(1 + e_0)\sigma'_{\text{average}}} = \frac{0.434 \times 1.11}{(1 + 1.939)(240)} = \mathbf{0.00068\ kPa^{-1}}$$

This is in excellent agreement with the value 0.67 MPa^{-1} computed in Example 8.2.
The bilinear $e - \log \sigma'_v$ plot can be seen as two straight line segments, where the one in the overconsolidated region is elastic and the one along the VCL is elastoplastic. In the overconsolidated state, the unloading and reloading paths coincide, which is typical of elastic behavior. In the normally consolidated state, the strains include both elastic and plastic components.

8.5 Field Corrections to e vs. log σ'_v Plot Developed in the Laboratory

In Sec. 8.4, we discussed the one-dimensional consolidation test carried out on an undisturbed clay specimen in the laboratory. When the sampling is carried out in the field, there can be some *mechanical disturbance* to the sample. In addition, when the sample is brought to the surface, its total overburden stress is reduced to zero, causing some *stress relief* to the sample. Therefore, the specimen may behave differently to the undisturbed clay in the field, which may give incorrect values of the preconsolidation pressure and the compression index.

Before using the $e - \log \sigma'_v$ plot developed in the laboratory to predict future settlements in the field, we have to make some adjustments to reflect the field conditions. The following two concerns are discussed separately in Secs. 8.5.1 and 8.5.2.

a. How do we define the preconsolidation pressure in the $e - \log \sigma'_v$ plot?
b. How do we define the *field* virgin consolidation line?

8.5.1 Defining the Preconsolidation Pressure σ'_p

Preconsolidation pressure σ'_p is an important parameter in the settlement computations, and is widely used in normalizing other geotechnical parameters for comparison. It can also be seen as the yield stress of the clay, which behaves elastically as long as σ'_v is less than σ'_p (i.e., when overconsolidated). In Example 8.2, we crudely estimated σ'_p. There are a few systematic ways to define the preconsolidation pressure in an $e - \log \sigma'_v$ plot. The most popular and the oldest of these is the method proposed by Casagrande (1936).

Casagrande's (1936) Method

Casagrande's (1936) graphical procedure to determine the preconsolidation pressure is summarized through the following steps.

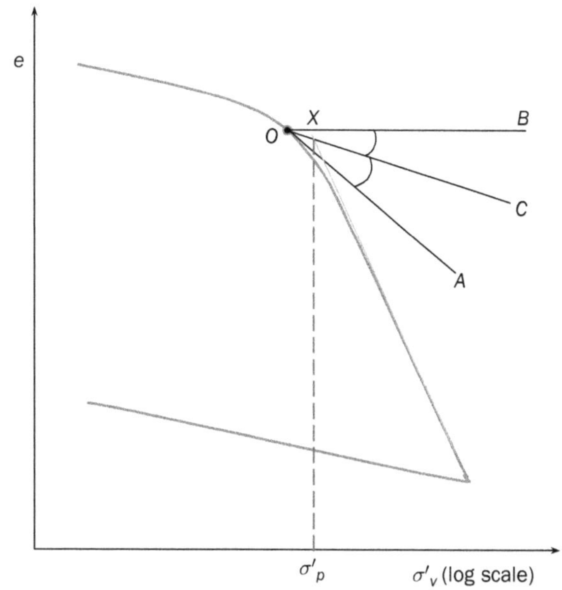

FIGURE 8.9 Casagrande's graphical procedure to determine σ'_p.

Step 1: In the laboratory e – log σ'_v plot, locate (eye estimate) the point where the minimum radius occurs (shown by O in Fig. 8.9).
Step 2: Draw the tangent OA.
Step 3: Draw the horizontal line OB.
Step 4: Bisect the angle AOB by line OC.
Step 5: Extend the straight line portion of the virgin consolidation line backward until it intersects the bisector OC at point X.
Step 6: The value of σ'_v at X is the preconsolidation pressure σ'_p.

There are methods other than Casagrande's for defining the preconsolidation pressure σ'_p. It has been reported in the literature that the σ'_p values predicted by Casagrande's method are unreliable (Boone, 2010; Clementino, 2005; Umar and Sandrekarimi, 2017). Clementino (2005) showed that the scales selected for the two axes can distort the e – log σ'_v plot and can influence the value of σ'_p predicted. He showed that Pacheco Silva's (1970) semilogarithmic method gives a realistic prediction of σ'_p. Umar and Sandrekarimi (2017) compared 11 methods (including Pacheco Silva's) and concluded that Oikawa's (1987) bilogarithmic method provides the most accurate prediction of σ'_p. They also noted that the traditional Casagrande's method gives poor predictions of σ'_p.

Pacheco Silva's (1970) Method

Pacheco Silva's (1970) method, illustrated through Fig. 8.10, is fairly simple and the steps are given below.

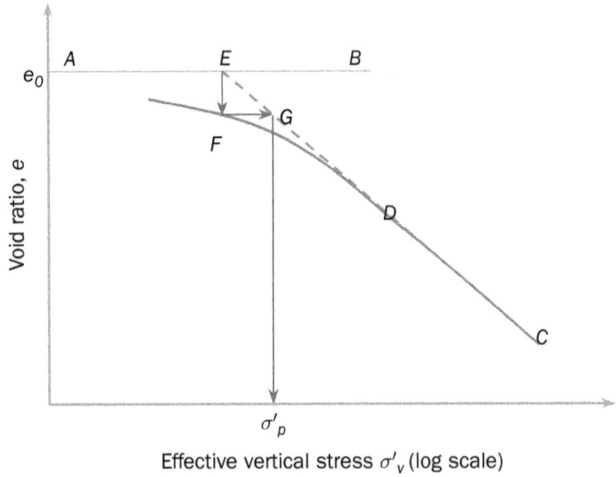

FIGURE 8.10 Pacheco Silva's graphical procedure to determine σ'_p.

Step 1. Determine the initial void ratio e_o and draw the horizontal line AB through e_0.

Step 2. Extend the straight line portion *CD* of the virgin consolidation line until it intersects the line *AB* at *E*.

Step 3. Drop a vertical line from *E* to intersect the $e - \log \sigma'_v$ plot at *F*.

Step 4. From *F*, draw a horizontal line to intersect the line *CDE* at *G*.

Step 5. The value of σ'_v at *G* is the preconsolidation pressure σ'_p.

Oikawa's (1987) Method

Oikawa (1987) studied several clays from Akita plane in Japan, with natural moisture content ranging 25–941% and void ratios ranging 0.7–17. He noted that for clays with natural moisture content greater than 200%, $e - \log \sigma'_v$ plot is no more a straight line in the normally consolidated stress range. He showed that plotting log $(1 + e)$ versus $\log \sigma'_v$ gives two straight line segments which intersect at the preconsolidation pressure. This method applies to clays at any natural moisture content.

Example 8.4

A consolidation test with an unloading-reloading cycle was carried out on a clay specimen obtained from 2.4 to 2.8 m depth at a site. The specimen was saturated and the initial void ratio e_0 was 2.20. The specific gravity of the soil grains was 2.70. The void ratio versus effective vertical stress data from the consolidation test are summarized below.

σ'_v (kN/m²)	e
1.4	2.14
6	2.08
13	2.03
26	1.95
38	1.88

58	1.81
86	1.70
130	1.55
194	1.45
110	1.47
26	1.53
52	1.52
104	1.49
208	1.43
416	1.22

Plot the data and determine the preconsolidation pressure σ'_p using

a. Casagrande's

b. Pacheco Silva's, and

c. Oikawa's methods.

Solution In the plots shown here, the unloading-reloading cycles are shown with different symbols. The three plots relevant to the three methods are shown in Figs. 8.11a, 8.11b, and 8.11c, respectively. The preconsolidation determined by the three different procedures are as follows:

Casagrande: $\sigma'_p = 40\ \text{kN/m}^2$

Pacheco Silva: $\sigma'_p = 33\ \text{kN/m}^2$

Oikawa: $\sigma'_p = 36\ \text{kN/m}^2$

8.5.2 Schmertmann's Procedure to Define the Field Virgin Consolidation Line

The mechanical disturbance and the stress relief in the field may distort the $e - \log \sigma'_v$ plot. Schmertmann (1955) suggested a graphical procedure to define the *field* virgin consolidation line from the laboratory $e - \log \sigma'_v$ plot. The slope of this field VCL is the true compression index, which must be used in the settlement calculations. The procedures for normally consolidated clays and overconsolidated clays are described separately here.

Normally Consolidated Clays (See Fig. 8.12a)

Step 1. Determine the preconsolidation pressure σ'_p using Casagrande's procedure.

Step 2. Determine the initial void ratio e_0.

Step 3. Mark e_0 and $0.42e_0$ on the vertical void ratio axis.

Step 4. Mark σ'_p on the horizontal σ'_v axis.

Step 5. Draw a vertical line through σ'_p and a horizontal line through e_0, to meet at A.

Step 6. Extend the straight line portion of the virgin consolidation line (laboratory) and the horizontal line through $0.42e_0$ to intersect at B.

Step 7. Join A and B, which is the field virgin consolidation line, the slope of which is the true C_c.

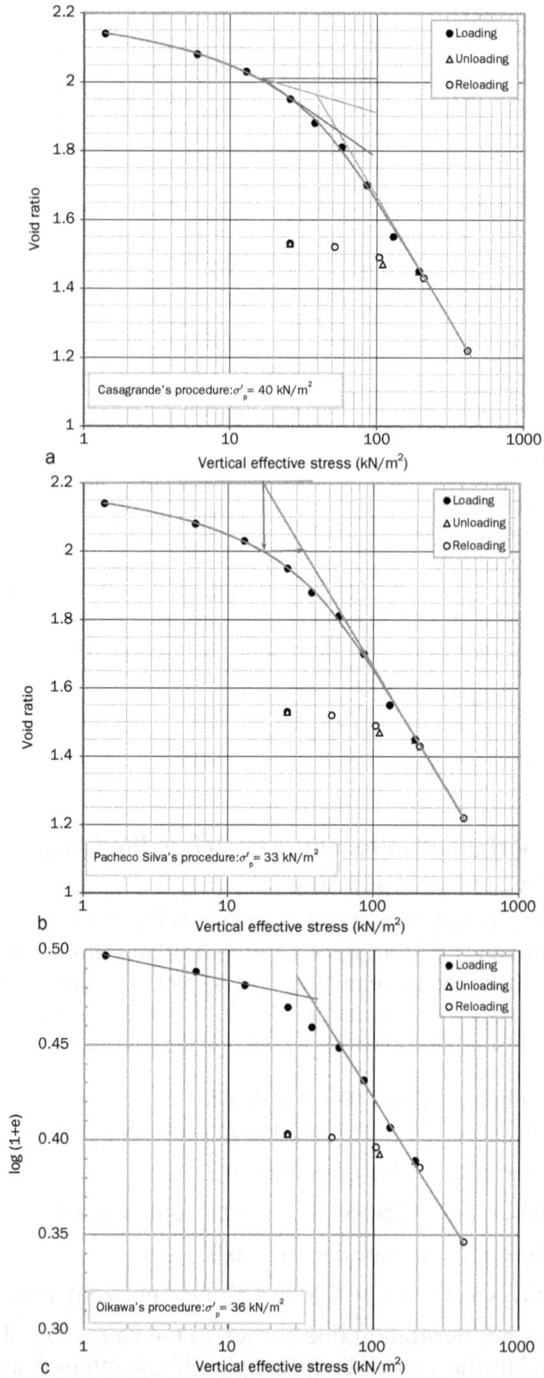

Figure 8.11 Determining preconsolidation pressure by the three methods for Example 8.4: (a) Casagrande, (b) Pacheco Silva, and (c) Oikawa.

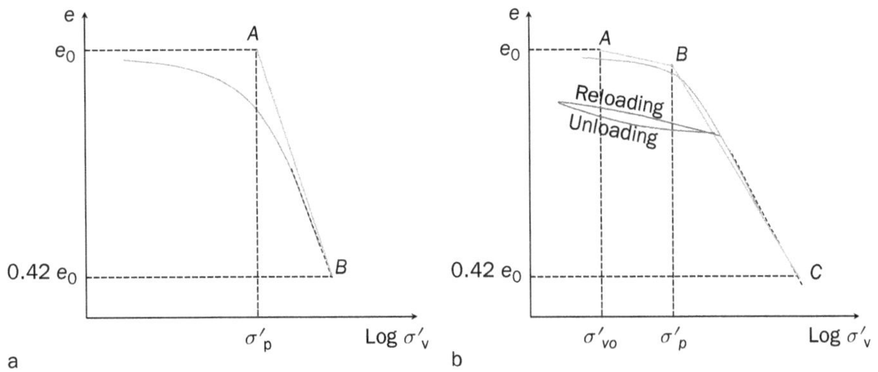

Figure 8.12 Schmertmann's procedure for determining virgin consolidation line: (a) Normally consolidated clays, (b) overconsolidated clays.

Overconsolidated Clays (See Fig. 8.12b)

Step 1. Determine the preconsolidation pressure σ'_p using Casagrande's procedure.

Step 2. Determine the initial void ratio e_0, and the initial effective overburden pressure σ'_{vo}.

Step 3. Mark e_0 and $0.42e_0$ on the vertical void ratio axis.

Step 4. Mark σ'_p and σ'_{vo} on the horizontal σ'_v axis.

Step 5. Draw a vertical line through σ'_{vo} and a horizontal line through e_0, to meet at A.

Step 6. Determine the recompression index C_r from an unload-reload cycle in the laboratory consolidation test.

Step 7. Draw a line through A at slope of C_r to intersect the vertical line through σ'_p at B.

Step 8. Extend the straight line portion of the virgin consolidation line (laboratory) and the horizontal line through $0.42e_0$ to intersect at C.

Step 9. Join B and C, which is the field virgin consolidation line.

8.6 Determination of Final Consolidation Settlement

Figure 8.13 shows the variation of the consolidation settlement with time. It can be seen that only after very long time (say $t = \infty$) the settlement becomes asymptotic to a value s_c, which is the *final consolidation settlement*. The final consolidation settlement s_c, can be computed using m_v or the $e - \log \sigma'_v$ plot. The two methods are discussed here.

8.6.1 Determination of Final Consolidation Settlement Using m_v

From Eqs. (8.10) and (8.11), it can be seen that

$$m_v = \frac{\left(\dfrac{\Delta H}{H_0}\right)}{\Delta\sigma'} = \frac{\left(\dfrac{s_c}{H_0}\right)}{q}$$

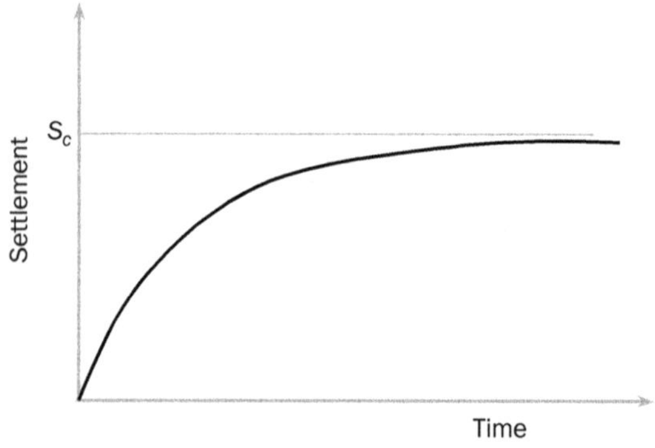

FIGURE 8.13 Consolidation settlement vs. time plot.

Thus,

$$s_c = m_v q H_0 \tag{8.12}$$

Equation (8.12) is fairly straightforward to apply. However, the problem is to find the appropriate value for m_v, which varies with the stress level. Incorrect choice of m_v can give an unrealistic estimate of the settlement.

8.6.2 Determination of Final Consolidation Settlement Using $e - \log \sigma'_v$ Plot

A better way to determine the consolidation settlement s_c is to use the parameters C_c, C_r, and σ'_p derived from the $e - \log \sigma'_v$ plot subjected to the field corrections discussed in Sec. 8.5. Unlike m_v, these values are independent of the stress levels, and can be easily determined from the $e - \log \sigma'_v$ plot. From Eq. (8.9), the final consolidation settlement can be determined as:

$$s_c = \Delta H = \left(\frac{\Delta e}{1 + e_0}\right) H_0 \tag{8.22}$$

where Δe is the change in void ratio during consolidation. If the clay is saturated, the degree of saturation S = 100%, and the initial void ratio e_0 can be determined as

$$e_0 = w_n G_s \tag{8.23}$$

where w_n = initial moisture content (*natural moisture content* in the field, or initial moisture content in the laboratory), G_s = specific gravity of the clay grains. The initial height of the clay layer H_0 is generally known from the bore logs. How does one determine the reduction in void ratio Δe due to consolidation? Let us look at this under three scenarios that cover all possibilities. In each case, σ'_{vo} = effective overburden stress at the *middle* of the clay layer, $\Delta\sigma'$ = effective stress increase at the *middle* of the clay layer (same as q in one-dimensional consolidation) which is causing the consolidation, σ'_p = preconsolidation pressure, e_0 = initial void ratio, and Δe = reduction in void ratio due to consolidation.

Scenario 1: Normally Consolidated Clays

In a normally consolidated clay (Fig. 8.14a), the initial state (e_0, σ'_{vo}) lies at point A on the virgin consolidation line. During consolidation due to an increase in effective stress of

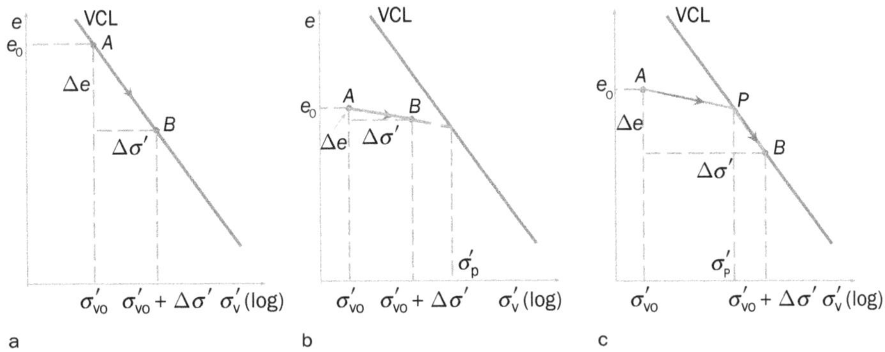

FIGURE 8.14 Determining Δe: (a) Normally consolidated clays, (b) overconsolidated clays where $\sigma'_{vo} + \Delta\sigma' < \sigma'_p$, (c) overconsolidated clays where $\sigma'_{vo} + \Delta\sigma' > \sigma'_p$.

$\Delta\sigma'$ the point moves to B, along the path AB, with a reduction in void ratio of Δe. The slope of the VCL is C_c. Therefore,

$$\Delta e = C_c \log\left(\frac{\sigma'_{vo} + \Delta\sigma'}{\sigma'_{vo}}\right) \tag{8.24}$$

Scenario 2: Overconsolidated Clays Where $\sigma'_{vo} + \Delta\sigma' < \sigma'_p$

In overconsolidated clays (Fig. 8.14b), the initial state (e_0, σ'_{vo}) lies at point A to the left of the VCL. During consolidation due to an increase in effective stress of $\Delta\sigma'$, the point moves to B along the path AB without reaching the preconsolidation pressure, as shown in Fig. 8.14b. The slope of line AB is C_r. Therefore,

$$\Delta e = C_r \log\left(\frac{\sigma'_{vo} + \Delta\sigma'}{\sigma'_{vo}}\right) \tag{8.25}$$

Scenario 3: Overconsolidated Clays Where $\sigma'_{vo} + \Delta\sigma' > \sigma'_p$

In overconsolidated clays (Fig. 8.14c), as in Scenario 2, the initial state (e_0, σ'_{vo}) lies at point A to the left of the VCL. During consolidation due to an increase in effective stress of $\Delta\sigma'$, the point moves to B along the path AB, passing the preconsolidation pressure at P, as shown in Fig. 8.14c. Part of the path AP is at a slope of C_r and the rest PB has a slope of C_c. The reduction in the void ratio during the two parts can be calculated as

$$\Delta e_{AP} = C_r \log\left(\frac{\sigma'_p}{\sigma'_{vo}}\right)$$

$$\Delta e_{PB} = C_c \log\left(\frac{\sigma'_{vo} + \Delta\sigma'}{\sigma'_p}\right)$$

Therefore, the reduction in the void ratio between A and B is

$$\Delta e = \Delta e_{AP} + \Delta e_{PB} = C_r \log\left(\frac{\sigma'_p}{\sigma'_{vo}}\right) + C_c \log\left(\frac{\sigma'_{vo} + \Delta\sigma'}{\sigma'_p}\right) \tag{8.26}$$

Example 8.5

The soil profile at a site is shown in Fig. 8.15. The bulk and saturated unit weights of the sand are 16.5 and 19.0 kN/m³, respectively. An undisturbed clay specimen, collected from the middle of the clay layer, has moisture content of 33.5% and the specific gravity of the soil grains is 2.73. A consolidation test is carried out on the clay specimen that gave the following properties: Compression index = 0.55, recompression index = 0.06, and preconsolidation pressure = 85.0 kN/m².

a. What is the overconsolidation ratio of the clay?
b. If a surcharge of 3.0 m high fill weighing 18.0 kN/m³ is placed at the ground level, what would be the consolidation settlement?

Solution

Part a For the clay specimen, $w = 33.5\%$, $S = 100\%$, $G_s = 2.73$
Therefore, $e_0 = 0.335 \times 2.73 = 0.915$

$$\gamma_{sat} = \left(\frac{G_s + e}{1 + e}\right)\gamma_w = \left(\frac{2.73 + 0.915}{1 + 0.915}\right)(9.81) = 18.7 \text{ kN/m}^3$$

The in situ effective overburden pressure on the specimen is

$$\sigma'_{vo} = (2)(16.5) + (1)(19 - 9.81) + (2)(18.7 - 9.81) = 60.0 \text{ kN/m}^2$$

Preconsolidation pressure $\sigma'_p = 85$ kN/m²

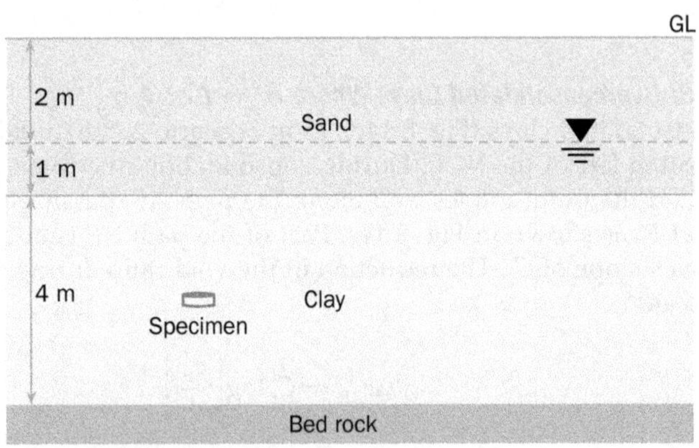

FIGURE 8.15 Soil profile for Example 8.5.

Therefore, the overconsolidation ratio is

$$OCR = \frac{\sigma'_p}{\sigma'_{vo}} = \frac{85}{60} = 1.42$$

Part b Due to surcharge by the fill, $\Delta\sigma' = 3 \times 18 = 54$ kN/m^2
$\sigma'_{vo} + \Delta\sigma' = 60.0 + 54.0 = 114.0$ kN/m^2, which is larger than σ'_p of 85 kN/m^2
This falls under scenario 3 and hence using Eq. (8.26)

$$\Delta e = C_r \log\left(\frac{\sigma'_p}{\sigma'_{vo}}\right) + C_c \log\left(\frac{\sigma'_{vo} + \Delta\sigma'}{\sigma'_p}\right)$$

$$\Delta e = 0.06 \log\left(\frac{85}{60}\right) + 0.55 \log\left(\frac{60 + 54}{85}\right) = 0.0792$$

Therefore, from Eq. (8.22)

$$s_c = \Delta H = \left(\frac{\Delta e}{1 + e_0}\right) H_0 = \left(\frac{0.0792}{1 + 0.915}\right)(4000)\ \text{mm} = \mathbf{165.4\,mm}$$

8.7 Preloading

When a normally consolidated clay is subjected to loading due to buildings or embankments, the consolidation occurs along the VCL which is at a slope of C_c. In overconsolidated clays, the consolidation occurs along the recompression line with a slope of C_r, which is significantly less than C_c. Therefore, for the same load, the reduction is void ratio and hence the settlement due to consolidation would be significantly less for overconsolidated clays than normally consolidated clays.

When higher than acceptable consolidation settlements are expected in normally consolidated clays, it is possible to make them slightly overconsolidated (e.g., *OCR* of 1 to 2) and hence settle significantly less. This process is known as *preloading* or *surcharging* and is illustrated through Fig. 8.16.

Figure 8.16a shows a normally consolidated clay which would settle excessively when loaded. The proposed structure is expected to impose a vertical stress of q. In Fig. 8.16b, the same NC clay is surcharged with earth fill or something similar, applying a vertical stress of q, which is left there for some time (e.g., 9 months) to consolidate the clay. Once substantial consolidation takes place, the load is removed in Fig. 8.16c. The three stages are illustrated in Fig. 8.16d with points *A*, *B*, and *C*, respectively. When the surcharge is removed (Point C), the clay becomes overconsolidated with *OCR* of $(\sigma'_{vo} + q)/\ \sigma'_{vo}$, with some reduction in the ground level which may have to be raised. When this overconsolidated clay is loaded with the proposed structure, applying vertical stress of q, the entire loading path stays on the recompression line, with significant reduction in the settlement.

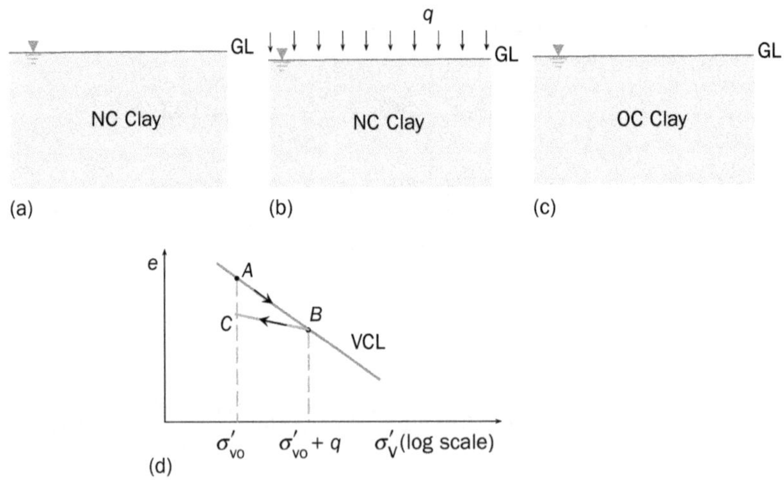

FIGURE 8.16 Preloading: (a) NC clay, (b) applying a surcharge q, (c) removal of surcharge, making the clay OC, (d) e vs. log σ'_v plot.

Generally, 90% or more of the primary consolidation settlement is removed by the preloading exercise. When the time is critical, lower values can be allowed. During the preloading period, how much settlement occurs depends on the coefficient of consolidation discussed in Sec. 8.8. The consolidation process can be accelerated by installing vertical drains within the clay, providing additional horizontal drainage, thus shortening the drainage path. The vertical drains in the form of wick drains (or prefabricated vertical drains) are generally installed at 1–3 m spacing in a triangular or rectangular grid pattern.

Example 8.6

The soil profile at a proposed construction site is shown in Fig. 8.17. The clay layer is normally consolidated. A clay specimen was collected from a depth of 5.0 m below the ground level and a consolidation test was carried out. The soil data are summarized below.

Unit weight of the sand above the water table = 18.0 kN/m^3

Unit weight of the sand below the water table = 20.5 kN/m^3

Moisture content of the clay = 45.5%

Specific gravity of the clay grains = 2.75

Compression index = 0.46

Recompression index = 0.06

It is proposed to carry out preloading, by applying a 60 kN/m^2 surcharge at the ground level and unloading after the consolidation is completed. After the surcharge is removed it is required to place additional fill and raise the ground level back to the original level. What should be the height of this additional fill?

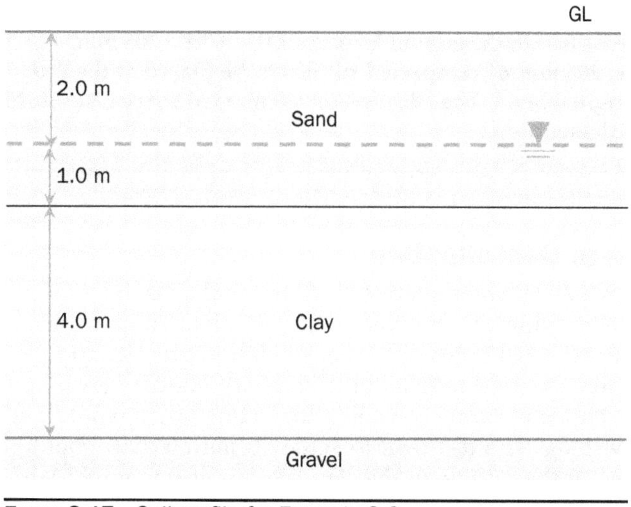

Figure 8.17 Soil profile for Example 8.6.

Solution For the clay, in situ $e_0 = 0.455 \times 2.75 = 1.251$

$$\gamma_{sat} = \left(\frac{G_s + e}{1 + e}\right)\gamma_w = \left(\frac{2.75 + 1.251}{1 + 1.251}\right)(9.81) = 17.4 \text{ kN/m}^3$$

The in situ effective overburden stress at the middle of the clay layer is

$$2(18.0) + 1(20.5 - 9.81) + 2(17.4 - 9.81) = 61.87 \text{ kN/m}^2$$

At the end of consolidation due to the surcharge,

$$\Delta e = C_c \log\left(\frac{\sigma'_{vo} + \Delta\sigma'}{\sigma'_{vo}}\right) = 0.46\log\left(\frac{61.87 + 60}{61.87}\right) = 0.135$$

$$\Delta H = \frac{\Delta e}{1 + e_0} H_0 = \frac{0.135}{1 + 1.251} \times 4000 = 240 \text{ mm}$$

The new void ratio and the clay layer height at the end of consolidation are

$$e = 1.251 - 0.135 = 1.116$$

$$H = 4000 - 240 = 3760 \text{ mm}$$

Let us use these as the initial values for the unloading of the surcharge.

Due to unloading,

$$\Delta e = 0.06 \times \log\left(\frac{121.87}{61.87}\right) = 0.018 \text{ (Increase)}$$

$$\Delta H = \frac{0.018}{1 + 1.116} \times 3760 = 32 \text{ mm (Heave)}$$

Therefore, the new layer thickness after the surcharge was removed is 3760 + 32 = 3792 mm.

The net reduction in the ground level is 4000 − 3792 = **208 mm**.

Therefore, 208 mm of compacted fill has to be placed at the ground level to bring it back to the original level. The addition of 208 mm fill will cause a further but small consolidation settlement.

8.8 Time Rate of Consolidation

From Secs 8.1–8.7, we were interested in the final consolidation settlement s_c, which was determined using m_v or e versus log σ'_v plot. These sections relate the initial ($t = 0$) and final ($t = \infty$) states of the consolidation process, and methods were developed to compute the void ratio and the height of the clay layer at the end of consolidation. What happens *during* consolidation? How much would the clay settle after 6 months? How long would it take to settle 50 mm? What would be the excess pore water pressure at the middle of the clay layer after 6 months? We will address these questions in this section.

Terzaghi (1925) proposed the one-dimensional consolidation theory based on the following assumptions.

a. Clay is fully saturated.

b. Darcy's law is valid.

c. The pore fluid (i.e., water) is incompressible.

d. Soil grains are incompressible.

e. Strains and drainage are both in one direction (generally vertical) and hence one-dimensional consolidation.

f. Small strain theory holds and hence hydraulic conductivity remains constant during consolidation.

Figure 8.18a shows a clay layer of thickness H sandwiched between two sand layers, consolidating under a stress increment $\Delta\sigma$, applied at the ground level. Recall the discussions we had in Sec. 8.2. Immediately upon loading, at time $t = 0^+$, the entire stress increment $\Delta\sigma$ is carried by the water with no increase in the effective stress (i.e., $\Delta u = \Delta\sigma$ and $\Delta\sigma' = 0$ at all depths). With time, the applied load is slowly transferred from the pore water to the soil skeleton, such that

$$\Delta\sigma = \Delta\sigma'(t) + \Delta u(t) \tag{8.4}$$

Figure 8.18b shows the variation of the excess pore water pressure with depth z at time t. The sand layers at the top and bottom are free draining and do not allow excess pore water pressure to be developed at any time at the top and bottom boundaries ($z = 0$ and $z = H$) of the clay layer. After a long time (at $t = \infty$), the entire excess pore water pressure has fully dissipated and the applied stress increment $\Delta\sigma$ is fully transferred to the soil skeleton (i.e., $\Delta u = 0$ and $\Delta\sigma' = \Delta\sigma$ at all depths).

Figure 8.18a also shows an infinitesimal element of dimensions, dx, dy, and dz, at depth z, which is also shown in Fig. 8.17c. The flow into the element is Q_{in} at velocity v, and flow out is Q_{out} with velocity $v + \frac{\partial v}{\partial z}dz$. Let the head at the element be $h(z)$.

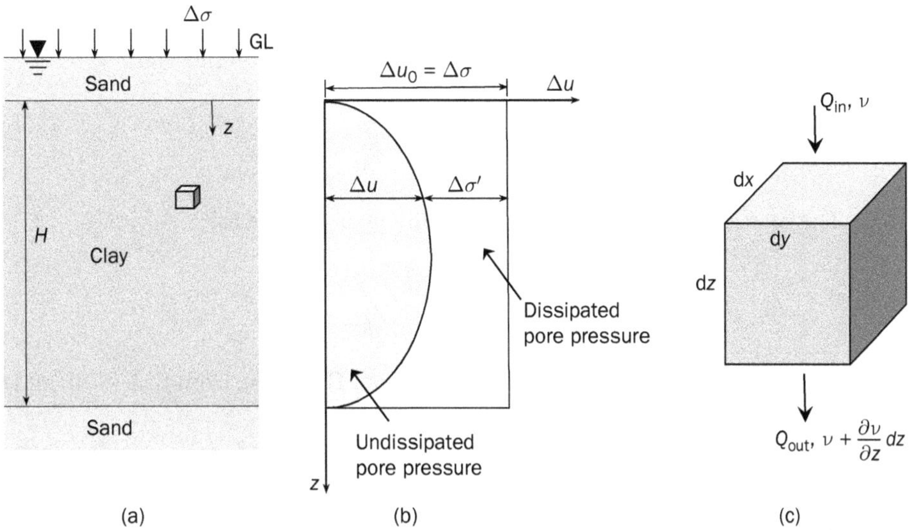

FIGURE 8.18 During consolidation: (a) clay layer, (b) excess pore water pressure distribution at time t, (c) an infinitesimal element.

$$\frac{\partial h}{\partial z} = \frac{1}{\gamma_w}\frac{\partial u}{\partial z} \tag{8.27}$$

From Darcy's law,

$$v = -k\frac{\partial h}{\partial z} = -\frac{k}{\gamma_w}\frac{\partial u}{\partial z} \tag{8.28}$$

The negative sign in Eq. (8.28) is to account for the reduction in h with z. The flow rate into the element is

$$Q_{\text{in}} = v\,dx\,dy$$

The flow rate out of the element is

$$Q_{\text{out}} = \left(v + \frac{\partial v}{\partial z}dz\right)dx\,dy$$

Therefore, the water squeezed out of the element is

$$Q_{\text{out}} - Q_{\text{in}} = \left(v + \frac{\partial v}{\partial z}dz\right)dx\,dy - v\,dx\,dy = \frac{\partial v}{\partial z}dz\,dx\,dy$$

Substituting for v from Eq. (8.28), the water squeezed out of the element is

$$Q_{\text{out}} - Q_{\text{in}} = -\frac{k}{\gamma_w}\frac{\partial^2 u}{\partial z^2}dx\,dy\,dz \tag{8.29}$$

During consolidation,

$$d[\Delta\sigma'(t)] = -d[\Delta u(t)] \tag{8.6}$$

$$\frac{\partial \sigma'}{\partial t} = -\frac{\partial u}{\partial t} \tag{8.30}$$

The coefficient of volume compressibility m_v is defined as

$$m_v = \frac{dV}{V} \Big/ d\sigma' \tag{8.31}$$

where V is the volume of the element $dx \cdot dy \cdot dz$. Therefore, from Eqs. (8.30) and (8.31),

$$\frac{\partial V}{\partial t} = m_v \frac{\partial \sigma'}{\partial t} dx\,dy\,dz = -m_v \frac{\partial u}{\partial t} dx\,dy\,dz \tag{8.32}$$

Equating the expressions on the right hand sides of Eqs. (8.29) and (8.32),

$$\frac{k}{\gamma_w}\frac{\partial^2 u}{\partial z^2} dx\,dy\,dz = m_v \frac{\partial u}{\partial t} dx\,dy\,dz$$

$$\frac{\partial u}{\partial t} = \frac{k}{m_v \gamma_w}\frac{\partial^2 u}{\partial z^2} = c_v \frac{\partial^2 u}{\partial z^2} \tag{8.33}$$

where c_v is known as the *coefficient of consolidation*, defined as

$$c_v = \frac{k}{m_v \gamma_w} \tag{8.34}$$

Equation (8.33) is the governing differential equation that can be solved with the following boundary conditions.

a. At $z = 0$, for all values of t, $\Delta u = 0$.
b. At $z = H$, for all values of t, $\Delta u = 0$.
c. At $z = H/2$, for all values of t, $\frac{\partial u}{\partial z} = 0$.
d. At $t = 0$, for all z, $\Delta u = \Delta u_0$.
e. At $t = \infty$, for all z, $\Delta u = 0$.

It is worth emphasizing that consolidation is driven by the dissipation of the *excess* pore water pressure, which was initially induced by the applied stress increment. Therefore, what is denoted by u in Eqs. (8.27)–(8.33) is really the excess pore water pressure Δu, which is in excess of the static pore water pressure u_0, which plays no role in the consolidation process.

Solving Eq. (8.33) with the above boundary conditions gives

$$\Delta u(z,t) = \Delta u_0 \sum_{m=0}^{\infty} \frac{2}{M} \sin(MZ) e^{-M^2 T} \tag{8.35}$$

where $M = \frac{\pi}{2}(2m+1)$. Z and T are the dimensionless *depth factor* and *time factor*, respectively, defined as

$$Z = \frac{z}{H_{dr}} \tag{8.36}$$

$$T = \frac{c_v t}{H_{dr}^2} \tag{8.37}$$

H_{dr} is the *average maximum length of drainage path during the consolidation process*. The maximum length of the drainage path varies during the consolidation process. It is suggested the average value of this maximum is used in the computations. In a situation similar to what is shown in Fig. 8.18a, the maximum length of the drainage path is from the middle of the clay to the top or bottom, which is $H/2$. Here the clay layer is known to be *doubly drained* (i.e., drained from top and bottom). When drainage is only from either top or bottom, $H_{dr} = H$. Here, the clay layer is known to be *singly drained*, which is the case when the clay layer is underlain by bedrock or an impervious stiff clay stratum. The drainage layer can be any granular soil that can be assumed to be free draining.

8.8.1 Degree of Consolidation *U*(*z*, *t*)

The degree of consolidation $U(z, t)$ is simply a measure of how much consolidation has occurred at a depth z at time t, expressed as a percentage. It varies in the range of 0–100%. Referring to Figs. 8.18a and 8.18b, the degree of consolidation at depth z and time t is the ratio of the dissipated excess pore water pressure to the initial excess pore water pressure. It can be written as

$$U(z,t) = \frac{\Delta u_0 - \Delta u(z,t)}{\Delta u_0} \tag{8.38}$$

From Eqs. (8.35) and (8.38),

$$U(z,t) = 1 - \sum_{m=0}^{\infty} \frac{2}{M} \sin(MZ) e^{-M^2 T} \tag{8.39}$$

The interrelationship among $U(z, t)$, Z and T is shown graphically in Fig. 8.19.

It can be seen from Fig. 8.19 that the degree of consolidation at any time is the minimum at the middle of the doubly drained clay layer or at the impervious boundary of the singly drained clay layer, where $Z = 1$. Figure 8.19 is valid for both doubly drained and singly drained clay layers. The degree of consolidation values for different depth and time factors are summarized in Table 8.5.

8.8.2 Average Degree of Consolidation $U_{avg}(t)$

At any time during consolidation, the degree of consolidation $U(z, t)$ varies with depth. We still can define an *average degree of consolidation* $U_{avg}(t)$ for the entire clay layer at a given time t. It is defined as the percentage of the overall dissipation of the excess pore water pressure for the entire clay layer at that time. It can be computed as the area of the dissipated excess pore water pressure distribution diagram (see Fig. 8.18b) divided by

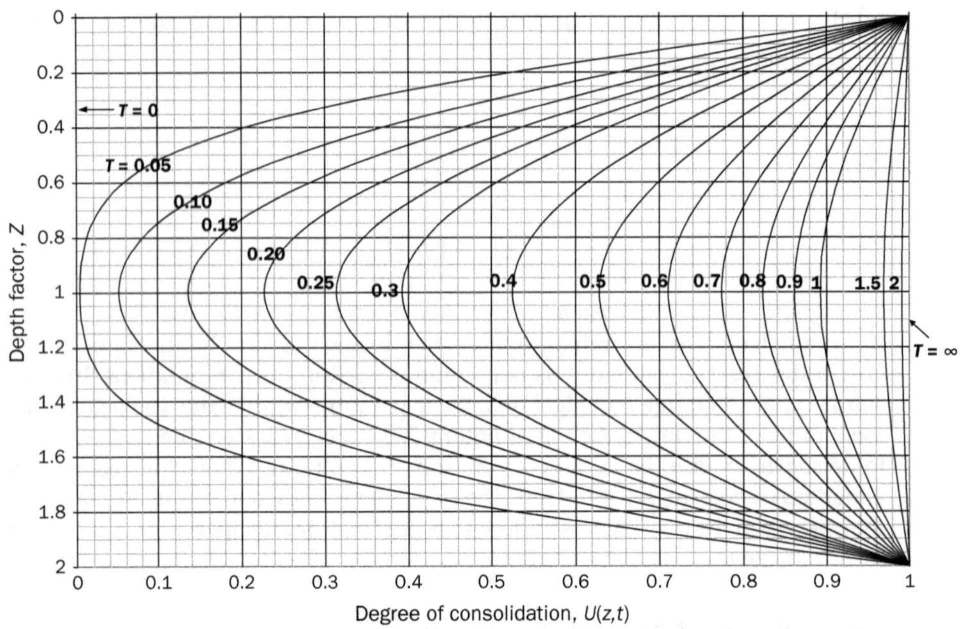

Figure 8.19 $U(z, t) - Z - T$ relationship.

the initial excess pore water pressure distribution diagram, expressed as a percentage. It is given by

$$U_{\text{avg}} = 1 - \sum_{m=0}^{\infty} \frac{2}{M^2} e^{-M^2 T} \tag{8.40}$$

$U_{\text{avg}} - T$ relationship satisfying Eq. (8.40) is shown in Fig. 8.20. $U_{\text{avg}}(t)$ is the same as the ratio of the consolidation settlement at that time to the final consolidation settlement. In other words, at 35% average degree of consolidation, 35% of the consolidation settlement has occurred.

When U_{avg} is expressed as percentage, $U_{\text{avg}} - T$ relationship can be approximated as

$$T = \frac{\pi}{4}\left(\frac{U_{\text{avg}}}{100}\right)^2 \qquad \text{for } U_{\text{avg}} \le 60\% \tag{8.41}$$

$$T = 1.781 - 0.933 \log\left(100 - U_{\text{avg}}\right) \qquad \text{for } U_{\text{avg}} \ge 60\% \tag{8.42}$$

$$U_{\text{avg}} = \sqrt{\frac{4T}{\pi}} \times 100\% \qquad \text{for } U_{\text{avg}} \le 60\% \tag{8.43}$$

$$U_{\text{avg}} = \left[1 - \frac{8}{\pi^2}\exp\left(-\frac{\pi^2}{4}T\right)\right] \times 100\% \qquad \text{for } U_{\text{avg}} \ge 60\% \tag{8.44}$$

Z	*T*														
	0.05	**0.1**	**0.15**	**0.2**	**0.25**	**0.3**	**0.4**	**0.5**	**0.6**	**0.7**	**0.8**	**0.9**	**1**	**1.5**	**2**
0	1.0000	1.0000	1.0000	1.0000	1.0000	1.0000	1.0000	1.0000	1.0000	1.0000	1.0000	1.0000	1.0000	1.0000	1.0000
0.125	0.6926	0.7799	0.8200	0.8456	0.8650	0.8812	0.9074	0.9277	0.9435	0.9558	0.9655	0.9730	0.9789	0.9939	0.9982
0.25	0.4292	0.5762	0.6494	0.6979	0.7355	0.7671	0.8183	0.8581	0.8891	0.9134	0.9323	0.9471	0.9587	0.9880	0.9965
0.375	0.2357	0.4020	0.4966	0.5632	0.6167	0.6620	0.7363	0.7940	0.8390	0.8742	0.9017	0.9232	0.9400	0.9825	0.9949
0.5	0.1139	0.2644	0.3675	0.4468	0.5130	0.5702	0.6644	0.7378	0.7951	0.8399	0.8749	0.9023	0.9236	0.9778	0.9935
0.625	0.0481	0.1644	0.2659	0.3527	0.4284	0.4949	0.6054	0.6917	0.7591	0.8118	0.8529	0.8851	0.9102	0.9739	0.9924
0.75	0.0178	0.0987	0.1934	0.2838	0.3658	0.4391	0.5616	0.6574	0.7323	0.7909	0.8366	0.8723	0.9002	0.9709	0.9915
0.875	0.0060	0.0623	0.1501	0.2418	0.3275	0.4048	0.5346	0.6363	0.7159	0.7780	0.8265	0.8645	0.8941	0.9692	0.9910
1	0.0031	0.0507	0.1358	0.2277	0.3146	0.3932	0.5255	0.6292	0.7103	0.7736	0.8231	0.8618	0.8920	0.9686	0.9908
1.125	0.0060	0.0623	0.1501	0.2418	0.3275	0.4048	0.5346	0.6363	0.7159	0.7780	0.8265	0.8645	0.8941	0.9692	0.9910
1.25	0.0178	0.0987	0.1934	0.2838	0.3658	0.4391	0.5616	0.6574	0.7323	0.7909	0.8366	0.8723	0.9002	0.9709	0.9915
1.375	0.0481	0.1644	0.2659	0.3527	0.4284	0.4949	0.6054	0.6917	0.7591	0.8118	0.8529	0.8851	0.9102	0.9739	0.9924
1.5	0.1139	0.2644	0.3675	0.4468	0.5130	0.5702	0.6644	0.7378	0.7951	0.8399	0.8749	0.9023	0.9237	0.9778	0.9935
1.625	0.2357	0.4020	0.4966	0.5633	0.6167	0.6620	0.7363	0.7940	0.8390	0.8742	0.9017	0.9232	0.9400	0.9825	0.9949
1.75	0.4292	0.5762	0.6494	0.6979	0.7355	0.7671	0.8183	0.8581	0.8891	0.9134	0.9323	0.9471	0.9587	0.9880	0.9965
1.875	0.6926	0.7799	0.8200	0.8456	0.8650	0.8812	0.9074	0.9277	0.9435	0.9558	0.9655	0.9730	0.9789	0.9939	0.9982
2	1.0000	1.0000	1.0000	1.0000	1.0000	1.0000	1.0000	1.0000	1.0000	1.0000	1.0000	1.0000	1.0000	1.0000	1.0000

Table 8.5 Degree of Consolidation $U(Z, T)$ Values

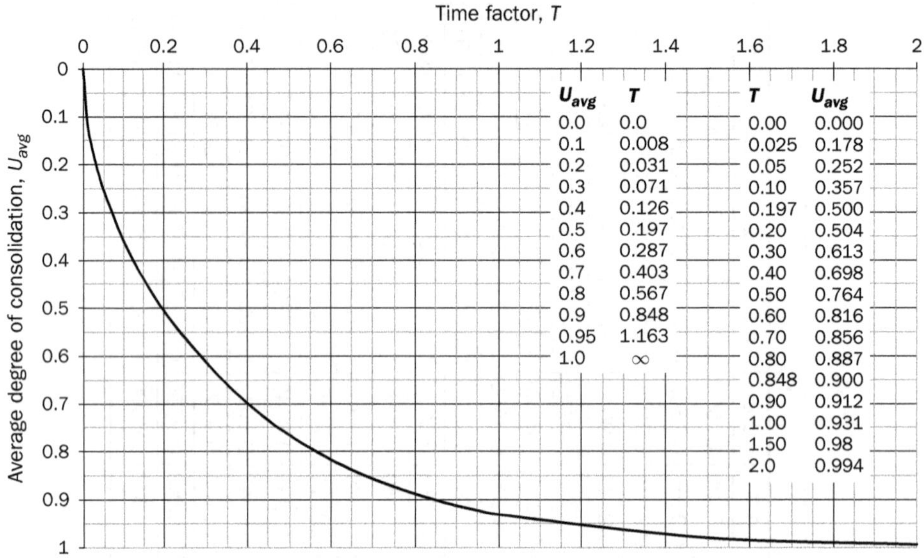

Figure 8.20 $U_{avg} - T$ relationship.

Example 8.7

The soil profile at a site consists of 2 m of sand at the ground level, underlain by 6 m of clay, followed by a very stiff clay stratum that can be assumed to be impervious and incompressible. The water table is at 1.5 m below the ground level. The moist and saturated unit weights of the sand are 17.0 and 18.5 kN/m^3, respectively. The clay has initial void ratio of 0.810, saturated unit weight of 19.0 kN/m^3, and coefficient of consolidation of 0.0014 cm^2/s.

a. When the ground is surcharged with 3-m-high compacted fill with moist unit weight of 19.0 kN/m^3, the settlement was 160 mm in the first year. What would be the consolidation settlement in the first 2 years?
b. If the clay is normally consolidated, what is the. compression index of the clay?
c. What would be the pore water pressure at the middle of the clay layer, after 1 year?

Solution

Part a The clay layer with one-way drainage has H_{dr} of 6.0 m.
After 1 year,

$$T = \frac{c_v t}{H_{dr}^2} = \frac{0.0014 \times 365 \times 24 \times 3600}{600^2} = 0.123$$

and the settlement s (1 year) = 160 mm
From Eq. (8.43), $U_{avg} = 39.6\%$
Therefore, the final consolidation settlement $s_c = 160/0.396 = 404$ mm

When $t = 2$ years, $T = 0.246$
From Eq. (8.43), $U_{avg} = 56.0\%$
Therefore, consolidation settlement during the first 2 years is $0.56 \times 404 =$ **226.2 mm**

Part b The initial effective overburden stress at the middle of the clay is

$$\sigma'_{vo} = 1.5 \times 17.0 + 0.5\,(18.5 - 9.81) + 3.0\,(19.0 - 9.81) = 57.4\ \text{kN/m}^2$$

$$\Delta\sigma' = 3 \times 19 = 57\ \text{kN/m}^2$$

$$\Delta e = \frac{\Delta H}{H_0}(1 + e_0) = \frac{404}{6000}(1 + 0.810) = 0.1219$$

The clay is normally consolidated. Therefore, from Eq. (8.24),

$$\Delta e = C_c \log\left(\frac{\sigma'_{vo} + \Delta\sigma'}{\sigma'_{vo}}\right)$$

$$0.1219 = C_c \log\left(\frac{57.4 + 57}{57.4}\right)$$

$$C_c = \mathbf{0.41}$$

Part c At the middle of the clay layer, the initial pore water pressure is

$$u_0 = 3.5 \times 9.81 = 34.3\ \text{kN/m}^2$$

The excess pore water pressure induced by the loading $\Delta u_0 = 57.0\ \text{kN/m}^2$
Out of this, a fraction has dissipated and the rest remains.

$$T = 0.123.$$

$$Z = 3/6 = 0.5$$

From Fig. 8.19, U ($z = 3$ m, $t = 1$ year) $= 0.316$

$$\Delta u = 57.0\,(1 - 0.316) = 39.0\ \text{kN/m}^2$$

Therefore, $u = u_0 + \Delta u = 34.3 + 39.0 =$ **73.3 kN/m²**

8.8.3 Coefficient of Consolidation

The coefficient of consolidation c_v is a measure of how fast the consolidation occurs. The larger the c_v, the faster the consolidation process. Figure 8.21 shows three scenarios where the final consolidation settlements are the same, but the *rate* of consolidation is different. The c_v is the highest for C and lowest for A. Terzaghi et al. (1996) suggested that clays with LL between 10 and100 have c_v in the range of 0.3–30 m²/year. Here, $LL = 10$ corresponds to $c_v = 30$ m²/year, and $LL = 100$ corresponds to $c_v = 0.3$ m²/year.

Intuitively one can see that the magnitudes of consolidation settlements are greater in soft clays than in stiff clays. What is not obvious is that the soft clays settle slower than the stiff clays as shown in Fig. 8.22. The coefficient of consolidation c_v comes into play here—it is generally significantly greater for stiff clays than soft clays, making the consolidation process a lot faster in stiff clays. Soft clays have larger m_v and hence lesser c_v [see Eq. (8.34)]. The rate of consolidation has little to do with the magnitude of the

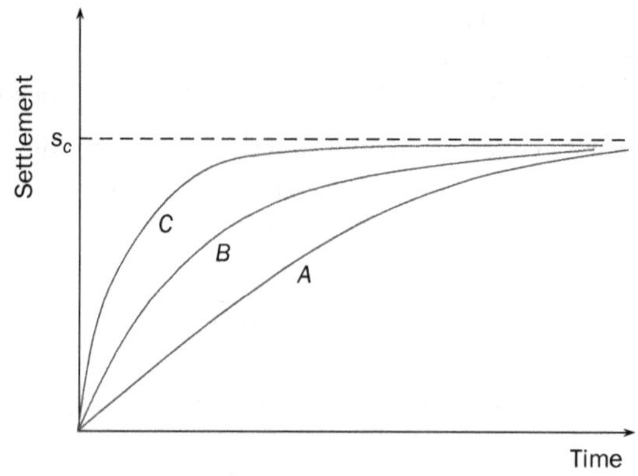

FIGURE 8.21 Effect of c_v on the rate of consolidation.

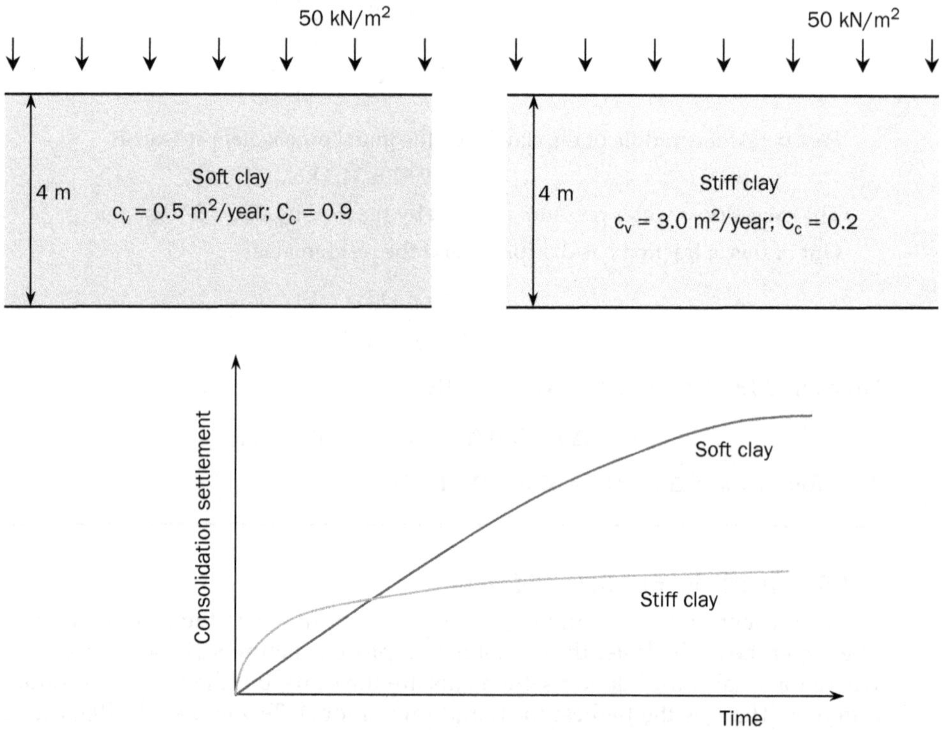

FIGURE 8.22 The difference between magnitude and rate of consolidation settlement.

consolidation settlement. Note that c_v depends on the compressibility of the clay skeleton and the permeability of the pore fluid, which is water.

Generally, the consolidation in the field is significantly faster than what is predicted from the c_v determined from the laboratory tests. This is due to the sand seams and fissures present in the field that make the drainage path shorter and hence make the consolidation quicker.

Generally, c_v of an overconsolidated clay is an order of magnitude larger than the c_v of the same clay if it is normally consolidated. Thus, overconsolidated clays consolidate faster than the normally consolidated clays. Coefficient of consolidation c_v is generally determined from laboratory consolidation tests or in situ tests such as piezocone dissipation tests. Approximate relationship between c_v and liquid limit, as suggested by the U. S. Navy (1971) is shown in Fig. 8.23.

While most of the consolidation parameters and the final consolidation settlement can be determined with good level of certainty, predicting the field coefficient of consolidation and the rate of settlement is relatively difficult due to the presence of sand seams and fissures in the clays, which make the consolidation faster.

In a laboratory consolidation test, stress increments are applied at 24-h intervals to allow sufficient time for the consolidation to take place during each increment. The time versus settlement data are recorded and used for determining c_v, which depends on the stress level. From the laboratory consolidation test data, the coefficient of consolidation is determined by one of the following two methods.

a. Casagrande's logarithm of time fitting method
b. Taylor's square root of time fitting method

Asaoka's (1978) semiempirical observational method is widely used in the field to determine c_v and the final consolidation settlement using the settlement records.

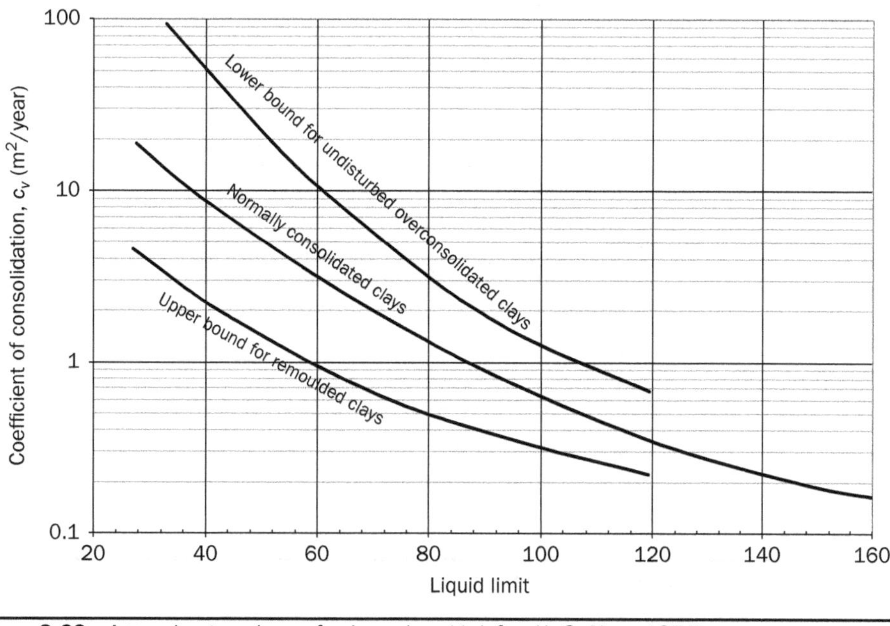

FIGURE 8.23 Approximate values of c_v based on *LL* (after U. S. Navy, 1971).

Casagrande's Logarithm of Time Fitting Method

Casagrande (1938) suggested plotting the dial gauge reading (or settlement) against the logarithm of time. He suggested defining d_0 and d_{100} corresponding to U_{avg} of 0% and 100%, respectively. From these two values, d_{50} is determined as

$$d_{50} = \frac{d_0 + d_{100}}{2}$$

The time corresponding to d_{50} is t_{50}, which is the time taken for 50% average consolidation of the specimen. From Fig. 8.20, it can be seen that when U_{avg} = 50%, T = 0.197. Therefore,

$$T_{50} = \frac{c_v t_{50}}{H_{dr}^2} = 0.197 \tag{8.45}$$

Since H_{dr} is known from the average specimen height during consolidation, c_v can be determined. Now, let us see how d_0 and d_{100} are determined.

In Fig. 8.24, the early part of the plot is a parabola. Times t and $4t$ are selected in the parabolic section of the plot. The corresponding dial gauge readings are noted in the plot (see Fig. 8.24). The difference in the dial gauge readings, denoted by x, is marked above the dial gauge reading corresponding to time t, to define d_0. Intersection of the two tangents to the tail end of the plot defines d_{100}.

Taylor's Square Root of Time Fitting Method

Taylor's (1948) method requires plotting the dial gauge reading (or settlement) against square root of time as shown in Fig. 8.25. The straight line segment of the plot is extended in both directions. The intersection with the vertical axis defines d_0. Another straight line is drawn through d_0 such that the abscissa is 1.15 times larger (see Fig. 8.25). The intersection of the second line with the laboratory curve defines the point corresponding to 90% average consolidation and hence $\sqrt{t_{90}}$ can be read off the figure.

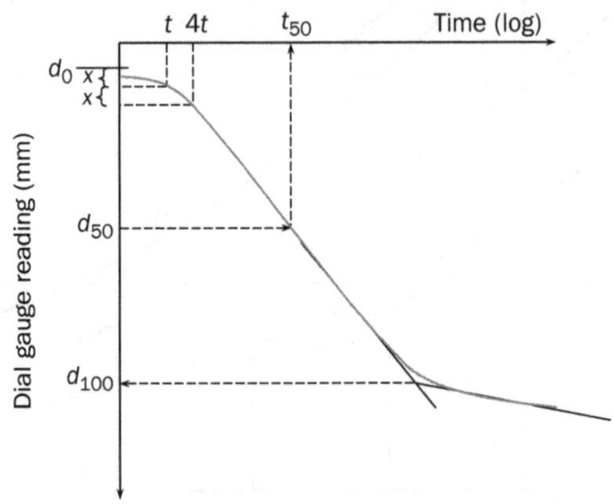

FIGURE 8.24 Dial gauge reading vs. log time plot.

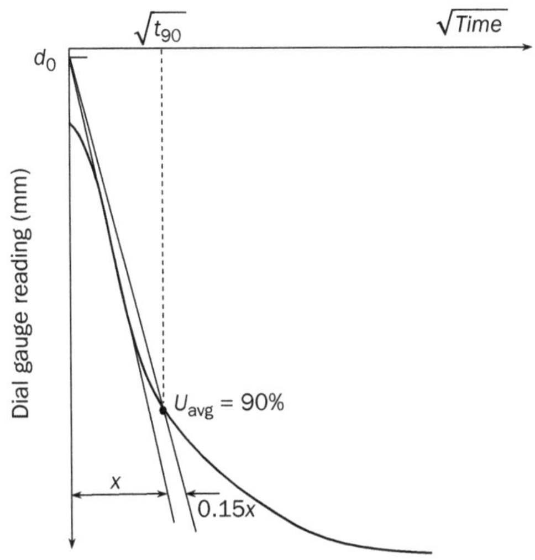

Figure 8.25 Dial gauge reading vs. square root of time.

From Fig. 8.20, it can be seen that when U_{avg} = 90%, T = 0.848. Therefore,

$$T_{90} = \frac{c_v t_{90}}{H_{dr}^2} = 0.848 \tag{8.46}$$

Since H_{dr} is known from the average specimen height during consolidation, c_v can be determined.

Asaoka's Observational Method

Asaoka (1978) proposed a semiempirical *observational method* to determine the *end-of-primary consolidation settlement*, herein denoted by s_{EoP} which is the same as the final consolidation s_c. From this, c_v can be back-calculated. This method was derived from Mikasa's (1965) and Terzaghi's (1925) consolidation theories. The method is quite popular among the practicing engineers in situations where the field monitoring data of settlement records are available.

The surface settlement is measured using settlement plates. It is required to have settlement records at successive *equal* time steps. Alternatively, the equal time step data can be generated from the settlement versus time plot. The settlement s_N at step N is plotted against the settlement s_{N-1} at the previous time step $N - 1$. As settlement progresses, the difference between s_N and s_{N-1} diminishes. Therefore, the best-fit line passing through these observational points will intersect the line of equality at a point where the EoP (end of primary consolidation) occurs (see Fig. 8.26).

The equation of the straight line of best fit is given by

$$s_N = \beta_0 + \beta_1 s_{N-1} \tag{8.47}$$

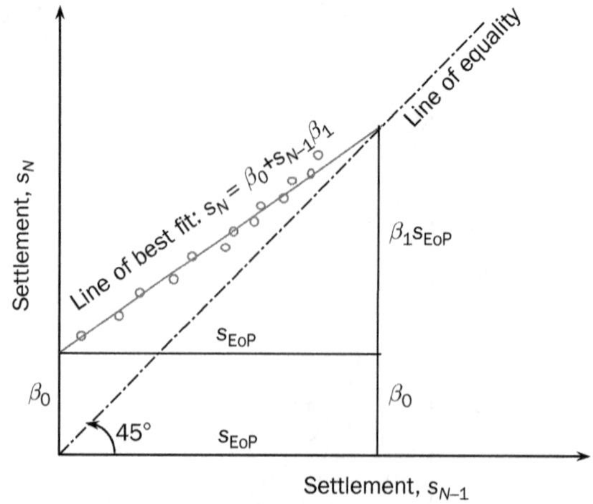

FIGURE 8.26 Asaoka's observational method to determine the final consolidation settlement.

where β_0 and β_1 are the intercept and slope of the straight line of best fit, shown in red. It can be seen from Fig. 8.26 that

$$s_{EoP} = \beta_0 + \beta_1 s_{EoP} \tag{8.48}$$

Therefore, s_{EoP} can be determined as

$$s_{EoP} = \frac{\beta_0}{1-\beta_1} \tag{8.49}$$

Balasubramaniam and Brenner (1981) showed that c_v can be determined as

$$c_v = -\frac{5}{12}\frac{H^2 \ln \beta_1}{\Delta t} \tag{8.50}$$

where H = average maximum length of drainage path, and Δt = time intervals between observations.

When the initial excess pore water pressure distribution is linear, the average degree of consolidation and the time factor are related by (Mesri and Huvaj-Sarihan, 2009)

$$U_{avg} = 1 - \frac{8}{\pi^2}\exp\left(-\frac{\pi^2}{4}T\right) \text{ for } U_{avg} > 40\% \tag{8.51}$$

Mesri and Huvaj-Sarihan (2009) showed that c_v can be determined as

$$c_v = -\frac{4}{\pi^2}\frac{H^2 \ln \beta_1}{\Delta t} \tag{8.52}$$

Observations over larger time intervals make the graphical interpretation clearer. However, larger time intervals and the number of observations required for the plot make the settlement monitoring spread over a longer period of time, which is not practical. The main attraction of Asaoka's method is that it can be used to identify when 90% consolidation would take place so that the preloading surcharge can be removed.

Example 8.8

A doubly drained 5.0-m-thick clay layer is undergoing consolidation under a fill load. The consolidation settlements are recorded monthly, and are as follows: 162, 193, 208, 230, 240, 257, 268, 271, 280, and 290 mm. Using Asaoka's method, find the end-of-primary consolidation settlement and the coefficient of consolidation.

Solution In Fig. 8.27, the equation of the line of best fit is

$$S_N = 0.8236 s_{N-1} + 55.567$$

From Eq. (8.49),

$$s_{\text{EoP}} = \frac{\beta_0}{1-\beta_1} = \frac{55.67}{1-0.8236} = \mathbf{315\,mm}$$

From Eq. (8.50),

s_{N-1} (mm)	s_N (mm)	
	162	
162	193	Plot
193	208	Plot
208	230	Plot
230	240	Plot
240	257	Plot
257	268	Plot
268	271	Plot
271	280	Plot
280	290	Plot
290		

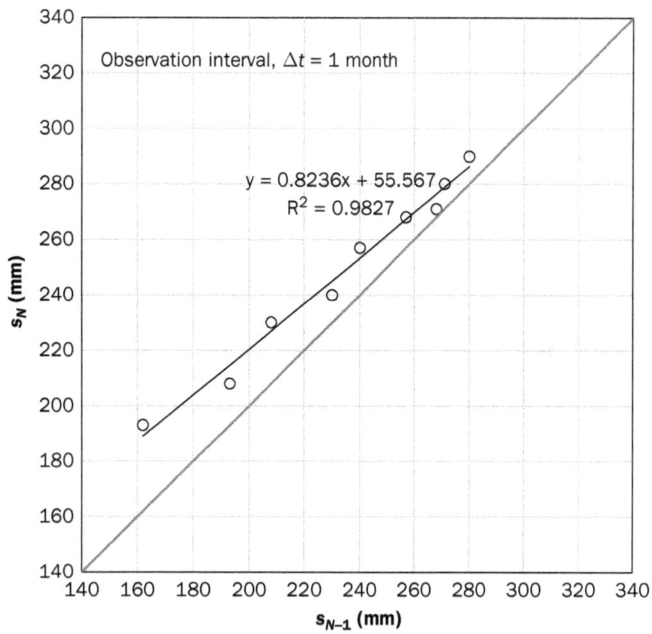

FIGURE 8.27 Asoka's method.

$$c_v = -\frac{5}{12}\frac{H^2 \ln \beta_1}{\Delta t} = -\frac{5}{12}\frac{(2.5)^2 \ln 0.8236}{\left(\frac{1}{12}\right)} = \mathbf{6.1\ m^2/year}$$

From Eq. (8.52),

$$c_v = -\frac{4}{\pi^2}\frac{H^2 \ln \beta_1}{\Delta t} = -\frac{4}{\pi^2}\frac{(2.5)^2 \ln 0.8236}{\left(\frac{1}{12}\right)} = \mathbf{5.9\ m^2/year}$$

8.8.4 Coefficient of Consolidation for Multilayered Clays

When a thick deposit of clay consists of layers of clay with varying values of c_v, U. S. Navy (1982) suggests a procedure to transform the clay layers into a single homogeneous layer with a specific value (assigned by the user) of coefficient c_v^* of consolidation c_v^*. Let there be n layers, where the thickness and the c_v of the ith layer are H_i and c_{vi}, respectively (Fig. 8.28). Here, the ith layer is transformed into a layer with thickness of H_i^* and coefficient of consolidation c_v^*. Here, H_i^* is given by

$$H_i^* = H_i\sqrt{\frac{c_v^*}{c_{vi}}} \tag{8.53}$$

Once all the layers are transformed, the total thickness of the consolidating clay deposit is given by

$$H^* = \sum_{i=1}^{n} H_i^* \tag{8.54}$$

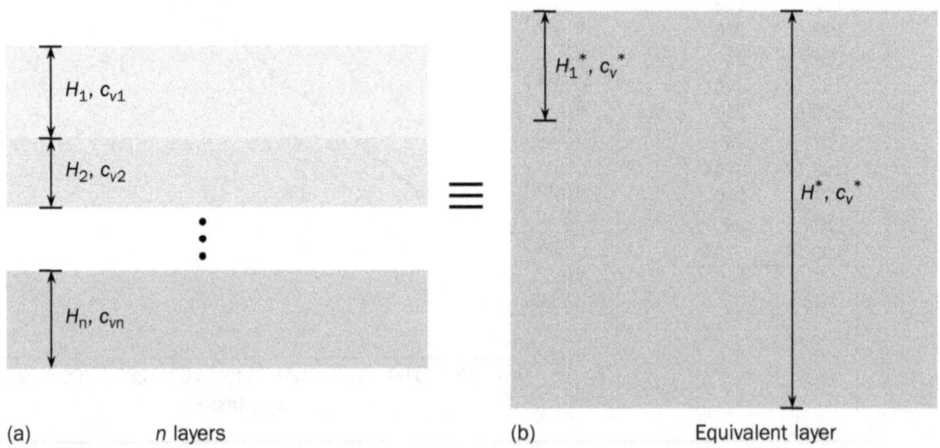

FIGURE 8.28 (a) n layers, (b) equivalent layer.

which has coefficient of consolidation of c_v^* throughout. The equivalent coefficient of consolidation $c_{v,\text{equivalent}}$ for a uniform layer of original layer thickness $H \left(= \sum_{i=1}^{n} H_i\right)$ can be calculated as

$$c_{v,\text{equivalent}} = c_v^* \left(\frac{H}{H^*}\right)^2 \tag{8.55}$$

Example 8.9

Determine the equivalent coefficient of consolidation c_v for the following layered clay system (Fig. 8.29). Maintain the same total thickness of 6 m for the equivalent clay.

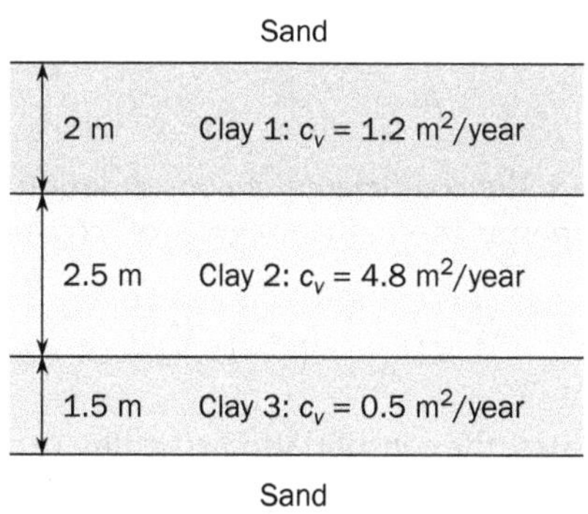

FIGURE 8.29 A three-layered clay.

Solution Let us assume a constant c_v^* of 1.5 m²/year and compute the clay layer thicknesses. Let the corresponding thicknesses of Clay 1, Clay 2, and Clay 3 be H_1^*, H_2^*, and H_3^*, respectively.

$$H_i^* = H_i \sqrt{\frac{c_v^*}{c_{vi}}}$$

$$H_1^* = 2\sqrt{\frac{1.5}{1.2}} = 2.236 \text{ m}$$

$$H_2^* = 2.5\sqrt{\frac{1.5}{4.8}} = 1.398 \text{ m}$$

$$H_3^* = 1.5\sqrt{\frac{1.5}{0.5}} = 2.598 \text{ m}$$

This gives the total thickness of a homogeneous clay of 6.232 m (2.236 + 1.398 + 2.598 m) with c_v^* of 1.5 m²/year. Let us convert this into a homogeneous clay of the same height 6.0 m, with $c_{v,\text{equivalent}}$.

$$c_{v,\text{equivalent}} = c_v^* \left(\frac{H}{H^*}\right)^2$$

$$c_{v,\text{equivalent}} = 1.5\left(\frac{6}{6.232}\right)^2 = 1.39 \text{ m}^2/\text{year}$$

8.8.5 Two- and Three-Dimensional Consolidation

In three-dimensional Cartesian coordinate system, the governing differential equation for consolidation can be written as

$$\frac{\partial u}{\partial t} = c_x \frac{\partial^2 u}{\partial x^2} + c_y \frac{\partial^2 u}{\partial y^2} + c_z \frac{\partial^2 u}{\partial z^2} \tag{8.56}$$

where c_x, c_y, and c_z are the coefficients of consolidation in x, y, and z directions. Using the principle of superposition, the degree of consolidation U_{xyz} can be obtained from

$$1 - U_{xyz} = (1 - U_x)(1 - U_y)(1 - U_z) \tag{8.57}$$

In polar coordinates, the consolidation equation can be written as

$$\frac{\partial u}{\partial t} = c_r \left(\frac{\partial^2 u}{\partial r^2} + \frac{1}{r}\frac{\partial u}{\partial r}\right) + c_z \frac{\partial^2 u}{\partial z^2} \tag{8.58}$$

The above equation consists of two separate components, where the first one represents radial flow and the second one represents vertical flow. The degree of consolidation U_{rv} can be derived from

$$1 - U_{rv} = (1 - U_r)(1 - U_v) \tag{8.59}$$

Here, U_r is given by (Barron, 1948)

$$U_r = 1 - \exp\left(-\frac{8T_r}{F(n)}\right) \tag{8.60}$$

where

$$F(n) = \frac{n^2}{n^2 - 1}\ln(n) - \frac{3n^2 - 1}{4n^2} \tag{8.61}$$

$$T_r = \frac{c_r t}{d_e^2} \tag{8.62}$$

and $n = d_e/d_w$. Here, d_e is the diameter of influence (i.e., twice the maximum drainage path length) and d_w is the diameter of the drain. The outer perimeter is acting as an impervious boundary.

8.9 Secondary Compression

The one-dimensional consolidation theory proposed by Terzaghi and discussed in the previous sections concludes that the consolidation process continues for ever and the final consolidation settlement s_c is reached only at $t = \infty$. In reality, consolidation process ends after some years where the excess pore water pressure has fully dissipated and the final consolidation settlement s_c is reached. The time at which consolidation ends is denoted by t_p (p for primary consolidation) or t_{100} in the literature. Beyond this time, there is no further excess pore water pressure dissipation, but the settlement continues indefinitely for a long time, and the process is called *secondary compression* or *creep*. The secondary compression occurs due to realignment of the clay grains which results in a change in the clay fabric and reduction in void ratio. There is no excess pore water pressure to dissipate, and the effective stress remains the same throughout the secondary compression process. The two processes are distinguished by the terms "primary consolidation" and "secondary consolidation," even though secondary compression is very different to consolidation.

While it is possible to have some overlap between primary consolidation and secondary compression, for simplicity, it is generally assumed that secondary compression starts when the consolidation ends at time t_p, as shown in Fig. 8.30.

As shown in Fig. 8.30, when time is plotted in logarithmic scale, the void ratio (or settlement) versus time plot is linear during secondary compression. The slope of this linear portion is known as the *secondary compression index*, denoted by C_α. It can be defined as the reduction in the void ratio per log cycle of time.

$$C_\alpha = \frac{\Delta e}{\Delta \log t} \tag{8.63}$$

It can be determined from the tail end of the void ratio versus log time plot used for determining c_v by Casagrande's (1938) method where the time is in logarithmic scale.

Instead of the void ratio, if the vertical strain is plotted against log time in Fig. 8.30, the slope of the linear segment is called the *modified secondary compression index* $C_{\alpha\varepsilon}$. Remember the way we separated C_c and C_r from $C_{c\varepsilon}$ (*CR*) and $C_{r\varepsilon}$ (*RR*), in Secs. 8.4.1 and 8.4.2, respectively. Here, we are doing similar.

$$C_{\alpha\varepsilon} = \frac{\varepsilon}{\Delta \log t} = \frac{C_a}{1+e_p} \tag{8.64}$$

where the vertical strain ε is

$$\varepsilon = \frac{\Delta e}{1+e_p}$$

Prior to consolidation, the clay layer has thickness of H_0, and void ratio of e_0. At the end of consolidation, at time t_p, they reduce to H_p and e_p, respectively. From Eq. (8.9) and

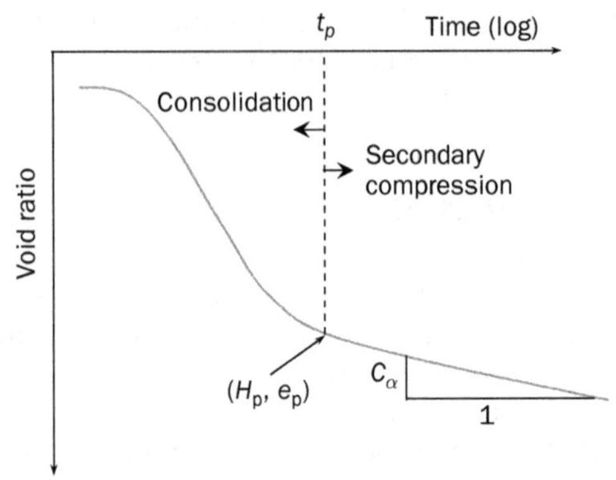

FIGURE 8.30 Consolidation and secondary compression.

assuming linear reduction in void ratio with logarithm of time, the void ratio reduction Δe and the secondary compression settlement s_s from time t_p to any time t ($>t_p$) are related by

$$\frac{s_s}{H_p} = \frac{\Delta e}{1 + e_p} \tag{8.65}$$

From Eqs. (8.63), (8.64), and (8.65),

$$s_s = \frac{H_p}{1 + e_p} C_\alpha \log\left(\frac{t}{t_p}\right) = H_p C_{\alpha\varepsilon} \log\left(\frac{t}{t_p}\right) \tag{8.66}$$

H_0 and e_0 are easily determined than H_p and e_p. From phase relationships, it can be shown that

$$\frac{H_0}{1 + e_0} = \frac{H_p}{1 + e_p}$$

Therefore, Eq. (8.66) can be written as

$$s_s = \frac{H_0}{1 + e_0} C_\alpha \log\left(\frac{t}{t_p}\right) \tag{8.67}$$

Mesri and Godlewski (1977) suggested that the ratio of C_α/C_c for cohesive soils generally varies in the range of 0.025–0.10, with an average value of about 0.05 and Table 8.6 gives the value ranges proposed by Mesri et al. (1994).

As a first approximation, $C_{\alpha\varepsilon}$ of normally consolidated clays can be estimated as (U. S. Navy, 1982)

$$C_{\alpha\varepsilon} = 0.0001\, w_n \quad \text{for } 10 < w_n(\%) < 3000 \tag{8.68}$$

where w_n is the natural moisture content of the clay in percentage. For most overconsolidated clays, $C_{\alpha\varepsilon}$ lies within the range of 0.0005–0.001 (Kulhawy and Mayne, 1990).

Material	C_α/C_c
Granular soils including rockfill	0.02 ± 0.01
Shale and mudstone	0.03 ± 0.01
Inorganic clays and silts	0.04 ± 0.01
Organic clays and silts	0.05 ± 0.01
Peats and muskeg	0.06 ± 0.01

Table 8.6 Some Typical C_α/C_c Values (After Mesri et al., 1994)

For overconsolidated clays, C_α and $C_{\alpha\varepsilon}$ can be significantly less than when they are normally consolidated. Alonso et al. (2000) suggested that the ratio of $C_{\alpha\varepsilon}$ for overconsolidated and normally consolidated clays can be written as

$$\frac{C_{\alpha\varepsilon,\,OC}}{C_{\alpha\varepsilon,\,NC}} = (1 - m)e^{-(\mathrm{OCR}-1)n} + m \tag{8.69}$$

where m and n are constants. The constant m is the minimum possible value for the ratio, which applies for very large OCR, and is similar to the C_r/C_c ratio. The magnitude of n controls the rate of decay in the ratio with OCR, with larger values of n giving a faster decay (see Fig. 8.31). Alonso et al. (2000) suggested $m = 0.1$ and $n = 12$ from limited data. In practice, smaller values for n are being used conservatively. Wong (2006) suggested $n = 6$ in organic clays for preliminary assessments.

8.10 A Note on Preloading

Some of the geotechnical instruments used in monitoring consolidation settlements in a preloading exercise include settlement plates, piezometers, stand pipes, and inclinometers (discussed in Sec. 11.10, Chap. 11). These instruments are used to assess any potential instability and settlements. The preloading is expected to achieve 90–95% of the primary consolidation, and the rate of consolidation is monitored through settlements and/or pore water pressures, which may not be always matching. Restrictions to the water flow through the piezometer due to entrapped air and slow flow rate in the low permeability clays can result in time lag in the pore water pressure measurements. Therefore, settlement is a better measure than pore water pressure dissipation for monitoring the rate of settlement.

The preload applied in the form of a fill is not always compacted to the same vigor as the traditional compacted fill. It is placed in thicker layers and compacted often using the machinery at the site such as dozers.

With the difficulty in determining a realistic value of the in situ c_v, the preloading may not always work as planned. In such circumstances, it may be necessary to increase the preload magnitude or delay its removal.

Preloading process delays the beginning of secondary compression. Since the clay becomes overconsolidated, the magnitude of the secondary compression settlement is also reduced.

When vertical drains are used in preloading to expedite the consolidation process, the drainage path between the drains is mainly horizontal and hence it is required to

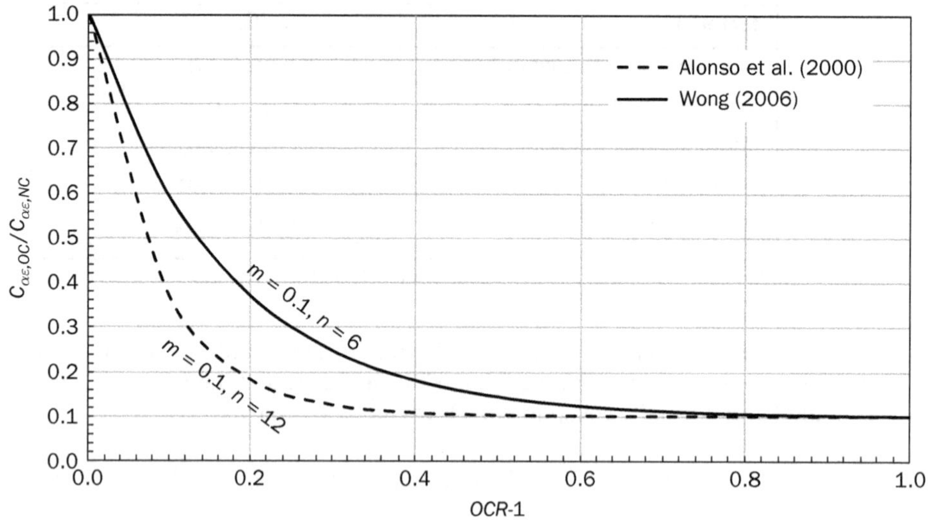

FIGURE 8.31 Variation of $C_{\alpha\varepsilon,OC}/C_{\alpha\varepsilon,NC}$ versus OCR – 1.

determine the horizontal coefficient of consolidation c_h. Typically, c_h/c_v is in the range of 1–4. While c_v is generally determined through laboratory oedometer tests, c_h can be determined by piezocone dissipation tests in the field.

Example 8.10

A 20-mm-thick clay sample at void ratio of 1.71 is subjected to a doubly drained consolidation test in an oedometer, where the dial gauge reading is set to 0.00 mm initially. The vertical pressure on the sample was increased from 0 to 272.6 kN/m² in few increments, each being applied for 24 h. During the next increment, where σ'_v was increased from 272.6 to 543 kN/m², the time-dial gauge readings are given below:

Time	Dial gauge reading (mm)
0^-	3.590 (just before applying the pressure increment)
1.5 s	3.676
15 s	3.690
30 s	3.718
1 min	3.756
2 min	3.806
4 min	3.884
8 min	3.983
16 min	4.130
32 min	4.330
60 min	4.562

141 min	4.853
296 min	5.027
429 min	5.086
459 min	5.095
680 min	5.141
1445 min	5.204
1583 min	5.212

a. Determine the coefficient of consolidation during this pressure increment using Casagrande's log time and Taylor's square root of time methods.

b. Determine the coefficient of volume compressibility during this increment.

c. Determine the coefficient of secondary compression during this increment.

d. Estimate the permeability during this increment.

Solution

Part a Casagrande's graphical construction for determining d_0, d_{100}, and t_{50} are shown in Fig. 8.32, where d_0 = 3.63 mm and d_{100} = 5.05 mm, from which d_{50} is determined as 4.34 mm and hence t_{50} is read off the plot as 33 min.

The average thickness of the sample during consolidation (i.e., at t_{50}) = 20 – 4.34 = 15.66 mm. Being doubly drained, H_{dr} = 15.66/2 = 7.83 mm

$$T_{50} = 0.197 = \frac{c_v t_{50}}{H_{dr}^2}$$

$$\therefore c_v = \frac{0.197 \times 7.83^2}{33} = \mathbf{0.37\ mm^2/min}$$

Taylor's graphical construction to determine t_{90} is shown in Fig. 8.33, from which $\sqrt{t_{90}}$ = 11.95 and hence t_{90} = 143 min.

$$d_0 = 3.60\text{ mm},\ d_{90} = 4.86\text{ mm} \rightarrow d_{50} = 3.60 + (4.86 - 3.60) \times 5/9 = 4.30\text{ mm}$$

∴ Average thickness of the sample during consolidation = 20 – 4.30 = 15.70 mm

$$\therefore H_{dr} = 15.70/2 = 7.85\text{ mm}$$

$$T_{90} = 0.848 = \frac{c_v t_{90}}{H_{dr}^2} \rightarrow c_v = \frac{0.848 \times 7.85^2}{143} = \mathbf{0.36\ mm^2/min}$$, very close to Casagrande's c_v.

Part b σ'_v has increased from 272.6 to 543 kN/m². Therefore,

$$\Delta\sigma' = 543.0 - 272.6 = 270.4\text{ kN/m}^2$$

$$\Delta H = d_{100} - d_0 = 5.05 - 3.63 = 1.42\text{ mm}$$

$$H_0 = 20.00 - 3.59 = 16.41\text{ mm}$$

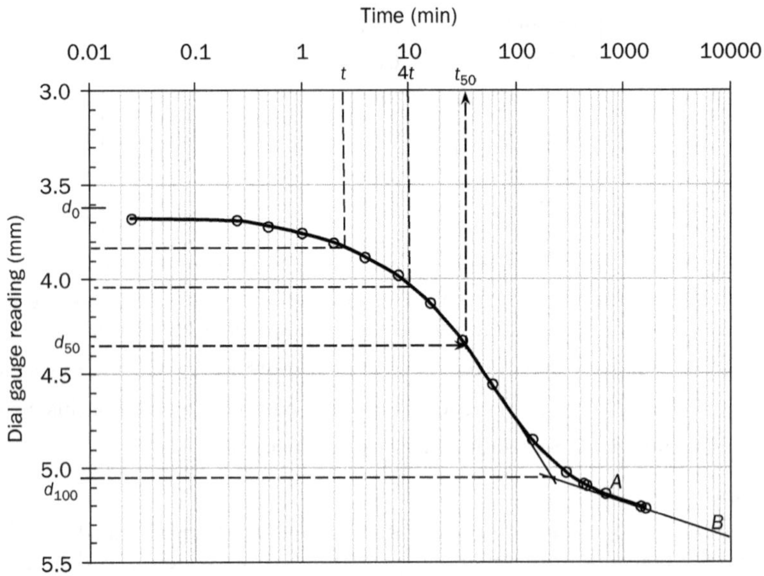

FIGURE 8.32 Casagrande's log time plot.

$$m_v = \frac{\left(\frac{\Delta H}{H_0}\right)}{\Delta\sigma'} = \frac{\left(\frac{1.42}{16.41}\right)}{270.4} = 0.32 \times 10^{-3}\ \text{kPa}^{-1}$$

Part c Let us consider the two points A and B on the tail of the Casagrande's plot (Fig. 8.32), and find the void ratios at these points.

From the very beginning of the consolidation test to A,

$$H_0 = 20.0\ \text{mm},\ e_0 = 1.71,\ \Delta H = 5.15\ \text{mm} \rightarrow \Delta e_A = \frac{5.15}{20.0} \times (1 + 1.71) = 0.698$$

From the very beginning of the consolidation test to B,

$$H_0 = 20.0\ \text{mm},\ e_0 = 1.71,\ \Delta H = 5.33\ \text{mm} \rightarrow \Delta e_B = \frac{5.33}{20.0} \times (1 + 1.71) = 0.722$$

$$\therefore e_A = 1.710 - 0.698 = 1.012;\ e_B = 1.710 - 0.722 = 0.988$$

$\therefore$ Change in void ratio between A and B = 1.012 – 0.988 = 0.024.

$$t_A = 680\ \text{min},\ t_B = 6000\ \text{min}$$

$$C_\alpha = \frac{0.024}{\log 6000 - \log 680} = \mathbf{0.025}$$

Part d From the definition of c_v as $c_v = \dfrac{k}{m_v \gamma_w}$,

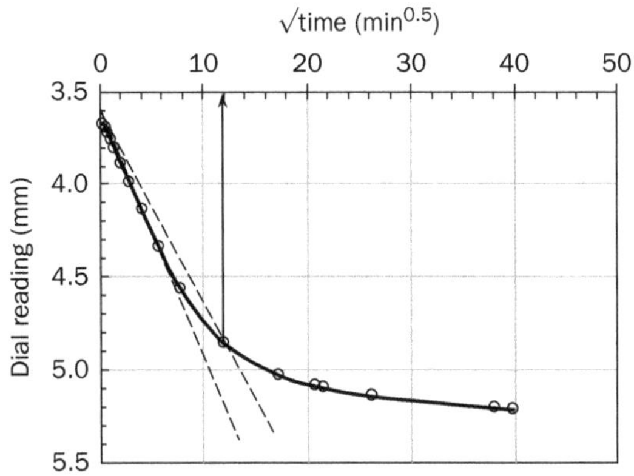

FIGURE 8.33 Taylor's square root of time plot.

$$k = \left(6.2 \times 10^{-9} \frac{\text{m}^2}{\text{s}}\right)(0.32 \times 10^{-6}\ \text{Pa}^{-1})\left(9810 \frac{\text{N}}{\text{m}^3}\right) = \mathbf{1.95 \times 10^{-11} m/s}$$

Example 8.11

The soil profile at a site is shown in Fig. 8.34. A 3-m-high compacted fill, with unit weight of 19.0 kN/m³, is placed at the ground level.

a. What would be the final consolidation settlement s_c?
b. What would be the consolidation settlement after 1 year?
c. How long will it take for 75 mm of consolidation settlement?
d. After 1 year, what would be the pore water pressure and effective vertical stress at the middle of the clay layer?

Solution

Part a From Eq. (8.12),

$$s_c = m_v q H_0 = (0.25)\left(\frac{3 \times 19}{1000}\right)(6)\ \text{m} = 85.5\ \mathbf{mm}$$

Part b The clay is doubly drained and hence $H_{dr} = 3.0$ m.

The time factor T after 1 year is

$$T = \frac{c_v t}{H_{dr}^2} = \frac{(4.9)(365 \times 24 \times 60)}{3000^2} = 0.286$$

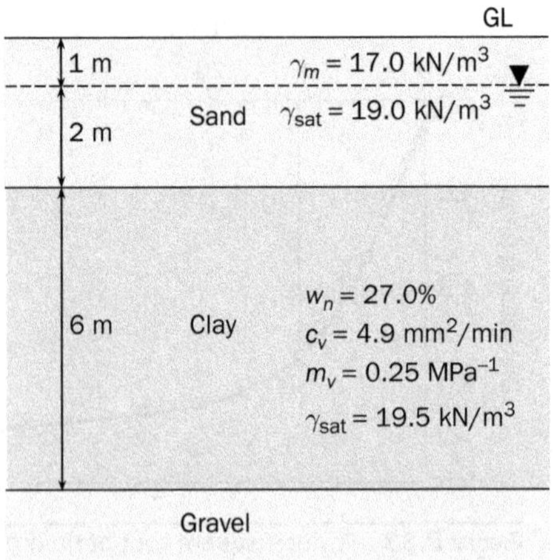

FIGURE 8.34 Soil profile for Example 8.11.

From Fig. 8.20 or Eq. (8.42), for $T = 0.286$, $U_{avg} = 60.3\%$. Therefore, the consolidation settlement after 1 year is

$$s \text{ (1 year)} = 0.603 \times 85.5 = \mathbf{51.6\ mm}$$

Part c When the consolidation settlement is 75 mm,

$$U_{avg} = \frac{75}{85.5} = 87.7\%$$

From Fig. 8.20 or Eq. (8.42), $T = 0.764$

$$T = 0.764 = \frac{c_v t}{H_{dr}^2} = \frac{(4.9)(t)}{3000^2}$$

The time t = 1,403,265 min = **2.67 years**

Part d At the middle of the clay layer, the initial effective overburden stress is given by

$$\sigma'_{vo} = (1)(17.0) + (2)(19.0 - 9.81) + (3)(19.5 - 9.81) = 64.5 \text{ kN/m}^2$$

The initial static pore water pressure at the middle of the clay layer is

$$u_0 = (5)(9.81) = 49.1 \text{ kN/m}^2$$

The initial excess pore water pressure induced by the fill is

$$\Delta u_0 = (3)(19) = 57 \text{ kN/m}^2$$

As consolidation progresses, a fraction of this 57 kN/m² is dissipated and is transferred as the increase in the effective stress $\Delta\sigma'$.

At the middle of the clay layer, with depth z being measured from the top of the clay layer,

$$Z = \frac{z}{H_{dr}} = \frac{3}{3} = 1$$

From Fig. 8.19, for $Z = 1$ and $T = 0.286$, $U(z, t) = 37.5\%$.

With 37.5% consolidation at the middle of the clay layer,

$$\Delta u = (1.0 - 0.375)\,(57) = 35.6\ \text{kN/m}^2$$

$$\Delta\sigma' = (0.375)\,(57) = 21.4\ \text{kN/m}^2$$

Therefore after 1 year, at the middle of the clay layer,

$$u = u_0 + \Delta u = 49.1 + 35.6 = \mathbf{84.7\ kN/m^2}$$

$$\sigma'_v = \sigma'_{vo} + \Delta\,\sigma' = 64.5 + 21.4 = \mathbf{85.9\ kN/m^2}$$

Example 8.12

The soil profile at a site consists of 3 m of sand below the ground level, followed by 5 m of clay. The water table is at the top of the clay layer. The bulk unit weight of the sand is 18.0 kN/m³ and the saturated unit weight of the clay is 19.5 kN/m³. The clay is normally consolidated, with $LL = 65$ and $PL = 31$. A 75 mm diameter undisturbed sample was collected from a depth of 6 m below the ground level. The natural (in situ) moisture content of the clay is 26.5%, and $G_s = 2.70$.

a. If a consolidation test is carried out on this sample, what would be the preconsolidation pressure?
b. Estimate the compression index from the liquid limit, using an empirical correlation (see Table 8.2).

A 50 kN/m² surcharge is placed at the above site, and piezometers were installed to monitor the decay of the excess pore water pressures. After 6 months, the piezometers indicated that the pore water pressure at the middle of the clay layer was 54.5 kN/m². The bore logs revealed that the clay was underlain by gravelly sand.

c. Find the coefficient of consolidation of the clay.
d. What would be the consolidation settlement after 6 months?

Solution

Part a The in situ effective vertical stress of the clay specimen is

$$3 \times 18 + 3(19.5 - 9.81) = 83.1\ \text{kN/m}^2$$

Since the clay is normally consolidated, the **preconsolidation pressure of this specimen would be 83.1 kN/m²**.

Part b The compression index of the normally consolidated clay can be estimated as

$$C_c = = 0.009\,(LL - 10) = 0.009\,(65 - 10) = \mathbf{0.495}$$

Part c

The initial pore water pressure at the middle of the clay layer $u_0 = 2.5 \times 9.81 = 24.5$ kN/m²

The initial excess pore water pressure $\Delta u_0 = 50$ kN/m²

The measured pore water pressure after 6 months is 54.5 kN/m², of which the excess pore water pressure $\Delta u = 54.5 - 24.5 = 30.0$ kN/m².

Therefore, the degree of consolidation at the middle of the clay layer after 6 months is

$$U\,(2.5\text{ m, 6 months}) = (50 - 30)/50 = 0.4$$

$$Z = z/H_{dr} = 2.5/2.5 = 1$$

From Fig. 8.19 or Table 8.5, $T = 0.307$

$$T = 0.307 = \frac{c_v t}{H_{dr}^2} = \frac{c_v \times 0.5}{2.5^2}$$

$$\mathbf{c_v = 3.8\ m^2/year}$$

Part d The in situ effective stress at the middle of the clay layer is

$$3 \times 18 + 2.5\,(19.5 - 9.81) = 78.2\ \text{kN/m}^2$$

Initial in situ void ratio $e_0 = 2.70 \times 0.265 = 0.716$

The change in void ratio during consolidation can be determined as

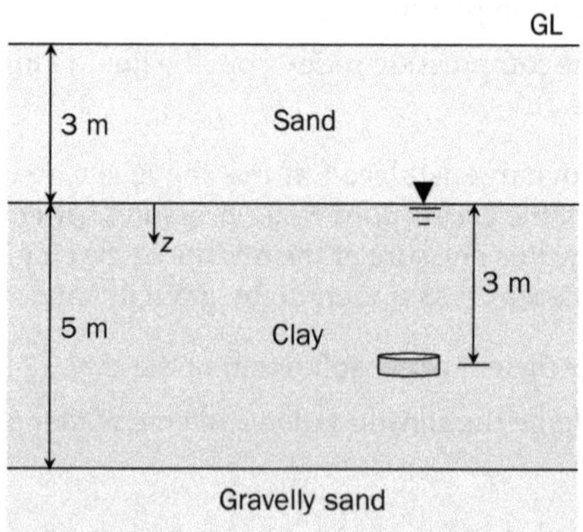

Figure 8.35 Soil profile for Example 8.12.

$$\Delta e = C_c \log\left(\frac{\sigma'_{vo} + \Delta\sigma'}{\sigma'_{vo}}\right) = 0.495\log\left(\frac{78.2 + 50}{78.2}\right) = 0.106$$

Therefore, the final consolidation settlement is

$$s_c = \frac{\Delta e}{1 + e_0} H_0 = \frac{0.106}{1 + 0.716} \times 5000 = 309 \text{ mm}$$

At t = 6 months, T = 0.307

$$U_{avg} = 62.0\%$$

The consolidation settlement in 6 months is 0.62 × 309 = **192 mm.**

Example 8.13

It is proposed to place a 3-m-high compacted fill, with unit weight of 19.0 kN/m^3 on the ground to preload the clay layer shown in Fig. 8.36. The coefficient of volume compressibility is 0.26 MPa^{-1}, and the coefficient of consolidation is 3.2 mm^2/min.

a. What is the final consolidation settlement of the clay layer?
b. What will be the consolidation settlement after 1 year?
c. How long will it take for 50 mm of consolidation settlement to occur?
d. Find the pore water pressures at 0, 1, 2, 2.5, 3, 4, and 5 m depths within the clay after 1 year and plot the variation with depth. In the same plot, show the variation of the initial in situ pore water pressures within the clay layer as well.

Solution

Part a

The applied surcharge q = 3 × 19 = 57 kN/m^2

The final consolidation settlement is

$$s_c = m_v\, q\, H_0 = 0.26 \times (57/1000) \times 5000 = \mathbf{74.1\ mm}$$

Part b The time factor after 1 year is

$$T = \frac{c_v t}{H_{dr}^2} = \frac{3.2 \times 365 \times 24 \times 60}{2500^2} = 0.269$$

$$U_{avg} = \sqrt{\frac{4T}{\pi}} \times 100 = \sqrt{\frac{4 \times 0.269}{\pi}} \times 100 = 58.5\%$$

Therefore, the consolidation settlement after 1 year is given by

$$= 0.585 \times 74.1 = \mathbf{43.3\ mm}$$

Part c At 50 mm settlement, the average degree of consolidation is given by

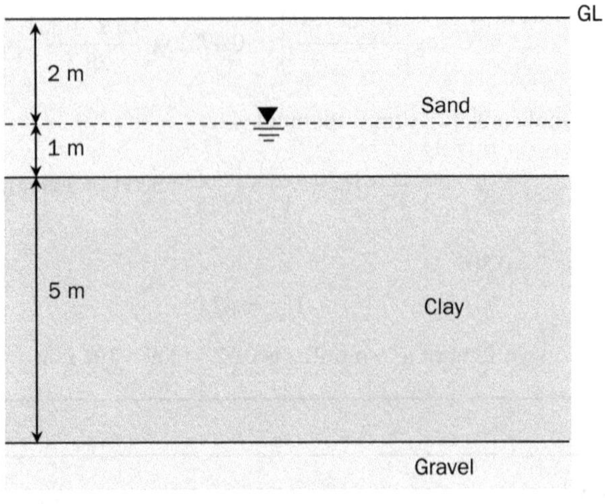

FIGURE 8.36 Soil profile for Example 8.13.

$$U_{avg} = (50/74.1) \times 100 = 67.5\%$$

The corresponding time factor can be determined as

$$T = 1.781 - 0.933 \log(100 - U_{avg}) = 1.781 - 0.933 \log(100 - 67.5) = 0.370$$

$$T = \frac{c_v t}{H_{dr}^2}$$

$$0.370 = \frac{3.2 \times t}{2500^2}$$

$t = 722{,}656$ min = **502 days or 1.37 years**

Part d

The initial excess pore water pressure $\Delta u_0 = 57$ kN/m^2

After 1 year, $T = 0.269$

The values of initial in situ pore water pressure u_0, degree of consolidation, and the pore water pressure after 1 year at the different depths are summarized below.

Z (m)	u_0 **(kN/m²)**	**Z**	**U (Z, T)**	Δu **(kN/m²)**	$u_0 + \Delta u$ **(kN/m²)**
0	9.8	0.00	1.000	0.00	9.81
1	19.6	0.40	0.614	22.00	41.62
2	29.4	0.80	0.377	35.51	64.94
2.5	34.3	1.00	0.346	37.28	71.61
3	39.2	1.20	0.377	35.51	74.75
4	49.1	1.60	0.614	22.00	71.05
5	58.9	2.00	1.000	0.00	58.86

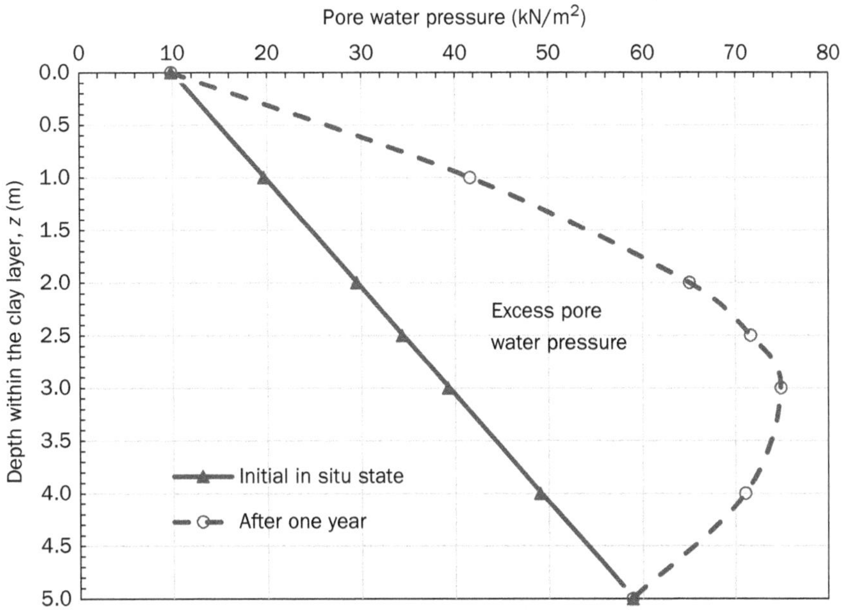

FIGURE 8.37 Variation of pore water pressure with depth.

8.11 Main Points

1. Consolidation occurs due to the dissipation of the excess pore water pressure from the voids.
2. During one-dimensional consolidation, both strains and drainage are vertical and hence the name one-dimensional consolidation.
3. Slope of the VCL is given by C_c, CR, and λ, depending on the axes used in the plot. Similarly, slope of the unloading/reloading path is given by C_r, RR, and κ.
4. Normally consolidated clays settle more than the overconsolidated clays.
5. Constrained or oedometer modulus D is the reciprocal of the coefficient of volume compressibility m_v.
6. Coefficient of consolidation c_v says how fast the consolidation will occur. It can be an order of magnitude larger for an overconsolidated clay than when it is normally consolidated.
7. It is assumed that the secondary compression begins when the primary consolidation ends.
8. The coefficient of secondary compression is significantly less for overconsolidated clays than normally consolidated clays.

Review Exercises

8.1 State whether the following are true or false.

a. During consolidation, effective stress increases.

b. Compression index C_c is larger than the recompression index C_r.

c. Young's modulus E is larger than the oedometer (constrained) modulus D.

d. Coefficient of consolidation increases with the liquid limit.

e. The larger the coefficient of consolidation c_v, the faster the consolidation process.

Answer: True, True, False, False, True

8.2 A 4-m-thick saturated clay layer is surcharged by a 3-m-high fill with unit weight of 18.5 kN/m^3 at the ground level. Assuming $m_v = 0.0002$ kPa^{-1}, estimate the final consolidation settlement.

Answer: 44.4 mm

8.3 The top 7 m at a site consists of sandy silt, having bulk unit weight of 16.5 kN/m^3 and saturated unit weight of 18.5 kN/m^3. The water table lies 1 m below the ground level. The sandy silt layer is underlain by a 2.5-m-thick clay layer (saturated unit weight 19.5 kN/m^3, average m_v = 0.5 MPa^{-1}), which is underlain by sands. If the water table is lowered by 3 m, what would be the consolidation settlement?

Answer: 29.3 mm

8.4 A site consists of 3 m of gravelly sand (γ_m = 17.0 kN/m^3) at the ground level underlain by 4 m of clay which is underlain by dense sand. Water table is at 3.0 m below the ground level. The properties of the clay are:

- moisture content = 26.0%;
- specific gravity of the soil grains = 2.74;
- compression index = 0.65;
- recompression index = 0.06; and
- overconsolidation ratio = 1.4.

If the ground level is raised by 2 m of compacted fill (γ_m = 19.0 kN/m^3) what would be the final consolidation settlement?

Answer: 81.3 mm

8.5 The soil profile at a site is shown in Fig. RE8.5. A consolidation test is carried out on an undisturbed clay specimen obtained from a depth of 5 m below the ground level. The natural moisture content of the clay is 30.0% and the specific gravity of the clay grains is 2.70. The bulk unit weight of the sand is 18.0 kN/m^3. The following parameters were determined from the consolidation test.

Preconsolidation pressure = 90.0 kN/m^2
Compression index = 0.45
Recompression index = 0.05

a. Is the clay normally consolidated or overconsolidated? What is the *OCR*?

b. It is proposed to build a warehouse on the ground, covering a large area in plan. The warehouse is expected to impose a uniform pressure of 35 kN/m^2 on the ground. What would be the final consolidation settlement of the warehouse?

Answer: Overconsolidated, 1.42, 84.1 mm

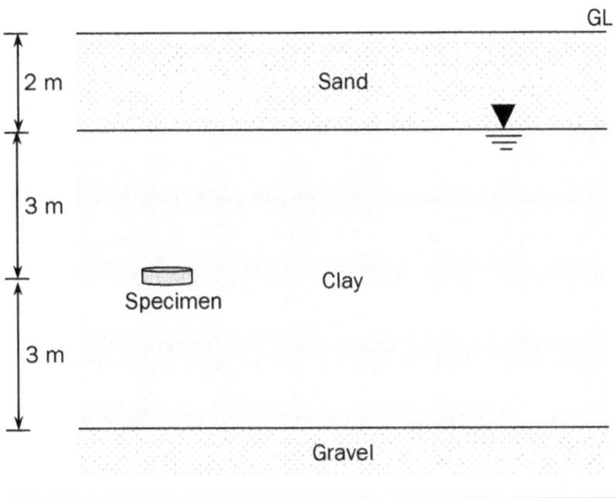

FIGURE RE8.5 Soil profile.

8.6 The soil profile at a site is shown in Fig. RE8.6. Assume the bulk and saturated unit weights of the two sands as 17.0 and 19.0 kN/m^3, respectively. The properties of the two clay layers are summarized below.

Property	Overconsolidated clay	Normally consolidated clay
Natural moisture content (%)	24.0	29.0
Specific gravity of the grains	2.74	2.74
Compression index	0.45	0.45
Recompression index	0.07	0.07
Overconsolidation ratio	1.5	1.0

If a 3.0 m high fill (unit weight = 18.5 kN/m^3) is placed on the ground covering a large extent, what would be the final consolidation settlement?

Answer: 289.7 mm

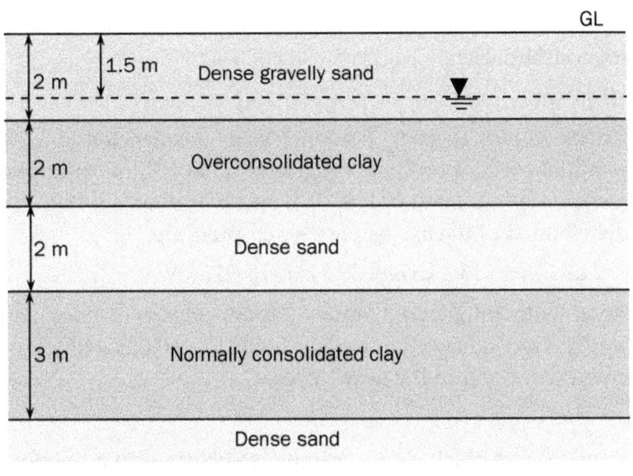

FIGURE RE8.6 Soil profile.

8.7 A saturated clay sample in an oedometer was under vertical stress of 120 kN/m^2 under normally consolidated state. The void ratio and the sample height at this stage are 1.21 and 18.40 mm, respectively. When the vertical stress was increased to 240 kN/m^2, at the end of consolidation, the thickness of the sample was reduced to 16.80 mm. When the vertical stress was reduced to the original value of 120 kN/m^2, the sample heaved to a thickness of 16.95 mm. Estimate the compression index and the recompression index of the clay.

What would be the reduction in thickness from now if the vertical stress is increased by 200 kN/m^2?

Answer: 0.64, 0.06; 0.82 mm

8.8 The equation of the virgin consolidation line for a clay deposit is given by

$$e = -0.64 \log \sigma_v' + 1.96$$

An undisturbed saturated specimen was recovered from the same clay, where the effective vertical overburden pressure is 50.0 kN/m^2 and the moisture content is 27.0%. Assuming G_s = 2.72 and recompression index to be 20% of the compression index, estimate the preconsolidation pressure of the specimen.

Answer: 93.3 kN/m^2

8.9 In Review Exercise 8.4, the coefficient of consolidation of the clay is 1.2 m^2/year. After 6 months, what would be the settlement and the excess pore water pressure that is yet to dissipate at the middle of the clay?

Answer: 35.5 mm, 32.9 kN/m^2

8.10 The soil profile at a site consists of 3 m of sand underlain by 6 m of clay (m_v = 0.32 MPa^{-1}, c_v = 4.9 mm^2/min) which is underlain by bedrock. The water table lies 1 m below the ground level. A 3-m-high compacted fill (γ_m = 20 kN/m^3) is placed on the ground in an attempt to raise the ground level.

a. What would be the final consolidation settlement of the clay layer?
b. How long will it take for 50 mm consolidation settlement to occur?
c. What would be the consolidation settlement in 1 year?

Answer: 115.2 mm, 2.07 years, 34.8 mm

8.11 A 2-m-thick singly drained clay layer is consolidating under a newly built embankment. The settlement (mm) records taken at 3 months intervals are: 265, 331, 396, 430, 465, 499, 520, and 537. Using Asaoka's method, find the end-of-primary consolidation settlement and the coefficient of consolidation.

Answer: 607 mm, 1.47 m^2/year

8.12 A 5-m-thick singly drained clay layer (c_v = 4.5 m^2/year) is subjected to a 50 kN/m^2 surcharge that induces initial excess pore water pressure of 50 kN/m^2 throughout the layer. Find the initial excess pore pressure at 0, 1, 2, 3, 4, and 5 m depths within the clay after 1 year, and plot the distribution of the excess pore water pressure.

Answer: 0, 13 kN/m^2, 24.3 kN/m^2, 33.3 kN/m^2, 38.5 kN/m^2, 40.5 kN/m^2

8.13 A 4-m-thick clay layer with initial void ratio of 0.98 appears to have consolidated in 4 years since loading. The secondary compression index C_α is 0.025. How much secondary compression settlement will occur in the next 15 years?

Answer: 34.2 mm

8.14 A 75-mm-diameter and 20-mm-thick clay specimen is subjected to a doubly drained consolidation test in an oedometer. After a few stress increments, the vertical stress is increased from 80 to 160 kN/m^2 and the resulting time versus dial gauge reading data are summarized below.

Time (min)	Dial reading (mm)
0.1	6.364
0.25	6.340
0.5	6.318
1	6.294
2	6.260
4	6.215
8	6.155
15	6.098
30	6.039
60	6.001
100	5.981
200	5.958
400	5.941
800	5.927
1600	5.915

The specimen thickness at the beginning of this increment was 16.24 mm.

a. Compute the coefficient of consolidation using
   i. Casagrande's logarithm of time method, and
   ii. Taylor's square root of time method.
b. Determine the coefficient of volume compressibility m_v.
c. What is the constrained modulus of this clay? Assuming Poisson's ratio of 0.2, estimate the Young's modulus.
d. What is the permeability of the clay during this stress increment?
e. Find the coefficient of secondary compression C_α if the void ratio at the beginning of this stress increment is 0.847.

Answers: 1.23 m²/year, 1.24 m²/year; 0.309 MPa⁻¹; 3.24 MPa, 2.92 MPa; 1.18 × 10⁻⁸ cm/s; 0.006

8.15 A clay deposit at a site consists of three layers as shown Fig. RE8.15. Find the equivalent height H^* assuming a uniform c_v^* of 2.0 m²/year. Find the equivalent c_v for the same soil profile height.

Answer: 0.77 m²/year

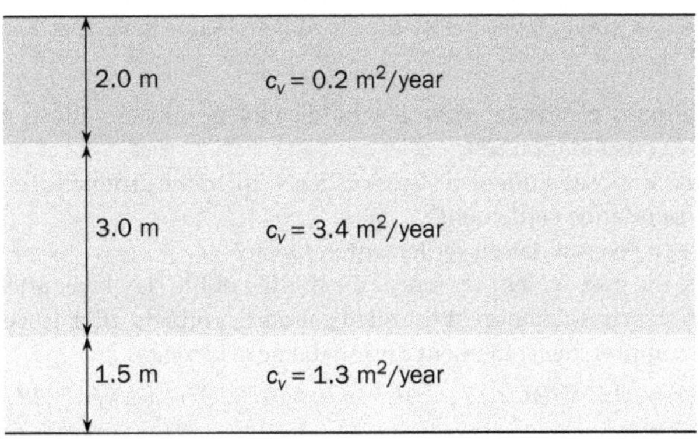

FIGURE RE8.15 Soil profile.

8.16 A 20.0-mm-thick clay specimen (G_s = 2.70, w = 35.0%) is subjected to a doubly drained consolidation test. After a few increments, when the vertical stress on the specimen was increased from 120 to 240 kN/m², the specimen thickness was plotted against time as shown in Fig. RE8.16. Determine the following during this stress increment.

a. Coefficient of consolidation
b. Coefficient of volume compressibility
c. Permeability (hydraulic conductivity)
d. Coefficient of secondary compression
e. Compression index of the clay assuming it is normally consolidated at this stress level

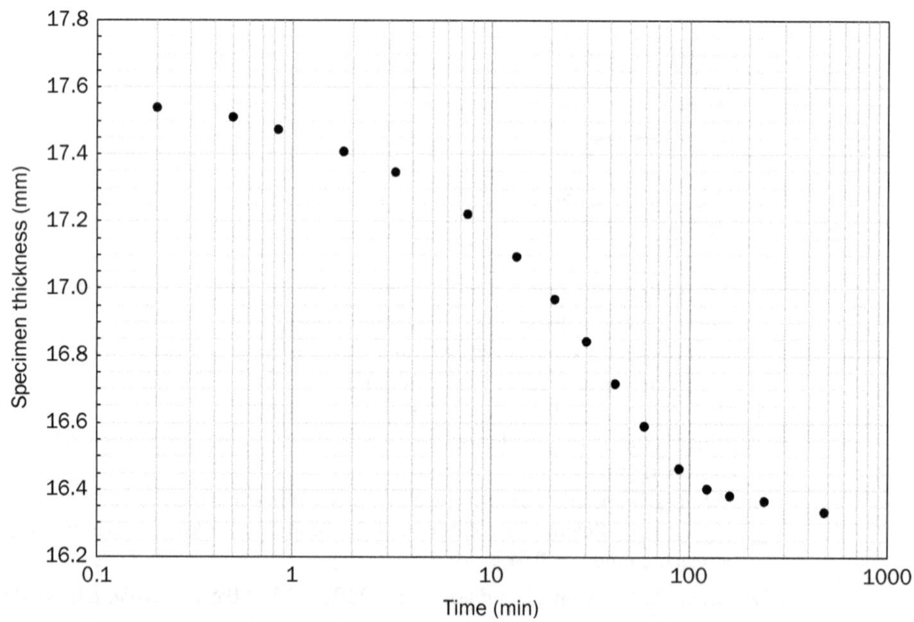

FIGURE RE8.16 Oedometer plot when σ'_v increased from 120 to 240 kN/m².

The above sample was taken from 5 m below the ground level at a site for a proposed construction of a warehouse with a large footprint. The soil profile at the site consists of 3.0 m of sand (γ_m = 17.5 kN/m³; γ_{sat} = 19.0 kN/m³) at the top, underlain by 4.0 m of clay (G_s = 2.72, w = 35.0%). There is a gravel layer below the clay layer. Water table is at 2.0 m below the ground level.

f. If the clay is normally consolidated, what would be the preconsolidation pressure of this specimen in the consolidation test?
g. If the warehouse imposes a uniform stress of 25 kN/m² at the ground level, what would be the final consolidation settlement?
h. What would be the consolidation settlement in 1 year?
i. What would be the pore water pressure at the middle of the clay layer after 1 year?
j. Assuming that the consolidation at the site is almost complete after 10 years, estimate the secondary compression settlement during the next 15 years.

Answer: 0.79 mm²/min; 0.57 MPa⁻¹; 0.74 × 10⁻⁸ cm/s; 0.01; 0.39; 61.6 kN/m²; 118 mm; 42.9 mm; 52.9 kN/m², 8.2 mm

8.17 A soil profile at a site is shown in Fig. RE8.17. The bulk unit weight of the sand and the saturated unit weight of the clay are 17.5 and 19.0 kN/m^3, respectively. A *standpipe piezometer* is installed to monitor the pore water pressure at the middle of the clay layer. A *settlement plate* is used to monitor the ongoing consolidation settlement of the clay layer. A 2.5 m compacted fill (bulk unit weight = 18.0 kN/m^3) was placed on the ground to raise the ground level. After 2 years, the piezometer reading shows that the pore water pressure at the middle of the clay is 40.0 kN/m^2, and the settlement plate reading shows that the clay has settled by 120 mm. Estimate the following:

a. Coefficient of consolidation
b. Final consolidation settlement
c. Coefficient of volume compressibility

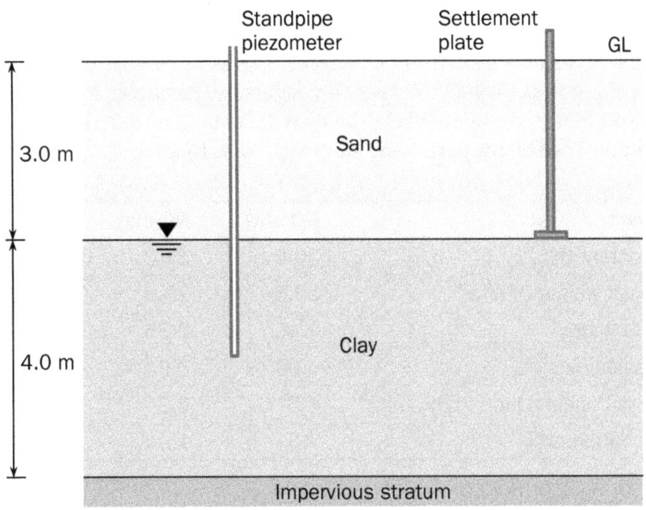

FIGURE RE8.17 Soil profile with settlement plate and piezometer.

Answer: 2.24 m^2/year, 201 mm, 1.117 MPa^{-1}

8.18 Which of the following two will consolidate *faster*, everything else being the same? Give your reasons very briefly in less than two sentences.

a. Larger c_v versus smaller c_v;
b. Stiff clay versus soft clay;
c. Higher permeability versus lower permeability; and
d. Doubly drained versus singly drained clay layer.

Answer: Larger c_v, Stiff clay, High permeability, Doubly drained

8.19 A soil profile at a site is shown in Fig. RE8.19. Transform the two saturated clay layers into a single 5.5-m-thick clay layer and determine the equivalent coefficient of consolidation and the average degree of consolidation after 3 years.

Answer: 0.945 m^2/year, 67.9%

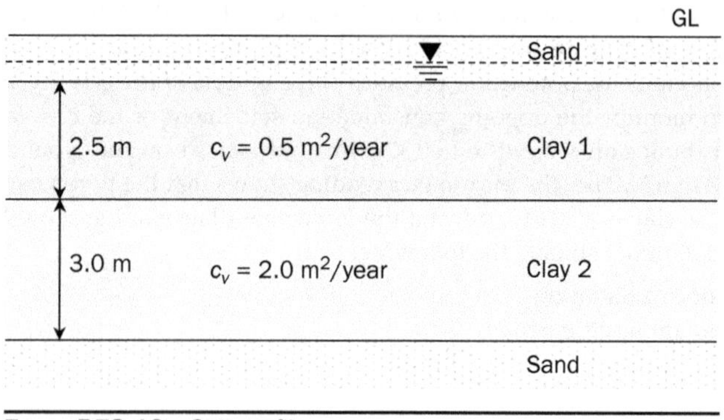

FIGURE RE8.19 Soil profile.

8.20 The soil profile at a site is shown in Fig. RE8.20. In this low-lying area, the water table is 1.00 m above the ground level. The two clay layers are separated by a 1.0-m-thick sand layer, and the normally consolidated clay layer at the bottom is underlain by an impervious stiff clay stratum. The soil properties are summarized as follows.

Soil parameter	OC clay	NC clay	Gravelly sand
Moisture content (%)	30.0	33.0	29.0
Saturated unit weight (kN/m^3)	20.5	20.0	19.7
Compression index	0.40	0.35	
Recompression index	0.04	0.04	
Coefficient of consolidation (mm^2/min)	4.5	2.3	
Overconsolidation ratio	2.0	1.0	

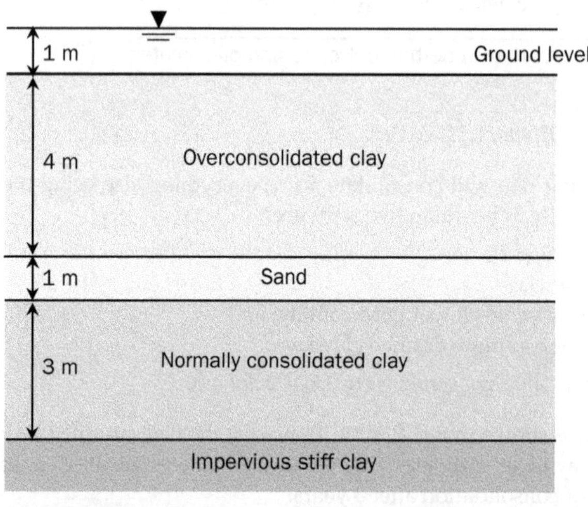

FIGURE RE8.20 Soil profile.

A 2.5 m fill (bulk unit weight = 18.0 kN/m^3 and saturated unit weight = 20 kN/m^3) was placed on the ground to raise the ground level.

a. Find the consolidation settlement, considering both clay layers.
b. What will be the consolidation settlement after 1 year?
c. Plot the variation of the pore water pressure with depth for the top 8 m of the soil,
   i. before placing the fill, and
   ii. 1 year after placing the fill.

Answer: 252.5 mm, 162.4 mm

8.21 The soil profile at a site is shown in Fig. RE8.21a. The normally consolidated clay has water content of 34% and specific gravity of 2.70. The dense gravel has bulk unit weight of 19 kN/m^3. A settlement plate was placed at the top of the clay layer to monitor the consolidation settlements over a period of 10 years, and a 2 m fill (γ_m = 20 kN/m^3) is placed on the ground in an attempt to raise the ground level. The consolidation settlement-time plot is shown in Fig. RE8.21b. Neglect the settlement in the dense gravel, and the secondary compression of clays (i.e., entire settlement is due to consolidation).

a. Estimate the compression index, coefficient of volume compressibility, and the coefficient of consolidation.
b. Sketch the variation of the pore water pressure with depth for the clay layer, after 12 months, showing necessary values in the sketch.

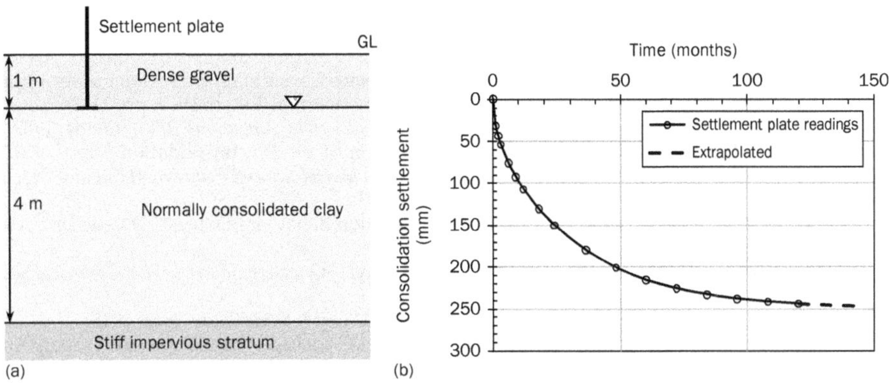

FIGURE RE8.21 (a) Soil profile and (b) settlement plot.

Answer: 0.37, 1.56 MPa^{-1}, 2.46 m^2/year

References

Al Khafaji, A. W. N. and O. B. Andersland. "Equations for Compression Index Approximation," *Journal of Geotechnical Engineering*, ASCE, 118(1):148–155, 1992.

Alonso, E. E., A. Gens, and A. Lloret. "Precompression Design for Secondary Settlement Reduction," *Geotechnique*, 50(6):645–656, 2000.

Asaoka, A. "Observational Procedure of Settlement Prediction," *Soils and Foundations*, 18(4):87–101, 1978.

ASTM D 2435-20. Standard Test Method for One-Dimensional Consolidation Properties of Soils using Incremental Loading.

Amertaunga, J., N. Sivakugan, and B. M. Das. *Correlations of Soil and Rock Properties in Geotechnical Engineering*, Springer, New Delhi, India, 2016.

Balasubramaniam, A. S., and P. Brenner. Consolidation and Settlement of Soft Clay, Soft Clay Engineering. In Brand, E. W., and R. P. Brenner (eds.), *Developments in Geotechnical Engineering* 20, Ch. 7, Elsevier, Amsterdam, the Netherlands, 1981, pp. 481–527.

Barron, R. A. "Consolidation of Fine Grained Soils by Drain Wells," *Transactions ASCE*, 113:718–724, 1948.

Bell, F. G. *Engineering Properties of Soils and Rocks*, 4th Edition, Blackwell science, UK, 2000.

Boone, S. J. "A Critical Reappraisal of Preconsolidation Pressure Interpretations using the Oedometer Test," *Canadian Geotechnical Journal*, 47:281–296, 2010.

Canadian Geotechnical Society. *Canadian Foundation Engineering Manual*, 3rd Edition, 1992.

Casagrande, A. "The Determination of the Pre-consolidation Load and Its Practical Significance," Discussion D-34, *Proceedings of the First International Conference on Soil Mechanics and Foundation Engineering*, Cambridge, III, 1936, pp. 60–64.

Casagrande, A. *Notes on Soil Mechanics—First Semester*, Harvard University (unpublished), 1938.

Clementino, R. V. Discussion of "An Oedometer Test Study on the Preconsolidation Stress of Glaciomarine Clays" by Grozic et al., *Canadian Geotechnical Journal*, 42(3):972–974, 2005.

Holtz, R. D., and W. D. Kovacs. *An Introduction to Geotechnical Engineering*, Prentice Hall, New Jersey, USA, 1981.

Kulhawy, F. H., and P. W. Mayne. *Manual on Estimating Soil Properties for Foundation Design*, Final report EL-6800, Electric Power Research Institute, Palo Alto, CA, USA, 1990.

Mesri, G., and M. Ajlouni. "Engineering Properties of Fibrous Peats," *Journal of the Geotechnical Engineering Division*, ASCE, 133(7):850–866, 2007.

Mesri, G., and P. M. Godlewski. "Time and Stress Compressibility Interrelationship," *Journal of Geotechnical Engineering Division*, ASCE, 103 (GT5):417–430, 1977.

Mesri, G., and N. Huvaj-Sarihan. "The Asaoka Method Revisited," *Proceedings of the 17th International Conference on Soil Mechanics and Geotechnical Engineering*, Alexandria, Egypt, IOS Press, 2009, pp. 131–134.

Mesri, G., D. O. K. Lo, and T. W. Feng. "Settlement of Embankments on Soft Clays," *Proceedings of Settlement '94, ASCE Specialty conference*, Geotechnical Special Publication, 40, 1, 1994, pp. 8–56.

Mesri, G., and B. Vardhanabhuti. "Compression of Granular Materials," *Canadian Geotechnical Journal*, 46:369–392, 2009.

Mikasa, M. "The Consolidation of Soft Clay—A New Consolidation Theory and Its Application," *Civil Engineering in Japan*, Japan Society of Civil Engineering, Tokyo, 1965, pp. 21–26.

Oikawa, H. "Compression Curves of Soft Soils," *Soils and Foundations*, 27(3):99–104, 1987.

Pacheco Silva, F. "A New Graphical Construction of the Preconsolidation Stress of a Soil Sample," *Proceedings of the 4th Brazilian Conference on Soil Mechanics and Foundation Engineering*, Rio de Janeiro, Brazil 2(1), 1970, pp. 225–232.

Schmertmann, J. H. "The Undisturbed Consolidation Behavior of Clays," *Transactions*, ASCE, 120:1201–1233, 1955.

Skempton, A. W. "Notes on Compressibility of Clays," *Quarterly Journal of the Geological Society of London*, 100:119–135, 1944.

Taylor, D. W. *Fundamentals of Soil Mechanics*, John Wiley & Sons, New York, 1948.

Terzaghi, K. *Erdbaumechanik auf Bodenphysikalischer Grundlage*, Franz Deuticke, Leipzig und Wein, 1925, p. 399.

Terzaghi, K., R. B. Peck, and G. Mesri. *Soil Mechanics in Engineering Practice*, 3rd Edition, Wiley, New York, 1996.

Umar, M., and A. Sandrekarimi. "Accuracy of Determining Pre-consolidation Pressure from Laboratory Tests," *Canadian Geotechnical Journal*, 54(3):441–450, 2017.

U. S. Navy. *Soil Mechanics, Foundations and Earth Structures*, NAVFAC Design Manual DM-7, Washington DC, USA, 1971.

U. S. Navy. *Soil Mechanics, Foundations and Earth Structures*, NAVFAC Design Manual DM-7, Washington DC, USA, 1982.

Wong, P. K. "Preload Design, Part 1: Review of Soil Compressibility Behavior in Relation to the Design of Preloads," Australian Geomechanics Society Sydney Chapter, *2006 Symposium*, 2006, pp. 27–32.

Wroth, C. P., and D. M. Wood "The Correlation of Index Properties with Some Basic Engineering Properties of Soils," *Canadian Geotechnical Journal*, 15(2):137–145, 1978.

Zheng, J., R. D. Hyrciw, and A. Ventola. "Compressibility of Sands of Various Geologic Origins at Pre-crushing Stress Levels," *Geotechnicall and Geological Engineering*, 35:2037–2051, 2017.

CHAPTER 9

Shear Strength

9.1 Introduction

Figure 9.1 shows an embankment. Depending on the slope angle, slope height, and the soil properties, the slope can fail along a *failure surface* as shown. When failure occurs, the soil enclosed within the failure surface slides downward. When the soil mass slides downward along the failure surface, the shear stress τ along the failure surface exceeds or becomes equal to the shear strength τ_f available there to resist the sliding. Here, the soil fails in *shear*. Along the failure surface, the soil grains slide over each other when shear failure occurs. There is no crushing of soil grains; it takes much larger stresses to crush the grains. Zheng et al. (2017) noted that the grain crushing occurs at stresses in the order of 0.3 MN/m^2 for biogenic carbonate sands to 30 MN/m^2 for quartz sands. In most geotechnical problems, failure occurs in shear—not in compression or tension. This is not the case in other engineering materials such as steel, concrete, or timber, where the failure can occur in compression, tension, or shear.

9.2 Mohr's Circles—A Review

Let us have a brief overview of the stress transformation and the application of Mohr's circles, which are generally covered in subjects such as Strength of Materials or Engineering Mechanics. Figure 9.2 shows the state of stress at a point (shown as an infinitesimal element) with respect to x and y axes in a Cartesian coordinate system. Here, the normal stresses on the x and y planes are σ_x and σ_y, respectively. The shear stresses are τ_{xy}, with the directions as shown. The sign convention used here is as follows:

- Normal stresses are positive when they are compressive.
- Shear stresses are positive when opposite faces produce a counterclockwise couple.

Suppose we know the values of σ_x, σ_y, and τ_{xy} with respect to the x and y axes. The maximum values of σ and τ can be in any direction. If the x and y axes are rotated counterclockwise by an angle θ as shown in the figure, what would be the values of $\sigma_{x'}$, $\sigma_{y'}$, and $\sigma_{x'y'}$ with respect to the new axes x' and y'? In what direction is the normal stress the maximum? In what direction is the shear stress the maximum? What are the magnitudes? Remember, we are still referring to the state of stress at the *same* point. An element is used to visualize the stresses at a point.

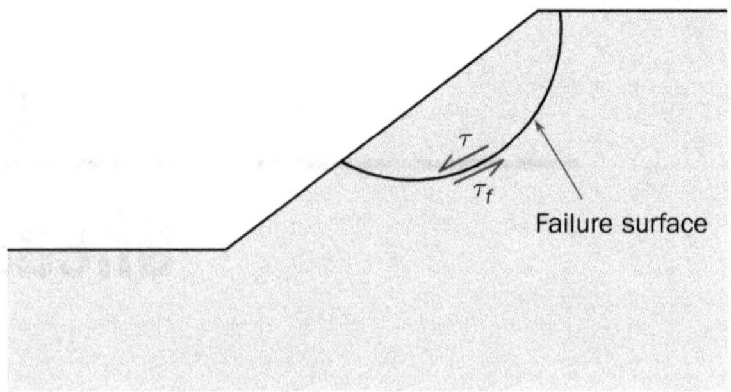

FIGURE 9.1 Shear failure of an embankment.

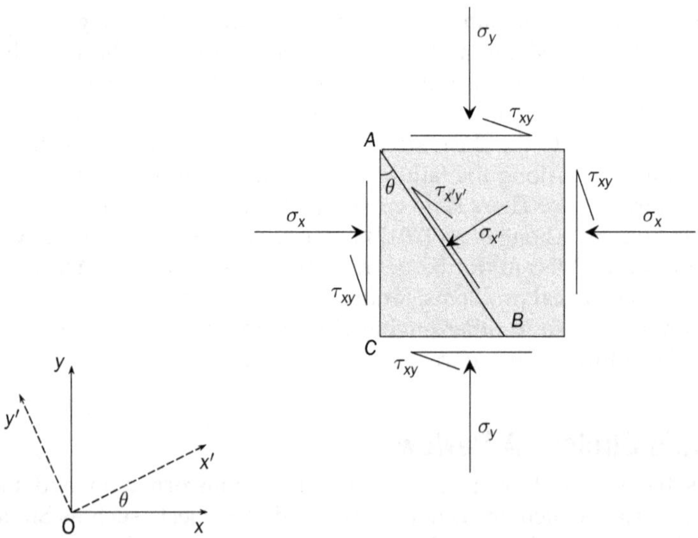

FIGURE 9.2 State of stress at a point.

9.2.1 Stress Transformation

The normal and shear stresses (both shown positive) acting on the plane *AB*, inclined at θ to the vertical, are $\sigma_{x'\sigma'}$ and $\tau_{x'y'}$, respectively. By considering the force equilibrium of the wedge *ABC*, it can be shown that

$$\sigma_{x'} = \left(\frac{\sigma_x + \sigma_y}{2}\right) + \left(\frac{\sigma_x - \sigma_y}{2}\right)\cos 2\theta - \tau_{xy}\sin 2\theta \tag{9.1}$$

$$\sigma_{y'} = \left(\frac{\sigma_x + \sigma_y}{2}\right) - \left(\frac{\sigma_x - \sigma_y}{2}\right)\cos 2\theta + \tau_{xy}\sin 2\theta \tag{9.2}$$

$$\tau_{x'y'} = \left(\frac{\sigma_x - \sigma_y}{2}\right)\sin 2\theta + \tau_{xy}\cos 2\theta \tag{9.3}$$

Here, $\sigma_{y'}$ is the normal stress in the y' direction. Equation (9.2) can be deduced by substituting $90 + \theta$ for θ in Eq. (9.1).

It can be seen from Eq. (9.1) that $\sigma_{x'}$ varies with the angle θ. For the maximum and minimum values for $\sigma_{x'}$,

$$\frac{d\sigma_{x'}}{d\theta} = 0$$

Differentiating Eq. (9.1) with respect to θ and equating it to zero, it can be shown that the maximum and minimum values for $\sigma_{x'}$, known as the *major* (largest) *and minor* (smallest) *principal stresses* σ_1 and σ_3, are given by

$$\sigma_{1,3} = \left(\frac{\sigma_x + \sigma_y}{2}\right) \pm R \tag{9.4}$$

where

$$R = \sqrt{\left(\frac{\sigma_x - \sigma_y}{2}\right)^2 + \tau_{xy}^{\ 2}} \tag{9.5}$$

Principal stresses occur on two mutually perpendicular planes known as *principal planes*, which have no shear stresses.

9.2.2 Mohr's Circles

From Eqs. (9.1) and (9.3),

$$\left[\sigma_{x'} - \left(\frac{\sigma_x + \sigma_y}{2}\right)\right]^2 = \left(\frac{\sigma_x - \sigma_y}{2}\right)^2 \cos^2 2\theta + \tau_{xy}^{\ 2} \sin^2 2\theta - 2\tau_{xy}\left(\frac{\sigma_x - \sigma_y}{2}\right) \sin 2\theta \cos 2\theta \tag{9.6}$$

$$\left[\tau_{x'y'} - 0\right]^2 = \left(\frac{\sigma_x - \sigma_y}{2}\right)^2 \sin^2 2\theta + \tau_{xy}^{\ 2} \cos^2 2\theta + 2\tau_{xy}\left(\frac{\sigma_x - \sigma_y}{2}\right) \sin 2\theta \cos 2\theta \tag{9.7}$$

Adding Eqs. (9.6) and (9.7),

$$\left[\sigma_{x'} - \left(\frac{\sigma_x + \sigma_y}{2}\right)\right]^2 + \left[\tau_{x'y'} - 0\right]^2 = \left(\frac{\sigma_x - \sigma_y}{2}\right)^2 + \tau_{xy}^{\ 2} = R^2 \tag{9.8}$$

Equation (9.8) describes a circle on $\tau_{x'y'}$ versus $\sigma_{x'}$ space, known as *Mohr's circle*, where σ_x, σ_y, τ_{xy}, and R are *known* constants. Every point on the circle represents the $\sigma_{x'}$, and $\tau_{x'y'}$ values on a specific plane in the element, inclined at a specific value of θ. It can be seen from Eq. (9.8) that the radius of the Mohr's circle is R and the coordinates of the center on $\tau_{x'y'}$ versus $\sigma_{x'}$ plane are $(\sigma_x + \sigma_y)/2$ and 0.

Figure 9.3 shows the Mohr's circle representing the state of stress at the same point in Fig. 9.2. The points C and D on the Mohr's circle represent the vertical and horizontal planes, respectively. They are diametrically opposite on the Mohr's circle. Both vertical (and horizontal) planes on the opposite side of the element are represented by

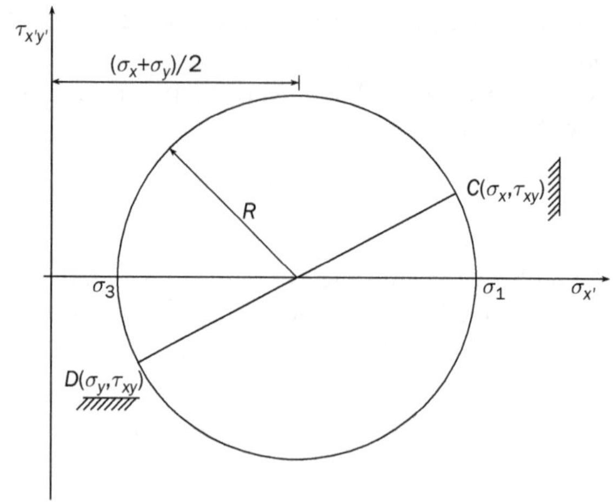

FIGURE 9.3 Mohr's circle for the point in Fig. 9.2.

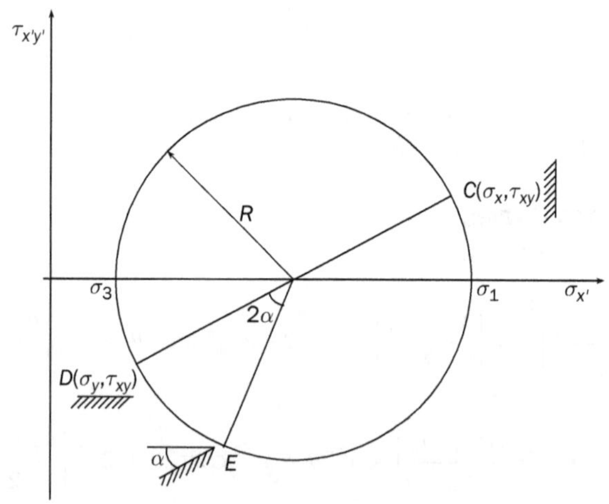

FIGURE 9.4 $\alpha°$ rotation of a plane means $2\alpha°$ rotation in the Mohr's circle.

the same point. The major and minor principal stresses are also shown in the figure. The maximum shear stress is R. Mohr's circle is a good way to represent the state of stress at a point. From the Mohr's circle, it is possible to deduce the $\sigma_{x'}$ and $\tau_{x'y'}$ values in any direction, the principal stresses, and the maximum shear stress.

An important point to note is that as the angle θ is varied from 0° to 180°, the point on the Mohr's circle moves by 360° along the perimeter. Therefore, if the angle θ is increased by $\alpha°$, the point on the Mohr's circle travels in the same direction (clockwise or anticlockwise) by $2\alpha°$. For example, the horizontal and vertical planes with 90° between them are 180° apart on the Mohr's circle and are diametrically opposite. This is illustrated in Fig. 9.4, where the point E on the Mohr's circle represents the plane inclined at angle of $\alpha°$ to the horizontal in counterclockwise direction.

Example 9.1

Figure 9.5a shows the state of stress at a point. Draw the Mohr's circle and find the following.

a. Principal stresses
b. Orientations of the principal planes
c. Maximum shear stress

Solution

Part a Radius of the Mohr's circle is given by

$$R = \sqrt{\left(\frac{\sigma_x - \sigma_y}{2}\right)^2 + \tau_{xy}^{\;2}} = \sqrt{\left(\frac{40 - 120}{2}\right)^2 + 30^2} = 50\,\text{kN/m}^2$$

The coordinates of the center are (40 + 120)/2, 0 (i.e., 80, 0).

From these, the Mohr's circle can be drawn as shown in Fig. 9.5b, and the principal stresses deduced as

$$\sigma_1 = 80 + 50 = \mathbf{130\ kN/m^2}, \textbf{ and } \sigma_3 = 80 - 50 = \mathbf{30\ kN/m^2}.$$

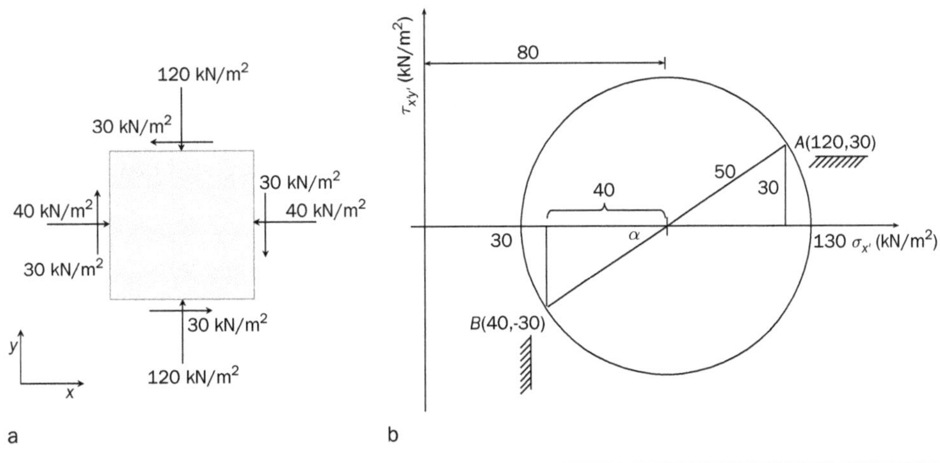

FIGURE 9.5 Example 9.1: (a) State of stress at a point, (b) Mohr's circle.

Part b The horizontal plane (120, 30) and the vertical plane (40, −30) are represented by the points *A* and *B*, respectively, on the Mohr's circle.

Tan $\alpha = 30/40$. Therefore, $\alpha = 36.87°$.

$$\alpha/2 = 18.44°$$

From the Mohr's circle, it can be seen that the vertical plane has to rotate clockwise by $\alpha/2$ (= 18.44°) to reach the minor principal plane. Similarly, the horizontal plane has to rotate clockwise by $\alpha/2$ (= 18.44°) to reach the major principal plane.

Major principal plane (where σ_1 acts) is at 18.44° clockwise to the horizontal.
Minor principal plane (where σ_3 acts) is at 18.44° clockwise to the vertical.

Part c

The maximum shear stress = 50 kN/m²

9.3 Mohr-Coulomb Failure Criterion

Mohr's circle fame Otto Mohr (1900) proposed a shear failure criterion for soils. He suggested that the shear failure takes place along a failure plane where the shear stress reaches the shear strength, and that the shear strength τ_f is a function of the normal stress σ_f on the same failure plane. Therefore,

$$\tau_f = f(\sigma_f) \tag{9.9}$$

The failure envelope suggested by Mohr is not necessarily a straight line. Every soil has a specific failure envelope in the σ_f versus τ_f plane, as shown in Fig. 9.6a. Due to symmetry, only the upper half of the Mohr's circle is shown here, which we will do in the future as well.

The state of stress at a point is fully defined by a Mohr's circle. When the Mohr's circle is within the failure envelope (*B* in Fig. 9.6a), the stresses in all directions are contained within the envelope and hence the point is stable. When the Mohr's circle touches the envelope (*A* in Fig. 9.6a), the point is at the verge of failure, where σ_f and τ_f are the normal and shear stresses on the failure plane as shown in Fig. 9.6b. Well before Mohr's suggestion of Eq. (9.9), Coulomb (1776) noted that shear strength has two separate components, coming from *friction* and *cohesion* and suggested that

$$\tau = \sigma \tan\phi + c \tag{9.10a}$$

where σ = normal stress, ϕ = *angle of internal friction* or simply *friction angle*, $\tan\phi$ = frictional coefficient, and c = *cohesion*. Here, c and ϕ are shear strength parameters that are specific to the soil and are constants. Combining Eqs. (9.9) and (9.10a), Mohr-Coulomb failure criterion can be written as

$$\tau_f = \sigma_f \tan\phi + c \tag{9.10b}$$

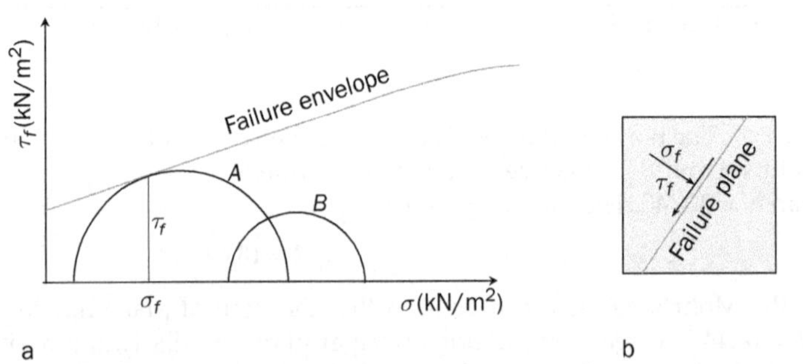

FIGURE 9.6 (a) Failure envelope, (b) failure plane at a point.

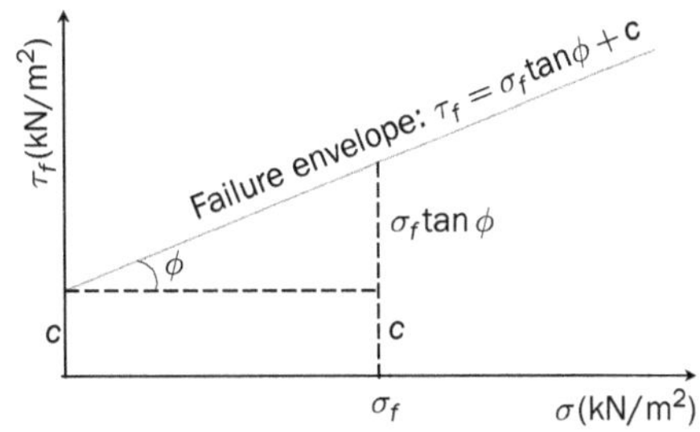

FIGURE 9.7 Mohr-Coulomb failure envelope.

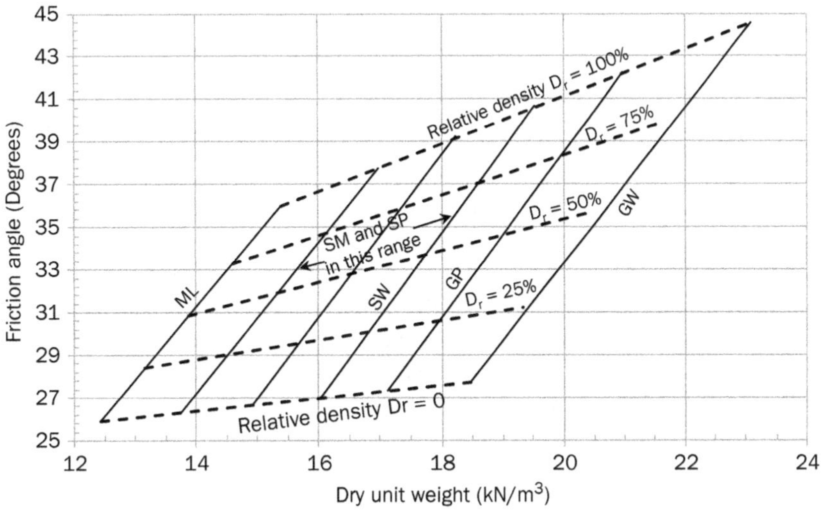

FIGURE 9.8 Friction angles for granular soils. (U. S. Navy, 1982.)

which is an equation of a straight line with slope of tan ϕ and intercept of c on the vertical τ_f axis as shown in Fig. 9.7.

The frictional component of the shear strength ($\sigma_f \tan\phi$) is proportional to the normal stress on the failure plane. The cohesive component of the shear strength (c) remains the same at all values of σ_f. Cohesion can be seen as the "stickiness" of the soil. Granular soils such as gravels and sands have no cohesion (i.e., $c = 0$), and are known as *cohesionless soils*. They are gritty but not sticky. For granular soils, Mohr-Coulomb failure envelope passes through the origin and becomes

$$\tau_f = \sigma_f \tan\phi \tag{9.11}$$

Friction angle of granular soils typically varies in the range of 28° for very loose sands to 45° for very dense gravels. It increases with relative density D_r, grain size, and angularity. Well-graded soils have larger friction angle than if they are poorly graded. Friction angles for granular soils suggested by U. S. Navy (1982) are shown in Fig. 9.8.

Duncan et al. (2014) noted that increasing the confining pressure can cause breakages of the soil grains that leads to a slight reduction in the friction angle and a curved failure envelope. In the case of Oroville dam shell material, consisting of well-graded hard amphibolite grains, they noted a 6.6° reduction for a 10-fold increase in the confining pressure. For rock fills, the friction angles can be large, especially at low confining pressures. Duncan et al. (2014) reported cases where the friction angle exceeded 50° at low confining pressures. Presence of water has no effect on the friction angle. Water only brings in surface tension and pore water pressures, without any measurable lubrication effects between the grains. Peck et al. (1974) noted that the magnitude of the friction angle is not changed appreciably by submergence. Holtz and Kovacs (1981) suggest that wet soils have 1–2° lower friction angle than dry soils.

9.4 A Simple Loading Scenario and Relevance of Mohr's Circle

Figure 9.9a shows an element that is initially under isotropic confining pressure σ_c acting in all directions. The Mohr's circle for this situation is a *point* represented by *A* in Fig. 9.9b. Now, if the vertical stress is increased by $\Delta\sigma$, while σ_3 remains the same at σ_c, σ_1 becomes $\sigma_c + \Delta\sigma$, and the Mohr's circle expands on the right. When $\Delta\sigma$ reaches $\Delta\sigma_f$, the Mohr's circle is tangent to the failure envelope and the element fails along a plane shown in Fig. 9.9c.

Let us have a close look at the Mohr's circle at failure, shown in Fig. 9.10a. The failure corresponds to point *F*, where the failure envelope is tangent to the Mohr's circle. The values of normal and shear stresses at failure on the failure plane are σ_f and τ_f, respectively. The radial line *OF* is perpendicular to the failure envelope. It can be seen from the figure that,

$$\tau_f = R\cos\phi \tag{9.12}$$

$$\sigma_f = \sigma_c + R - R\sin\phi \tag{9.13}$$

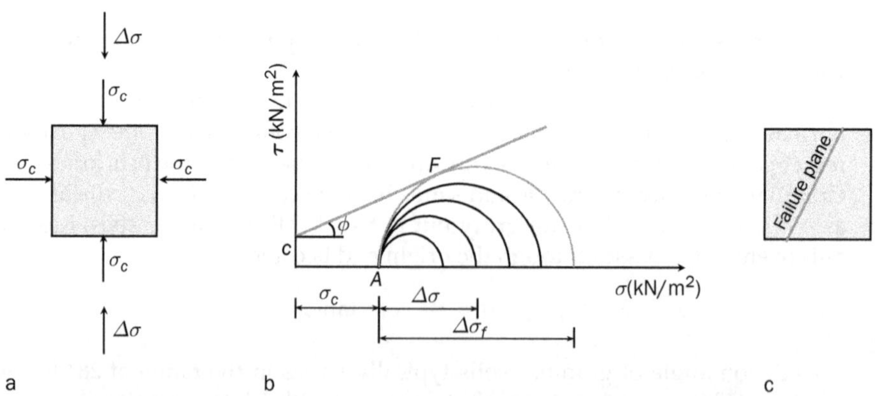

FIGURE 9.9 A simple loading: (a) state of stress at a point, (b) Mohr's circle, (c) failure plane.

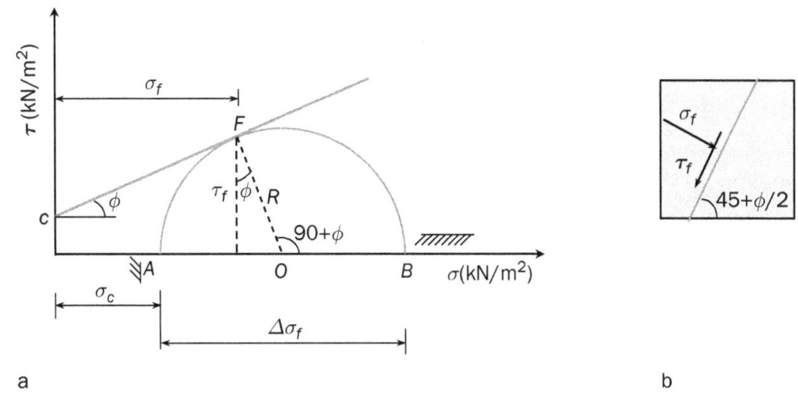

FIGURE 9.10 (a) Mohr's circle at failure, (b) element at failure showing the failure plane.

where R is the radius of the Mohr's circle, given by $R = \Delta\sigma_f/2$. The maximum shear stress within the element at failure is given by

$$\tau_{\max} = R \tag{9.14}$$

The principal stress difference at any time during the loading is $\Delta\sigma$, which is sometimes called the *deviator stress*. The principal stress difference being the Mohr's circle's diameter, is also twice the maximum shear stress within the element. Therefore, the diameter of the Mohr's circle at failure is a clear measure of the shear strength of the soil. It can also be seen that the larger the confining pressure σ_c, the larger the Mohr's circle and the larger the principal stress difference at failure $\Delta\sigma_f$. A soil element located at larger depth in the ground and hence consolidated *in situ* under a larger overburden pressure will have a larger shear strength.

The points A and B on the Mohr's circle represent the vertical and horizontal planes, respectively. The angle BOF is $90 + \phi$. Therefore, the horizontal plane has to rotate by an angle $(90 + \phi)/2$ counterclockwise to coincide with the failure plane. Thus, the failure plane in Fig. 9.10b is inclined at $45 + \phi/2$ to the horizontal.

Example 9.2

A sand specimen is initially under an isotropic confining pressure of 100 kN/m². Later, the specimen is subjected to an additional vertical stress, which was increased from 0 to 240 kN/m² when failure occurred.

a. What is the friction angle of the sand?
b. What are the normal and shear stresses on the failure plane?
c. What is the orientation of the failure plane?
d. When failure occurs, what is the maximum shear stress within the specimen?

Solution

Part a From Fig. 9.11a,

$$\sin\phi = \frac{OF}{OD} = \frac{120}{100+120} = 0.546$$

Therefore, $\phi = \mathbf{33.1°}$

Part b From Eq. (9.13) or Fig. 9.11a,

$$\sigma_f = \sigma_c + R - R\sin\phi = 100 + 120 - 120\sin 33.1 = \mathbf{154.5\ kN/m^2}$$

From Eq. (9.12) or Fig. 9.11a,

$$\tau_f = R\cos\phi = 120\cos 33.1 = \mathbf{100.5\ kN/m^2}$$

Part c

$$45 + \frac{33.1}{2} = 61.6$$

The failure plane is inclined to the horizontal at 61.6°.

Part d Maximum shear stress within the element is **120 kN/m²**.

Note that the failure does not occur on the plane inclined at 45° to horizontal where the shear stress is the maximum. The reason is that, on this plane the shear strength available is more than the shear stress (see the Mohr's circle).

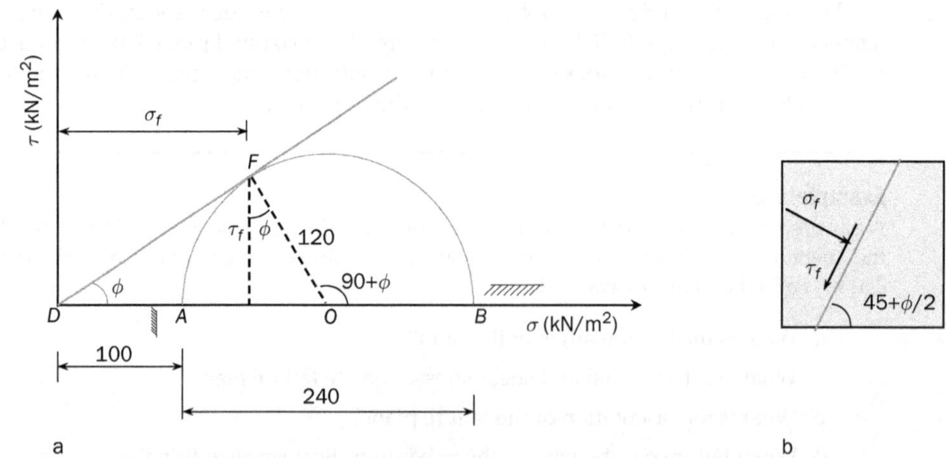

FIGURE 9.11 Example 9.2: (a) Mohr's circle, (b) failure plane.

9.5 Mohr's Circles and Failure Envelopes in Terms of Total and Effective Stresses

Let us consider the total stresses, effective stresses and the pore water pressure at a point in a saturated soil mass as shown in Fig. 9.12. Here, the vertical and horizontal normal stresses are taken as the major and minor principal stresses, respectively. There are no shear stresses on these two planes, which are principal planes. Using the major and minor principal stresses, the Mohr's circles in terms of total or effective stresses can be drawn as shown in the figure.

$$\sigma_1 = \sigma_1' + u$$

$$\sigma_3 = \sigma_3' + u$$

$$\sigma_1 - \sigma_3 = \sigma_3' - \sigma_3' \tag{9.15}$$

From Eq. (9.15), the two Mohr's circles have the same diameter. As shown in the figure, they are separated by the pore water pressure "u." This is true at any point within the soil mass. When the pore water pressure is negative, the effective stresses are larger than the total stresses.

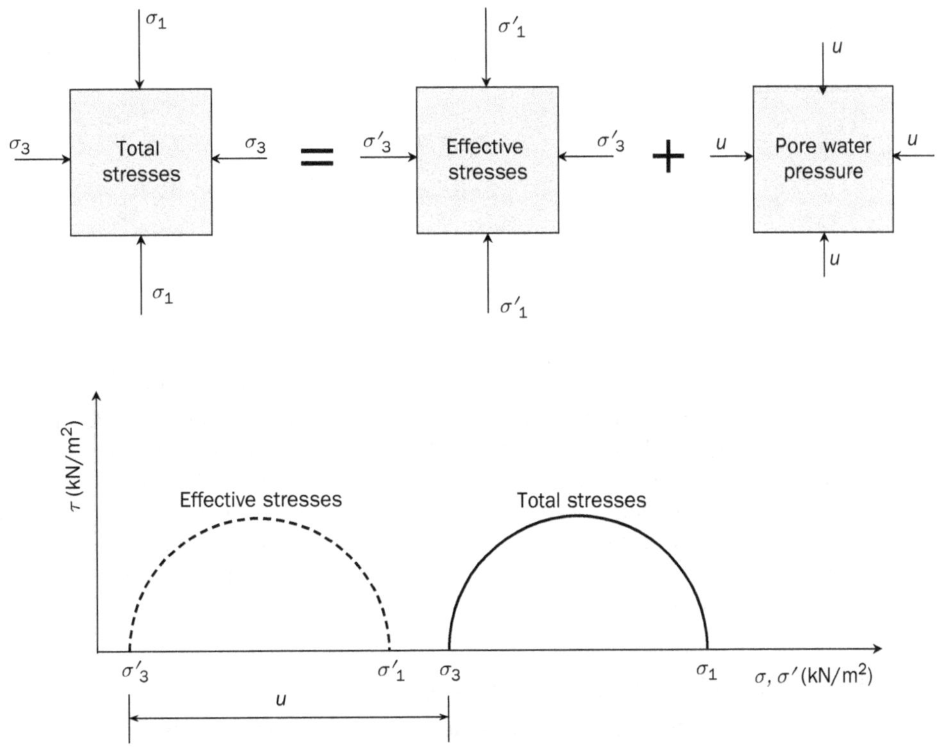

FIGURE 9.12 Mohr's circles in terms of total and effective stresses.

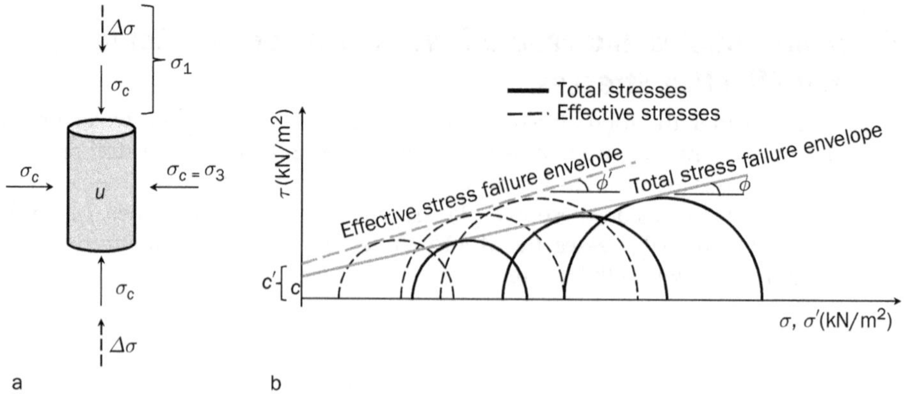

FIGURE 9.13 (a) Specimen and the stresses at failure, (b) failure envelopes in terms of total and effective stresses.

Figure 9.13a shows a cylindrical soil specimen subjected to an initial isotropic confining pressure of σ_c, which is later subjected to an additional vertical stress $\Delta\sigma$, which is increased steadily from zero till failure occurs. This is the same loading scenario discussed in Sec. 9.4. The pore water pressure u within the specimen varies during loading. The confining pressure σ_c remains constant throughout the loading. At any time during the loading,

$$\sigma_3 = \sigma_c$$

$$\sigma_1 = \sigma_c + \Delta\sigma$$

$$\sigma'_c = \sigma_c - u$$

$$\sigma'_c = \sigma_c + \Delta\sigma - u$$

When failure occurs, the stresses are distinguished with a subscript "f" as σ_{1f}, σ_{3f}, σ'_{1f}, σ'_{3f}, $\Delta\sigma_f$, and u_f. Figure 9.13b shows the Mohr's circles at failure for three identical samples but at different confining pressures and hence failing at different deviator stresses. The figure shows two sets of Mohr's circles. The ones in solid lines are the Mohr's circles in terms of the total stresses, and the ones in dashed lines are the Mohr's circles in terms of effective stresses. Separate failure envelopes, tangent to the Mohr's circles, are drawn in terms of total and effective stresses, defining c, ϕ, c', and ϕ'. The first two are the shear strength parameters in terms of total stresses, and the last two are shear strength parameters in terms of effective stresses.

9.6 Drained and Undrained Loadings

When a saturated soil element is subjected to external loads, excess pore water pressure develops. In granular soils, the excess pore water pressure dissipates almost immediately. In the case of clays, the dissipation can take a long time. In the *short term*, i.e., immediately after loading, the clay does not allow any water to be squeezed out, and is said to be *undrained*. In the *long term*, i.e., after a very long time, the *excess* pore water pressure has fully dissipated and the clay is said to have *drained* over this long period, with the pore water pressure reaching the initial in situ value. In addition, if the loading is very slow such that there is no excess pore water pressure developing, the situation can be treated as drained loading.

When analyzing the long-term stability (i.e., drained loading) of a foundation or an embankment on a clay, an *effective stress analysis* is commonly carried out using c' and ϕ'. In the case of short-term or undrained loading, the common practice is to carry out a *total stress analysis*, using total stress shear strength parameters. Here, no attempt is made to separate the effective stresses and the pore water pressures; soil is treated as a saturated homogeneous continuum that does not undergo volume change. Silts can behave drained or undrained and hence it is a good practice to test both conditions. Granular soils are always analyzed in terms of effective stresses and considered fully drained in all circumstances.

The undrained tests in the laboratory and field are carried out at a faster rate where the clay is sheared within few minutes. In reality, the field loading takes several days and hence the failure occurs over a long time and at a slower strain rate. Kulhawy and Mayne (1990) suggest that undrained strength increases by 10% for 10-fold increase in strain rate.

In deciding whether a soil is loaded under drained or undrained conditions, a close look at the dimensionless time factor $T = c_v t / H_{dr}^2$ is useful. Here, c_v is the coefficient of consolidation, t is the loading duration, and H_{dr} is the maximum length of the drainage path. When the time factor is less than 0.1, the loading can be treated as undrained, and when it is larger than 2, it can be treated as drained.

9.7 Triaxial Test

To simulate the loading scenario discussed in Sec. 9.4 on multiple soil specimens subjected to different confining pressures, the triaxial test apparatus shown in Fig. 9.14a is widely used. Here, it is also possible to simulate drained or undrained loadings to determine the different parameters. The concept is fairly straightforward. As discussed with Fig. 9.13, three or more soil specimens are tested at different confining pressures and the failure values are used to define the Mohr's circles in terms of total and effective stresses, and hence the two failure envelopes.

Typically, the soil specimens are 38–50 mm in diameter, twice that in length. On special occasions, larger specimens can be tested. The specimen is enclosed in an impermeable membrane, with porous stones at the top and bottom to facilitate drainage and pore water pressure measurements. The two o-rings at the top and bottom provide the soil specimen a watertight seal, allowing drainage only from the top and bottom of the specimen. The specimen is surrounded by pressurized water within the cylindrical cell, applying the isotropic cell pressure σ_c. The additional vertical stress (i.e., deviator stress or the principal stress difference) is applied through a piston at the top, in the form of an axial load. Figure 9.14b shows a photograph of a soil specimen enclosed within a membrane being setup for a triaxial test.

As discussed in Secs. 9.4 and 9.5, each soil specimen is loaded in two separate stages:

a. Isotropic cell pressure σ_c is applied right around the specimen in all directions. During the application of σ_c, the drainage valve can be open to consolidate the

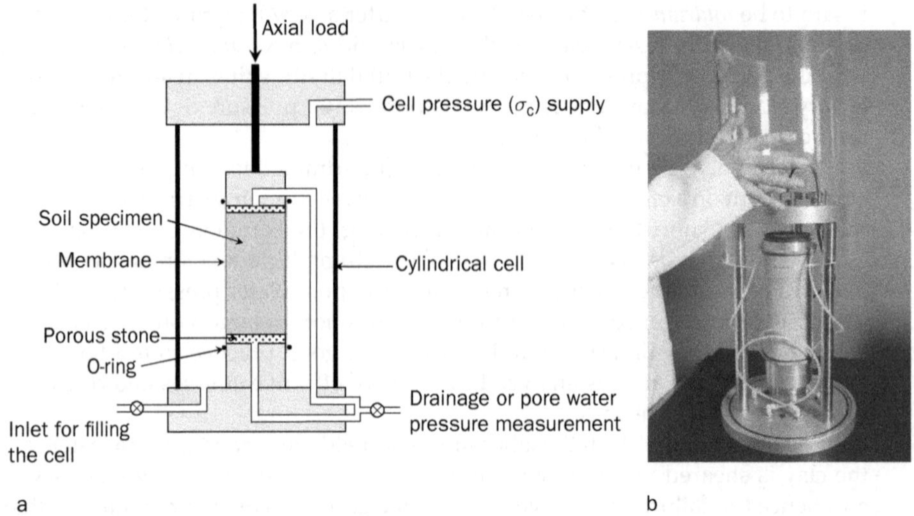

Figure 9.14 Triaxial apparatus: (a) schematic diagram, (b) photograph. (Courtesy of Mr. Rodrigo Palma, Business Development and Technical Manager, Geo-Con Products Pty Ltd., Australia.)

specimen or it can be closed so that the saturated soil specimen does not consolidate at any confining pressure. Depending on this (i.e., drainage conditions), the specimen is consolidated or unconsolidated.

b. The additional vertical stress $\Delta\sigma$ is applied and increased from zero till failure occurs at $\Delta\sigma_f$. During this period, the drainage valve can be opened to allow drainage and hence permit drained loading or it can be closed to ensure undrained loading. Therefore, depending on the drainage conditions, the specimen is loaded drained or undrained. The loading rate is relatively slow for drained loading, and fast for undrained loading. This second stage of loading is "shearing" where the shear stress is increased from zero till failure occurs.

Depending on the drainage conditions during the two loading stages, three possible combinations are relevant to the field situations. They are:

i. Consolidated drained (CD) triaxial test
ii. Consolidated undrained (CU) triaxial test (ASTM D 4767-11)
iii. Unconsolidated undrained (UU) triaxial test (ASTM D 2850-15)

9.7.1 Consolidated Drained Triaxial Test

In a consolidated drained triaxial test, with the drainage valve open, the soil specimen is isotropically consolidated under a specific confining pressure σ_c. Once consolidation is completed, with the drainage valve still open, the specimen is subjected to an additional vertical stress $\Delta\sigma$, which is increased slowly until failure occurs, such that there is no excess pore water pressure build-up during the loading. During shearing, since

$u = 0$, $\sigma = \sigma'$. Therefore, the Mohr's circles at failure are the same in terms of total stresses and effective stresses, and hence there is only one envelope in terms of both total and effective stresses, defining c' and ϕ'.

When the soil specimen being tested contains entrapped air pockets or it is unsaturated, the applied load is partially carried by the compressed air bubbles, in the form of pore air pressure, which makes the analysis complex. Therefore, the common technique is to saturate the specimen by pressurizing the water within the voids, which dissolves the entrapped air bubbles. This is often accomplished by the use of *back pressure* u_0, which is applied into the pore water in the soil specimen through the drainage valve. In a triaxial setup, the back pressure and the confining pressure are increased simultaneously in small steps such that the cell pressure is always greater than the back pressure. The difference is the effective consolidation pressure σ_c', which has to be positive to ensure that the specimen does not collapse. It is the net or effective confining pressure σ_c' that causes consolidation.

$$\sigma_c' = \sigma_c - u \tag{9.16}$$

In a CD triaxial test, the back pressure is maintained during consolidation and also while applying the deviator stress to shear the specimen.

9.7.2 Consolidated Undrained Triaxial Test

In a consolidated undrained triaxial test, firstly the soil specimen is isotropically consolidated under a specific confining pressure σ_c. When the consolidation is complete, the drainage valve is closed and the specimen is subjected to an additional vertical stress $\Delta\sigma$, which is increased quickly under undrained condition until failure occurs, with pore water pressure building up during the undrained loading. The total and effective stresses at failure are different and hence there are two sets of Mohr's circles from which c, ϕ, c', and ϕ' can be determined.

The back pressure u_0 applied throughout the consolidation process is locked into the specimen when carrying out the undrained loading. Therefore, any excess pore water pressure developed during the undrained loading is in addition to the back pressure. Therefore, the pore water pressure at failure u_f is given by,

$$u_f = u_0 + \Delta u_f \tag{9.17}$$

where Δu_f is the excess pore water pressure *developed during the application* of $\Delta\sigma_f$.

Either CD or CU triaxial test can be carried out to determine the effective stress shear strength parameters c' and ϕ'. CU test is quicker and cheaper than a CD test for this purpose. Any effective stress analysis has to be carried out using appropriate values of c' and ϕ'. For normally consolidated clays and granular soils, $c' = 0$. For compacted clays and overconsolidated clays, c' can be in the range of 0–25 kN/m^2. The friction angle ϕ' for granular soils can be in the range of 28–45°. It is less for clays. For natural intact normally consolidated clays, ϕ' can vary from less than 20° to little more than 30°. For clays, ϕ' decreases with increasing plasticity. For compacted clays, ϕ' is typically in the range of 25–30°, but can be slightly higher.

Australian Standards for retaining walls (AS 4678) suggests the values for c' and ϕ' in Table 9.1.

		Soil parameters	
Soil group	Typical soils in group	c' (kN/m^2)	ϕ' (deg.)
Poor	Soft and firm clay of medium to high plasticity; silty clays; loose variable clayey fills; loose sandy silts	0–5	17–25
Average	Stiff sandy clays; gravelly clays; compact clayey sands and sandy silts; compacted clay fills	0–10	26–32
Good	Gravelly sands, compacted sands, controlled crushed sandstone and gravel fills, dense well-graded sands	0–5	32–37
Very good	Weak weathered rock, controlled fills of road base, gravel and recycled concrete	0–25	36–43

TABLE 9.1 Typical Values of c' and ϕ' (After AS 4678, 2002)

Material	Poisson's ratio, ν
Saturated clays (undrained)	0.5
Saturated clays (drained)	0.2–0.4
Dense sand	0.3–0.4
Loose sand	0.1–0.3
Loess	0.1–0.3
Rocks	0.1–0.3
Ice	0.36
Aluminum	0.35
Steel	0.29
Concrete	0.15

TABLE 9.2 Typical Values of Poisson's Ratio

Kulhawy and Mayne (1990) suggested that the drained Poisson's ratio ν_d of a granular soil can be estimated using the ϕ' determined from a triaxial compression test as

$$\nu_d = 0.1 + 0.3\left(\frac{\phi' - 25}{20}\right) \text{ for } 25^\circ < \phi' < 45^\circ \tag{9.18}$$

Equation (9.18) suggests that higher the friction angle, higher the drained Poisson's ratio. From the theory of elasticity, without considering whether drained or undrained, it can be shown that

$$\nu = \frac{1 - \sin\phi'}{2 - \sin\phi'} \tag{9.19}$$

Table 9.2 summarizes some typical values of Poisson's ratios for different materials.

9.7.3 Unconsolidated Undrained Triaxial Test

Unconsolidated undrained triaxial tests are generally carried out on saturated clay specimens at natural (in situ) moisture content to simulate undrained loading. The specimens taken from below the water table are generally saturated. The drainage valve is closed from the start and hence the specimen does not consolidate when subjected to the isotropic confining pressure σ_c, which is mainly transferred to the pore water if the specimen is saturated. In the second stage, an additional vertical stress $\Delta\sigma$ is applied and increased quickly under undrained condition until failure occurs, with pore water pressure building up during the loading. The pore water pressure is never measured at any stage and hence the effective stresses are not known during the test and at failure. The Mohr's circles are drawn in terms of total stresses only. For saturated soil specimens, irrespective of the magnitude of the confining pressure, the additional stress required to cause failure $\Delta\sigma_f$ remains the same. Thus, the failure Mohr's circles are of the same diameter for any confining pressure as shown in Fig. 9.15. For all three total stress circles, the effective stresses are the same and hence there is only one effective stress circle, which remains unknown in a UU test. The total stress and effective stress Mohr's circles are separated by the pore water pressures at failure, denoted by u_1, u_2, and u_3, for the three tests.

As shown in Fig. 9.15, the failure envelope in terms of the total stresses is horizontal with the shear strength parameters c_u and ϕ_u (= 0). The subscript "u" denotes undrained loading. c_u is known as *undrained shear strength* or *undrained cohesion*, which is related to the initial moisture content of the clay specimens. A short-term analysis assuming undrained loading is generally carried out in terms of total stresses with appropriate value for c_u, with $\phi_u = 0$.

When UU tests are carried out on unsaturated specimens, higher confining pressures cause more consolidation and hence larger Mohr's circle, making the envelope no more horizontal.

Based on the Danish code of practice for foundations, Sorensen and Okkels (2013) suggest that a conservative estimate of c' for overconsolidated clays can be obtained from

$$c' = 0.1c_u \tag{9.20}$$

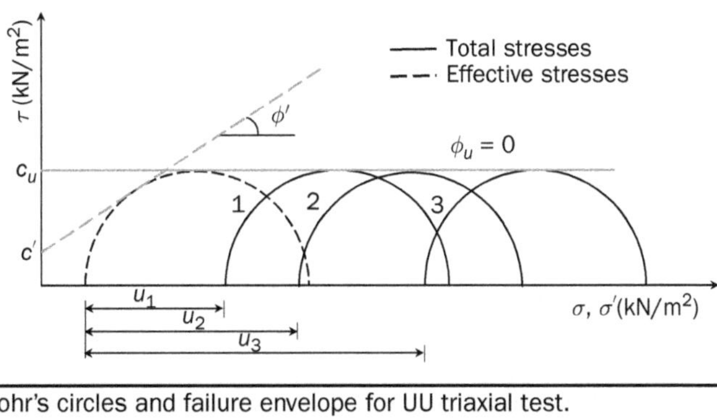

FIGURE 9.15 Mohr's circles and failure envelope for UU triaxial test.

9.7.4 Unconfined Compression Test

Unconfined compression test (ASTM D 2166-16), sometimes called *uniaxial compression test*, is a special case of a UU triaxial test. It can only be carried out on clay specimens that can stand unsupported. The test does not require the triaxial setup and hence is very much simpler than a UU test. There is no confining pressure applied (i.e., $\sigma_3 = \sigma_c = 0$) and the vertical stress is increased till failure occurs at $\Delta\sigma_f$, which is known as *unconfined compressive strength* q_u. With $\sigma_3 = 0$ and $\sigma_1 = q_u$, it can be seen from Fig. 9.16 that unconfined compressive strength q_u and undrained shear strength (also called *undrained cohesion*) c_u are related by

$$c_u = \frac{1}{2} q_u \tag{9.21}$$

The undrained cohesion c_u is a necessary parameter for a short-term stability analysis. It is very much easier to determine than the effective stress parameters c' and ϕ'. There are simple empirical correlations that can be used for a first estimate of c_u. Skempton (1957) suggested that for normally consolidated clays,

$$\left(\frac{c_u}{\sigma'_{\mathrm{vo}}}\right)_{NC} = 0.0037\,PI + 0.11 \tag{9.22}$$

where σ'_{vo} is the effective vertical overburden stress, and *PI* is the plasticity index of the clay. In overconsolidated clays (Ladd et al., 1977),

$$\left(\frac{c_u}{\sigma'_{\mathrm{vo}}}\right)_{OC} = \left(\frac{c_u}{\sigma'_{\mathrm{vo}}}\right)_{NC} OCR^{0.8} \tag{9.23}$$

where *OCR* is the overconsolidation ratio. For clays of low to moderate plasticity (Jamiolkowski et al., 1985)

$$\left(\frac{c_u}{\sigma'_{\mathrm{vo}}}\right)_{OC} = (0.23 \pm 0.04)OCR^{0.8} \tag{9.24}$$

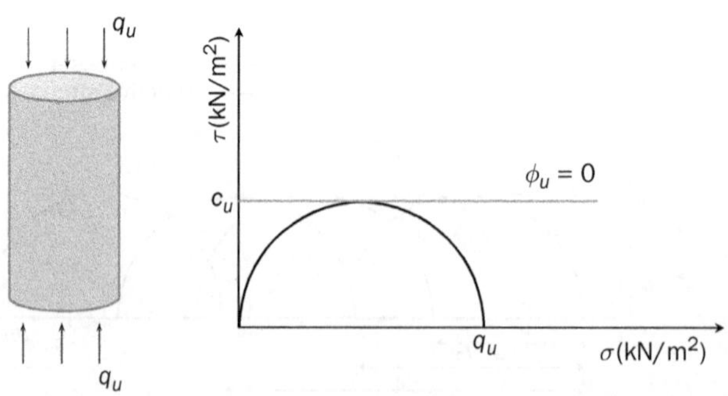

Figure 9.16 Unconfined compression test.

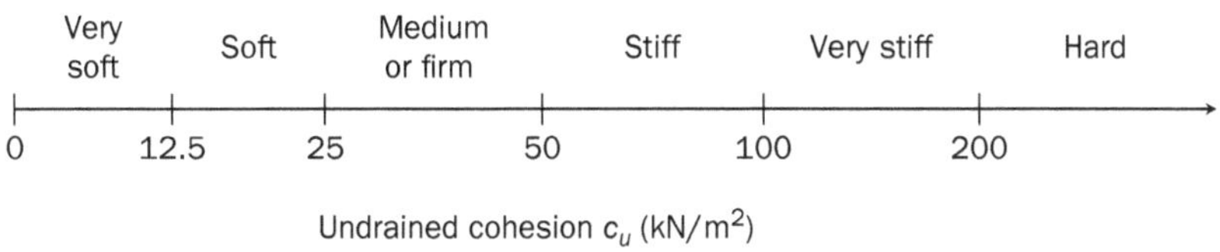

FIGURE 9.17 Consistency terms for clays based on c_u.

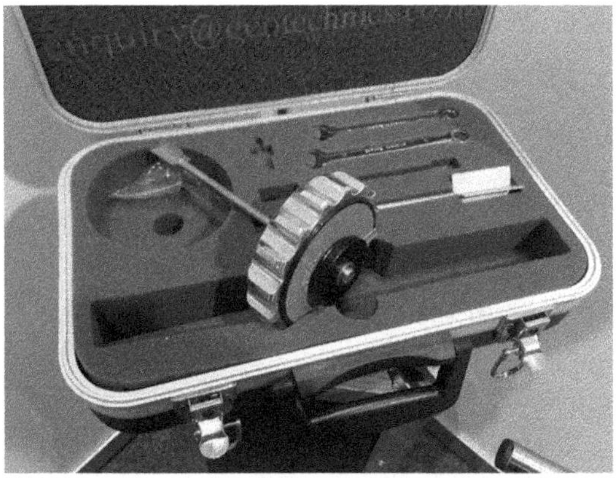

FIGURE 9.18 Laboratory vane to measure undrained shear strength.

Mesri (1989) suggested that the preconsolidation pressure σ'_p and undrained cohesion c_u are related by

$$\frac{c_u}{\sigma'_p} = 0.22 \tag{9.25}$$

Figure 9.17 shows the terms that can be used to describe the consistency of a clay based on the undrained cohesion c_u.

Very soft clay is of toothpaste or jelly-like consistency. If a shovel of clay is dug out from the ground, the hole is slowly filled by the clay flowing in from the surrounding area. With increasing c_u, the clays can be soft, medium (or firm), stiff, very stiff, or hard.

Unconfined compressive strength q_u is often estimated at the site or from the clay samples, using pocket penetrometers. The pocket penetrometer is pushed into the walls of the excavations or the sides of the clay samples, and the unconfined compressive read off the scale directly. These measurements are very approximate only. Similarly, handheld torvane, or laboratory miniature vane (Fig. 9.18) can be used for this purpose. The torque required to turn the vane pushed into the clay is converted into the undrained shear strength. The undrained shear strength is proportional to the torque required to turn the vane.

Example 9.3

The soil profile at a site consists of 3 m of sand ($\gamma_m = 17.0$ kN/m^3) at the top, underlain by a thick clay deposit ($\gamma_{sat} = 20.0$ kN/m^3; $PI = 32$). The water table is at the top of the clay layer. What would be the undrained shear strength c_u of the clay at 5 m below the water table if

a. the clay is normally consolidated?

b. the clay is overconsolidated with $OCR = 2$?

Solution

Part a At 5 m below the water table, the effective overburden stress is given by

$$\sigma'_{vo} = (3)\,(17) + (5)\,(20 - 9.81) = 101.95 \text{ kN/m}^2$$

From Skempton (1957),

$$\left(\frac{c_u}{\sigma'_{vo}}\right)_{NC} = 0.0037\,PI + 0.11 = 0.0037 \times 32 + 0.11 = 0.228$$

Therefore, $c_u = (0.228)\,(101.95) =$ **23.2 kN/m²**

Part b From Ladd et al. (1977),

$$\left(\frac{c_u}{\sigma'_{vo}}\right)_{OC} = \left(\frac{c_u}{\sigma'_{vo}}\right)_{NC} OCR^{0.8} = (0.228)\left(2^{0.8}\right) = 0.397$$

Therefore, $c_u = (0.397)\,(101.95) =$ **40.5 kN/m²**

Example 9.4

The data from a series of consolidated undrained triaxial test are summarized below. Draw the three Mohr's circles and plot the failure envelope in terms of effective stresses, and find c' and ϕ'.

Sample number	Confining pressure (kN/m²)	Deviator stress at failure (kN/m²)	Pore water pressure at failure (kN/m²)
1	100	88.2	57.4
2	200	138.5	123.7
3	300	232.1	208.8

Solution The table on the next page can be prepared from the given data, and the Mohr's circles are plotted as shown in Fig. 9.19.

The failure envelope is drawn tangent to the Mohr's circles in the figure and from measurements, $c' = 10$ kPa and $\phi' = 24.7°$.

Sample no.	σ_3 (kN/m²)	$\Delta\sigma_f$ (kN/m²)	u_f (kN/m²)	σ_1 (kN/m²)	σ_3' (kN/m²)	σ_1' (kN/m²)
1	100	88.2	57.4	188.2	42.6	130.8
2	200	138.5	123.7	338.5	76.3	214.8
3	350	232.1	208.8	582.1	141.2	373.3

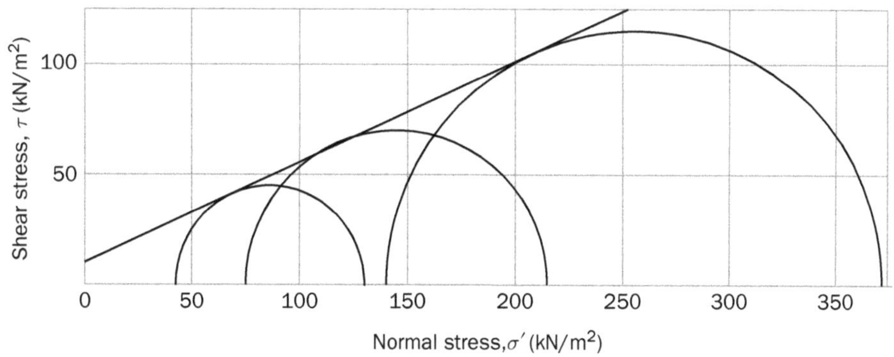

FIGURE 9.19 Mohr circles and the failure envelope.

9.8 Direct Shear Test

Direct shear test (ASTM D 3080-11) is mostly carried out on granular soils and sometimes on cohesive soils too. The problem with cohesive soils is in controlling the strain rates to achieve drained or undrained loading. In the case of granular soils, loading is always assumed drained. Schematic diagram of the shear box is shown in Fig. 9.20a. The soil specimen is placed in a square (sometimes circular) box, approximately 60 mm × 60 mm in plan, which is split into upper and lower halves as shown. One of the halves (lower in the figure) is fixed and the other is pushed or pulled horizontally relative to the other half, thus *forcing* the soil sample to shear (fail) along the horizontal plane separating the two halves. Under a specific normal load (N), the shear load (S) is increased from zero until the specimen is fully sheared. During the test, the horizontal (δ_{hor}) and vertical (δ_{ver}) deformations of the specimen are recorded continuously using two separate dial gauges, along with the shear load using a load cell or proving ring. Normal stress and shear stress on the horizontal failure plane are calculated as $\sigma = N/A$ and $\tau = S/A$, where A is the plan area of the sample, which decreases slightly with the horizontal deformation, requiring a correction. A photograph of the direct shear test apparatus is shown in Fig. 9.20b.

While the normal stress remains constant, the shear stress changes during the loading. The shear stress is plotted against the horizontal displacement to define the shear

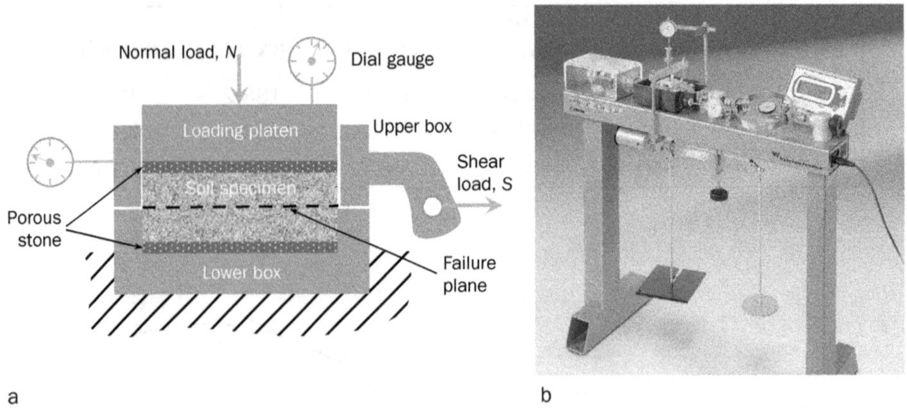

FIGURE 9.20 Direct shear test apparatus: (a) schematic diagram, (b) photograph. (Courtesy of Mr. Rodrigo Palma, Business Development and Technical Manager, Geo-Con Products Pty Ltd., Australia.)

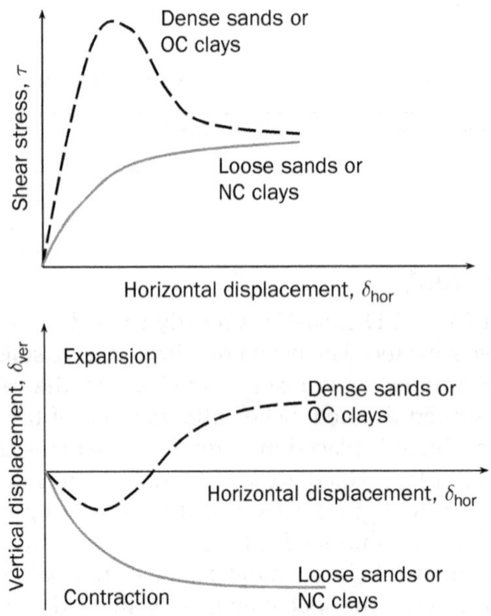

FIGURE 9.21 Typical variations of τ and δ_{ver} against δ_{hor}.

strength at the specific normal stress. By repeating this for three different values of normal stress, the failure envelope can be defined and the cohesion and friction angle be determined. Figure 9.21 shows the typical τ versus δ_{hor} and δ_{ver} versus δ_{hor} plots for different soils. It can be seen that loose sands and normally consolidated clays behave similarly; dense sands and overconsolidated clays behave similarly. In the direct shear test, there is a slight reduction in the cross-sectional area of the specimen during the shear, which can be corrected in computing the stresses.

In dense sands and overconsolidated clays, the specimen contracts initially, and once it reaches a stage where it cannot contract any more, it begins to expand which is called *dilation*. The shear stress increases to a peak value and decreases to a constant value at large strain. How do we define failure here—the peak or the final value? This is discussed in Sec. 9.9. In loose sands and normally consolidated clays, the specimen contracts throughout the loading, and the shear stress increases to an asymptotic value at large strains, which is also the peak.

9.9 Peak, Residual, and Critical States

Figure 9.22 shows typical stress-strain plots of two different types of soils: (a) normally consolidated clays or loose sands, and (b) overconsolidated clays or dense sands. The first one exhibits contractive (i.e., volume reduction) behavior throughout the shearing process, and the shear stress peaks at *critical state* which occurs at relatively large strains. At critical state, the soil deforms under constant volume or void ratio. The second type of soil exhibits some initial contraction followed by dilative (i.e., volume increase) behavior during the shear, and the peak value of the shear stress is reached at a relatively small strain in the order of 1–2%. The shear stress reduces on further straining, and reaches the critical state value, which occurs at strains in the order of 10–20%. The friction angle and cohesion derived based on the peak shear stress values are denoted by ϕ'_{peak} and c'_{peak}, respectively.

Irrespective of the initial state (i.e., normally consolidated or overconsolidated clay; or loose or dense sand), the shear stress and the void ratio reach constant values at *critical state*. The friction angle and the void ratio at the critical state are known as the *critical state friction angle* ϕ'_{cv} and *critical state void ratio* e_{cv}, respectively, where "cv" stands for constant volume or critical state void ratio. ϕ'_{cv} largely depends on the mineralogy of the soil grains.

At very large strains, clay soils would go from critical state to *residual state*, where the friction angle is known as the *residual friction angle* ϕ'_{res}. In clays, strains in the order of 100% or more are required to remold the clay and reach residual state (Kulhawy and Mayne, 1990). In granular soils, residual state is not far from the critical state, where it can be assumed that $\phi'_{res} \approx \phi'_{cv}$. For cohesive soils ϕ'_{res} can be several degrees lower than ϕ'_{cv}. For all soils in general, $\phi'_{res} < \phi'_{cv} < \phi'_{peak}$. ϕ'_{res} and hence the residual shear strength depends largely on the mineralogy of the grains, grain sizes and shapes, clay fraction, and plasticity. At large strains, the fabric of the clay is destroyed and there is no cohesion (i.e., $c'_{res} = c'_{cv} = 0$).

One can define failure in terms of the shear stresses at peak, critical, or residual state. The corresponding shear strength parameters c' and ϕ' derived will also be different. Which of these should we use in the designs? It depends on the strain level—i.e., where we are on the stress-strain diagram. In most geotechnical problems such as foundations, strains are relatively small, and well within the peak. Therefore, peak shear strength parameters ϕ'_{peak} and c'_{peak} are commonly used in geotechnical engineering. Only in large strain problems such as slope stability or landslides, the strains are large and the soil has strained well beyond the peak. Here, it is appropriate to use the residual or critical state shear strength parameters. In this book and in most literature, unless stated otherwise, the default cohesion and friction angle used in empirical equations and design charts are the *peak* values c'_{peak} and ϕ'_{peak} [e.g., Eq. (9.20), Table 9.1, and Fig. 9.8].

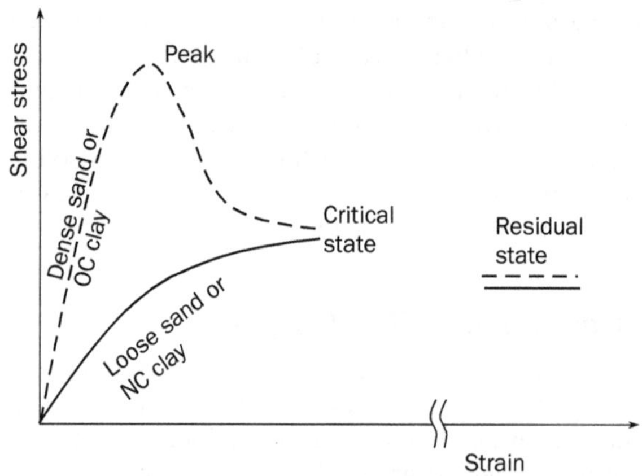

Figure 9.22 Definition of peak, residual, and critical states.

When a dense sand undergoes shearing, its volume increases and this behavior is known as dilatancy. The dilatancy angle ψ is defined as

$$\psi = \tan^{-1}\frac{d\varepsilon_v}{d\gamma} \tag{9.26}$$

where ε_v = volumetric strain, and γ = shear strain. The peak friction angle, the critical state friction angle, and the dilatancy angle are related by

$$\phi'_{\text{peak}} = \phi'_{\text{cv}} + \psi \tag{9.27}$$

The dilatancy angle ψ (pronounced psi) is an important input parameter in most numerical modeling work, in defining the constitutive model to describe the soil behavior. It typically varies in the range of 0–15° for granular soils. Except for heavily overconsolidated clays, clays in general show little dilatancy.

9.9.1 Typical Values

The effective friction angle generally decreases with increasing plasticity index. For normally consolidated clays, based on the data from Bjerrum and Simmons (1960), Duncan et al. (2014) suggest the values given in Table 9.3.

Based on decades of triaxial compression test data on clays from Denmark, Sorensen, and Okkels (2013) suggested the following three Eqs. (9.28)–(9.30).

$$\phi'_{\text{peak}} = 43 - 10\log PI \quad \text{for NC clays} \tag{9.28}$$

$$\phi'_{\text{peak}} = 45 - 14\log PI \quad \text{for OC clays with } 4 < PI < 50 \tag{9.29}$$

$$\phi'_{\text{peak}} = 26 - 3\log PI \quad \text{for OC clays with } 50 < PI < 150 \tag{9.30}$$

Equation (9.20) can be used to estimate c'_{peak} of overconsolidated clays. For normally consolidated clays, and all granular soils, $c'_{\text{peak}} = 0$.

Plasticity index	ϕ'_{peak} (degrees)
10	33 ± 5
20	31 ± 5
30	29 ± 5
40	27 ± 5
60	24 ± 5
80	22 ± 5

TABLE 9.3 Suggested Values of ϕ'_{peak} for Normally Consolidated Clays

USCS symbol	σ'_{peak} (kN/m^2)	ϕ'_{peak} (degrees)
SM-SC	14.4	33
SC	12.0	31
ML	9.6	32
CL-ML	21.5	32
CL	14.4	28
MH	21.5	23
CH	12.0	19

TABLE 9.4 Typical Values for c'_{peak} and ϕ'_{peak} of Compacted Cohesive Soils

Duncan et al. (2014) suggested values of c'_{peak} and ϕ'_{peak} for cohesive soils compacted to relative compaction of 100% with respected to standard Proctor compaction effort, shown in Table 9.4. For lower compactive effort, ϕ'_{peak} remains the same but c'_{peak} decreases. At relative compaction of 90%, c'_{peak} should be 50% of what is given in Table 9.4.

The critical state friction angle is influenced by the mineralogy, grain compressibility, and the grain shape; and ψ is the dilatancy angle that reflects the density of the grain packing. The value of ϕ'_{cv} is about 32° for quartz sands, 33° for silty quartz sands with up to 20% fines, 34° for quartz-feldspar siliceous sands, 39° for calcareous sands, and 40° for fledspathic sands (Mayne, 2014). For most natural sands it is in the range from 30° to 33°. Mitchell (1976) suggested that the critical state friction angle is related the plasticity index by

$$\sin\phi'_{cv} = 0.8 - 0.094 \ln PI \tag{9.31}$$

The peak friction angle of a granular soil can be written as (BS 8002, 1994; AS 4678, 2002)

$$\phi'_{peak} = 30 + k_A + k_B + k_C \tag{9.32}$$

where k_A, k_B, and k_C account for the angularity of the grains (0–4°), grain size distribution (0–4°), and relative density expressed in terms of blow counts from the standard penetration test (0–9°), respectively. These values are given in Table 9.5.

		k value
k_A	Rounded grains	0
	Subangular grains	2
	Angular grains	4
k_B	Uniformly graded ($C_u < 2$)	0
	Moderately graded ($2 < C_u < 6$)	2
	Well graded ($C_u > 6$)	4
k_C	$N_{60} < 10$	0
	$N_{60} = 20$	2
	$N_{60} = 40$	6
	$N_{60} = 60$	9

Table 9.5 k_A, k_B, and k_C Values for Eqs. (9.32) and (9.33) (After AS 4678, 2002)

The critical state friction angle ϕ'_{cv}, which is independent of the relative density, can be estimated as

$$\phi'_{cv} = 30 + k_A + k_B \tag{9.33}$$

The residual friction angle for kaolinite, illite, and montmorillonite are about 15°, 10°, and 5°, respectively. For clay mineral grains that are not plate-like (e.g., gibbsite, hematite, bauxite), the residual friction angle can be larger than 25° (Fell et al., 2005).

9.9.2 Sensitivity

Clays have a fabric, which comes from the particle orientations and the bonds between them. Two extreme situations are flocculated and dispersed fabrics (see Sec. 3.8). Most of the time, it is in between these two. When a clay is *remolded* (i.e., highly disturbed), some of the bonds are broken and the fabric is partly destroyed. This leads to reduction in strength and stiffness. *Sensitivity* (S_t) is defined as the ratio of the undisturbed to remolded shear strength. At very large strains, clay becomes remolded and therefore, the ratio of peak to residual shear strength is approximately equal to the sensitivity. Highly sensitive clays have flocculated fabric. In highly sensitive clays, sensitivity can be as high as 10 or even more, where the clay will lose most of its strength when remolded. Some clays would regain their strength after some time since remolding. The process is known to be *thixotropy*, which occurs at constant volume and composition. There are different scales to classify sensitive soils, and the one that is most popular, suggested by Skempton and Northey (1952) is shown in Fig. 9.23. In soft clays where it is difficult to obtain good-quality undisturbed samples, laboratory or field vane shear test (see Sec. 11.8) is a good alternative for measuring sensitivity.

Remolded clays are often used in earthworks. In addition, in situations where the clay is known to be remolded, it is prudent to use the remolded shear strength than that of the undisturbed clay, which can be significantly higher. Wroth and Wood (1978) suggested that the remolded undrained shear strength of a clay is about 1.7 kN/m^2 at *LL* and increases by 100 times over the plastic range to 170 kN/m^2 at *PL*. Based on this

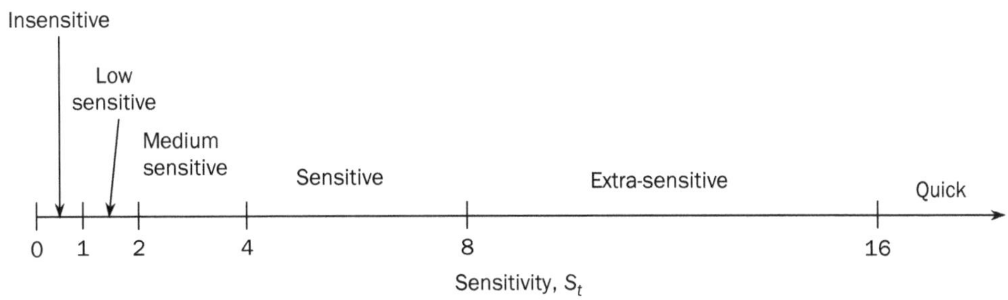

FIGURE 9.23 Sensitivity classification (after Skempton and Northey, 1952).

several equations are proposed in the literature relating the remolded undrained shear strength to the liquidity index, which are summarized by O'Kelly (2013).

9.10 Skempton's Pore Pressure Coefficients for Undrained Loading

In 1954, Sir Alec Skempton, a professor at Imperial College, U.K., introduced a simple concept to estimate the change in pore water pressure (Δu) in a soil element due to the *changes* in major and minor *total* principal stresses ($\Delta\sigma_1$ and $\Delta\sigma_3$) during undrained loading (Skempton, 1954). This is widely used in engineering practice due to its simplicity and for its practical value, especially in clays.

Figure 9.24a shows the major ($\Delta\sigma_1$) and minor ($\Delta\sigma_3$) total principal stress increments applied on an element which results in a pore water pressure change of Δu which has to be determined. This can be separated into two scenarios shown in Figs. 9.24b and 9.24c. First is an isotropic loading where $\Delta\sigma_3$ is applied in all directions, leading to a pore water pressure change of $\Delta u_{\text{isotropic}}$, and the second is deviator stress of $\Delta\sigma_1 - \Delta\sigma_3$ applied only in the vertical direction which changes the pore water pressure by $\Delta u_{\text{deviator}}$. Therefore, $\Delta u = \Delta u_{\text{isotropic}} + \Delta u_{\text{deviator}}$. Skempton (1954) expressed the change is pore water pressure due to $\Delta\sigma_1$ and $\Delta\sigma_3$ as:

$$\Delta u = \Delta u_{\text{isotropic}} + \Delta u_{\text{deviator}} = B\left[\Delta\sigma_3 + A\left(\Delta\sigma_1 - \Delta\sigma_3\right)\right] \tag{9.34}$$

where A and B are known as Skempton's pore pressure coefficients, $\Delta u_{\text{isotropic}} = B\Delta\sigma_3$ and $\Delta u_{\text{deviator}} = AB(\Delta\sigma_1 - \Delta\sigma_3)$. AB is sometimes denoted by $\bar{A}$. Here, $\Delta\sigma_1$ and $\Delta\sigma_3$ are not necessarily the changes in σ_1 and σ_3, respectively; they are the *algebraically* larger and smaller values of the principal stress changes, noting that compressive stresses are positive.

Skempton (1954) defined B as

$$B = \frac{1}{1 + n\dfrac{C_v}{C_{\text{sk}}}} \tag{9.35}$$

where n = porosity, C_v = compressibility of the pore fluid, and C_{sk} = compressibility of the soil skeleton. In saturated soils, the pore fluid is water that is relatively incompressible and hence C_v/C_{sk} is very small, making B approximately 1.

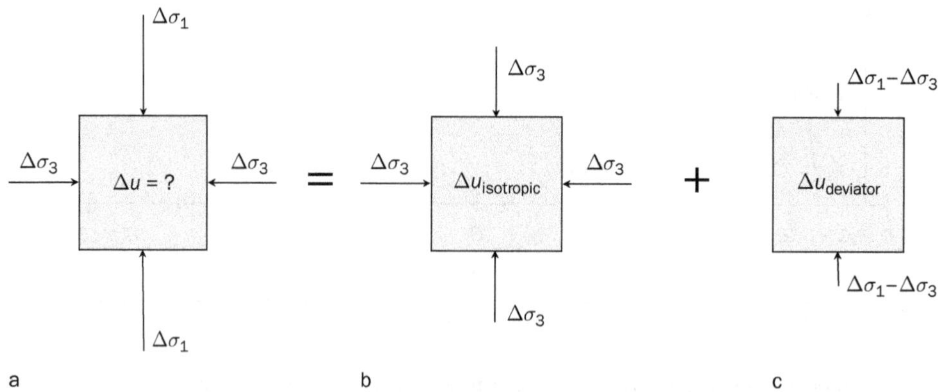

FIGURE 9.24 Pore pressure increase due to (a) applied loadings, (b) isotropic stresses, (c) deviator stresses.

Skempton's coefficient B is a measure of saturation and increases rapidly with the degree of saturation S. In a saturated clay, $B \approx 1$. It is the ratio of the pore water pressure increase to the increase in the isotropic confining pressure during undrained loading. For example, in a saturated clay, a 50 kN/m^2 increase in the confining pressure should increase the pore water pressure by approximately 50 kN/m^2. The coefficient B larger than about 0.98 is a good indication that the clay is saturated.

The coefficient A is a function of overconsolidation ratio, stress path, anisotropy, and strain rate. It is a measure of the pore pressure developed during the application of deviator stress. The coefficient A can vary during shearing, and at failure it is denoted by A_f. For normally consolidated clays A_f lies in the range of 0.5–1.0, generally closer to the upper end of the range. In highly overconsolidated clays, it can be negative, lying in the range of −0.5 to 0. In sensitive clays, A_f can exceed 1.0.

Example 9.5

A clay specimen is consolidated under a cell pressure of 200 kN/m^2 and a back pressure of 50 kN/m^2 to ensure full saturation of the specimen. When the consolidation is completed after 24 h, the drainage valve was closed (locking in the back pressure) to carry out some undrained loading quickly.

Under undrained conditions, the cell pressure was increased to 300 kN/m^2, and the pore water pressure increased to 145 kN/m^2. What is Skempton's pore pressure coefficient B? Note: Generally, we do this (B-Check) in a triaxial test to check whether the sample is saturated.

After the above, the deviator (axial) stress was increased from 0 to 75 kN/m^2 (no failure yet) and the pore water pressure was measured as 160 kN/m^2. Find Skempton's pore pressure coefficient A.

Solution The changes in σ_3, σ_1, and u are shown in Fig. 9.25. From the changes during the isotropic loading, applying Skempton's pore pressure equation

$$\Delta u = B[\Delta\sigma_3 + A(\Delta\sigma_1 - \Delta\sigma_3)]$$

$$95 = B[100 + A(100 - 100)]$$

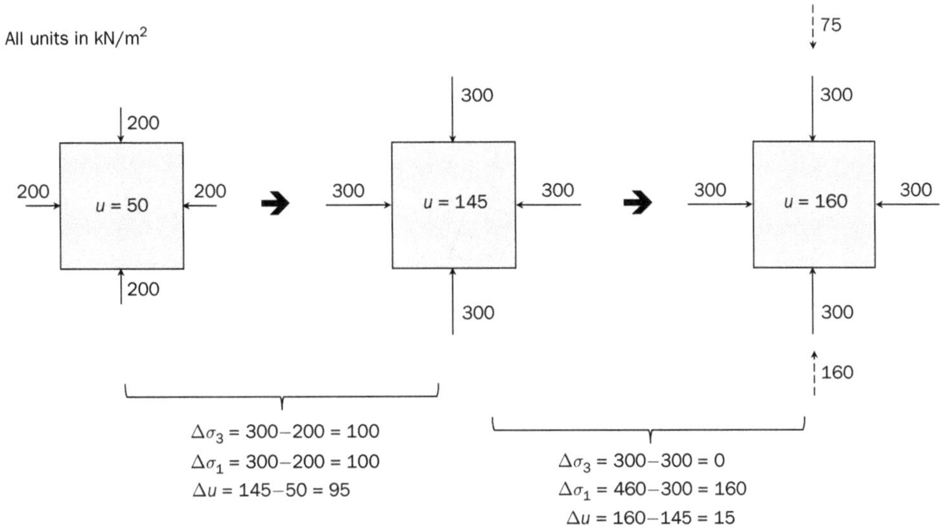

FIGURE 9.25 Changes in σ_3, σ_1, and u.

Therefore, $B = \mathbf{0.95}$.

During the application of the deviator stress,

$$\Delta u = B[\Delta\sigma_3 + A(\Delta\sigma_1 - \Delta\sigma_3)]$$

$$15 = 0.95[0 + A(75 - 0)]$$

Therefore, $A = \mathbf{0.21}$.

Example 9.6

A soil specimen is subjected to a consolidated undrained (CU) triaxial test, with the confining pressure of 250 kN/m^2. Previous tests on the same soil suggest that for a CU triaxial test the total and effective stress shear strength parameters are: $c = 5.0$ kN/m^2, $c' = 4.0$ kN/m^2, $\phi = 15°$, and $\phi' = 26°$. What would be the pore water pressure at failure of this soil specimen? Assuming the specimen is saturated, estimate Skempton's pore pressure coefficient A at failure.

Solution As shown in Fig. 9.26, both failure Mohr's circles, in terms of total and effective stresses have the same diameter, taken as $2x$ here.

$$\sigma_{3f} = 250 \text{ kN/m}^2$$

In terms of total stresses,

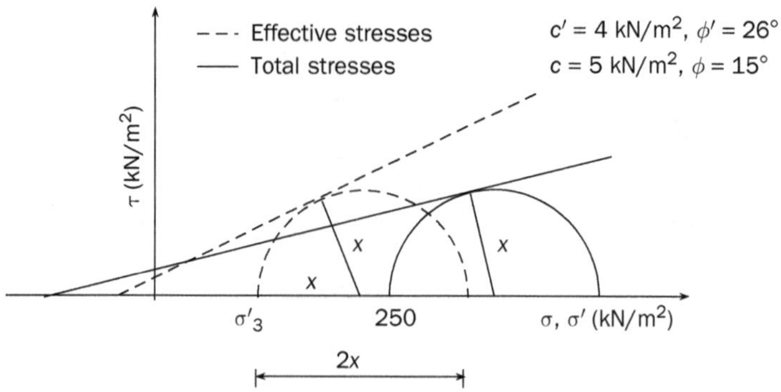

FIGURE 9.26 Mohr's circles and failure envelopes.

$$\sin\phi = \frac{x}{c\cot\phi + 250 + x}$$

$$\sin 15 = \frac{x}{5\cot 15 + 250 + x}$$

Therefore, $x = 93.8\ \text{kN/m}^2$

In terms of effective stresses,

$$\sin\phi' = \frac{x}{c'\cot\phi' + \sigma'_{3f} + x}$$

$$\sin 26 = \frac{93.8}{4\cot 26 + \sigma'_{3f} + 93.8}$$

Therefore, $\sigma'_{3f} = 112.0\ \text{kN/m}^2$

Therefore, pore water pressure at failure = 250 − 112 = **138.0 kN/m²**

Assuming $B = 1$,

$$\text{Skempton's } A_f = 138/(2 \times 93.8) = \mathbf{0.73}$$

9.11 Relationship between σ_1 and σ_3 at Failure

Figure 9.27 shows the Mohr's circle at failure, where σ_{1f} and σ_{3f} are the major and minor principal stresses at failure. The radial line *OP* is perpendicular to the tangent *AP*.

$$\sin\phi = \frac{OP}{AO} = \frac{(\sigma_{1f} - \sigma_{3f})/2}{c\cot\phi + (\sigma_{1f} + \sigma_{3f})/2}$$

$$\frac{\sigma_{1f} - \sigma_{3f}}{2} = \left(\frac{\sigma_{1f} + \sigma_{3f}}{2}\right)\sin\phi + c\cos\phi$$

$$\sigma_{1f} = \left(\frac{1+\sin\phi}{1-\sin\phi}\right)\sigma_{3f} + 2c\sqrt{\left(\frac{1+\sin\phi}{1-\sin\phi}\right)} \tag{9.36}$$

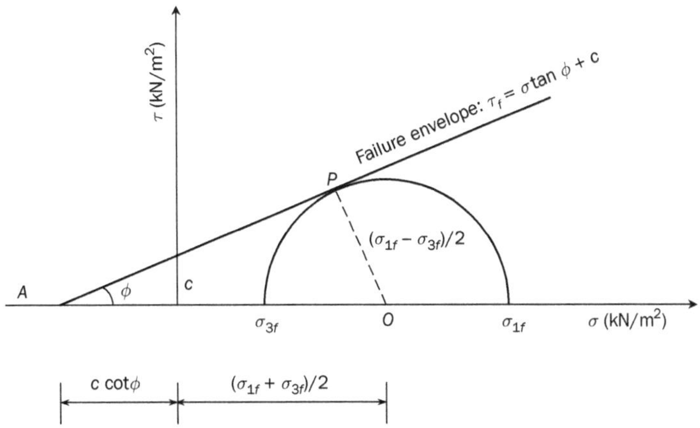

Figure 9.27 Mohr's circle at failure.

$$\sigma_{3f} = \left(\frac{1-\sin\phi}{1+\sin\phi}\right)\sigma_{1f} - 2c\sqrt{\left(\frac{1-\sin\phi}{1+\sin\phi}\right)} \tag{9.37}$$

$$\frac{1+\sin\phi}{1-\sin\phi} = \tan^2\left(45+\frac{\phi}{2}\right) = N_\phi$$

$$\frac{1-\sin\phi}{1+\sin\phi} = \tan^2\left(45-\frac{\phi}{2}\right) = \frac{1}{N_\phi}$$

Equations (9.36) and (9.37) are valid in terms of effective stresses or total stresses.

9.12 Stress Paths

In the triaxial testing in the laboratory or in some geotechnical application in the field, we often want to keep track of the progress in the loading. In most geotechnical engineering applications, the vertical (σ_v) and horizontal (σ_h) normal stresses are assumed to be the principal stresses. With each pair of σ_v and σ_h values, it is possible to draw the Mohr's circle. When there are hundreds of values recorded, a series of Mohr's circles can be a way to present the data, although with some clutter. Stress path discussed in this section is an effective way to display these results without any clutter, and clearly show the progress in loading at any time.

For now we will assume that σ_v and σ_h are the major and minor principal stresses, respectively, i.e., $\sigma_v = \sigma_1$ and $\sigma_h = \sigma_3$. Figure 9.28a shows the Mohr's circle for a point subjected to these stresses. The top of the Mohr's circle is known as the *stress point* (see Fig. 9.28a) where

$$\sigma = \frac{\sigma_v + \sigma_h}{2} = s \tag{9.38}$$

$$\tau = \frac{\sigma_v - \sigma_h}{2} = t \tag{9.39}$$

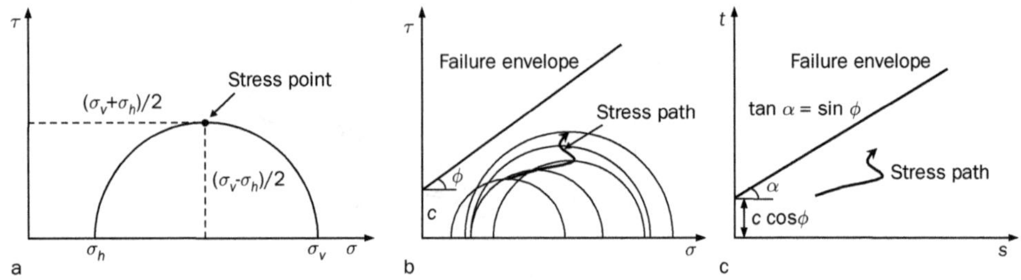

Figure 9.28 Stress path concept: (a) stress point, (b) Mohr circles and stress path, (c) stress path and failure envelope in s-*t* plane.

We will call the above values from Eqs. (9.38) and (9.39) as s and t, respectively. Defining the top of the Mohr's circle (i.e., s and t), clearly defines the entire Mohr's circle. Therefore, instead of drawing a series of Mohr's circles, we can draw the stress points, which is a neater way of presenting the progress in loading. The path followed by the stress points, or the locus of the stress points, is called the *stress path*. Figure 9.28b shows the Mohr's circles and the stress paths. Figure 9.28c shows the stress path in the s-t plane.

Instead of plotting Mohr's circles in σ-τ plane, it is better to plot the stress path on s-t plane. Here, s and t are defined by Eqs. (9.38) and (9.39). In σ-τ plane, the failure envelope is

$$\tau_f = c + \sigma_f \tan\phi \tag{9.10b}$$

When the Mohr's circles get closer to the failure envelope, failure is imminent. What would be the failure envelope in s-t plane?

From the Mohr's circle at failure, it can be shown that

$$\sin\phi = \frac{\left(\dfrac{\sigma_{vf} - \sigma_{hf}}{2}\right)}{c\cot\phi + \left(\dfrac{\sigma_{vf} + \sigma_{hf}}{2}\right)} \tag{9.40}$$

where σ_{vf} and σ_{hf} are the values of σ_v and σ_h at failure. The values of s and t at failure are given by

$$s_f = \frac{\sigma_{vf} + \sigma_{hf}}{2} \tag{9.41}$$

$$t_f = \frac{\sigma_{vf} - \sigma_{hf}}{2} \tag{9.42}$$

Therefore, Eq. (9.40) can be written as

$$\sin\phi = \frac{\left(\dfrac{\sigma_{vf} - \sigma_{hf}}{2}\right)}{c\cot\phi + \left(\dfrac{\sigma_{vf} + \sigma_{hf}}{2}\right)} = \frac{t_f}{c\cot\phi + s_f}$$

$$t_f = s_f \sin\phi + c\cos\phi \tag{9.43}$$

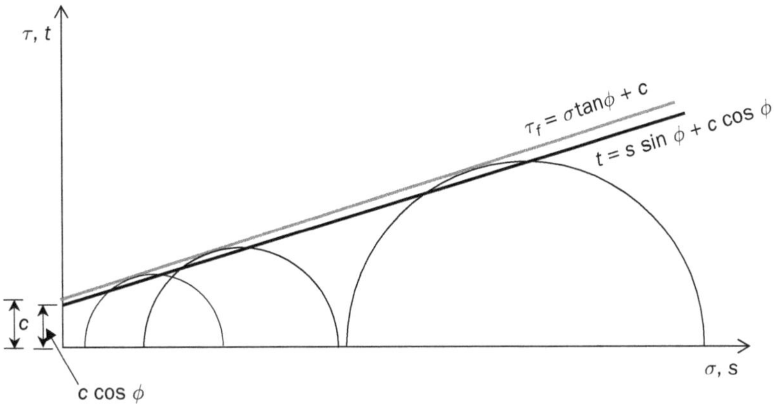

FIGURE 9.29 Failure envelopes in σ-τ and s-t planes.

When the failure values s_f and t_f are plotted, the points lie along the failure envelope on the *s-t* plane, which is a straight line. The slope of the envelope is $\sin\phi$ and the intercept on the *t* axis is $c \cos\phi$ as shown in Fig. 9.28c. When the stress path meets the failure envelope on *s-t* plane, failure takes place. Figure 9.29 shows the two envelopes and the Mohr's circles at failure on the same plot for comparison. The upper line showing the envelope in the σ-τ plane, tangent to the Mohr's circles, has a larger intercept *c* on the vertical axis and a larger slope $\tan\phi$. The failure envelope on the *s-t* plane, passing through the top of the Mohr's circles (dashed line), has a smaller intercept $c \cos\phi$ on the vertical axis and a smaller slope $\sin\phi$.

As in the case of Mohr's circles, stress paths can also be drawn in terms of effective stresses, where *s* can be replaced with s', where $s' = (\sigma'_v + \sigma'_h)/2$. There is no t', since *t* is the same as t'.

$$s = \frac{\sigma_v + \sigma_h}{2}$$

$$s' = \frac{\sigma'_v + \sigma'_h}{2}$$

Therefore,

$$s - s' = \frac{\sigma_v - \sigma'_v}{2} + \frac{\sigma_h - \sigma'_h}{2} = u$$

The effective stress principle $\sigma = \sigma' + u$ applies here too. At all times, at all points within the soil mass,

$$s = s' + u \tag{9.44}$$

Generally total and effective stress paths are plotted on the same graph, where both *s* and s' are shown on the horizontal axis, preferably using the same scale for all s, s', and *t*. From the failure envelope on s'-*t* plane, c' and ϕ' can be determined.

In some literature, s and t are replaced by p and q, respectively, which are also defined slightly differently as follows.

$$p = \frac{\sigma_1 + \sigma_2 + \sigma_3}{3} \tag{9.45}$$

$$q = \frac{1}{\sqrt{2}}\sqrt{\left[(\sigma_1 - \sigma_2)^2 + (\sigma_2 - \sigma_3)^2 + (\sigma_3 - \sigma_1)^2\right]} \tag{9.46}$$

Under axisymmetric loading such as in a triaxial test, Eq. (9.46) becomes

$$q = \sigma_1 - \sigma_3 \tag{9.47}$$

From the above discussions, it can be seen that s or p are similar to the normal stress σ and t is similar to the shear stress τ.

Example 9.7

A consolidated undrained triaxial test on a specimen of normally consolidated saturated clay ($c' = 0$) was carried out under a confining pressure of 500 kN/m². Consolidation took place against a back pressure of 100 kN/m². After the consolidation was completed, the specimen was subjected to undrained loading, where the additional vertical stress $\Delta\sigma$ was increased until failure occurred ($\Delta\sigma_f$) and the test data are summarized below.

$\Delta\sigma$ (kN/m²)	0	68	134	182	237	272*
u (kN/m²)	100	129	177	218	288	333*

*Failure

a. Draw total and effective stress paths.

b. Find the effective friction angle ϕ'.

c. What is Skempton's A-parameter at failure?

Another specimen of the same clay, consolidated under 500 kN/m² and back pressure of 100 kN/m² is subjected to a drained loading to failure, where the vertical stress was increased slowly to failure without inducing excess pore water pressure.

d. Draw the effective stress path in the above plot.

e. Find the total horizontal and vertical stresses, and the principal stress difference at failure.

Solution

Part a The total and effective stress paths are drawn in Fig. 9.30, by plotting s versus t and s' versus t.

σ_h (kN/m²)	$\Delta\sigma$ (kN/m²)	σ_v (kN/m²)	u (kN/m²)	s (kN/m²)	s′ (kN/m²)	t (kN/m²)
500	0	500	100	500	400	0
500	68	568	129	534	405	34
500	134	634	177	567	390	67
500	182	682	218	591	373	91
500	237	737	288	618.5	330.5	118.5
500	272	772	333	636	303	136

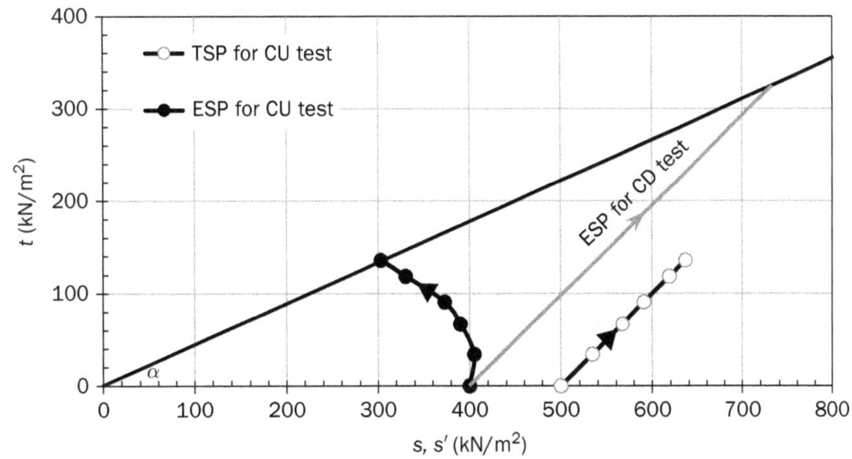

Figure 9.30

Part b

$$\sin\phi' = \tan\alpha = \frac{136}{303} = 0.449$$

Therefore, $\phi' = \mathbf{26.7°}$

Part c

$$A_f = \frac{\Delta u_f}{\Delta\sigma_f} = \frac{233}{272} = \mathbf{0.86}$$

Part d **The total stress path is the same as in CU test. The effective stress path remains parallel at all times, separated by the back pressure of 100 kN/m².**

Part e For the drained loading, the stress path (with 45° to s' axis) intersects the failure envelope at

$$s' = 735 \text{ kN/m}^2 \text{ and } t = 325 \text{ kN/m}^2$$

$$s' = \frac{\sigma'_v + \sigma'_h}{2} = 735 \text{ kN/m}^2 \quad (1)$$

$$t = \frac{\sigma'_v - \sigma'_h}{2} = 325 \text{ kN/m}^2 \quad (2)$$

Solving these two equations, at failure, $\sigma'_{hf} = 410$ kN/m² and $\sigma'_{vf} = 1050$ kN/m².

$u_f = 100$ kN/m², $\boldsymbol{\sigma_{hf}}$ **= 510 kN/m², $\boldsymbol{\sigma_{vf}}$ = 1150 kN/m², and $\boldsymbol{\Delta\sigma_f}$ = 640 kN/m².**

9.13 Critical State Soil Mechanics

Critical state soil mechanics (CSSM) was developed at University of Cambridge in 1960s (Roscoe et al., 1958; Schofield and Wroth, 1968) for normally consolidated or slightly overconsolidated soils, both granular and cohesive. It is an effective stress framework for describing soil behavior during loading. It is assumed that the soil is isotropic, elastoplastic, and deforms as a continuum, unaffected by creep.

In critical state soil mechanics, the state of stress at a point is represented by three variables, namely, *mean effective normal stress* (p'), *deviator stress* (q), and *specific volume* (v). They are defined as

$$p' = \frac{1}{3}(\sigma'_1 + \sigma'_2 + \sigma'_3) \quad (9.48)$$

$$q = \sqrt{\frac{(\sigma'_1 - \sigma'_2)^2 + (\sigma'_2 - \sigma'_3)^2 + (\sigma'_3 - \sigma'_1)^2}{2}} \quad (9.49)$$

$$v = 1 + e \quad (9.50)$$

where σ'_1, σ'_2, and σ'_3 are the major, intermediate, and minor effective principal stresses, and e is the void ratio. In the case of axisymmetric situations such as triaxial loading,

$$p' = \frac{1}{3}(\sigma'_1 + 2\sigma'_3) \quad (9.51)$$

$$q = \sigma'_1 - \sigma'_3 \quad (9.52)$$

The three variables (p', q, and v) are represented by three mutually perpendicular axes as shown in Fig. 9.31, which shows the *state boundary surface* which is curved in all directions. The state boundary surface consists of:

a. *Roscoe surface* bound by the *isotropic virgin consolidation line* (*ICL*) and the *critical state line* (*CSL*), for normally and slightly overconsolidated soils,

b. *Hvorslev surface* for heavily overconsolidated soils, and

c. *No tension cut-off surface*.

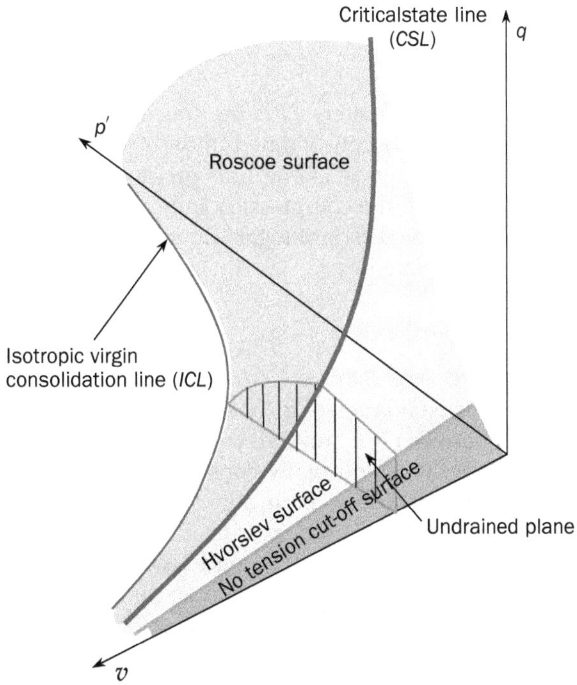

FIGURE 9.31 State boundary surface in p'-q-v space.

The state boundary surface separates the states at which soils can exist from the states that are not possible. When the state is represented by a point below the surface, the soil behaves elastically. If the point is on the state boundary surface (either Roscoe or Hvorslev), the soil undergoes plastic deformation. In this section, we will be dealing mainly with the Roscoe surface, which is relevant to normally consolidated and slightly overconsolidated clays. Critical state soil mechanics postulates that when the soil is loaded to failure, it moves in the p'-q-v space until it reaches the critical state (i.e., failure) represented by the critical state line where the specific volume and effective stresses remain unchanged at the critical state values. For normally consolidated clays, any drained or undrained loading starts from *ICL* and traverses along the Roscoe surface until the critical state is reached on *CSL*.

When a normally consolidated soil is isotropically consolidated, it follows the *ICL* in Fig. 9.31, which is the isotropic virgin consolidation line lying entirely on the p'-v plane. As shown in Fig. 9.32, the projections of *ICL* and *CSL* on the v-ln p' plane are two parallel straight lines, with a slope of λ. Along the *ICL*,

$$q = 0 \tag{9.53}$$

and

$$v = N - \ln p' \tag{9.54}$$

Along CSL, $q > 0$ and

$$v = \Gamma - \lambda \ln p' \tag{9.55}$$

where N and λ are soil parameters. N is the specific volume of the normally consolidated soil when isotropically consolidated under p' of one unit pressure. Γ is the specific volume of the soil when p' is one unit pressure and the soil is at critical state. The constant λ is similar to the compression index C_c introduced in consolidation, the difference coming from the different logarithmic axes—one base 10 and other natural logarithm. C_c and λ are related by

$$\lambda = \frac{C_c}{\ln 10} \tag{9.56}$$

Any anisotropic consolidation (e.g., one-dimensional consolidation) line will lie on the Roscoe surface, between ICL and CSL with the same slope λ on v-ln p' plane.

A soil lying on the ICL is normally consolidated. Slightly overconsolidated soils, with overconsolidation ratio less than about 2.5, lie between the ICL and VCL in Fig. 9.32. They are called "wet" soils in critical state soil mechanics, developing positive excess pore water pressure during undrained shear. Dense sands and heavily overconsolidated clays, called "dry" soils, lie on the Hvorslev surface and plot below the CSL and develop negative excess pore water pressures during undrained shear. When loaded, wet soils will undergo compression and strain hardening; dry soils will undergo expansion and strain softening.

Figure 9.33 shows the stress paths followed by two identical normally consolidated specimens isotropically consolidated under p'_0 subjected to (a) undrained, and (b) drained loadings. The initial state, at the end of isotropic consolidation, is represented by the point O, where the specific volume is v_0. When the specimen is sheared undrained, the entire stress path OF remains on the vertical constant volume v_0 plane shown in Fig. 9.33a, while moving along the Roscoe surface until it reaches the critical state at F.

In a conventional consolidated drained triaxial loading, p' and q increase at a ratio of 1:3. The stress path OG in Fig. 9.33b remains on this shaded plane while moving along

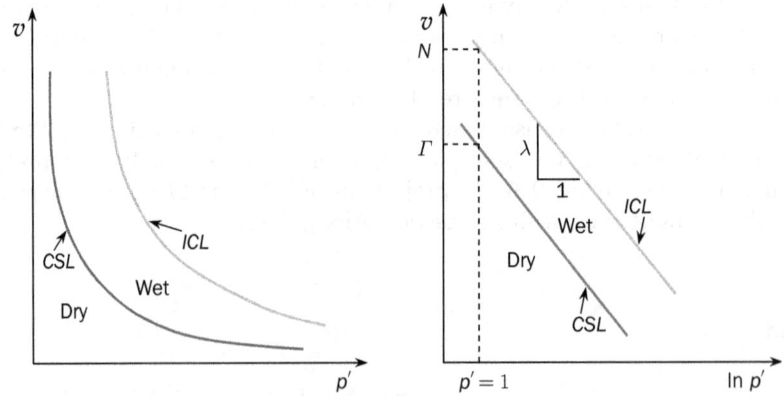

FIGURE 9.32 On v-p' and v-ln p' planes.

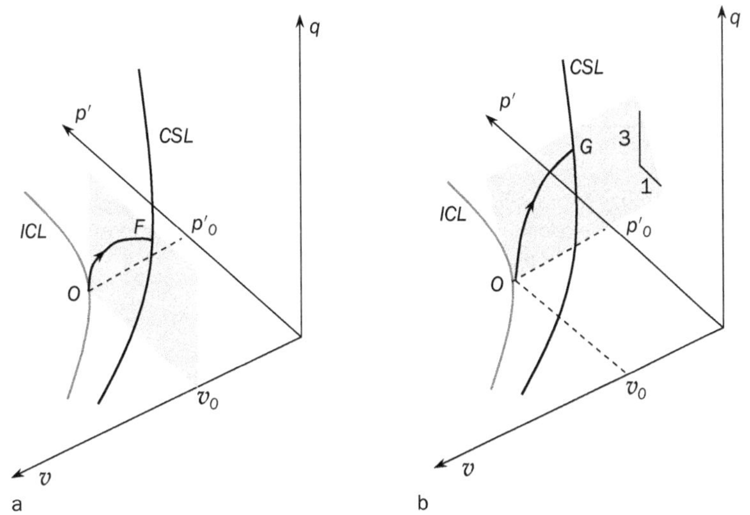

FIGURE 9.33 Triaxial stress paths: (a) undrained, (b) drained loadings.

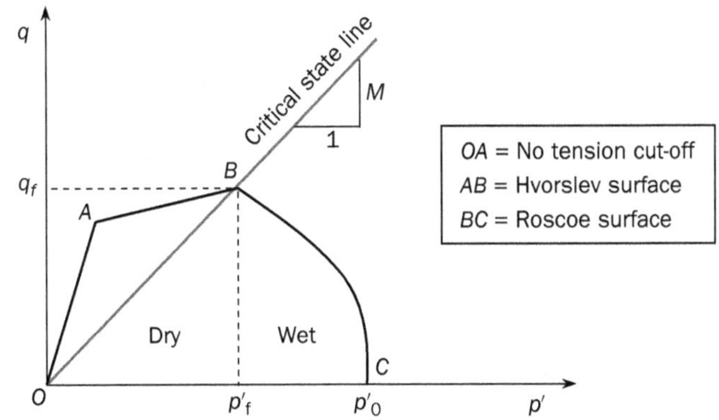

FIGURE 9.34 Projection on p'-q plane.

Roscoe surface. The stress path is defined by the intersection of the Roscoe surface and the inclined plane shown in the figure. During drained loading, along the stress path *OG*, all three parameters p', q, and v will vary.

Figure 9.34 shows the projection of *CSL* on p'-q plane, where the slope of the *CSL* is *M*. The equation of the *CSL*'s projection on p'-q plane is given by

$$q = Mp' \tag{9.57}$$

The projection of isotropic virgin consolidation line lies along the p' axis on this p'-q plane. The initial isotropic consolidation pressure is p'_0. The projection of the state boundary surface *ABC* is also shown in the figure. The CSL intersects the state boundary surface at *B* where the deviatoric stress q is the maximum at q_f. Hvorslev surface is a straight line (*AB*) on p'-q plane. A no tension cut-off is introduced to ensure $q = 0$ when

$p' = 0$. From Eqs. (9.51) and (9.52), it can be seen that in a conventional isotropically consolidated drained triaxial test, with $\sigma_3' = 0$, the slope of the stress path on q-p' plane is 3.0 as well. Therefore, the slope of the no-tension cut-off line OA is also taken as 3.0.

Example 9.8

A normal clay has $M = 0.85$, $N = 2.15$, $\Gamma = 2.02$, and $\lambda = 0.12$. If a clay specimen is subjected to isotropic effective consolidation pressure of 150 kN/m² prior to undrained triaxial loading, determine the values of p', q, and v at failure.

Solution From Eq. (9.54),

$$v = N - \lambda \ln p'$$

$$v = 2.15 - 0.12 \ln 150 = 1.549$$

During undrained loading, v remains the same at 1.549.

At failure (i.e., critical state)

$$v = \Gamma - \lambda \ln p'$$

$$1.549 = 2.02 - 0.12 \ln p_f'$$

Therefore, $p_f' = 50.65$ kN/m²

$$q_f = Mp_f' = 0.85 \times 50.65 = 40.5 \text{ kN/m}^2$$

Therefore, at failure, $p' =$ **50.7 kN/m²**, $q =$ **40.5 kN/m²**, and $v =$ **1.549**.

Example 9.9

A normally clay has $M = 0.85$, $N = 2.15$, $\Gamma = 2.02$, and $\lambda = 0.12$. If a clay specimen is subjected to isotropic effective consolidation pressure of 150 kN/m² prior to drained triaxial loading, determine the values of p', q, and v at failure.

Solution

$$p_0' = 150 \text{ kN/m}^2$$

Let the values of v, p', and q at failure be v_f, p_f', and q_f.

$$\frac{p_f' - p_0'}{q_f} = \frac{1}{3}$$

$$\frac{p_f' - 150}{q_f} = \frac{1}{3} \tag{1}$$

$$q_f = Mp_f' = 0.85p_f' \tag{2}$$

From Eqs. (1) and (2), $p_f' =$ **209.3 kN/m²**, and $q_f =$ **177.9 kN/m²**.

$$v_f = \Gamma - \lambda \ln p_f' = 2.02 - 0.12 \ln 209.3 = \mathbf{1.379}$$

Example 9.10

In Example 9.9, what is the reduction in the void ratio of the specimen during the shear?

Solution At the end of consolidation,

$$1 + e_0 = v_0 = 2.15 = 0.12 \ln 150 = 1.549$$
$$1 + e_f = v_f = 1.379$$

$$\text{Reduction in void ratio} = e_0 - e_f = 0.549 - 0.379 = \mathbf{0.17}$$

Form the normally consolidated state, unloading takes place from *ICL* along a straight line in v-ln p' plane, at a slope of κ, as shown in Fig. 9.35. This is similar to the void ratio versus log of the effective stress plot in consolidation. Reloading takes place along the same swelling line, which is defined as

$$v = v_s - \kappa \ln p' \tag{9.58}$$

Similar to Eq. (9.56),

$$\kappa = \frac{C_r}{\ln 10} \tag{9.59}$$

where C_r is the recompression index, and v_s is the specific volume when p' is one unit pressure.

Along any swelling path, the strains are purely elastic. It is assumed that, on the curved vertical surface above this swelling path on p'-v plane, the strains remain elastic until the state boundary surface is reached when plastic strains occur. This vertical surface is known as an "elastic wall."

For triaxial compression, it can be shown that the slope of the CSL on p'-q space is given by

$$M_c = \frac{6 \sin \phi'_{cv}}{3 - \sin \phi'_{cv}} \tag{9.60}$$

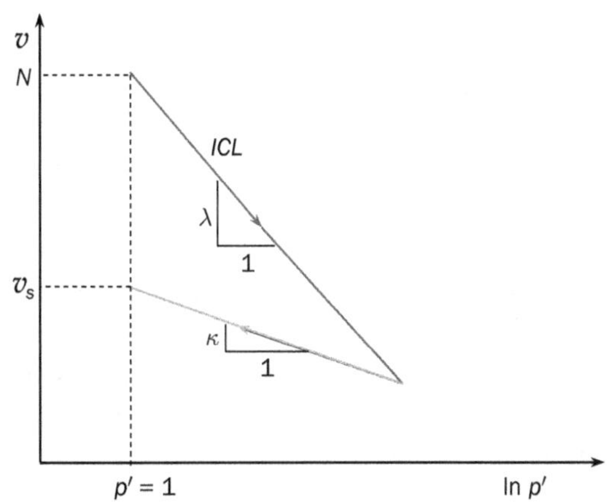

FIGURE 9.35 ICL and unloading-reloading line.

For triaxial extension,

$$M_e = \frac{6\sin\phi'_{cv}}{3+\sin\phi'_{cv}} \tag{9.61}$$

In Fig. 9.34, the projection of the Roscoe surface on the p'-q plane defines the yield curve for the wet soils. It can be seen that the larger the initial isotropic consolidation pressure p'_0, the larger the yield surface. Cam-clay model and modified Cam-clay model are two different constitutive models that define the yield curve slightly differently. The yield curve for the Cam-clay model takes the shape of a bullet; the one for the modified Cam-clay takes the shape of an ellipse.

Example 9.11

A normally consolidated clay specimen is subjected to isotropic consolidation under a cell pressure of 100 kN/m². When it is subjected to drained axial compression, it failed under principal stress difference of 110 kN/m². Find M and the friction angle ϕ'.

Solution At the end of consolidation, $p'_0 = 100.0$ kN/m²
Let the values of p' and q at failure be p'_f and q_f.

$$\frac{p'_f - p'_0}{q_f} = \frac{1}{3}$$

$$\frac{p'_f - 100}{110} = \frac{1}{3}$$

Therefore, $p'_f = 136.67$ kN/m²

$$M = q_f/p'_f = 110/136.67 = \mathbf{0.805}$$

$$M_c = 0.805 = \frac{6\sin\phi'_{cv}}{3-\sin\phi'_{cv}}$$

$$\sin\phi'_{cv} = 0.355$$

$$\phi'_{cv} = \mathbf{20.8^\circ}$$

9.14 Main Points

1. Soils fail in shear, where shear stress along the failure plane exceeds or becomes equal to the shear strength.
2. Shear strength τ_f has two separate components: friction and cohesion. The frictional component is proportional to the normal stress, and the cohesive component is a constant.

3. Shear strength parameters cohesion and friction angle are denoted by c and ϕ. They can be defined in terms of total or effective stresses.
4. Triaxial or direct shear test can be carried out to determine the shear strength parameters.
5. For normally consolidated clays and granular soils $c' = 0$.
6. CD or CU triaxial test gives c' and ϕ'. UU triaxial test gives c_u.
7. Short-term or undrained analysis is carried out in terms of total stresses using c_u and $\phi_u = 0$. Long-term or drained analysis is carried out in terms of effective stresses, using c' and ϕ'.
8. Skempton's pore pressure coefficients are used for determining pore water pressure changes in clays during undrained loading.
9. Stress path is a simple way of keeping track of the progress in loading.

Review Exercises

9.1 State whether the following are true or false.

a. Total stress is always larger than the effective stress.
b. An unconsolidated undrained triaxial test gives c' and ϕ'.
c. A CD triaxial test is faster than a CU triaxial test.
d. The unconfined compressive strength is twice the undrained cohesion.
e. A clay with unconfined compressive strength of 40 kN/m^2 is classified as a soft clay.
f. Residual friction angle of a clay is larger than the critical state friction angle.
g. The dilatancy angle ψ is increases with the relative density of a dense sand.

Answer: False, False, False, True, True, False, True

9.2 Draw the Mohr's circles and find the principal stresses, orientations of the principal planes, and the maximum shear stress at the point for the following situations shown in Fig. RE9.2.

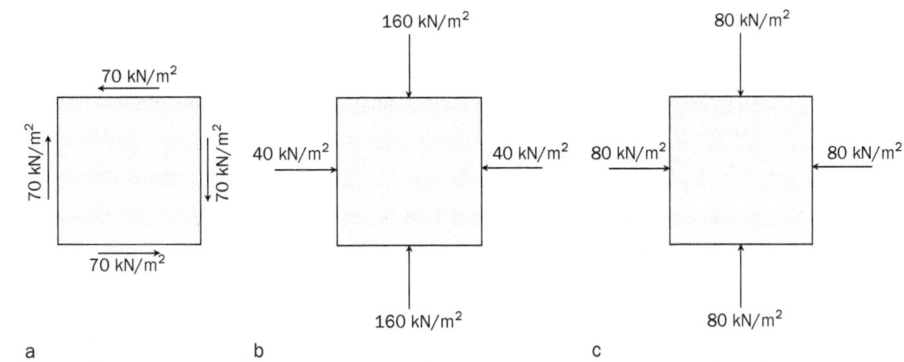

FIGURE RE9.2 State of stress.

Answer: 70 kN/m^2, −70 kN/m^2, 45° to horizontal, 70 kN/m^2; 160 kN/m^2, 40 kN/m^2, horizontal and vertical, 60 kN/m^2; 80 kN/m^2, 80 kN/m^2, Mohr's circle is a point, 0.

9.3 A sand specimen is initially under an isotropic confining pressure of 150 kN/m^2. Later, the specimen is subjected to an additional vertical stress, which was increased from 0 to 370 kN/m^2 when failure occurred.

a. What is the friction angle of the sand?
b. What are the normal and shear stresses on the failure plane?
c. What is the orientation of the failure plane?
d. When failure occurs, what is the maximum shear stress within the specimen?

Answer: 33.5°; 232.9 kN/m^2, 154.3 kN/m^2; 61.8° to horizontal; 185 kN/m^2

9.4 A saturated clay sample was consolidated in the triaxial cell under a cell pressure of 150 kN/m^2 without any back pressure. The drainage valve was then closed and the deviator stress was quickly increased from 0 to 200 kN/m^2 when failure occurred. If $c' = 15$ kN/m^2 and $\phi' = 20°$, find the pore water pressure and Skempton's pore pressure coefficient A at failure.

Answer: −1.2 kN/m^2, −0.006

9.5 A soil sample is consolidated under a cell pressure of 250 kN/m^2 with back pressure of 100 kN/m^2. It is given that the effective shear strength parameters of the soil are: c' = 10 kN/m^2 and $\phi' = 25°$. The sample is subjected to an axial load that is increased very slowly from zero to failure, maintaining the same back pressure of 100 kN/m^2, such that there is no build-up of excess pore water pressure. What would be the deviator stress at failure?

Answer: 251 kN/m^2

9.6 A series of consolidated undrained triaxial tests were carried out on three identical saturated clay specimens. The results are:

Specimen no.	Cell pressure (kN/m^2)	At failure (kN/m^2)	
		Deviator stress	**Pore water pressure**
1	100	170	40
2	200	260	95
3	300	360	135

Determine c, ϕ, $c_,'$ and ϕ' using (a) Mohr's circles, and (b) stress points at failure.

Answer: 26 kN/m^2, 19°; 19 kN/m^2, 28°

9.7 A 50-mm-diameter normally consolidated clay specimen is subjected to an unconfined compression test where it failed under an axial load of 140 N. What is the undrained shear strength?

Answer: 35.7 kN/m^2

9.8 In Review Exercise 9.7, if the friction angle in terms of effective stresses ϕ' is 24°, what would have been the pore water pressure within the specimen at failure?

Answer: −52.1 kN/m^2

9.9 A direct shear test is carried out on a dense sand at three different normal loads and the shearing was continued well beyond the peak until the residual state was reached. The loads at failure are as follows.

Normal load (N)	100	200	400	500
Peak shear load (N)	80	156	365	388
Residual shear load (N)	55	112	215	260

The specimen size is 60 mm ×60 mm. Determine the peak and residual friction angles.

Answer: 39.5°, 27.9°

9.10 Figure RE9.10 shows the total and effective stress paths from an isotropically consolidated undrained traixial test.

a. What is the back pressure in the test?
b. Find c' and ϕ'.
c. What is the deviator stress applied at failure?
d. Does Skempton's pore pressure coefficient A remain constant during the application of the deviator stress? What is the value at failure? Assume saturation and hence $B = 1$.

Answer: 50 kN/m², 13.3 kN/m², 19.4°; 100 kN/m²; 0.95

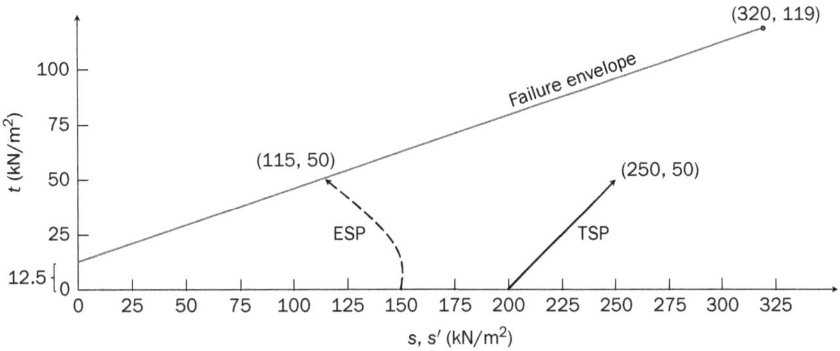

FIGURE RE9.10 Stress paths.

9.11 In a direct shear test on a sand specimen, the normal and shear stresses at failure are 450 and 310 kN/m², respectively.

a. What is the friction angle of the sand?
b. Assuming the failure plane is horizontal, what are the major and minor principal stresses at failure? What is the maximum shear stress at failure?

Answer: 34.6°; 1040.7 kN/m², 286.9 kN/m²; 376.9 kN/m²

9.12 The current state of stress at a point O within a saturated clay in the ground is as follows:

Total vertical stress = 110 kPa
Total horizontal stress = 90 kPa
Pore water pressure = 25 kPa

The friction angle and cohesion, in terms of effective stresses, are 24° and 5 kPa, respectively. Skempton's pore pressure coefficients A and B are 0.3 and 1.0, respectively. The loadings at the ground level are changed such that the increase in the total vertical stress at point O is twice that of the increase in total horizontal stress at O. Show the directions of the total and effective stress paths.

If the loading at the ground level is continued in the same way, what is the maximum vertical stress that can be applied at point O before failure? What would be the pore water pressure at failure?

Answer: 119.5 kN/m², 102.7 kN/m²

9.13 Redo RE 9.12 using Mohr's circles and find the maximum vertical stress that can be applied at point O before failure. Also, find the pore water pressure at failure.

Answer: 119.7 kN/m², 102.8 kN/m²

9.14 A clay sample was consolidated in a triaxial apparatus under a cell pressure of 500 kN/m², and back pressure of 100 kN/m². Then, the drainage valve was closed and the cell pressure was increased to 650 kN/m² where the pore water pressure increased to 242 kN/m².

a. Find Skempton's *B*-parameter.

Keeping the drainage valve closed, the vertical deviator stress was increased until failure occurred at axial strain of 10%. The axial strain, deviator stress, and pore water pressure values during the test are summarized below:

Axial strain, %	0	2	4	6	8	10
Deviator stress, kN/m²	0	68	132	182	236	272
Pore water pressure, kN/m²	242	271	319	360	430	475

b. Plot the total and effective stress paths.
c. Find Skempton's *A*-parameter at failure.
d. Is the clay normally consolidated? Why?
e. Find the effective friction angle.

Answer: 0.95, 0.90, NC, 25.9°

9.15 A normal clay has $M = 0.90$, $N = 2.85$, $\Gamma = 2.73$, and $\lambda = 0.14$. If a clay specimen is subjected to isotropic effective consolidation pressure of 250 kN/m² prior to undrained triaxial loading, determine the values of p', q, and v at failure.

Answer: 106.1 kN/m², 95.5 kN/m², 2.077

9.16 A normal clay has $M = 0.90$, $N = 2.85$, $\Gamma = 2.73$, and $\lambda = 0.14$. If a clay specimen is subjected to isotropic effective consolidation pressure of 250 kN/m² prior to drained triaxial loading, determine the values of p', q, and v at failure.

Answer: 357.1 kN/m², 321.4 kN/m², 1.315

References

AS 4678 (2002). Earth-Retaining Structures, Standards Australia.

ASTM D 2166-16. Standard Test Method for Unconfined Compressive Strength of Cohesive Soils.

ASTM D 2850-15. Standard Test Method for Unconsolidated Undrained Triaxial Compression Test on Cohesive Soils.

ASTM D 3080-11. Standard Test Method for Direct Shear Test of Soils under Consolidated Drained Conditions.

ASTM D 4767-11. Standard Test Method for Consolidated Undrained Triaixal Compression Test for Cohesive Soils.

BS 8002. "Code of Practice for Earth Retaining Structures," British Standards Institution, 1994.

Bjerrum, L., and N. E. Simons. "Comparison of Shear Strength Characteristics of Normally Consolidated Clays," *Proceedings of the ASCE Research Conference on Shear Strength of Cohesive Soils*, Boulder, 7111-726, 1960.

Coulomb, C. A. "Essai sur une Application des Règles de Maximus et Minimis à Quelques Problèmes de Statique, Relatifs à L'architecture," *Mémoires de Mathématique et de Physique, Présentés a l'Académie Royale des Sciences, par divers Savans, et lûs dans ses Assemblées*, Paris, 7 (Volume for 1773 published in 1776), 1776, pp. 343–382.

Duncan, J. M., S. G. Wright, and T. L. Brandon. *Soil Strength and Slope Stability*, 2nd Edition, Wiley, New York, 2014.

Fell, R., P. MacGregor, D. Stapledon, and G. Bell. *Geotechnical Engineering of Dams*, A.A. Balkema Publishers, UK, 2005.

Holtz, R. D., and W. D. Kovacs. *An Introduction to Geotechnical Engineering*, Prentice-Hall, New Jersey, US, 1981.

Jamiolkowski, M., C. C. Ladd, J. T. Germaine, and R. Lancellotta. "New Developments in Field and Laboratory Testing of Soils," *Proceedings of the 11th International Conference on Soil Mechanics and Foundation Engineering*, San Francisco, 1, 1985, pp. 57–154.

Kulhawy, F. H., and P. W. Mayne. *Manual on Estimating Soil Properties for Foundation Design*, Report EL-6800, Electric Power Research Institute, Palo Alto, US, 1990.

Ladd, C. C., R. Foott, K. Ishihara, F. Schlosser, and H. G. Poulos. Stress-Deformation and Strength Characteristics, *Proceedings of 9th International Conference on Soil Mechanics and Foundation Engineering*, Tokyo, 2, 1977, pp. 421–494.

Mayne, P. W. "Interpretation of Geotechnical Parameters from Seismic Piezocone Tests," *Proceedings of 3rd International Symposium on Cone Penetration Testing (CPT'14)*, Robertson, P. K., and K. I. Cabal (eds.), Las Vegas, Nevada, 2014, pp. 47–73.

Mesri, G. "A Reevaluation of $s_{u(mob)} \approx 0.22\,\sigma'_p$ using Laboratory Shear Tests," *Canadian Geotechnical Journal*, 26(1):162–164, 1989.

Mitchell, J. K. *Fundamentals of Soil Behaviour*, 1st Edition, Wiley, New York, 1976.

Mohr, O. "Welche Umstände Bedingen die Elastizitätsgrenze und den Brunch Eines Materiales," *Zeitschrift des Vereines Deutscher Ingenieure*, 44:1524–1530; 1572–1577, 1900.

O'Kelly, B. C. "Atterberg Limits and Remolded Shear Strength—Water Content Relationships," *Geotechnical Testing Journal*, ASTM, 36(6):939–947, 2013.

Peck, R. B., W. E. Hansen, and T. H. Thornburn. *Foundation Engineering*, 2nd Edition, Wiley, New York, 1974.

Roscoe, K. H., A. N. Schofield, and C. P. Wroth. "On the Yielding of Soils," *Geotechnique*, 8(1):22–53, 1958.

Schofield, A. N., and C. P. Wroth. *Critical State Soil Mechanics*, McGraw Hill, New York, 1968.

Skempton, A. W. "The Pore-Pressure Coefficients A and B," *Geotechnique*, 4:143–147, 1954.

Skempton, A. W. Discussion on "The Planning and Design of the New Hong Kong Airport," *Proceedings of the Institution of Civil Engineers*, London, 7, 1957, pp. 305–307.

Skempton, A. W., and R. D. Northey. "The Sensitivity of Clays," *Geotechnique*, 3(1):30–53, 1952.

Sorensen, K. K., and N. Okkels. "Correlation between Drained Shear Strength and Plasticity Index of Undisturbed Overconsolidated Clays," *Proceedings of the 18th International Conference on Soil Mechanics and Geotechnical Engineering*, Paris, Presses des Ponts, 1, 2013, pp. 423–428.

U. S. Navy. *Soil Mechanics—Design Manual 7.1*, Department of the Navy, Naval Facilities Engineering Command, U.S. Government Printing Office, Washington, DC, 1982.

Wroth, C. P., and D. M. Wood, "The Correlation of Index Properties with Some Basic Engineering Properties of Soils," *Canadian Geotechnical Journal*, 15(2):137–145, 1978.

Zheng, J., R. D. Hyrciw, and A. Ventola. "Compressibility of Sands of Various Geologic Origins at Pre-crushing Stress Levels," *Geotechnicall and Geological Engineering*, 35:2037–2051, 2017.

CHAPTER 10

Lateral Earth Pressures

10.1 Introduction

The pressure at a point within water is hydrostatic and is the same in all directions. This is not the case in soils, where the normal stress varies with direction. Until now, we were dealing mainly with vertical overburden stresses and vertical stress increases due to surface loads. The methods to compute horizontal earth pressures are discussed in this chapter. It is necessary to determine the lateral earth pressures for the design and analysis of retaining walls, sheet pile walls, braced excavations, and tunnels.

Figure 10.1a shows a gravity boulder retaining wall in Kyoto, Japan. Figure 10.1b shows a deep-braced excavation at a busy road Bukit Bintang in the heart of Kuala Lumpur, Malaysia. Figure 10.1c shows a tunnel for the underground railway in Malaysia.

10.2 At-Rest State and K_0

In a homogenous soil mass where the ground level is horizontal and is without any other loads coming from other structures as shown in Fig. 10.2a, the soil is said to be "at-rest." The ratio of the horizontal to vertical *effective* stresses is known as the *coefficient of earth pressure at rest*, denoted by K_0 (pronounced K-nought), which is a constant throughout the soil. Thus,

$$K_0 = \frac{\sigma'_h}{\sigma'_v} \tag{10.1}$$

In at-rest state, also known as K_0 *state*, there are no lateral strains within the soil mass. The soil is just being compressed vertically while it is restrained from any lateral deformation. One-dimensional consolidation is a special case of K_0 state, where the clay is loaded such that there is no lateral deformation. In normally consolidated clays and sands, K_0 is typically in the range of 0.3–0.6. Figure 10.2b shows a soil profile, with specific values of K_0 for each layer. In heavily overconsolidated soils and rocks, K_0 can be larger than 1, implying that σ'_h is larger than σ'_v. Considering the soil mass as a linear elastic continuum, it can be shown that

$$K_0 = \frac{\nu}{1-\nu} \tag{10.2}$$

FIGURE 10.1 Applications: (a) retaining wall in Kyoto, Japan, (b) braced excavation in Kuala Lumpur, Malaysia, (c) tunnel in Malaysia.

where ν is the Poisson's ratio. Jaky (1944, 1948) showed that for normally consolidated sands

$$K_0 = (1 - \sin\phi')\left(\frac{1 + \frac{2}{3}\sin\phi'}{1 + \sin\phi'}\right) \tag{10.3}$$

which was approximated and widely used as

$$K_0 = 1 - \sin\phi' \tag{10.4}$$

Equation (10.4) works well for normally consolidated clays too. Massarsch (1979) showed that for normally consolidated clays

$$K_0 = 0.44 + 0.0042\,PI \tag{10.5}$$

where *PI* is the plasticity index of the clay. Alpan (1967) showed that for normally consolidated clays,

$$K_0 = 0.19 + 0.233\log PI \tag{10.6}$$

Equation (10.4) suggests that K_0 decreases with increasing ϕ'. Equations (10.5) and (10.6) suggest that in normally consolidated clays, K_0 increases with *PI*.

In overconsolidated soils, K_0 increases with the overconsolidation ratio. It can be estimated as

$$K_{0,\text{overconsolidated}} = K_{0,\text{normally consolidated}} OCR^n \tag{10.7}$$

where n is often taken as 0.5. Mayne and Kulhawy (1982) suggested $n = \sin\phi'$. Ladd et al. (1977) suggested $n = 0.42$ for low plastic clays and 0.32 for high plastic clays.

Example 10.1

In a normally consolidated sand water table is at 5 m below the ground level. The unit weights of the sand above and below the water table are 16.5 and 19.0 kN/m³, respectively. The friction angle of the sand is 34°. Find the total horizontal stress at 8 m depth.

Solution At 8 m depth,

$$\sigma'_{vo} = (5)(16.5) + (3)(19.0 - 9.81) = 110.1 \text{ kN/m}^2$$
$$K_0 = 1 - \sin\phi' = 1 - \sin 34 = 0.441$$

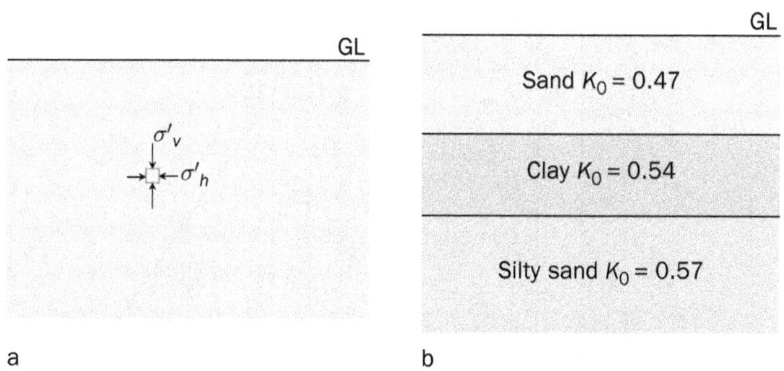

Figure 10.2 (a) K_0 in a homogeneous soil mass, (b) K_0 values with a layered soil.

Therefore,

$$\sigma'_{ho} = K_0 \times \sigma'_{vo} = (0.441)\,(110.1) = 48.5\ \text{kN/m}^2$$

$$\sigma_{ho} = \sigma'_{ho} + u = 48.5 + (3)\,(9.81) = \mathbf{78.0\ kN/m^2}$$

Example 10.2

Figure 10.3a shows a rigid basement wall that will not allow lateral movement and hence K_0 condition can be assumed for the sand it retains. The unit weights of the sand above and below the water table are 16.5 and 19.0 kN/m³, respectively. $K_0 = 0.40$.

a. Plot the variation of the total horizontal stress with depth.

b. Find the magnitude and location of the lateral thrust on the wall.

Solution

Part a At the water table,

$$\sigma'_v = (2)\,(16.5) = 33.0\ \text{kN/m}^2$$

$$\sigma'_h = K_0\,\sigma'_v = (0.40\,(33.0) = 13.2\ \text{kN/m}^2$$

$$u = 0 \text{ and hence } \sigma_h = 13.2\ \text{kN/m}^2$$

At the bottom of the sand,

$$\sigma'_v = (2)\,(16.5) + (4)\,(19.0 - 9.81) = 69.8\ \text{kN/m}^2$$

$$\sigma'_h = K_0\,\sigma'_v = (0.40\,(69.8) = 27.9\ \text{kN/m}^2$$

$$u = (4)\,(9.81) = 39.2\ \text{kN/m}^2$$

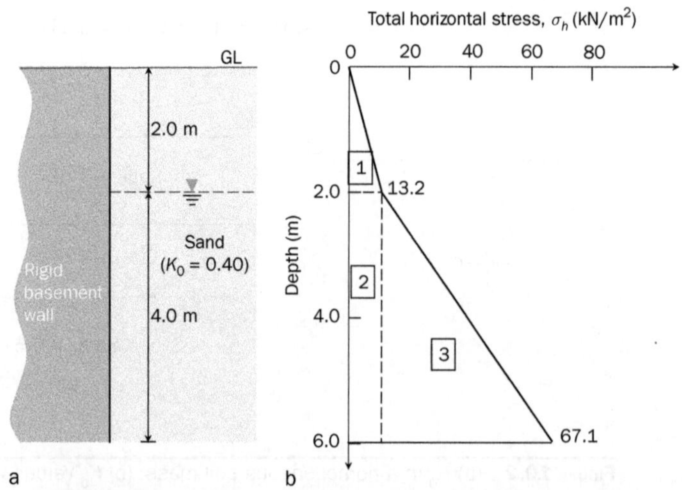

FIGURE 10.3 (a) Rigid basement wall, (b) variation of σ_h with depth.

Therefore, $\sigma_h = 27.9 + 39.2 = 67.1\ \text{kN/m}^2$

Figure 10.3b shows the plot of σ_h with depth.

Part b

Block	Horizontal load per m width (kN/m)	Height above the base (m)	Moment (kN·m/m)
1	(0.5) (13.2) (2) = 13.2	4.667	61.6
2	(4) (13.2) = 52.8	2.0	105.6
3	(0.5) (67.1 − 13.2) (4) = 107.8	1.333	143.7
Sum	173.8		310.9

From the table above,

The total horizontal load on the wall is 173.8 kN/m, which acts at a height of 310.9/173.8 = 1.79 m above the bottom of the basement wall.

Note that if there are different soils with different values for K_0, there will be breaks in the σ_h versus depth profile, occurring at the interfaces between the different soils.

How does K_0 vary with time when a clay is aging? This question was raised first by Schmertmann (1983). Mesri and Hayat (1993) suggested that, K_0 of a normally consolidated clay during secondary compression at time t can be estimated as

$$K_{0,\text{during secondary compression}} = K_{0p}\left(\frac{t}{t_p}\right)^{C_\alpha/C_c} \tag{10.8}$$

where K_{0p} is the K_0 at the end of primary consolidation, which they suggest can be estimated using ϕ'_{cv} in Jaky's Eq. (10.4). C_α = coefficient of secondary compression, C_c = compression index, and t_p = time taken for the primary consolidation. Mitchell and Soga (2005) suggest that K_0 of a normally consolidated soil should increase toward 1.0, which is supported by limited data from Kavazanjian and Mitchell (1984) on San Francisco Bay mud. In heavily overconsolidated soil K_0 should decrease with time toward 1.0.

10.3 Active and Passive States

Figure 10.4a shows a smooth vertical wall OA, retaining a backfill where σ'_v and σ'_h are vertical and horizontal effective stresses at a point X, depth z below the ground level. Therefore, assuming no lateral strains, initially at X, $\sigma'_h = K_0\sigma'_v = K_0\gamma z$. Since the wall is smooth, σ'_v and σ'_h are principal stresses (i.e., no shear stresses on the horizontal and vertical planes).

If the wall rotates about point O and moves away from the soil, σ'_h would decrease while σ'_v remains constant at γz. As the wall moves further away, at some stage when A reaches A', shear failure occurs within the soil mass along the failure plane shown in Fig. 10.4b. Here, the soil on the left of the failure plane slides down. It will be shown later that the failure plane is inclined at $45 + \phi'/2$ to the horizontal. When the soil fails this way (i.e., being allowed to stretch laterally), it is known to be in *active state*.

If the wall is rotated about point O in the other direction and moved toward the soil, σ'_h would increase while σ'_v remains constant at γz. As the wall moves further toward

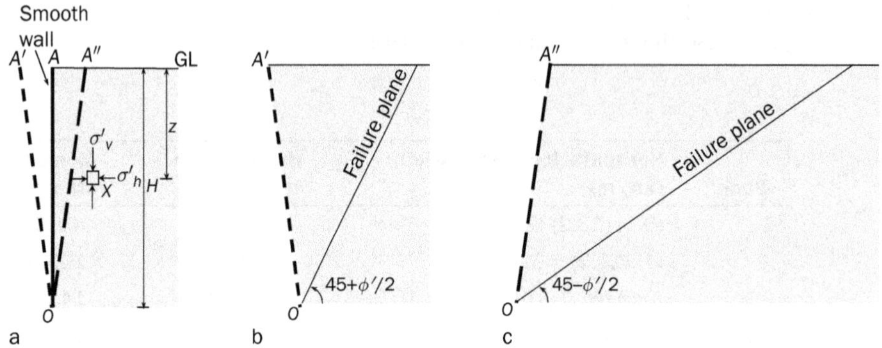

FIGURE 10.4 (a) Wall positions, (b) wall moves away from the soil to reach active state, (c) wall moves toward the soil to reach passive state.

Soil	Active	Passive
Loose sand	0.1–0.2% of *H*	1% of *H*
Dense sand	0.05–0.1% of *H*	0.5% of *H*
Soft clay	2% of *H*	4% of *H*
Stiff clay	1% of *H*	2% of *H*

TABLE 10.1 Horizontal Displacements Required to Induce Active or Passive States in Soils

the soil, at some stage when A reaches A'', and shear failure occurs within the soil mass along the failure plane shown in Fig. 10.4c. Here, the soil on the left of the failure plane slides upward. It will be shown later that the failure plane is inclined at $45 - \phi'/2$ to the horizontal. When the soil fails this way (i.e., being compressed laterally), it is known to be in *passive state.*

For the active or passive state to take place within a soil, the wall movement does not have to be only *rotational,* it can also be *translational.* Table 10.1 lists approximate horizontal displacements required to induce active or passive states in some soils. Following observations can be made from Table 10.1.

a. Weaker soils require larger displacements to reach active or passive states.
b. It requires larger displacements to induce passive state than active state.
c. It requires larger displacements to induce active or passive states in clays than sands.
d. Displacement required to induce active or passive state is proportional to the wall height *H.*

Figure 10.5 summarizes the variation in σ'_h with the horizontal displacement of the wall. The effective horizontal stresses at active and passive states are denoted by σ'_a and σ'_p, respectively. It can be seen that the lateral earth pressure lies within the range from σ'_a to σ'_p.

We only discussed the state of stress at a point X during the wall movement. Failure occurs at every point within the soil mass where active or passive state has been

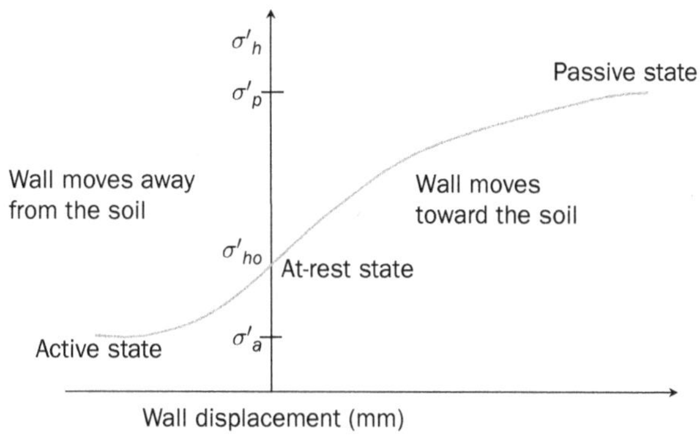

Figure 10.5 Wall movement versus effective horizontal stress.

reached—not just along the plane passing through point *O*. There will be hundreds of parallel failure planes within the soil mass.

10.4 Rankine's Earth Pressure Theory

Lateral earth pressure theories proposed by Rankine (1857) and Coulomb (1776) are discussed in this chapter. Between the two, Rankine's theory is simpler and is widely used in the designs and analysis of retaining walls, sheet piles, and braced excavations. It assumes that the wall retaining the soil is *smooth and vertical*, and hence σ'_v and σ'_h are principal stresses during the wall movement either way.

10.4.1 Active State

Let us consider a smooth vertical wall retaining a *granular* soil such as sand (Fig. 10.4a), where the friction angle and unit weight are ϕ' and γ, respectively. Initially at *X*,

$$\sigma'_v = \gamma z$$

With no wall movement initially, the soil is at rest and hence the initial value of σ'_h is given by

$$\sigma'_{h0} = K_0 \gamma z$$

The Mohr's circle corresponding to the initial at-rest state is shown in Fig. 10.6a with dashed line. When the wall moves away from the soil, there is no change in σ'_v while σ'_h decreases. This makes the Mohr's circle become larger and expand on the left as shown in Fig. 10.6a, until it becomes tangent to the failure envelope at active state, with σ'_h becoming σ'_a.

Figure 10.6b shows only the Mohr's circle at failure (active state), along with the failure envelope. The Mohr's circle is tangent to the failure envelope at *F*, which is joined with the center *C*.

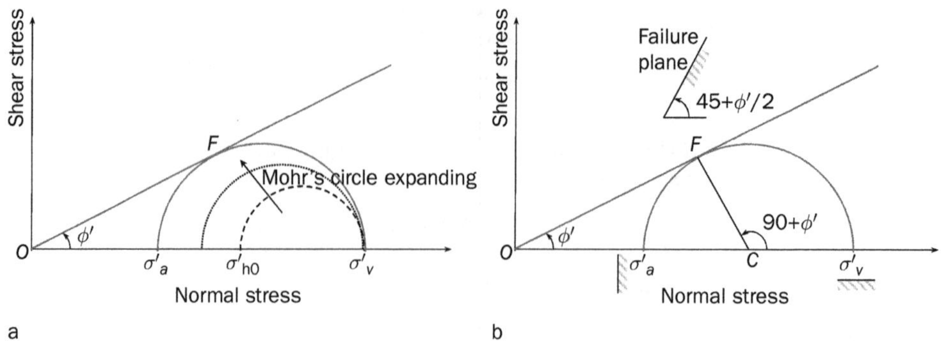

FIGURE 10.6 Wall moving away from the soil: (a) expansion of Mohr's circles, (b) at active state.

$$OC = \frac{\sigma'_v + \sigma'_a}{2}$$

$$FC = \frac{\sigma'_v - \sigma'_a}{2}$$

Therefore,

$$\sin\phi' = \frac{FC}{OC} = \frac{\sigma'_v - \sigma'_a}{\sigma'_v + \sigma'_a}$$

$$\sigma'_a = \frac{1 - \sin\phi'}{1 + \sin\phi'}\sigma'_v = K_a\sigma'_v \tag{10.9}$$

where

$$K_a = \frac{1 - \sin\phi'}{1 + \sin\phi'} = \tan^2\left(45 - \frac{\phi'}{2}\right) \tag{10.10}$$

Here, K_a is known as *Rankine's active earth pressure coefficient*.

σ'_v and σ'_a are acting on the horizontal and vertical planes, respectively. From Fig. 10.6b, it can be deduced that the failure plane, which corresponds to point *F* on the Mohr's circle, is inclined at $45 + \phi'/2$ to the horizontal.

In cohesive soils, Eq. (10.9) becomes

$$\sigma'_a = K_a\sigma'_v - 2c'\sqrt{K_a} \tag{10.11}$$

10.4.2 Passive State

Let us consider a smooth wall retaining a granular soil with friction angle and unit weight of ϕ' and γ, respectively. Initially, when the wall has not moved, the soil is at-rest and hence $\sigma'_v = \gamma z$ and $\sigma'_{h0} = K_0\gamma z$. The Mohr's circle corresponding to the initial at-rest state is shown in Fig. 10.7a with dashed line on the left. When the wall moves *toward* the soil, there is no change in σ'_v while σ'_h increases. At one stage σ'_h becomes equal to σ'_v and the Mohr's circle becomes a point. With further wall movement, σ'_h becomes larger than σ'_v and the Mohr's circle expands on the right side as shown with the dotted circles. Finally, when the passive state is reached, the Mohr's circle is tangent to the failure envelope.

Figure 10.7b shows only the Mohr's circle at passive state, along with the failure envelope.

$$OC = \frac{\sigma'_p + \sigma'_v}{2}$$

$$FC = \frac{\sigma'_p - \sigma'_v}{2}$$

Therefore,

$$\sin\phi' = \frac{FC}{OC} = \frac{\sigma'_p - \sigma'_v}{\sigma'_p + \sigma'_v}$$

$$\sigma'_p = \frac{1+\sin\phi'}{1-\sin\phi'}\sigma'_v = K_p\sigma'_v \tag{10.12}$$

where

$$K_p = \frac{1+\sin\phi'}{1-\sin\phi'} = \tan^2\left(45 + \frac{\phi'}{2}\right) \tag{10.13}$$

Here, K_p is known as *Rankine's passive earth pressure coefficient*.
σ'_v and σ'_p are acting on the horizontal and vertical planes, respectively. From Fig. 10.7b, it can be deduced that the failure plane, which corresponds to point F on the Mohr's circle, is inclined at $45 - \phi'/2$ to the horizontal.

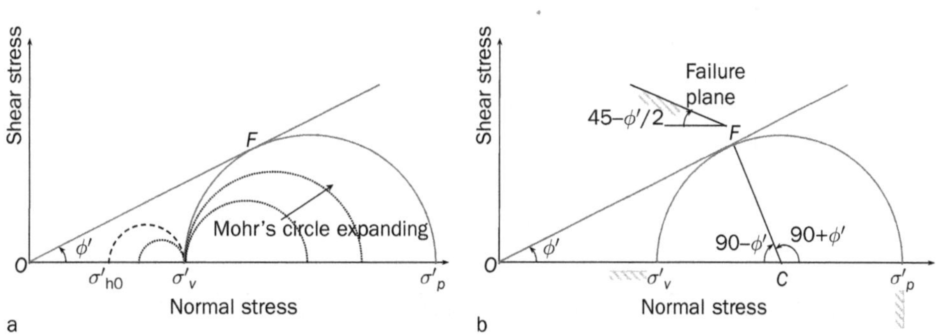

FIGURE 10.7 Wall moving toward the soil: (a) expansion of Mohr's circles, (b) at passive state.

In cohesive soils, Eq. (10.12) becomes

$$\sigma'_p = K_p\sigma'_v + 2c'\sqrt{K_p} \tag{10.14}$$

For a typical friction angle $\phi' = 30°$, $K_a = 0.333$ and $K_0 = 0.5$ and $K_p = 3.0$. As in this case, generally, the passive earth pressure coefficient K_p is an order of magnitude larger than the active earth pressure coefficient K_a.

10.4.3 Lateral Earth Pressure Distribution

Figure 10.8a shows a sandy soil of height H, retained by a smooth vertical wall. At depth z below the ground level, $\sigma'_v = \gamma z$. If the sand is in active state, σ'_a varies linearly with depth and is given by $\sigma'_a = K_a\gamma z$. At the bottom of the wall, $\sigma'_a = K_a\gamma H$ as shown in Fig. 10.8b. The resultant force of the lateral earth pressure distribution on the wall is P_a, acting at a height of $H/3$ from the bottom. P_a is given by the area of the triangular pressure distribution and hence

$$P_a = 0.5K_a\gamma H^2 \tag{10.15}$$

If the sand is in passive state, the σ'_p varies linearly with depth and the lateral pressures are much larger. The resultant P_p, acting at $H/3$ above the base, is given by (Fig. 10.8c).

$$P_p = 0.5K_p\gamma H^2 \tag{10.16}$$

In the case of a smooth wall retaining a cohesive soil (Fig. 10.9a),

$$\sigma'_a = K_a\sigma'_v - 2c'\sqrt{K_a} \tag{10.11}$$

$$\sigma'_a = K_a\gamma z - 2c'\sqrt{K_a}$$

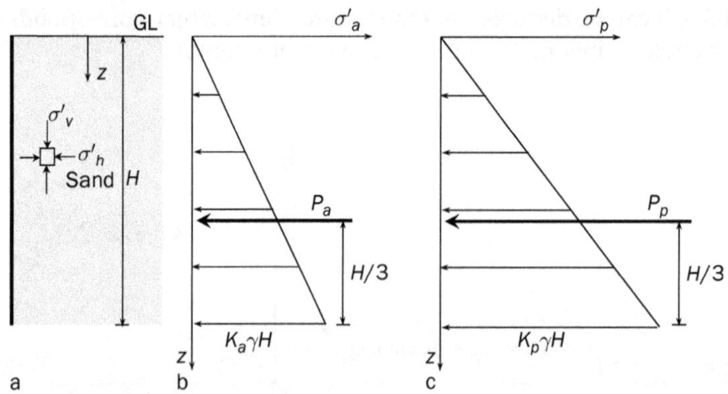

Figure 10.8 Variation of σ'_h with depth in granular soils: (a) smooth wall retaining the sand, (b) active state, (c) passive state.

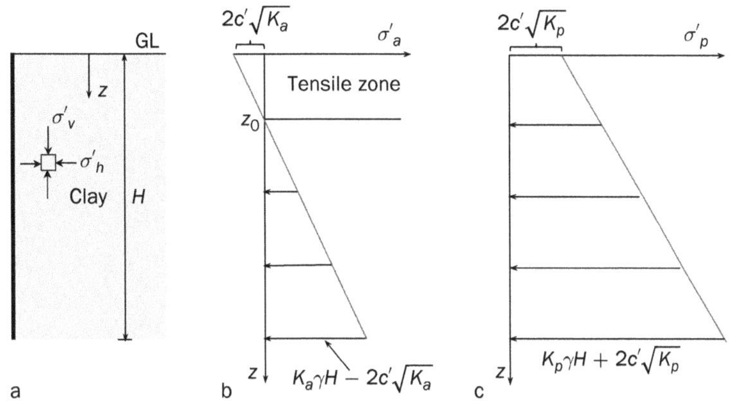

FIGURE 10.9 Variation of σ'_h with depth in cohesive soils: (a) smooth wall retaining the clay, (b) active state, (c) passive state.

This suggests a linear variation of σ'_a with depth as shown in Fig. 10.9b. At $z = 0$, $\sigma'_a = -2c'\sqrt{K_a}$. The clay is in tension (i.e., $\sigma'_a < 0$) up to a depth of z_0, which can be computed as follows.

At depth z_0, $\sigma'_a = 0$. Therefore,

$$\sigma'_a = 0 = K_a\gamma z_0 - 2c'\sqrt{K_a}$$

$$z_0 = \frac{2c'}{\gamma\sqrt{K_a}} \tag{10.17}$$

At the bottom of the wall, from Eq. (10.11),

$$\sigma'_a = K_a\gamma H - 2c'\sqrt{K_a} \tag{10.18}$$

From Eqs. (10.17) and (10.18), σ'_a at the bottom of the wall is given by

$$\sigma'_a = K_a\gamma(H - z_0)$$

Within the depth z_0, the clay is in tension and may crack, resulting in little or no contact between the wall and the clay. In the designs of earth retaining structures, active thrust P_a is seen as a load and the passive thrust P_p as resistance. Therefore, when computing the active thrust, it is unwise to rely on the tensile stresses within the top z_0, which would theoretically reduce the load. Conservatively, the tensile zone can be neglected in computing the active thrust as

$$P_a = 0.5K_a\gamma(H - z_0)^2 \tag{10.19}$$

In passive state in clays, the variation of σ'_p with depth is linear as shown in Fig. 10.9c. For clays under undrained conditions, c_u and $\phi_u = 0$ must be used in computing

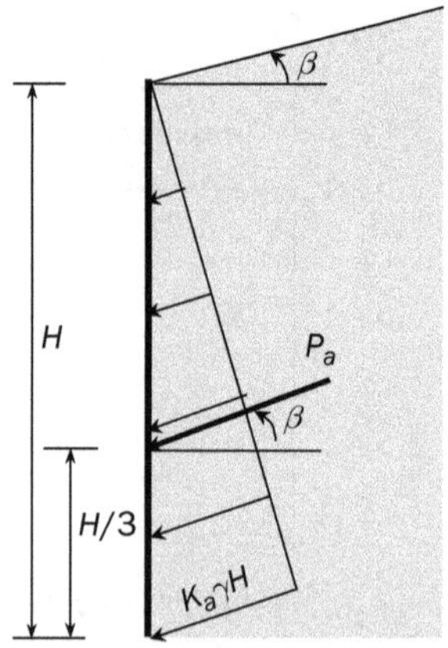

FIGURE 10.10 Active state in granular soil with sloping backfill.

the lateral earth pressures in active or passive state, but in terms of total stresses. Here, $K_a = K_p = 1$. For active state in clays under undrained conditions, $z_0 = 2c_u/\gamma$.

10.4.4 Inclined Granular Backfills in Active and Passive States

Rankine's lateral earth pressure equations (10.9), (10.11), (10.12), and (10.14) are derived for smooth vertical walls that retain backfills where the *ground level is horizontal.* For a smooth vertical wall retaining a granular backfill where the ground is inclined at an angle β as shown in Fig. 10.10, the active earth pressure σ'_a at any depth is also inclined at the same angle as shown, and is given by

$$\sigma'_a = K_a \gamma z$$

where

$$K_a = \cos\beta \frac{\cos\beta - \sqrt{\cos^2\beta - \cos^2\phi'}}{\cos\beta + \sqrt{\cos^2\beta - \cos^2\phi'}} \tag{10.20}$$

The resultant active thrust P_a, acts at $H/3$ from the bottom of the wall, inclined at β to the horizontal. Its magnitude is given by

$$P_a = 0.5 K_a \gamma H^2 \tag{10.15}$$

The equations are similar for sloping granular backfills in passive state. Here,

$$K_p = \cos\beta \frac{\cos\beta + \sqrt{\cos^2\beta - \cos^2\phi'}}{\cos\beta - \sqrt{\cos^2\beta - \cos^2\phi'}} \tag{10.21}$$

$$P_p = 0.5 K_p \gamma H^2 \tag{10.16}$$

Rankine's K_a and K_p values for the inclined backfills, determined from Eqs. (10.20) and (10.21) are given in Tables 10.2 and 10.3, respectively.

	Friction angle ϕ'(°)												
β(°)	**28**	**29**	**30**	**31**	**32**	**33**	**34**	**35**	**36**	**37**	**38**	**39**	**40**
0	0.3610	0.3470	0.3333	0.3201	0.3073	0.2948	0.2827	0.2710	0.2596	0.2486	0.2379	0.2275	0.2174
1	0.3612	0.3471	0.3335	0.3202	0.3074	0.2949	0.2828	0.2711	0.2597	0.2487	0.2380	0.2276	0.2175
2	0.3618	0.3476	0.3339	0.3207	0.3078	0.2953	0.2832	0.2714	0.2600	0.2489	0.2382	0.2278	0.2177
3	0.3627	0.3485	0.3347	0.3214	0.3084	0.2959	0.2837	0.2719	0.2605	0.2494	0.2386	0.2282	0.2181
4	0.3639	0.3496	0.3358	0.3224	0.3094	0.2967	0.2845	0.2726	0.2611	0.2500	0.2392	0.2287	0.2186
5	0.3656	0.3512	0.3372	0.3237	0.3105	0.2978	0.2855	0.2736	0.2620	0.2508	0.2399	0.2294	0.2192
6	0.3676	0.3531	0.3389	0.3253	0.3120	0.2992	0.2868	0.2747	0.2631	0.2518	0.2409	0.2303	0.2200
7	0.3701	0.3553	0.3410	0.3272	0.3138	0.3008	0.2883	0.2761	0.2644	0.2530	0.2420	0.2313	0.2209
8	0.3730	0.3580	0.3435	0.3294	0.3159	0.3027	0.2900	0.2778	0.2659	0.2544	0.2432	0.2325	0.2220
9	0.3764	0.3611	0.3463	0.3320	0.3182	0.3049	0.2921	0.2796	0.2676	0.2560	0.2447	0.2338	0.2233
10	0.3802	0.3646	0.3495	0.3350	0.3210	0.3074	0.2944	0.2818	0.2696	0.2578	0.2464	0.2354	0.2247
11	0.3846	0.3686	0.3532	0.3383	0.3241	0.3103	0.2970	0.2841	0.2718	0.2598	0.2482	0.2371	0.2263
12	0.3896	0.3731	0.3573	0.3421	0.3275	0.3134	0.2999	0.2868	0.2742	0.2621	0.2503	0.2390	0.2281
13	0.3952	0.3782	0.3620	0.3464	0.3314	0.3170	0.3031	0.2898	0.2770	0.2646	0.2527	0.2412	0.2301
14	0.4015	0.3839	0.3671	0.3511	0.3357	0.3209	0.3068	0.2931	0.2800	0.2674	0.2552	0.2435	0.2322
15	0.4086	0.3903	0.3729	0.3564	0.3405	0.3253	0.3108	0.2968	0.2834	0.2705	0.2581	0.2461	0.2346
16	0.4165	0.3975	0.3794	0.3622	0.3458	0.3302	0.3152	0.3008	0.2871	0.2739	0.2612	0.2490	0.2373
17	0.4255	0.4056	0.3867	0.3688	0.3518	0.3356	0.3201	0.3053	0.2911	0.2776	0.2646	0.2521	0.2401
18	0.4357	0.4146	0.3948	0.3761	0.3584	0.3415	0.3255	0.3102	0.2953	0.2817	0.2683	0.2555	0.2433
19	0.4473	0.4249	0.4039	0.3842	0.3657	0.3481	0.3315	0.3156	0.3006	0.2862	0.2724	0.2593	0.2467
20	0.4605	0.4365	0.4142	0.3934	0.3739	0.3555	0.3381	0.3216	0.3060	0.2911	0.2769	0.2634	0.2504
21	0.4758	0.4498	0.4259	0.4037	0.3830	0.3637	0.3455	0.3283	0.3120	0.2965	0.2818	0.2678	0.2545
22	0.4936	0.4651	0.4392	0.4154	0.3934	0.3729	0.3537	0.3356	0.3186	0.3025	0.2872	0.2727	0.2590
23	0.5147	0.4829	0.4545	0.4287	0.4050	0.3832	0.3628	0.3438	0.3259	0.3091	0.2932	0.2781	0.2638
24	0.5404	0.5041	0.4724	0.4440	0.4183	0.3948	0.3731	0.3529	0.3341	0.3164	0.2997	0.2840	0.2692
25	0.5727	0.5299	0.4936	0.4619	0.4336	0.4081	0.3847	0.3631	0.3431	0.3245	0.3070	0.2905	0.2750

TABLE 10.2 Rankine's K_a for Inclined Backfills [Eq. (10.20)]

	Friction angle ϕ'(°)												
β(°)	28	29	30	31	32	33	34	35	36	37	38	39	40
0	2.7698	2.8821	3.0000	3.1240	3.2546	3.3921	3.5371	3.6902	3.8518	4.0228	4.2037	4.3955	4.5989
1	2.7676	2.8798	2.9977	3.1217	3.2522	3.3897	3.5347	3.6877	3.8493	4.0201	4.2010	4.3927	4.5960
2	2.7610	2.8731	2.9909	3.1147	3.2451	3.3825	3.5273	3.6801	3.8415	4.0122	4.1929	4.3843	4.5874
3	2.7499	2.8618	2.9795	3.1032	3.2333	3.3704	3.5150	3.6675	3.8286	3.9990	4.1793	4.3704	4.5730
4	2.7344	2.8461	2.9635	3.0870	3.2168	3.3536	3.4978	3.6499	3.8106	3.9805	4.1603	4.3509	4.5530
5	2.7145	2.8260	2.9431	3.0662	3.1957	3.3320	3.4758	3.6274	3.7875	3.9569	4.1360	4.3259	4.5272
6	2.6903	2.8015	2.9182	3.0408	3.1699	3.3057	3.4489	3.5999	3.7594	3.9280	4.1064	4.2954	4.4959
7	2.6618	2.7726	2.8889	3.0110	3.1395	3.2748	3.4173	3.5676	3.7263	3.8941	4.0716	4.2596	4.4590
8	2.6290	2.7393	2.8551	2.9768	3.1046	3.2392	3.3810	3.5305	3.6884	3.8552	4.0316	4.2185	4.4167
9	2.5919	2.7018	2.8171	2.9381	3.0653	3.1991	3.3401	3.4887	3.6455	3.8112	3.9865	4.1722	4.3690
10	2.5507	2.6601	2.7748	2.8952	3.0216	3.1546	3.2946	3.4422	3.5980	3.7625	3.9365	4.1207	4.3161
11	2.5054	2.6142	2.7283	2.8479	2.9736	3.1057	3.2447	3.3912	3.5457	3.7090	3.8816	4.0643	4.2580
12	2.4559	2.5643	2.6777	2.7966	2.9213	3.0525	3.1904	3.3357	3.4890	3.6508	3.8219	4.0029	4.1948
13	2.4025	2.5102	2.6230	2.7411	2.8649	2.9950	3.1318	3.2759	3.4278	3.5881	3.7575	3.9368	4.1268
14	2.3450	2.4523	2.5643	2.6816	2.8045	2.9335	3.0691	3.2119	3.3623	3.5210	3.6887	3.8661	4.0540
15	2.2836	2.3903	2.5017	2.6182	2.7401	2.8680	3.0024	3.1437	3.2926	3.4496	3.6154	3.7908	3.9766
16	2.2183	2.3246	2.4353	2.5509	2.6718	2.7986	2.9316	3.0715	3.2188	3.3740	3.5379	3.7112	3.8947
17	2.1491	2.2549	2.3650	2.4798	2.5998	2.7254	2.8571	2.9955	3.1411	3.2945	3.4564	3.6275	3.8086
18	2.0759	2.1815	2.2910	2.4051	2.5240	2.6485	2.7788	2.9157	3.0596	3.2111	3.3709	3.5397	3.7183
19	1.9988	2.1042	2.2133	2.3266	2.4447	2.5680	2.6970	2.8323	2.9744	3.1240	3.2816	3.4481	3.6241
20	1.9176	2.0230	2.1318	2.2446	2.3618	2.4840	2.6116	2.7454	2.8857	3.0333	3.1888	3.3528	3.5262
21	1.8320	1.9379	2.0466	2.1589	2.2754	2.3965	2.5229	2.6551	2.7936	2.9392	3.0924	3.2540	3.4247
22	1.7417	1.8485	1.9575	2.0696	2.1855	2.3057	2.4308	2.5615	2.6983	2.8419	2.9928	3.1519	3.3198
23	1.6462	1.7546	1.8643	1.9766	2.0920	2.2115	2.3355	2.4647	2.5998	2.7414	2.8901	3.0466	3.2117
24	1.5443	1.6555	1.7667	1.8796	1.9950	2.1139	2.2370	2.3649	2.4983	2.6379	2.7844	2.9383	3.1006
25	1.4343	1.5502	1.6641	1.7783	1.8942	2.0129	2.1352	2.2620	2.3938	2.5316	2.6758	2.8273	2.9867

TABLE 10.3 Rankine's K_p for Inclined Backfills [Eq. (10.21)]

Example 10.3

A gravity retaining wall is being designed to retain 9.0 m of soil consisting of 7.0 m of sand and 2.0 m of clay as shown in Fig. 10.11a. Assuming the entire soil is in active state, determine the location and magnitude of the *total* thrust on the wall. The soil properties are as follows.

Soil property	Sand	Clay
Bulk unit weight (kN/m³)	17.0	
Saturated unit weight (kN/m³)	19.0	20.0
Effective friction angle (°)	34	25
Effective cohesion (kN/m²)	0	10

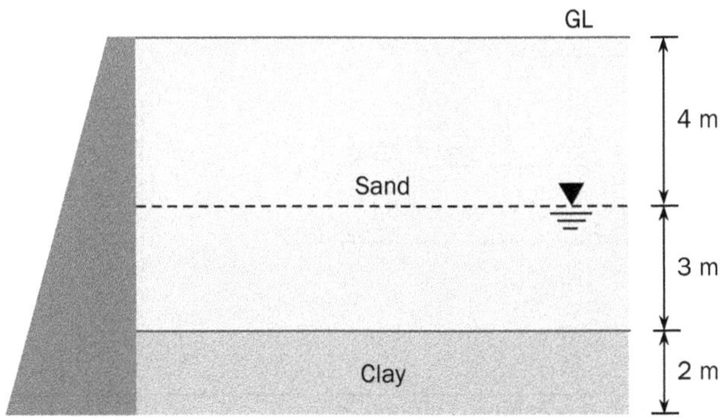

FIGURE 10.11a Gravity retaining wall for Example 10.3.

Solution For sand, $K_a = \tan^2(45 - 34/2) = 0.283$

For clay, $K_a = \tan^2(45 - 25/2) = 0.406$ and $\sqrt{K_a} = 0.637$

At the ground level, $\sigma'_v = 0$ and $\sigma'_a = 0$.

At the water table level,

$$\sigma'_v = (4)(17) = 68.0 \text{ kN/m}^2$$
$$\sigma'_a = (0.283)(68.0) = 19.2 \text{ kN/m}^2$$

At the bottom of the sand,

$$\sigma'_v = (4)(17) + (3)(19.0 - 9.81) = 95.6 \text{ kN/m}^2$$
$$\sigma'_a = (0.283)(95.6) = 27.0 \text{ kN/m}^2$$

At the top of the clay,

$$\sigma'_v = 95.6 \text{ kN/m}^2$$
$$\sigma'_a = K_a\sigma'_v - 2c'\sqrt{K_a} = (0.406)(95.6) - (2)(10)(0.637) = 26.1 \text{ kN/m}^2$$

At the bottom of the clay,

$$\sigma'_v = (4)(17.0) + (3)(19.0 - 9.81) + (2)(20.0 - 9.81) = 116.0 \text{ kN/m}^2$$

$$\sigma_a' = K_a\sigma_v' - 2c'\sqrt{K_a} = (0.406)(116.0) - (2)(10)(0.637) = 34.4 \text{ kN/m}^2$$

Pore water pressure at the bottom of the clay = (5) (9.81) = 49.1 kN/m²

The variation of σ_a' and u with depth are plotted *separately* as shown in Fig. 10.11b. The total active thrust P_a is given by the sum of the loads coming from the soil grains and water, shown by the six blocks in the figure. It is also possible to plot the total stress σ_a with depth.

From Fig. 10.11b, the following table can be prepared.

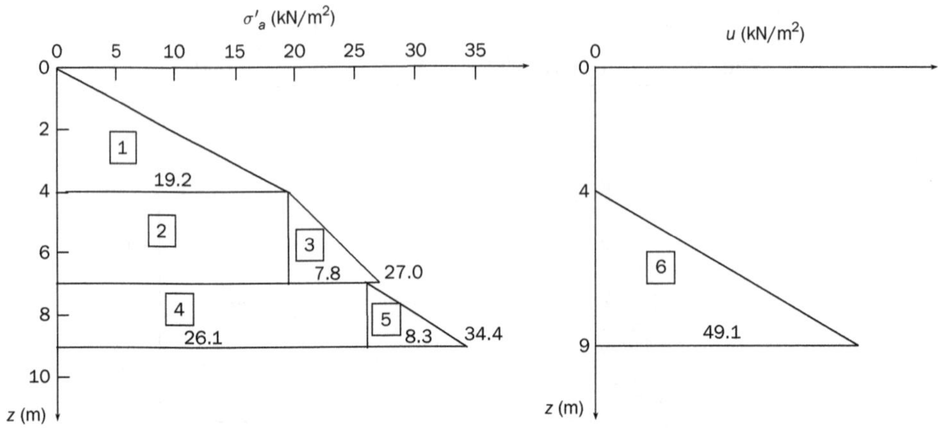

FIGURE 10.11b Lateral pressure distribution.

Block	Horizontal load (kN/m)	Height above base (m)	Moment (kN·m/m)
1	(0.5) (19.2)(4) = 38.4	6.33	243.2
2	(19.2) (3) = 57.6	3.50	201.6
3	(0.5) (7.8) (3) = 11.7	3.0	35.1
4	(26.1) (2) = 52.2	1.0	52.2
5	(0.5) (8.3) (2) = 8.3	0.667	5.47
6	(0.5) (49.1) (5) = 122.7	1.667	204.6
SUM	290.9		742.2

The total horizontal active thrust P_a = 290.9 kN/m. It acts at a height of 742.2/290.9 = 2.55 m above the bottom of the wall.

10.5 Coulomb's Earth Pressure Theory

Coulomb's (1776) theory was proposed about 80 years before Rankine's and the solution is a little more complex. However, Coulomb's theory allows the wall to be inclined and frictional, making the solutions more realistic.

Let us discuss Coulomb's active earth pressure theory for granular soils. Figure 10.12a shows a sloping granular backfill that is retained by a wall of height H, inclined at α to the horizontal. The slope angle of the backfill is β to the horizontal. It is assumed that the wall moves laterally to the left to allow the wedge ABC to slide down the failure plane AC, which is yet to be located. Let us consider a potential failure plane AC inclined at θ to the horizontal.

On the potential failure plane AC, which is a soil-soil interface, the reaction R is inclined at ϕ' to the normal to the failure plane as shown. At the soil-wall interface AB, the active thrust P_a is inclined at δ to the normal. The *wall friction angle* δ is in the range of 0 to ϕ', depending on the frictional characteristics of the wall-soil interface. Lower end of the range applies when the soil is in contact with smoother surfaces such as steel, timber, or precast concrete, and the upper end applies for rougher surfaces such as cast-in-place concrete. For very smooth walls $\delta = 0$ and for very rough walls $\delta = \phi'$. For concrete-soil interface δ is commonly taken as $^2/_3\,\phi'$, while it can be in the range of 0.5–0.8 ϕ'.

The soil wedge ABC is in equilibrium under three forces:

a. Self-weight W—magnitude known and is vertical in direction.
b. Reaction R—magnitude unknown and direction known.
c. Active thrust P_a—magnitude unknown and direction known.

From these, a force triangle can be drawn as shown in Fig. 10.12b.

From the force triangle,

$$\frac{W}{\sin(\alpha - \delta + \theta - \phi')} = \frac{P_a}{\sin(\theta - \phi')} \tag{10.22}$$

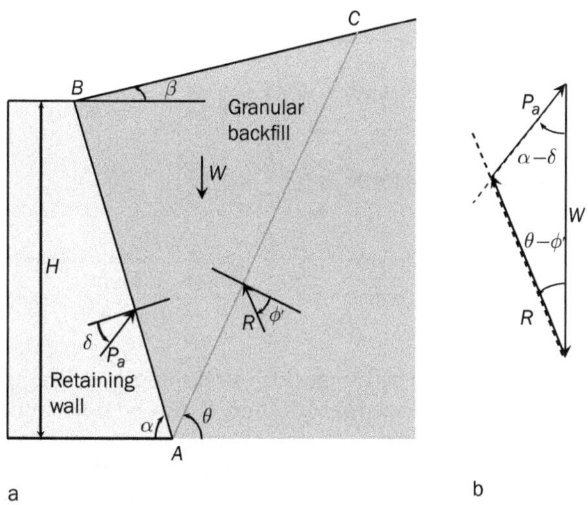

FIGURE 10.12 Coulomb's analysis in granular soils: (a) failure wedge, (b) force triangle.

The length AC is given by

$$AC = \frac{H}{\sin\alpha}\frac{\sin(\alpha+\beta)}{\sin(\theta-\beta)}$$

The self-weight W is given by

$$W = \frac{1}{2}\gamma H^2\left[\frac{\sin(\alpha+\beta)\sin(\alpha+\theta)}{\sin^2\alpha\sin(\theta-\beta)}\right] \tag{10.23}$$

From Eqs. (10.22) and (10.23),

$$\begin{aligned} P_a &= \frac{W\sin(\theta-\phi')}{\sin(\alpha-\delta+\theta-\phi')} \\ &= \frac{1}{2}\gamma H^2\left[\frac{\sin(\alpha+\beta)\sin(\alpha+\theta)}{\sin^2\alpha\sin(\theta-\beta)}\right]\frac{\sin(\theta-\phi')}{\sin(\alpha-\delta+\theta-\phi')} \end{aligned} \tag{10.24}$$

Varying the value of θ, different values of P_a are obtained. The maximum value of P_a occurs when

$$\frac{dP_a}{d\theta} = 0 \tag{10.25}$$

Solving Eq. (10.25) gives the value of θ which defines the failure plane. Substituting this value in Eq. (10.24) gives

$$P_a = 0.5K_a\gamma H^2$$

where

$$K_a = \frac{\sin^2(\alpha+\phi')}{\sin^2\alpha\sin(\alpha-\delta)\left[1+\sqrt{\dfrac{\sin(\phi'+\delta)\sin(\phi'-\beta)}{\sin(\alpha-\delta)\sin(\alpha+\beta)}}\right]^2} \tag{10.26a}$$

An alternate expression for K_a is

$$K_a = \left(\frac{\sin(\alpha+\phi')\,/\,\sin\alpha}{\sqrt{\sin(\alpha-\delta)}+\sqrt{\dfrac{\sin(\phi'+\delta)\sin(\phi'-\beta)}{\sin(\alpha+\beta)}}}\right)^2 \tag{10.26b}$$

When the ground level of the backfill is sloping and irregular, W has to be determined from the area of the wedge ABC and P_a can be determined from Eq. (10.22) for a set of values for θ. The plot of P_a versus θ will be convex upward and the maximum value of P_a can be determined from the plot. For $\alpha = 90°$, $\delta = 0$, and $\beta = 0$, K_a from Eq. (10.26) reduces to what is given by Rankine's theory for smooth vertical walls with horizontal backfills. For smooth vertical walls retaining a sloping backfill, K_a and K_p values determined from Rankine's and Coulomb's expressions are slightly different.

The directions of the thrusts are different too; Rankine's one is parallel to the backfill slope, and Coulomb's is normal to the wall and hence horizontal.

In situations where the granular soil is in passive state,

$$P_p = 0.5K_p\gamma H^2$$

where

$$K_p = \frac{\sin^2(\alpha - \phi')}{\sin^2\alpha\sin(\alpha+\delta)\left[1+\sqrt{\dfrac{\sin(\phi'+\delta)\sin(\phi'+\beta)}{\sin(\alpha+\delta)\sin(\alpha+\beta)}}\right]^2} \tag{10.27a}$$

An alternate expression for K_p is

$$K_p = \left(\frac{\sin(\alpha-\phi')\,/\,\sin\alpha}{\sqrt{\sin(\alpha+\delta)}-\sqrt{\dfrac{\sin(\phi'+\delta)\sin(\phi'+\beta)}{\sin(\alpha+\beta)}}}\right)^2 \tag{10.27b}$$

Values of Coulomb's K_a from Eq. (10.26a) or (10.26b) for $\delta = {}^2/_3\,\phi'$ are given in Table 10.4.

Coulomb's theory does not give the location of the active or passive thrust. It can be assumed acting at $H/3$ from the bottom of the wall, inclined at angle δ to the normal to the wall.

Allowing friction along the soil-wall interface leads to a reduction in P_a and increase in P_p, compared to what is expected when the wall is smooth. In reality, the failure planes (or more appropriately surfaces) are curved (approximately a log spiral) near the bottom of the wall, which leads to a slight underestimation of the active thrust. The error is more significant on the passive side, especially when $\delta > \phi'/3$, grossly overestimating the passive thrust. More realistic estimates of P_p can be obtained by neglecting the wall friction (i.e., $\delta = 0$) or by using Rankine's theory.

In granular soils, the soil wedge ABC is in equilibrium under three forces W, P_a, and R. In cohesive soils, two additional forces have to be considered: (a) adhesive resistance C_a along the soil-wall interface AB, and (b) cohesive resistance C along the failure plane AC. The cohesive resistance C per unit width of the wall can be determined by multiplying the length AC by the cohesion c. The adhesive resistance C_a per unit width of the wall can be determined by multiplying the length AB by the adhesion c_a along the wall. Typically, c_a is taken as 0.5–0.7 of the cohesion c. The soil wedge ABC is in equilibrium under five forces W, R, P_a, C, and C_a. With known magnitudes and directions for W, C, and C_a (and known directions and unknown magnitudes for P_a and R), a force polygon can be drawn to determine P_a for a specific value of θ. By repeating this for different values of θ, the maximum value of P_a can be determined.

For a vertical wall, retaining a sloping backfill, Coulomb's K_a from Eq. (10.26a) or (10.26b) becomes

$$K_a = \left(\frac{\cos\phi'}{\sqrt{\cos\delta}+\sqrt{\dfrac{\sin(\phi'+\delta)\sin(\phi'-\beta)}{\cos\beta}}}\right)^2 \tag{10.28}$$

		$\alpha(°)$						
$\beta(°)$	$\phi'(°)$	**60**	**65**	**70**	**75**	**80**	**85**	**90**
0	28	0.6417	0.5662	0.5026	0.4481	0.4007	0.3588	0.3213
0	29	0.6306	0.5547	0.4908	0.4362	0.3886	0.3467	0.3091
0	30	0.6198	0.5435	0.4794	0.4245	0.3769	0.3349	0.2973
0	31	0.6093	0.5326	0.4682	0.4133	0.3655	0.3235	0.2860
0	32	0.5991	0.5220	0.4574	0.4023	0.3545	0.3125	0.2750
0	33	0.5891	0.5117	0.4469	0.3917	0.3439	0.3019	0.2645
0	34	0.5795	0.5017	0.4367	0.3813	0.3335	0.2916	0.2543
0	35	0.5701	0.4919	0.4267	0.3713	0.3235	0.2816	0.2444
0	36	0.5610	0.4824	0.4170	0.3615	0.3137	0.2719	0.2349
0	37	0.5521	0.4732	0.4076	0.3520	0.3042	0.2626	0.2257
0	38	0.5434	0.4642	0.3984	0.3428	0.2950	0.2535	0.2168
0	39	0.5350	0.4554	0.3894	0.3337	0.2861	0.2447	0.2082
0	40	0.5268	0.4468	0.3806	0.3250	0.2774	0.2361	0.1999
0	41	0.5188	0.4384	0.3721	0.3164	0.2989	0.2278	0.1918
0	42	0.5111	0.4303	0.3638	0.3081	0.2607	0.2197	0.1840
5	28	0.7068	0.6191	0.5461	0.4843	0.4311	0.3845	0.3431
5	29	0.6936	0.6056	0.5325	0.4707	0.4175	0.3709	0.3295
5	30	0.6809	0.5926	0.5194	0.4575	0.4043	0.3578	0.3165
5	31	0.6685	0.5800	0.5067	0.4448	0.3916	0.3451	0.3039
5	32	0.6565	0.5678	0.4943	0.4324	0.3793	0.3329	0.2919
5	33	0.6448	0.5559	0.4823	0.4204	0.3673	0.3211	0.2803
5	34	0.6335	0.5443	0.4707	0.4088	0.3558	0.3097	0.2691
5	35	0.6225	0.5331	0.4594	0.3975	0.3446	0.2987	0.2583
5	36	0.6118	0.5222	0.4484	0.3866	0.3338	0.2881	0.2479
5	37	0.6015	0.5116	0.4377	0.3759	0.3233	0.2778	0.2379
5	38	0.5914	0.5012	0.4273	0.3656	0.3132	0.2679	0.2282
5	39	0.5816	0.4912	0.4172	0.3556	0.3033	0.2582	0.2188
5	40	0.5720	0.4814	0.4074	0.3458	0.2937	0.2489	0.2098
5	41	0.5628	0.4718	0.3978	0.3363	0.2844	0.2398	0.2011
5	42	0.5538	0.4626	0.3885	0.3271	0.2753	0.2311	0.1927
10	28	0.7859	0.6834	0.5992	0.5287	0.4686	0.4163	0.3702
10	29	0.7697	0.6672	0.5831	0.5128	0.4528	0.4007	0.3548
10	30	0.7541	0.6516	0.5676	0.4974	0.4376	0.3857	0.3400
10	31	0.7390	0.6365	0.5527	0.4826	0.4230	0.3713	0.3259
10	32	0.7244	0.6220	0.5382	0.4683	0.4089	0.3575	0.3123
10	33	0.7103	0.6079	0.5242	0.4545	0.3953	0.3442	0.2993
10	34	0.6967	0.5942	0.5107	0.4412	0.3822	0.3314	0.2868
10	35	0.6835	0.5810	0.4976	0.4283	0.3696	0.3190	0.2748
10	36	0.6707	0.5682	0.4849	0.4158	0.3574	0.3072	0.2633

TABLE 10.4 Coulomb's K_a from Eq. (10.26a) or (10.26b) for $\delta = \frac{2}{3}\phi'$

		$\alpha(°)$						
$\beta(°)$	$\phi'(°)$	60	65	70	75	80	85	90
10	37	0.6583	0.5558	0.4726	0.4037	0.3456	0.2957	0.2522
10	38	0.6464	0.5438	0.4607	0.3920	0.3342	0.2846	0.2415
10	39	0.6347	0.5321	0.4492	0.3807	0.3232	0.2740	0.2313
10	40	0.6235	0.5208	0.4379	0.3697	0.3125	0.2637	0.2214
10	41	0.6125	0.5097	0.4270	0.3590	0.3021	0.2537	0.2119
10	42	0.6019	0.4990	0.4164	0.3487	0.2921	0.2441	0.2027
15	28	0.8886	0.7671	0.6685	0.5869	0.5179	0.4585	0.4065
15	29	0.8672	0.7463	0.6483	0.5672	0.4987	0.4397	0.3881
15	30	0.8469	0.7265	0.6291	0.5484	0.4804	0.4219	0.3707
15	31	0.8275	0.7076	0.6107	0.5305	0.4629	0.4049	0.3541
15	32	0.8089	0.6895	0.5930	0.5134	0.4462	0.3887	0.3384
15	33	0.7910	0.6721	0.5761	0.4969	0.4303	0.3732	0.3234
15	34	0.7739	0.6554	0.5599	0.4811	0.4150	0.3583	0.3091
15	35	0.7575	0.6393	0.5442	0.4660	0.4003	0.3442	0.2954
15	36	0.7417	0.6239	0.5292	0.4514	0.3862	0.3306	0.2823
15	37	0.7265	0.6089	0.5146	0.4373	0.3726	0.3175	0.2698
15	38	0.7118	0.5945	0.5006	0.4238	0.3596	0.3050	0.2578
15	39	0.6976	0.5806	0.4871	0.4107	0.3470	0.2929	0.2463
15	40	0.6839	0.5672	0.4740	0.3980	0.3348	0.2813	0.2353
15	41	0.6707	0.5541	0.4614	0.3858	0.3231	0.2702	0.2248
15	42	0.6580	0.5415	0.4491	0.3740	0.3118	0.2594	0.2146
20	28	1.0370	0.8881	0.7689	0.6715	0.5900	0.5205	0.4602
20	29	1.0050	0.8581	0.7407	0.6446	0.5643	0.4958	0.4364
20	30	0.9753	0.8303	0.7144	0.6196	0.5403	0.4728	0.4142
20	31	0.9477	0.8044	0.6898	0.5961	0.5179	0.4513	0.3935
20	32	0.9219	0.7800	0.6667	0.5741	0.4968	0.4311	0.3742
20	33	0.8976	0.7570	0.6448	0.5533	0.4770	0.4121	0.3559
20	34	0.8746	0.7352	0.6241	0.5335	0.4581	0.3941	0.3388
20	35	0.8528	0.7145	0.6044	0.5148	0.4403	0.3771	0.3225
20	36	0.8321	0.6948	0.5857	0.4970	0.4233	0.3609	0.3071
20	37	0.8124	0.6760	0.5678	0.4799	0.4071	0.3455	0.2925
20	38	0.7935	0.6580	0.5507	0.4637	0.3916	0.3308	0.2787
20	39	0.7755	0.6408	0.5343	0.4481	0.3768	0.3168	0.2654
20	40	0.7583	0.6242	0.5185	0.4331	0.3627	0.3034	0.2529
20	41	0.7417	0.6084	0.5034	0.4188	0.3491	0.2906	0.2408
20	42	0.7259	0.5931	0.4888	0.4049	0.3360	0.2784	0.2294

Table 10.4 Coulomb's K_a from Eq. (10.26a) or (10.26b) for $\delta = \frac{2}{3}\phi'$ *(Continued)*

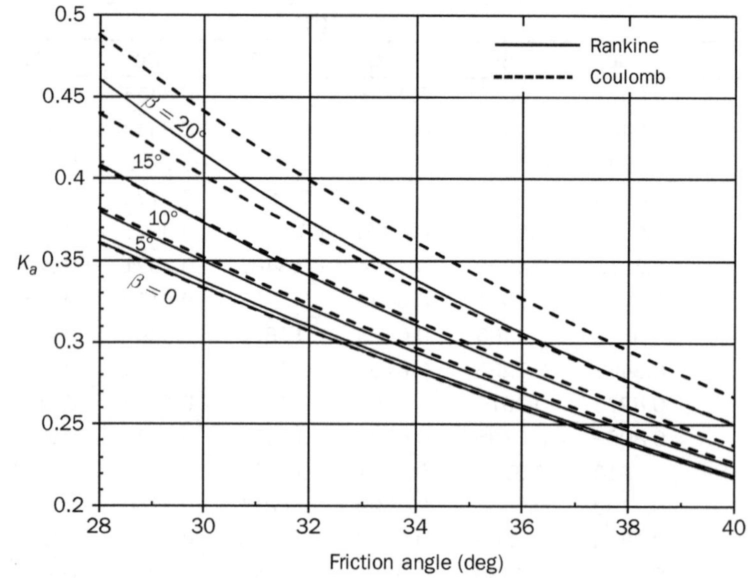

FIGURE 10.13 Rankine's and Coulomb's K_a values for smooth vertical walls with sloping backfills.

For a vertical wall, retaining a horizontal backfill, Eq. (10.28) becomes

$$K_a = \left(\frac{\cos\phi'}{\sqrt{\cos\delta} + \sqrt{\sin(\phi' + \delta)\sin\phi'}} \right)^2 \tag{10.29}$$

For smooth vertical wall, retaining a sloping backfill, Coulomb's K_a from Eq. (10.26a) or (10.26b) becomes

$$K_a = \left(\frac{\cos\phi'}{1 + \sqrt{\frac{\sin\phi' \sin(\phi' - \beta)}{\cos\beta}}} \right)^2 \tag{10.30}$$

Similarly, the expressions for K_p can be deduced for these different scenarios.

Figure 10.13 compares Rankine's and Coulomb's K_a for smooth vertical walls with sloping granular backfills. Both Rankine's and Coulomb's theories suggest increasing K_a values with increasing slope angle β. Coulomb's values are slightly larger than Rankine's. Only for horizontal backfills they are the same.

10.6 Lateral Earth Pressures Based on Elastic Analysis

In this section, the soil is treated as a linear elastic continuum without worrying about active or passive states. From the theory of elasticity, and modified by some laboratory test results, expressions have been developed for some special cases.

10.6.1 Line Load at the Ground Level

Figure 10.14 shows a line load Q kN/m acting at a horizontal distance aH from the vertical wall of height H. At a depth $z = bH$, the horizontal earth pressure σ'_h acting on the wall, the resultant load P_h and its height from the bottom $\bar{z}$ are as follows.

For $a > 0.4$,

$$\sigma'_h = \frac{4Q}{\pi H}\frac{a^2 b}{(a^2 + b^2)^2} \tag{10.31}$$

$$P_h = \frac{0.64Q}{1 + a^2} \tag{10.32}$$

$$\bar{z} \approx 0.5H \tag{10.33}$$

For $a \leq 0.4$,

$$\sigma'_h = \frac{0.203\,Q}{H}\frac{b}{(0.16 + b^2)^2} \tag{10.34}$$

$$P_h = 0.55Q \tag{10.35}$$

$$\bar{z} \approx 0.6H \tag{10.36}$$

Note that for $a \leq 0.4$, σ'_h distribution is independent of a.

10.6.2 A Uniform Strip Load at the Ground Level

Highways and railways are good examples of strip loads. When they run parallel to sheet pile walls, they can impose lateral loads on the sheet piles. From elastic analysis, due to a uniformly loaded strip load of intensity q as shown in Fig. 10.15, the lateral earth pressure σ'_h acting on the wall is given by (Jarquio, 1981)

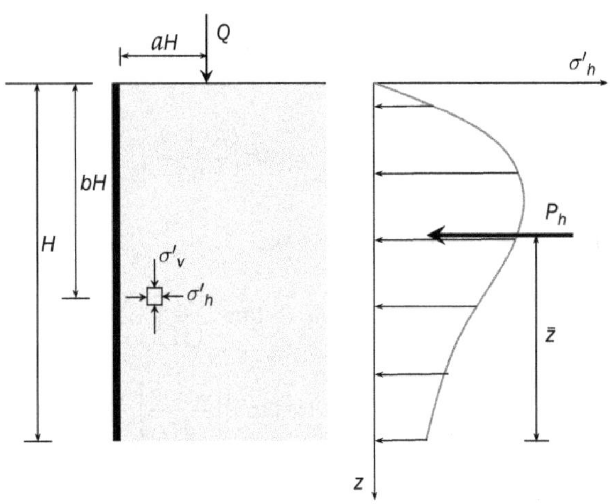

FIGURE 10.14 σ'_h due to a line load.

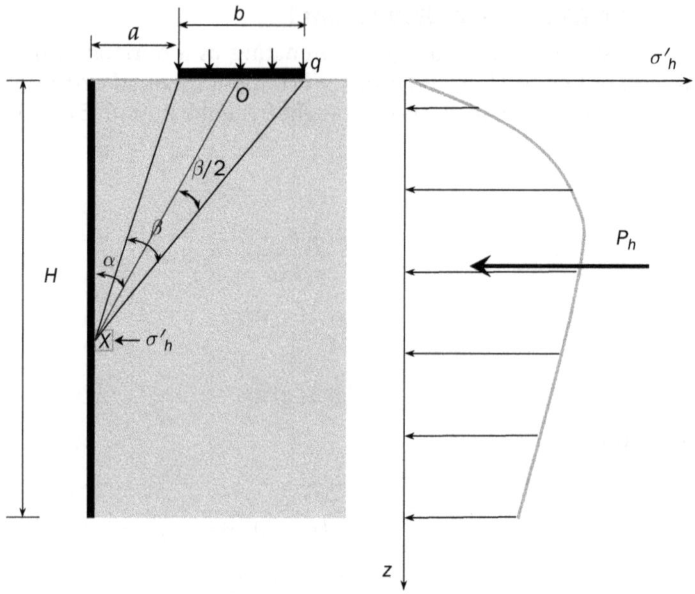

FIGURE 10.15 σ'_h due to a uniform strip load.

$$\sigma'_h = \frac{q}{\pi}(\beta - \sin\beta\cos 2\alpha) \tag{10.37}$$

where the angles α and β are in radians. In reality, the soil mass yields, and σ'_h is better approximated as

$$\sigma'_h = \frac{2q}{\pi}(\beta - \sin\beta\cos 2\alpha) \tag{10.38}$$

Note that the point O is not the mid-point of the strip load; it is the point where the bisector of the angle β intersects the strip load. The resultant lateral load P_h is given by (Jarquio, 1981)

$$P_h = qH\left(\frac{\theta_2 - \theta_1}{90}\right) \tag{10.39}$$

where θ_1 and θ_2 are in degrees defined as

$$\theta_1 = \tan^{-1}\left(\frac{a}{H}\right) \tag{10.40}$$

$$\theta_2 = \tan^{-1}\left(\frac{a+b}{H}\right) \tag{10.41}$$

Example 10.4

A 10-m-high gravity retaining wall retains a granular backfill ($\phi' = 33°$, $\delta = 22°$, and $\gamma = 18.5$ kN/m^3) as shown in Fig. 10.16. Using the graphical procedure discussed in Sec. 10.5 under Coulomb's method, assuming potential failure planes with $\theta = 55–80°$, determine the active thrust P_a.

Compare the above value of P_a with the one computed using K_a from Eq. (10.26a).

Solution The values of W, computed from Eq. (10.23) and P_a computed from Eq. (10.24) for the different values of θ are summarized below. The plot of P_a versus θ is shown in Fig. 10.17, where it can be seen that the maximum value for P_a is 562 kN/m, which should be taken as the active thrust P_a.

θ (deg)	W (kN/m)	P_a (kN/m)
55	1332.2	550.7
56	1296.2	554.4
57	1261.3	557.3
58	1227.6	559.5
59	1194.9	561.1
60	1163.2	562.0
61	1132.4	562.2
62	1102.4	562.0
63	1073.3	561.2
64	1044.9	559.9
65	1017.2	558.1
66	990.2	555.8
67	963.9	553.2
68	938.1	550.1
69	912.9	546.6
70	888.1	542.8
71	863.9	538.5
72	840.2	534.0
73	816.9	529.0
74	794.0	523.8
75	771.5	518.2
76	749.3	512.3
77	727.5	506.1
78	706.0	499.6
79	684.9	492.7
80	664.0	485.6

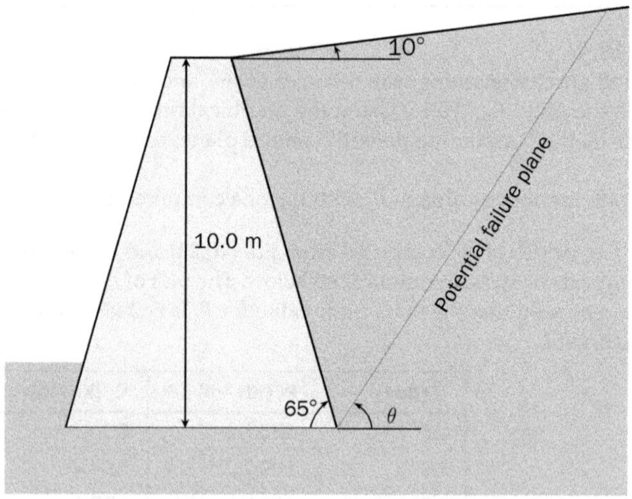

FIGURE 10.16 Gravity retaining wall for Example 10.4.

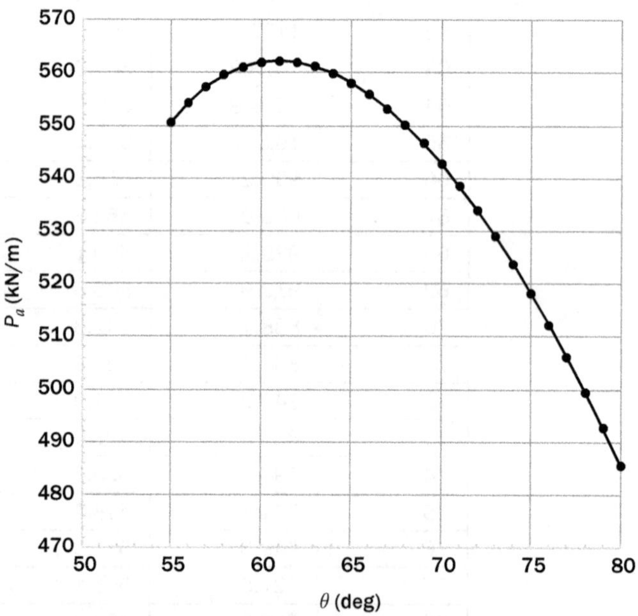

FIGURE 10.17 Plot of P_a vs. θ.

From Eq. (10.26a)

$$K_a = \frac{\sin^2(\alpha + \phi')}{\sin^2\alpha \sin(\alpha - \delta)\left[1 + \sqrt{\dfrac{\sin(\phi' + \delta)\sin(\phi' - \beta)}{\sin(\alpha - \delta)\sin(\alpha + \beta)}}\right]^2}$$

$$K_a = \frac{\sin^2(65 + 33)}{\sin^2 65 \sin(65 - 22)\left[1 + \sqrt{\dfrac{\sin(33 + 22)\sin(33 - 10)}{\sin(65 - 22)\sin(65 + 10)}}\right]^2} = 0.6078$$

Therefore,

$$P_a = 0.5\, K_a \gamma H^2 = 0.5 \times 0.6078 \times 18.5 \times 10^2 = \mathbf{562\ kN/m}$$

10.7 Main Points

1. When the soil is at rest (K_0 state), there are no lateral strains and $\sigma'_h / \sigma'_v = K_0$ at any point within the soil.
2. In clays, K_0 increases with the plasticity index. In granular soils, K_0 decreases with increasing friction angles.
3. At active state, $\sigma'_a = K_a\sigma'_v - 2c'\sqrt{K_a}$ where $K_a = \dfrac{1 - \sin\phi'}{1 + \sin\phi'} = \tan^2\left(45 - \dfrac{\phi'}{2}\right)$. The failure plane is inclined at $45 + \phi'/2$ to horizontal.
4. At passive state, $\sigma'_p = K_p\sigma'_v + 2c'\sqrt{K_p}$ where $K_p = \dfrac{1 + \sin\phi'}{1 - \sin\phi'} = \tan^2\left(45 + \dfrac{\phi'}{2}\right)$. The failure plane is inclined at $45 - \phi'/2$ to the horizontal.
5. Elastic analysis does not allow for wall movement and hence does not distinguish active, passive, or K_0 states.

Review Exercises

10.1 Circle the correct answer in the following five questions.

a. In normally consolidated soils, which of the following has the largest lateral earth pressure coefficient?
(i) Active state (ii) Passive state (iii) At-rest state

b. In which of the following states is the soil not at failure?
(i) Active state (ii) Passive state (iii) At-rest state

c. In which of the following states must the lateral earth pressure coefficient be always less than one?
(i) Active state (ii) Passive state (iii) At-rest state

d. What is the angle between the failure plane and the horizontal plane at active state?
(i) 45 degrees (ii) $45 + \phi'/2$ degrees (iii) $45 - \phi'/2$ degrees

Answer: Passive state, At-rest state, Active state, 45 + $\phi'/2$ degrees

10.2 The backfill retained by a basement wall shown in Fig. RE10.2 consists of two sand layers, compacted at different densities. The properties of the sand are as follows.
Sand 1: Unit weight = 17.5 kN/m^3, friction angle = 32°
Sand 2: Unit weight = 18.5 kN/m^3, friction angle = 36°
Assuming K_0 condition, determine the magnitude and location of the thrust on the wall.

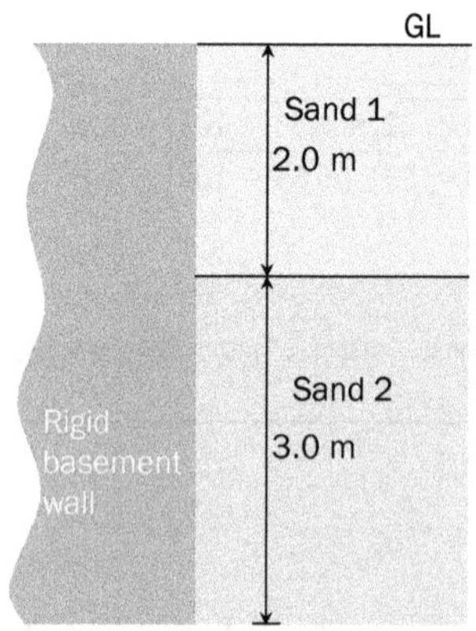

Figure RE10.2 Basement wall.

Answer: 94.1 kN/m at 1.70 m above the bottom of the wall

10.3 A 7-m-high gravity retaining wall retains 4 m of sand underlain by 3 m of clayey sand (Fig. RE10.3). The strength parameters and unit weights of the backfill soils are shown in the figure. Assuming that the entire backfill is in active state, determine the location and magnitude of the thrust on the wall.

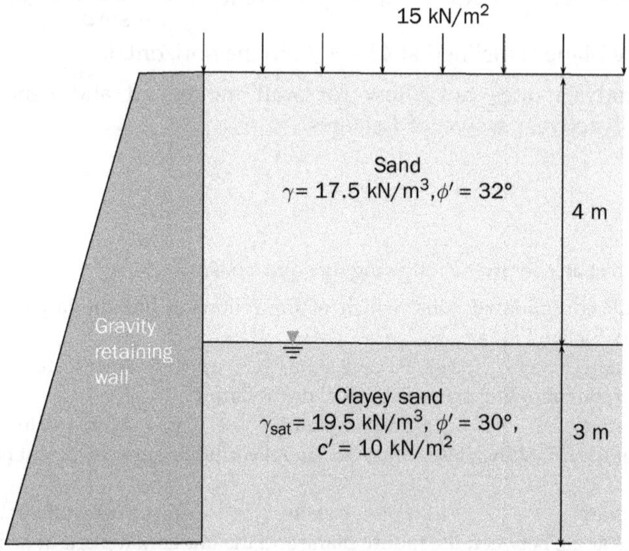

Figure RE10.3 Gravity retaining wall.

Answer: 170.3 kN/m acting at 2.42 m above the bottom of the wall

10.4 A smooth vertical wall retains a clay backfill with $c' = 15$ kN/m^2, $\phi' = 24°$, and $\gamma = 19.0$ kN/m^3. If the clay is in active state, determine the following:

a. The maximum tensile stress within the clay.
b. The depth of the tensile cracks.

Answer: 19.5 kN/m^2, 2.43 m

10.5 A 6-m-high smooth vertical wall retains a sandy back fill ($\gamma = 18.0$ kN/m^3, $\phi' = 33°$) which is sloping at an angle of 15° to the horizontal.

a. If the wall is allowed to move away and allow the sand to reach the active state, what would be the thrust on the wall?
b. If the wall is pushed toward the sand to make the sand reach the passive state, what would be the thrust on the wall?

Answer: 105.4 kN/m, which acts at 15° to the horizontal, at 2 m from the bottom of the wall, 969.4 kN/m, which acts at 15° to the horizontal, at 2 m from the bottom of the wall.

10.6 A concrete gravity retaining wall retains a sandy back fill of 5 m height as shown in Fig. RE10.6. The sand has the following properties:

- Friction angle = 34°
- Moisture content = 22.5%
- Specific gravity of the soil grains = 2.65
- Hydraulic conductivity = 2×10^{-2} cm/s

a. Determine the flow rate beneath the wall in m^3/day per meter.
b. Determine the safety factor with respect to piping.
c. Determine the uplift on the retaining wall from the pore water pressures at E and F. Assess the possibility of the retaining wall being lifted. Assume unit weight of concrete at 24 kN/m^3.

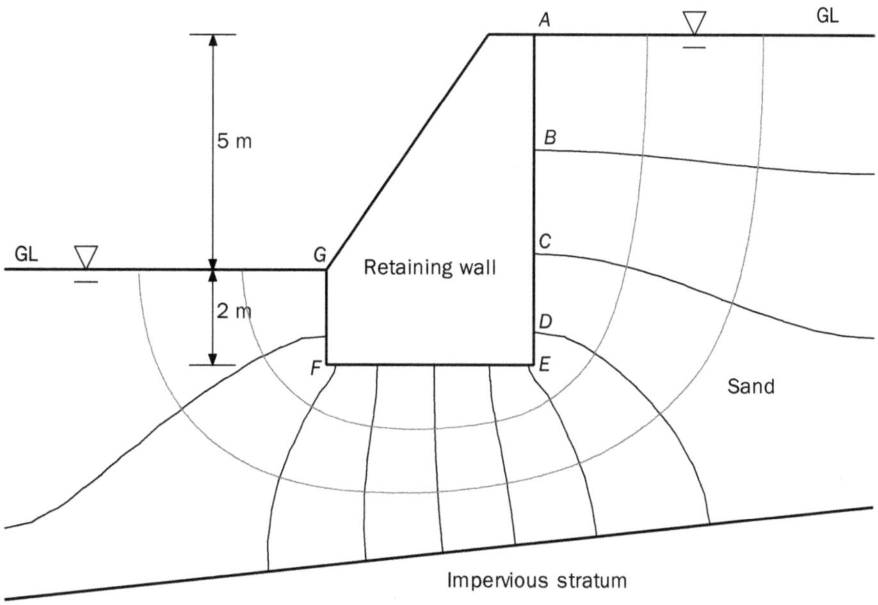

FIGURE RE10.6 Concrete gravity retaining wall and the flow net.

d. Assuming that the backfill retained is in active state, determine the total lateral earth pressure at the bottom of the wall E.

Note: The crest and base widths of the wall are scaled from Fig. RE10.6 as 1.0 and 4.5 m, respectively.

Answer: 25.9 m³/day per m width; 2.91; 50.5 kN/m², 28.0 kN/m², 176.6 kN/m; 75.8 kN/m²

10.7 Refer to Fig. 10.12, where $\alpha = 60°$, $\beta = 10°$, $\phi' = 36°$, $\delta = 24°$, $\gamma = 19.0\ \text{kN/m}^3$, and $H = 8$ m.

a. Find the load P_a using Eq. (10.24) assuming different failure planes. Select the maximum value as the active thrust.

b. How does it compare with the active load computed using the K_a determined from Eq. (10.26)?

Answer: 407.6 kN/m; 407.7 kN/m, compares well

10.8 Refer to Fig. 10.14, where a 6-m-high vertical wall retains an elastic soil mass. If a line load of 10 kN/m is applied at a horizontal distance of (a) 1 m and (b) 4 m from the wall, plot the distribution of lateral pressures on the wall. What are the maximum values for σ'_h?

Answer: 1.708 kN/m², 1.029 kN/m²

10.9 Refer to Fig. 10.15, where a 6-m-high vertical wall retains an elastic soil mass. If a strip load applies a uniform pressure at the ground level such that $a = 3$ m, $b = 2$ m, $H = 6$ m, $q = 50$ kN/m², plot the variation of lateral earth pressure on the wall with depth. Find the magnitude and depth for the maximum value of σ'_h.

Answer: 9.5 kN/m² at 2.0 m depth

10.10 A 10-m-high rough vertical wall retains a granular backfill with a slope of 15°. The parameters for the granular backfill are: $\phi' = 34°$, $\delta = 22°$, and $\gamma = 18.5\ \text{kN/m}^3$.

a. Carry out Coulomb's graphical procedure and determine the active load P_a.

b. Determine P_a using the value of K_a obtained from Eq. (10.28).

Answer: 286 kN/m

10.11 A 10-m-high rough inclined wall retains a granular backfill as shown in Fig. RE10.11. The parameters for the granular backfill are: $\phi' = 35°$, $\delta = 24°$, and $\gamma = 18.5\ \text{kN/m}^3$.

a. Carry out Coulomb's graphical procedure and determine the active load P_a.

b. Determine P_a using the value of K_a obtained from Eq. (10.26).

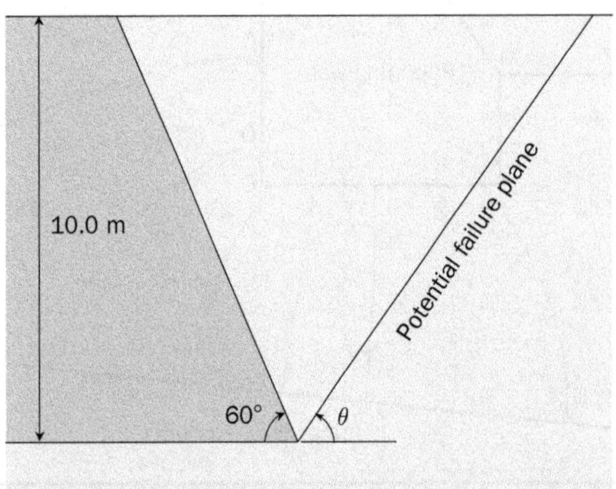

FIGURE RE10.11 Retaining wall.

Answer: 530 kN/m

10.12 A 7.0-m-high retaining wall shown in Fig. RE10.12 retains a granular backfill where $\phi' = 30°$, $\delta = 20°$, and $\gamma = 18.5$ kN/m^3.

a. Considering a vertical plane shown by the dashed line *AB* passing through the heel, determine the location, direction, and magnitude of the active thrust P_{aR} using Rankine's equation.
b. From the force triangle for the soil wedge *ABC* between the wall and the vertical plane, determine the magnitude and direction of the thrust P_{a*} on the wall *AC*.
c. Determine the active thrust P_{aC} on the wall using Coulomb's method.
d. Compare the two solutions.

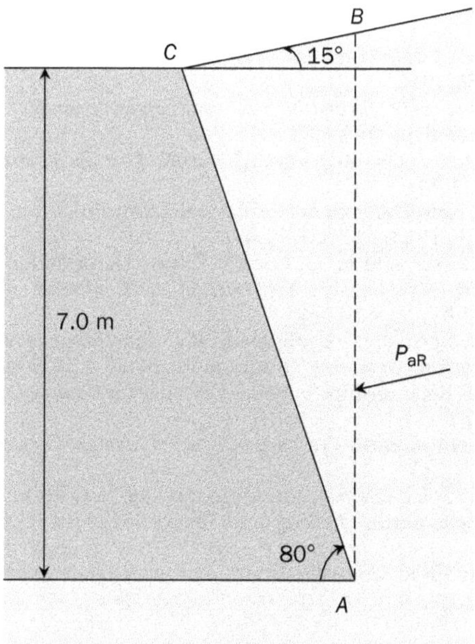

FIGURE RE10.12 Retaining wall.

Answer: 185.3 kN/m, 2.44 m above A, inclined at 15° to horizontal; 222.2 kN/m inclined at 53.6° to the vertical; 217.7 kN/m inclined at 60° to the vertical

10.13 A 5.0-m-high vertical wall shown in Fig. RE10.13 retains a granular backfill where $\phi' = 32°$, $\delta = 21°$, and $\gamma = 18.5$ kN/m^3. A line load of 70 kN/m has to be located at the ground level, at distance *a* from the wall. Using Coulomb's approach, determine the minimum value of

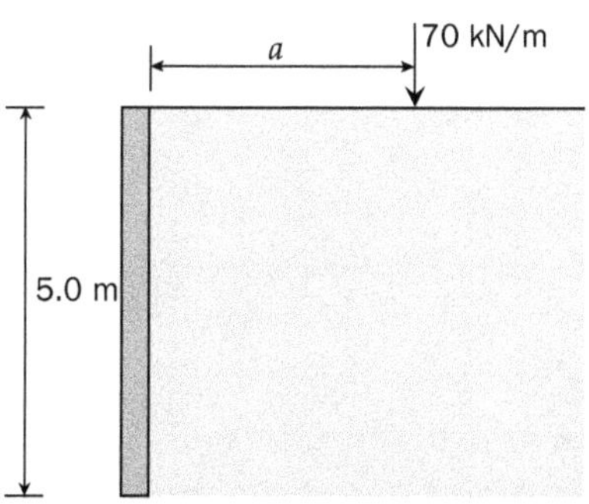

FIGURE RE10.13 Vertical wall.

a such that the active thrust on the wall does not exceed the value determined without the line load.

Hint: Consider trial failure planes meeting the ground surface at both sides of the line load.

Answer: 5.27 m

References

Alpan, I. "The Empirical Evaluation of the Coefficient K_0 and K_{0r}," *Soils and Foundations*, 7(1):31–40, 1967.

Coulomb, C. A. "Essai sur une Application des Règles de Maximus et Minimis à Quelques Problèmes de Statique, Relatifs à l'Architecture," Mémoires de Mathématique et de Physique, Présentés a l'Académie Royale des Sciences, par divers Savans, et lûs dans ses Assemblées, Paris, 7 (Volume for 1773 published in 1776):343–382, 1776.

Jaky, J. "The Coefficient of Earth Pressure at Rest" (in Hungarian), *Journal of the Society of Hungarian Architects and Engineers*, 78(22):355–358, 1944.

Jaky, J. "Earth Pressures in Silos," *Proceedings of the Second International Conference on Soil Mechanics and Foundation Engineering*, Rotterdam, I, 1948, pp. 103–107.

Jarquio, R. "Total Lateral Surcharge Pressures due to Strip Loads," *Journal of the Geotechnical Engineering Division*, ASCE, 107(GT10):1424–1428, 1981.

Kavazanjian, E., Jr., and J. K. Mitchell. "Time Dependence of Lateral Earth Pressure," *Journal of the Geotechnical Engineering Division*, ASCE, 110(4):530–533, 1984.

Ladd, C. C., R. Foote, K. Ishihara, F. Schlosser, and H. G. Poulos. "Stress Deformation and Strength Characteristics," State-of-the-art Report, *Proceedings of the 9th International Conference on Soil Mechanics and Foundation Engineering*, Tokyo, 2, 1977, pp. 421–494.

Massarsch, K. R. "Lateral Earth Pressure in Normally Consolidated Clay," *Proceedings of the 7th European Conference on Soil Mechanics and Foundation Engineering*, Brighton, England, 2, 1979, pp. 245–250.

Mayne, P. W., and F. H. Kulhawy. "K_0-OCR Relationships in Soil," *Journal of the Geotechnical Engineering Division*, ASCE, 108(GT6):851–872, 1982.

Mesri, G., and T. M. Hayat. "The Coefficient of Earth Pressure at Rest," *Canadian Geotechnical Journal*, 30(4):647–666, 1993.

Mitchell, J. K., and K. Soga. *Fundamentals of Soil Behaviour*, 3rd Edition, Wiley, New York, 2005.

Rankine, W. J. M. "On the Stability of Loose Earth," *Philosophical Transaction of the Royal Society*, 147, London, 1857.

Schmertmann, J. H. "A Simple Question about Consolidation," *Journal of Geotechnical Engineering*, 109(1):119–122, 1983.

PART 2

Applications

CHAPTER 11

Site Investigation

11.1 Introduction

When it is proposed to construct a building, embankment, or any other structure on the ground, it is necessary to fully understand the ground conditions so that the relevant soil parameters can be derived and the designs can be carried out. This is accomplished through a well-planned *site investigation* program, which consists of the following.

a. *Identifying the data needs:* The engineer must have a big picture and foresee the analysis required and the geotechnical parameters that are necessary for the computations. In other words, we have to know in advance what we are looking for in the site investigation program.
b. *A desk study to gather all available information:* The engineer must look for all available information about the site including geological or soil maps, contours, utility (electricity, gas, telecommunication, sewer, water) locations, topography maps, history of the site (e.g., previous construction, filling, dewatering), and site investigation reports of nearby locations. This exercise should not cost anything. A quick look at the Google Maps or Google Earth can be a good start.
c. *A site visit:* The engineer must visit the site with a camera and look for topography, geological features (e.g., rock outcrops, water), soil/rock type, nearby structures (foundations, conditions), utility location, and site access. In a difficult terrain, site access is not always straightforward (see Fig. 11.1).
d. *Developing a preliminary site model:* From *b* and *c*, the engineer should have a rough idea on what lies beneath, including the stratigraphy and expected ground water level. This preliminary site model will obviously change with a more detailed investigation.
e. *Planning a site investigation program:* Now, the engineer must decide on the number of boreholes and their depths, sampling locations for disturbed and undisturbed samples, and the required in situ tests (e.g., SPT, CPT, or both—more on these in the following sections). At this point, the engineer can prepare a realistic estimated cost of the site investigation exercise, which may be submitted to the client.
f. *In situ testing and sampling:* The in situ testing and sampling can take a few days.
g. *Laboratory testing:* The laboratory tests include index tests (e.g., w, LL, PL, G_s) for classifying the soils at the different locations, to more involved triaxial or consolidation tests to define the strength and compressibility parameters. While

FIGURE 11.1 Mobilization of drilling rig and site investigation equipment.

the triaxial or consolidation tests require undisturbed samples, the classification and index tests can be conducted on disturbed samples or trimmings.

h. *Preparation of the site investigation report:* All the in situ and laboratory tests carried out in the site investigation program are presented as the site investigation report, which serves as the basis for the foundation and substructure design. The stratigraphy pertaining to a borehole or trial pit, along with some soil test data, is summarized in a *bore log*. A site investigation report will have many bore logs.

The boring, sampling, in situ tests, and laboratory tests can be supplemented by geophysical exploration (e.g., seismic refraction, electrical resistivity, ground penetrating radar, SASW, and MASW) if necessary. Geophysical methods are discussed briefly in Sec. 11.12.

The quality of the analysis or design can only be as good as the quality of the input parameters. If the input parameters are of poor quality, no matter how sophisticated the analysis is the results are of little value. The soil parameters determined in a site investigation program are the basis of the design and analysis. A thorough and thoughtful site investigation program can make the problem well defined, and enable a realistic solution with less uncertainty.

Figure 11.2 shows a simple soil profile where a drill rig is used to make the boreholes and the backhoe digs the trial pits. The trial pits are typically less than 4 m in depth and enable a clear view of what the soil looks like near the surface and good-quality block samples be taken. A backhoe can dig about eight trial pits in an 8-h day. Trial pits cost significantly less than the boreholes. However, boreholes are required to understand the soil conditions at depths greater than 4 m. They can extend to depths in excess of 100 m. Before any digging, it is necessary to locate the service lines (e.g., telephone, gas, water, electricity, etc.) to avoid any damage to them.

In clayey soils, it is possible to push in metal tubes at the bottom of the boreholes and obtain undisturbed (or intact) samples for laboratory tests such as triaxial or consolidation tests, which require good-quality undisturbed samples. In sands, where undisturbed sampling is difficult, generally we rely on in situ tests. Alternatively, it is possible

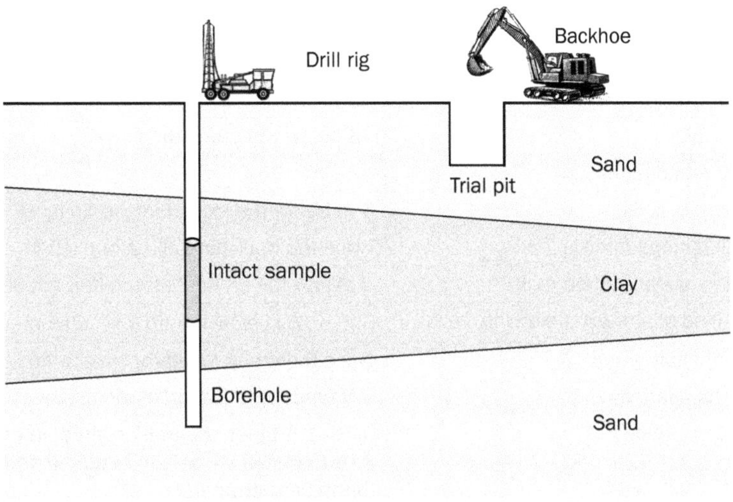

Figure 11.2 Soil profile determined through boring.

to carry out laboratory tests on *reconstituted* sand samples, which are packed to the in situ density. When the boreholes or test pits are no longer required, they have to be backfilled with compacted soil, locally available soils or in some special cases with grout.

The competitive pricing of the tenders and the tight schedules make it difficult for one to appreciate the value of the site investigation exercise. Cost cutting now by reducing the scope of the site investigation can cost dearly later. The false economy can result in overdesign, variation claims, and time delays due to unforeseen or different site conditions and significant cost overruns.

11.2 Spacing and Depth of Investigation

11.2.1 Spacing

The spacing of the boreholes and depth of exploration are governed by the type of structure and the budget. While there is no hard and fast rule for deciding on the number of boreholes, some guidelines given by the Eurocode 7 Part 2 are as follows.

- 15 m to 40 m grid spacing for industrial and high-rise buildings.
- Grid with less than 60 m spacing for structures with very large footprints.
- 20 m to 200 m spacing for linear structures (e.g., roads, railways, channels, pipelines, retaining walls, tunnels, and dikes).
- Two to six locations for special structures such as bridges, stacks, machine foundations, etc.
- 25 m to 75 m spacing along the vertical section for dams and weirs.

11.2.2 Depth

American Society of Civil Engineers (1972) suggests that the boreholes be extended to the *lesser* of the following two.

Structure	Minimum depth
Spread footings:	
$L < 2B$	$2B$ below the foundation level
$L > 5B$	$4B$ below the foundation level
Deep foundations:	
Piles in soils	6 m below the expected pile tip level
Pile groups in soils	Twice the maximum pile group length
Piles supported on rock	3 m into the rock to ensure it is not a boulder
Drilled shafts extending into rocks	3 m or $2D$ below the tip (D = diameter)
Drill shaft group	Twice the group length below the tip
Raft foundations	1.5 times B (the smaller dimension)
Retaining walls	0.75–1.5 times the wall height below the ground level. If deep instability or high settlements expected, extend to the hard stratum
Roads	2 m below the proposed subgrade level
Excavations	5 m below the excavation level
Embankments	Twice the embankment heights if no hard stratum encountered before.
Culverts	Similar to embankments

TABLE 11.1 Minimum Depths of Boreholes (Adopted from Mayne et al., 2001)

a. The depth where the net vertical stress increase $\Delta\sigma'$ is equal to 10% of the vertical stress applied by the structures

b. The depth where the net vertical stress increase $\Delta\sigma'$ is equal to 5% of the effective vertical overburden stress

More detailed guidelines specific to different structures are given in Table 11.1, where B and L are the breadth and length of the structure.

11.3 Boring and Sampling

Borings are generally advanced using hollow or solid stem *continuous flight augers*, wash boring, percussion drilling, or by rotary drilling. Hollow stem auger has a large hole at the center, through which sampler can be sent when required. The hole is plugged while advancing into the ground. They are used in sands and clays. Solid stem augers are used mainly in stiff clays, when no sampling is required. They are rotated and advanced into the ground with a downward thrust. There are also *discontinuous flight augers* (e.g., bucket augers) which do not extend to the full length of the borehole. They have to be removed from the borehole to remove the soil periodically. They are mainly used for shallow boreholes that do not require a casing.

In *wash boring*, a relatively old labor intensive method, the wall of the borehole is supported by a casing (for the top 3 m) and/or drilling fluid such as bentonite slurry. The cuttings at the bottom of the borehole, produced by the drill bit, are washed upward along the borehole to the surface. The water forced downward through the pipe, comes at high velocity to clean the bottom of the borehole and brings the cuttings up to the surface through the annular zone between the pipe and the casing. The stratification can be deduced from the reaction of the chopping tool, the color of the wash water and the sediment carried in the suspension, collected in a sump.

Percussion drilling is commonly carried out on hard rocks. It works like a jackhammer within a borehole and the cuttings are brought to the surface by air. The cutting tool (e.g., a chisel), suspended from a cable or drill rod, is raised and dropped to give repeated blows to the bottom of the borehole. The cutting tool used in sands is a heavy metal tube called a shell (or baler), and in clays it is a clay cutter, a metal tube with a cutting edge.

Rotary drilling is very common in rocks or stiff clays. A rotating drill bit is pushed into the rock to grind the material, with one or more tubes (e.g., double or triple tube) within the sampler to catch the core. A coolant can be circulated which also brings up the cuttings.

The high strength of the rock makes it necessary to use thick-walled core barrels (tubes or pipes) with tips made of some of the hardest materials such as diamond or tungsten carbide (Fig. 11.3a). The rotary drill grinds away an annular zone around the specimen and advances into the ground while the cuttings are washed out by circulating water, in a manner similar to wash borings in soils. The central rock core is collected within the core barrel, which can typically retain cores of 0.5–3.0 m in length. The coring process subjects the cores to some torsion and significant mechanical disturbance. In addition, the core can undergo swelling and get contaminated by the drilling fluid, especially if the rock is weak or heavily fractured. These disturbances can be minimized by using *double-tube* (Fig. 11.3b) or *triple-tube core barrels.*

The drill rod, core barrel, and casing are slightly different in diameter. The early drill holes had diameters of 38 (1½ in.), 51 (2 in.), 63.5 (2½ in.), and 76 mm (3 in.), matching the standard steel pipes available at that time, and they were given designations of

a b

Figure 11.3 (a) Diamond-tipped coring tool, (b) double- and single-tube core barrels.

E, A, B, and N, respectively. With some standardization worldwide in 1930, an "X" was added. H and P are larger sizes that were introduced later.

Some of the common core sizes and their standard designations are given in Table 11.2. The first letter of the symbol (e.g., E, A, B, N, H, P) identifies the core diameter. The second letter Q signifies *wire line drilling*, a technique widely used for deep drilling to minimize the time lost in removing and re-inserting the entire length of drill rods and core barrel for recovering the cores. Instead, the core barrel is lowered down a wire line inside the outer barrel that extends to the full depth of the hole. Upon reaching the bottom of the hole, the core barrel is latched inside the outer barrel, and drilling proceeds.

Single-tube core barrels are the most rugged and least expensive. They are used in the beginning of the drilling operation and are adequate in homogeneous hard intact rock mass or in situations when very good-quality sampling is not required. Double-tube core barrels are the most common, and are often used with NX cores. While the outer barrel moves with the cutting bit, the inner barrel retains the core. In fractured or highly weathered rocks, to minimize the disturbance, triple tube core barrels are preferred. They are also effective on brittle rocks with low strength. The outer barrel does the first cutting, while the middle one does the finer cutting. The third and the innermost barrel retains the core. This process reduces the heat generated at the cutting edge that can otherwise damage the core. A "3" or "TT" is added to the two-letter symbol given in Table 11.2 for triple tube core barrels (i.e., PQ3).

In rocks, the standard sample size is NX, with 54 mm ($2\frac{1}{8}$ in.) diameter, with typical core length of 1.5 m. Longer core lengths are prone to get damaged.

Soil samples in clays are obtained by pushing sampling tubes through the bottom of the boreholes. The mechanical disturbance of the sample and hence its quality can be affected significantly by the wall thickness. Higher the wall thickness, higher is the degree of disturbance. The mechanical disturbance produces a disturbed annular region near the wall, the thickness of which increases with wall thickness. For good-quality intact samples, it is suggested to use *thin-walled sampling tubes* (ASTM D 1587-15) sometimes called Shelby™ tubes. In softer clays and relatively shallow depths, it is possible to push the tubes directly into the ground using some mechanical effort (see Fig. 10.4). The thin-walled samplers are made using stainless steel, carbon steel, galvanized

	Nominal core diameter		Nominal hole diameter	
Symbol	**(mm)**	**(in.)**	**(mm)**	**(in.)**
AQ	27.0	$1\frac{1}{16}$	48.0	$1\frac{57}{64}$
BQ	36.5	$1\frac{7}{16}$	60.0	$2\frac{23}{64}$
NQ	47.6	$1\frac{7}{8}$	75.8	$2\frac{63}{64}$
HQ	63.5	$2\frac{1}{2}$	96.0	$3\frac{25}{32}$
PQ	85.0	$3\frac{11}{32}$	122.6	$4\frac{53}{64}$
EX	22.2	$\frac{7}{8}$	36.5	$1\frac{7}{16}$
AX	30.2	$1\frac{3}{16}$	47.6	$1\frac{7}{8}$
BX	41.3	$1\frac{5}{8}$	58.7	$2\frac{5}{16}$
NX	54.0	$2\frac{1}{8}$	74.6	$2\frac{15}{16}$

Table 11.2 Core Size Designations and Nominal Diameters

FIGURE 11.4 Thin-walled sampling tube being pushed into the ground. (Courtesy of Dr. S. Prapaharan, Haran Engineering, New Jersey.)

coated carbon steel, or brass. They vary in diameter between 50 (2 in.) and 125 mm (5 in.) and come in 700–1000 mm (28–39 in.) lengths. Hvorslev (1949) suggested that when sampling in clays, the tube length should be limited to 10–20 times the diameter.

The intact soil samples collected in tubes are generally sealed at the ends with wax and/or wrapped in polythene sheets for storage so that the moisture is preserved till testing in the laboratory. Abdellaziz et al. (2020) tested two clay samples stored over 27 years in Canada, which showed very little change in the physical and mechanics properties.

Inside clearance ratio and the *area ratio* of a sampling tube are defined as

$$\text{Inside clearance ratio,} \quad C_r(\%) = \frac{D_i - D_e}{D_e} \times 100 \tag{11.1}$$

$$\text{Area ratio,} \quad A_r(\%) = \frac{D_o^2 - D_e^2}{D_e^2} \times 100 \tag{11.2}$$

where D_o = outside diameter of the tube, D_i = inside diameter of the tube, and D_e = inside diameter of the cutting edge, and $D_o > D_i > D_e$. It can be seen that the area ratio is the ratio of the cross-sectional area of the tube wall to the cross-sectional area of the sample, expressed as a percentage.

For good-quality samples, the inside clearance ratio must be less than about 2% and area ratio less than 10%. Longer samples may justify larger C_r. Typical thin-walled sampling tubes are manufactured to C_r of 0.5–1.0%. Tubes with larger A_r are acceptable for stiffer clays to avoid buckling of the tube. The split-spoon sampler used in the standard penetration test (see Sec. 11.6) has area ratio in the order of 100% and hence the samples obtained are highly disturbed.

Example 11.1
A thin-walled tube is manufactured to $D_o = 76.30$ mm, wall thickness = 1.63 mm, and $D_e = 72.30$ mm. Determine the inside clearance ratio and the area ratio.

Solution

$$\text{Inside diameter} \quad D_i = 76.30 - 2 \times 1.63 = 73.04 \text{ mm}$$

$$\text{Inside clearance ratio} \quad C_r(\%) = \frac{D_i - D_e}{D_e} \times 100 = \frac{73.04 - 72.30}{72.30} \times 100 = \mathbf{1.02\%}$$

$$\text{Area ratio} \quad A_r(\%) = \frac{76.3^2 - 72.30^2}{72.30^2} \times 100 = \mathbf{11.4\%}$$

Piston samplers (e.g., Hvorslev, Osterberg) were developed during 1900–1940 in Europe and the United States, for obtaining better quality samples than those obtained from the thin-walled samplers. They use a piston contained within the sampling tube, which is pulled upward to facilitate sampling. Piston samplers with small area ratios are very effective in sampling soft and sensitive clays, which are otherwise prone to high disturbance.

11.4 Laboratory versus In Situ Tests

While triaxial and consolidation tests require the best possible quality undisturbed samples, highly disturbed samples and trimmings can be used for the determination of index properties such as moisture content, Atterberg limits, and specific gravity of the soil grains, and visual classification.

It is possible to determine some of the design parameters from either in situ tests or laboratory tests. It is the responsibility of the geotechnical engineer to strike the right balance between the two. They both have their advantages and disadvantages, as summarized in Table 11.3. Laboratory and in situ tests should complement each other; one should not be at the expense of the other.

Laboratory tests on specimens	In situ tests
Test results are not available immediately after the field work.	Test results are available immediately.
More rational interpretation.	Empirical and semiempirical interpretation.
Smaller volumes tested.	Larger volumes tested in relatively shorter period. More representative of the soil mass.
Well-defined boundary conditions.	Poorly defined boundary conditions.
Better control of the boundary conditions.	Little control of the boundary conditions.
Stress relief and mechanical disturbance in sampling.	Testing in situ; no stress relief and limited sample disturbance.
Better control of strain rate and stress path.	Difficult to control strain rate or stress path.
Very few points tested along the depth.	Continuous profiling possible.

Table 11.3 Advantages and Disadvantages of Laboratory and In Situ Tests

Standard	Purpose
D 1195	Repetitive static plate load tests of soils and flexible pavement components, for airport and highway pavements
D 1196	Nonrepetitive static plate load tests of soils and flexible pavement components, for use in evaluation and design of airport and highway pavements
D 1452	Soil exploration and sampling by auger borings
D 1586	Standard penetration test and split-barrel sampling of soils
D 1587	Thin-walled tube sampling of fine-grained soils
D 2113	Rock core drilling and sampling of rock for site exploration
D 2487	Classification of soils for engineering purposes—USCS
D 2488	Description and identification of soils (Visual-manual procedure)
D 2573	Field vane shear test in saturated fine-grained soil
D 3441	Mechanical cone penetration testing of soils
D 3550	Thick wall ring-lined split barrel drive sampling of soils
D 4220	Preserving and transporting soil samples
D 4428	Crosshole seismic test
D 4429	In situ CBR
D 4633	Energy measurements for dynamic penetrometers
D 4700	Soil sampling
D 5092	Design and installation of ground water monitoring wells
D 5777	Guide for seismic refraction method for subsurface investigation
D 5778	Electronic friction cone and piezocone penetration testing of soils
D 6429	Selecting surface geophysical methods
D 6431	Resistivity method for subsurface exploration
D 6635	Flat plate dilatometer testing
D 7128	Seismic reflection method for shallow subsurface investigation
G 57	Field measurement of soil resistivity using Wenner array

TABLE 11.4 ASTM Standards Relevant to Site Investigations

The ASTM standards covering the major in situ tests and sampling procedures are listed in Table 11.4.

11.5 In Situ Testing

A significant fraction of the site investigation budget is allocated to in situ testing and sampling, which go hand in hand. Not all in situ tests are carried out within boreholes. When in situ tests such as standard penetration test or vane shear test are carried out in boreholes, undisturbed samples are also collected from various depths, especially in clays. Some of the in situ tests are as follows.

a. Standard penetration test

b. Cone penetration test

c. Vane shear test
d. Dilatometer test
e. Pressuremeter test
f. Plate loading test
g. Borehole shear test
h. K_0 stepped-blade test

Out of the above in situ tests, most of the site investigation for foundations are carried out using the first two—standard penetration test and cone penetration test. Vane shear test has a special place in soft clays as an effective tool for determining undrained shear strength c_u. These three tests are discussed in detail in Secs. 11.6, 11.7, and 11.8. Most in situ tests require a probe be pushed or driven into the soil, and the resistance is seen as a measure of the soil strength and stiffness.

11.6 Standard Penetration Test

Penetration resistance testing using an open-ended pipe started in early 1900s. The Raymond Concrete Pile Company developed the standard penetration test in 1927, with Terzaghi and Peck (1948) introducing this in soil exploration and ASTM standardizing it in 1950s. The test involves a 63.5 kg (140 lb) hammer that is repeatedly lifted by 760 mm (30 in.) and dropped on an anvil on top of the drill rod, to drive an open-ended *split-spoon sampler* into the soil at the bottom of the borehole. The number of blows, known as *N-value, penetration number* or *blow count*, required to penetrate 300 mm into the ground is determined. It is denoted by N, which is typically in the range of 0–50 for most soils. Only in very hard materials, it can exceed 50. The test is discussed in ASTM D 1586-11, and the procedure for determining the *normalized penetration resistance* N_1 is discussed in ASTM D 6066-11.

Figure 11.5a shows the complete split spoon sampler. It consists of a longitudinally split thick-walled sampling tube, held between the driving shoe and the coupling head, by threaded pipes at both ends (see Fig. 11.5b). The coupling head is connected to a series of drill rods to reach the desired depth before being struck by the hammer, sometimes called a monkey.

The outer diameter of the sampler is 50.8 mm (2.0 in.), and the inner diameter is 34.9 mm (1.375 in.) to match the inner diameter of the cutting shoe. Sometimes, the inner diameter is kept at 38.1 mm (1.5 in.), allowing for liners to be used. The inner diameter of the driving shoe is 34.9 mm (1.375 in.) and the penetrating end is slightly tapered and rounded. The length of the sampler varies from 457 (18.0 in.) to 762 mm (30.0 in.).

The test is carried out in boreholes, generally at 1.0–1.5-m-depth intervals, and the blow count is plotted against depth where the points are connected by straight lines. When clays are encountered while advancing boreholes, it is possible to obtain thin-walled Shelby™ tube samples between SPTs.

There are different types of hammers. In early days, a pin-guided donut hammer was lifted by a person using a rope over a pulley, wound 1¾–2¼ turns around a rotating drum called *cathead* (Fig. 11.6a). The friction between the rope and the cathead leads to high energy losses. Due to poor efficiency and safety hazard, the old pin-guided *donut hammers* are nowadays replaced by *safety hammers* where the moving parts are enclosed,

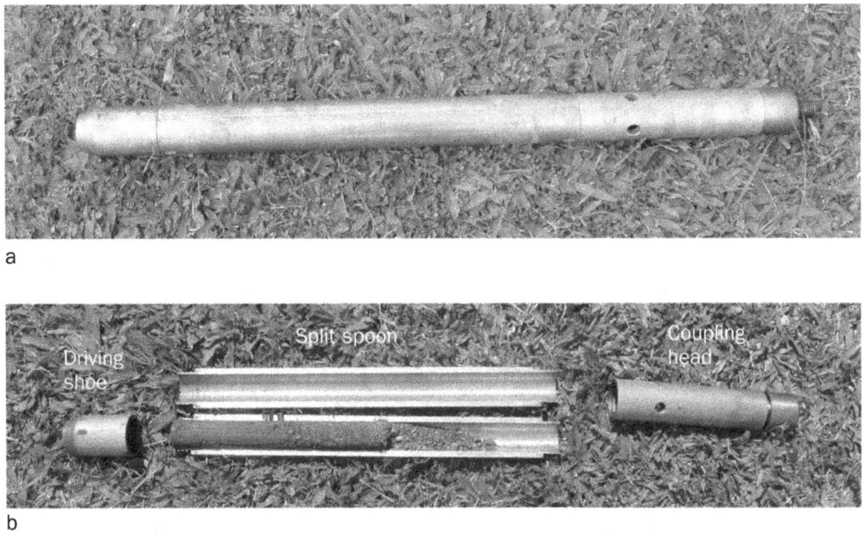

Figure 11.5 Split-spoon sampler in: (a) closed, (b) open positions. (Courtesy of N. Sivakugan, James Cook University, Australia.)

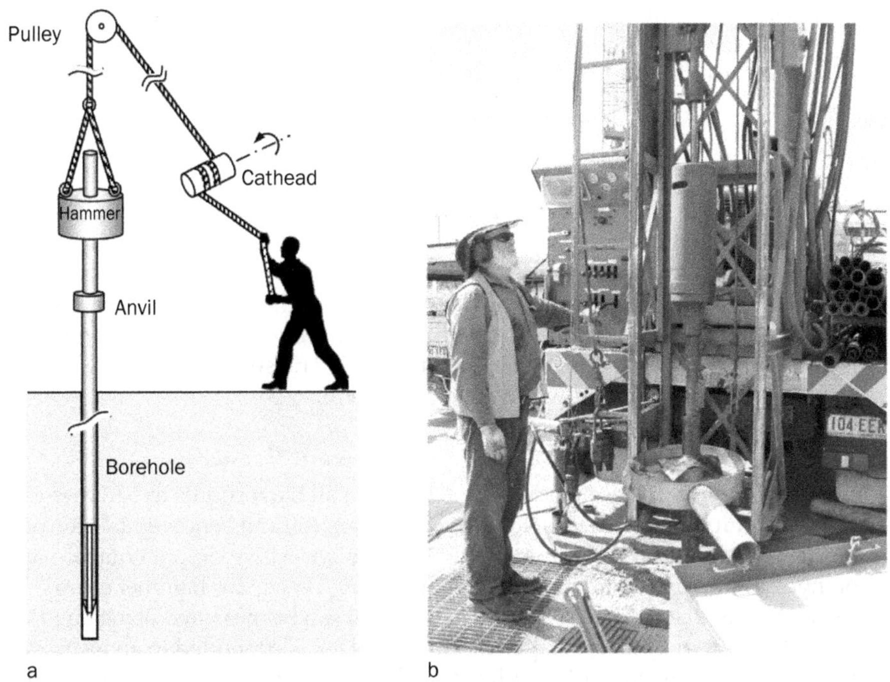

Figure 11.6 (a) Schematic diagram, (b) safety hammer. (Courtesy of N. Sivakugan, James Cook University, Australia.)

or *automatic hammers* that use a chain cam driven by a hydraulic motor and reduce the energy losses. Donut hammers were the oldest, followed by safety hammers and lately the automatic hammers. Figure 11.6b shows a photograph of a standard penetration test rig, which uses a safety hammer.

When recording the blow count, the standard practice is to record the blow counts for *three* consecutive 150 mm (6 in.) intervals, and discard the first record (called seating drive) for possible end effects. The sum of the blows for the *last two* 150 mm (6 in.) intervals is taken as the blow count. After each test, the split spoon sampler is withdrawn from the borehole and the collected sample is visually classified and taken to the laboratory for index property tests. In clays, the sample would be highly disturbed and should not be used for triaxial tests or similar, where good-quality intact specimens are required.

SPT is better suited for granular soils. In cohesive soils, pore water pressures develop while driving the sampler into the ground, which can reduce the effective stresses near the sampler, making it difficult to derive reliable parameters. In the 20th century, SPT was probably the most used in situ test for soil exploration in geotechnical engineering. However, in the 21st century, cone penetration test is becoming more popular for the better rationale behind it. There has been always criticism for the crudeness of SPT. However, it is a simple test that can be carried out in most soils with a rugged equipment. There is a large database, correlations and decades of experience behind its use.

11.6.1 SPT Corrections

Noting the crudeness of the test, we expect to have some corrections to the blow count. This would improve the rationale in the test and enable better predictions of the soil parameters using the blow count. The most important correction is for the *energy rating* (or *efficiency*) of the hammer. Theoretically, the energy delivered by a blow can be calculated as follows.

$$\begin{aligned}\text{Energy/blow} &= \text{Weight of hammer} \times \text{drop height}\\ &= (63.5 \times 9.81\ \text{N})\ (0.76\ \text{m}) = 473.4\ \text{N}\cdot\text{m (or Joules)}\end{aligned}$$

There can be significant energy loss in driving the sampler with the hammer, depending on the way the hammer is lifted and released, and other factors associated with the test. The energy transfer efficiency (η_H) can vary in a wide range of 30–90%, depending on the hammer type used. Therefore, for the same location at a site, the measured N value can vary widely depending on the hammer used. The hammer efficiency is inversely proportional to the blow count. Table 11.5 summarizes the efficiency values of different hammers used worldwide. In general,

$$\eta_{H,\ \text{automatic hammer}} > \eta_{H,\ \text{safety hammer}} > \eta_{H,\ \text{donut hammer}}$$

The standard practice worldwide is to adjust all blow counts to hammer efficiency of 60% by multiplying the measured blow count by $\eta_H/60$ and hence remove the dependence of the blow count on the hammer efficiency. This way, we do not compare apples with oranges. The adjusted blow count is denoted by N_{60}. When the hammer energy transferred to the drill rod (i.e., energy rating) is unknown, it can be measured using an *SPT Analyzer*. The SPT analyzer, similar to a pile driving analyzer, is connected to an instrumented SPT rod, attached to the top of the drill rods to take the hammer blows. The instrumentation with accelerometers and strain gauges enables force and wave velocity measurements, from which the hammer efficiency can be determined, as described in ASTM D 4633-16.

Country	Hammer type	Hammer release	η_H (%)
Japan	Donut	Free fall	78
	Donut	Rope and pulley	67
United States	Safety	Rope and pulley	60
	Donut	Rope and pulley	45
	Automatic	Trip	80–90
Argentina	Donut	Rope and pulley	45
China	Donut	Free fall	60
	Donut	Rope and pulley	50
UK	Automatic	Trip	73

TABLE 11.5 Hammer Efficiencies (η_H)

Borehole diameter correction η_B		
Diameter (mm)	**Diameter (in.)**	η_B
60–120	2.4–4.7	1
150	6	1.05
200	8	1.10
Sampler liner correction η_S		
Sampler		η_S
Standard sampler (no liner)		1.0
Sampler with liner in dense sand or clay		0.8
Sampler with liner in loose sand		0.9
Rod length correction η_R		
Rod length (m)	**Rod length (ft)**	η_R
10	>33	1.0
6–10	20–33	0.95
4–6	13–20	0.85
0–4	0–13	0.75

TABLE 11.6 Suggested Values of η_B, η_S, and η_R

The other corrections are for (a) larger borehole diameter—η_B, (b) use of liner in the sampler—η_S, and (c) very short rod length—η_R. When the test is carried out in boreholes of large diameter, the measured blow count is lower. When a liner is used within the sampler, the resulting friction increases the blow count. When the drill rod is short, considerable amount of hammer energy is reflected back into the rod, thus reducing the energy available to drive the sampler into the ground and increasing the blow count. The values recommended by Seed et al. (1985) and Skempton (1986) for these corrections are summarized in Table 11.6.

The corrected blow count N_{60} is given by

$$N_{60} = N\frac{\eta_H}{60}\eta_B\eta_S\eta_R \tag{11.3}$$

where N is the measured blow count, without any corrections. Most of the time, the last three correction factors η_B, η_S, and η_R can be taken as unity, and it is only required to correct for the hammer efficiency.

11.6.2 Normalized Penetration Resistance

The overburden pressure has significant influence on the N_{60} value determined. The blow count increases with the overburden pressure. A *normalized blow count*, adjusted for the overburden pressure and denoted by $(N_1)_{60}$, is widely used in general geotechnical engineering, foundations designs, and assessing liquefaction potential. It is defined as

$$(N_1)_{60} = C_N N_{60} \tag{11.4}$$

where C_N is the *overburden correction factor*, which decreases with depth. This correction adjusts N_{60} to a standard overburden pressure σ'_{vo} of approximately100 kN/m² (2000 lb/ft²). While ASTM D 1586-11 discusses the test procedure for determining the blow count N, ASTM D 6066-11 discusses the determination of the blow count corrected for overburden (N_1). ASTM D 6066-11 suggests C_N in the form of

$$C_N = \left(\frac{\sigma'_{v,\mathrm{ref}}}{\sigma'_{vo}}\right)^n \tag{11.5}$$

where $\sigma'_{v,\mathrm{ref}}$ is the reference vertical overburden stress level, generally taken as 100 kN/m² ($\approx$ 1 kg/cm² $\approx$ 1 bar $\approx$ 1 atm $\approx$ 1 tsf), n is the exponent in the range of 0.45–0.6 for normally consolidated sands. Several suggestions have been made for the overburden correction factor C_N over the years and they are listed in Table 11.7, and are shown in Fig. 11.7 for comparison. Here, p_a is the atmospheric pressure (101.3 kN/m²). It can be seen from Fig. 11.7 that there is good agreement among all methods, except for very shallow depths where $\sigma'_{vo}/p_a < 0.5$. It is suggested that C_N be limited to 1.6 at very shallow depths. While Skempton's C_N is mainly based on laboratory test data, the others are obtained from field measurements.

Liao and Whitman (1986) method gives very large values for C_N at shallow depths. It has good agreement with Skempton's (1986) equation for normally consolidated fine sands for $\sigma'_{vo}/p_a > 0.5$. Kulhawy and Mayne (1990) noted that either of these two equations are adequate for normally consolidated and overconsolidated sands for $\sigma'_{vo}/p_a > 0.5$.

In dense or very dense fine or silty sands below the water table, driving the sampler into the ground induces negative excess pore water pressure which temporarily increases the effective stresses and the blow count is overestimated. To account for this dilatancy effect, the corrected blow count can be computed as (Terzaghi and Peck, 1948)

$$(N_1)_{60,\mathrm{corrected}} = 15 + 0.5[(N_1)_{60} - 15] \tag{11.6}$$

This correction is generally applied after adjusting the blow count for the energy rating and overburden pressure.

Authors	Proposed correction factor
Bazaraa (1967)	$C_N = \dfrac{4}{1+4\left(\dfrac{\sigma'_{vo}}{p_a}\right)}$ for $\dfrac{\sigma'_{vo}}{p_a} \le 0.75$ $C_N = \dfrac{4}{3.25+\left(\dfrac{\sigma'_{vo}}{p_a}\right)}$ for $\dfrac{\sigma'_{vo}}{p_a} > 0.75$
Peck et al. (1974)	$C_N = 0.77\log\left(\dfrac{20}{\sigma'_{vo}/p_a}\right)$ for $\dfrac{\sigma'_{vo}}{p_a} \ge 0.25$
Seed et al. (1975)	$C_N = 1 - 1.25\log\left(\dfrac{20}{\sigma'_{vo}/p_a}\right)$
Skempton (1986)	$C_N = \dfrac{2}{1+\left(\dfrac{\sigma'_{vo}}{p_a}\right)}$ for normally consolidated fine sand $C_N = \dfrac{3}{2+\left(\dfrac{\sigma'_{vo}}{p_a}\right)}$ for normally consolidated coarse sand $C_N = \dfrac{1.7}{0.7+\left(\dfrac{\sigma'_{vo}}{p_a}\right)}$ for overconsolidated sand
Liao and Whitman (1986)	$C_N = \sqrt{\dfrac{p_a}{\sigma'_{vo}}}$

TABLE 11.7 Overburden Correction Factor C_N

11.6.3 Correlations for Relative Density of Granular Soils

The two major factors influencing the penetration resistance in a granular soil are the relative density and the overburden pressure. Meyerhof (1957) related them as

$$\frac{N_{60}}{D_r^2} = 17 + 24\left(\frac{\sigma'_{vo}}{p_a}\right) \tag{11.7}$$

where D_r is the relative density as a decimal number less than 1, and p_a is the atmospheric pressure. Cubrinovski and Ishihara (1999) proposed that

$$\frac{N_{60}}{D_r^2} = 9\frac{\left(\dfrac{\sigma'_{vo}}{p_a}\right)}{\left(0.23 + \dfrac{0.06}{D_{50}}\right)^{1.7}} \tag{11.8}$$

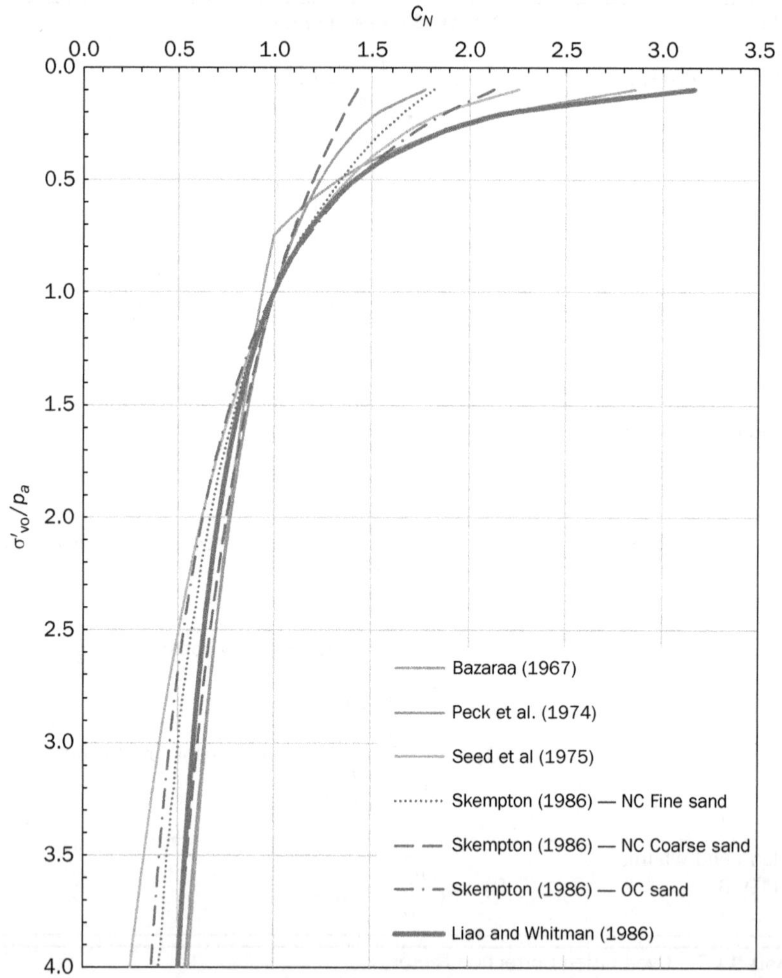

FIGURE 11.7 Overburden correction factor C_N.

where D_{50} is in mm.

The three parameters N_{60}, σ'_{vo}, and D_r are often lumped into a single parameter $(N_1)_{60}/D_r^2$, which varies in a narrow range of 55–70. Note that σ'_{vo} is incorporated into $(N_1)_{60}$.

Skempton (1986) suggested that for normally consolidated and recently deposited sands, it is reasonable to assume that $(N_1)_{60} \,/\, D_r^2 = 60$. From the extensive work of Skempton (1986), it is evident that $(N_1)_{60}/D_r^2$ increases slightly with the median grain size D_{50}, the age of the granular soil deposit, and the overconsolidation ratio. Kulhawy and Mayne (1990) summarized this connection as

$$\frac{(N_1)_{60}}{D_r^2} = (60 + 25 \log D_{50})\left(1.2 + 0.05 \log \frac{t}{100}\right)(OCR^{0.18}) \tag{11.9}$$

	Very loose		Loose		Medium		Dense		Very dense	

D_r	0	15	35	50	65	85	100
$(N_1)_{60}$	0	3(2)	8(5)	15	25(20)	42(35)	58
$(N_1)_{60}/D_r^2$			65	60	59	58	58

FIGURE 11.8 Correlations between $(N_1)_{60}$ and D_r for granular soils.

where D_{50} is in mm, and t is the time since deposition in years. The second term in the right-hand side of the equation is not very sensitive to t, which can be assumed as 100 years if no data are available on the age of the soil deposit.

Figure 11.8 shows the classification of granular soils based on relative density as suggested by Lambe and Whitman (1979). The borderline $(N_1)_{60}$ values proposed by Skempton (1986) are also shown, along with those suggested by Tokimatsu and Seed (1987) in parentheses. $(N_1)_{60}/D_r^2$ ratios given by Skempton (1986) are also given in the figure.

11.6.4 Correlations for Drained Friction Angle of Granular Soils

The drained friction angle of a granular soil ϕ' is better related to $(N_1)_{60}$ than N_{60}, since the overburden pressure σ'_{vo} has some influence. The few empirical correlations relating ϕ' to N_{60}, also include σ'_{vo} in the equations. Peck et al. (1974) gave a correlation between $(N_1)_{60}$ and ϕ' graphically, which was approximated by Wolff (1989) as

$$\phi'(\text{deg}) = 27.1 + 0.3(N_1)_{60} - 0.00054(N_1)_{60}^2 \tag{11.10}$$

Schmertmann's (1975) graphical correlation relating ϕ', N_{60}, and σ'_{vo} was presented by Kulhawy and Mayne (1990) as

$$\phi'(\text{deg}) = \tan^{-1}\left[\frac{N_{60}}{12.2 + 20.3\left(\dfrac{\sigma'_{vo}}{p_a}\right)}\right]^{0.34} \tag{11.11}$$

Equation (11.11) estimates ϕ' conservatively. It should not be used for shallow depths less than 2 m.

Hatanaka and Uchida (1996) proposed that

$$\phi'(\text{deg}) = \sqrt{20(N_1)_{78}} + 20 \tag{11.12}$$

In terms of $(N_1)_{60}$, the above equation becomes

$$\phi'(\text{deg}) = \sqrt{15.4(N_1)_{60}} + 20 \tag{11.13}$$

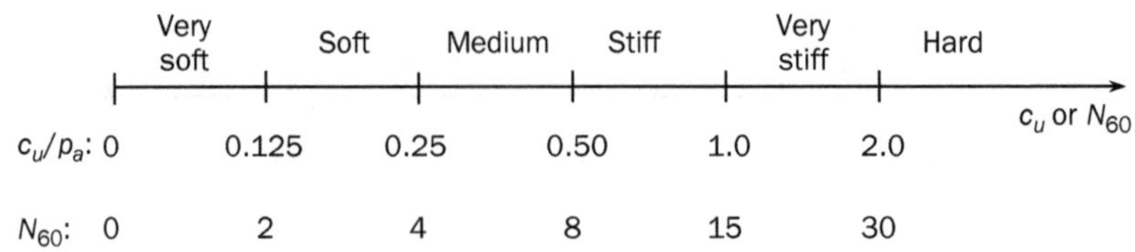

FIGURE 11.9 Interrelationship among N_{60}, c_u, and consistency of clays.

Kulhawy and Chen (2007) suggested that a first-order estimation of the triaxial compression friction angle can be obtained from

$$\phi'(\text{deg}) = 27.5 + 9.2\log(N_1)_{60} \tag{11.14}$$

11.6.5 Correlations for Drained Young's Modulus of Granular Soils

Young's modulus E is a measure of the soil stiffness. It is a necessary parameter for computing soil deformations. Kulhawy and Mayne (1990) suggest that for granular soils,

$$\frac{E}{p_a} = \alpha\, N_{60} \tag{11.15}$$

where α is 5 for sands with fines, 10 for clean normally consolidated sands, and 15 for clean overconsolidated sands.

The blow count N_{60} from a standard penetration test and the dynamic cone penetration index $DCPI$ (see Sec. 4.8) are related by

$$\log DCPI = -1.05 + 2.03 \log N_{60} \tag{11.16}$$

Here, $DCPI$ is the penetration in mm per blow in a dynamic cone penetration test carried out in road works.

11.6.6 Correlations in Cohesive Soils

Standard penetration test induces excess pore water pressures in cohesive soils and hence reduces the effective stress near the sampler, which can underestimate the blow count and make the penetration results unreliable when used in estimating the soil parameters. In addition, there is significant reduction in the blow count with increasing sensitivity. Terzaghi and Peck (1967) suggested the relationship among N_{60}, c_u, and the consistency of the clay as shown in Fig. 11.9. The correlations suggested in Table 11.8 are very approximate and should be used with caution. Here, some of the correlations are based on N values at different hammer energies and hence the scatter is large.

Example 11.2

During a standard penetration test in a 75-mm-diameter borehole in clean dry normally consolidated sands ($D_{50} = 1.0$ mm), at 12 m depth, the blow count was measured as 20. The hammer efficiency is 75%. The unit weight of the sand is 18.0 kN/m^3. No liner was used within the sampler.

Parameter	Equation	Comments
c_u	$\dfrac{c_u}{p_a} = \alpha N_{60}$ $\dfrac{c_u}{p_a} = 0.29N_{60}^{0.72}$	Terzaghi and Peck (1967) table approximated by Kulhaway and Mayne (1990) as $\alpha = 0.06$. Values in the range 0.04–0.15 reported in the literature. Low end is more reliable. From Hara et al. (1974) for 25 clay sites in Japan—Overestimates c_u.
σ'_p	$\dfrac{\sigma'_p}{p_a} = 0.47N_{60}^m$ $\dfrac{\sigma'_p}{p_a} = 0.15N_{60}$	Mayne (2007). $m = 0.6$ for clean quartz sand, 0.8 for silty sand to sandy silt, 1.0 for intact clays and clayey silts. Kulhawy and Chen (2007)
OCR	$OCR = 0.58N\left(\dfrac{p_a}{\sigma'_o}\right)$	Kulhawy and Mayne (1990). Uncorrected N.
K_o	$K_o = 0.073N\left(\dfrac{p_a}{\sigma'_o}\right)$	Kulhawy and Mayne (1990). Uncorrected N.

Table 11.8 Some Approximate SPT Correlations for Clays

a. Find N_{60}.
b. Using Liao and Whitman (1986) overburden correction find $(N_1)_{60}$.
c. Determine the relative density, friction angle, and the Young's modulus of the sand, using all possible relationships.

Solution

Part a

$$\eta_B = 1;\ \eta_S = 1;\ \eta_R = 1$$

$$N_{60} = \frac{\eta_H}{60}\eta_B\eta_S\eta_R = (20)\frac{75}{60}(1.0)(1.0(1.0) = \mathbf{25.0}$$

Part b From Liao and Whitman (1986),

$$C_N = \sqrt{\frac{p_a}{\sigma'_{vo}}} = \sqrt{\frac{101.3}{12 \times 18}} = 0.68$$

$$(N_1)_{60} = C_N N_{60} = 0.68 \times 25.0 = \mathbf{17.0}$$

Part c From Meyerhof (1974),

$$\frac{N_{60}}{D_r^2} = 17 + 24\left(\frac{\sigma'_{vo}}{p_a}\right) = 17 + 24\left(\frac{12 \times 18}{101.3}\right) = 68.2$$

$$D_r = \sqrt{\frac{25.0}{68.2}} = \mathbf{0.605} \quad \text{or} \quad \mathbf{60.5\%}$$

From Cubrinovski and Ishihara (1999),

$$\frac{N_{60}}{D_r^2} = 9\frac{\left(\frac{\sigma'_{\text{vo}}}{p_a}\right)}{\left(0.23 + \frac{0.06}{D_{50}}\right)^{1.7}} = 9\frac{\left(\frac{12 \times 18}{101.3}\right)}{\left(0.23 + \frac{0.06}{1}\right)^{1.7}} = 157.4$$

$$D_r = \sqrt{\frac{25.0}{157.4}} = \mathbf{0.399} \quad \text{or} \quad \mathbf{39.9\%}$$

From Kulhawy and Mayne (1990),

$$\frac{(N_1)_{60}}{D_r^2} = (60 + 25\log D_{50})\left(1.2 + 0.05\log\frac{t}{100}\right)(OCR^{0.18})$$

$$\frac{(N_1)_{60}}{D_r^2} = (60 + 25\log 1)\left(1.2 + 0.05\log\frac{100}{100}\right)(1^{0.18}) = 72$$

$$D_r = \sqrt{\frac{17.0}{72}} = \mathbf{0.486 \text{ or } 48.6\%}$$

The calculated values of D_r vary in the range 40–60%.

From Peck et al. (1974),

$$\phi'(\text{deg}) = 27.1 + 0.3(N_1)_{60} - 0.00054(N_1)_{60}^2 = 27.1 + 0.3(17) - 0.00054(17)^2 = \mathbf{32.0^\circ}$$

From Schmertmann (1975)

$$\phi'(\text{deg}) = \tan^{-1}\left[\frac{N_{60}}{12.2 + 20.3\left(\frac{\sigma'_{\text{vo}}}{p_a}\right)}\right]^{0.34} = \tan^{-1}\left[\frac{25}{12.2 + 20.3\left(\frac{12 \times 18}{101.3}\right)}\right]^{0.34} = \mathbf{37.3^\circ}$$

From Hatanaka and Uchida (1996),

$$\phi'(\text{deg}) = \sqrt{15.4(N_1)_{60}} + 20 = \sqrt{15.4(17)} + 20 = \mathbf{36.2^\circ}$$

Peck et al. (1974) equations appears to be slightly conservative for determining ϕ', which probably is in the range of 36–37°.

From Kulhawy and Mayne (1990), $\frac{E}{p_a} = \alpha\, N_{60}$

For clean normally consolidated sand, $\alpha = 10$.

$$E = 10 \times 101.3 \times 25 \text{ kN/m}^2 = \mathbf{25.3 \text{ MN/m}^2}$$

11.7 Cone Penetration Test

The *cone penetration test* (CPT), originally known as the *Dutch cone penetration test,* is a versatile in situ test that can be carried out in most soil conditions, for determining the soil profile and various soil parameters. The interpretation of the test is more rational than in the standard penetration test. SPT and CPT are used in more than 90% of the soil exploration programs worldwide.

The original Dutch cone developed in Holland in1932 was a 60° *mechanical cone* (ASTM D 3441-16) with 35 mm outside diameter, with approximately 10 cm^2 (1.55 $in.^2$) projected area. The cone is pushed into the ground at a rate of 20 mm/s, and the resistance at the cone and the sleeve, known as *cone resistance* (q_c) and the *sleeve resistance* (f_s), respectively, are measured separately. The measurements are recorded at intervals in the order of 200 mm (8 in.). These mechanical cones are labor intensive and are limited in accuracy. They are currently used mainly in developing countries.

Fugro developed the first commercially available *electric cone* (ASTM D 5778-20) in 1965 as an improvement of the mechanical cone. In the electric cone, the measurements were made using strain gauges and transducers, generally for every 20 mm (0.8 in.), thereby almost giving continuous measurements with depth. The modern cones also have the pore water pressure measurement facility and are known as *piezocones* (CPTu). The standard cone of 10 cm^2 (1.55 $in.^2$) cross section in plan, has friction sleeve area of 150 cm^2 (23.3 $in.^2$). The 133.7-mm-long (5.3-in.) frictional sleeve has an outer diameter of 35.7 mm (1.4 in.). Piezocones are the modern electric friction cones. One of the major advantages of piezocone is its ability to provide three independent parameters q_c, f_s, and u simultaneously.

Figure 11.10a shows a piezocone with a filter (for pore pressure measurement) located between the cone and the shaft. In piezocones, pore water pressures are

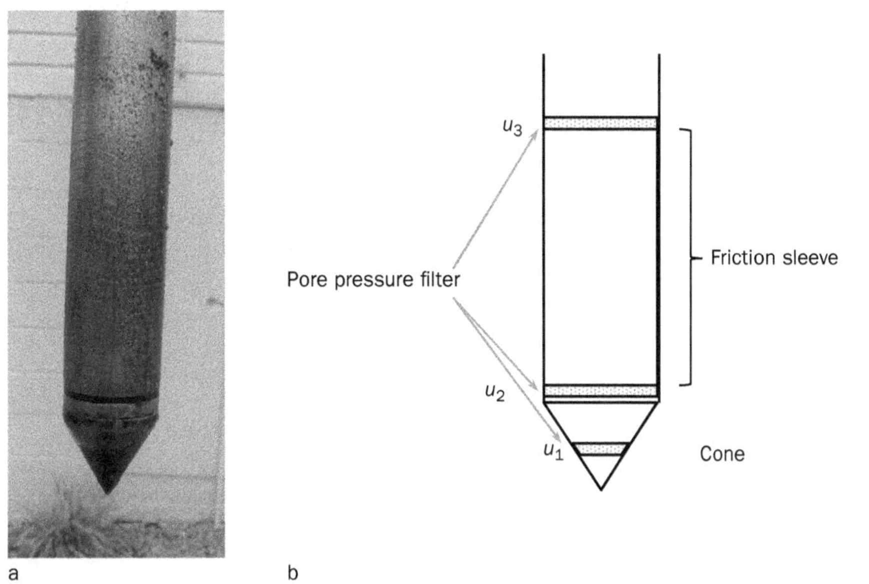

FIGURE 11.10 (a) Piezocone with one filter, (b) possible pore pressure measurement locations.

measured from one or more of the three possible locations shown in Fig. 11.10b. They are denoted by u_1, u_2, and u_3. Out of the three, u_2 (see Figs. 11.10a and 11.10b) is used the most in computations and hence is the most common location, followed by u_1. When profiling stiff clays, $u_2 \approx 0$, and it is recommended to use u_1. With an addition of geophone approximately 500 mm above the cone, *seismic cone penetration tests* (SCPT or SCPTu) are nowadays possible, which can also measure shear wave velocity v_s (and compression wave velocity v_p) and hence the shear modulus G. There are also *resistivity piezocones* (RCPTu), which can determine the electrical resistivity profile of the soil.

The extension rods can be added in 1 m segments. The cone is pushed at the rate of 20 mm/s in 1 m increments, with a rod break every meter. During these breaks, dissipation test or shear wave velocity measurement can be carried out. A typical cone can be advanced by about 60 m/day, which is faster than advancing boreholes typically at the rate of 15–30 m/day.

11.7.1 Corrected Cone Resistance

When the pore water pressure filter is located at the bottom of the friction sleeve (Fig. 11.10a), ambient pore water pressure acts on the shoulder area between the cone and the friction sleeve. It is suggested to correct the cone resistance q_c to account for this pore water pressure u_2. The *cone resistance corrected for the pore water effects* q_t is defined as

$$q_t = q_c + u_2(1 - a) \tag{11.17}$$

where a is the *net area ratio*, approximately equal to the ratio of the shaft cross-sectional area (A_n) to the cone cross-sectional area (A_c) as shown in Fig. 11.11. It is determined from laboratory calibration, with typical values in the range of 0.70–0.85. This correction becomes particularly important in soft clays where $u_2\,(1 - a)$ can be significant compared to q_c. In sandy soils $q_c = q_t$. For simplicity, we will neglect the correction for now and treat q_c as q_t.

11.7.2 Soil Behavior Type Chart

One of the most attractive features of the cone is in profiling the soil, where it enables identification of the soil type in real time, based on q_t and f_s.

Friction ratio (R_f) is defined as

$$R_f(\%) = \frac{\text{Sleeve friction, } f_s}{\text{Corrected cone resistance,} q_t} \times 100 \tag{11.18}$$

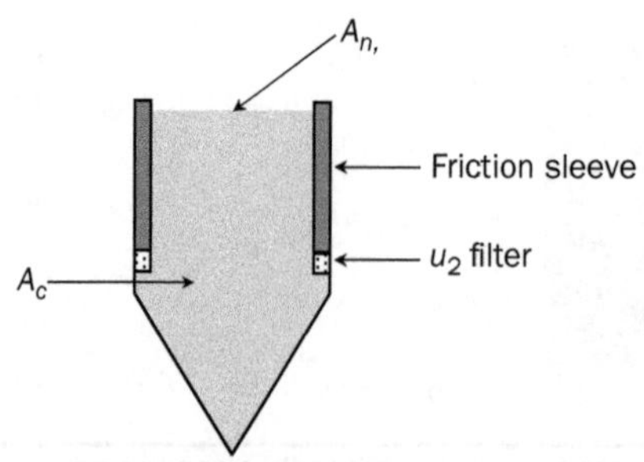

FIGURE 11.11 A_c and A_n used in defining the net area ratio a in Eq. (11.17).

R_f varies in the range of 0–10%, taking lower end of the range for granular soils and the upper end for cohesive soils. Typically, the corrected cone resistance q_t is high in sands and low in clays. The friction ratio R_f is low in sands and high in clays. This distinction makes it possible to identify the soil type based on the two values of q_t and R_f. The most commonly used *soil behavior type* (SBT) chart for cone penetration test is the one suggested by Robertson et al. (1986), consisting of 12 zones. This was updated by Robertson (1990, 2010) with nine zones, as shown in Fig. 11.12. The asterisks in two of the regions at the top right (zones 8 and 9) suggest that the soil is heavily overconsolidated or cemented. The values of $(q_c/p_a)/N_{60}$ are shown within parentheses in Fig. 11.12.

The bulk unit weight of the soil can be estimated by (Robertson, 2010)

$$\gamma / \gamma_w = 0.27[\log R_f] + 0.36[\log(q_t/p_a)] + 1.236 \tag{11.19}$$

where p_a is the atmospheric pressure.

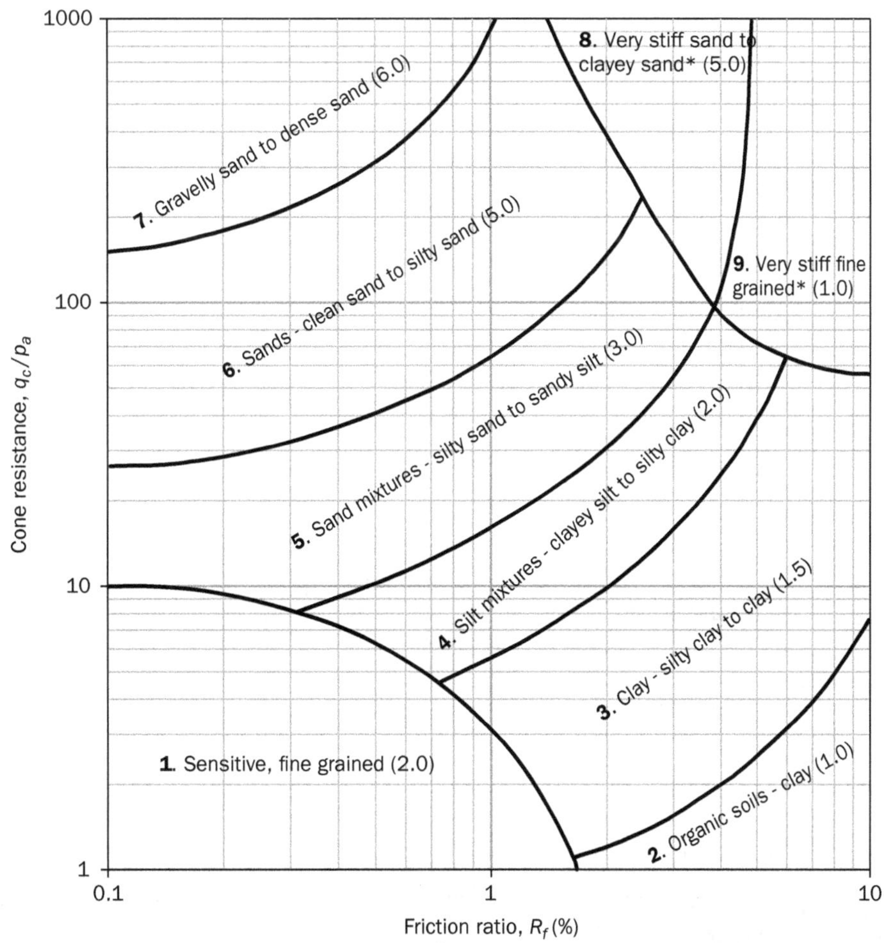

FIGURE 11.12 Soil behavior type chart. (Adapted from Robertson et al., 1986 and Robertson, 2010.)

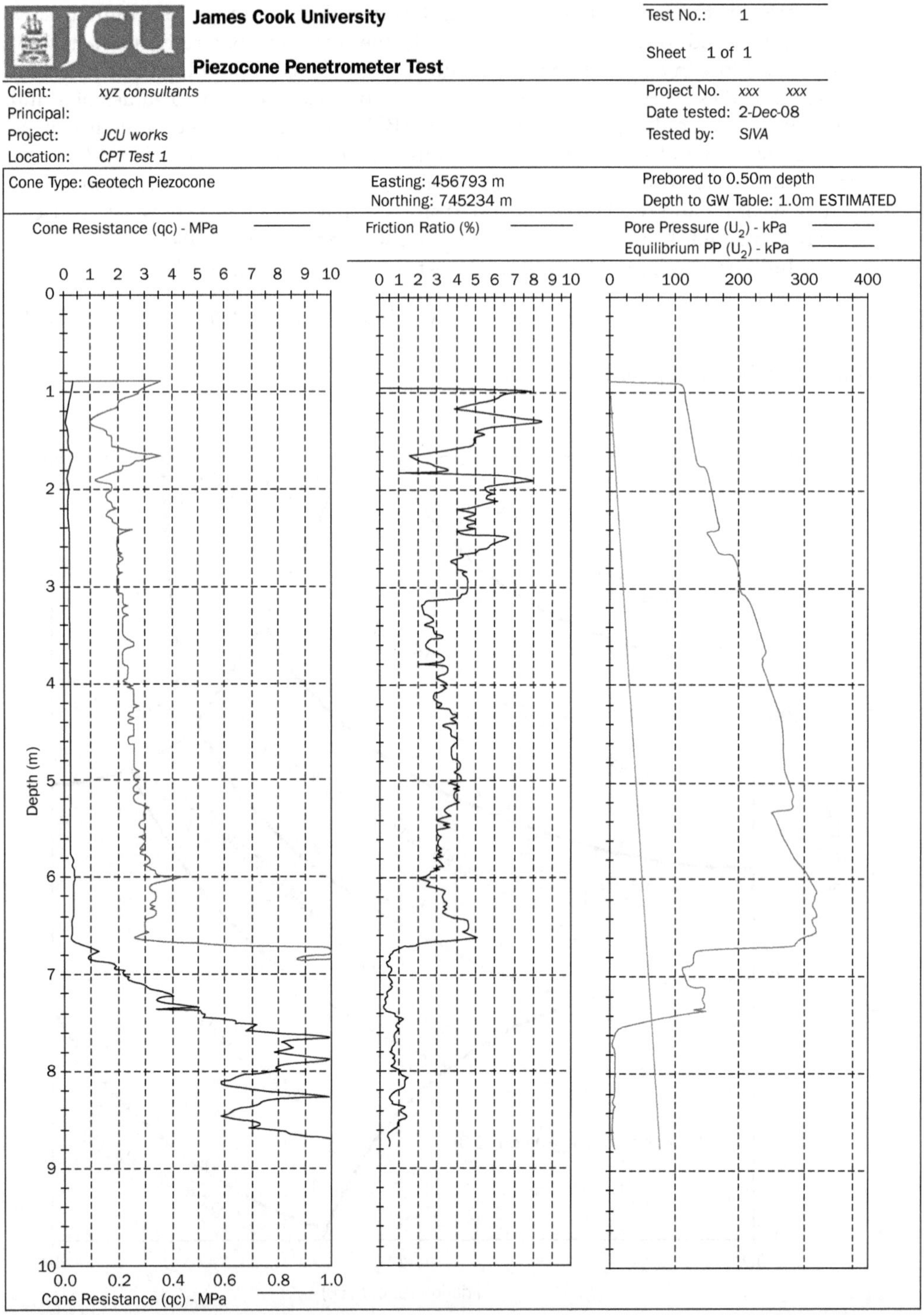

FIGURE 11.13 Cone penetration test data from a site.

Figure 11.13 shows the data logged from a piezocone test at a site where the water table is at 1.0 m below the ground level. The values of q_c, R_f, and u_2 are plotted separately. The equilibrium values of the pore pressure u_0 are also shown along with u_2. As shown in the figure, the cone resistance q_c is often plotted in two separate scales, one is 10 times the other, for ease of reading.

11.7.3 Empirical Relationships in Granular Soils

Geotechnical engineers do not always have the luxury of having both SPT and CPT data from the site. With so many empirical correlations available for both tests, there is a need to go from one to the other (i.e., from q_c to N_{60} or from N_{60} to q_c).

For granular soils, many researchers have noted that the ratio $(q_c/p_a)/N_{60}$ increases with the median grain size D_{50}. The relationships proposed can be approximated as

$$\left(\frac{q_c \,/\, p_a}{N_{60}}\right) = cD_{50}^{a} \tag{11.20}$$

The trend lines and the values of a and c are shown in Fig. 11.14. While Burland and Burbidge (1985) and Robertson et al. (1983) provide a range, the Canadian Foundation Engineering Manual (1992), Kulhawy and Mayne (1990), and Anagnostopoulos et al. (2003) suggest specific values of $(q_c/p_a)/N_{60}$ for a given D_{50}.

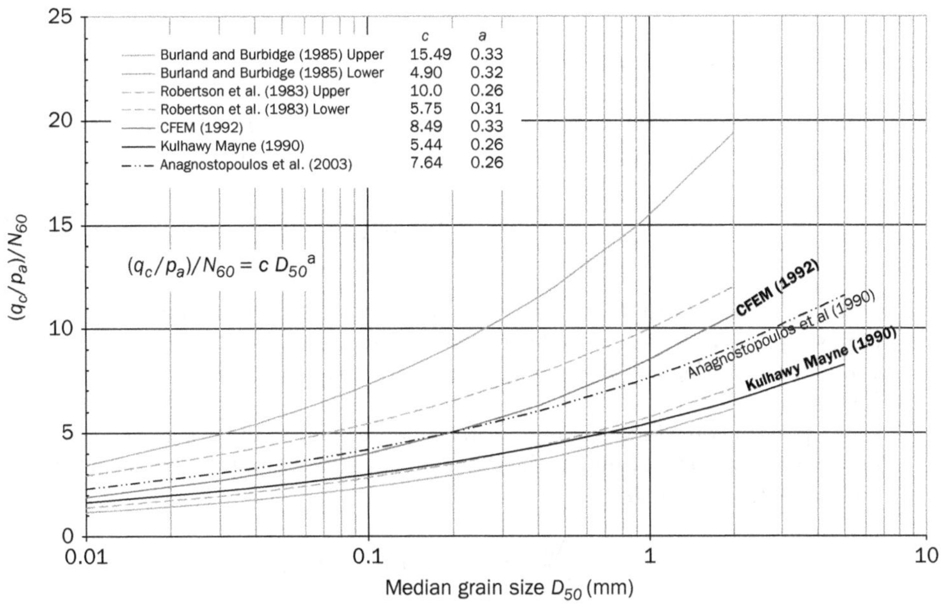

FIGURE 11.14 $(q_c/p_a)/N_{60}$ versus D_{50} variation.

Kulhawy and Mayne (1990) suggested from an extensive data set for clean, rounded, uncemented normally consolidated and overconsolidated quartz sands that the effective friction angle from a triaxial compression test is given by

$$\phi' = 17.6 + 11.0 \log \left[\frac{\left(\dfrac{q_t}{p_a} \right)}{\left(\dfrac{\sigma'_{vo}}{p_a} \right)^{0.5}} \right] \tag{11.21}$$

Robertson and Campanella (1988) showed that the peak friction angle of uncemented, unaged, moderately compressible quartz sands can be given as

$$\tan \phi' = 0.38 \log \left(\frac{q_t}{\sigma'_{vo}} \right) + 0.1 \tag{11.22}$$

The Young's modulus of granular soils varies in the range of two to four times q_c.

11.7.4 Empirical Relationships in Cohesive Soils

The undrained shear strength of clays can be estimated from

$$c_u = \frac{q_t - \sigma_{vo}}{N_{kt}} \tag{11.23}$$

where σ_{vo} is the total overburden stress, and N_{kt} is the total stress *cone factor*, varying in the range of 10–18, with average value of 14. N_{kt} increases with plasticity. N_{kt} is better be determined for the site through calibration against corrected vane shear strength data.

One of the many advantages of piezocone is that it can be used to measure the horizontal coefficient of consolidation c_h through a *dissipation test*. When the cone is paused at a certain depth in the clay, the excess pore water pressure $\Delta u = u_2 - u_0$ starts dissipating with time. Here, u_0 is the static pore water pressure. By monitoring the time t_{50} taken for 50% dissipation of the excess pore water pressure, c_h can be determined from Eq. (11.24):

$$c_h = \left(\frac{T_{50}}{t_{50}} \right) r_0^2 \sqrt{I_R} \tag{11.24}$$

where T_{50} = time factor for radial consolidation at 50% consolidation, r_0 = penetrometer radius, $I_R = G/c_u$ = rigidity index of the soil, and G = shear modulus. The time factor values for other degrees of consolidation, based on u_1 and u_2 readings are summarized in Table 11.9.

Depending on whether u_1 or u_2 readings were taken in the dissipation test, the excess pore water pressure at time t to the initial excess pore water pressure are approximately related to the time factor by

$$\frac{\Delta u_1}{\Delta u_{1,0}} \approx (1 + 27T)^{-0.4} - 0.06 \tag{11.25}$$

$$\frac{\Delta u_2}{\Delta u_{2,0}} \approx (0.85 + 10T)^{-0.45} - 0.08 \tag{11.26}$$

Degree of consolidation, U (%)	Time factor T based on u_1	Time factor T based on u_2
20	0.014	0.038
30	0.032	0.078
40	0.063	0.142
50	0.118	0.245
60	0.226	0.439
70	0.463	0.804
80	1.04	1.60

Table 11.9 *U-T* Values for Piezocone Dissipation Tests (After Teh and Houlsby, 1991)

Therefore, the degree of consolidation determined from u_1 and u_2 are related to the time factor by

$$1 - U_1 \approx (1 + 27T)^{-0.4} - 0.06 \tag{11.27}$$

$$1 - U_2 \approx (0.85 + 10T)^{-0.45} - 0.08 \tag{11.28}$$

where

$$T = \frac{c_h t}{r_0^2 \sqrt{I_R}} \tag{11.29}$$

Example 11.3

In Fig. 11.13, what are the values of q_c, f_s, u_2, and u_0 at 6 m depth? What is the soil type?

Solution At 6 m depth,

$$q_c = \mathbf{0.45\ MN/m^2},\ R_f = \mathbf{2.2\%},\ \text{and hence}\ f_s = \mathbf{9.9\ kN/m^2}$$

$$u_2 = \mathbf{320\ kN/m^2},\ \text{and}\ u_0 = \mathbf{50\ kN/m^2}$$

$$q_c/p_a = 450/100 = 4.5$$

From Fig. 11.12, the point lies in zone 3 which is "**Clay—silty clay to clay**."

Example 11.4

A piezocone pushed into the ground gives the following readings at 6 m depth: $q_c = 14.5$ MN/m², and $f_s = 205$ kN/m². Using the SBT chart, identify the soil at this depth.

Solution

$$R_f = \frac{f_s}{q_c} \times 100 = \frac{205}{14.5 \times 10^3} \times 100 = 1.41\%$$

$$\frac{q_c}{p_a} = \frac{14500}{100} = 145$$

From, Fig. 11.12, the point lies in zone 6, and hence the soil is "**Sands—Clean sand to silty sand.**"

Example 11.5

In Example 11.4, assuming the soil up to 6 m depth is dry with average unit weight of 17.5 kN/m^3, estimate the friction angle of the sand.

Solution

$$q_c = 14.5 \text{ MN/m}^2$$
$$p_a = 100 \text{ kN/m}^2$$
$$\sigma'_{vo} = 6 \times 17.5 = 105 \text{ kN/m}^2$$

From Eq. (11.21),

$$\phi' = 17.6 + 11.0 \log\left[\frac{\left(\dfrac{q_t}{p_a}\right)}{\left(\dfrac{\sigma'_{vo}}{p_a}\right)^{0.5}}\right] = 17.6 + 11.0 \log\left[\frac{\left(\dfrac{14{,}500}{100}\right)}{\left(\dfrac{105}{100}\right)^{0.5}}\right] = \mathbf{41.3°}$$

From Eq. (11.22),

$$\tan\phi = 0.38 \log\left(\frac{q_t}{\sigma'_{vo}}\right) + 0.1 = 0.38 \log\left(\frac{14{,}500}{105}\right) + 0.1 = 0.913$$

Therefore, $\phi' = \mathbf{42.4°}$.

Similar to the overburden correction in SPT in deriving the normalized penetration resistance (see Sec. 11.6.2), a normalized cone tip resistance q_{c1} can be defined as (Robertson and Wride, 1998; Robertson, 2012)

$$q_{c1} = q_c\left(\frac{p_a}{\sigma'_v}\right)^{0.5} \tag{11.30}$$

It can be deduced that

$$\frac{q_c}{N_{60}} = \frac{q_{c1}}{(N_1)_{60}} \tag{11.31}$$

11.8 Vane Shear Test

Vane shear test (ASTM D 2573) is used on soft to medium stiff clays to measure the undrained shear strength c_u. The standard rectangular vane consists of two rectangular metal blades at 90° to each other (see Fig. 11.15). The vane, at the end of the vane rod, is pushed into the clay through the bottom of the borehole (Fig. 11.15a) or right through the overburden using a vane housing. A torque is applied at the top of the vane rod, at the ground level, rotating it at 0.1 degrees per second, to shear the vane through the clay, failing the clay along the vertical cylindrical surface and the two horizontal (top and bottom) circular surfaces. The torque T applied to shear the clay is measured by the torque meter, to compute the undrained shear strength as

$$c_u = \frac{2T}{\pi d^2\left(h + \dfrac{d}{3}\right)} \tag{11.32}$$

where d = width of the vane blade, and h = height of the vane blade as shown in Fig. 11.15b (see Example 11.6 for proof). The undrained shear strength thus determined is the peak value. To determine the remolded value, the vane should be rotated quickly 5–10 revolutions to remold the clay where the torque will be lower. Using this lower remolded value of the torque in Eq. (11.32) gives the remolded undrained shear strength. The ratio of the peak to remolded shear strength gives the sensitivity (see Sec. 9.9.2). The vane shear test is generally carried out at 0.5- to 1.0-m-depth intervals or at depths suggested by the engineer. Figure 11.15c shows the photograph of a full size field vane and the torquemeter.

Noting that the test is of no value in sandy layers, sometimes it is necessary to carry out a cone penetration test to locate the clay layers where vane shear test can be carried out.

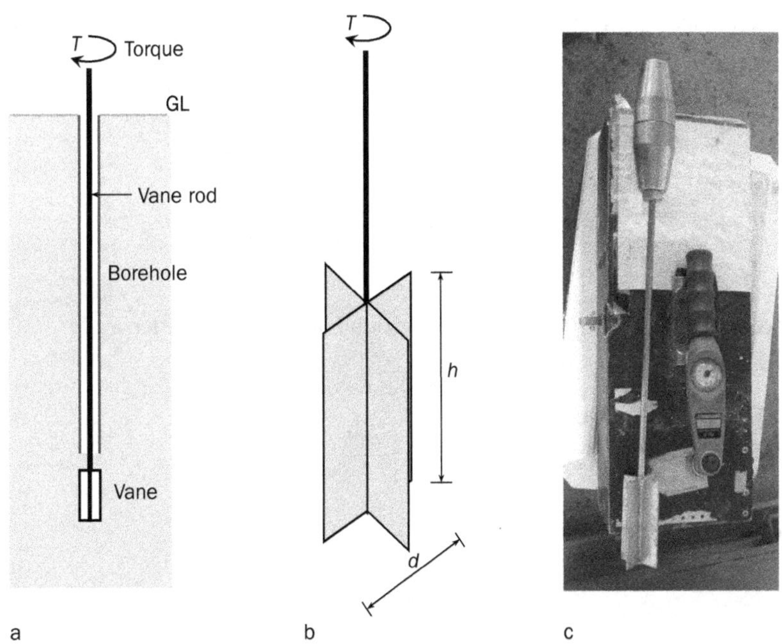

FIGURE 11.15 Vane shear test: (a) vane in borehole, (b) dimensions, (c) field vane and the torquemeter.

In carrying out a vane shear test in clay, the vane is rotated and the clay is sheared very quick and hence undrained conditions can be assumed. Thus, the shear stresses along the failure surfaces of the clay are the same as c_u.

Some of the vanes have the top and bottom ends tapered and Eq. (11.32) is modified as follows.

$$c_u = \frac{12T}{\pi d^2 \left(\dfrac{d}{\cos i_T} + \dfrac{d}{\cos i_B} + 6h \right)} \tag{11.33}$$

where i_T = angle of taper at the top with respect to horizontal, i_B = angle of taper at the bottom with respect to horizontal, and h = length of the straight section of the vane.

An advantage of vane shear test is that it avoids the problems associated with sampling and laboratory testing and works well in soft clays. The test is not suitable in fibrous peats, granular soils, clays with sand seams, or clays where c_u exceeds 75 kN/m^2.

Example 11.6

From the first principles, show that in a vane shear test c_u and T are related by Eq. (11.32).

Solution The applied torque T is resisted by the shear stress τ_f developed along the failure surface, which consists of two horizontal circular areas and one vertical cylindrical wall as shown in Fig. 10.16. The torque from one of the horizontal circular areas is given by

$$T_{\text{circle}} = \int_{r=0}^{r=0.5d} \tau_f \, 2\pi r dr \; r = 2\pi\tau_f \int_{r=0}^{r=0.5d} r^2 dr = \frac{\pi d^3}{12}\tau_f$$

Torque from the vertical cylindrical surface is given by

$$T_{\text{cylinder}} = \pi d h \tau_f \frac{d}{2} = \frac{\pi d^2 h}{2}\tau_f$$

Therefore, the torque T applied at the ground level can be written as

$$T = T_{\text{cylinder}} + 2T_{\text{circle}}$$

$$T = \tau_f \left(\frac{\pi d^2 h}{2} + \frac{\pi d^3}{6} \right)$$

Noting that $\tau_f = c_u$,

$$\tau_f = c_u = \frac{2T}{\pi d^2 \left(h + \dfrac{d}{3} \right)}$$

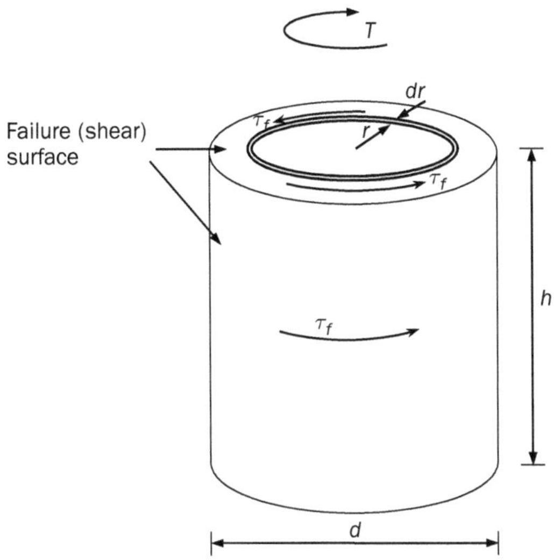

FIGURE 11.16 Failure surface of the clay sheared by the vane.

The vane width d is generally in the range of 40–100 mm and the h/d is maintained as 2.0. Typical vane widths are 38.1, 50.8, 63.5, and 92.1 mm. The diameter of the vane rod is 12.7 mm.

Miniature vanes are used in laboratory samples using the same principles. Figure 9.18 in Chap. 9 shows a laboratory vane shear device. It can be pushed into the walls of excavations or into the clay samples and rotated to determine the undrained shear strength c_u.

Back analysis of failed embankments, excavations, and foundations shows that the undrained shear strength is overestimated in a vane shear test, and the degree of overestimation is strongly related to PI. Bjerrum (1972) corrected the undrained shear strength as

$$c_{u(\text{corrected})} = \mu c_{u(\text{vane})} \tag{11.34}$$

where μ is a correction factor recommended by Bjerrum (1972) as shown in Fig. 11.17. The correction factor μ can be approximated as

$$\mu = 1.7 - 0.54 \log PI \tag{11.35}$$

Skempton (1957) suggested that the ratio of the uncorrected undrained shear strength of a normally consolidated clay determined by vane to the effective overburden pressure at the test depth is a constant that depends on the plasticity index. This can be expressed as

$$\left(\frac{c_{u(\text{VST})}}{\sigma'_{\text{vo}}}\right)_{NC} = 0.0037\, PI + 0.11 \tag{11.36}$$

Equation (11.36) is the same as Eq. (9.22), and the equation works better for c_u derived from a vane shear tests. The predictions from this equation are generally within ± 25%.

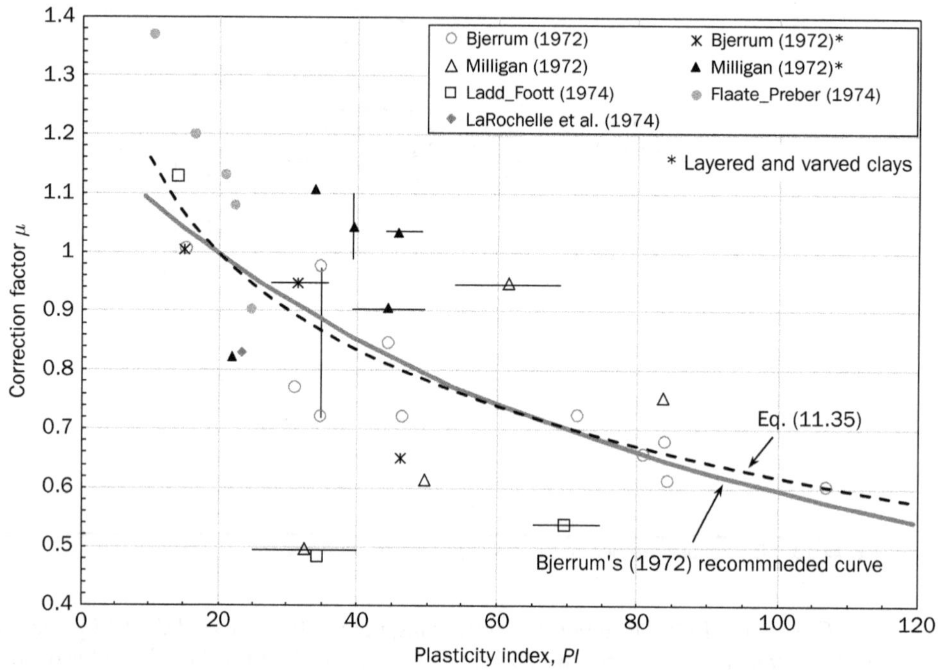

Figure 11.17 Bjerrum's correction factor μ.

Example 11.7

A 63.5-mm-diameter and 127-mm-high NX vane is pushed into a clay to 10 m depth. Torque was applied at the ground level and increased to shear the clay. The peak and remolded torques are 25.0 and 15.8 N·m, respectively.

a. Find the peak and remolded undrained shear strength of the clay.
b. What is the sensitivity of the clay?
c. If the plasticity index of the clay is 30, what is the corrected peak undrained shear strength?

Solution

Part a Substituting in Eq. (11.32),

$$c_{u,\text{peak}} = \frac{2T}{\pi d^2\left(h + \frac{d}{3}\right)} = \frac{2(25.0)}{\pi(0.0635)^2\left(0.127 + \frac{0.0635}{3}\right)} = \mathbf{26{,}639\ N/m^2\ or\ 26.6\ kN/m^2}$$

$$c_{u,\text{remolded}} = \frac{2T}{\pi d^2\left(h + \frac{d}{3}\right)} = \frac{2(15.8)}{\pi(0.0635)^2\left(0.127 + \frac{0.0635}{3}\right)} = \mathbf{16{,}836\ N/m^2\ or\ 16.8\ kN/m^2}$$

Part b

$$\text{Sensitivity} = 26.6/16.8 = \mathbf{1.58}$$

Part c From Eq. (11.35),

$$\mu = 1.7 - 0.54 \log 30 = 0.90$$

Therefore,

$$c_{u,\text{peak(corrected)}} = \mu c_{u,\text{peak(vane)}} = 0.9 \times 26.6 = \mathbf{23.9\ kN/m^2}$$

11.9 Other In Situ Tests

The pressuremeter test (ASTM D 4719-20) involves a cylindrical probe of 32–74 mm in diameter and 400–800 mm in length, which is inserted inside the borehole and expanded laterally using compressed gas (N_2 or CO_2) and water, pushing against the borehole wall. The applied pressure is plotted against the volume of the cell and from this, the soil parameters are derived. The probe consists of the measuring cell at the middle with guard cells at the top and bottom to minimize the end effect and thus ensure plane strain conditions.

The pressuremeter was originally developed by Menard (1956). It was meant for predrilled boreholes. To overcome the soil disturbance within the borehole and the stress relief, self-boring pressuremeter was developed in 1970s, simultaneously in France and the United Kingdom. The self-boring pressuremeter has a cutting tool at the bottom and does not require a prebored hole. It is a very effective tool for measuring in situ lateral earth pressure and hence K_0. In addition, other soil parameters such as friction angle of sand, undrained shear strength c_u and horizontal coefficient of consolidation c_h of clay, and soil stiffness can be determined.

The flat blade dilatometer (ASTM D 6635-15) was developed by Dr. Sylvano Marchetti in Italy in 1975, and has become increasingly popular over the past few decades. It consists of a 240-mm-long, 95-mm-wide, and 15-mm-thick stainless steel blade with a thin expandable 60 mm diameter circular steel membrane that is mounted flush with one of the two faces. The bottom 50 mm of the blade is tapered to provide a sharp cutting edge when penetrating the soil. The dilatometer blade is advanced into the soil at a rate of 20 mm/s. A gas pressure unit at the surface is used to inflate the membrane pushing against the soil laterally. The dilatometer pressure measurements can be interpreted to determine the lateral earth pressure, K_0, OCR, soil stiffness, undrained shear strength of clay, and the friction angle of a granular soil. Seismic dilatometer was introduced in 2006 for measuring the shear wave velocity, in addition to the other parameters. Here, two geophones located above the blade, 500 mm apart, measure the shear wave velocities from the signals produced at the ground level by striking a plate.

Borehole shear test, developed by Dr. Richard Handy at Iowa State University in the 1960s involves carrying out a direct shear test on the borehole wall. The device is advanced into the borehole to the desired depth where the normal stress is applied to the borehole wall through two serrated shear plates. Under this normal stress, a shear stress is applied to the plate by pulling it upward until it moves enough to shear the soil. From three test points, the failure envelope can be determined and the cohesion and friction angle determined.

K_0 stepped blade test was also developed by Dr. Richard Handy at Iowa State University, to determine the lateral in situ stress and hence K_0. The device uses a long blade that is made of four steps, 100 mm apart, with thicknesses of 3.0, 4.5, 6.0, and 7.5 mm, decreasing in thickness downward. Each step carries a pneumatic pressure cell flush with the flat surface that comes into contact with the soil when pushed into the ground. The test is carried out in a borehole, where the first (3.0 mm) blade is pushed into the soil and the pressure $P_{3.0}$ is measured. Then, the blade is pushed down further sending the second step (4.5 mm) into the soil and the two pressure readings $P_{3.0}$ and $P_{4.5}$ are taken. The third (6.0 mm) and fourth (7.5 mm) steps are then pushed into the soil with pressure readings taken from the pressure cells that are in the ground. The fifth step has the same thickness as the fourth, but has no pressure cells. The test gives 14 $(1 + 2 + 3 + 4 + 4)$ pressure readings, which are plotted as pressure (log scale) versus blade thickness (mm). The line of best fit is extrapolated to zero thickness and the reading gives the in situ lateral earth pressure. From this, K_0 can be determined.

Plate loading test (ASTM D 1196-16) is generally carried out to simulate the loading on a prototype foundation or pavement. A 300–500-mm-wide square or circular plate is placed on the ground and is loaded in increments with the settlements measured. The settlement is plotted against the applied pressure, from which the modulus of subgrade reaction is determined.

11.10 Bore Logs

Figure 11.18 shows the locations for the 10 boreholes, 3 piezocone tests, and 4 trial pits in a site investigation program required for a proposed high-rise building. The standard penetration tests will be carried out in each borehole and undisturbed samples may be collected from the boreholes at some depths where clays are encountered. Once the in situ testing and sampling are completed, the samples are sent to the geotechnical laboratory for necessary tests, which include the simple index property tests (e.g., moisture content, Atterberg limits) and more sophisticated consolidation and strength tests. On completion of all tests, the results are summarized in a *site investigation report*, which is sent to the client and used as the basis for the design parameters.

All field and laboratory soil data pertaining to a borehole or trial pit are summarized graphically in a *bore log* similar to the one shown in Fig. 11.19. When the boreholes are deeper, the bore log can run into few pages. A bore log shows the soil profile, delineating the different soil layers, water table depth, standard penetration test data, and any simple field estimates of undrained shear strength of the clays based on pocket penetrometers or miniature vanes in an attempt to define the consistency of the clays encountered. Pocket penetrometer is a simple and inexpensive handheld device that can be pushed into the walls of an excavation or the clay samples to read off the unconfined compressive strength q_u, which is $2c_u$.

11.11 Geotechnical Instrumentation

Limited site investigation data, variabilities in the loadings and the soil parameters, and the approximations in the theoretical models make it necessary to use relatively large factors of safety in the geotechnical designs. An effective method to minimize the risk or

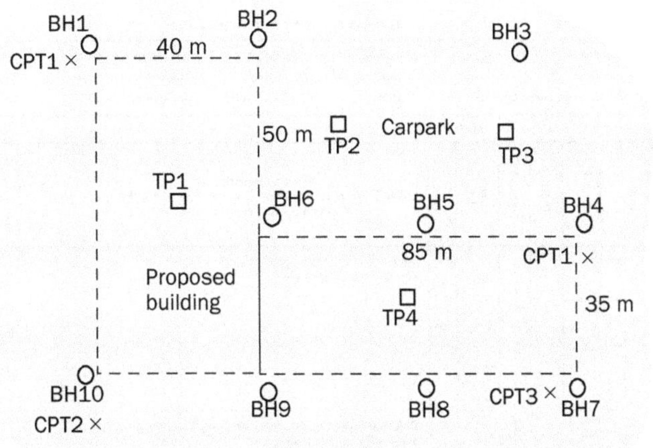

FIGURE 11.18 Locations of boreholes, trial pits, and CPTs.

verify the design is to instrument the structure and monitor its performance as long as it is necessary. The most common geotechnical instruments used on a structure include settlement plates, piezometers, inclinometers, total earth pressure cells, load cells, and strain gauges.

Settlement plates (Fig. 11.20) located at different depths can measure the settlements within specific layers. Here, the elevations are monitored over time, with respect to a fixed survey marker. Piezometers are used to measure the pore water pressures. They are available in different forms, from the old-fashioned standpipe piezometers to the modern vibrating wire piezometers (Fig. 11.21a). Inclinometer consists of a probe that is sent through a casing installed vertically or horizontally, to detect the deviation in alignment over time. The grooves in the casing (see Fig. 11.21b) maintain the orientation of the probe.

Load cells (Fig. 11.22a) are used for measuring the loads, and strain gauges are used for measuring strains. Total earth pressure cells (Fig. 11.22b) can be embedded within the soil to monitor the total earth pressures.

Geotechnical instrumentation is increasingly becoming popular for projects with new design or construction methods, where the instrumentation data are used to verify that the performance of the structure is as expected. These measurements can be a strong evidence in legal disputes. Geotechnical instrumentation can be used for early warnings against natural hazards such as landslides, tsunamis, or earthquakes.

In situ testing and geotechnical instrumentation have some similarities. In situ tests are for one-off measurements at specific locations within the soil. In geotechnical instrumentation, the measurements are often continuous over a long period. In situ tests are carried out in the very early stages of the project. Geotechnical instrumentation is often carried out during construction or to monitor the postconstruction performance.

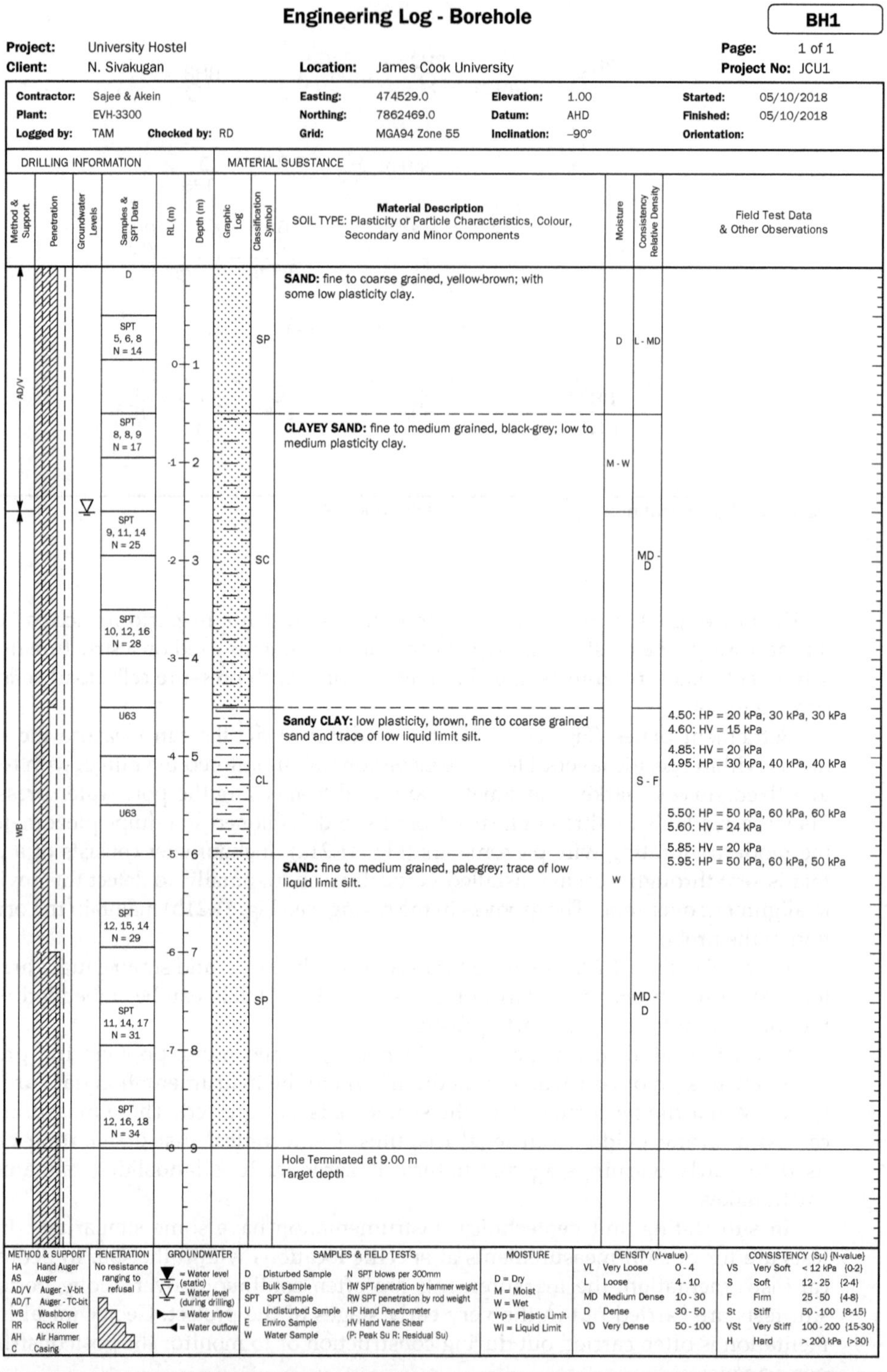

FIGURE 11.19 A typical bore log.

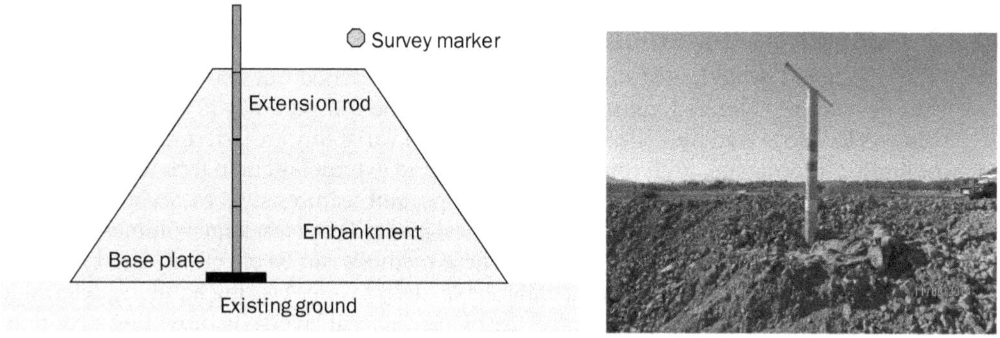

a b

Figure 11.20 Settlement plate: (a) schematic diagram, (b) photograph.

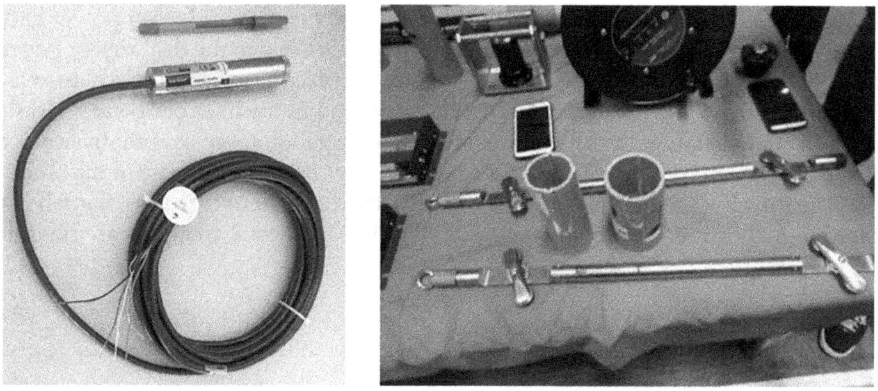

a b

Figure 11.21 (a) Vibrating wire piezometer, (b) inclinometer probe and casing.

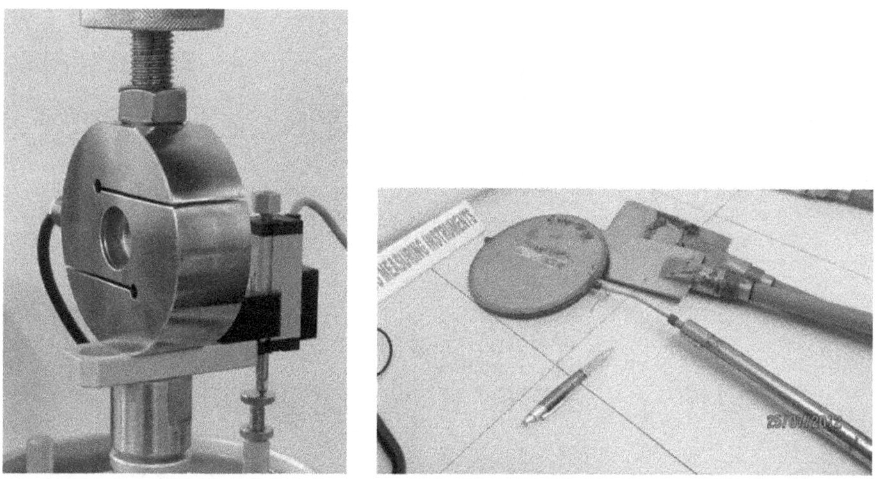

a b

Figure 11.22 (a) Load cell, (b) earth pressure cell.

11.12 Geophysical Methods

The traditional geotechnical site investigation work, carried out using boreholes and in situ devices, tests limited volumes of soils. Here, the soil parameters are determined at discreet locations, and the soil profile and its spatial variability are determined on the basis of limited information, with some interpolation and extrapolation. In fact, when the boreholes are far apart, they can miss out some important features such as cavities or buried service lines. On the other hand, the geophysical methods can test large volumes of soils in a relatively short time and economically. These methods can be effectively used to determine the depths of the bedrock and water table, detect water-bearing aquifers, determine the soil profile, and the dynamic properties of the different layers. By providing a continuous stream of data, these methods can give a two- or three-dimensional view of the subsoil profile. With the ability to cover large terrains in relatively short time, the limitation of geophysical exploration is that the interpreted results (e.g., soil profile) are often not clear cut.

The geophysical test equipment are portable and are relatively easier to manage when it comes to site access. For example, they can be used beneath bridges or power lines, contaminated sites and urban areas that may not be accessible to the drill rigs or penetration test rigs. They can be applied in all types of soils. Surface based geophysical tests are noninvasive, with no threat of encountering underground utility lines or operator risks. When used in conjunction with the traditional site investigation exercise, geophysical exploration can prove to be very effective. One should never be at the expense of the other.

The geophysical methods can be broadly divided into two groups, namely, the surface methods and the borehole methods (Fig. 11.23). Surface methods are nonintrusive

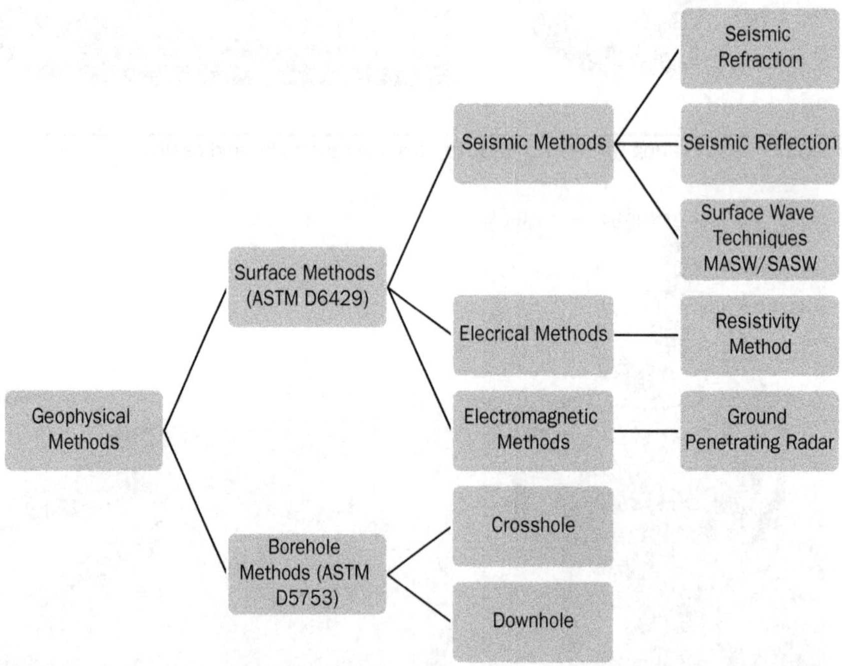

FIGURE 11.23 Geophysical methods.

(ASTM D 6429-20), and include seismic methods, electrical methods, and electromagnetic methods. They include seismic refraction, seismic reflection, multichannel analysis of surface waves (MASW), spectral analysis of surface waves (SASW), electrical resistivity method, and ground penetrating radar. Borehole methods (ASTM D 5753-18) require boreholes for placing the energy source or the geophone receivers, and are hence intrusive. They include crosshole and downhole methods. Geophones or accelerometers are simple *seismic sensors* that are used for detecting ground movement by converting the wave velocity into voltage.

11.12.1 Dynamic Properties of Soils and Rocks

Machine vibrations, blasting, and earthquake loading are some of the situations where the *dynamic properties* of the surrounding soil and/or rock are required. These are obtained under low-strain (less than 0.001%) environments where the soil/rock deformation is not large enough to induce nonlinear nonelastic stress-strain behavior. Such tests include seismic refraction/reflection, suspension logging, steady-state vibration, downhole, crosshole, MASW, and SASW. The traditional in situ tests such as SPT, CPT, dilatometer, and pressuremeter tests are large strain tests.

The dynamic soil properties that are commonly required include shear wave velocity (V_s), shear modulus (G), maximum shear modulus (G_{max}), damping ratio (D), and Poisson's ratio (ν). These dynamic properties are also used in nondynamic problems. For medium or large strain problems, geophysical properties are of little use, where other tests such as cyclic triaxial shear test may be useful. Other laboratory test devices used for measuring the dynamic properties of soils and rocks include resonant column, torsional shear, and bender elements. Seismic cone (ASTM D 5778-20) and seismic dilatometer (ASTM D 6635-15) are two of the commonly used in situ devices that can measure dynamic soil/rock properties.

The shear modulus G decays with the shear strain γ. It is common to plot G, often normalized with respect to G_{max}, versus log γ, which is known as the *modulus reduction curve* showing the decay of shear modulus with shear strain. Damping ratio is the ratio of damping to the critical damping coefficient. It is a measure of the motion attenuation and reflects the ability of the material to dissipate the dynamic load and dampen the system.

When a seismic event (e.g., earthquake) occurs, waves propagate outward from the source. In geophysical methods, we deal with four types of waves: Compression (P), Shear (S), Rayleigh (R), and Love (L) waves, with $V_p > V_s > V_R$. V denotes wave velocity. The first two are *body waves* and the last two are *surface waves*. Body waves propagate through the soil mass, whereas the surface waves travel along the ground surface. The surface waves have larger amplitudes and longer periods. Waves travel faster through denser materials. Shear wave does not travel through water. Shear wave velocity of soils and rocks can vary in the range of 500–5000 m/s. P waves in the air are the sound waves, travelling at 330 m/s. In water, they travel at 1475 m/s.

Compression waves are also called longitudinal waves, plane waves, primary waves, or P-waves. Shear waves are also known as transverse waves, distortional waves, secondary waves, or S-waves. Rayleigh wave is a kind of hybrid P-S wave at the ground surface. It is slower than the body waves, and V_R is approximately 0.90 V_s to 0.94 V_s. Love waves, are slightly faster than Rayleigh waves (i.e., $V_L > V_R$). Table 11.10 shows some typical values of P- and S-wave velocities in different types of soils and rocks.

Material	V_p (m/s)	V_s (m/s)	Density (kg/m³)
Dry sand/gravel	750	200	1,800
Clay	900	300	2,000
Saturated sand	1,500	350	2,100
Saturated clay	1,800	400	2,200
Shale	3,500	1,500	2,500
Sandstone	2,850	1,400	2,100
Limestone	4,000	2,200	2,600
Granite	6,000	3,500	2,600
Water	1,475	0	1,000

TABLE 11.10 Typical Values for the Wave Velocities (After ASTM D 7128-05)

To eliminate the dependence of the shear wave velocity on the effective overburden stress, the measured shear wave velocity V_s is often normalized and presented as

$$V_{s1} = V_s \left(\frac{p_a}{\sigma'_{vo}} \right)^{0.25} \tag{11.37}$$

where p_a = atmospheric pressure and σ'_{vo} = effective overburden stress. The shear modulus can be determined from the shear wave velocity as

$$G = \rho V_s^2 \tag{11.38}$$

where ρ is the bulk density of the soil. For geophysical testing, where the shear strains are very small, G is denoted by G_{max} or G_0, which is the shear modulus at zero strain. The shear modulus is the maximum at zero strain and decays with strain.

The compression wave velocity V_p is related to the constrained modulus D (see Sec. 8.3.2) by

$$D = \rho V_p^2 \tag{11.39}$$

11.12.2 Seismic Refraction/Reflection Method

Seismic refraction method (ASTM D 5777-18) is more popular than seismic reflection method. As shown in Fig. 11.24, a sledgehammer is used to produce seismic energy source at the ground level. A series of interconnected geophone receivers are placed at certain distances and the times taken for the different waves to reach the geophones are recorded through a seismograph. When the depth of interest is large (e.g., >20 m), explosive charge may be required to produce a larger energy source. Seismic waves travel in straight lines, and when they come in contact with a medium of different density they can refract or reflect. The seismic wave velocities are determined and compared with those soils/rocks of known velocities.

The seismic refraction method requires that each successive layer has a higher velocity, the layers to be thick enough, and have enough velocity contrast. Since the bedrock is generally denser than the overburden soils, the method is very effective in

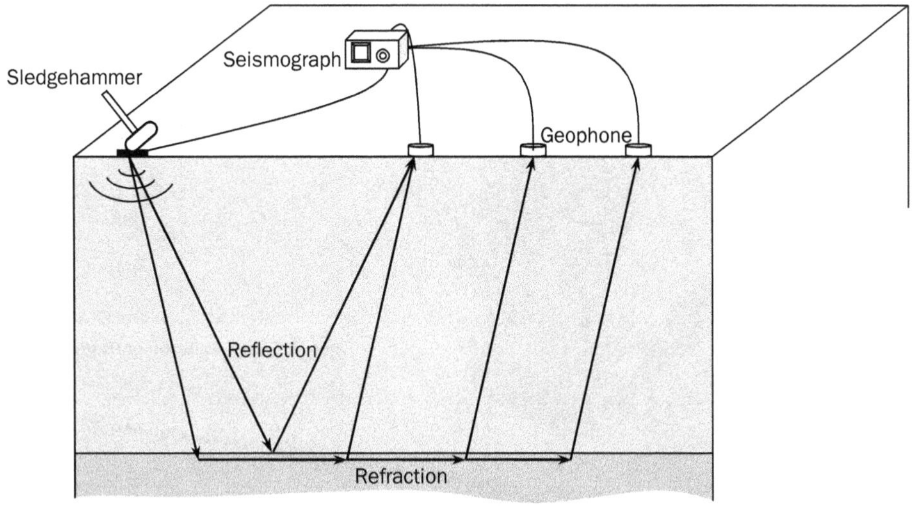

Figure 11.24 Seismic refraction/reflection method.

determining the bedrock depth, irrespective of the soil conditions above. Presence of softer material beneath a stiffer material may not be detected correctly. The seismic refraction method will not detect thin layers.

Seismic refraction method relies on the waves that are refracted at the interface between two layers as per Snell's law. The refracted wave travels along the interface while it is continuously radiates seismic waves back to the ground surface where they are detected by the geophones as shown in Fig. 11.24. It is quite effective in determination of shear and compression wave velocities in the bedrock. Compression waves are faster and hence reach the geophones first.

11.12.3 MASW and SASW Methods

Multichannel analysis of surface waves (MASW) and spectral analysis of surface waves (SASW) are two surface wave techniques that are noninvasive tests for soil exploration. Seismic waves are created by a sledgehammer (e.g., 5 kg) striking a plate (e.g., 300 mm square) as shown in Fig. 11.25, which are captured by geophones. They determine the surface wave (Rayleigh wave) velocity profile relatively quickly. The tests are useful when dealing with large extent of soils such as for seismic microzonation work. SASW was developed in early 1980s, followed by MASW in late 1990s.

In MASW, a series of (generally more than 24) geophones are used which are equally spaced at 1–2 m intervals (see Fig. 11.26). The shear wave velocity can be determined from the Rayleigh wave velocity ($V_R/V_S = 0.90$–0.94). Shear velocity versus depth profiles are generated which are used to interpret the soil profile.

MASW tests are very fast and economical. They are good for near surface soil characterization, and are generally used for 30–80 m depths. With special provisions, they can be extended to 250–300 m depths. MASW is simpler than SASW to conduct. Multichannel recording allows better quality data processing, where the noise is eliminated.

Figure 11.25 Sledgehammer striking a pad to produce the signal for an MASW test.

Figure 11.26 MASW test setup with an array of geophones.

SASW uses only two geophone receivers, and measures the property of the exposed layer. The measurements are made in few seconds and the equipment is portable. The shear wave velocity V_s is used for determining the shear modulus G from Eq. (11.38). SASW gives an overall average of the subsurface velocities, which may be more appropriate than the detailed measurements of successive layers.

In geophysical tests such as MASW or SASW, where the strains are very low, the measured shear modulus is generally G_{max}, the maximum shear modulus, which is also denoted by G_0.

The velocity of the Rayleigh wave, frequency (f), and wave length (γ) are related by

$$V_R = f\lambda \tag{11.40}$$

Higher frequency (i.e., shorter wavelength) waves propagate only at shallow depths, whereas, lower frequency (i.e., longer wavelengths) waves travel through deeper layers. For shallow explorations, the receivers are placed at closer spacing and only a light seismic source of low amplitude and high frequency is required. This can be often achieved through a sledgehammer. For deeper exploration, the receivers are placed at larger intervals and a heavy seismic source of high amplitude and low frequency, in the form of a bulldozer or a heavy weight dropped from a large height can be used. The depth of exploration is approximately taken as $\lambda/3$ to $\lambda/2$.

11.12.4 Electrical Resistivity Method

Electrical resistivity is the reciprocal of the *electric conductivity* of the soil. The SI units of the two are, Ω·m (ohm·m) and S/m (Siemens per m), respectively. In electrical resistivity method (ASTM D 6431-18), current is passed into the ground through two electrodes, and the potential difference is measured between two electrodes, as shown in Fig. 11.27. The electrodes are generally placed in a linear array. When the electrodes are equally spaced (e.g., Wenner array), the resistivity of the ground is given by:

$$\rho = \frac{2\pi s V}{I} \tag{11.41}$$

where s = spacing (see Fig. 11.27), V = voltage drop, and I = current.

The test gives the bulk resistivity of the soils and rocks that influences the flow of the current, multiplied by a geometric factor that depends on the electrode layout. The test is simple and quick, where 70 m depth of soil can be tested in less than an hour. Typical values of resistivity of some soils are given in Table 11.11. It can be seen that the resistivity values for the geomaterials can vary in a very wide range.

Larger depths can be tested by increasing the electrode spacing. Generally, the separation between the current electrodes should be about three times the depth of exploration. Unlike seismic refraction method, electrical resistivity method works well in soil profiles, where the resistivity of successive layers can increase or decrease. The method is good for subsurface hydrogeology characterization, determination of overburden

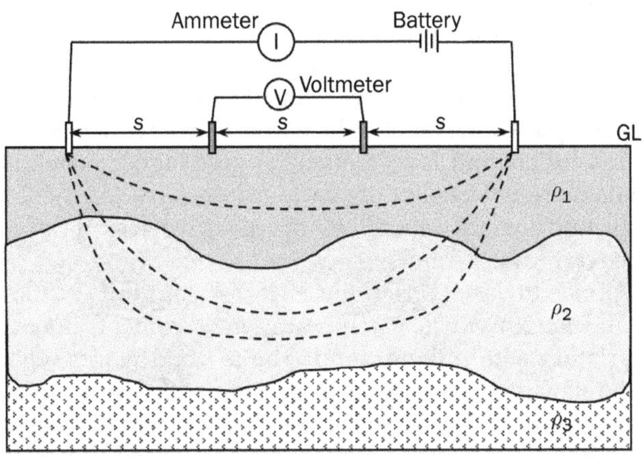

FIGURE 11.27 Electrical resistivity method.

Material	Resistivity (Ω·m)
Saltwater	0.2
Clay	1–20
Clayey sand	30–100
Groundwater (fresh)	10–100
Moist to wet sand	20–200
Shale	1–500
Sandstone	100–8000
Porous limestone	100–1000
Dense limestone	1,000–1,000,000
Metamorphic rocks	50–1,000,000
Igneous rocks	100–1,000,000

Table 11.11 Typical Values of Electrical Resistivity

thickness or ground water depth, mapping salt water intrusion, mapping karst features (e.g., sinkholes) measuring landfill thickness, and determining extents of contaminant plumes.

Electrical resistivity imaging (ERI) is a technique in which many individual resistivity measurements are combined to produce a two- or three-dimensional resistivity cross section of the subsurface.

11.12.5 Ground Penetrating Radar

Ground penetrating radar (ASTM D 6432) is one of the common geophysical exploration devices used by engineers, archaeologists, and others, for detecting the soil profile, cavities, buried artefacts, service lines, and underground storage tanks. It is a nonintrusive electromagnetic method that works on the basis of the dielectric constant of the material through which the electromagnetic pulse goes through. It is a relatively fast method, and presents results as the survey progresses.

The equipment consists of a recorder, and a transmitting and receiving antenna. The GPR unit, similar to a lawn mower, is wheeled across the ground with the radar antenna pointing downward, generating well-timed bursts of high frequency (25–1500 MHz) electromagnetic pulses that probe the earth. High frequencies give lower penetration but higher resolution, and low frequencies give higher penetration but lower resolution. The pulses are reflected at the various interfaces (e.g., soil horizons, water table) within the ground and are detected by the radar receiver. It can be vehicle mounted for fast use, or covering larger areas.

GPR responds to variations in the dielectric constant (Water = 80, dry sand = 3–5, saturated sand 20–30) which varies with water content, it can effectively detect the moisture variation within the ground. The method works well in sands up to 30 m depth, but the signals attenuate in clays and limit the depth of penetration. It can also be used for assessing the integrity of concrete and masonry structures and scour near the bridge foundations. Considerable expertise is required to design, conduct, and interpret the survey.

11.12.6 Seismic Downhole, Uphole, and Crosshole

Seismic downhole, uphole, and crosshole methods are geophysical borehole methods used commonly for soil exploration. Similar to seismic refraction/reflection, the methods require a seismic energy source and one or more geophone receivers.

In a downhole test, the seismic source (e.g., a sledgehammer hitting a plank) is on the ground surface and an array of geophone receivers, at 1–3 m intervals, is lowered down a borehole. From the wave velocities measured, the soil profile and the dynamic properties can be determined. The test is similar to a seismic cone penetration test. In an *uphole test*, the seismic source is placed in the borehole and the geophone receivers are placed at the ground level. Uphole or downhole tests require only one borehole and are hence economical than the crosshole ones which require two or more boreholes. These tests give shear and compression wave velocity profiles that can be related to the shear (G) and constrained (D) moduli by Eqs. (11.38) and (11.39).

Crosshole test (ASTM D 4428-14) requires at least two, and preferably three or more boreholes, 3–6 m apart. One borehole contains the seismic energy source and geophone receivers are placed within the other boreholes, generally at the same depths. The seismic energy source is generally in the form of a vertical impulse created at the bottom of the borehole. The boreholes are 75–100 mm in diameter, and are often lined in soils using PVC or inclinometer casings, and grouted to ensure good transmission of the wave energy. No linings are required in rocks. The energy source should be rich in shearing energy (S-wave or shear wave), and poor in compression energy (P-wave or compression wave), to ensure that the arrival of S-waves at the geophones can be detected easily. The test can give shear and compression wave velocity profiles within soils and rocks, and the dynamic soil properties such as stiffness, Poisson's ratio, damping, etc. Vertical impact or shallow explosives are effective in creating P-waves; a horizontal sledgehammer impact on a horizontal wood plank is effective in producing S-waves. The need for multiple boreholes makes the crosshole test slow and expensive. The major advantage is that the test can be done at a certain depth so that the waves pass only through the desired material.

11.13 Main Points

1. The complete site investigation program includes in situ testing, sampling, and laboratory testing.
2. Wash boring, percussion drilling, and rotary drilling are different ways of drilling a borehole.
3. It is difficult to obtain undisturbed samples from granular soils. Better test them in situ.
4. From the measured blow count N in an SPT, N_{60} and $(N_1)_{60}$ are derived. From these, D_r, ϕ', and E can be estimated for a granular soil.
5. Piezocone gives three independent simultaneous measurements—q_c, f_s, and u. It can be used in granular and cohesive soils.
6. Vane shear test is mainly for clays where it gives the undrained shear strength c_u.
7. A bore log summarizes the field data pertaining to a borehole or trial pit.
8. Geophysical methods can effectively complement the traditional site investigation methods.

Review Exercises

11.1 State whether the following are true or false.

a. Higher wall thickness in a sampling tube gives higher area ratio A_R.
b. In a standard penetration test, higher the energy rating, higher is the blow count.
c. Standard penetration test is more suitable for clays than sands.
d. In a cone penetration test, the friction ratio R_f is high in sands and low in clays.
e. Vane shear test is not for sands.

Answer: True, False, False, False, True

11.2 What is the maximum wall thickness of a sampling tube that can produce 75 mm diameter clay samples with area ratio less than 10%?

Answer: 1.83 mm

11.3 What is the area ratio for the split spoon sampler (D_o = 50.8, D_i = 34.9 mm) used in the standard penetration test?

Answer: 113.6%

11.4 In a sandy soil, standard penetration test blow count was measured as 15 at a depth of 10 m. The energy rating of the test setup is 80%. The water table is at 4 m below the ground level and the unit weight of the sand above and below the water table are 17.0 and 19.0 kN/m^3, respectively.

a. Find N_{60} and $(N_1)_{60}$. Apply Liao and Whitman (1986) overburden correction.
b. Estimate the friction angle of the sand.

Answer: 20, 18: 37–39°

11.5 A piezocone pushed into the ground gives the following readings at 6 m depth: q_c = 5.3 MN/m^2, and f_s = 85 kN/m^2. Using the SBT chart, identify the soil at this depth.

Answer: Sand mixtures—silty sand to sandy silt

11.6 In RE11.5, if the effective overburden pressure is 80.5 kN/m^2, find the friction angle.

Answer: 37–38°

11.7 In the CPT login Fig. 11.13, what is the soil at 8 m depth? What would be N_{60} at this depth if a standard penetration test is carried out?

Answer: Sands—clean sands too silty sands; 16

11.8 A BX vane of 50.8 mm diameter and 101.6 mm is pushed into a clay to 6 m depth. Torque was applied at the ground level and increased to shear the clay. The peak and remolded torques are 18.2 and 13.5 Nm, respectively. The plasticity index of the clay is 34. Find the corrected peak and residual undrained shear strengths of the clay. What is the sensitivity?

Answer: 30.9 kN/m^2, 22.9 kN/m^2; 1.35

References

ASTM D 1196-16. Standard Test Method for Nonrepetitive Static Plate Load Tests of Soils and Flexible Pavement Components, for Use in Evaluation and Design of Airport and Highway Pavements.
ASTM D1586-11. Standard Test Method for Standard Penetration Test (SPT) and Split-barrel Sampling of Soils.

ASTM D 1587-15. Standard Practice for Thin-walled Tube Sampling of Fine Grained Soils for Geotechnical Purposes.
ASTM D 3441-16. Standard Test Method for Mechanical Cone Penetration Testing of Soils.
ASTM D 4428-14. Standard Test Method for Crosshole Seismic Testing.
ASTM D 4633-16. Standard Test Method for Energy Measurement for Dynamic Penetrometers.
ASTM D 4719-20. Standard Test Methods for Prebored Pressuremeter Testing in Soils.
ASTM D 5753-18. Standard Guide for Planning and Conducting Geotechnical Borehole Geophysical Logging.
ASTM D 5777-18. Standard Guide for Using the Seismic Refraction Method for Subsurface Investigation.
ASTM D 5778-20. Standard Test Method for Electronic Friction Cone and Piezocone Penetration Testing of Soils.
ASTM D 6066-11. Standard Practice for Determining the Normalized Penetration Resistance in Sands for Evaluation of Liquefaction Potential.
ASTM D 6429-20. Standard Guide for Selecting Surface Geophysical Methods.
ASTM D 6431-18. Standard Guide for Using the Direct Current Resistivity Method for Subsurface Site Characterization.
ASTM D 6635-15. Standard Test Method for Performing the Flat Plate Dilatometer.
ASTM D 7128-05. Standard Guide for Using the Seismic Refraction Method for Shallow Subsurface Investigation.
Abdellaziz, M., M. N. Hussein, M. Chekired, and M. Karray. "Does Long-term Storage of Clay Samples Influence Their Mechanical Characteristics?" *Canadian Geotechnical Journal*, 57(2):304–309, 2020.
Bazaraa, A. *Use of the Standard Penetration Test for Estimating Settlements of Shallow Foundations on Sand*, PhD Dissertation, University of Illinois, Champaign-Urbana, Illinois, US, 1967.
Bjerrum, L. "Embankments on Soft Ground," *Proceedings of ASCE Specialty Conference on Performance on Earth and Earth Supported Structures*, Purdue University, Indiana, USA, 2, 1972, pp. 1–54.
Cubrinovski, M., and K. Ishihara. "Empirical Correlations between SPT N-values and Relative Density for Sandy Soil," *Soils and Foundations*, 39(5):61–92, 1999.
Eurocode 7—*Geotechnical design—Part 2*: Ground Investigation and testing, 2007.
Hara, A., T. Ohta, M. Niwa, S. Tanaka, and T. Banno. "Shear Modulus and Shear Strength of Cohesive Soils," *Soils and Foundations*, 14(3):1–12, 1974.
Hatanaka, M., and A. Uchida. "Empirical Correlation between Penetration Resistance and Internal Friction Angle of Sandy Soil," *Soils and Foundations*, 36(4):1–10, 1996.
Houlsby, G. T., and C. I. Teh. "Analysis of the Piezocone in Clay," *Proceedings of the International Symposium on Penetration Testing*, ISOPT-1, Orlando, Rotterdam, AA Balkema, 2, 1988, pp. 777–783.
Hvorslev, M. J. *Subsurface Exploration and Sampling of Soils for Civil Engineering Purposes*, U.S. Army Engineer Waterways Experiment Station, Vicksburg, MS, 1949.
Kulhawy, F. H., and J.-R. Chen. Discussion of "Drilled Shaft Side Resistance in Gravelly Soils," by K. M. Rollins, R. J. Clayton, R. C. Miskell and B. C. Blaise, *Journal of Geotechnical and Geoenvironmental Engineering*, ASCE, 133(10):1325–1328, 2007.
Kulhawy, F. H., and P. W. Mayne. *Manual on Estimating Soil Properties for Foundation Design*, Electric Power Research Institute, Palo Alto, CA, 1990.
Liao, S. S. C., and R. V. Whitman. "Overburden Correction Factors for SPT in Sand," *Journal of Geotechnical Engineering*, ASCE, 112(3):373–377, 1986.
Mayne, P. W. "In Situ Test Calibrations for Evaluating Soil Parameters," Tan, T. S., K. K. Phoon, D. W. Hight, and S. Leroueil (eds.), *Proceedings of Singapore Workshop: Characterisation and Engineering Properties of Natural Soils II*, London, Taylor and Francis Group, 2007, pp. 1601–1652.
Mayne, P. W., B. R. Christopher, and J. DeJong. *Geotechnical Site Characterization*, Manual on Subsurface Investigations, FHWA NH1-01-031, Federal Highway Administration, Washington, DC, July 2001.
Menard, L. An Apparatus for Measuring the Strength of Soil in Place, Master's thesis, University of Illinois, Urbana-Champaign, IL, 1956.
Peck, R. B., W. E. Hanson, and T. H. Thornburn. *Foundation Engineering*, 2nd Edition, New York, Wiley, 1974.
Robertson, P. K. "Soil Classification Using Cone Penetration Test," *Canadian Geotechnical Journal*, 27(1):151–158, 1990.
Robertson, P. K. "Soil Behaviour Type from the CPT: An Update," *The 2nd International Symposium on Cone Penetration Testing*, CPT'10, Huntington Beach, CA, USA, 2010. www.cpt10.com.
Robertson, P. K. Discussion of "Influence of Particle Size on the Correlation between Shear Wave Velocity and Cone Tip Resistance," *Canadian Geotechnical Journal*, 49(1):121–123, 2012.
Robertson, P. K., R. G. Campanella, D. Gillespie, and J. Greig. "Use of Piezometer Cone Data," *Use of In Situ Tests in Geotechnical Engineering*, Geotechnical Special Publication, No. 6, ASCE, 1986, pp. 1263–1280.

Robertson, P. K., and R. G. Campanella. Guidelines for Geotechnical Design Using CPT and CPTU, University of British Columbia, Vancouver, 1988.

Robertson, P. K., and C. E. Wride). "Evaluating Cyclic Liquefaction Potential Using the Cone Penetration Test." *Canadian Geotechnical Journal*, 35(3):442–459, 1998.

Schmertmann, J. H. "Measurement of In Situ Strength," *Proceedings of Specialty Conference on In Situ Measurement of Soil Properties*, ASCE, 2:57–138, 1975.

Seed, H. B., I. Arango, and C. K. Chan. *Evaluation of Soil Liquefaction Potential during Earthquakes*, Report No. EERC 75-28, Earthquake Engineering Research Centre, University of California, Berkeley, CA, 1975.

Seed, H. B., K. Tokimatsu, L. F. Harder, and R. M. Chung. "Influence of SPT Procedures in Soil Liquefaction Resistance Evaluations," *Journal of Geotechnical Engineering*, ASCE, 111(12):1425–1445, 1985.

Skempton, A. W. "Discussion: The Planning and Design of New Hong Kong Airport," *Proceedings of the Institute of Civil Engineers*, London, 7, 1957, pp. 305–307.

Skempton, A. W. "Standard Penetration Test Procedures and the Effects in Sands of Over-Burden Pressure, Relative Density, Particle Size, Ageing and Overconsolidation," *Geotechnique*, 36(3):425–447, 1986.

Teh, C. I., and G. T. Houlsby. "An Analytical Study of the Cone Penetration Test in Clay," *Geotechnique*, 41(1):17–34, 1991.

Terzaghi, K., and R. Peck. *Soil Mechanics in Engineering Practice*, New York, Wiley, 1948.

Terzaghi, K., and R. B. Peck. *Soil Mechanics in Engineering Practice*, 2nd Edition, New York, John Wiley & Sons, 1967.

Tokimatsu, K., and H.B. Seed. "Evaluation of Settlements in Sands due to Earthquake Shaking," *Journal of Geotechnical Engineering*, ASCE, 113(8):861–878, 1987.

Wolff, T.F. "Pile capacity predictions using parameter functions," *Proceedings of Predicted and Observed Axial Behaviour of Piles*, Results of a pile prediction symposium, ASCE, Evanston, Geotechnical Special Publication, 23, 1989, pp. 96–106.

CHAPTER 12

Shallow Foundations

12.1 Introduction

Foundations are structural elements like beams, columns, or slabs, often made of reinforced concrete. The main function of the foundation is to transfer the building loads coming from the roof, slabs, and beams to the underlying ground, safely and economically. A typical high-rise building contributes 10–15 kN/m^2 per floor, over the footprint of the building. The foundations have to satisfy the following two major considerations.

a. *Bearing capacity*: The load on the foundation must be limited such that the underlying soil does not fail.

b. *Settlement*: The settlement of the foundation must be within tolerable limits.

Many building codes suggest conservative values of presumed bearing capacity, which are the maximum foundation pressures that are permissible, based on visual classification of the soil. A proper design requires a more rational analysis of bearing capacity and settlement, discussed in this chapter.

Figure 12.1 shows a foundation carrying a load Q, where the base dimensions are B (breadth) and L (length), and the depth of the base below the ground level is D_f. When D_f is less than B (the shorter dimension), it is a *shallow foundation*. When D_f is greater than B, it is a *deep foundation*. Generally, for deep foundations, D_f is much larger than B. This chapter is limited to bearing capacity considerations of shallow foundations. Deep foundations are discussed in Chap. 13.

Figure 12.2 shows the three major types of shallow foundations, namely pad, strip, and raft foundations. Pad foundations (Fig. 12.2a), also known as *pad footings* or *spread footings*, are used to support point loads coming down from columns. They spread the column load evenly to the underlying soil such that the contact pressure is not excessive. A pad footing of base dimensions B and L, supporting a column load Q, applies a pressure q to the underlying ground, given by

$$q = \frac{Q}{BL} \tag{12.1}$$

A *strip foundation* (Fig. 12.2b), also known as strip footing or continuous foundation, is generally used to support line loads such as wall loads. A strip foundation of width B,

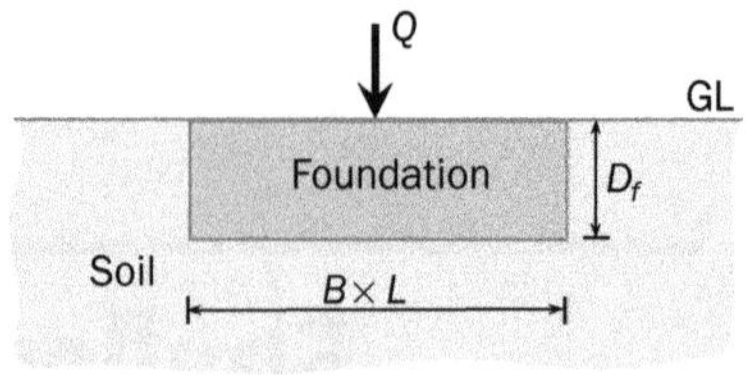

FIGURE 12.1 A shallow foundation.

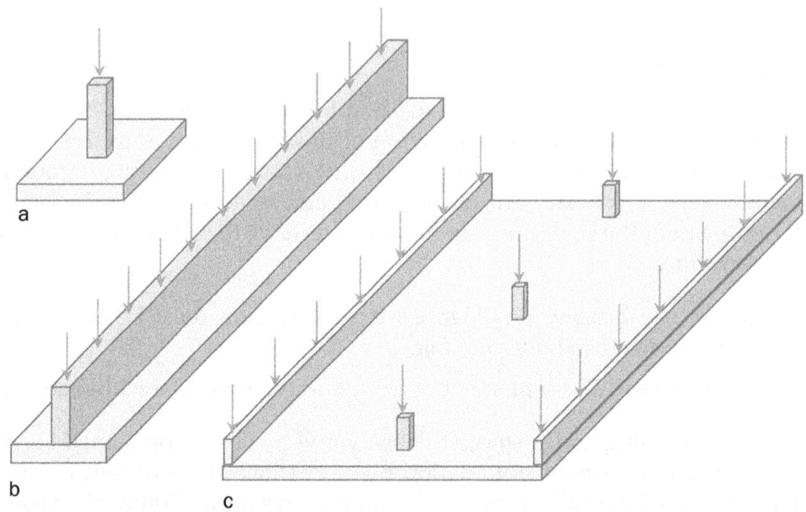

FIGURE 12.2 Types of shallow foundations: (a) pad, (b) strip, (c) raft.

carrying a wall load Q (unit of kN/m or kip/ft), applies a pressure q to the underlying ground, given by

$$q = \frac{Q}{B} \tag{12.2}$$

Raft or *mat foundations* (Fig. 12.2c), can support several column loads and wall loads. When the plan area of the building requires substantial area for pad and strip footings, it may be economical to consider a raft covering the entire footprint.

"Footing" is a term widely used for pad and strip foundations. Figure 12.3 shows some pad footings to support column loads in a multistoreyed building.

12.2 General, Local, and Punching Shear Failure Modes

Depending on how dense the soil is, the bearing capacity failure of a shallow foundation can be in one of the three slightly different modes: *General shear failure, local shear failure,* and *punching shear failure* (Terzaghi, 1943). Figure 12.4a shows the general shear failure

FIGURE 12.3 Pad footings with the column reinforcement cages.

mode within a dense soil beneath the foundation. Here, when the failure load is reached, the soil enclosed between the ground level and the failure surface is pushed sideways and upward, resulting in significant heave above the ground level near the foundation. The pressure versus settlement plot shows a well-defined peak value where the failure occurs (Fig. 12.4b). This peak value of pressure is known as the *ultimate bearing capacity* q_{ult}, which is the maximum pressure that can be applied before failure. Considering the uncertainty in the estimates of the loads and soil parameters, and simplifications in the bearing capacity theory discussed in Sec 12.3, we cautiously allow only a fraction of the ultimate bearing capacity as the design pressure or *allowable bearing capacity* q_{all}, defined as

$$q_{all} = \frac{q_{ult}}{F} \tag{12.3}$$

where F is the factor of safety typically taken as 3.0. This is a relatively larger factor of safety compared to the ones used in structural designs with less uncertainty.

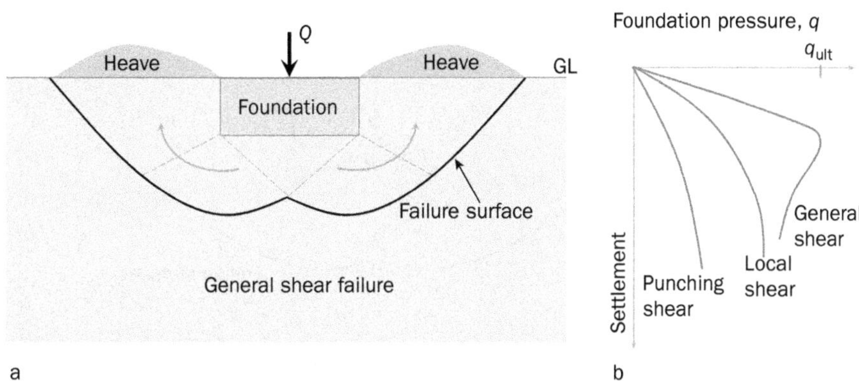

FIGURE 12.4 (a) General shear failure, (b) pressure versus settlement plots for the three failure modes.

When the soil is moderately dense, there is no pronounced peak value in the pressure versus settlement plot (see Fig. 12.4b), and local shear failure occurs with minor heave above the ground level. In soils that are loosely packed, the failure surface will not extend to the ground level. The foundation simply punches into the underlying soil, with no pronounced peak value in the pressure versus settlement plot. This is known as punching shear failure. In granular soils, with relative density greater than 70%, shallow foundations ($D_f < B$) fail mainly in general shear failure mode (Vesic, 1963). Foundations are generally placed on ground that is well compacted and hence we will deal mainly with general shear failure mode in this chapter.

12.3 Terzaghi's Bearing Capacity Theory

Terzaghi (1943) studied the bearing capacity of a strip footing of width B at a depth of D_f within a soil with unit weight of γ. He assumed the footing-soil interface to be rough and a failure surface *GEDFH* as shown in Fig. 12.5. He treated the soil within the depth D_f as a surcharge pressure γD_f, ignoring the shear resistance provided along the dashed lines *GI* and *HJ*. The soil within the failure surface consists of three distinct zones:

a. Triangular zone ACD, immediately below the foundation, with the angle α assumed as ϕ'.
b. Radial shear zones *AED* and *CFD*, where the curves *DE* and *DF* are logarithmic spirals.
c. Rankine passive zones *GAE* and *CHF*.

Terzaghi (1943) suggested that the bearing capacity of the strip footing is given by

$$q_{\text{ult}} = cN_c + \gamma D_f N_q + 0.5B\gamma N_\gamma \tag{12.4}$$

where c = cohesion, γ = unit weight of the soil, and N_c, N_q, and N_γ are three bearing capacity factors which are functions of the friction angle. The three components in the right hand side of Eq. (12.4) reflect the contribution from cohesion, surcharge, and the

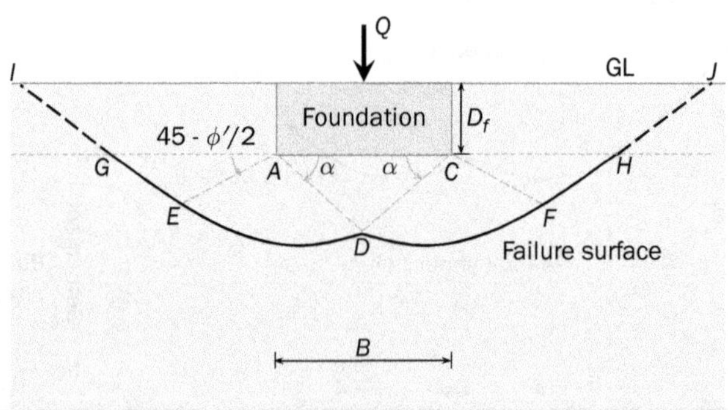

FIGURE 12.5 General shear bearing capacity failure.

self-weight of the soil, respectively. For square and circular footings, Eq. (12.4) can be modified as

$$q_{\text{ult}} = 1.3cN_c + \gamma D_f N_q + 0.4B\gamma N_\gamma \text{ for square} \tag{12.5}$$

$$q_{\text{ult}} = 1.3cN_c + \gamma D_f N_q + 0.3B\gamma N_\gamma \text{ for circle} \tag{12.6}$$

The bearing capacity equations can be used in terms of effective or total stresses depending on whether drained or undrained conditions apply.

Terzaghi neglected the shear resistance within the overburden soil. The angle α in Fig. 12.5 appears to be closer to $45 + \phi'/2$ than his assumed ϕ'. Subsequent research by several others noted that the bearing capacity factors N_q and N_c are better given by

$$N_q = e^{\pi \tan\phi} \tan^2\left(45 + \frac{\phi}{2}\right) \tag{12.7}$$

$$N_c = \left(N_q - 1\right)\cot\phi \tag{12.8}$$

Equation (12.7) is different to what was originally proposed by Terzaghi, but is the same as the one given by Reissner (1924). Equation (12.8) is the same equation given by Terzaghi (1943) and Prandtl (1921). The two expressions for N_q and N_c given by Eqs. (12.7) and (12.8) are mathematically rigorous and are universally accepted.

There are no direct analytical solutions for N_γ. Kumbhojkar (1993) provided numerical values for Terzaghi's N_γ. These values can be approximated as (Diaz-Segura, 2013).

$$N_\gamma = \left(N_q + 3\right)\tan(1.34\phi) \tag{12.9}$$

There are several expressions suggested for N_γ in the literature, based on different assumptions, giving a wide range of values for a specific friction angle. Motra et al. (2016) and Diaz-Segura (2013) listed more than 60 equations for N_γ, all in terms of ϕ' Table 12.1 summarizes some of the popular expressions for N_γ, which are compared in Fig. 12.6. In spite of the wide variation expected in N_γ for the different equations reported in the literature, there is good agreement among these selected equations, especially for

Expression	Reference
$(N_q - 1)\tan(1.4\phi)$	Meyerhof (1963)
$1.5\,(N_q - 1)\tan\phi$	Brinch Hansen (1970)
$2.0\,(N_q + 1)\tan\phi$	Vesic (1973)
$1.1\,(N_q - 1)\tan(1.3\phi)$	Spangler and Handy (1982)
$2.0\,(N_q - 1)\tan\phi$	Eurocode 7 and DIN 4017; Chen (1975)
$(N_q - 1)\tan(1.338\phi)$	Martin (2005)

TABLE 12.1 Expressions for N_γ

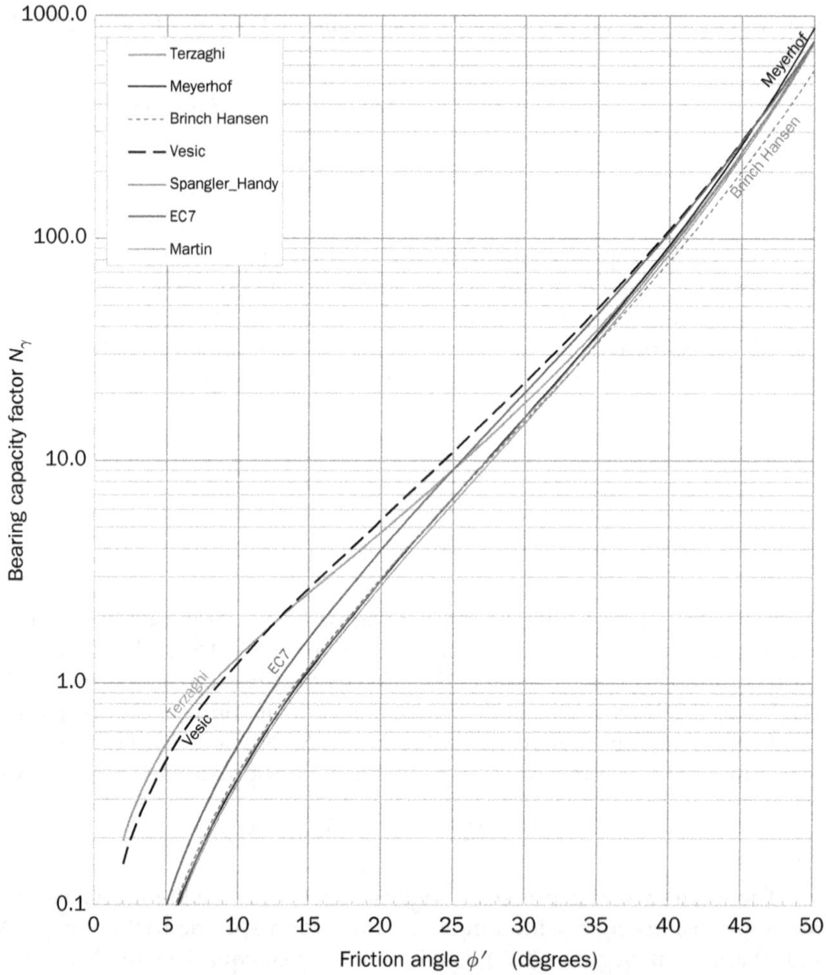

FIGURE 12.6 Bearing capacity factor N_γ.

$\phi' = 20°–50°$. The differences are mainly for the low-friction angles. Meyerhof (1963), Brinch Hansen (1970), Spangler and Handy (1982), and Martin (2005) give values that are very similar, with only a slight variation at higher friction angles. Vesic (1973) gives higher values. Chen (1975) solution is in very good agreement with the one proposed by Zhu and Michalowski (2005).

The N_γ values suggested by Brinch Hansen are widely used in Europe, and the ones suggested by Vesic and Meyerhof are popular in North America. Values of N_q [Eq. (12.7)], N_c [Eq. 12.8)] and N_γ from Meyerhof, Brinch Hansen, and Vesic are summarized in Table 12.2.

When the water table is at the bottom of the foundation, in computing the bearing capacity in terms of effective stresses, submerged unit weight should be used as γ in the third bearing capacity term [see Eqs. (12.4), (12.5), and (12.6)]. With $\alpha = 45 + \phi'/2$, the failure surface in Fig. 12.5 extends approximately to a depth of B below the foundation level.

$\phi(°)$	N_q	N_c	$N_{\gamma(M)}$	$N_{\gamma(BH)}$	$N_{\gamma(V)}$
0	1.0	5.14	0.0	0.0	0.0
5	1.6	6.5	0.1	0.1	0.4
10	2.5	8.4	0.4	0.4	1.2
15	3.9	11.0	1.1	1.2	2.7
20	6.4	14.9	2.9	3.0	5.4
21	7.1	15.9	3.4	3.5	6.2
22	7.8	16.9	4.1	4.1	7.1
23	8.7	18.1	4.8	4.9	8.2
24	9.6	19.4	5.7	5.8	9.5
25	10.7	20.8	6.8	6.8	10.9
26	11.9	22.3	8.0	8.0	12.6
27	13.2	24.0	9.5	9.4	14.5
28	14.8	25.9	11.2	11.0	16.8
29	16.5	27.9	13.3	12.9	19.4
30	18.5	30.2	15.7	15.1	22.5
31	20.7	32.8	18.6	17.8	26.1
32	23.3	35.6	22.1	20.9	30.3
33	26.2	38.8	26.3	24.6	35.3
34	29.6	42.3	31.3	28.9	41.2
35	33.4	46.3	37.3	34.1	48.2
36	37.9	50.8	44.7	40.3	56.6
37	43.1	55.9	53.6	47.6	66.5
38	49.2	61.6	64.4	56.5	78.4
39	56.2	68.2	77.8	67.1	92.7
40	64.5	75.6	94.2	80.0	110.0
41	74.3	84.2	114.7	95.6	131.0
42	85.8	94.2	140.2	114.6	156.5
43	99.6	105.6	172.3	137.9	187.7
44	116.0	119.0	212.9	166.6	226.1
45	135.7	134.6	264.6	202.1	273.5
46	159.5	152.9	331.2	246.3	332.6
47	188.4	174.6	417.5	301.7	406.5
48	223.8	200.5	530.7	371.4	499.6
49	267.3	231.4	680.6	459.9	617.8
50	321.4	268.6	881.6	573.1	768.9

TABLE 12.2 Bearing Capacity Factors

Therefore, γ for the third bearing capacity term [see Eqs. (12.4), (12.5), and (12.6)] should be the weighted average value of the unit weights within this depth B below the base of the foundation.

12.4 Gross and Net Pressures

The ultimate bearing capacities computed using Eqs. (12.4), (12.5), and (12.6) are all *gross* ultimate bearing capacities. There is already an overburden pressure of γD_f acting at the foundation level. The *net* ultimate bearing capacity is the maximum *additional* soil pressure that can be sustained before failure. Therefore, net ultimate bearing capacity is obtained by subtracting the overburden pressure from the gross ultimate bearing capacity. Similarly, the net applied pressure is the additional pressure applied at the foundation level in excess of the existing overburden pressure. The safety factor with respect to bearing capacity failure is generally defined in terms of the *net* values as:

$$F = \frac{q_{\text{ult,net}}}{q_{\text{app,net}}} = \frac{q_{\text{ult,gross}} - D_f}{q_{\text{app,gross}} - D_f} \tag{12.10}$$

In most spread footing designs, the gross pressures are significantly larger than the overburden pressures γD_f. In other words, the gross and net pressures are not very different, as seen in most of the review problems in this chapter. Only in problems involving removal of large overburden pressures, such as buildings with basements, gross and net pressures can be quite different. The difference can be substantial when D_f is large as in the case of excavations for deep basements, and rafts. In *compensated* or *floating foundations*, the applied net pressure is reduced substantially, almost to the extent of making it negligible, by increasing D_f. The factor of safety against bearing capacity failure for such foundations would be very high. Here, the design is governed by the settlement criterion.

We generally carry out an effective stress analysis for assessing the long-term stability, assuming drained conditions. Here, c' and ϕ' are used in the bearing capacity equation. To assess the short-term stability (e.g., immediately after loading) undrained conditions are assumed, and the analysis is carried out in terms of total stresses using c_u and $\phi_u = 0$. For $\phi_u = 0$, $N_q = 1$, $N_\gamma = 0$, and $N_c = 5.14$ (or $\pi + 2$).

The bearing capacity theory discussed above assumes that the footing-soil interface is rough. Concrete footings are made by pouring concrete directly on the ground and therefore the soil-footing interface is rough. Schultze and Horn (1967) noted that the way the concrete footings are cast in place, there is adequate friction at the base which mobilizes friction angle equal to ϕ. Even the bottom of a metal storage tank is not smooth since the base is always treated with paint or asphalt to resist corrosion (Bowles, 1996). Therefore, the assumption of rough base is more realistic than a smooth one. Based on experimental studies, Vesic (1975) stated that foundation roughness has little effect on the ultimate bearing capacity, provided the footing load is vertical.

Example 12.1

A 2.0-m-wide square footing shown in Fig. 12.7 carries a column load of 500 kN. Unit weight of concrete is 24.0 kN/m^3. The properties of the clay are: $c' = 5$ kN/m^2, $\phi' = 25°$, and $\gamma = 18.0$ kN/m^3. What is the factor of safety with respect to bearing capacity failure? Use Vesic's N_γ.

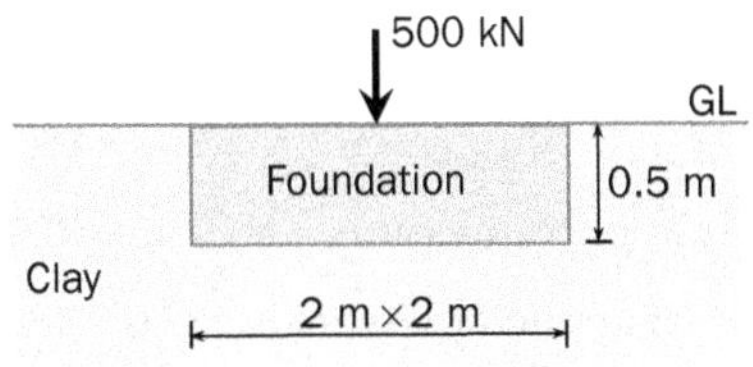

FIGURE 12.7 Square footing for Example 12.1.

Solution From Eq. (12.7), for $\phi' = 25°$,

$$N_q = e^{\pi \tan\phi} \tan^2\left(45 + \frac{\phi}{2}\right) = 10.7$$

From Eq. (12.8),

$$N_c = \left(N_q - 1\right)\cot\phi = 20.8$$

Vesic's N_γ is given by

$$N_\gamma = 2\,(N_q + 1)\tan\phi' = 2(10.7 + 1)\tan 25 = 10.9$$

These values can also be read off from Table 12.2.

From Eq. (12.5),

$$q_{\text{ult}} = 1.3cN_c + \gamma D_f N_q + 0.4B\gamma N_\gamma$$

$$q_{\text{ult,gross}} = 1.3(5)(20.8) + 18(0.5)(10.7) + 0.4(2)(18)(10.9) = 388.5\ \text{kN/m}^2$$

$$q_{\text{ult,net}} = q_{\text{ult,gross}} - \gamma D_f = 388.5 - 18(0.5) = 379.5\ \text{kN/m}^2$$

$$\text{Gross applied pressure} = 500/(2 \times 2) + 24(0.5) = 137.0\ \text{kN/m}^2$$

$$\text{Net applied pressure} = 137.0 - 18(0.5) = 128.0\ \text{kN/m}^2$$

Therefore, the factor of safety with respect to bearing capacity failure is given by

$$F = 379.5/128.0 = \mathbf{2.96}$$

Example 12.2

Assess the short-term stability of the same square footing in Example 12.1 if it is given that $c_u = 40\ \text{kN/m}^2$. Assume undrained conditions.

Solution For undrained condition, $\phi_u = 0$. $N_q = 1$, $N_\gamma = 0$, and $N_c = 5.14$.

From Eq. (12.5),

$$q_{\text{ult}} = 1.3cN_c + \gamma D_f N_q + 0.4B\gamma N_\gamma$$

$$q_{\text{ult,gross}} = 1.3(40)(5.14) + 18(0.5)(1.0) + 0.4(2)(18)(0) = 276.3 \text{ kN/m}^2$$

$$q_{\text{ult,net}} = q_{\text{ult,gross}} - \gamma D_f = 276.3 - 18(0.5) = 267.3 \text{ kN/m}^2$$

$$\text{Gross applied pressure} = 500/(2 \times 2) + 24(0.5) = 137.0 \text{ kN/m}^2$$

$$\text{Net applied pressure} = 137.0 - 18(0.5) = 128.0 \text{ kN/m}^2$$

Therefore, the factor of safety with respect to bearing capacity failure is given by

$$F = 267.3/128.0 = \mathbf{2.09}$$

Note: The short-term stability is more critical than the long-term stability.

12.5 The General Bearing Capacity Equation

Noting that the original bearing capacity equation [Eq. (12.4)] is mainly for a strip footing and that the equation has to be modified for square, circular, and rectangular footings, it is desirable to have one equation that covers all shapes. Further, the original bearing capacity equation does not adequately account for the following factors:

a. Shape of the footing (Meyerhof, 1963; Brinch Hansen, 1970; Vesic, 1973), *s*
b. Depth of the footing (Meyerhof, 1963; Brinch Hansen, 1970; Vesic, 1973), *d*
c. Inclination of the applied load to the vertical (Meyerhof, 1963; Brinch Hansen, 1970; Vesic, 1973), *i*
d. Ground inclination (Brinch Hansen, 1970; Vesic, 1973), *g*
e. Base inclination (Brinch Hansen, 1970; Vesic, 1973), *b*
f. Soil compressibility (Vesic, 1973), *C*

A general bearing capacity equation, covering all footing shapes, was proposed by Meyerhof (1963), Brinch Hansen (1970), and Vesic (1973), along with their own correction factors discussed below. In this book, we will use the one proposed by Meyerhof (1963) which is simpler and accounts for the first three factors (shape, depth, and inclination) only. In rare situations where there is ground or base inclination, it is suggested to use Brinch Hansen's or Vesic's equation along with their correction factors. They should never be mixed. These equations are meant to work well with the specific set of factors. Vesic's (1973) correction factors are summarized in App. B.

In the most generalized form, the bearing capacity equation can be written as

$$q_{\text{ult}} = s_c d_c i_c g_c b_c C_c \mathbf{cN_c} + s_q d_q i_q g_q b_q C_q \boldsymbol{\gamma D_f N_q} + s_\gamma d_\gamma i_\gamma g_\gamma b_\gamma C_\gamma \mathbf{0.5B}\boldsymbol{\gamma}\mathbf{N_\gamma} \tag{12.11}$$

Equation (12.11) is of the same form as the original bearing capacity equation [Eq. (12.4)] shown in bold, but includes the six factors listed above (*s*, *d*, *i*, *g*, *b*, and *C*), applied to the three bearing capacity components coming from cohesion (c), surcharge (q), and self-weight (γ) of the soil. With only the first three correction factors accounting for shape, depth, and inclination, Meyerhof's bearing capacity equation becomes

$$q_{\text{ult}} = s_c d_c i_c\, \mathbf{cN_c} + s_q d_q i_q\, \boldsymbol{\gamma D_f N_q} + s_\gamma d_\gamma i_\gamma\, \mathbf{0.5B}\boldsymbol{\gamma}\mathbf{N_\gamma} \tag{12.12}$$

Shape factors
$s_c = 1 + 0.2\frac{B}{L}\tan^2(45 + \phi/2)$
$s_q = s_\gamma = 1 + 0.1\frac{B}{L}\tan^2(45 + \phi/2)$ for $\phi > 10°$
$s_q = s_\gamma = 1$ for $\phi = 0$
Depth factors
$d_c = 1 + 0.2\frac{D_f}{B}\tan(45 + \phi/2)$
$d_q = d_\gamma = 1 + 0.1\frac{D_f}{B}\tan(45 + \phi/2)$ for $\phi > 10°$
$d_q = d_\gamma = 1$ for $\phi = 0$
Inclination factors
$i_c = i_q = \left(1 - \frac{\alpha°}{90}\right)^2$
$i_\gamma = \left(1 - \frac{\alpha°}{\phi}\right)^2$ for $\phi > 0$
$i_\gamma = 1$ for $\phi = 0$

Note: α is the inclination of the load to the vertical

TABLE 12.3 Meyerhof's Shape, Depth, and Inclination Factors

The expressions to compute Meyerhof's shape, depth, and inclination factors are summarized in Table 12.3. For strip footings, $L >> B$ and hence B/L can be taken as zero.

Example 12.3

Redo Example 12.1 using Meyerhof's general equation for the bearing capacity.

Solution For $\phi' = 25°$, from Table 12.2, $N_q = 10.7$, $N_c = 20.8$, and $N_\gamma = 6.8$.

$$s_c = 1 + 0.2\frac{2}{2}\tan^2(45 + 25/2) = 1.49$$

$$s_q = s_\gamma = 1 + 0.1\frac{2}{2}\tan^2(45 + 25/2) = 1.25$$

$$d_c = 1 + 0.2\frac{0.5}{2}\tan(45 + 25/2) = 1.08$$

$$d_q = d_\gamma = 1 + 0.1\frac{0.5}{2}\tan(45 + 25/2) = 1.04$$

Since the load is vertical, all three inclination factors are unity.

$$q_{\text{ult,gross}} = s_c d_c i_c\, cN_c + s_q d_q i_q \gamma D_f N_q + s_\gamma d_\gamma i_\gamma 0.5B\gamma N_\gamma$$

$$q_{\text{ult,gross}} = (1.49)(1.08)(1)(5)(20.8) + (1.25)(1.04)(1)(18)(0.5)(10.7) + (1.25)(1.04)(1)(0.5)(2.0)(18)(6.8) = 451.7 \text{ kN/m}^2$$

$$q_{\text{ult,net}} = 451.7 - (18)(0.5) = 442.7 \text{ kN/m}^2$$

From Example 12.1, the net applied pressure = 128.0 kN/m²

Therefore, $F = 442.7/128.0 = \mathbf{3.46}$

As seen from Examples 12.1 and 12.3, Terzaghi's original bearing capacity equation without the additional correction factors is conservative and often underestimates the allowable loads.

12.5.1 Plane Strain Correction

Figure 12.8a shows a circular footing where the soil along the vertical centerline is subjected to axi-symmetric loading. Square footings can be approximated as circular and hence the loading is approximately axi-symmetric. The loading is quite different in a strip footing as shown in Fig. 12.8b. Due to the wall loading, and the footing being very long, the strains in the soil are confined to the transverse planes with no strains occurring in the longitudinal direction. Such loading is known as plane strain loading. In geotechnical engineering, we come across plane strain loadings often in situations such as retaining walls, embankments, buried pipes in trenches, and roadwork.

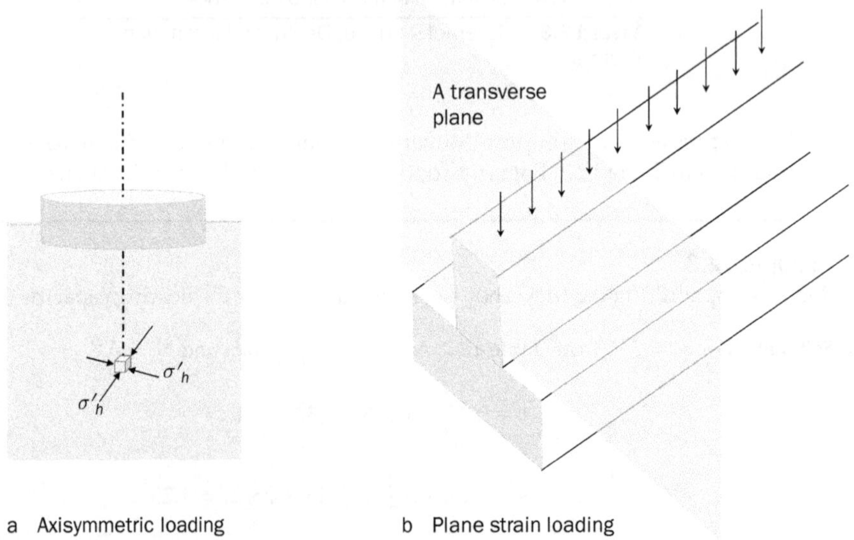

FIGURE 12.8 (a) Axisymmetric, (b) plane strain loading.

It has been observed by many researchers that the friction angle obtained from a triaxial compression test ϕ'_{tx}, is smaller than that obtained from a plane strain compression test ϕ'_{ps}, by 4°–9° in dense sands and 2°–4° in loose sands (Ladd et al., 1977). This is due to the difference in axi-symmetric to plane strain loading. The loading along the vertical centerline of a circular or square footing is axi-symmetric, whereas plane strain loading exists beneath a strip footing. Therefore, the friction angle has to be adjusted depending on how it was measured and the shape of the footing.

Based on the suggestions from Bishop (1961), Bjerrum and Kummeneje (1961), and Lee (1970) that the plane strain angle is approximately 10% larger than the triaxial ones, Meyerhof (1963) suggested that the friction angle for a rectangular footing be adjusted by the following equation.

$$\phi'_{\text{rectangular}} = \left(1.1 - 0.1\frac{B}{L}\right)\phi'_{tx} \tag{12.13}$$

Equation (12.13) essentially interpolates between ϕ'_{tx} (for $B/L = 1$) and ϕ'_{ps} (for $B/L = 0$), which differ by 10%.

12.5.2 Bearing Capacity Reduction due to Eccentric Loading

Foundation loads are not always applied at the center. When there are lateral loads or moments acting on the structure, the column loads may act with some eccentricities. A foundation near the property boundary may also experience some eccentricity with the column not placed at the center. Eccentricity reduces the bearing capacity.

Figure 12.9 shows a rectangular footing with dimensions B and L, with center O. The load is applied at O', with eccentricities of e_B and e_L along the x and y directions, respectively. Meyerhof (1963) suggested neglecting the hatched area shown in Fig. 12.9,

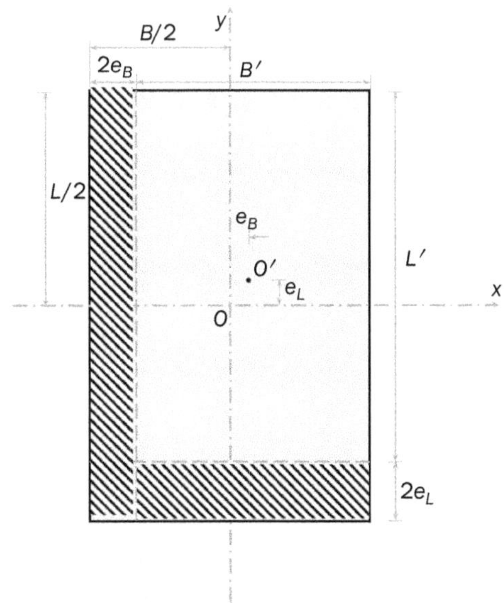

FIGURE 12.9 Meyerhof's eccentricity correction.

which consists of two strips of width $2e_B$ and $2e_L$. For the remaining area, the load is acting at its center O'. The effective area of the foundation in computing the load is $B' \times L'$, where $B' = B - 2e_B$ and $L' = L - 2e_L$. In computing the shape factors and in the bearing capacity equation, B' and L' must be used. In depth factor computations, B must be used.

12.6 Pressure Distributions beneath Eccentrically Loaded Foundations

Sometimes foundations are subjected to horizontal loads, which lead to eccentricity in the applied load. There can also be moments that induce eccentricity. Eccentricity makes the pressure distribution beneath the footing uneven.

Figure 12.10a shows a strip footing of width B subjected to an eccentric line load Q per unit length, with eccentricity of e. The vertical pressure beneath the footing will be the maximum at the right edge and minimum at the left edge. The eccentric load shown in Fig. 12.10a is replaced by a centric load Q and a moment M (= Qe) in Fig. 12.9b. Considering the vertical pressure induced by Q and M separately, and applying the principle of superposition, the applied pressure $q(x)$ at horizontal distance x from the centerline is given by

$$q(x) = \frac{Q}{B} + \frac{M}{I}x \tag{12.14}$$

where I is the moment of inertia about the horizontal centerline for a unit length of the strip, and is given by $B^3/12$. Equation (12.14) can be written as

$$q(x) = \frac{Q}{B} + \frac{12Qe}{B^3}x = \frac{Q}{B}\left(1 + 12\frac{e}{B^2}x\right) \tag{12.15}$$

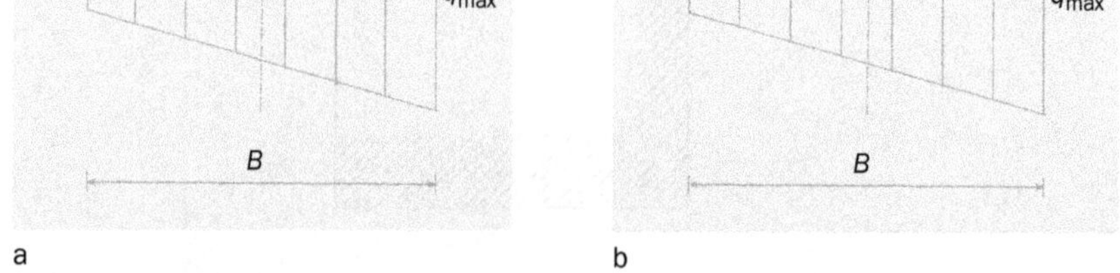

FIGURE 12.10 Pressure distribution beneath an eccentrically loaded strip footing: (a) eccentric load, (b) equivalent centric load with a moment.

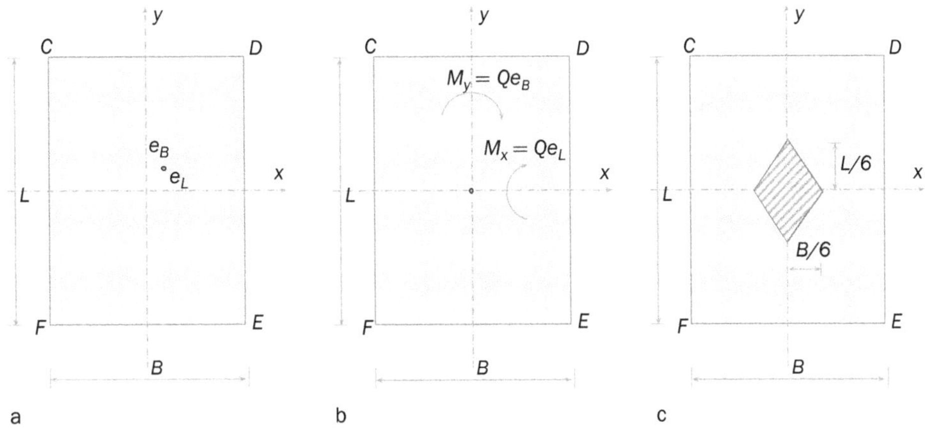

FIGURE 12.11 Two-way eccentricity in a rectangular footing: (a) eccentric loading, (b) equivalent centric load and moments, (c) kern.

It can be deduced from Eq. (12.15) that $q(x)$ varies linearly, with minimum value occurring at $x = -B/2$ and maximum at $x = B/2$. These values are given by

$$q_{\min} = \frac{Q}{B}\left(1 - 6\frac{e}{B}\right) \tag{12.16}$$

$$q_{\max} = \frac{Q}{B}\left(1 + 6\frac{e}{B}\right) \tag{12.17}$$

Soils are weak in tension, and therefore it is desirable to ensure the entire footprint of the foundation is in compression. To ensure this, $q_{\min} > 0$. In other words, $e < B/6$. It is a common practice to limit the eccentricity to $B/6$. When e exceeds $B/6$, there will be redistribution of the soil pressures.

Figure 12.11a shows a rectangular footing of dimensions B and L, subjected to a vertical load Q with eccentricities e_B and e_L along the x and y directions, respectively. Figure 12.11b shows an equivalent loading system including a point load Q at the center and the moments M_x and M_y about the x and y axes. Considering each of the three components Q, M_x, and M_y, and applying the principle of superposition, the pressure $q(x,y)$ at any point under the footing is given by

$$q(x,y) = \frac{Q}{BL} + \frac{M_y}{I_y}x + \frac{M_x}{I_x}y$$

$$q(x,y) = \frac{Q}{BL} + \frac{Qe_B}{\left(\frac{1}{12}LB^3\right)}x + \frac{Qe_L}{\left(\frac{1}{12}BL^3\right)}y$$

$$q(x,y) = \frac{Q}{BL}\left(1 + 12\frac{e_B}{B^2}x + 12\frac{e_L}{L^2}y\right) \tag{12.18}$$

Substituting the values of x and y in Eq. (12.18), the soil pressure can be determined at any point beneath the foundation. In Figs. 12.11a and 12.11b, the maximum value q_{max} occurs at D ($x = B/2$ and $y = L/2$) and the minimum value q_{min} ($x = -B/2$ and $y = -L/2$) occurs at F. They are given by

$$q_{max} = \frac{Q}{BL}\left(1 + 6\frac{e_B}{B} + 6\frac{e_L}{L}\right) \tag{12.19a}$$

$$q_{min} = \frac{Q}{BL}\left(1 - 6\frac{e_B}{B} - 6\frac{e_L}{L}\right) \tag{12.19b}$$

For q_{min} to be positive,

$$1 - 6\frac{e_B}{B} - 6\frac{e_L}{L} > 0$$

$$\frac{e_B}{B} + \frac{e_L}{L} < \frac{1}{6}$$

Figure 12.11c shows the hatched area at the center, a rhombus known as *kern*, within which the load has to be applied to ensure q_{min} is positive so that the soil beneath the footing is in compression at all locations.

Example 12.4

A $B \times L$ rectangular footing carries a vertical load Q with eccentricities $B/24$ and $L/10$ along the B and L directions, respectively. Determine the vertical pressure distribution along the perimeter of the footing.

Solution

$$\begin{aligned} q(x,y) &= \frac{Q}{BL}\left(1 + 12\frac{e_B}{B^2}x + 12\frac{e_L}{L^2}y\right) \\ &= \frac{Q}{BL}\left(1 + 12\frac{e_B}{B}\frac{x}{B} + 12\frac{e_L}{L}\frac{y}{L}\right) \\ &= \frac{Q}{BL}\left(1 + 12\frac{1}{24}\frac{x}{B} + 12\frac{1}{10}\frac{y}{L}\right) \\ &= \frac{Q}{BL}\left(1 + 0.5\frac{x}{B} + 1.2\frac{y}{L}\right) \end{aligned}$$

$$q_C(-B/2, L/2) = \frac{Q}{BL}\left(1 - 0.5\frac{1}{2} + 1.2\frac{1}{2}\right) = 1.35\frac{Q}{BL}$$

$$q_D(B/2, L/2) = \frac{Q}{BL}\left(1 + 0.5\frac{1}{2} + 1.2\frac{1}{2}\right) = 1.85\frac{Q}{BL}$$

$$q_E(B/2, -L/2) = \frac{Q}{BL}\left(1 + 0.5\frac{1}{2} - 1.2\frac{1}{2}\right) = 0.65\frac{Q}{BL}$$

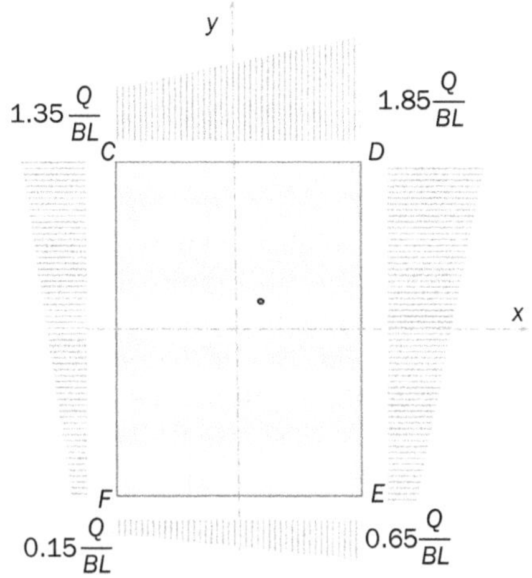

FIGURE 12.12 Pressure distribution along the perimeter.

$$q_F(-B/2,-L/2) = \frac{Q}{BL}\left(1 - 0.5\frac{1}{2} - 1.2\frac{1}{2}\right) = 0.15\frac{Q}{BL}$$

The pressure distribution along the perimeter is shown in Fig. 12.12.

Since the applied pressure is positive at all locations, the load is applied within the kern.

12.7 Raft Foundations

Raft or mat foundation is a thick reinforced concrete slab that supports several column loads and/or wall loads. It can also support silos, chimneys, towers, and machinery. When proposed pad and strip footings cover more than 50% of the footprint of the building, it may be preferable to consider a raft foundation covering the entire area.

Bearing capacity computations for raft foundations are similar to those of pad or strip footings discussed above. In clays under undrained conditions ($\phi_u = 0$; $N_q = 1$, $N_\gamma = 0$, and $N_c = 5.14$), with $s_c = 1 + 0.2B/L$, $d_c = 1 + 0.2D_f/B$, $s_q = 1$, and $d_q = 1$ (see Table 12.3), the net ultimate bearing capacity of a raft can be estimated as

$$q_{\text{ult,gross}} = 5.14c_u\left(1 + 0.2\frac{B}{L}\right)\left(1 + 0.2\frac{D_f}{B}\right) + \gamma D_f$$

$$q_{\text{ult,net}} = 5.14c_u\left(1 + 0.2\frac{B}{L}\right)\left(1 + 0.2\frac{D_f}{B}\right) \tag{12.20}$$

With large value of B, the ultimate bearing capacity of the raft in sands is very high, thus giving a very high factor of safety with respect to bearing capacity failure. The design is controlled by the allowable settlements. Extending Meyerhof's (1956) work,

Bowles (1988) proposed an empirical relation for estimating the net allowable bearing capacity of a raft foundation in sands as:

$$q_{\text{all,net}}\left(\text{kN/m}^2\right) = 12.5N_{60}\left(\frac{B+0.3}{B}\right)^2\left(1+\frac{1}{3}\frac{D_f}{B}\right)\left(\frac{\text{maximum settlement (mm)}}{25}\right) \quad (12.21)$$

In rafts, total settlements as high as 50 mm or more can be allowed while differential settlements are still within tolerable limits. This is about twice the total settlement allowed for isolated footings in granular soils.

The structural design of a raft can be carried out by considering the raft as rigid or flexible. The simplified method outlined below considers the raft as rigid, which can translate or rotate. Figure 12.13 shows the column layout for a proposed building

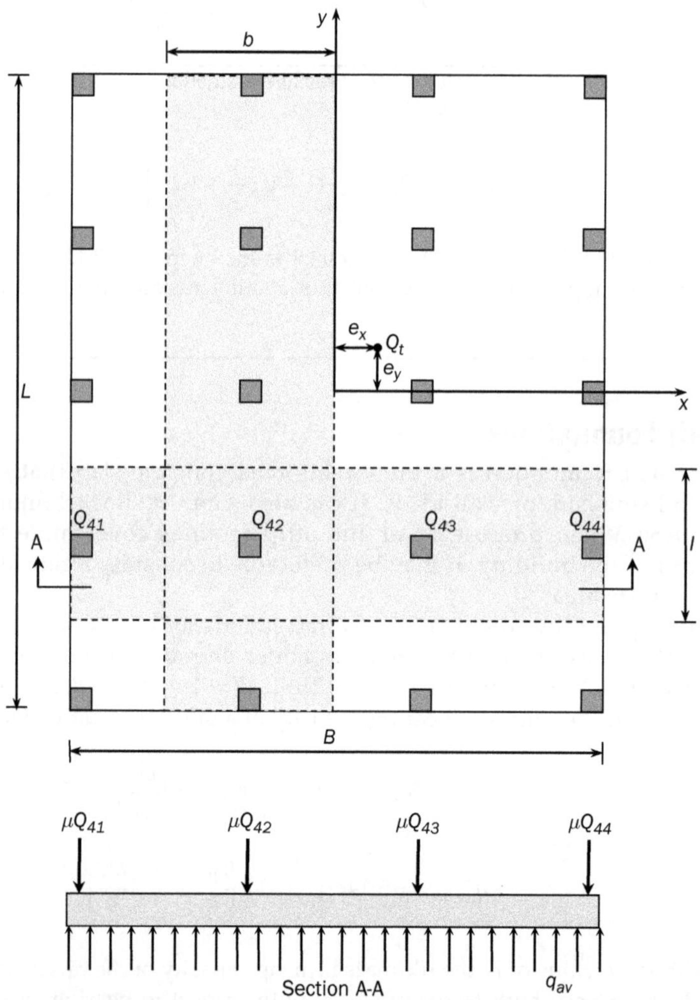

FIGURE 12.13 Raft design as a two-way slab.

with loads Q_{ij}. The coordinates of the column positions are defined with respect to a Cartesian coordinate system as shown in the figure.

The first step in the design is to find the equivalent point load Q_t and the eccentricities e_x and e_y based the column loads and their locations.

$$Q_t = \sum_{i=1,j=1}^{i=m,j=n} Q_{ij} \tag{12.22}$$

$$Q_t e_x = \sum_{i=1,j=1}^{i=m,j=n} Q_{ij} x_{ij} \tag{12.23}$$

$$Q_t e_y = \sum_{i=1,j=1}^{i=m,j=n} Q_{ij} y_{ij} \tag{12.24}$$

Here, x_{ij} and y_{ij} are the x and y coordinates of the column load Q_{ij}. From Eqs. (12.22), (12.23), and (12.24), Q_t, e_x, and e_y can be determined. With respect to the same Cartesian coordinate system, the foundation pressure at a point beneath the raft can be written as

$$q(x,y) = \frac{Q_t}{BL} + \frac{M_y}{I_y}x + \frac{M_x}{I_x}y \tag{12.25}$$

where $M_y = Q_t e_x$, $M_x = Q_t e_y$, $I_x = BL^3/12$, and $I_y = LB^3/12$. The maximum net contact pressure computed from Eq. (12.25) must be less than the net allowable bearing capacity of the raft.

The next step is to divide the raft into several strips along x and y directions, as shown by the shaded areas in the figure, similar to the structural design of a two-way slab considering the tributary areas. Since we are isolating the strips in the analysis and neglecting the bending moments and shear forces between the strips, the summation of the column loads within a strip will not match the upward load obtained by integration the contact pressure $q(x, y)$ over the strip area. This simplification violates the force equilibrium and requires an adjustment to the column loads. If the average value of $q(x,y)$ over the strip is q_{av}, it can be determined by averaging the values at the four corners. For the $B \times l$ strip shown in the figure, the adjustment factor μ can be determined from

$$\mu(Q_{41} + Q_{42} + Q_{43} + Q_{44}) = q_{av}(B)(l) \tag{12.26}$$

With the adjusted loads, the bending moment and shear force diagrams can be drawn and the reinforcements and slab thickness can be determined.

Example 12.5

The raft shown in Fig. 12.14 supports 20 columns. The columns are spaced at 5 m and 4 m intervals in x and y directions, respectively. The column loads are shown in the figure. It is desired to keep the contact pressure within 75 kN/m². Check whether this is satisfied. Also compute the contact pressures at A, B, C, and D. Treating it as a rigid raft, determine the modified column loads within the strip $ABCD$ shown hatched.

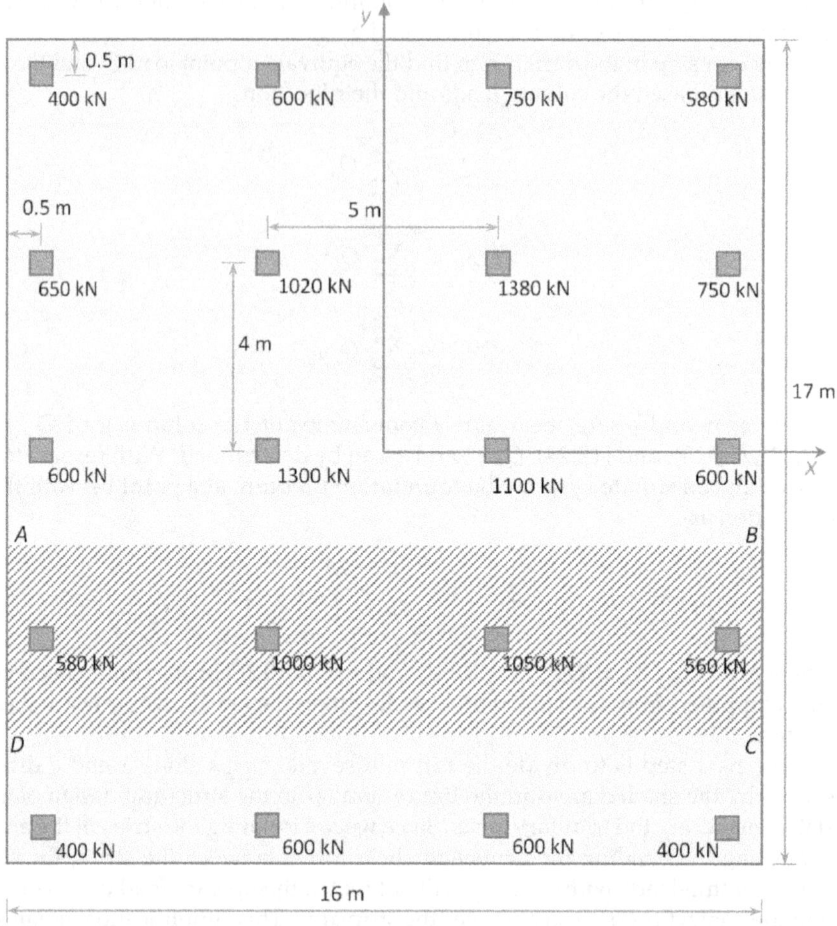

Figure 12.14 Raft supporting 20 columns.

Solution The total load coming from all the columns is

$$\sum Q_{ij} = 14{,}920 \text{ kN}$$

Let this equivalent load be acting with eccentricities of e_x and e_y, respectively.

Taking moments about y-axis,

$$\begin{aligned} 14{,}920\, e_x &= (580 + 750 + 600 + 560 + 400)(7.5) + (750 + 1380 + 1100 + 1050 + 600)(2.5) \\ &\quad - (600 + 1020 + 1300 + 1000 + 600)(2.5) - (400 + 650 + 600 + 580 + 400)(7.5) \\ &= 2850 \text{ kN·m} \end{aligned}$$

Therefore, $e_x = 0.1910$ m $= 191.0$ mm.

Taking moments about x axis,

$$\begin{aligned} 14{,}920\, e_y &= (400 + 600 + 750 + 580)(8) + (650 + 1020 + 1380 + 750)(4) \\ &\quad - (580 + 1000 + 1050 + 560)(4) - (400 + 600 + 600 + 400)(8) = 5080 \text{ kN·m} \end{aligned}$$

Therefore, $e_y = 0.3405$ m = 340.5 mm.

The soil pressure distribution beneath the raft, as a function of the coordinates, is given by

$$q(x,y) = \frac{Q}{BL} + \frac{M_y}{I_y}x + \frac{M_x}{I_x}y$$

where $Q = 14{,}920$ kN, $B = 16$ m, $L = 17$ m

$$M_y = 14{,}920 \times 0.191 = 2850 \text{ kN·m}$$

$$I_y = \frac{17 \times 16^3}{12} = 5802.7 \text{ m}^4$$

$$M_x = 14{,}920 \times 0.3405 = 5080 \text{ kN·m}$$

$$I_x = \frac{16 \times 17^3}{12} = 6550.7 \text{ m}^4$$

$$q(x,y) = \frac{14{,}920}{16 \times 17} + \frac{2850}{5802.7}x + \frac{5080}{6550.7}y$$

$$q(x,y) = 54.85 + 0.4912x + 0.7755y$$

The maximum contact pressure occurs at the top right corner where $x = 8.0$ m and $y = 8.5$ m. This value is given by

$$q(x,y) = 54.85 + 0.4912(8.0) + 0.7755(8.5) = 65.37 \text{ kN/m}^2 < 75.0 \text{ kN/m}^2$$

$$\text{At } A,\ q(-8,-2) = 54.85 + 0.4912(-8) + 0.7755(-2) = 49.37 \text{ kN/m}^2$$

$$\text{At } B,\ q(8,-2) = 54.85 + 0.4912(8) + 0.7755(-2) = 57.23 \text{ kN/m}^2$$

$$\text{At } C,\ q(8,-6) = 54.85 + 0.4912(8) + 0.7755(-6) = 54.13 \text{ kN/m}^2$$

$$\text{At } D,\ q(-8,-6) = 54.85 + 0.4912(-8) + 0.7755(-6) = 46.27 \text{ kN/m}^2$$

The average pressure in the hatched area *ABCD* is

$$q_{\text{avg}} = \frac{49.37 + 57.23 + 54.13 + 46.27}{4} = 51.75 \text{ kN/m}^2$$

The average pressure computed using the four column loads is

$$q_{\text{avg}} = \frac{580 + 1000 + 1050 + 560}{4 \times 16} = 49.84 \text{ kN/m}^2$$

Let us use the average of these two values (50.8 kN/m^2) as the average applied pressure in *ABCD*. The column loads have to be multiplied by a factor F to match this pressure.

$$50.8 = \frac{F \times 580 + F \times 1000 + F \times 1050 + F \times 560}{4 \times 16}$$

Therefore, $F = 1.02$.

The new loads for the four columns are 592, 1020, 1071, and 571 kN.

12.8 Total and Differential Settlements

The extent of settlement beneath a foundation depends on the soil conditions and the loadings. The settlement of an isolated pad or strip footing is called the *total settlement*. The absolute value of the difference between the settlements of the two adjacent foundations is known as the *differential settlement* between them. The ratio of the differential settlement to the span length is known as the *angular distortion*. Rigid body rotation is called *tilt*.

A building consists of a framework of slabs, beams, columns, and foundations, all of which are structural elements made of engineering materials such as concrete, steel, timber, etc. When the entire building settles equally at every location, the magnitude of settlement is of little concern. Here, the total settlements only affect the aesthetics, drainage, utility lines, and alignments. Palace of Fine Arts, built in early 1900s in Mexico City, settled more than 3.5 m, but is still in use. The Leaning Tower of Pisa in Italy experienced a tilt in the order of 1/10. There are structures that have undergone large total settlements, but with insignificant differential settlements, making them still functional. It is often the *differential settlement* that is a worry. Differential settlements can induce angular distortion, which can cause structural damages and should be limited.

Limiting values for the settlements have been discussed in the literature since 1950s [e.g., Polshin and Tokar (1957) and Skempton and MacDonald (1956)]. The settlements in clays are slower than in sands, occurring over many years and hence the structural system can gradually adjust to the settlements and sustain less damage. Therefore, larger settlements are generally allowed in clays than in sands. Similarly, raft foundations are subject to relatively lower differential settlements due to their rigidity and hence can tolerate larger total settlements than pad footings. Typically, angular distortion in the order of 1/500 to 1/300 is expected to cause architectural damage (e.g., cracking of the plaster and partitions) and problems with alignments and services lines. Structural damages can be expected when the angular distortion exceeds a limiting value in the order of 1/300 to 1/150. By limiting the total settlements, it is possible to limit the differential settlements and angular distortions. Terzaghi and Peck (1967) suggested limiting the total settlement of isolated footings in sand to 25 mm (1.0 in) would limit the differential settlement to 19 mm (¾ in).

Various codes and standards impose limits on total and differential settlements for the different types of structures. O'Brien (2013) suggests the following limiting values for total settlements in routine designs of shallow foundations.

Pad footings in clays	65 mm
Pad footings in sands	40 mm
Raft foundations in clays	100 mm
Raft foundations in sands	65 mm

12.9 Settlement Computation Based on Elastic Analysis (Drained Soils)

In this section, we will treat the soil as a linear elastic continuum with Young's modulus of E and Poisson's ratio of ν. The Young's modulus of soils can vary in the range of 1–70 MN/m^2, which is often estimated from in situ tests such as standard penetration or cone penetration test. Poisson's ratio for all materials, including soils and rocks, vary in the range of 0–0.5. For saturated clays, under undrained conditions without any volume change, theoretically it can be shown to be 0.5.

12.9.1 Surface Foundation on Semi-Infinite Elastic Half Space

For a uniformly loaded flexible rectangular foundation of dimensions $B \times L$ lying on top of an elastic half space, the settlement under a point on the foundation is given by

$$s = \frac{qB}{E}(1 - \nu^2)I \tag{12.27}$$

where I is the influence factor that depends on the location of the point of interest on the foundation, and L/B, q = pressure applied by the foundation to the underlying soil, B = width of the foundation (always the shorter dimension), E = Young's modulus of the soil, and ν = Poisson's ratio.

Schleicher (1926) expressed the influence factor for the *corner of a flexible footing* as

$$I_{\text{corner}} = \frac{1}{\pi}\left[m' \ln\left(\frac{1 + \sqrt{m'^2 + 1}}{m'}\right) + \ln\left(m' + \sqrt{m'^2 + 1}\right)\right] \tag{12.28}$$

where $m' = L/B$. The influence factors for the other locations on the foundation can be determined by dividing the foundation into four rectangles and using the principle of super position. It can be deduced that the influence factor for the center is twice that of the corner. The influence factors for the corner, center, mid-point of the short side, and mid-point of the long side, determined from Eq. (12.28), are shown in Fig. 12.15 for $1 \leq m' \leq 1000$. The values computed here are the same as those reported by Poulos and Davis (1974) who refer the work of Giroud (1968). These values can be approximated as follows.

$$\text{Corner:} \quad I = 0.7283 \log m' + 0.5469 \tag{12.29}$$

$$\text{Center:} \quad I = 1.4566 \log m' + 1.0939 \tag{12.30}$$

$$\text{Mid-point short side:} \quad I = 0.7318 \log m' + 0.7617 \tag{12.31}$$

$$\text{Mid-point long side:} \quad I = 1.4357 \log m' + 0.6894 \tag{12.32}$$

The settlements under the center and the perimeter of a uniformly loaded flexible circular foundation of diameter D are given by:

$$\text{Center:} \; s = \frac{qD}{E}(1 - \nu^2) \tag{12.33}$$

$$\text{Perimeter:} \; s = \frac{qD}{E}(1 - \nu^2)\frac{2}{\pi} \tag{12.34}$$

Comparing Eqs. (12.33) and (12.34) with Eq. (13.27), it can be seen that the influence factors for the center and a point on the perimeter of a flexible circular foundation are 1.00 and $2/\pi$, respectively.

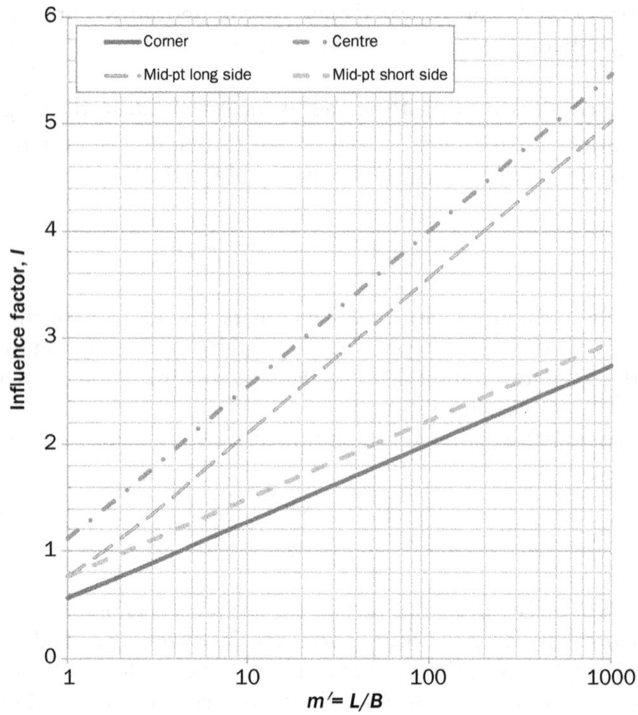

FIGURE 12.15 Variation of I versus m' for a flexible foundation.

$m' = L/B$	Flexible	Rigid
Circle	0.85	0.79 ($\pi/4$)
1	0.95	0.82
1.5	1.20	1.07
2	1.30	1.21
3	1.52	1.42
5	1.82	1.60
10	2.24	2.00
100	2.96	3.40

TABLE 12.4 Influence Factors to Compute Average Settlements of Rigid and Flexible Foundations

Flexible foundations apply uniform pressure to the underlying ground but settle nonuniformly. Rigid foundations apply nonuniform pressure but settle uniformly. Some influence factors for estimating the average values of the settlements of flexible and rigid footings, as reported in the literature, are given in Table 12.4.

The average values of I reported by Giroud (1968) and Poulos and Davis (1974) suggest that the average value of the settlement of a flexible foundation can be computed by using 84–88% of the I value for the center. Bowles (1982) suggests that the settlement of the rigid foundation can be estimated as 93% of the average settlement computed for a flexible foundation.

The ratio of the foundation stiffness to the soil stiffness is known as the *relative stiffness* K. It is a measure widely used to quantify the raft stiffness in relation to the soil stiffness and decide whether to treat the foundation as rigid or flexible. Clancy (1993) defined the relative stiffness of a circular raft as

$$K = \frac{E_r\,(1-\nu_s^2)}{E_s\,(1-\nu_r^2)}\left(\frac{t}{a}\right)^3 \tag{12.35}$$

where E_r = Young's modulus of the raft, E_s = Young's modulus of the soil, t = raft thickness, ν_s = Poisson's ratio of the soil, ν_r = Poisson's ratio of the raft and a = raft radius. Neglecting the effects of Poisson's ratios, the relative stiffness of a circular raft (K) can be approximated as (Brown 1969)

$$K \approx \frac{E_r}{E_s}\left(\frac{t}{a}\right)^3 \tag{12.36}$$

The relative stiffness of the raft influences the differential settlement within the raft. Higher the relative stiffness, lower is the differential settlement. The foundation can be assumed perfectly rigid for $K > 5$ and perfectly flexible for $K < 0.1$. For any other value of K, the influence factor I for the center of a circular foundation can be interpolated between $\pi/4$ (rigid) and 1.0 (flexible) as follows (Mayne and Poulos, 1999)

$$I_{\text{centre}} \approx \frac{\pi}{4} + \frac{1}{4.6 + 10K} \tag{12.37}$$

For circular foundations, the ratio of edge to center settlement is given by (Mayne and Poulos, 1999)

$$\frac{s_{\text{edge}}}{s_{\text{center}}} \approx 1 - \frac{1.533}{4.6 + 10K} \tag{12.38}$$

For rectangular foundations, the ratio of corner to center settlement is given by (Mayne and Poulos, 1999)

$$\frac{s_{\text{corner}}}{s_{\text{center}}} \approx 1 - \frac{2.3}{4.6 + 10K} \tag{12.39}$$

In the case of rectangular foundations, slightly different equations have been reported in the literature for the relative stiffness K.

Fraser and Wardle (1976):

$$K = \frac{4\,E_r\,(1-\nu_s^2)}{3\,E_s\,(1-\nu_r^2)}\left(\frac{t}{B}\right)^3 \tag{12.40}$$

Hain and Lee (1978):

$$K = \frac{4}{3\pi}\left(\frac{E_r}{E_s}\right)\left(1-\nu_s^2\right)\left(\frac{B}{L}\right)\left(\frac{t}{L}\right)^3 \tag{12.41}$$

Horikoshi and Randolph (1997):

$$K = 5.57\frac{E_r(1-\nu_s^2)}{E_s(1-\nu_r^2)}\left(\frac{B}{L}\right)^{0.5}\left(\frac{t}{L}\right)^3 \tag{12.42}$$

For a circular foundation where $B = L = a\sqrt{\pi}$, neglecting the Poisson's ratio effect, Eq. (12.42) gives the same value as Eq. (12.36).

Example 12.6

A 4-m-diameter and 200-mm-thick concrete footing applies 80 kN/m^2 pressure to the underlying ground. The Poisson's ratio of the concrete and soil are 0.16 and 0.20, respectively. The Young's modulus of concrete and the soil are 30 GPa and 20 MPa, respectively.

a. Determine the relative stiffness of the footing and check whether it can be considered flexible or rigid.
b. Determine the settlements beneath the center.
c. Determine the settlement beneath the perimeter.

Solution

Part a From Eq. (12.35),

$$K = \frac{E_r}{E_s}\frac{(1-\nu_s^2)}{(1-\nu_r^2)}\left(\frac{t}{a}\right)^3 = \left(\frac{30\times10^9}{20\times10^6}\right)\left(\frac{1-0.2^2}{1-0.16^2}\right)\left(\frac{0.2}{2.0}\right)^3 = \mathbf{1.48}$$

The footing is neither flexible nor rigid, somewhere in the middle.

Part b From Eq. (12.37), the influence factor for the center of the footing can be interpolated between the extremes perfectly rigid and perfectly flexible as

$$I_{\text{centre}} \approx \frac{\pi}{4} + \frac{1}{4.6+10K} \approx \frac{\pi}{4} + \frac{1}{4.6+10\times1.48} \approx 0.84$$

From Eq. (12.27),

$$s = \frac{qB}{E_s}(1-\nu_s^2)I = \frac{(80\times10^3)(4.0)}{20\times10^6}(1-0.2^2)(0.84) = 12.9\times10^{-3}\text{m} = \mathbf{12.9\ mm}$$

Part c From Eq. (12.38),

$$\frac{s_{\text{perimeter}}}{s_{\text{center}}} = 1 - \frac{1.533}{4.6+10K} = 1 - \frac{1.533}{4.6+10\times1.48} = 0.92$$

Therefore, $s_{\text{perimeter}} = 0.92\times12.9 = \mathbf{11.9\ mm}$

12.9.2 Surface Foundation on Elastic Continuum of Limited Depth

When the soil is underlain by a rigid stratum and hence the depth of soil below the foundation is finite, the settlement computed using Eq. (12.27) with Schleicher's (1926) influence factor has to be reduced. Steinbrenner (1934) suggested the following expression for the influence factor for the corner of a rectangular flexible surface foundation, taking into consideration the presence of the rigid layer, at depth of H below the foundation and $n' = H/B$ and $m' = L/B$.

$$I_{\text{corner}} = F_1 + \left(\frac{1-2\nu}{1-\nu}\right)F_2 \tag{12.43}$$

where

$$F_1 = \frac{1}{\pi}(A_0 + A_1) \tag{12.44}$$

$$F_2 = \frac{n'}{2\pi}\tan^{-1} A_2 \tag{12.45}$$

A_0, A_1, and A_2 are given by

$$A_0 = m'\ln\left[\frac{\left(1+\sqrt{m'^2+1}\right)\sqrt{\left(m'^2+n'^2\right)}}{m'\left(1+\sqrt{m'^2+n'^2+1}\right)}\right] \tag{12.46}$$

$$A_1 = \ln\left[\frac{\left(m'+\sqrt{m'^2+1}\right)\sqrt{1+n'^2}}{m'+\sqrt{m'^2+n'^2+1}}\right] \tag{12.47}$$

$$A_2 = \frac{m'}{n'\sqrt{m'^2+n'^2+1}} \tag{12.48}$$

The influence factor determined using Eqs. (12.43)–(12.48) is for the corner of a uniformly loaded flexible rectangular surface foundation. The influence factor for any other point on the foundation can be determined using the principle of superposition as before. For $H = \infty$ and hence $n' = \infty$, and Eqs. (12.46)–(12.48) become

$$A_0 = m'\ln\frac{1+\sqrt{m'^2+1}}{m'} \tag{12.49}$$

$$A_1 = \ln\left(m' + \sqrt{m'^2+1}\right) \tag{12.50}$$

$$A_2 = 0 \tag{12.51}$$

Substituting these in Eqs. (12.44) and (12.45),

$$F_1 = \frac{1}{\pi}\left[m'\ln\left(\frac{1+\sqrt{m'^2+1}}{m'}\right) + \ln\left(m' + \sqrt{m'^2+1}\right)\right] \tag{12.52}$$

$$F_2 = 0 \tag{12.53}$$

Substituting Eqs. (12.52) and (12.53) in Eq. (12.43) gives the same expression we had from Schleicher (1926) in Eq. (12.28). The values of F_1 and F_2 for different values of m' and n' are given in Tables 12.5 and 12.6.

	m'									
n'	1	1.2	1.4	1.6	1.8	2	2.5	3	5	10
0.25	0.014	0.013	0.012	0.011	0.011	0.011	0.010	0.010	0.010	0.010
0.50	0.049	0.046	0.044	0.042	0.041	0.040	0.038	0.038	0.036	0.036
0.75	0.095	0.090	0.087	0.084	0.082	0.080	0.077	0.076	0.073	0.071
1.00	0.142	0.138	0.134	0.130	0.127	0.125	0.121	0.118	0.113	0.111
1.25	0.186	0.183	0.179	0.176	0.173	0.170	0.165	0.161	0.154	0.151
1.50	0.224	0.224	0.222	0.219	0.216	0.213	0.207	0.203	0.194	0.189
1.75	0.257	0.259	0.259	0.258	0.255	0.253	0.247	0.242	0.232	0.225
2.00	0.285	0.290	0.292	0.292	0.291	0.289	0.284	0.279	0.267	0.259
2.25	0.309	0.317	0.321	0.323	0.323	0.322	0.317	0.313	0.300	0.291
2.50	0.330	0.341	0.347	0.350	0.351	0.351	0.348	0.344	0.331	0.320
2.75	0.348	0.361	0.369	0.374	0.377	0.378	0.377	0.373	0.359	0.347
3.00	0.363	0.379	0.389	0.396	0.400	0.402	0.402	0.400	0.386	0.373
3.25	0.376	0.394	0.406	0.415	0.420	0.423	0.426	0.424	0.412	0.397
3.50	0.388	0.408	0.422	0.431	0.438	0.442	0.447	0.447	0.435	0.420
3.75	0.399	0.420	0.436	0.447	0.454	0.460	0.467	0.468	0.458	0.441
4.00	0.408	0.431	0.448	0.460	0.469	0.476	0.484	0.487	0.479	0.462
4.25	0.417	0.441	0.459	0.473	0.483	0.490	0.501	0.505	0.499	0.481
4.50	0.424	0.450	0.469	0.484	0.495	0.503	0.516	0.521	0.517	0.499
4.75	0.431	0.458	0.478	0.494	0.506	0.515	0.530	0.536	0.535	0.517
5.00	0.437	0.465	0.487	0.503	0.516	0.526	0.543	0.551	0.552	0.534
5.50	0.448	0.478	0.501	0.520	0.534	0.546	0.566	0.576	0.588	0.565
6.00	0.457	0.489	0.514	0.534	0.550	0.563	0.585	0.598	0.610	0.595
6.50	0.465	0.498	0.524	0.546	0.563	0.577	0.603	0.618	0.635	0.622
7.00	0.471	0.506	0.533	0.556	0.575	0.590	0.618	0.635	0.658	0.647
7.50	0.477	0.513	0.541	0.565	0.585	0.601	0.631	0.650	0.679	0.670
8.00	0.482	0.519	0.549	0.573	0.594	0.611	0.643	0.664	0.697	0.692
8.50	0.487	0.524	0.555	0.580	0.601	0.619	0.653	0.676	0.714	0.713
9.00	0.491	0.529	0.560	0.587	0.609	0.627	0.663	0.687	0.730	0.732
9.50	0.495	0.533	0.565	0.592	0.615	0.634	0.671	0.697	0.744	0.751
10.0	0.498	0.537	0.570	0.597	0.621	0641	0.679	0.707	0.758	0.768
20.0	0.529	0.575	0.614	0.647	0.677	0.702	0.756	0.797	0.896	0.977
50.0	0.548	0.598	0.640	0.678	0.711	0.740	0.803	0.853	0.989	1.146

TABLE 12.5 F_1 Values

	m′									
n′	**1**	**1.2**	**1.4**	**1.6**	**1.8**	**2**	**2.5**	**3**	**5**	**10**
0.25	0.049	0.050	0.051	0.051	0.051	0.052	0.052	0.052	0.053	0.053
0.50	0.074	0.077	0.080	0.081	0.083	0.084	0.085	0.086	0.087	0.088
0.75	0.083	0.089	0.093	0.097	0.099	0.101	0.104	0.106	0.109	0.110
1.00	0.083	0.091	0.098	0.102	0.106	0.109	0.114	0.117	0.122	0.124
1.25	0.080	0.089	0.096	0.102	0.107	0.111	0.118	0.122	0.130	0.133
1.50	0.075	0.084	0.093	0.099	0.105	0.110	0.118	0.124	0.134	0.139
1.75	0.069	0.079	0.088	0.095	0.101	0.107	0.117	0.123	0.136	0.142
2.00	0.064	0.074	0.083	0.090	0.097	0.102	0.114	0.121	0.136	0.145
2.25	0.059	0.069	0.077	0.085	0.092	0.098	0.110	0.119	0.136	0.146
2.50	0.055	0.064	0.073	0.080	0.087	0.093	0.106	0.115	0.135	0.147
2.75	0.051	0.060	0.068	0.076	0.082	0.089	0.102	0.111	0.133	0.147
3.00	0.048	0.056	0.064	0.071	0.078	0.084	0.097	0.108	0.131	0.147
3.25	0.045	0.053	0.060	0.067	0.074	0.080	0.093	0.104	0.129	0.147
3.50	0.042	0.050	0.057	0.064	0.070	0.076	0.089	0.100	0.126	0.146
3.75	0.040	0.047	0.054	0.060	0.067	0.073	0.086	0.096	0.124	0.145
4.00	0.037	0.044	0.051	0.057	0.063	0.069	0.082	0.093	0.121	0.144
4.25	0.036	0.042	0.049	0.055	0.061	0.066	0.079	0.090	0.119	0.144
4.50	0.034	0.040	0.046	0.052	0.058	0.063	0.076	0.086	0.116	0.143
4.75	0.032	0.038	0.044	0.050	0.055	0.061	0.073	0.083	0.113	0.142
5.00	0.031	0.036	0.042	0.048	0.053	0.058	0.070	0.080	0.111	0.140
5.50	0.028	0.033	0.039	0.044	0.049	0.054	0.065	0.075	0.106	0.138
6.00	0.026	0.031	0.036	0.040	0.045	0.050	0.060	0.070	0.101	0.135
6.50	0.024	0.029	0.033	0.038	0.042	0.046	0.056	0.066	0.096	0.132
7.00	0.022	0.027	0.031	0.035	0.039	0.043	0.053	0.062	0.092	0.129
7.50	0.021	0.025	0.029	0.033	0.039	0.041	0.050	0.059	0.088	0.126
8.00	0.020	0.023	0.027	0.031	0.035	0.038	0.047	0.055	0.084	0.124
8.50	0.018	0.022	0.026	0.029	0.033	0.036	0.045	0.053	0.080	0.121
9.00	0.017	0.021	0.024	0.028	0.031	0.034	0.042	0.050	0.077	0.118
9.50	0.017	0.020	0.023	0.026	0.029	0.033	0.040	0.048	0.074	0.115
10.0	0.016	0.019	0.022	0.025	0.028	0.031	0.038	0.046	0.071	0.112
20.0	0.008	0.010	0.011	0.013	0.014	0.016	0.020	0.024	0.039	0.071
50.0	0.003	0.004	0.004	0.005	0.006	0.006	0.008	0.010	0.016	0.031

TABLE 12.6 F_2 Values

12.9.3 Effect of Embedment

The expressions developed in Secs. 12.9.1 and 12.9.2 were for foundations placed at the ground level with no embedment. When the foundation base is located at a depth D_f beneath the ground level, the embedment reduces the settlement. Fox (1948) proposed reduction factor I_f to account for this reduction in settlement. The settlement computed using Schleicher (1926) or Steinbrenner (1934) influence factors have to be multiplied by I_f to account for the settlement reduction due to embedment. I_f values for different D_f/B, L/B, and ν are given in Fig. 12.16.

Example 12.7

A 2 m × 3 m pad footing applies a pressure of 80 kN/m² to the underlying ground. The Young's modulus and Poisson's ratio of the soil are 30 MN/m² and 0.3, respectively. Determine the settlement beneath the center for the following cases.

a. Surface foundation ($D_f = 0$)
b. Bedrock at 2 m depth
c. Bedrock at 2 m depth and $D_f = 1$ m.

Solution

Part a Using the approximation [Eq. (12.30)] to Schleicher (1926) equation

$$\text{For } m' = L/B = 3/2 = 1.5, I = 1.35 \text{ (for the center)}$$

From Eq. (12.27),

$$s = \frac{qB}{E}(1 - \nu^2)I$$

$$s = \frac{(80)(2.0)}{30{,}000}(1 - 0.3^2)(1.35) = 0.0066 \text{ m} = \mathbf{6.6 \text{ mm}}$$

Part b With limited soil depth, let us use Steinbrenner's solution, which is for the corners. Let us divide the area into four quarters where the center becomes a corner for each quarter.

With bedrock (rigid stratum) at 2 m depth, and $B = 1$ m (for a quarter)

$$n' = H/B = 2/1 = 2$$
$$m' = 1.5$$

For Steinbrenner equation, from Tables 12.5 and 12.6, $F_1 = 0.292$ and $F_2 = 0.087$.

From Eq. (12.35),

$$I_{\text{corner}} = F_1 + \left(\frac{1 - 2\nu}{1 -}\right)F_2 = 0.292 + \left(\frac{1 - 2 \times 0.3}{1 - 0.3}\right)(0.087) = 0.342$$

From Eq. (12.27), and considering the four quarters,

$$s = \frac{(80)(1.0)}{30{,}000}(1 - 0.3^2)(4\ 0.343) = 0.0033 \text{ m} = \mathbf{3.3 \text{ mm}}$$

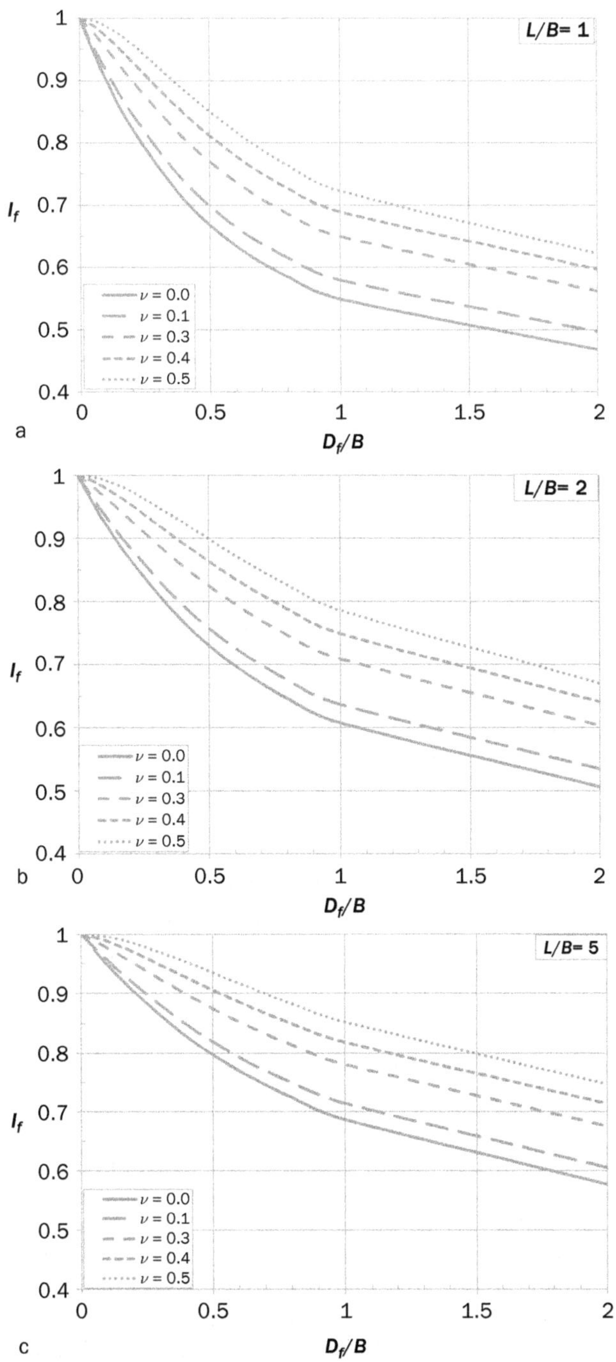

Figure 12.16 Variation of I_f with D_f/B: (a) $L/B = 1$, (b) $L/B = 2$, (c) $L/B = 5$.

Part c

$$D_f/B = 1/2 = 0.5 \quad \text{and} \quad \nu = 0.3$$

From Fig. 12.16,

for $L/B = 1$, $I_f = 0.77$
for $L/B = 2$, $I_f = 0.82$

Therefore, for $L/B = 1.5$, $I_f = 0.80$.
Therefore the settlement = 0.80 × 3.3 = **2.6 mm.**

12.10 Settlement Computations in Granular Soils

The majority of the settlements of foundations in granular soils are almost instantaneous, with a slight increase over time due to creep. There are more than 40 empirical and semiempirical methods to estimate the expected settlements of shallow foundations in granular soils. However, the quality of predictions of these methods remains poor, as demonstrated by the settlement prediction symposium in Texas in 1994 where five prototype square footings were loaded to failure and the data were compared with the predictions from 31 experts (Briaud and Gibbens, 1994). This is discussed briefly in Sec. 12.10.5.

The five major factors that influence the magnitude of the settlement of a shallow foundation in a granular soil are applied pressure (q), soil stiffness (E), foundation width (B), foundation shape (B/L), and foundation depth (D_f/B). Most of the settlement prediction methods take these factors into account. The soil stiffness or Young's modulus E is generally estimated from the in situ tests such as standard penetration or cone penetration tests.

12.10.1 Terzaghi and Peck's (1967) Method

Terzaghi and Peck (1967) proposed the first method for predicting settlement of a shallow foundation in a granular soil. They related the settlement of a square footing (s_{footing}) of width B to the settlement of a 300-mm-wide (1-ft) square plate (s_{plate}) under the same pressure by

$$s_{\text{footing}} = s_{\text{plate}} \left(\frac{2B}{B + 0.3} \right)^2 \left(1 - \frac{1}{4} \frac{D_f}{B} \right) \tag{12.54}$$

where B is in meters.

The last term in Eq. (12.54) accounts for the reduction in settlement with the increase in foundation depth. Leonards (1986) suggested replacing 1/4 by 1/3, based on additional load test data. The values of s_{plate} can be obtained from Fig. 12.17 that summarizes the plate loading test data, given by Terzaghi and Peck (1967). These load tests were carried out on thick deposits of drained normally consolidated sands. This method was originally proposed for square foundations, but can be applied to rectangular and strip foundations with caution. The deeper influence zone and increase in the stresses within the underlying soil mass in the case of rectangular or strip foundations are compensated by the increase in the soil stiffness. The method does not account for creep effects.

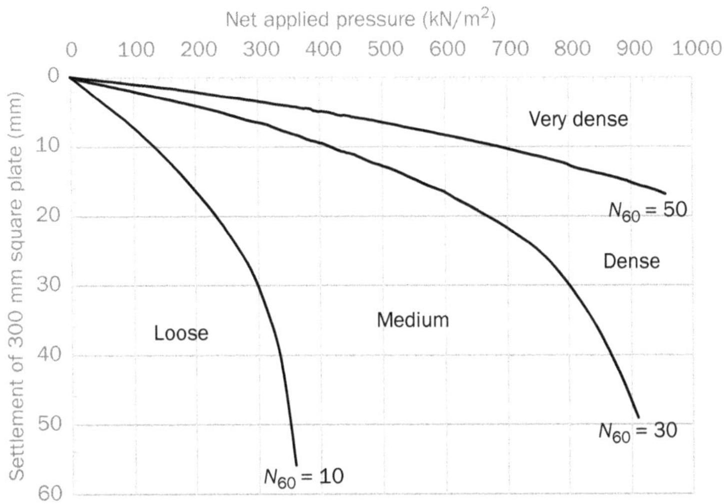

FIGURE 12.17 Settlement versus applied pressure plots for 300 mm × 300 mm plate load test. (Data from late Professor G. A. Leonards from Purdue University.)

Example 12.8

A 3.0-m-wide square foundation, placed at a depth of 1.0 m within a sand ($\gamma = 18.0$ kN/m³, $N_{60} = 28$), applies a net pressure of 100 kN/m² to the underlying sand. What would be the settlement?

Solution From Fig. 12.17, for the net applied pressure of 100 kN/m² and $N_{60} = 28$, $s_{\text{plate}} = 3.0$ mm.

$$s_{\text{footing}} = 3.0\left(\frac{2 \times 3}{3 + 0.3}\right)^2\left(1 - \frac{1}{3}\frac{1}{3}\right) = \mathbf{8.8\ mm}$$

12.10.2 Schmertmann et al.'s (1970, 1978) Method

When an elastic half space is subjected to a uniform pressure q over a large extent, the vertical strain induced at depth z is given by

$$\varepsilon_z = \frac{q}{E_z}$$

where E_z is the Young's modulus at depth z, where the vertical normal stress increases by q. When this pressure q is applied only over a limited area as shown in Fig. 12.18, the vertical normal strain along the vertical centerline can be very much less, depending on the depth z. It can be written as

$$\varepsilon_z = \frac{q}{E_z} I_z \tag{12.55}$$

where I_z is known as the *strain influence factor*, which must be less than unity. It varies with the depth below the foundation z. Thus, by discretizing the compressible soils beneath the foundation, computing the settlements within each sublayer and summing them, the total settlement along the vertical centerline can be determined as

$$s = \sum_{i=1}^{n} \frac{q}{E_z} I_z dz \tag{12.56}$$

Figure 12.19a shows a shallow foundation of breadth B, at depth D_f in an elastic medium. Schmertmann (1970) proposed that the strain influence factor varies as shown in Fig. 12.19b, which is known as *2B-0.6 distribution*. The influence factor increases linearly with depth, peaking at 0.6 at a depth of 0.5B and decaying to zero at depth of 2B. The diagram does not differentiate the foundation shape, and relies only on the width B. He also suggested deriving the Young's modulus from a cone penetration test as $E = 2q_c$ at any depth. To account for the reduction in settlement due to the embedment depth D_f, and the increase in settlement due to creep, he introduced two correction factors C_1 and C_2 and modified Eq. (12.56) as

$$s = C_1 C_2 q_{\text{net}} \sum_{i=1}^{n} \frac{I_z}{E_z} dz \tag{12.57}$$

where

$$C_1 = 1 - 0.5 \frac{\sigma'_{\text{vo}}}{q_{\text{net}}} \geq 0.5 \tag{12.58}$$

$$C_2 = 1 + 0.2 \log\left(\frac{t}{0.1}\right) \tag{12.59}$$

Here, σ'_{vo} = vertical overburden pressure at the foundation level, q_{net} = net applied pressure, and t = time since loading (years). Leonards (1986) and Holtz et al. (1991) noted that Schmertmann's data are based on foundations in Florida sands, which are interbedded with clays and silts that produce the time-dependent settlements. They suggest using $C_2 = 1$ in sands that do not contain any fines.

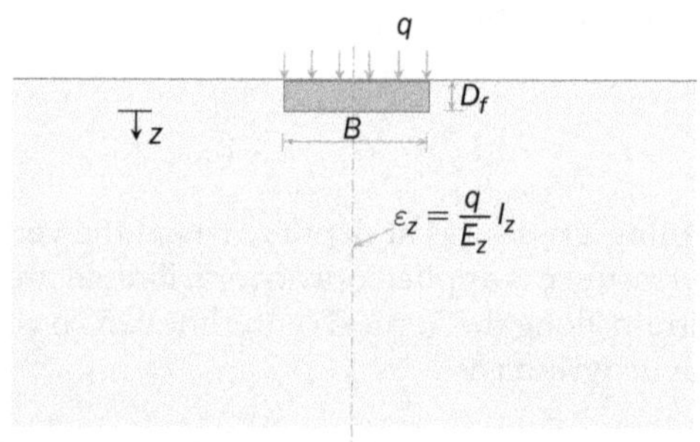

FIGURE 12.18 A foundation on an elastic half space.

Realizing the need to account for the foundation shape, 8 years later, Schmertmann et al. (1978) proposed the modified strain influence factor diagram shown in Fig. 12.19c, which distinguishes the square and strip foundations. They suggested $E = 2.5q_c$ for square foundations reflecting axi-symmetric loading and $E = 3.5q_c$ strip foundations reflecting plane strain loading. This is in agreement with the observation by Lee (1970) that the Young's modulus is 40% greater under plane strain loading compared to axi-symmetric loading. The I_z value peaks at a depth of $0.5B$ below the square foundation and B below the strip foundation, and the value of $I_{z,peak}$ is given by

$$I_{z,\text{peak}} = 0.5 + 0.1\sqrt{\frac{q_{\text{net}}}{\sigma'_{\text{vo}}}} \tag{12.60}$$

where σ'_{vo} is the initial effective overburden pressure at $0.5B$ below the square foundation and B below the strip foundation, where the peak value occurs. For rectangular foundations, the settlements can be computed for strip ($B/L = 0$) and square foundations ($B/L = 1$), and interpolated on the basis of B/L.

Terzaghi et al. (1996) simplified Schmertmann et al.'s (1978) method with a simpler I_z versus z diagram shown in Fig. 12.19d. For rectangular foundations, the strain influences factor I_z becomes zero at depth z_r, which is given by

$$z_r = 2B\left(1 + \log\frac{L}{B}\right) \quad \text{for} \frac{L}{B} \le 10 \tag{12.61}$$

Terzaghi et al. (1996) suggested increasing E to $3.5q_c$ for square foundations and the following expression for rectangular and strip foundations.

$$E_{\text{rectangular}} = 3.5\left(1 + 0.4\log\frac{L}{B}\right)q_c \quad \text{for} \frac{L}{B} \le 10 \tag{12.62}$$

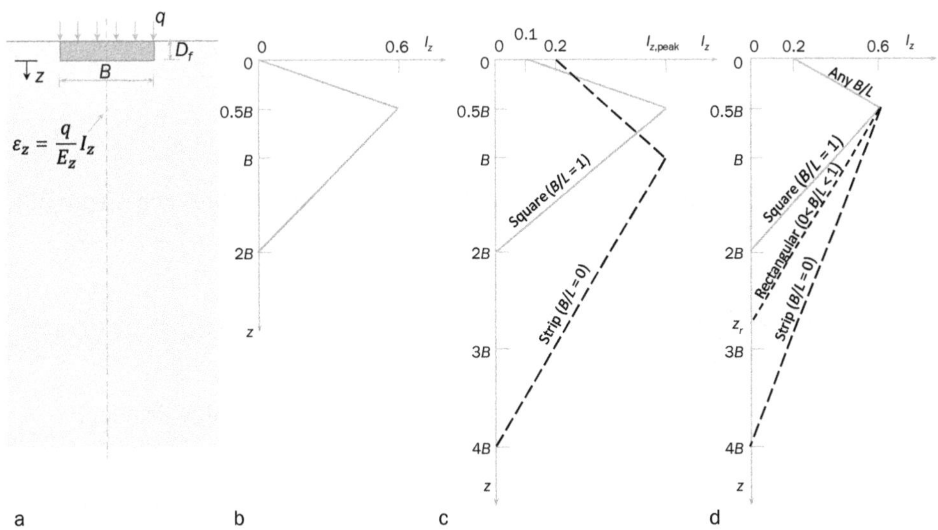

FIGURE 12.19 (a) Foundation on elastic medium, (b) I_z from Schmertmann (1970), (c) I_z from Schmertmann et al. (1978), (d) I_z from Terzaghi et al. (1996).

For strip foundations, L/B can be taken as 10. A critical review of Schmertmann's method and its modifications was presented by Pantelidis (2020).

Example 12.9

A 2-m-wide strip foundation is placed at 0.5 m depth in a sand (q_c = 12 MN/m², γ = 18.5 kN/m³), applying a net pressure of 80 kN/m². Estimate the settlement in the 50 years of its design life using Schmertmann's (1970) method.

Solution

$$\int_{z=0}^{2B} \frac{I_z dz}{E_z} = \frac{(0.5)(2 \times 2)(0.6)}{2 \times 12} = 0.05\,\text{m/MPa}$$

$$C_1 = 1 - 0.5\frac{\sigma'_{vo}}{q_{net}} = 1 - 0.5\frac{0.5(18.5)}{80} = 0.94$$

$$C_2 = 1 + 0.2\log\left(\frac{t}{0.1}\right) = 1 + 0.2\log\frac{50}{0.1} = 1.54$$

$$s = C_1 C_2 q_{net} \int_{i=1}^{n} \frac{I_z}{E_z} dz = (0.94)(1.54)\left(\frac{80}{1000}\right)(0.05)\ \text{m} = \mathbf{5.8\ mm}$$

Example 12.10

A 2.5-m-wide strip foundation placed at 1.0 m depth in a sand as shown in Fig. 12.20a, applies a net pressure of 110 kN/m² to the underlying ground. What would be the settlement in 25 years? Use Schmertmann et al.'s (1978) method.

Solution

$$I_{z,peak} = 0.5 + 0.1\sqrt{\frac{q_{net}}{\sigma'_{vo}}} = 0.5 + 0.1\sqrt{\frac{110}{18.0 \times 3.5}} = 0.63$$

The I_z versus z diagram shown in Fig. 12.20b is divided into three zones with different E values. Using $E = 3.5\, q_c$,

Upper layer: E = 28.0 MN/m²

Middle layer: E = 35.0 MN/m²

Bedrock: $E = \infty$

Let us compute the $\int \frac{I_z dz}{E_z}$ values within the different soils and sum them up.

$$\int_{z=0}^{10\,\text{m}} \frac{I_z}{E_z} dz = \frac{(0.2 + 0.63)(2.5)}{2(28)} + \frac{(0.34 + 0.63)(3.5)}{2(35)} + 0 = 0.0856\,\text{m/MPa}$$

$$C_1 = 1 - 0.5\frac{\sigma'_{vo}}{q_{net}} = 1 - 0.5\frac{(1.0)(18.0)}{110} = 0.92$$

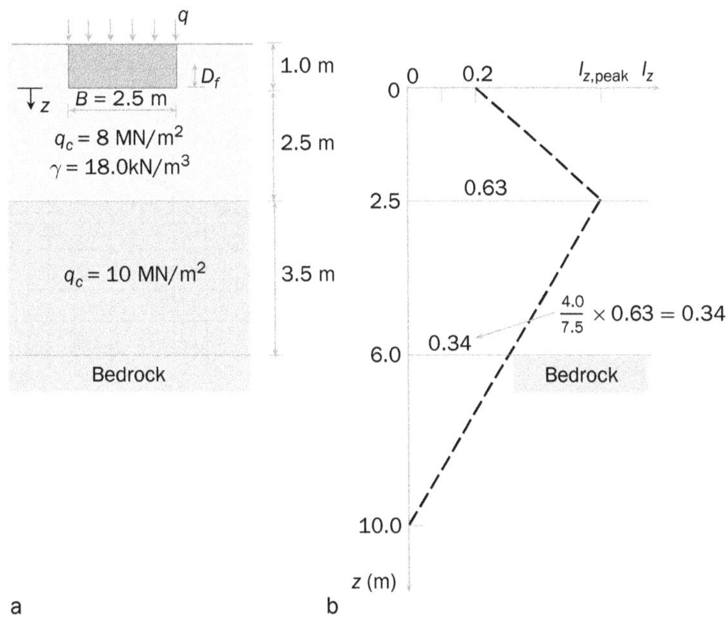

FIGURE 12.20 Soil profile and the strain influence factor diagram.

$$C_2 = 1 + 0.2\log\left(\frac{t}{0.1}\right) = 1 + 0.2\log\frac{25}{0.1} = 1.48$$

$$s = C_1 C_2 q_{\text{net}} \int_{i=1}^{n} \frac{I_z}{E_z} dz = (0.92)(1.48)\left(\frac{110}{1000}\right)(0.0856)\text{m} = \mathbf{12.8\ mm}$$

12.10.3 Burland and Burbidge's (1985) Method

Burland et al. (1977) suggested the possible upper limits for the settlement per unit pressure, as shown in Fig. 12.20, based on a large database of more than 200 settlement records of shallow foundations in granular soils. They noted that the probable settlement is about 50% of the upper limit shown in the figure and that maximum settlement is unlikely to exceed 75% of this upper limit. Figure 12.21 can be used as a good check even when the settlement is estimated by other methods.

Burland and Burbidge (1985) proposed an empirical settlement prediction methods based on the above settlement records and N_{60} value from the standard penetration test. They noted that the depth of influence z_I that affects the settlement beneath the foundation is related to the foundation width B by

$$z_I = B^{0.7} \tag{12.63}$$

where both z_I and B are in meters. They expressed the compressibility of the soil by compressibility index (I_c), which is similar to the coefficient of volume compressibility

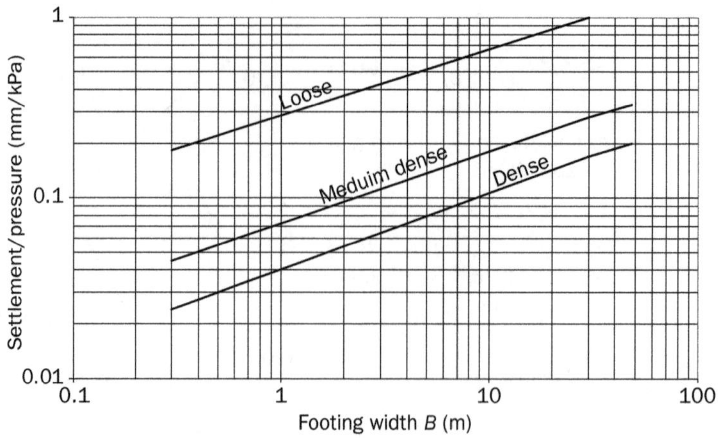

FIGURE 12.21 Upper limit of settlement per unit pressure (after Burland et al., 1977).

(m_v) used in consolidation of saturated clays. For normally consolidated granular soils, I_c was related to the average blow count within the influence depth $\overline{N}_{60}$ by:

$$I_c = \frac{1.71}{\overline{N}_{60}^{1.4}} \tag{12.64}$$

where I_c is in MPa^{-1}. For overconsolidated granular soils I_c is 1/3 of the value given by Eq. (12.64). Burland and Burbidge (1985) suggested estimating settlements using

$$s = qI_c z_I \tag{12.65}$$

For normally consolidated granular soils, the settlement is given by

$$s = q\frac{1.71}{\overline{N}_{60}^{1.4}}B^{0.7} \tag{12.66}$$

For overconsolidated soils where the applied pressure q is less than the preconsolidation pressure σ'_p, the soil remains overconsolidated at the end of loading and hence,

$$s = \frac{1}{3}q\frac{1.71}{\overline{N}_{60}^{1.4}}B^{0.7} \tag{12.67}$$

In overconsolidated granular soils, when q exceeds σ'_p, the soil becomes normally consolidated. Here, the settlement is given by

$$s = \frac{1}{3}\sigma'_p\frac{1.71}{\overline{N}_{60}^{1.4}}B^{0.7} + (q - \sigma'_p)\frac{1.71}{\overline{N}_{60}^{1.4}}B^{0.7}$$

$$s = \left(q - \frac{2}{3}\sigma'_p\right)\frac{1.71}{\overline{N}_{60}^{1.4}}B^{0.7} \tag{12.68}$$

They suggested that in gravel or sandy gravel, N_{60} must be increased by 25%. For fine sand and silty sand below the water table where $N_{60} > 15$, driving the sampler dilates the sand, producing negative pore water pressure and increased effective stresses which result in an overestimated N_{60}. This can be adjusted as

$$N_{60,\ \text{corrected}} = 15 + 0.5(N_{60} - 15) \tag{12.69}$$

Equations (12.66)–(12.68) work well for square foundations. For rectangular or strip foundations, the settlement has to be multiplied by a shape factor f_s given by

$$f_s = \left(\frac{1.25L/B}{0.25 + L/B}\right)^2 \tag{12.70}$$

For strip foundations with $L/B = \infty$, f_s becomes 1.56. The settlements computed in Eqs. (12.66)–(12.68) are based on the assumption that there is no hard stratum in the vicinity of the foundation and the thickness of the granular soil bed is greater than the influence depth z_I. When this is not the case, with the granular soil thickness of H_s is underlain by a hard stratum, the settlement computed above has to be reduced by multiplying by a factor f_l given by

$$f_l = \frac{H_s}{z_I}\left(2 - \frac{H_s}{z_I}\right) \tag{12.71}$$

Recognizing some time-dependent settlements in granular soils, Burland and Burbidge (1985) suggested a multiplication factor f_t given by

$$f_t = 1 + R_3 + R_t \log\frac{t}{3} \tag{12.72}$$

The factor R_3 accounts for the time-dependent settlements taking place within the first 3 years, and lies in the range 0.3–0.7. $R_t \log t/3$ accounts for the time-dependent settlement taking place after the first 3 years, and R_t is in the range 0.2–0.8. The lower end of the range applies for static loads, and the upper end for fluctuation loads (e.g., bridge abutments, silos, chimneys).

12.10.4 Berardi and Lancellotta's (1991) Method

In most settlement prediction methods, the stiffness is determined from the penetration tests such as SPT or CPT, and a constant value is assumed for the Young's modulus (e.g., $E = 2.5\ q_c$). Berardi and Lancellotta (1991) recognized that the stiffness varies with the stress level, with Young's modulus decreasing with increasing strain. They proposed a new method for predicting settlement, incorporating the stress level in determining the soil stiffness. They suggested that the influence depth H beneath the foundation is B for square foundation and $2B$ for strip foundation. For rectangular foundation, it can be interpolated on the basis of B/L as

$$H = \left(1 + \log\frac{L}{B}\right)B \quad \text{for } L/B \le 10 \tag{12.73}$$

When $L/B > 10$, it should be taken as 10 in Eq. (12.73). From the theory of elasticity, the settlement of a foundation with dimensions B and L, resting on a linear elastic continuum is given by

$$s = \frac{qB}{E}(1 - \nu^2)I \tag{12.27}$$

where q = applied pressure, B = footing width, E = Young's modulus, ν = Poisson's ratio, and I = influence factor that depends on the footing dimensions. Assuming Poisson's ratio of 0.15 and the footing to be rigid, values of $(1 - \nu^2)\,I$ are given in Table 12.7 (after Berardi et al., 1991). Here, H^* is the depth of the compressible stratum below the foundation.

When the depth of the compressible layer H^* is more than $2B$ in thickness, it can be taken as $2B$ in computing the value of $(1 - \nu^2)I$ in Table 12.7. The depth of the granular soil below $2B$ has no effect on the settlement.

The foundation load increases the soil stiffness, and Janbu (1963) suggested expressing Young's modulus as

$$E = K_E p_a \left(\frac{\sigma'_{vo} + 0.5\Delta\sigma'_v}{p_a} \right)^{0.5} \tag{12.74}$$

where K_E is a dimensionless *modulus number* for the sand, which depends on the relative density, strain level, and the influence depth H; σ'_{vo} = initial effective overburden pressure at the center of the influence zone; $\Delta\sigma'$= the vertical effective stress increase due to the foundation load; and p_a = atmospheric pressure (=101.3 kN/m²).

The value of modulus number at s/B of 0.1%, denoted by $K_{E,0.1\%}$ is related to the relative density D_r, as shown in Fig. 12.22 for $H = B$ and $H = 2B$. Here, the vertical strain is related to the ratio of the settlement s to the foundation width B. When using Fig. 12.22 to determine $K_{E,0.1\%}$, it is suggested that D_r be obtained from Skempton's (1986) empirical correlation given by

$$\frac{(N_1)_{60}}{D_r^2} \approx 60 \tag{12.75}$$

Berardi et al. (1991) presented the data shown in Fig. 12.23, and suggested that the modulus number at any other strain level can be determined from the settlement as

	L/B					
H*/B	**1**	**2**	**3**	**5**	**10**	**Circle**
0.5	0.35	0.39	0.40	0.41	0.42	0.34
1.0	0.56	0.65	0.67	0.68	0.71	0.52
1.5	0.63	0.76	0.81	0.84	0.89	0.59
> 2.0	0.69	0.88	0.96	0.99	1.06	0.63

TABLE 12.7 $(1 - \nu^2)I$ Values for a Rigid Footing

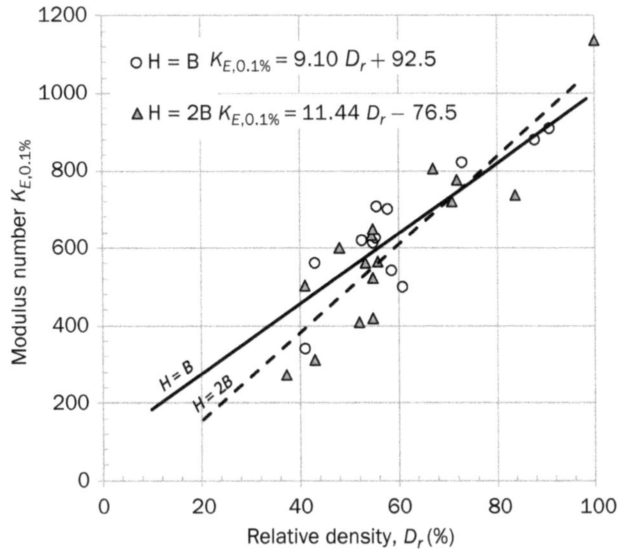

FIGURE 12.22 Variation of $K_{E,0.1\%}$ with the relative density D_r (based on Lancellotta, 2009).

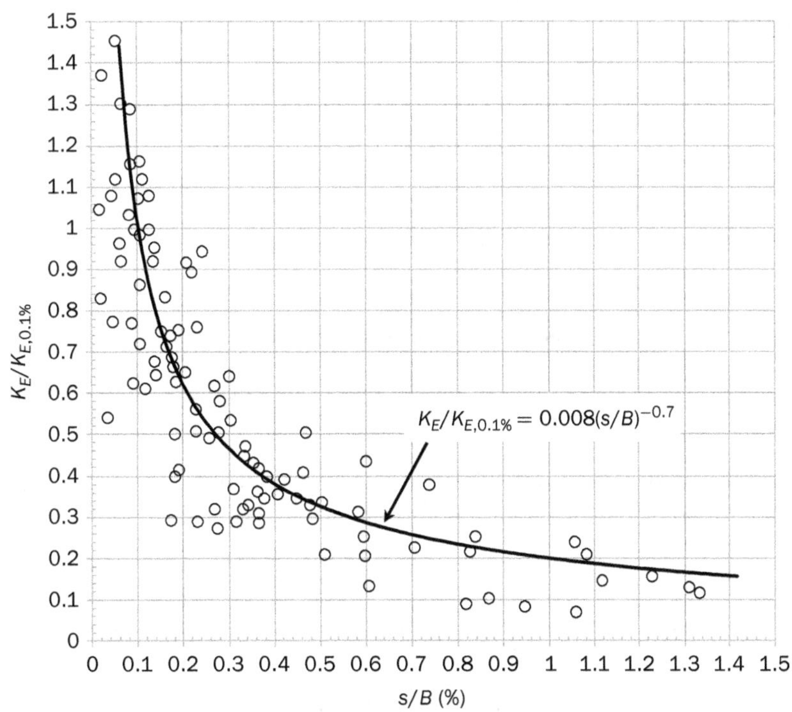

FIGURE 12.23 $K_E/K_{E,0.1\%}$ vs. s/B (based on Berardi et al., 1991).

$$\frac{K_E}{K_{E,0.1\%}} = 0.008\left(\frac{s}{B}\right)^{-0.7} \tag{12.76}$$

From the work of Berardi (1991) and Eq. (12.76), Lancellotta (2009) suggested that

$$\frac{E}{E_{0.1\%}} = 0.008\left(\frac{s}{B}\right)^{-0.7} \tag{12.77}$$

where $E_{0.1\%}$ is the Young's modulus when $s/B = 0.1\%$. The value of $K_{E,0.1\%}$ determined from Fig. 12.22 can be substituted in Eq. (12.74) for determining $E_{0.1\%}$. From Eqs. (12.27) and (12.77),

$$\left(\frac{s}{B}\right)^{0.3} = \frac{125q(1-\nu^2)I}{E_{0.1\%}} \tag{12.78}$$

The settlement can be determined form Eq. (12.78).

Example 12.11

A 2-m-wide strip foundation carrying a 260 kN/m wall load is placed at a depth of 1.0 m in sand where the unit weight is 19.0 kN/m³ and $(N_1)_{60}$ is 32. Assuming Poisson's ratio of 0.15, estimate the settlement of the foundation.

Solution

The applied pressure at the footing level = 260 ÷ 2 = 130 kN/m²

From Skempton (1986),

$$D_r = \sqrt{\frac{(N_1)_{60}}{60}} = \sqrt{\frac{32}{60}} = 0.730 \text{ or } 73.0\%$$

The influence zone extends to 2*B* below the foundation. Therefore, the in situ effective overburden stress at the center of the influence zone is computed as

$$\sigma'_{vo} = 3.0 \times 19 = 57.0 \text{ kN/m}^2$$

The vertical stress increase at this depth can be estimated as

$$\Delta\sigma'_v = \frac{(130)(2)}{(2+2)} = 65.0 \text{ kN/m}^2$$

The influence depth $H = 2B = 4.0$ m (below the foundation)

From Table 12.7, for $H^*/B > 2$ and $L/B = 10$, $(1-\nu^2)I = 1.06$

$$p_a = 101.3 \text{ kN/m}^2$$

$$\left(\frac{\sigma'_{vo} + 0.5\Delta\sigma'_v}{p_a}\right)^{0.5} = \left(\frac{57.0 + (0.5)(65.0)}{101.3}\right)^{0.5} = 0.940$$

For $D_r = 73.0\%$ and $H = 2B$, from Fig. 12.22 → $K_{E,0.1\%} = 759$

From Eq. (12.74),

$$E_{0.1\%} = K_{E,0.1\%} p_a \left(\frac{\sigma'_{vo} + 0.5\Delta\sigma'_v}{p_a} \right)^{0.5} = (759)(101.3)(0.94) = 72{,}273 \text{ kN/m}^2$$

Substituting these in Eq. (12.78),

$$\left(\frac{s}{B} \right)^{0.3} = \frac{(125)(130)(1.06)}{72{,}273} = 0.2383$$

Therefore, $s/B = 0.0084$, and hence settlement $= 0.0084 \times 2000 =$ **16.8 mm**

12.10.5 Reliability of the Settlement Calculation Methods

It is widely documented in the literature that the design of shallow foundations in granular soils is governed by settlements than bearing capacity. Therefore, more care is required in the settlement computations.

Briaud and Gibbens (1994) documented a class A settlement prediction session held at Texas A&M University, College Station, Texas in 1994, where 16 academics and 15 geotechnical professional engineers participated. An extensive site investigation with six standard penetration tests, seven cone penetration tests, four pressuremeter tests, four dilatometer tests, four crosshole tests, three borehole shear tests, and a step blade tests were carried out in a 12 m × 18 m site, where five different square pad footings of widths ranging from 1.0 to 2.5 m were to be constructed at specific locations and load tested at a future date. Laboratory test data including maximum/minimum densities, specific gravity of the grains, natural water content, void ratio, densities, and friction angles were also available for sand samples taken from 0.6 to 3.0 m depths. The soil profile consisted predominantly of sands. The entire soil data were available to all participants, who were asked to predict the loads, Q_{25} and Q_{150}, which would make the five footings settle by 25 and 150 mm, respectively. Q_{25} is the allowable footing load satisfying the settlement criterion, and Q_{150} is more or less the failure load corresponding to ultimate bearing capacity. A total of 22 different methods were used by the 31 participants, with Schmertmann's (1970, 1978), Burland and Burbidge's (1985) methods, and finite element analysis being more popular. The predictions from the participants were compared with the load test data which clearly showed the deficiencies in the current state-of-the-art, where the predictions from the 31 international experts varied in a wide range. In spite of having access to the full suite of data from a rigorous site investigation program, their predictions of Q_{25}, the load required to produce 25 mm settlement, were significantly less than what was measured, implying that the settlements were overestimated in general. In reality, the geotechnical engineer has access to very limited data from the field, and the quality of predictions can only be worse. For a 12 m × 18 m site with shallow foundations, one would not expect to have more than two or three SPTs or CPTs.

Tan and Duncan (1991) defined two parameters for comparing settlement prediction methods: *accuracy* and *reliability*. Accuracy is how close the predictions by a specific method are to the measured values, and is defined as the average value of the ratio of the calculated to measured settlements. Reliability is the probability that the actual settlement would be less than that computed by a specific method. It is a measure of conservativeness of a settlement prediction method. A good method should have accuracy closer to 1 and reliability closer to 100%. Tan and Duncan (1991) found that there is generally a trade-off between accuracy and reliability within all methods. Terzaghi and Peck's (1948) and Schmertmann et al.'s (1978) methods appear to have high reliability and poor accuracy, reflecting their conservativeness. On the other hand, Burland and Burbidge's (1985) and Berardi and Lancellotta's (1991) methods have good accuracy, with values close to unity, but low reliability.

Sivakugan and Johnson (2004) proposed a probabilistic design charts shown in Fig. 12.24, which was developed from a large database of settlement records. The charts enable one to evaluate the uncertainty associated with the four settlement prediction methods. For example, when Terzaghi and Peck's (1967) method predicts 25 mm settlement, the probability of the actual settlement exceeding 25 mm is only 0.26. When the settlement is predicted as 25 mm by Berardi and Lancellotta's (1991) method, the probability that the actual settlement will exceed 25 mm is 0.59.

Example 12.12

For a footing in sand, when 30 mm of settlement is predicted by the following methods, what is the probability that the actual settlement will exceed 25 mm?

a. Terzaghi and Peck's (1967) method
b. Schmertmann et al.'s (1978) method
c. Burland and Burbidge's (1985) method
d. Berardi and Lancellotta's (1991) method

Solution From Fig. 12.24,

Terzaghi and Peck's (1967) method: 0.31
Schmertmann et al.'s (1978) method: 0.33
Burland and Burbidge's (1985) method: 0.48
Berardi and Lancellotta's (1991) method: 0.66

12.10.6 Design Charts for Shallow Foundations

Pad and strip foundations in granular soils are designed to limit the settlement to 25 mm. On the basis of this, allowable soil pressures estimated by Burland and Burbidge (1985) and Terzaghi and Peck (1967) are suggested in Fig. 12.25. Here it is assumed that the sand is normally consolidated.

12.11 Settlement Computations in Cohesive Soils

The settlement of a shallow foundation in a saturated cohesive soil has three components as shown in Fig. 12.26. They are *immediate settlement* s_i, *consolidation settlement* s_c, and the *secondary compression settlement* s_s. At a time $t > t_p$, all three components would be present and they have to be computed separately and the total settlement becomes

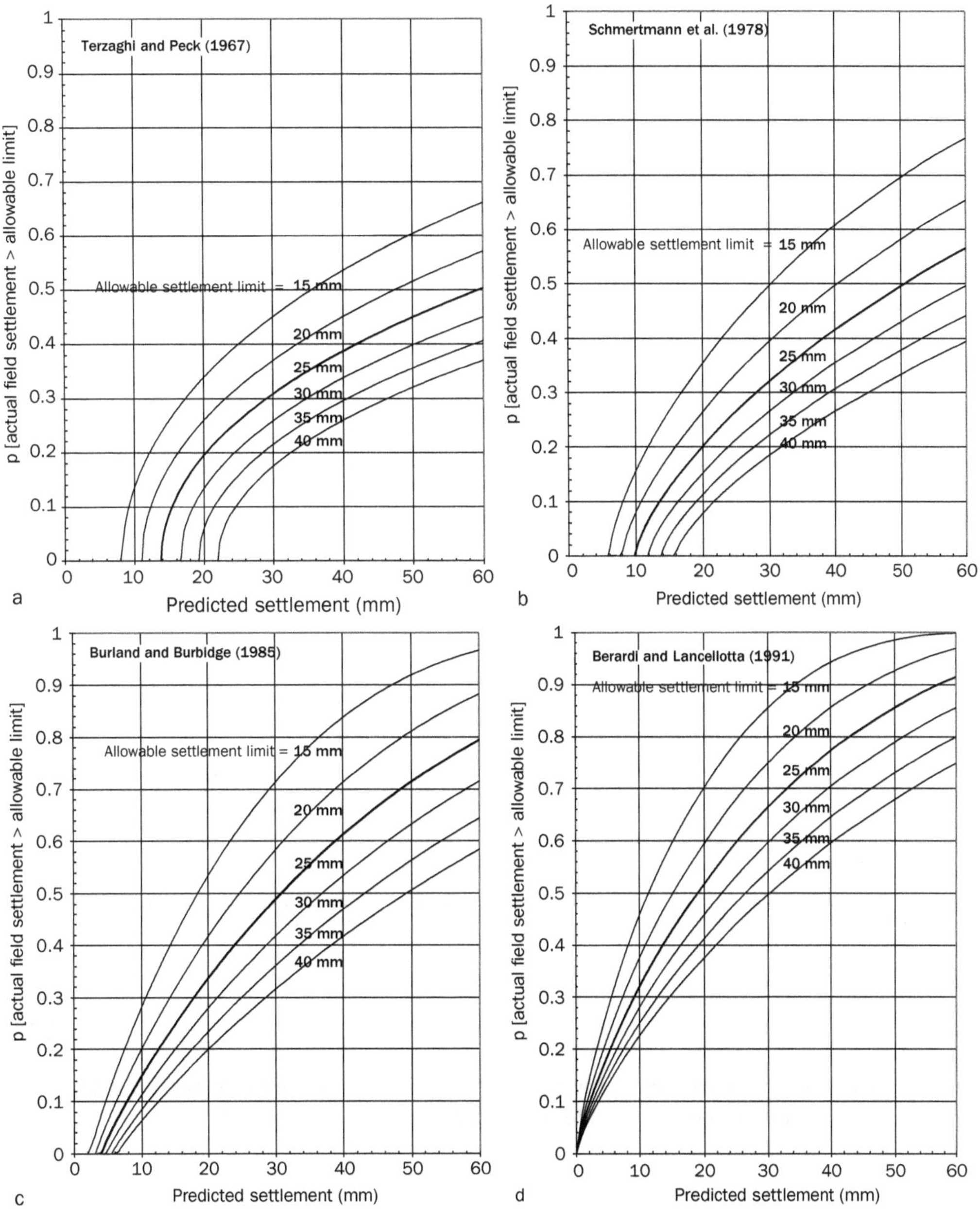

FIGURE 12.24 Probabilistic design charts: (a) Terzaghi and Peck's method, (b) Schmertmann et al.'s method, (c) Burland and Burbidge's method, (d) Berardi and Lancellotta's method.

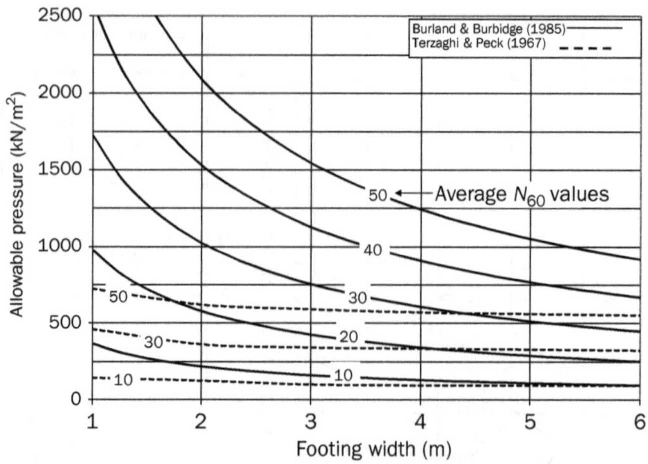

FIGURE 12.25 Allowable pressures for shallow foundations on sands with 25 mm maximum settlement.

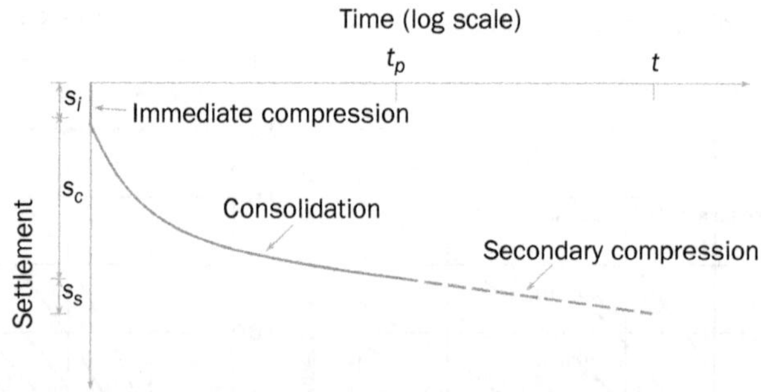

FIGURE 12.26 Settlement of a shallow foundation in a cohesive soil.

$$s = s_i + s_c + s_s \tag{12.79}$$

At $t < t_p$, the secondary compression component is not present, and only a fraction of consolidation has occurred. Here, the total settlement is given by

$$s = s_i + U_{avg}(t)s_c \tag{12.80}$$

where $U_{avg}(t)$ = average degree of consolidation at time t, t_p = time to complete consolidation, and s_c = final consolidation settlement.

12.11.1 Immediate Settlement s_i

Immediate settlement, also known as distortion, initial or elastic settlement, occurs immediately upon loading, due to lateral distortion of the soil beneath the foundation. In saturated cohesive soils where drainage is poor, it is reasonable to assume undrained

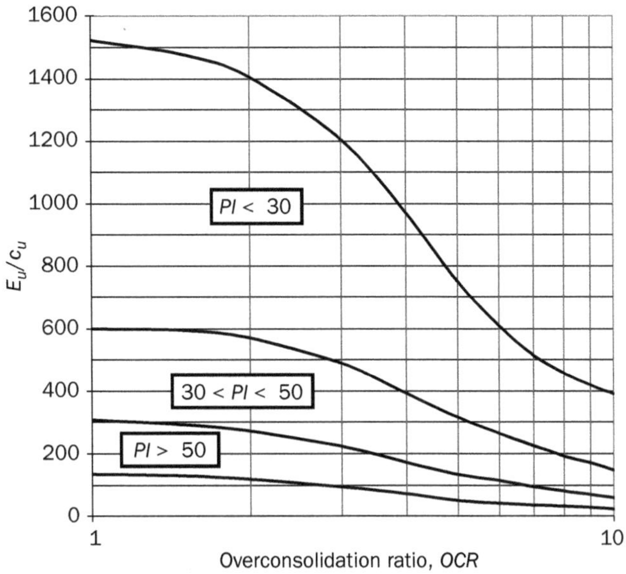

FIGURE 12.27 Variation of E_u/c_u with PI and OCR.

conditions (i.e., $\nu = 0.5$) in computing the settlements. From the theory of elasticity, Janbu et al. (1956) suggested that

$$s_i = \frac{qB}{E_u}\mu_0\mu_1 \tag{12.81}$$

where E_u = undrained Young's modulus, which can be obtained from Fig. 12.27 suggested by Duncan and Buchignani (1976). The values of the coefficients μ_0 and μ_1, as suggested by Christian and Carrier (1978), are given in Fig. 12.28.

Example 12.13

A 2 m × 4 m foundation is placed at 1 m depth, applying 120 kN/m² to the underlying clay where c_u = 40 kN/m², PI = 25, and OCR = 2. Bedrock lies at 6 m below the foundation. Estimate the immediate settlement.

Solution For $PI = 25$ and $OCR = 2$, from Duncan and Buchignani's (1976) Fig. 12.27, $E_u/c_u = 700$

Therefore, $E_u = (700)(40) = 28{,}000$ kN/m²

$D_f/B = 1/2 = 0.5$. From Fig. 12.28, $\mu_0 = 0.95$

$B/L = 2/4 = 0.5$; $H/B = 6/2 = 3.0$. From Fig. 12.28, $\mu_1 = 0.73$

$$s_i = \frac{qB}{E_u}\mu_0\mu_1 = \frac{(120)(2)}{(28{,}000)}(0.95)(0.73)\text{ m} = 6.2\text{ mm}$$

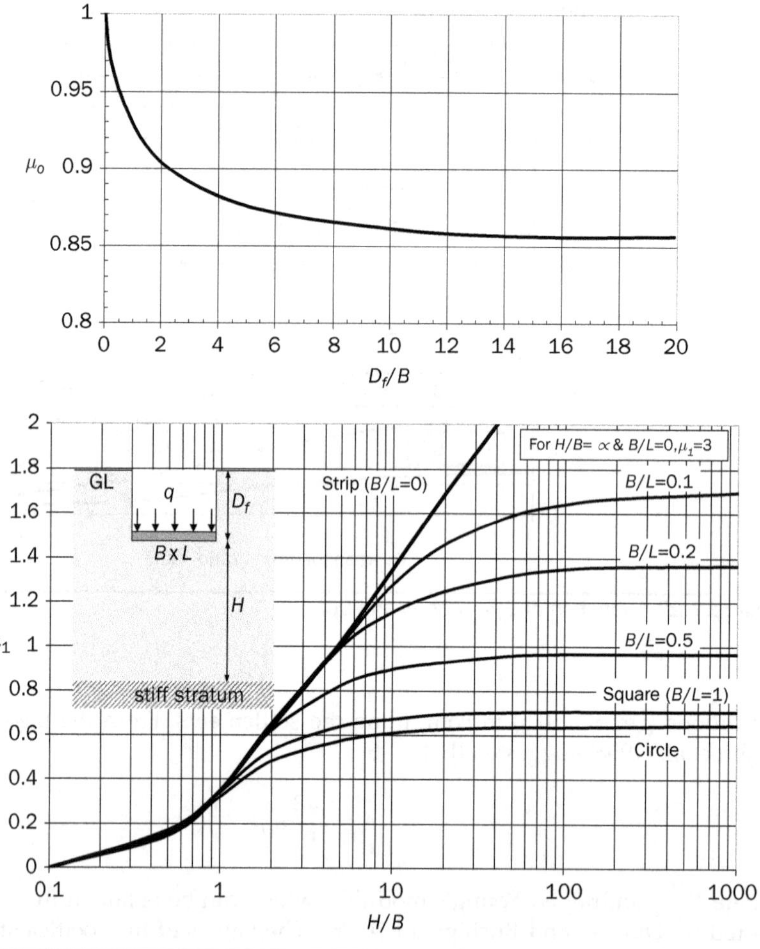

FIGURE 12.28 The coefficients μ_0 and μ_1.

12.11.2 Consolidation Settlement s_c

Consolidation settlement takes place right after the immediate settlement, and can go on for several years, depending on the thickness and consolidation characteristics of the clay layer, and the drainage conditions. In Fig. 12.26, the time to complete the consolidation is denoted by t_p.

Methods of computing the consolidation settlements are discussed in detail in Chap. 8 for one-dimensional consolidation, where the lateral deformations are constrained. When the lateral deformations are allowed, as in the case of two- or three-dimensional consolidation, the vertical consolidation settlement is reduced. When the loading is two or three dimensional, the computations can still be made using the same principles assuming one-dimensional consolidation, with a correction factor μ proposed by Skempton and Bjerrum (1957) to account for the two- or three-dimensional effects. Figure 12.29 gives the values of μ as a function of Skempton's pore pressure coefficient A, for square and strip foundations. The settlement computed based on one-dimensional consolidation has to be multiplied by this factor to derive the actual settlement. Here, H is the depth of clay to the rigid stratum.

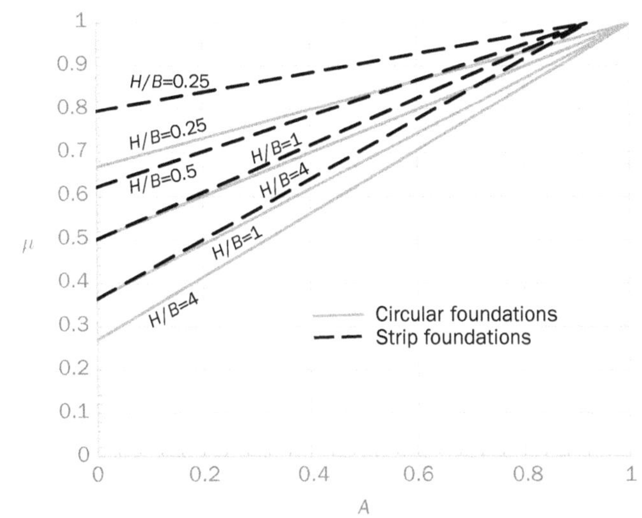

FIGURE 12.29 Skempton and Bjerrum correction factor μ.

In one-dimensional consolidation, the normal strains and drainage are assumed to be taking place only in the vertical direction. This situation arises when the applied pressure at the ground level is uniform and is of very large lateral extent, as we saw in most of the questions in Chap. 8. Subsequently, the vertical stress increase ($\Delta\sigma'$) is also the same at any depth within the clay layer. In reality, when the foundations and the applied pressures are of limited lateral extent, the consolidation is not one dimensional. The vertical stress increase ($\Delta\sigma'$) will be decaying with depth, and we will use the value of $\Delta\sigma'$ at the middle of the clay layer. This value can be computed using the methods discussed in Chap. 7, and it can be significantly less than what is applied at the ground level. When the clay layer is thick, it is a good practice to divide it into several sublayers and to use the appropriate values of σ'_{vo}, $\Delta\sigma'$, etc. for each sublayer in computing the changes in void ratios Δe the consolidation settlements within the sublayers. These settlements are added up to give the total consolidation settlement of the clay layer.

12.11.3 Secondary Compression Settlement s_s

Secondary compression and computation of secondary compression settlement are discussed in detail in Sec. 8.9. As a simplification, it is generally assumed that secondary compression begins at the completion of consolidation and continues indefinitely (see Fig. 12.26). In reality, some secondary compression can take place during consolidation.

12.12 Main Points

1. General, local and punching shear failures take place in dense, moderately dense, and loose soils, respectively. The bearing capacity equation discussed in this chapter assumes general shear failure.
2. Factor of safety for a shallow foundation against bearing capacity failure is typically taken as 3.0, defined in terms of net pressures.
3. The general bearing capacity equation takes the form of the original Terzaghi's bearing capacity equation, and in addition, it includes correction factors for shape, depth, and load inclination.

4. For short-term stability use c_u and ϕ_u and analyze in terms of total stresses; for long-term stability use c' and ϕ' and analyze in terms of effective stresses.
5. For foundations in clays under undrained conditions $\phi_u = 0$ and hence $N_q = 1$, $N_\gamma = 0$, and $N_c = 5.14$.
6. Most settlement calculation methods for granular soils overestimate the settlements. It is better to overestimate than underestimate the settlements and thus be conservative.
7. In clays, the settlement has three components, namely, the immediate, consolidation, and secondary compression settlements, which have to be computed separately.

Review Exercises

12.1 A 2.5-m-wide strip footing carrying a wall load of 800 kN/m is placed at a depth of 1.0 m below the ground level, and the soil parameters are $c' = 5$ kN/m², $\phi' = 32°$ (from CU triaxial test), and $\gamma = 19$ kN/m³. What is the factor of safety with respect to bearing capacity failure?

Answer: 6.1

12.2 A 2.0-m-wide square footing is placed in sand, at 1.0 m depth below the ground level. The water table lies at the base of the footing, with unit weight of the soil above and below the footing being 17.0 and 19.0 kN/m³, respectively. The friction angle of the sand is 33°. What is the maximum load that can be allowed on the footing with factor of safety of 3?

Answer: 1338 kN

12.3 A 2.5 m × 2.5 m square footing, placed at 1.5 m depth, carries a vertical column load with eccentricities of $e_B = 0.1$ m and $e_L = 0.1$ m. The soil properties are $c' = 10$ kN/m², $\phi' = 25°$, and $\gamma =$ 19 kN/m³. What is the maximum column load that can be applied with factor of safety of 3.0?

Answer: 1750 kN

12.4 A 2.5-m-wide strip footing carries a wall load in a clay where short-term stability appears to be critical. The footing depth is 1.25 m, unit weight of the clay is 19 kN/m³, and the undrained shear strength is 35 kN/m². Find the maximum load that can be allowed for the wall load, with $F = 3$.

Answer: 185 kN/m

12.5 Figure RE12.5 shows a 3.0-m-wide strip footing subjected to a vertical load of 450 kN/m and horizontal load of 50 kN/m, as shown. What are the values you would use for the inclination angle α and the eccentricity e in computing the ultimate bearing capacity?

Answer: 6.34°, 0.167 m

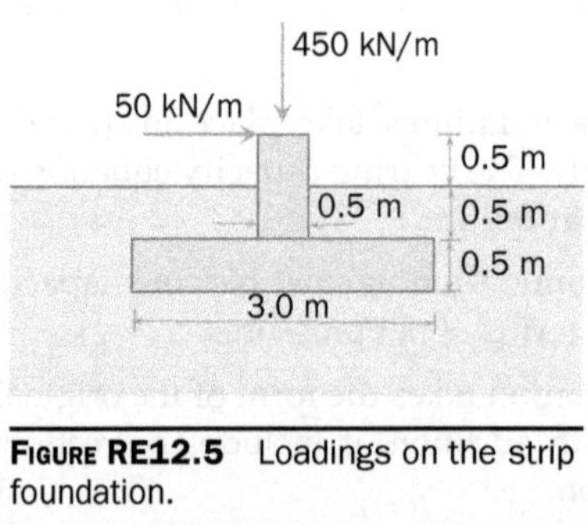

FIGURE RE12.5 Loadings on the strip foundation.

12.6 A $B \times L$ rectangular footing carries a vertical load Q with eccentricities $B/12$ and $L/20$ along the B and L directions, respectively. Determine the vertical pressure distribution along the perimeter of the footing. Give the values at the four corners and the center.

Answer: 0.8Q/BL, 1.8Q/BL, 1.2Q/BL, 0.2Q/BL, Q/BL

12.7 It is proposed to construct a 12 m × 15 m raft at 2 m depth within a sand where N_{60} is 24. What would be the net allowable bearing capacity of the raft if a maximum settlement of 50 mm is allowed?

Answer: 665.4 kN/m²

12.8 It is proposed to construct a 12 m × 15 m raft at 2 m depth within a clay where $c_u = 45$ kN/m². What would be the net allowable bearing capacity of the raft foundation?

Answer: 92.4 kN/m²

12.9 A 2 m × 4 m flexible loaded surface foundation applies a pressure of 100 kN/m² to the underlying ground. The Young's modulus and Poisson's ratio of the soil are 25 MN/m², and 0.2, respectively. Determine the settlement beneath the center using

a. Schleicher's method, and
b. Steinbrenner's method.

Answer: 11.8 mm, 11.4 mm

12.10 A 2-m-wide square foundation is placed at 1.0 depth in a sand where N_{60} = 33, and $\gamma = 18.0$ kN/m³. What would be the settlement at 250 kN/m² contact pressure?

Answer: 13.9 mm

12.11 A 2.5-m-square foundation is placed at a depth of 0.5 m within a sand as shown in Fig. RE12.11. The top 3.5 m of sand is underlain by a deep bed of denser sand, and the q_c values from a CPT are given in the figure. What would be the settlement expected in 50 years if the foundation applies a net pressure of 125 kN/m²? Use Schmertmann's (1970) method.

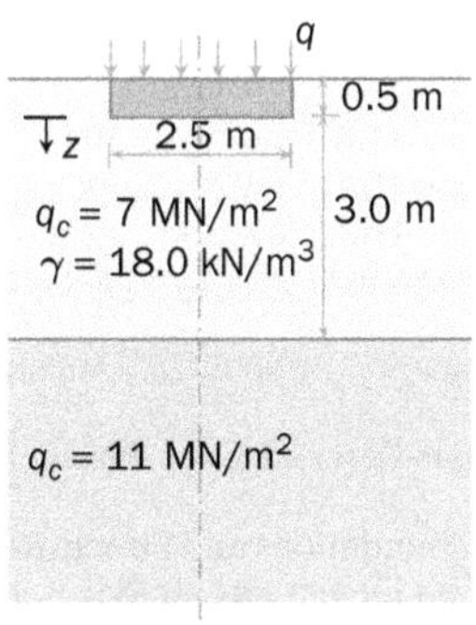

FIGURE RE12.11 Soil profile.

Answer: 18.3 mm

12.12 Redo Problem 12.11 using Schmertmann et al.'s (1978) method.

Answer: 17.7 mm

12.13 A 2.0-m-square footing placed on a sand at depth of 0.5 m carries a column load of 600 kN. The sand has unit weight of 19 kN/m³ and $(N_1)_{60}$ of 28. Estimate the settlement using Berardi and Lancellotta's method. Assume Poisson's ratio of the sand as 0.15.

Answer: 7.4 mm

12.14 A 3 m × 4 m foundation shown in Fig. RE12.14 applies a net pressure of 150 kN/m^2 to the underlying ground. Determine the immediate settlement.

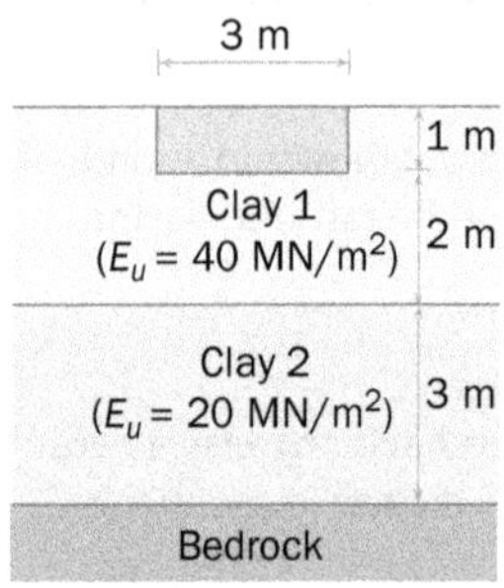

FIGURE RE12.14 Soil profile.

Answer: 9.4 mm

References

Berardi, R. "Nonlinear Elastic Approaches in Foundation Design," 2nd International Symposium on Pre-Failure Deformation Characteristics of Geomaterials, IS Torino 99, Jamiolkowski, M., R. Lancellotta, and D. LoPresti (eds.), Torino, Italy, September 28–30, 1999.

Berardi, R., and R. Lancellotta. "Stiffness of Granular Soil from Field Performance," *Geotechnique*, 1:149–157, 1991.

Berardi, R., M. Jamiolkowski, and R. Lancellotta. "Settlement of Shallow Foundations in Sands Selection of Stiffness on the Basis of Penetration Resistance," *Geotechnical Engineering Congress*, Colorado, ASCE Geotechnical Special Publication No. 27, 1, 1991, pp. 185–200.

Bishop, A. W. Discussion on "Soil Properties and Their Measurement," *Proceedings of International Conference on Soil Mechanics*, 3:97, 1961.

Bjerrum, L., and O. Kummeneje. "Shearing Resistance of Sand Samples with Circular and Rectangular Cross Sections," *Norwegian Geotechnical Institute Publication No. 44*, 1, 1961.

Bowles, J. E. "Elastic Foundation Settlements on Sand Deposits," *Journal of Geotechnical Engineering*, 113(8):846–860, 1987.

Bowles, J. E. *Foundation Analysis and Design*, 3rd Edition, McGraw-Hill, New York, 1982.

Bowles, J. E. *Foundation Analysis and Design*, 4th Edition, McGraw-Hill, New York, 1988.

Bowles, J. E. *Foundation Analysis and Design*, 5th Edition, McGraw-Hill, New York, 1996.

Briaud, J.-L., and R. M. Gibbens. *Predicted and Measured Behavior of Five Spread Footings on Sand*, Geotechnical Special Publication 41, ASCE, Texas, 1994.

Brinch Hansen, J. *A Revised and Extended Formula for Bearing Capacity*, Bulletin 28, Danish Geotechnical Society, Copenhagen, 1970.

Brown, P. T. "Numerical Analysis of Uniformly Loaded Circular Rafts on Deep Elastic Foundations," *Geotechnique*, 19(3):399–404, 1969.

Burland, J. B., B. B. Broms, and V. F. B. De Mello. "Behaviour of Foundations and Structures," *Proceedings of 9th International Conference on Soil Mechanics and Foundation Engineering*, Tokyo, 2, 1977, 495–538.

Burland, J. B., and M. C. Burbidge. "Settlements of Foundations on Sand and Gravel," *Proceedings of Institution of Civil Engineers*, 78(1):1325–1381, 1985.

Chen, W. F. *Limit Analysis of Soil Plasticity*, Elsevier, New York, 1975.

Christian, J. T., and Carrier III, W. D. "Janbu, Bjerrum and Kjaernsli's Chart Reinterpreted," *Canadian Geotechnical Journal*, 15:123–128, 1978.

Clancy, P. *Numerical Analysis of Piled Raft Foundations*, PhD Thesis, University of Western Australia, Australia, 1993.

Diaz-Segura, E. G. "Assessment of the Range of Variation of N_γ from 60 Estimation Methods for Footing on Sand," *Canadian Geotechnical Journal*, 50:793–800, 2013.

Duncan, J. M., and A. L. Buchignani. *An Engineering Manual for Settlement Studies*, Geotechnical Engineering Report, Dept. of Civil Engineering, University of California, Berkeley, USA, 1976, p. 94.

European Committee for Standardisation. *Eurocode 7: Geotechnical Design—Part 1*, Brussles, 1994.

Fox, E. N. "The Mean Elastic Settlement of a Uniformly Loaded Area at a Depth below the Ground Surface," *Proceedings of the 2nd International Conference on Soil Mechanics and Foundation Engineering*, Rotterdam, 1, 1948, pp. 129–132.

Fraser, R. A., and L. J. Wardle. "Numerical Analysis of Rectangular Rafts on Layered Foundations," *Geotechnique*, 26(4):613–630, 1976.

Giroud, J. P. "Settlement of Linearly Loaded Rectangular Area," *Journal of Soil Mechanics and Foundations Division*, ASCE, 94 (SM4):813–831, 1968.

Hain, S. J., and I. K. Lee. "The Analysis of Flexible Raft-Pile System," *Geotechnique*, 28(1):65–83, 1978.

Horikoshi, K., and M. F. Randolph. "On the Definition of Raft-Soil Stiffness Ratio," *Geotechnique*, 47(5):1055–1061, 1997.

Holtz, R. D. Stress Distribution and Settlement of Shallow Foundations, Chapter 5, *Foundation Engineering Handbook*, Fang H.-Y. (ed.), 2nd Edition, van Nostrand Reinhold, New York, 1991, pp. 166–222.

Janbu, N., L. Bjerrum, and B. Kjaernsli. *Veiledning ved Losning av Fundamenteringsoppgaver*, Norwegian Geotechnical Institute Publication 16, Oslo, 1956, pp. 30–32.

Janbu, N. "Soil Compressibility as Determined by Oedometer and Triaxial Tests," *Proceedings of the ECSMFE*, Wiesbaden, Germany, 1, 1963, pp. 19–25.

Kumbhojkar, A. S. "Numerical Evaluation of Terzaghi's N_γ," *Journal of Geotechnical Engineering*, ASCE, 119(3):598–607, 1993.

Ladd, C. C., R. Foote, K. Ishihara, F. Schlosser, and H. G. Poulos. "Stress Deformation and Strength Characteristics," State-of-the-art Report, *Proceedings of the 9th International Conference on Soil Mechanics and Foundation Engineering (ICSMFE)*, Tokyo, 2, 1977, pp. 421–494.

Lancellotta, R. *Geotechnical Engineering*, 2nd Edition, Taylor & Francis, London, 2009, p. 499.

Lee, K. L. "Comparison of Plane Stain and Triaxial Tests on Sand," *Journal of Soil Mechanics and Foundations Division*, ASCE, 96(SM3):901–923, 1970.

Leonards, G. A. CE683-Advanced Foundation Engineering, Lecture Notes, Purdue University, West Lafayette, IN, 1986.

Mayne, P. W., and H. G. Poulos. "Approximate Displacement Influence Factors for Elastic Shallow Foundations," *Journal of Geotechnical and Geoenvironmental Engineering*, 125(6):453–460, 1999.

Martin, C. M. "Exact Bearing Capacity Calculations Using the Method of Characteristics," *Proceedings of 11th International Assn for Computer Methods and Advances in Geomechanics (IACMAG)*, Turin, 4, 2005, pp. 441–450.

Meyerhof, G. G. "Penetration Tests and Bearing Capacity of Cohesionless Soils," *Journal of the Soil Mechanics and Foundations Division*, ASCE, 82(SM1):1–19, 1956.

Meyerhof, G. G. "Some Recent Research on Bearing Capacity of Foundations," *Canadian Geotechnical Journal*, 1(1):16–26, 1963.

Motra, H. B., H. Stutz, and F. Wuttke. "Quality Assessment of Soil Bearing Capacity Factor Models of Shallow Foundations," *Soils and Foundations*, 56(2):265–276, 2016.

O'Brien, A. S. Foundation Types and Conceptual Design Principles, Chapter 52, *ICE Manual of Geotechnical Engineering*, Burland, J. B., T. Chapman, H. Skinner, and M. Brown (eds.), ICE Publishing, London, II, 2013, pp. 733–764.

Pantelidis, L. "A Critical Review of Schmertmann's Strain Influence Factor Method for Immediate Settlement Analysis," *Geotechnical and Geological Engineering*, 38(1):1–18, 2020.

Polshin, D. E., and R. A. Tokar. "Maximum Allowable Nonuniform Settlement of Structures," *Proceedings of 4th International Conference of Soil Mechanics and Foundation Engineering*, London, 1957, pp. 402–406.

Poulos, H. G., and E. H. Davis. *Elastic Solutions for Soil and Rock Mechanics*, John Wiley & Sons, New York, 1974.

Prandtl, L. "Uber die Eindringungsfestigkeit Plastischer Baustoffe und die Festigkeit von Schneiden," *Zeitschrift für Angewandte Mathematik und Mechanik*, Basel, Switzerland, 1(1):15–20, 1921.

Reissner, H. "Zum Erddruck-Problem," *Proceedings of First International Congress on Applied Mechanics*, Delft, 1924, pp. 295–311.

Schleicher, F. "Zur Theorie des Baurundes." *Bauingenieur*, 7(48):931–935, 1926.

Schmertmann, J. H. "Static Cone to Compute Static Settlement Over Sand," *Journal of the Soil Mechanics and Foundations Division*, ASCE, 96(SM3):1011–1043, 1970.

Schmertmann, J. H. *Guidelines for Cone Penetration Test Performance and Design*, Federal Highway Administration, Report FHWA-TS-78-209, Washington DC, 1978, p. 145.

Schultze, E., and A. Horn. "The Base Friction for Horizontally Loaded Footings in Sand and Gravel," *Geotechnique*, 17(4):329–347, 1967.

Sivakugan, N., and K. Johnson. "Settlement Predictions in Granular Soils: A Probabilistic Approach." *Geotechnique*, 54(7):499–502, 2004.

Skempton, A. W., and L. Bjerrum. "A Contribution to Settlement Analysis of Foundations on Clay," *Geotechnique*, 7:168–178, 1957.

Skempton, A. W., and D. H. MacDonald. "The Allowable Settlement of Buildings," *Proceedings of the Institution of Civil Engineers*, 3(5):727–768, 1956.

Spangler, M. G., and R. L. Handy. *Soil Engineering*, 4th Edition, Harper and Row Publishers, New York, 1982.

Steinbrenner, W. "Tafeln zur Setzungsberechnung," *Die Strasse*, 1:121–124, 1934.

Tan, C. K., and J. M. Duncan. "Settlement of Footings on Sands: Accuracy and Reliability." *Geotechnical Engineering Congress 1991*, Colorado, 1, 1991, pp. 446–455.

Terzaghi, K. *Theoretical Soil Mechanics*, Wiley & Sons, New York, 1943.

Terzaghi, K., and R. B. Peck. *Soil Mechanics in Engineering Practice*, 2nd Edition, John Wiley & Sons, New York, 1967.

Terzaghi, K., R. B. Peck, and G. Mesri. *Soil Mechanics in Engineering Practice*, 3rd Edition, John Wiley & Sons, New York, 1996.

Vesic, A. S. "Bearing Capacity of Deep Foundations in Sand," Highway Research Record, 39, Highway Research Board, Washington, DC, 1963.

Vesic, A. S. Bearing Capacity of Shallow Foundations, Chapter 3, *Foundation Engineering Handbook*, Winterkorn, H. F. and H.-Y. Fang (eds.), Van Nostrand Reinhold Co, New York, 1975.

Zhu, M., and R. L. Michalowski. "Shape Factors for Limit Loads on Square and Rectangular Footings," *Journal of Geotechnical and Geoenvironmental Engineering*, ASCE, 131(2):223–231, 2005.

CHAPTER 13
Deep Foundations

13.1 Introduction

In Chap. 12, shallow foundations were defined as ones where the depth D_f is less than the breadth B. In deep foundations, D_f is greater (in fact significantly) than B. Wherever shallow foundations can serve the purpose, they are fine; only when they are inadequate, deep foundations can be the way to go. Compared to shallow foundations, deep foundation cost more and require specialized equipment and skill.

Deep foundations include single piles, pile groups, and drilled shafts. Piles are slender structural elements that are used to transfer the loads from the structure to the underlying ground. Figure 13.1 shows some applications of deep foundations. When the column load is high for a traditional pad footing, a pile can be a good alternative (Fig. 13.1a). When the column load is significantly high for a single pile, instead of providing a very large diameter pile, a pile group can be provided (Fig. 13.1b). Here, the reinforced concrete pile cap connects the piles and the column. In situations where the top soil is weak, shallow foundations cannot carry the required column loads. Piles can bypass the weak layer and transfer the column load to the stronger stratum lying below (Fig. 13.1c).

Underreamed piles are cast-in-place piles where the base is enlarged forming a bulb as shown in Fig. 13.1d. A special underreaming tool is used to enlarge the cavity at the base, which is filled with concrete. They provide a good anchorage against any uplift which can take place in expansive clays or structures like pump houses located in ground with high water tables (Fig. 13.1d). Some underreamed piles have more than one bulb at different heights. Franki pile also has an enlarged base which is formed slightly differently. Here, a concrete plug at the bottom of the hole is rammed into the soil below, forming the enlarged base.

Piles are also effective in resisting lateral loads. When the piles are installed in an inclined position, they are called *battered piles* or *raked piles*. Figure 13.1e shows an anchored sheet pile supporting an excavation. The sheet pile is held in place by a concrete anchor (known as dead man) supported by raked piles. They are used in offshore platforms, wharves, and jetties, resisting lateral impact loads coming from the berthing vessels.

Figure 13.1f shows a transmission tower subjected to lateral loads, which is supported by underreamed piles providing resistance against uplift. Pad footings may not be able to resist the lateral loads and the uplift. Figure 13.1g shows 50–150-mm-diameter micropiles, also known as *root piles* or *pali radice*, used in *underpinning* an old foundation that has settled excessively. Underpinning is the process of strengthening existing foundations.

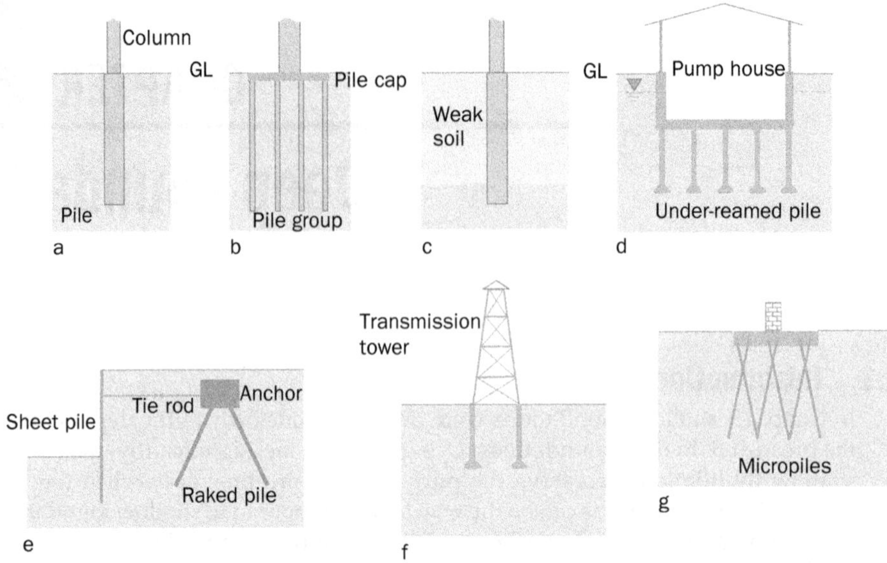

FIGURE 13.1 Applications of deep foundations: (a) single pile, (b) pile group, (c) piling through a weak soil, (d) under-reamed piles, (e) raked piles, (f) piles for transmission tower, and (g) micropiles used in underpinning.

Piles have been driven into loose sands to densify them and provide reinforcement. They are called compaction piles.

13.2 Pile Materials

Piles are generally made of timber, steel or concrete. They have their specific advantages and disadvantages.

13.2.1 Timber Piles

Timber is cheaper than concrete or steel, but it is limited in length, diameter, and load-carrying capacity. Timber piles are tree trunks that have the branches and barks removed, with circular cross section and a natural taper, with the top end (butt or head) larger in diameter than the bottom end (tip). The load-carrying capacity of a timber pile is limited to about 500 kN. Timber piles are available up to 20 m and they can be spliced to form longer piles. They are installed by driving and used as friction piles than in end bearing. Very high driving stresses can cause brooming or splitting at the head, or can damage the tip. The damage can be minimized by providing a cushion at the head and a steel shoe at the tip.

In marine environment with fluctuating water levels, timber deteriorates fast when attacked by marine organisms such as marine borers and fungi. It performs better when permanently submerged. Timber can be pressure impregnated by creosotes and oil to make it more durable. While timber piles were widely used in the ancient times, due to scarcity of timber at present and their low load-carrying capacities, they are no more popular.

13.2.2 Concrete Piles

The concrete piles can be precast or cast-in-place. Precast piles, which are often installed by driving or jetting (with the help of a water jet to loosen and displace soil at the tip, mostly in sands), can be conventionally reinforced or prestressed. One of the major drawbacks with precast piles is that they require casting yards, storage space, transport to the site, and special care in handling to avoid damage due to bending stresses. They have to be lifted from the right locations to avoid high bending stresses, which can damage the pile. Figure 13.2a shows octagonal precast concrete piles with tapered ends to facilitate driving. Figure 13.2b shows a precast concrete battered pile being driven into the ground by a *pile-driving hammer*.

Drilled shafts or *bored piles* are made by drilling a vertical hole and filling it with concrete, thus forming a cast-in-place pile. A reinforcement cage is provided mainly for the upper section of the pile which may be subjected to tensile stresses. The base can also be underreamed to increase the load-carrying capacity and/or provide resistance to uplift. Figure 13.3a shows the reinforcement cage being placed for a 3-m-diameter bored pile, where the top few meters are cased. Continuous flight auger piles, known as CFA piles, are cast-in-place piles where the concrete or grout is pumped into the hole while the auger is withdrawn (Fig. 13.3b).

13.2.3 Steel Piles

Steel piles are often in the form of pipes or H-sections. They are installed by driving. High load-carrying capacity, resistance to driving, and relative ease of splicing make them an attractive option. In marine environment they can corrode and require special treatments such as *cathodic protection*. Here, the steel pile (cathode) is connected to a sacrificial metal (anode), thus forming an electrochemical cell.

13.2.4 Composite Piles

Composite piles are made of two different materials, exploiting the advantages of the two. A steel pipe filled with concrete is a composite pile. A timber pile in a marine

a b

FIGURE 13.2 (a) Octagonal precast concrete piles, (b) raked pile driven into the ground.

a b

FIGURE 13.3 (a) A 3-m-diameter bored pile, (b) a CFA pile.

environment, where a concrete cover is used around the head to protect the timber from the fluctuating water table is also a composite pile. Lately, piles for marine environments are also made of polymer composites, PVC, and fiber-reinforced plastic/polymer (FRP).

13.3 Pile Installation

While most piles are installed into the ground by driving or boring, some are installed by jetting or screwing. Figure 13.4a shows a pile-driving rig at work, consisting of a crane, leader, and pile hammer. The leader is a stiff tubular member mounted on the crane, and supports the hammer and the pile. By repeatedly lifting and dropping a weight over a certain height, the pile is driven into the ground. There are also vibratory hammers, which are different from the impact hammers. They are effective in granular soils. Figure 13.4b shows the vertical hole for an uncased bored pile, with the reinforcement cage for the upper length. When the walls of the hole are not firm, a casing is required.

Jetted piles are installed using a pressurized water jet to weaken the soil which makes it possible to advance the pile. Jetting is more effective in granular soils than in clays. *Screw piles*, also known as *helical piles*, can have one or more helices, and are wound into the ground applying a torque. They provide good anchorage and hence resistance to uplift.

When the pile is installed into the ground, nearby soil may be displaced laterally. Depending on the extent of soil displacement, the piles can be categorized as nondisplacement, low-displacement, medium-displacement, or high-displacement piles. The larger the displacement, the larger the pile capacity, especially the fraction derived from the shaft (see Sec. 13.4). Bored piles and micropiles are nondisplacement piles. CFA piles are very low-displacement piles. An open-ended-pipe pile, H-pile, or screw pile

a b

FIGURE 13.4 (a) Pile driving, (b) reinforcement cage for a bored pile.

are low-displacement piles. Large-diameter driven piles and closed-end-pipe piles are high-displacement piles.

13.4 Shaft and Tip Loads

As shown in Fig. 13.5a, when a vertical load Q is applied on the *pile head*, a fraction of it is transferred to the ground through the pile shaft (Q_s) and the remainder is transferred to the ground through the pile tip (Q_t). Therefore,

$$Q = Q_s + Q_t \tag{13.1}$$

The vertical load should ideally include the self-weight of the pile, which is small in comparison and is neglected here. When the soil near the ground surface is very weak, and is underlain by a stiffer stratum at some depth, the pile is driven past the weak layer into the stiffer stratum. Here, most of the load is transferred to the stiff stratum through the pile tip and hence, $Q_s \approx 0$ and $Q_t \approx Q$. These piles are called *end-bearing piles*. When most of the load is transferred to the soil through the pile shaft with $Q_s \approx Q$ and $Q_t \approx 0$, the piles are called *friction piles*.

The shaft load Q_s and the tip load Q_t increase with the load Q applied at the pile head, and would reach their maximum capacities $Q_{s,\text{ult}}$ and $Q_{t,\text{ult}}$, respectively, when they are fully mobilized. The *ultimate load-carrying capacity* of the pile Q_{ult} is this given by

$$Q_{\text{ult}} = Q_{s,\text{ult}} + Q_{t,\text{ult}} \tag{13.2}$$

The *allowable load-carrying capacity* of the pile Q_{all} is given by

$$Q_{\text{all}} = \frac{Q_{\text{ult}}}{F} \tag{13.3}$$

where F is the factor of safety in the range of 2–3. Depending on whether the pile load capacity is verified by a proper pile load test, or a pile-driving analyzer (PDA), or not verified at all, the factor of safety is taken as 2.0, 2.5, or 3.0, respectively.

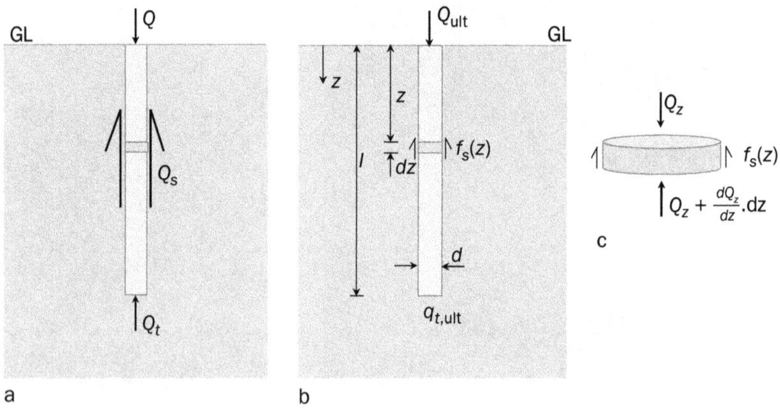

FIGURE 13.5 (a) Shaft and tip loads, (b) ultimate bearing capacities of the shaft and tip, (c) the disc element.

Figure 13.5b shows a pile of length l and diameter d, carrying the ultimate load Q_{ult}. Let us consider a small disc element of thickness dz at depth z. The *ultimate skin friction* $f_s(z)$ (unit of stress) varies along the length of the pile. The shaft load carried by the disc element is

$$dQ_{s,ult} = p(z)\, f_s(z)\, dz$$

where $p(z)$ is the perimeter of the pile at depth z. For nontapered piles with circular cross sections, the perimeter is a constant for the entire pile length, given by πd.

Therefore, the total shaft load carried by the pile at failure is given by

$$Q_{s,ult} = \int_{z=0}^{z=l} p(z) f_s(z) dz \tag{13.4}$$

The load carried by the pile tip at failure is given by

$$Q_{t,ult} = q_{t,ult} A_t \tag{13.5}$$

where $q_{t,ult}$ = ultimate bearing capacity at the pile tip and A_t = cross-sectional area of the pile tip. To compute the allowable load on a pile Q_{all}, it is necessary to determine Q_{ult}, which requires $f_s(z)$ and $q_{t,ult}$. The methods to compute these two parameters are discussed in Sec. 13.6.

13.5 Pile Load Transfer Mechanism

Figure 13.5c shows the disc element from the pile located at a depth z below the ground level. Due to the skin friction acting along the pile shaft, the pile load Q_z acting on the cross section of the pile, decreases with depth. Considering the equilibrium of the disc element,

$$Q_z + \frac{dQ_z}{dz} dz - Q_z + f_s(z) p(z) dz = 0$$

$$f_s(z) = -\frac{1}{p(z)}\frac{dQ_z}{dz} \tag{13.6a}$$

$$\frac{dQ_z}{dz} = -f_s(z)p(z) \tag{13.6b}$$

Equation (13.6) is valid at any stage of loading, not necessarily at the ultimate loads.

From the variation of $f_s(z)$ with depth, it is possible to determine the variation of pile load Q_z with length using Eq. (13.6a) or (13.6b). Figure 13.6 shows four possible distributions of the ultimate skin friction $f_s(z)$ with depth. This includes, constant, linearly increasing, linearly decreasing, and parabolic distributions. Also shown in the figure are the corresponding variations of the pile load Q_z with depth, derived from Eq. (13.6b). You can see how the load Q applied at the top of the pile is separated into Q_s and Q_t, for each case.

When a pile is subjected to uplift or a tensile load, the entire load is carried by the pile shaft, with no contribution from the pile tip. Under a compressive load, the load Q applied at the head is shared by the tip and the shaft.

Figure 13.7 shows the variations of the shaft load Q_s and tip load Q_t with settlement when the load Q ($Q = Q_s + Q_t$) at the pile head is increased from zero to failure. During the very early stages of loading, most of the load is carried by the shaft. The entire shaft load is mobilized when it becomes equal to $Q_{s,ult}$, which occurs at a relatively lower value of settlement, in the order of 0.5–1.0% of the pile diameter. For the tip load to be

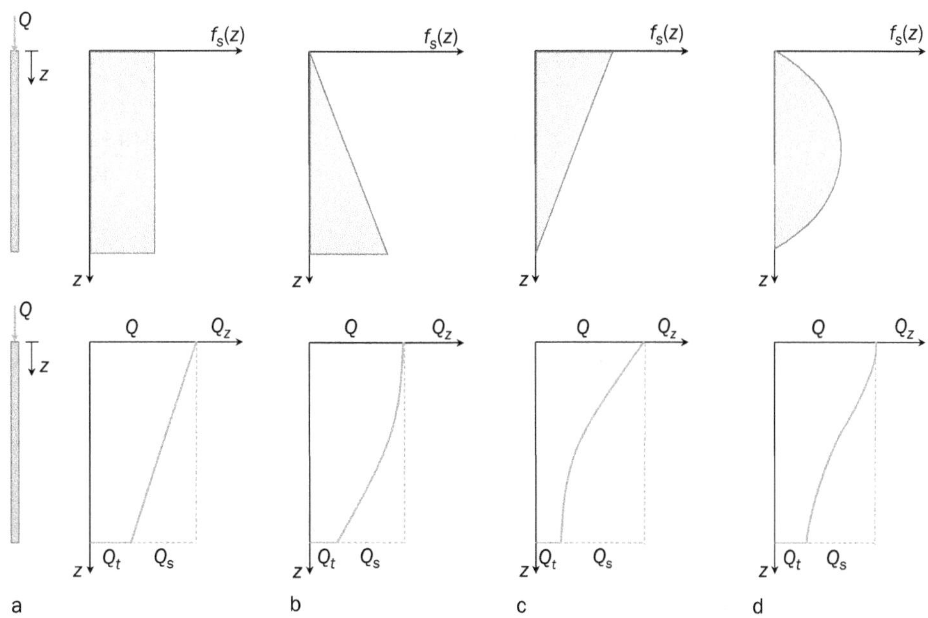

FIGURE 13.6 Variation of the skin friction $f_s(z)$ and the pile load Q_t with depth z: (a) constant, (b) linearly increasing, (c) linearly decreasing, and (d) parabolic skin friction.

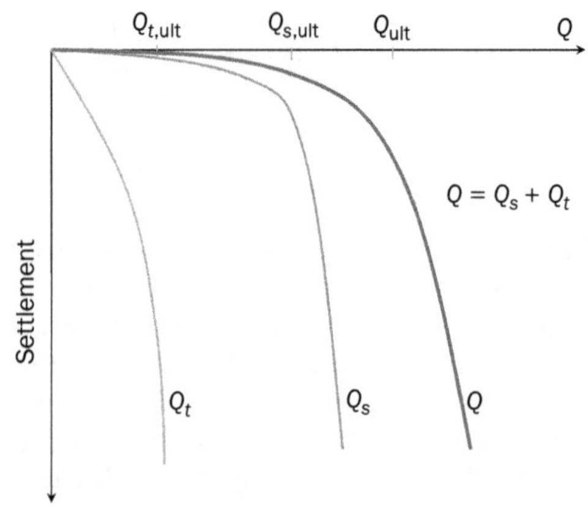

Figure 13.7 Variations of Q_s and Q_t with settlement.

fully mobilized, it requires significantly larger settlement, in the order of 10% of the pile diameter. The shaft load is fully mobilized well before the tip load.

Example 13.1

A 16-m-long and 500-mm-diameter driven concrete pile carries a 1000-kN load (Fig. 13.8a). The variation of skin friction with depth is shown in Fig. 13.8b. Plot the variation of the pile load with depth.

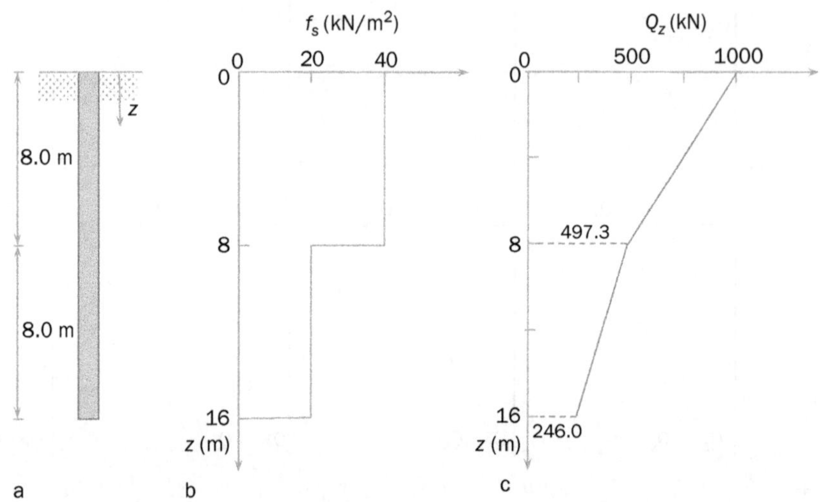

Figure 13.8 (a) Pile, (b) variation of $f_s(z)$ with depth, (c) Pile load Q_z versus depth plot.

Solution

$$\frac{dQ}{dz} = -\pi d\, f_s(z)$$

Thus, when $f_s(z)$ is constant, the pile load Q_z decreases linearly with depth.
Shaft load for $z = 0–8$ m is given by

$$Q_{s1} = (\pi \times 0.5)(8 \times 40) = 502.7 \text{ kN}$$

The pile load Q_z at $z = 8$ m is

$$Q_z = 1000 - 502.7 = 497.3 \text{ kN}$$

Shaft load for $z = 8–16$ m is given by

$$Q_{s2} = (\pi \times 0.5)(8 \times 20) = 251.3 \text{ kN}$$

$$\text{Total load carried by the pile} = 1000 \text{ kN}$$

Therefore, the load carried by the tip is

$$Q_t = 1000 - (502.7 + 251.3) = 246.0 \text{ kN}$$

Example 13.2

The ultimate load-carrying capacities of the tip and the shaft of a 500-mm-diameter pile were determined as 1200 and 1500 kN, respectively. Figure 13.9a shows the expected load versus settlement (pile head) diagram for the pile shaft and the pile tip. The shaft load is fully mobilized at settlement of 7.5 mm, and the tip load is fully mobilized only at settlement of 50 mm.

a. Plot the load versus settlement for the pile.
b. What is the load that can be allowed on the pile with safety factor of 2.5? What would be the settlement at this load?
c. Of this allowable load, what fraction is carried by the shaft?

Solution

Part a At 7.5 mm settlement,

$$Q_s = 1500 \text{ kN}$$

$$Q_t = 1200(7.5/50) = 180 \text{ kN}$$

Therefore, $Q = 1500 + 180 = 1680$ kN
At 50 mm settlement,

$$Q_s = 1500 \text{ kN}$$

$$Q_t = 1200 \text{ kN}$$

Therefore, $Q = 1500 + 1200 = 2700$ kN
From these, the load versus settlement plot can be developed as shown in Fig. 13.9b.

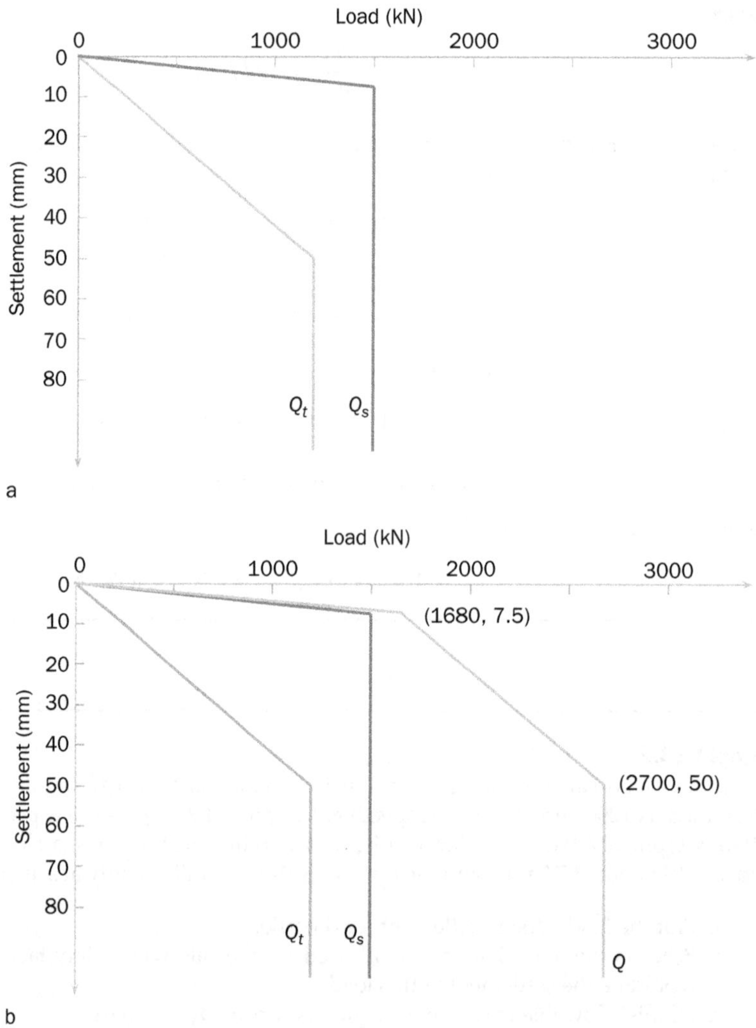

FIGURE 13.9 (a) Variation of Q_s and Q_t with settlement, (b) Load Q versus settlement.

Part b From the plot, $Q_{ult} = 2700$ kN
Therefore, $Q_{all} = 2700/2.5 =$ **1080 kN**

Part c At 1080 kN load, settlement = 1080(7.5/1680) = 4.82 mm
At this settlement of 4.82 mm,

$$Q_s = 1500(4.82/7.50) = 964 \text{ kN}$$

$$Q_t = 1200(4.82/50) = 116 \text{ kN}$$

Percentage of the shaft load = (964/1080)(100) = **89.3%**

13.6 Load-Carrying Capacity of a Single Pile

The allowable load on a pile Q_{all} is derived from the ultimate load Q_{ult}, applying an appropriate factor of safety [see Eq. (13.3)]. The ultimate load consists of two components, the ultimate shaft load $Q_{s,ult}$ and the ultimate tip load $Q_{t,ult}$. To determine these two components, it is necessary to evaluate the variation of the ultimate skin friction $f_s(z)$ with depth z, and the ultimate bearing capacity at the tip of the pile $q_{t,ult}$. The methods to compute these two parameters in granular soils and cohesive soils, are discussed separately in Secs. 13.6.1, 13.6.2, 13.6.3, and 13.6.4.

13.6.1 Ultimate Skin Friction in Granular Soils

Next to the pile shaft, at depth z below the ground level, the effective vertical stress is σ'_v. The effective horizontal stress σ'_h can be written as

$$\sigma'_h = K_s \sigma'_v$$

where K_s is a lateral earth pressure coefficient, which depends on factors such as the method of installation, pile material, and the pile diameter. The ultimate skin friction $f_s(z)$ is given by

$$f_s(z) = K_s \sigma'_v \tan\delta \tag{13.7}$$

where δ is the *interface friction angle between the soil and the pile*, which depends on the friction angle ϕ' of the granular soil and the pile material. This is similar to the wall friction angle between the granular soil and the concrete wall discussed in Sec. 10.5. Table 13.1 gives some guidance in estimating K_s, where K_0 is the coefficient of earth pressure at rest that can be determined as

$$K_0 = 1 - \sin\phi' \tag{13.8}$$

in normally consolidated soils. Canadian Foundation Engineering Manual (CGS, 2006) suggests $K_s = 2K_0$ for driven displacement piles, which is slightly higher than the values suggested in Table 13.1. For tapered piles, it suggests increasing the values by further 30–50%.

Based on field tests on 24 driven piles in sands, Coyle and Castello (1981) suggested values for K_s as a function of the friction angle and the embedment ratio l/d, as shown in Fig. 13.10. It was assumed that $\delta = 0.8\phi'$. This K_s is generally used for computing the average skin friction along the pile, using the average overburden pressure $(\overline{\sigma'_{vo}})$ determined at the mid-height of the pile. Thus,

Pile type	K_s
Bored pile	K_0
Jetted pile	K_0
Low-displacement driven pile	1.0–1.4 times K_0
High-displacement driven pile	1.4–1.8 times K_0

TABLE 13.1 Values of K_s

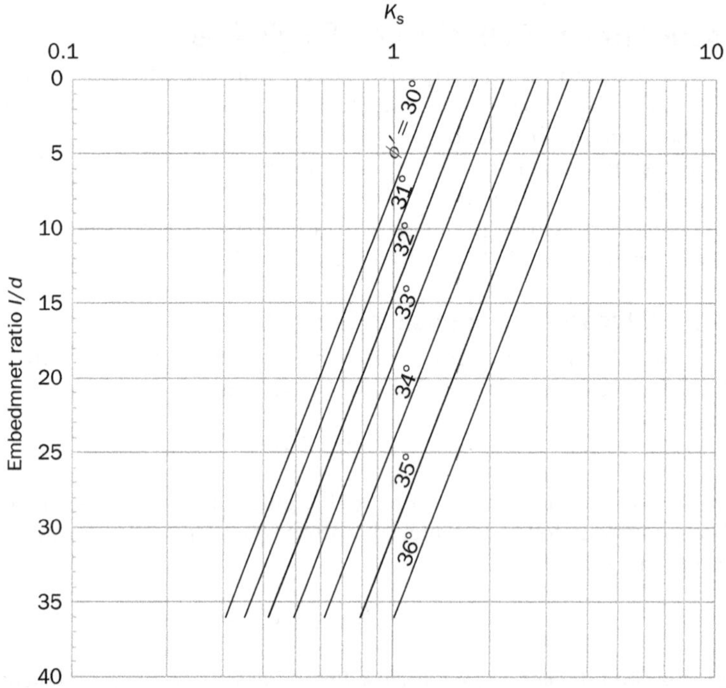

FIGURE 13.10 K_s values from Coyle and Castello (1981).

Pile material	Kulhawy (1984)	US Army (1993)
Timber	0.8–0.9	0.8–1.0
Smooth steel	0.5–0.7	0.67–0.8
Rough (corrugated) steel	0.7–0.9	
Precast concrete	0.8–1.0	0.8–1.0
Cast-in-place concrete	1.0	

TABLE 13.2 Values of δ/ϕ' in Sand Suggested by Kulhawy (1984) and US Army (1994)

Soil	Driven piles	Cast-in-place piles
Silts	0.3–0.5	0.2–0.3
Loose sands	0.3–0.8	0.2–0.4
Medium sands	0.6–1.0	0.3–0.5
Dense sands	0.8–1.2	0.4–0.6
Gravels	0.8–1.5	0.4–0.7

TABLE 13.3 β-coefficient (After CGS, 2006)

$$Q_{s,\text{ult}} = K_s \overline{\sigma'_{vo}}(\tan 0.8'\phi)pl \tag{13.9}$$

where p is the average perimeter of the pile, and l is the pile length.

The pile-soil interface friction angle δ lies in the range of $\phi'/2$ to ϕ'. Canadian Foundation Engineering manual (CGS, 2006) suggests taking tan δ as 0.7–1.0 times tan ϕ'. Some values of δ/ϕ' suggested in the literature are given in Table 13.2.

The term K_s in Eq. (13.7) is called the β-coefficient, and its values suggested by Canadian Foundation Engineering Manual (CGS, 2006) are given in Table 13.3, for driven piles and cast-in-place piles. Bhushan (1982) suggested that for high-displacement piles in sands

$$\beta = 0.18 + 0.65D_r \tag{13.10}$$

From Eqs. (13.4) and (13.7), the ultimate shaft load $Q_{s,\text{ult}}$ can be determined as

$$Q_{s,\text{ult}} = \int_{z=0}^{z=l} p(z) K_s \sigma'_v \tan\delta \, dz \tag{13.11}$$

For piles which are not tapered, with circular cross section and diameter of d,

$$Q_{s,\text{ult}} = K_s \pi d \tan\delta \int_{z=0}^{z=l} \sigma'_v dz \tag{13.12}$$

The last term within the integral in Eq. (13.12) is the area enclosed within the σ'_v versus z diagram.

The vertical stress σ'_v adjacent to a driven pile in a granular soil does not increase indefinitely with depth. Probably due to arching of the soil adjacent to the pile, it does not increase beyond a certain depth. This *critical depth* d_c depends on the relative density, and lies in the range of 10–20 times the pile diameter d, with the lower and upper ends of the range for very loose and very dense sands, respectively (US Army, 1993).

13.6.2 Ultimate Bearing Capacity at the Tip in Granular Soils

The ultimate bearing capacity at the pile tip $q_{t,\text{ult}}$ can be computed by treating the pile as a footing of width d and depth l. In granular soils where $c' = 0$, neglecting the shape and depth factors,

$$q_{t,\text{ult}} = \gamma l N_q + 0.5 d \gamma N_\gamma \tag{13.13}$$

The failure surface and the failure mechanism for the deep foundations are different to the ones for shallow foundations discussed in Chap. 12. While N_q and N_γ for the piles are different to those used in Chap. 12, they are of the same order of magnitude. Since $l >> 0.5d$, Eq. (13.13) can be written as

$$q_{t,\text{ult}} \approx \gamma l N_q \approx \sigma'_{v,\text{tip}} N_q \tag{13.14}$$

where $\sigma'_{v,\text{tip}}$ is the vertical stress adjacent to the pile tip. A wide range of values have been proposed for the pile bearing capacity factor N_q by Terzaghi (1943), Meyerhof

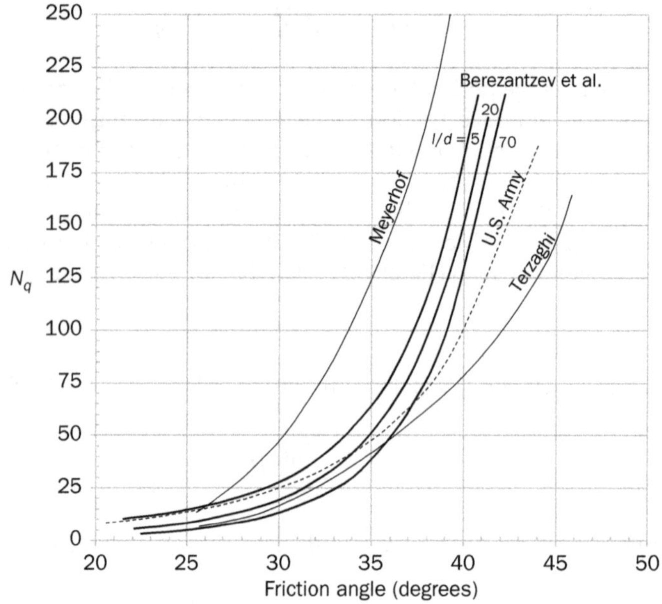

Figure 13.11 N_q values for driven piles.

(1976), Berezantzev et al. (1961), and US Army (1993). They are shown in Fig. 13.11. The values suggested by Berezantzev et al. (1961) appear to be popular. US Navy (1986) suggested N_q values for driven and cast-in-place piles. The ones suggested for the driven piles are very close to Berezantzev et al.'s (1961) values for $l/d = 20$. The values suggested by the US Navy (1986) for the cast-in-place piles are 50% of the ones for the driven piles. The N_q values suggested by Coyle and Castello (1981) are shown in Fig. 13.12. The N_q values suggested by the Canadian Foundation Engineering Manual (CGS, 2006) for cast-in-place and driven piles are given in Table 13.4.

Due to arching, the ultimate bearing capacity of the pile tip does not increase with the pile length indefinitely. At critical depth d_c, σ'_v reaches its maximum value and hence $q_{t,\text{ult}}$ can be taken as a constant for any further increase in the pile length. The vertical stress at the critical depth can be used in Eq. (13.14) as a substitute for $\sigma'_{v,\text{tip}}$. From Eq. (13.5),

$$Q_{t,\text{ult}} = q_{t,\text{ult}} A_t = \sigma'_{v,\text{tip}} N_q A_t \tag{13.15}$$

The computed values of the ultimate bearing capacity near the tip $q_{t,\text{ult}}$ are generally very high. They must not be taken as anything more than the limiting value, typically in the order of 10–15 MN/m^2 in silica sand and 3–5 MN/m^2 in calcareous sand.

Table 13.5 summarizes the empirical equations reported in the literature relating f_s or $q_{t,\text{ult}}$ with the penetration resistance from the *in situ* tests such as standard penetration test or cone penetration test.

Example 13.3

A 500-mm-diameter and 25-m-long concrete pile is driven into a sand where $\gamma = 18.0$ kN/m^3, $\phi' = 30°$, and $\delta = 21°$. With a factor of safety of 2.5, and considering critical depth, what would be the maximum load that can be allowed on the pile?

Solution Let us take the critical depth as $11d = 5.5$ m.

The vertical overburden stress near the pile shaft at the critical depth $= 5.5 \times 18.0 = 99.0$ kN/m^2, which is assumed to remain the same for $z > 5.5$ m, as shown in Fig. 13.13.

Shaft load The area of the σ'_v versus z diagram can be determined as

$$\frac{(25.0 + 19.5)}{2}(99.0) = 2202.8 \text{ kN/m}$$

$$K_0 = 1 - \sin\phi' = 1 - \sin 30 = 0.5$$

Let's take K_s as 1.5 K_0 (see Table 13.1)

$$K_s = 1.5 \times 0.5 = 0.75$$

From Eq. (13.12),

$$Q_{s,\text{ult}} = K_s \tan\delta\pi d \int_{z=0}^{z=l} \sigma'_v dz$$

$$Q_{s,\text{ult}} = (0.75)(\tan 21)(\pi)(0.5)(2202.8) = 996.2 \text{ kN}$$

Tip load

$$l/d = 25/0.5 = 50$$

From Fig. 13.11, Berezantzev et al.'s $N_q = 16$.

From Eq. (13.14),

$$q_{t,\text{ult}} \approx \sigma'_{v,\text{tip}} N_q$$

$$q_{t,\text{ult}} \approx (99.0)(16) = 1584 \text{ kN/m}^2$$

From Eq. (13.15),

$$Q_{t,\text{ult}} = (1584)\left(\frac{\pi \times 0.5^2}{4}\right) = 311.0 \text{ kN}$$

The ultimate load the pile can carry is

$$Q_{\text{ult}} = Q_{s,\text{ult}} + Q_{t,\text{ult}}$$

$$Q_{\text{ult}} = 996.2 + 311.0 = 1307.2 \text{ kN}$$

The allowable load, $Q_{\text{all}} = Q_{\text{ult}}/F = 1307.2/2.5 =$ **522.9 kN**

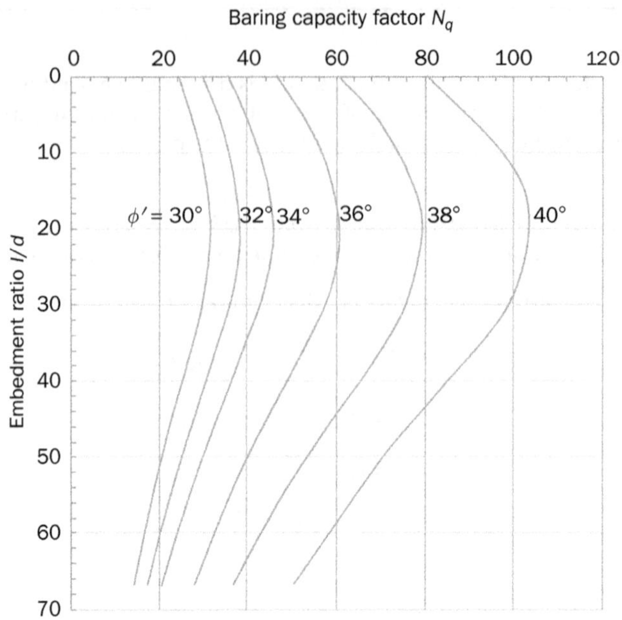

FIGURE 13.12 N_q values suggested by Coyle and Castello (1981) for driven piles in granular soils.

Soil type	Driven piles	Cast-in-place piles
Silt	20–40	10–30
Loose sand	30–80	20–30
Medium sand	50–120	30–60
Dense sand	100–120	50–100
Gravel	150–300	80–150

TABLE 13.4 N_q Values from Canadian Foundation Engineering Manual (CGS, 2006)

Expression	Comments	Reference
$f_s = (0.028N_{60} + 0.1)p_a$	Displacement piles	Decourt (1995)
$f_s = 0.55(0.028N_{60} + 0.1)p_a$	Nondisplacement piles	Decourt (1995)
$f_s = 0.02p_a\overline{N_{60}}$	High-displacement piles	Meyerhof (1976)
$f_s = 0.01p_a\overline{N_{60}}$	Low-displacement piles	Meyerhof (1976)
$f_s = 0.224p_a\left(\overline{N_{60}}\right)^{0.29}$	Driven piles	Briaud et al. (1985)

TABLE 13.5 Relationships for f_s and $q_{t,ult}$ in Granular Soils

Expression	Comments	Reference
$f_s = q_c / 150$	Driven concrete piles; < 120 kN/m²	MELT (1993)
$f_s = q_c / 200$	Drilled; < 120 kN/m²	MELT (1993)
$q_{t,\text{ult}} = 0.4 p_a N_{60} \frac{l}{d}$	N_{60} at pile tip. Driven piles. Not to exceed $4p_a N_{60}$ in sands and gravels; $3p_a N_{60}$ in nonplastic silts	Meyerhof (1976)
$q_{t,\text{ult}} = 19.7 p_a (N_{60})^{0.36}$	Driven piles	Briaud et al. (1985)
$q_{t,\text{ult}} = 0.14 p_a N_{60} \frac{l}{d}$	Cast-in-place piles	Meyerhof (1976)
$q_{t,\text{ult}} = 3.25 p_a N_{60,\text{tip}}$	Displacement piles in sand	Decourt (1995)
$q_{t,\text{ult}} = 2.05 p_a N_{60,\text{tip}}$	Displacement piles in sandy silt	Decourt (1995)
$q_{t,\text{ult}} = 1.65 p_a N_{60,\text{tip}}$	Nondisplacement piles in sand	Decourt (1995)
$q_{t,\text{ult}} = 1.15 p_a N_{60,\text{tip}}$	Nondisplacement piles in sandy silt	Decourt (1995)
$q_{t,\text{ult}} = 0.6 p_a N_{60,\text{tip}}$	Drilled shafts in sands	Reese and O'Neill (1989)
$q_{t,\text{ult}} = 0.15\, q_c$	Nondisplacement piles in granular soils	MELT (1993)
$q_{t,\text{ult}} = 0.5\, q_c$	Displacement piles in granular soils	MELT (1993)

Notes: $\overline{N_{60}}$ = average N_{60} value along the pile length, $N_{60,\text{tip}} = N_{60}$ at the pile tip, q_c = cone penetration resistance, p_a = atmospheric pressure = 101.3 kN/m².

Table 13.5 Relationships for f_s and $q_{t,\text{ult}}$ in Granular Soils (*Continued*)

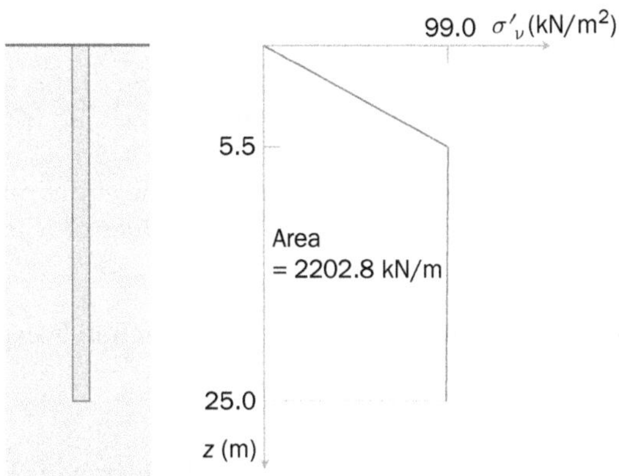

Figure 13.13 Variation of σ'_v with depth.

Example 13.4

A 650-mm-diameter and 20-m-long high-displacement precast concrete pile is driven into a sand where the average N_{60} along the pile is 22 and 24 at the pile tip. Estimate the maximum load that can be allowed on the pile with factor of safety of 2.5. Use Meyerhof's (1976) empirical equations in Table 13.5.

Solution

Shaft load

$$f_s = 0.02p_a\overline{N_{60}} = 0.02 \times 101.3 \times 22 = 44.6\ \text{kN/m}^2$$

$$Q_{s,\text{ult}} = (\pi \times 0.65 \times 20)(44.6) = 1820.4\ \text{kN}$$

Tip load

$$q_{t,\text{ult}} = 0.4p_aN_{60}\frac{l}{d} = 0.4 \times 101.3 \times 24 \times \frac{20}{0.65} = 29{,}922.5\ \text{kN/m}^2$$

The limiting value of $q_{t,\text{ult}} = 4\,p_a\,N_{60} = 4 \times 101.3 \times 24 = 9724.8\ \text{kN/m}^2$

$$Q_{t,\text{ult}} = \left(\frac{\pi}{4} \times 0.65^2\right)(9724.8) = 3227.0\ \text{kN}$$

$$Q_{\text{ult}} = 1820.4 + 3227.0 = 5047.4\ \text{kN}$$

$$Q_{\text{all}} = 5047.4/2.5 = \mathbf{2019\ kN}$$

13.6.3 Ultimate Skin Friction in Clays

There are different ways of determining the skin friction (or better adhesion) along the pile shaft in clays. They are discussed here.

λ Method

Vijayvergia and Focht (1972) proposed this total stress method, assuming that the pile driving induces passive lateral earth pressures. They suggested that the average skin friction is given by

$$f_s(z) = \lambda\left(\overline{\sigma'_v} + 2c_u\right) \tag{13.16}$$

where $\overline{\sigma'_v}$ is the mean effective vertical stress for the entire pile embedment length, and c_u is the average undrained shear strength. The values of λ are given in Table 13.6.

α Method

This is also a total stress method where the skin friction is expressed as

$$f_s(z) = \alpha c_u \tag{13.17}$$

The coefficient α is the *adhesion factor*, lying in the range of 0.3–1.0, depending on the overconsolidation ratio and the undrained shear strength c_u. Soft clays and normally

l (m)	0	5	10	15	20	25	30
λ	0.5	0.336	0.245	0.200	0.173	0.150	0.136
l (m)	35	40	50	60	70	80	90
λ	0.132	0.127	0.118	0.113	0.110	0.110	0.110

Table 13.6 Variation of λ with Pile Embedment Length *l*

consolidated clays take the upper end of the range, and stiff clays and heavily overconsolidated clays take the lower end.

From the pile load test data from Tomlinson (1957) on 56 driven piles and Kulhawy and Jackson (1989) on 106 bored piles, for piles in compression and uplift,

$$\alpha = 0.21 + 0.26\frac{p_a}{c_u} \leq 1 \quad \text{for} \frac{c_u}{p_a} < 3 \tag{13.18}$$

Canadian Foundation Engineering Manual (CGS, 2006) recommends Eq. (13.18) but suggests verification through load tests. Based on the work of Randolph and Murphy's (1985), American Petroleum Institute (API, 1987) suggested computing α as

$$\alpha = 0.5\left(\frac{c_u}{\sigma'_{vo}}\right)^{-0.5} \quad \text{for} \frac{c_u}{\sigma'_{vo}} \leq 1 \tag{13.19a}$$

$$\alpha = 0.5\left(\frac{c_u}{\sigma'_{vo}}\right)^{-0.25} \quad \text{for} \frac{c_u}{\sigma'_{vo}} > 1 \tag{13.19b}$$

where $\alpha \leq 1$. Equation (13.19) was later refined in API (1993), where it is suggested that the average skin friction must be taken as the larger of

$$\left[f_s(z)\right]_{avg} = 0.5(c_u)^{0.5}\left(\overline{\sigma'_{vo}}\right)^{0.5} \tag{13.20a}$$

$$\left[f_s(z)\right]_{avg} = 0.5(c_u)^{0.75}\left(\overline{\sigma'_{vo}}\right)^{0.25} \tag{13.20b}$$

where $\overline{\sigma'_{vo}}$ is the average effective vertical stress along the pile length. Kolk and van der Velde (1996) suggested that for driven piles in clays,

$$\alpha = 0.55\left(\frac{40d}{l}\right)^{0.2}\left(\frac{\sigma'_{vo}}{c_u}\right)^{0.3} \tag{13.21}$$

Karlsrud et al. (2005) proposed the α-values shown in Fig. 13.14 in their design method known as NGI-99.

β Method

For clays under drained conditions, an effective stress method called β-method can be applied. Here, the skin friction is determined as

$$f_s(z) = \beta\sigma'_{vo} \tag{13.22}$$

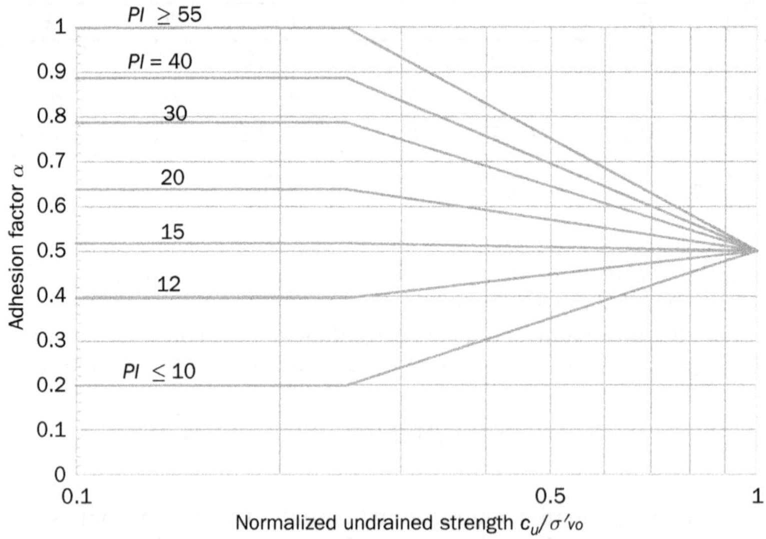

FIGURE 13.14 Variation of α with c_u/σ'_{vo} after Karlsrud et al. (2005).

Burland's (1993) data for all types of piles suggest that β can be estimated as

$$\beta = 0.52\left(\frac{c_u}{\sigma'_{vo}}\right) + 0.11 \tag{13.23}$$

Noting that $\beta = K_s \tan\delta$, and assuming $\delta = \phi'_{cv}$

$$K_s = K_0 = (1 - \sin\phi'_{cv})\sqrt{OCR} \tag{13.24}$$

$$\beta = (1 - \sin\phi'_{cv})\sqrt{OCR}\tan\phi'_{cv} \tag{13.25}$$

The skin friction in clays has also been related to the penetration resistance from SPT and CPT. For example, it can be determined from the sleeve friction f_s from a cone penetration test as

$$f_s(z) = \alpha' f_{s,\mathrm{CPT}} \tag{13.26}$$

where α' varies in the range of 0.2–1.2. French practice, based on an extensive database of bored and driven piles, is to use q_c rather than f_s from the CPT for determining the skin friction of piles. Almeida et al. (1996) suggested that

$$f_s(z) = \frac{q_c - \sigma_{vo}}{\alpha_{\mathrm{CPT}}} \tag{13.27}$$

where α_{CPT} is a reduction factor given by

$$\alpha_{\mathrm{CPT}} = 11.8 + 14\log\left(\frac{q_c - \sigma_{vo}}{\sigma'_{vo}}\right) \tag{13.28}$$

The skin friction $f_s(z)$ computed by any of these methods is generally limited to 120 kN/m^2.

13.6.4 Ultimate Bearing Capacity at the Tip in Clays

In Chap. 12, we saw that for shallow foundations in clays under undrained conditions, $\phi_u = 0$, which gives $N_q = 1$, $N_\gamma = 0$, and $N_c = 5.14$. The net ultimate bearing capacity $q_{\text{ult,net}}$ becomes $c_u N_c$.

For piles in saturated clays under undrained conditions, the ultimate bearing capacity near the tip is given by

$$q_{t,\text{ult}} = c_u N_c \tag{13.29}$$

Skempton (1951) studied the bearing capacity of foundations at different depths in clays under undrained conditions, and his findings are summarized in Fig. 13.15. Here, the bearing capacity factor N_c incorporates the shape factor as well. For strip, N_c increases from 5.14 (or $2 + \pi$) at $D_f/B = 0$ to an asymptotic value of 7.5 at $D_f/B > 5$. For square or circular foundations, N_c increases from 6.2 (or 2π) at $D_f/B = 0$ to a maximum of 9 at $D_f/B > 5$. For rectangular foundations, N_c can be obtained by multiplying the value for square foundation by $0.84 + 0.16\, B/L$. In pile foundations, where l/d is generally significantly larger than 5, N_c can be taken as 9.

In this chapter, it is assumed that the load at the ground level, applied on the pile head is shared by the shaft and the tip. The shaft load is carried in the form of friction or adhesion. Here, the pile moves downward relative to the soil and hence the shaft load acts upward on the shaft. In some soils such as soft clays or compressible fills that are undergoing settlements, the soil can move down relative to the pile and exert a down drag on the pile. This is called *negative skin friction*, which has to be accounted in the design. It reduces the load-carrying capacity of the shaft and the tip has to carry a load greater than what is applied at the top.

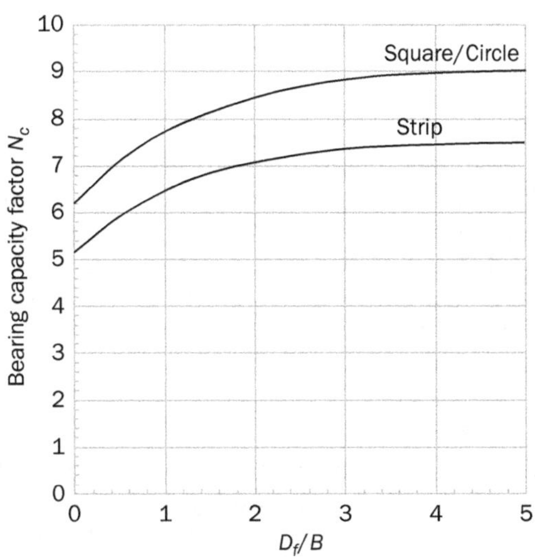

Figure 13.15 Bearing capacity factor N_c in undrained clays.

Example 13.5
The soil profile at a site consists of 10 m of soft clay (c_u = 20 kN/m^2) underlain by a deep bed of stiff clay (c_u = 80 kN/m^2). What should be the length of a 600-mm-diameter driven concrete pile for carrying a 750 kN load with a factor of safety of 2.5? Use α-method to determine the shaft load, with α from Eq. (13.18).

Solution From Eq. (13.18),

$$\alpha = 0.21 + 0.26\frac{p_a}{c_u} < 1$$

For soft clay,

$$\alpha = 0.21 + 0.26\frac{101.3}{20} = 1.0$$

For stiff clay,

$$\alpha = 0.21 + 0.26\frac{101.3}{80} = 0.54$$

Let the pile length be l (m).

$$Q_{s,\text{ult}} = \pi(0.6)[(10 \times 1.0 \times 20 + (l - 10) \times 0.54 \times 80] = 81.4l - 437 \text{ kN}$$

$$Q_{t,\text{ult}} = \left(\frac{\pi}{4} \times 0.6^2\right)(80 \times 9) = 203.6 \text{ kN}$$

$$81.4\,l - 437.0 + 203.6 = 2.5 \times 750$$

$$l = \mathbf{25.9\ m}$$

13.7 Pile Driving

Precast or prestressed concrete, timber, or steel piles are generally installed by driving. Figures 13.2b and 13.4a show piling rigs at work, driving a concrete pile into the ground. The main component of a pile-driving hammer is a ram, which moves up and down to hit the pile head, making it penetrate the ground. The penetration into the ground per blow is called the *set*. This section discusses the different types of hammers and their components, PDA, and the use of pile-driving formulas in estimating the load-carrying capacity of a driven pile.

13.7.1 Types of Hammers

Drop hammers, air or steam hammers, diesel hammers, hydraulic hammers, and vibratory hammers are used in pile driving. Drop hammers, used from the ancient times, are raised and lifted by a cable running around a pulley as shown in Fig. 13.16a. In air or steam hammers, compressed air or steam is used to raise the ram as shown in Fig. 13.16b. Depending on whether the ram is allowed to fall freely under gravity

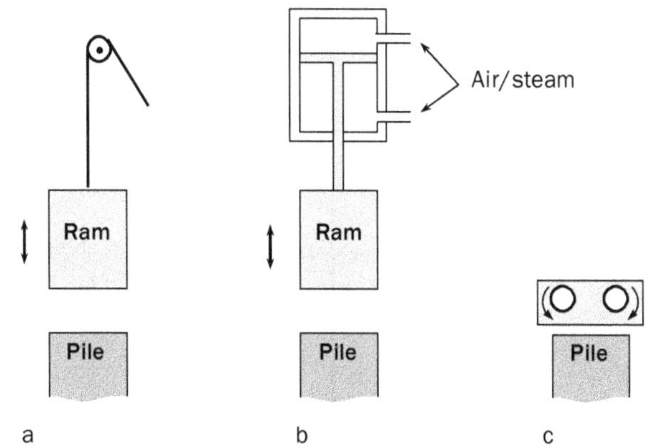

FIGURE 13.16 Type of pile-driving hammers: (a) drop hammer, (b) air or steam hammer, (c) vibratory hammer.

or pushed down by air or steam, it is called a *single-acting* or a *double-acting hammer*. In a single-acting hammer, the ram is pushed up by air or steam, followed by a free fall under gravity. In a double-acting hammer, the fall is further assisted by air or steam pushing the ram down in addition to the gravity. In the vibrating hammer shown in Fig. 13.16c, two counterrotating masses produce a sinusoidal exciting force that drives the pile into the ground. The horizontal components of the centrifugal forces cancel each other. These hammers are used for sheet piles and for driving piles in locations where noise is not desirable. When the ram of a diesel hammer strikes the anvil, a fuel-air mixture is ignited which raises the ram. Hydraulic hammers use pressurized oil to lift the ram.

Some of the pile hammer manufacturers are Delmag, MKT, Vulcan, Mitsubishi, Kobelco, BSP, Junttan, Conmaco, ICE, and Menck. The pile hammer is selected based on the soil characteristics and the pile weight. Two simple but crude rules of thumb in selecting the ram weight are:

a. Ram weight is better be in the range of 25% to 200% of the pile weight, and

b. Ram weight is about 2% of the ultimate load-carrying capacity of the pile.

The pile-driving equipment consists of a *ram* (the falling weight), *anvil* (striker plate), *hammer cushion* (also called cap block or dolly), *helmet* (pile cap), and *pile cushion* (packing). Figure 13.17 shows these components in a single-acting air or steam hammer.

Pile cushion, placed between the pile and the helmet to protect the pile, is made of plywood or similar material. It is used mainly in concrete piles. It is replaced for each pile as it gets damaged. Hammer cushion is placed between the helmet and the anvil, to protect the hammer ram from excessive driving stresses. It is made of durable, heat-resistant materials such as aluminum, micarta, or polymers. The helmet holds the pile in position, aligned under the ram to distribute the impact energy evenly.

Figure 13.18 separates the different impact hammers into groups. Table 13.7 shows the ram weight, drop height, make and the model of some commercially used pile-driving hammers.

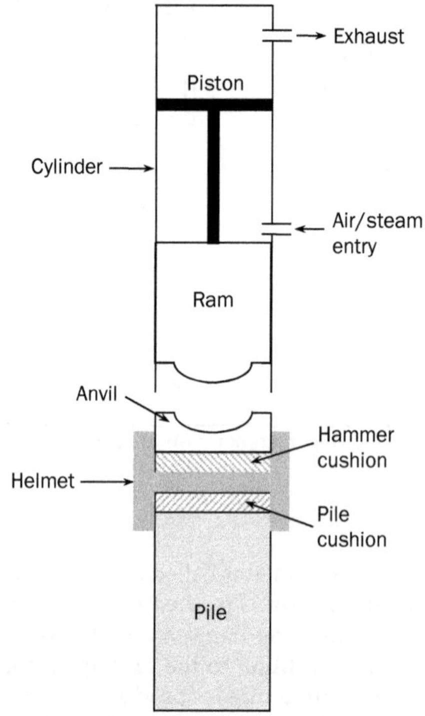

FIGURE 13.17 Components of a single-acting air or steam hammer.

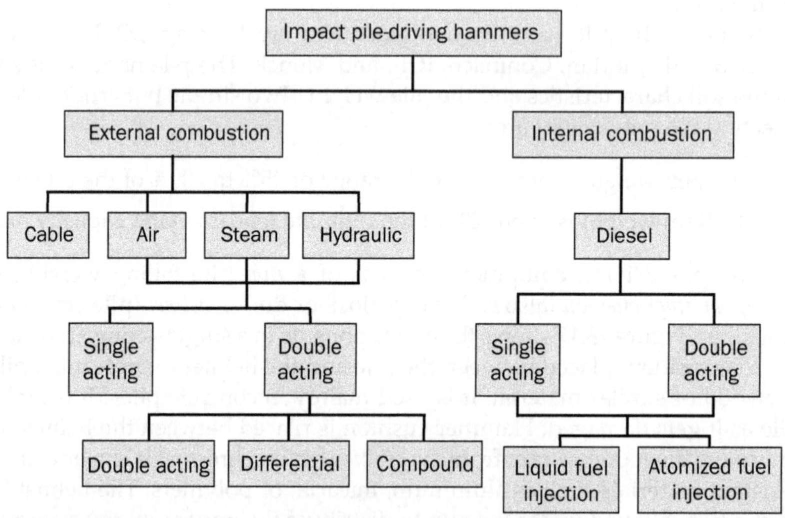

FIGURE 13.18 Hammer categories (adopted from Rausche et al., 1986).

Rated energy		Hammer				Ram weight		Nominal stroke	
kN·m	kip·ft	Make	Model	Type	Blows/min	kN	kip	m	in.
11.9	8.75	MKT	9B3	DA (A/S)	145	7.1	1.6	0.432	17
17.8	13.1	MKT	10B3	DA (A/S)	105	13.3	3.0	0.483	19
20.3	15.0	Vulcan	305	SA (A/S)	60	22.2	5.0	0.914	36
26.0	19.15	MKT	11B3	DA (A/S)	95	22.2	5.0	0.483	19
26.44	19.5	Raymond	65C	DA/Differential (S)	100–110	28.93	6.5	0.406	16
33.2	24.5	Raymond	80C	DA/Differential (S)	95–105	35.6	8.0	0.419	16.5
44.1	32.5	Vulcan	010	SA (A/S)	50	44.5	10.0	0.991	39
44.6	32.9	Vulcan	100C	DA/Differential (A/S)	103	44.5	10.0	0.419	16.5
48.8	36.0	Vulcan	140C	DA/Differential (A/S)	103	62.3	14.0	0.394	15.5
66.1	48.75	Raymond	150C	DA/Differential (A/S)	95–105	66.75	15.0	0.457	18.0
68.1	50.2	Vulcan	200C	DA/Differential (A/S)	98	89.0	20.0	0.394	15.5
77.1	56.9	Raymond	5/0	SA (A/S)	44	77.9	17.5	0.991	39.0
81.3	60.0	Vulcan	320	SA (A/S)	55	89.0	20.0	0.914	36
107.1	79.0	Kobe	K-42	SA Diesel	45–60	41.2	9.26	2.600	102.4
110.2	81.25	Raymond	8/0	SA (A/S)	33–38	111.3	25.0	0.991	39
122.3	90.2	Delmag	D-44	SA Diesel	37–56	42.27	9.50	2.893	113.9
132.0	97.3	Junttan	HHK 9S	Hydraulic	30–100	88.29	19.84	1.500	59.1
135.6	100.0	Vulcan	520	SA (A/S)	42	89.0	20.0	1.524	60
141.3	104.2	Junttan	HHK 12A	Hydraulic	40–100	117.72	26.46	1.200	47.2
153.9	113.5	Vulcan	400C	DA (A/S)	100	178.0	40.0	0.419	16.5
165.0	121.6	Junttan	HHK 14A	Hydraulic	40–100	137.3	30.9	1.200	47.2
203.4	150.0	MKT	DE150	SA Diesel	40–50	66.75	15.0	3.05	120
204	150.0	Vulcan	530	SA (A/S)	42	133.5	30.0	1.524	60

Table 13.7 Specifications of Some Commercial Pile-driving Hammers

13.7.2 Pile-Driving Analyzer

Nowadays, it is common to see a PDA used for monitoring pile driving. It determines the energy delivered to the pile using strain gauges and accelerometers. The technique was originally developed for driven piles, but later extended to cast-in-situ piles as well. The impulse generated by the pile hammer in the case of driven piles, or separately using a drop weight in bored piles, is analyzed for determining the integrity of the pile and to estimate its load-carrying capacity (ASTM D 4945-17). The analysis is carried out using a signal matching program CAPWAP (Case Pile Wave Analysis Program), developed at Case Western Reserve University in the United States. These days, it is possible to have real-time monitoring from remote locations using internet and mobile phones. Figure 13.19 shows a PDA in use.

A hammer blow is applied to the pile head, with sufficient energy to mobilize at least 150% of the design load. The pile head is instrumented with accelerometers and strain gauges, and the data are interpreted using CAPWAP or other wave equation analysis methods. The method is widely used in quality control to check the pile integrity and design verification works. The test gives the load-settlement behavior obtained by a best-fit curve matching estimate. In the case of high capacity piles of large diameters, even the heaviest hammer may not be able to mobilize the required 150% of the design load.

These are mainly for driven piles. In the case of bored piles, it is called a drop hammer test, where a steel mass is dropped from a height to impart a force on the pile head.

13.7.3 Pile-Driving Formulas

Pile-driving formulae can be used to estimate, however crudely, the ultimate load-carrying capacity of a driven pile. When the ram of weight W_R falls over a height of h, the work done is $W_R h$, which is imparted to the pile tip in driving it into the ground. The pile has to reach the ultimate bearing capacity Q_{ult} to penetrate the soil by a distance s, known as set. Here the work done is $Q_{ult}s$. Allowing for energy loss during pile driving and hence with an efficiency of η,

$$W_R h\eta = Q_{ult}s \tag{13.30}$$

The term $W_R h$ can be replaced by the *energy rating of the hammer*. Equation (13.30) is the basis of all pile-driving equations. There are dozens of pile-driving formulas reported

Figure 13.19 Pile-driving analyzer.

in the literature. Some of the popular ones are discussed in this section. Because of the crudeness of the equation, a factor of safety as high as six is generally applied in arriving at the allowable pile load Q_{all}. A hydraulic or diesel hammer has higher efficiency than the air or steam hammer.

ENR Formula

Engineering News Record (ENR) formula is one of the oldest and simplest pile-driving formulas, which was developed for timber piles driven by drop or single-acting hammers. The ultimate pile load is given by (Wellington, 1888)

$$Q_{ult} = \frac{W_R h \eta}{s + c} \tag{13.31}$$

where W_R = ram weight, h = ram drop height, s = set (penetration per blow), and c is a correction factor, which is taken as 2.54 mm for steam hammers and 25.4 mm for drop hammers. With 5–10 blows/minute, drop hammers are very slow in operation and are only used for small jobs. For other types of hammers, c can be taken as 2.54 mm.

The modified ENR formula was proposed later as

$$Q_{ult} = \frac{W_R h \eta}{(s + c)}\left(\frac{W_R + n^2 W_P}{W_R + W_P}\right) \tag{13.32}$$

where W_P is the weight of the pile, and n is the *coefficient of restitution*, which is the measure of the bounciness of the hammer, typically in the range of 0.25 (timber) to 0.5 (steel). It is the ratio of the relative velocity between the pile and the hammer after and before the impact. For a perfectly elastic blow, $n = 1$; for a perfectly inelastic blow $n = 0$. The constant c is taken as 2.54 mm. η is the efficiency of the pile-driving setup, which is in the range of 0.75–1.0.

Hiley Formula

Hiley formula (Olson and Flaate, 1967), widely used in Europe, is similar to the modified ENR formula [Eq. (13.32)], where c is replaced by $(C_1 + C_2 + C_3)/2$. Here,

- C_1 = elastic compression of the pile cushion and helmet, typically in the range 1–5 mm.
- C_2 = temporary elastic compression of the pile $Q_{ult} l/AE$, which can be taken as 0.2–1.0 times the pile length l. Here, l is in m and C_2 in mm.
- C_3 = elastic compression of the soil surrounding the pile and that lies below the pile, typically in the range 1–6 mm.

Navy-McKay Formula

$$Q_{ult} = \frac{\eta \times \text{Energy rating}}{s\left(1 + 0.3\dfrac{W_P}{W_R}\right)} \tag{13.33}$$

Danish Formula (Olson and Flaate, 1967)

$$Q_{\text{ult}} = \frac{\eta \times \text{Energy rating}}{s + \sqrt{\dfrac{\eta \times \text{Energy rating} \times l}{2A_pE_p}}} \tag{13.34}$$

where η = efficiency, A_p = cross-sectional area of the pile, E_p = Young's modulus of the pile, and l = pile length.

Gates Formula (Gates, 1957)

$$Q_{\text{ult}} = 104.5\sqrt{\eta \times \text{Energy rating}}\,(2.4 - \log s) \tag{13.35}$$

where the energy rating is in kN.m, set s is in mm, and Q_{ult} is in kN. It is suggested that the efficiency $\eta = 0.75$ for drop hammers, and 0.85 for all others. Equation (13.35) is strictly empirical, based on a statistical analysis of about 100 load tests.

There is a widespread understanding among geotechnical engineers that pile-driving formulas are very rough and unreliable. For Hiley and Gates formulas, which are more reliable than the others, a lower factor of safety in the order of three can be used. For all others, factor of safety of six is suggested.

Example 13.6

A Delmag D-22 diesel hammer with the maximum energy rating of 53.8 kN·m, efficiency of 0.9, and ram weight of 21.6 kN is used to drive a 300-mm-diameter and 20-m-long concrete pile into the ground. What should be the blows per 100 mm, to ensure that the pile can carry 400 kN, based on (a) Danish formula, and (b) Navy-McKay formula? $E_{\text{concrete}} = 25$ GPa; $\gamma_{\text{concrete}} = 24$ kN/m^3. Assume factor of safety of 6 for both methods.

Solution

Part a: Danish formula

Cross-sectional area of the pile $A_p = (\pi/4)(0.3^2) = 0.0707$ m^2

Allowing factor of safety of 6, $Q_{\text{ult}} = 6 \times 400 = 2400$ kN

$$Q_{\text{ult}} = \frac{\eta \times \text{Energy rating}}{s + \sqrt{\dfrac{\eta \times \text{Energy rating} \times l}{2A_pE_p}}}$$

$$2400 = \frac{0.9 \times 53.8}{s + \sqrt{\dfrac{0.9 \times 53.8 \times 20}{2 \times 0.0707 \times 25 \times 10^6}}}$$

$$s = 0.0036 \text{ m or } 3.6 \text{ mm}$$

Blows required for 100 mm penetration = 100/3.6 = **28**

Part b: Navy-McKay formula

Weight of the pile $W_p = 0.0707 \times 20 \times 24 = 33.9$ kN

$$Q_{\text{ult}} = \frac{\eta \times \text{Energy rating}}{s\left(1 + 0.3\dfrac{W_P}{W_R}\right)}$$

$$2400 = \frac{0.9 \times 53.8}{s\left(1 + 0.3\frac{33.9}{21.6}\right)}$$

$$s = 0.0137 \text{ m} = 13.7 \text{ mm}$$

Number of blows required for 100 mm penetration = 100/13.7 = **7.3** which is significantly lower than the 28 blows estimated in part a. That is the nature of the pile-driving formulas.

13.8 Pile Load Test

Proof of the pudding is in the eating. This is true for pile design too. With all the approximations and assumptions we have made in "determining" the static load-carrying capacity of the pile, how realistic is the calculated pile load capacity? The best way to assess our design methods is to carry out a pile load test (ASTM D 1143-20) and study the performance of the pile. In a pile load test, the pile is loaded to well beyond the design load and the load versus settlement plot is developed. From the plot, the ultimate pile load Q_{ult} can be determined. The general understanding is that the ultimate load is reached when the tip (not the head) has settled 10% of the pile diameter.

13.8.1 Net and Gross Settlements

When a pile is loaded, the pile itself compresses and undergoes an elastic shortening s_e (or Δl) which can be determined as

$$s_e = \Delta l = \frac{Q_{z,avg}\, l}{A_p E_p} \tag{11.36}$$

where $Q_{z,avg}$ = average pile load at the cross section, A_p = cross-sectional area of the pile, and E_p = Young's modulus of the pile. Due to the skin friction, Q_z varies with depth and $Q_{z,avg}$ can be significantly less than the load Q applied at the pile head. Due to the elastic shortening of the pile, the pile head settles more than the pile tip, and the settlements can be related by

$$s_{head} = s_{tip} + s_e \tag{13.37}$$

where s_{tip} is the settlement of the pile tip, known as the *net settlement* or s_{net}, and s_{head} is the settlement of the pile head, known as the *gross settlement* or s_{gross}. For an incompressible pile ($E_p = \infty$), there is no elastic shortening and hence the net and gross settlements are the same.

13.8.2 Pile Load Test and Load-Settlement Plot

Figure 13.20 shows the schematic diagrams for the two common pile load test arrangements. The main difference between the two is the way the horizontal reaction beam is held in place. In Fig. 13.20a, a *kentledge*, consisting of heavy weights, is used to hold the reaction beam in place and the hydraulic jack is jacked against the beam to apply the pile load. Kentledges can be used to apply loads as high as 5000 kN.

In Fig. 13.20b, the horizontal reaction beam is held in place by two reaction piles located far away. The loads are applied in increments as specified by the relevant codes

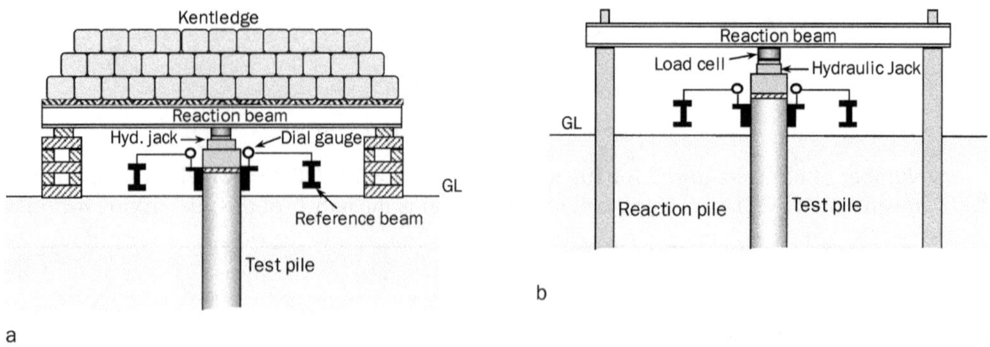

FIGURE 13.20 Pile load test: (a) using kentledge, (b) using reaction beam.

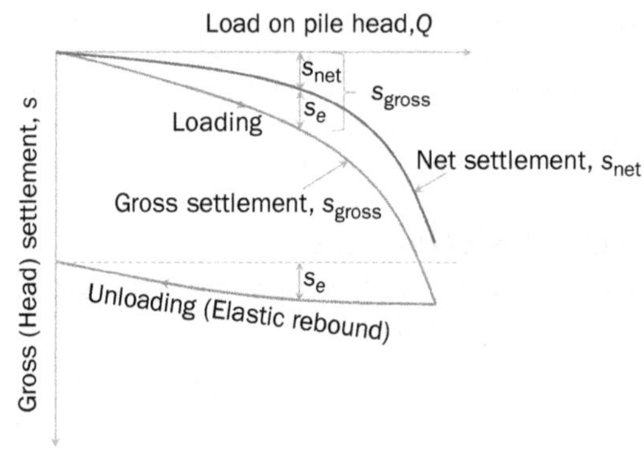

FIGURE 13.21 Load-settlement plot.

or standards, with sufficient time between the increments. The piles are generally loaded well beyond the working loads (e.g., twice the working load), and upon reaching the maximum load, the pile is unloaded in steps.

Static pile load tests shown in Fig. 13.20 are expensive to carry out. In a large project, about 1% of the piles, selected randomly, are subjected to static pile load tests. Up to 10% of the piles are tested using a PDA, which is less reliable but costs significantly less.

During a pile load test, dial gauges or other displacement measurement devices are placed on top of the pile to measure the settlement of the pile head s_{head} (or s_{gross}). When the maximum load is reached, the pile is unloaded where the elastic shortening s_e is fully recovered. The pile head settlement is plotted against the applied load as shown in Fig. 13.21, which also defines the elastic rebound (settlement) s_e. At any load, subtracting the elastic settlement from gross settlement gives the net settlement, as shown in the figure. Therefore, it is possible to derive the load versus *net* settlement plot from the pile load test data where only gross settlements are recorded.

13.8.3 Determining the Ultimate Load Q_{ult}

Now that we have the two load-settlement plots, we will use them to derive the ultimate pile load Q_{ult}. Should this be based on the gross settlement or net settlement? The full capacity of the pile load is reached when the *tip* has penetrated about 10% of pile diameter. Therefore, it appears sensible to use net (tip) settlement in determining the pile load. Some of the methods reported in the literature are discussed below.

Some of the codes define the ultimate pile load as the one that corresponds to gross (head) settlement of 10% of the pile diameter, 25 mm, or $\frac{Ql}{A_pE_p} + \frac{d}{30}$ mm. DeBeer (1968) suggested plotting both gross settlement and load on logarithmic scale, where the intersection of the two straight line segments defines the ultimate pile load.

Davisson (1973) proposed that the ultimate pile load can be taken as the one that corresponds to a gross settlement of $4.0\,(\text{mm}) + \frac{Q_{\text{ult}}l}{A_pE_p}(\text{mm}) + \frac{d(\text{mm})}{120}$. This method is widely used in practice. The procedure to determine the ultimate pile load from the load versus gross settlement plot is shown in Fig. 13.22. Firstly, $4.0 + d(\text{mm})/120$ is marked on the settlement axis, defining the point A. From the point A, a straight line with a slope of $1(\text{V}){:}A_pE_p/l(\text{H})$ is drawn, which meets the load versus gross settlement plot at B. The load at point B is the ultimate pile load Q_{ult}.

US Army (1993) procedure is based on net settlements. It is suggested to determine the ultimate pile load by the following three methods and to average them.

a. Load corresponding to net settlement of 6 mm (0.25 in.).
b. Load at the intersection of the two tangents drawn at the start and end of the net settlement plot.
c. Load when the slope of the load versus net settlement plot becomes 0.0254 mm/kN (0.01 in./ton).

13.8.4 Osterberg Cell® (O-Cell) Test

O-Cell, also known as bidirectional test for its ability to apply loads both upward and downward simultaneously, was developed by Professor Jorj Osterberg in 1980s. It works mainly with bored piles where, a hydraulically driven sacrificial pressure cell is built within the pile, often near the bottom. The cell is jacked against the pile so that the shaft is

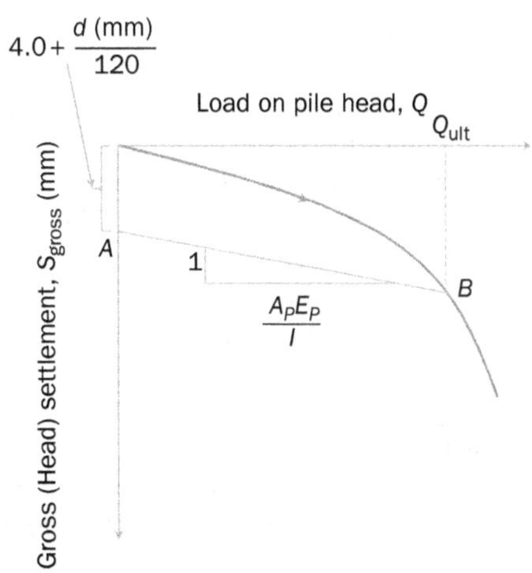

FIGURE 13.22 Davisson's method.

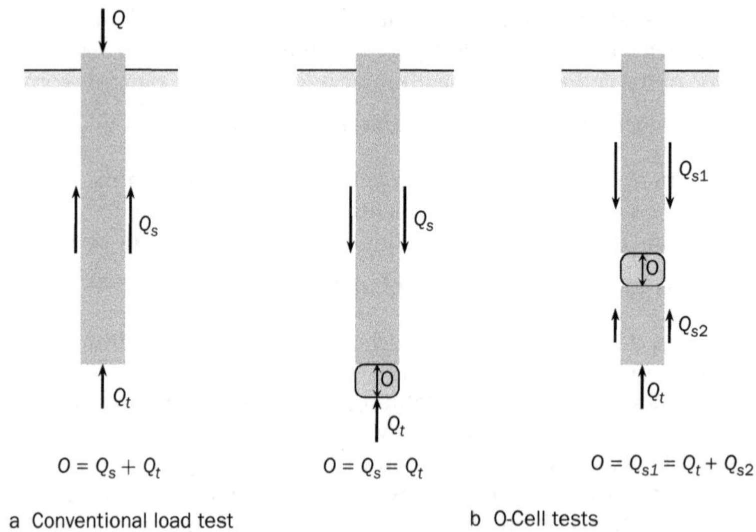

FIGURE 13.23 O-Cell principle.

pushed upward and the toe is pushed downward, allowing bidirectional measurements and hence separation of the shaft and end-bearing loads as shown in Fig. 13.23. The loading continues till the toe or the shaft reaches the capacity. An attractive feature of O-Cell is that it does not require a kentledge or reaction piles as it is jacked against the pile itself. O-Cells have been used to measure pile loads in excess of 200 MN. Often O-Cells are installed at two levels to better define the shaft and the base capacities of the pile.

13.8.5 Statnamic™ Test

The word Statnamic comes from static and dynamic, as the method attempts to exploit the advantages of static and dynamic pile load tests. It was developed in the late 1980s in a collaboration between Canada and Netherlands, and has been widely used since the early 1990s. It is a rapid compressive pile load test, where the load test takes place over a fraction of a second.

The combustion of a fast burning fuel within a chamber propels a reaction mass upward at a very high acceleration in the order of 20 g. This in turn applies a downward force of 20 times the reaction mass on the pile head, over a short duration of 120–200 milliseconds. Within this short time, the load is applied in a linearly increasing manner, followed by unloading which is achieved by controlled venting of the pressure. The load and the pile head displacement are measured at frequent intervals during this time, thus developing the load-settlement plot.

13.9 Settlement of a Pile

Settlements of pile foundations are relatively small compared to the shallow foundations. Settlements under working loads (i.e., $F \approx 2.5$) can be crudely estimated as 1% of the pile diameter.

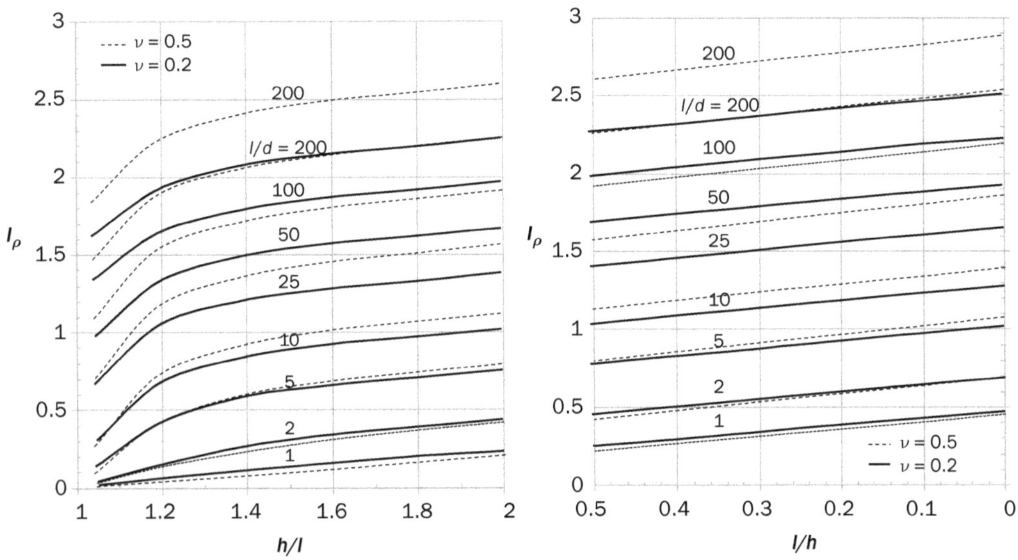

FIGURE 13.24 Influence factor I_ρ for incompressible piles.

13.9.1 Poulos and Davis Method

A simple approach based on the elastic solutions proposed by Poulos and Davis (1974) is presented here. For *incompressible piles*, where the net and gross settlements are the same, the settlement is given by

$$\text{Settlement} = \frac{Q}{lE_s} I_\rho \tag{13.38}$$

where Q = pile load on the head, l = pile length, d = pile diameter, h = depth of the soil bed from the ground level to the stiff stratum, E_s = Young's modulus of the soil, and $I\rho$ = influence factor from Fig. 13.24.

To account for the compressibility of the pile, the elastic shortening s_e can be computed using the average pile load Q_{avg} on the cross section, and hence

$$s_{gross} = \frac{Q}{lE_s} I_\rho + s_e \tag{13.39}$$

where $s_e = Q_{avg} l/(A_p E_p)$. The average pile load can be taken as the average of the pile load at the head and the tip.

Example 13.7

Determine the gross settlement of a 500-mm-diameter and 25-m-long concrete pile, subjected to a 1200 kN vertical load on its head. Assume it is a friction pile where the load carried by the tip is negligible. The bedrock is at a depth of 50 m. Young's modulus of the soil and concrete are 25 MPa and 30 GPa, respectively. Poisson's ratio of the soil = 0.2.

Solution

$$l/d = 25/0.5 = 50$$

$$l/h = 25/50 = 0.5$$

From Fig. 13.24, $I_\rho = 1.7$.

Average load on the pile cross section = (1200 + 0)/2 = 600 kN

$$s_{\text{gross}} = \frac{Q}{lE_s} I_\rho + s_e$$

$$s_{\text{gross}} = \frac{1200 \times 10^3}{25 \times (25 \times 10^6)}(1.7) + \frac{600 \times 10^3 \times 25}{\left(\frac{\pi}{4} \times 0.5^2\right)(30 \times 10^9)} \text{ m}$$

$$= 0.00326 + 0.00255 \text{ m} = 0.00581 \text{ m} = \mathbf{5.81\ mm}$$

13.9.2 Vesic's Method

Vesic (1977) proposed elastic solutions to determine the settlement of the pile head. He separated the gross (head) settlement into three components:

a. Elastic shortening of the pile s_1;
b. Settlement s_2 due to the working load at the tip (Q_{wt}); and
c. Settlement s_3 due to the working load along the pile shaft (Q_{ws}).

Therefore,

$$s_{\text{gross}} = s_1 + s_2 + s_3 \tag{13.40}$$

Settlement s_1 due to Elastic Shortening

The pile load at the cross section Q_z varies along the pile, and the variation depends on the skin friction. It can be shown analytically that the elastic shortening of the pile is given by

$$s_1 = \frac{(Q_{\text{wt}} + \xi Q_{\text{ws}})l}{A_p E_p} \tag{13.41}$$

where ξ = a constant in the range of 0.50–0.67, which depends on the skin friction distribution; for situations where the skin friction is uniform or parabolic, $\xi = 1/2$, and for situation where the skin friction increases or decreases linearly $\xi = 2/3$ (see Fig. 13.6).

Q_{wt} = working load at the pile tip,
Q_{ws} = working load along the pile shaft,
A_p = cross sectional area of the pile tip,
E_p = Young's modulus of the pile, and
l = pile length.

Soil type	Driven piles	Bored piles
Sand (dense to loose)	0.02–0.04	0.09–0.18
Clay (stiff to soft)	0.02–0.03	0.03–0.06
Silt (Dense to loose)	0.03–0.05	0.09–0.12

Table 13.8 C_t Values for Driven and Bored Piles

Settlement s_2 due to Q_{wt}

Based on elastic solutions, Vesic suggested that the settlement due to the tip load is given by

$$s_2 = \frac{Q_{\mathrm{wt}} d}{A_p E_s}\left(1-\nu^2\right) I_{\mathrm{wt}} \tag{13.42}$$

where E_s and ν are the Young's modulus and Poisson's ratio of the soil, respectively. I_{wt} is an influence factor which can be taken as 0.85. Vesic (1977) also proposed an empirical equation to determine s_2 as

$$s_2 = \frac{Q_{\mathrm{wt}} C_t}{d q_{t,\mathrm{ult}}} \tag{13.43}$$

where $q_{\mathrm{t,ult}}$ is the ultimate bearing capacity at the pile tip, and C_t is an empirical coefficient, given in Table 13.8.

Settlement s_3 due to Q_{ws}

Settlement due to the shaft load can be computed as (Vesic, 1977)

$$s_3 = \left(\frac{Q_{\mathrm{ws}}}{pl}\right)\frac{d}{E_s}\left(1-\nu^2\right) I_{\mathrm{ws}} \tag{13.44}$$

where p is the pile perimeter, and I_{ws} is an influence factor given by

$$I_{\mathrm{ws}} = 2 + 0.35\sqrt{\frac{l}{d}} \tag{13.45}$$

Vesic (1977) also suggested an alternate empirical equation to determine s_3 as

$$s_3 = \frac{Q_{\mathrm{ws}} C_s}{l q_{t,\mathrm{ult}}} \tag{13.46}$$

where the empirical coefficient C_s is given by

$$C_s = \left(0.96 + 0.16\sqrt{\frac{l}{d}}\right) C_t \tag{13.47}$$

Knowing that the shaft load is mobilized well before the tip load (see Fig. 13.7), it requires some good judgment to separate the working loads between the shaft and the tip.

Example 13.8

In a concrete pile installed in a soil, the elastic shortening of the pile is given by (Vesic, 1977)

$$s_1 = \frac{(Q_{wt} + Q_{ws})l}{A_p E_p}$$

Show from the first principles that when the skin friction is uniform, $\xi = 1/2$.

Solution In Fig. 13.25a, it is assumed that out of the working load Q_w acting on the pile head, fraction kQ_w is carried by the shaft and $(1 - k)Q_w$ is carried by the pile tip.

Let us consider a disc element of thickness dz at depth z below the ground (Fig. 13.25b). Let us find the pile load Q_z acting at this depth. For the pile segment above this depth (Fig. 13.24c), the shaft load is $k(z/l)Q_w$. Therefore, the pile load Q_z at depth z is

$$Q_z = Q_w - \frac{kz}{l}Q_w = Q_w\left(1 - \frac{kz}{l}\right)$$

With this load acting on the disc element, the elastic shortening of the disc element is

$$ds_1 = \frac{Q_z dz}{A_p E_p} = \frac{Q_w}{A_p E_p}\left(1 - \frac{kz}{l}\right)dz$$

Therefore, the elastic compression of the pile is

$$s_1 = \int_{z=0}^{z=l} \frac{Q_w}{A_p E_p}\left(1 - \frac{kz}{l}\right)dz = \frac{Q_w}{A_p E_p}\int_{z=0}^{z=l}\left(1 - \frac{kz}{l}\right)dz$$

$$s_1 = \frac{Q_w}{A_p E_p}\left(1 - \frac{k}{2}\right)l$$

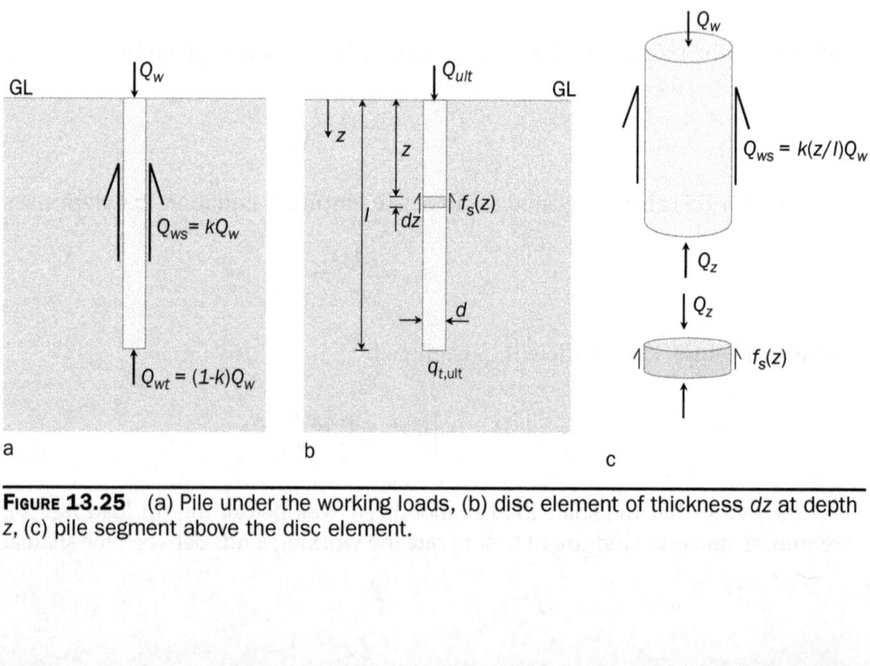

FIGURE 13.25 (a) Pile under the working loads, (b) disc element of thickness dz at depth z, (c) pile segment above the disc element.

$$s_1 = \frac{(Q_{wt} + \xi Q_{ws})l}{A_p E_p} = \frac{((1-k)Q_w + \xi k Q_w)l}{A_p E_p}$$

Equating the two expressions for s_1,

$$\frac{Q_w}{A_p E_p}\left(1 - \frac{k}{2}\right)l = \frac{((1-k)Q_w + \xi k Q_w)l}{A_p E_p}$$

$$\left(1 - \frac{k}{2}\right) = (1-k) + \xi k$$

$$\boldsymbol{\xi = 1/2}$$

Example 13.9

Redo Example 13.7 using Vesic's method, assuming that the working load of 1200 kN on the pile head is distributed between the shaft and the tip as $Q_{ws} = 900$ kN and $Q_{wt} = 300$ kN, and the skin friction remains the same at all depths.

Solution

$$s_1 = \frac{(Q_{wt} + \xi Q_{ws})l}{A_p E_p} = \frac{(300\times10^3 + 0.5\times900\times10^3)(25)}{\left(\frac{\pi}{4}\times0.5^2\right)(30\times10^9)} = 0.00318\,\text{m}$$

$$s_2 = \frac{Q_{wt}d}{A_p E_s}(1 -^{2})I_{wt} = \frac{300\times10^3(0.5)}{\left(\frac{\pi}{4}\times0.5^2\right)(25\times10^6)}(1-0.2^2)(0.85) = 0.0249\text{ m}$$

$$I_{ws} = 2 + 0.35\sqrt{\frac{l}{d}} = 2 + 0.35\sqrt{\frac{25}{0.5}} = 4.47$$

$$s_3 = \left(\frac{Q_{ws}}{pl}\right)\frac{d}{E_s}(1-\nu^2)I_{ws} = \left(\frac{900\times10^3}{\pi\times0.5\times25}\right)\frac{0.5}{25\times10^6}(1-0.2^2)(4.47) = 0.00197\text{ m}$$

$$s_1 + s_2 + s_3 = 3.18 + 24.9 + 1.97\text{ mm} = \mathbf{30.0\ mm}$$

Here, it appears that s_2 is too high. The empirical equation 13.43 may give a more realistic estimate of s_2.

13.10 Pile Groups

When the loads are very large, instead of providing a very large diameter pile, it is possible to use several small diameter piles, as a *pile group*. All the piles are connected by a rigid reinforced concrete pile cap. Figure 13.26a shows a 2 × 9 pile group, consisting of

FIGURE 13.26 (a) A 2 × 9 pile group, (b) pile cap for a 2 × 2 pile group.

18 piles driven in two rows. This particular pile group is used to support a heavy bridge abutment. The pile segments above the cap will be sawn off. Figure 13.26b shows the reinforcement cage for the pile cap of a smaller 4 × 4 pile group.

When the piles are installed too close, their influence zones overlap, which reduces the load-carrying capacity of the pile group. Piles are generally placed at 2.5d–3.5d center-to-center spacing both ways, where d is the pile diameter. The efficiency of a pile group η is defined as

$$\eta = \frac{\text{Ultimate load carrying capacity of the pile group}}{\text{No. of piles in the group} \times \text{Ultimate load carrying capacity of a single pile}} \tag{13.48}$$

Determining the pile group efficiency is a difficult task. In loose sands, pile driving densifies the sand which increases the efficiency to greater than 1. In dense sands, the pile driving can loosen the sand and therefore the efficiency of the group is less than 1. Generally, η is taken as 1 in driven pile groups, and reduced slightly for bored pile groups where it can be as low as 2/3.

13.10.1 Efficiency of a Friction Pile Group

Let us consider a friction pile group, where the load carried by the bottom of the pile group is negligible compared to the load carried by the four sides. There are m piles in one direction, at center-to-center spacing of s, and n piles in the perpendicular direction at the same spacing. The pile diameter is d. Let B and L be the width and length of the pile group. Therefore,

$$B = (m - 1)s + d$$
$$L = (n - 1)s + d$$

Let us assume a constant skin friction of f_s.

Considering the pile group, where the piles and the soil between the piles act as a block, the ultimate load-carrying capacity of the pile group is given by

$$Q_{\text{ult,group}} = 2(B+L)lf_s = \left[2(m+n-2)s+4d\right]lf_s$$

The ultimate load-carrying capacity of a single pile is

$$Q_{\text{ult,pile}} = \pi d l f_s$$

Therefore, the efficiency can be determined as

$$\eta = \frac{\left[2(m+n-2)s+4d\right]lf_s}{mn \times \pi d l f_s} = \frac{2(m+n-2)s+4d}{\pi dmn} \tag{13.49}$$

One of the earliest equations used to calculate the group efficiency of a friction pile group was proposed by Converse-Labarre (Bolin, 1941) as

$$\eta = 1 - \frac{\varepsilon}{90}\left(\frac{(m-1)n+(n-1)m}{mn}\right) \tag{13.50}$$

where $\varepsilon = \tan^{-1}(d/s)$ expressed in degrees, and s is the center-to-center spacing of the piles.

According to Los Angeles Group Action equation,

$$\eta = 1 - \frac{d}{\pi smn}\left[m(n-1)+n(m-1)+\sqrt{2}(m-1)(n-1)\right] \tag{13.51}$$

13.10.2 Ultimate Load-Carrying Capacity of a Pile Group in Clay

In a pile group in clays, the piles and the clay between the piles can act as a single monolithic block. For pile groups in clays, it is suggested to compute the ultimate load-carrying capacity of the group in two methods discussed below using α-method, and to use the lower value.

Method 1: As a Collection of Single Piles

The ultimate load-carrying capacity of a single pile is

$$Q_{\text{ult,pile}} = A_t 9c_{u,\text{base}} + \pi dl(\alpha c_{u,\text{side}})$$

where α = adhesion factor, $c_{\text{u,base}}$ = undrained cohesion at the pile tip, and $c_{u,\text{side}}$ = average undrained cohesion along the pile shaft. Here, the bearing capacity factor N_c is taken as 9. Assuming $\eta = 1$, the load-carrying capacity of the pile group is computed as

$$Q_{\text{ult,group}} = mnQ_{\text{ult,pile}} = mn\left[(A_t)9c_{u,\text{base}} + \pi dl(\alpha c_{u,\text{side}})\right] \tag{13.52}$$

Here, m and n are the number of piles in the two perpendicular directions within the pile group.

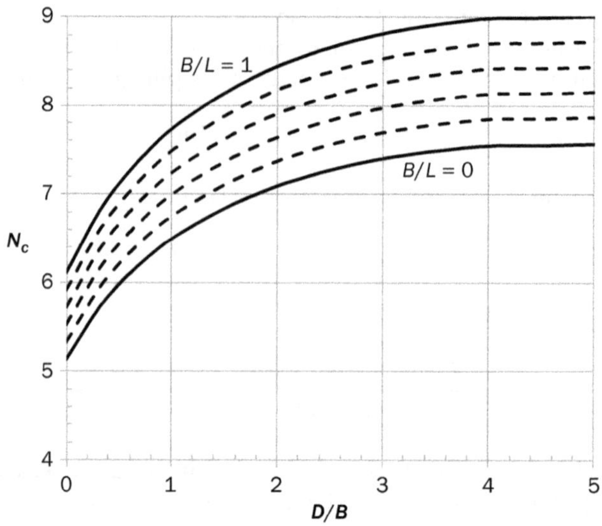

FIGURE 13.27 Bearing capacity factor N_c for a pile group in clay.

Method 2: Block Theory

Here, the piles and the clay within the pile group are treated as a block, for which the ultimate load-carrying capacity is computed. Let the breadth, length, and depth of the block be B, L, and D, respectively.

$$Q_{\text{ult,group}} = Q_{\text{ult,base}} + Q_{\text{ult,side}}$$

$$Q_{\text{ult,group}} = BLc_{u,\text{base}}N_c + 2(B + L)Dc_{u,\text{side}} \tag{13.53}$$

Here, $c_{u,\text{base}}$ and $c_{u,\text{side}}$ are the undrained cohesion at the base and the sides, respectively. The N_c values for pile groups, as suggested by Skempton (1951), are shown in Fig. 13.27.

The lower value calculated for $Q_{\text{ult,group}}$ should be used.

Example 13.10

A pile group consists of 3 × 4 square piles of 300 mm width and 20 m length, at 800 mm center-to-center spacing. The pile group is placed in a homogeneous saturated clay deposit where $c_u = 25$ kN/m². What is the ultimate load-carrying capacity of the pile group?

Solution

Method 1: Considering Single Piles

$$\alpha = 0.21 + 0.26\frac{p_a}{c_u} < 1$$

$$\alpha = 0.21 + 0.26\frac{101.3}{25} = 1.0$$

$$Q_{\text{ult,group}} = mn\left[(A_t)9c_{u,\text{base}} + \pi dl(\alpha c_{u,\text{side}})\right]$$

$$Q_{\text{ult,group}} = (3 \times 4)\left[(0.3 \times 0.3)(9)(25.0) + (\pi \times 0.3)(20)(1 \times 25.0)\right]$$

$$Q_{\text{ult,group}} = 5898 \text{ kN}$$

Method 2: Block Theory

$$B = 2 \times 0.8 + 0.3 = 1.9 \text{ m}$$

$$L = 3 \times 0.8 + 0.3 = 2.7 \text{ m}$$

$$D = l = 20 \text{ m}$$

For $D/B = 20/1.9 = 10.5$ and $B/L = 1.9/2.7 = 0.7$, from Fig. 13.27, $N_c = 8.7$ (Extrapolated)

$$Q_{\text{ult,group}} = BLc_{u,\text{base}}N_c + 2(B + L)Dc_{u,\text{side}}$$

$$Q_{\text{ult,group}} = (1.9)(2.7)(25)(8.7) + 2(1.9 + 2.7)(20)(25) = 5716 \text{ kN}$$

Let us take the lower of the two values and hence $Q_{\text{ult,group}} = \mathbf{5716\ kN}$.

13.10.3 Settlements of Pile Groups

A simplified approach is suggested in the literature for computing settlements of pile groups. For a friction pile group shown in Fig. 13.28a, it is assumed that the pile cap with the load is treated as an equivalent raft placed at $\frac{2}{3}l$ depth. From here, the load is assumed to spread in 2V:1H distribution, and the equivalent raft is treated as a shallow foundation in computing settlements using the methods discussed in Chap. 12. For the end bearing pile group shown in Fig. 13.28b, the equivalent raft is at the bottom of the pile group, located in the firm ground. From here, the load is spread in 2V:1H distribution.

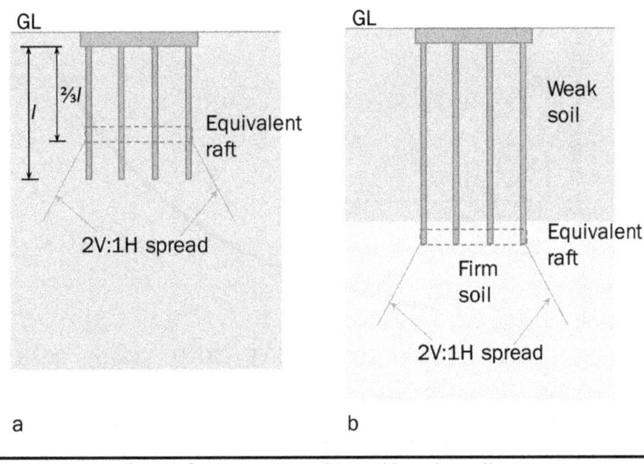

Figure 13.28 Equivalent raft: (a) friction piles, (b) end-bearing piles.

13.11 Foundations for Super-Tall Buildings

Figure 13.29 suggests that the number of tall buildings in excess of 300 m in height are increasing worldwide at a fast rate. Such buildings, known as *super-tall buildings*, have some special challenges that were not present for the traditional tall buildings. They are (Poulos, 2017):

- Building weight increasing rapidly and nonlinearly with height.
- Large lateral wind loads inducing high moments.
- Cyclic nature of the wind loads can degrade the ultimate bearing capacity and increase the settlements.
- The main high-rise structure surrounded by low-rise podium structure can induce differential settlements.
- Consideration of seismic loads and possibility of resonance.

13.11.1 Piled Rafts

Piled rafts are relatively new in foundation engineering where we have seen piles and rafts separately. For super-tall building foundations, piled-rafts are becoming very popular. The lateral extent of the raft is so large, there is never a problem with bearing capacity. However, differential settlements are often a problem with rafts. Piles are introduced to reduce the settlements. In a piled raft, the piles provide the majority of the foundation stiffness while the raft contributes toward the bearing capacity. Rafts alone will satisfy bearing capacity considerations, but may undergo very high differential settlements. By providing piles of different lengths and diameters as necessary, at strategic locations under the raft, these settlements can be controlled. In addition, the raft thickness can also be varied across the raft.

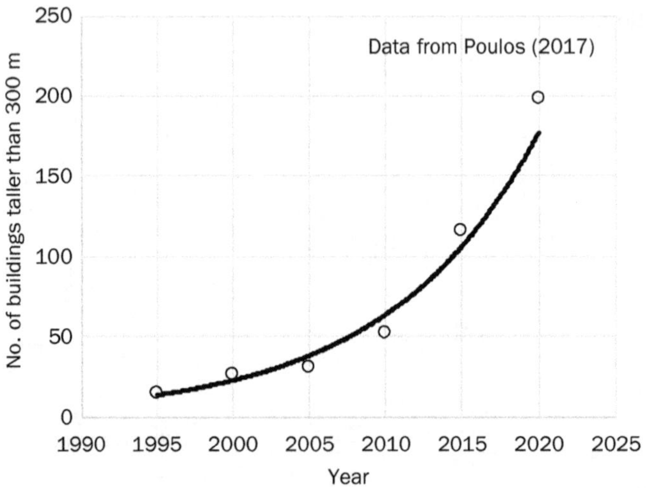

FIGURE 13.29 Number of super-tall buildings in the world.

13.11.2 Barrettes

Barrette foundations have become increasingly popular worldwide since the 1980s for carrying very large column loads in super-tall buildings. A barrette is a cast-in-place reinforced concrete column, generally rectangular in cross section, having widths of 0.5–1.5 m, lengths of 1.5–7.0 m, and depths as high as 100 m. The construction of a barrette is similar to that of a diaphragm wall. A trench is excavated, supported by bentonite or polymer slurry, using a clamshell grab-bucket or a hydrofraise that acts similar to a chainsaw cutting into the ground. Then, the reinforcement cage is placed, and the trench is filled by concrete replacing the slurry. Barrettes can be constructed in various shapes such as I, H, T, L, X, T, or Y.

Rectangular barrettes have higher circumference and hence higher specific surface compared to circular shafts of the same cross-sectional area. This increases the load-carrying capacity along the shaft. With large cross sections, they can carry large loads both vertically and horizontally. They can carry vertical loads of 5000–50,000 kN. Their strong axis can be aligned in the wind direction to resist the high wind loads expected in tall buildings.

The same equipment used for diaphragm walls can be used for barrettes, thus avoiding mobilization of additional machinery for the foundation work. Design methods of barrettes are similar to those of drilled shafts.

13.12 Rock-Socketed Piles

Rock-socketed piles or shafts are the preferred option when there is possibility of transferring the loads from the superstructure to the competent bearing stratum at a certain depth beneath the ground level. They also limit the settlements. They transfer the load to the sockets through the side shear resistance and end bearing resistance. The ultimate load-carrying capacity of the rock socket can be determined as

$$Q_{\text{ult}} = Q_{\text{side,ult}} + Q_{\text{base,ult}} \tag{13.54}$$

where $Q_{\text{side,ult}}$ and $Q_{\text{base,ult}}$ are the ultimate load-carrying capacities of the side and the base, respectively, of the rock socket. They can be determined as

$$Q_{\text{side,ult}} = \pi B D_s f_s \tag{13.55}$$

$$Q_{\text{base,ult}} = \pi \frac{B^2}{4} q_b \tag{13.56}$$

where B = shaft diameter, D_s = socket length, f_s = ultimate unit side friction, and q_b = ultimate end bearing capacity (see Fig. 13.30).

In its simplest form, the ultimate side friction f_s is expressed generally as

$$\frac{f_s}{p_a} = a\left(\frac{\sigma_c}{p_a}\right)^b \tag{13.57}$$

where a and b are constants, σ_c = unconfined compressive strength (UCS) of the *intact rock*, and p_a = atmospheric pressure (101.3 kN/m^2). A range of values are reported in the

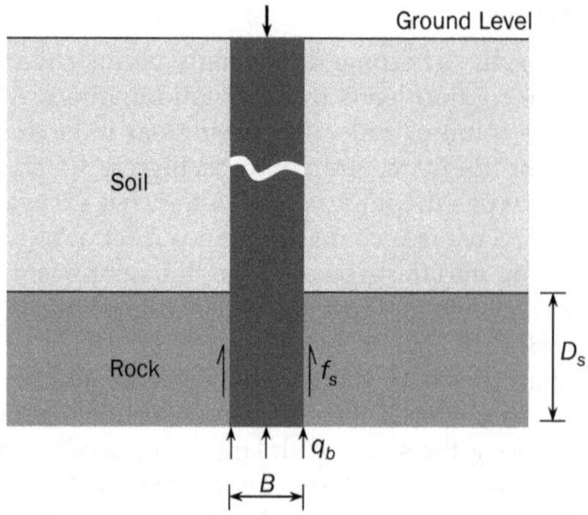

FIGURE 13.30 Schematic diagram of a rock-socketed pile.

literature for a and b. The Canadian Foundation Engineering Manual (CGS, 1992) suggests $a = 1.42$ and $b = 0.5$. Carter and Kulhawy (1988) also suggested that in the absence of load test data, in situ tests or local experience, f_s should be computed as

$$f_s = 0.15\sigma_c \tag{13.58}$$

Equation (13.58) is generally preferred for situations where σ_c/p_a is less than 20. Further, f_s should not exceed 5% of the compressive strength of concrete f_c', to avoid bond failure at the interface, unless there is enough data to justify higher values.

The end bearing of socketed shafts is less understood than the side shear resistance. In its simplest form, the ultimate end bearing capacity of the rock socket can be written as

$$q_b = N_{cr}\,\sigma_c \tag{13.59}$$

where N_{cr} is the bearing capacity factor for rock. While values for N_{cr} in the range of 2.7–8.0 have been suggested in the literature, a lower bound value of 2.5 is suggested.

Some researchers have suggested that in good-quality rock, the ultimate end bearing capacity of the pile base q_b and σ_c are related by

$$q_b(\text{MPa}) = c\sigma_c^d \tag{13.60}$$

where σ_c is the UCS in MPa. Some values reported in the literature for c and d are summarized in Table 13.9.

Authors	c	d	Comments
Zhang and Einstein (1998)	4.83	0.51	39 pile load tests
Vipulanandan et al. (2007)	4.66	0.56	21 pile load tests
Zhang (2008)	4.93	0.5	50 pile load tests
Rowe and Armitage (1987)	2.7	1	
Coates (1967)	3	1	
ARGEMA (1992)	4.5	1	$q_b \leq 10$ MPa

TABLE 13.9 Values of c and d for Eq. (13.60)

13.13 Main Points

1. Piles are effective in carrying compressive, tensile, and lateral loads.
2. The shaft load is fully mobilized well before the tip load.
3. The load-carrying capacity of a cast-in-place pile is about 50% of that of a driven pile of same size.
4. For piles in clays under undrained conditions, $N_c = 9$.
5. The load-carrying capacity of piles estimated from pile-driving formulas are very approximate.
6. Due to elastic shortening of piles, the head settles more than the tip.
7. Treat the pile group as an equivalent raft and hence a shallow foundation in determining settlement of the pile group.

Review Exercises

13.1 Surf the Web and discuss the following.

a. Cathodic protection of steel piles against corrosion.
b. Use of PDA.
c. Some real examples of foundations for super-tall buildings.
d. Specifications of five pile-driving hammers other than the ones listed in Table 13.7.

13.2 Write a one-page case study of a foundation system for a super-tall building, giving the information on the loadings, foundations, and the soil conditions.

13.3 A 20-m-long and 600-mm-diameter driven concrete pile carries a 1500-kN load (Fig. RE13.3a). The variation of skin friction with depth is shown in Fig. RE13.3b. Plot the variation of the pile load with depth.

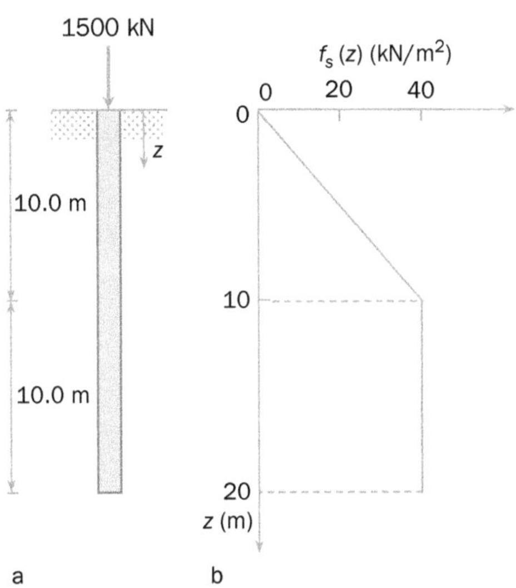

FIGURE RE13.3 (a) Pile, (b) variation of $f_s(z)$.

13.4 A 600-mm-diameter and 18-m-long precast concrete pile is driven into a sand where $\gamma = 18.0$ kN/m^3 and $\phi' = 32°$. Assume $\delta = 0.8\phi'$. Using Coyle and Castello's (1981) method determine the maximum load that can be allowed on the pile, with a factor of safety of 2.5. Neglect the effect of critical depth.

Answer: 2040 kN

13.5 A 500-mm-diameter and 15-m-long precast concrete pile is driven into a sand where $\phi' = 34°$ and $\delta = 22°$. The water table is at 3 m below the ground level. The unit weights of the sand above and below the water table are 17.5 and 19.8 kN/m^3, respectively. Determine the maximum load that can be allowed on the pile with factor of safety of 3.0. Use the β coefficients from Table 13.3 for computing the shaft load and N_q values from Table 13.4 for computing the tip load.

Answer: 905 kN

13.6 A site consists of clay ($PI = 30$, $c_u = 40$ kN/m^2, $\gamma_{sat} = 19.0$ kN/m^3) where the water table is at the ground level. A 20-m-long and 600-mm-diameter pile is driven into the ground. Find the average value of adhesion along the pile shaft using the different options discussed under the α method in Sec. 13.6.3.2.

Answer: 27–35 kN/m²

13.7 The soil profile shown in Fig. RE13.7 is fully submerged. It consists of two different clays where the saturated unit weights and the undrained shear strengths are shown in the figure. Determine the maximum load a 600-mm-diameter and 20-m-long cast-in-place pile can carry with a factor of safety of 2.5. Use α method with Kulhawy and Jackson (1989) equation for determining the skin friction.

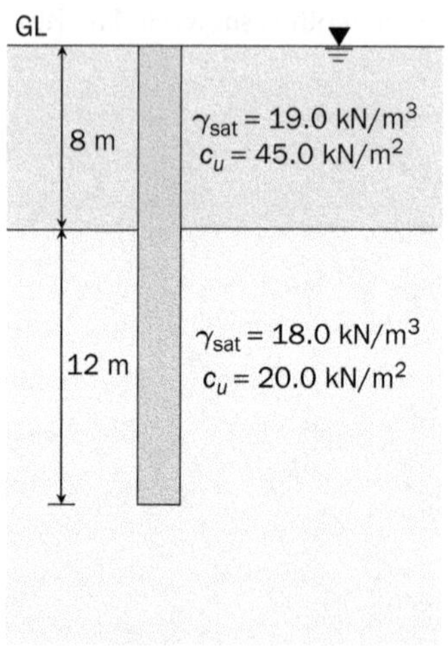

FIGURE RE13.7

Answer: 417 kN

13.8 Redo Example 13.6 using Gates formula and factor of safety of three.

Answer: 18

13.9 A single-acting Vulcan 320 hammer (see Table 13.7) is used to drive 650-mm-diameter and 25.0-m-long concrete pile into the ground where the set was measured as 13.2 mm. Determine the ultimate pile load Q_{ult}, using the different pile-driving formulas. Assume $n = 0.4$, $\eta = 0.8$, $E_p = 30$ GPa, and $\gamma_{concrete} = 24$ kN/m³. Using appropriate factors of safety, estimate the allowable loads for the pile.

Answer: ENR = 689 kN, modified ENR = 289 kN, Hiley = 379 kN, Navy-McKay = 491 kN, Danish = 488 kN, Gates = 359 kN

13.10 A 20-m-long 350-mm diameter bored concrete pile is load tested and the data are given below:

Load (kN)	0	250	500	750	1000	1250	1000	750	500	250	0
Settlement (mm)	0	2.0	5.0	8.2	15.0	39.6	39.5	38.2	37.5	36.1	35.5

$E_{concrete} = 25$ GPa. Estimate the allowable pile load using Davisson's (1973) method, with $F = 2.0$.

At the maximum load of 1250 kN, what are the total and net settlements?

Answer: 500 kN; 39.6 mm, 35.5 mm

13.11 A 400 mm × 400 mm cross section and 20-m-long concrete pile carries a vertical load of 800 kN. The tip and shaft loads are expected to be about 200 and 600 kN, respectively. The average Young's modulus and Poisson's ratio of the soil are 20 MPa and 0.2. Young's modulus of concrete is 30 GPa. The depth to the stiff stratum is very large.

a. Determine the settlement of the pile head.
b. What is the settlement of the pile tip?

Answer: 5.9 mm, 3.8 mm

13.12 In a concrete pile installed in a soil, the elastic shortening of the pile is given by (Vesic, 1977)

$$s_1 = \frac{(Q_{wt} + \xi Q_{ws})l}{A_p E_p}$$

Show from the first principles that when the skin friction is increasing linearly, from zero at the ground level, $\xi = 2/3$.

13.13 Determine the efficiency of a 4 × 5 friction pile group where the piles are 400 mm in diameter, at center-to-center spacing of 1200 mm.

Answer: 75%

13.14 A site consists of 15 m of very weak soil, underlain by good-quality rock where the uniaxial compressive strength σ_c is 40 MN/m². It is proposed to install a 650-mm-diameter rock-socketed pile with 1.5-m-long socket into the rock. The compressive strength of concrete is 35 MN/m². With factor of safety of 3.0, what would be the maximum load that can be allowed on top of the pile? Use Rowe and Armitage (1987) values of c and d in Table 13.9 for computing qb.

Answer: 14.9 MN

13.15 Surf the Web for some interesting case histories on foundations for high-rise buildings and discuss the soil conditions, loadings, foundation types, and expected settlements. Some examples you may consider are:

- Burj Khalifa Tower, Dubai, UAE
- Jeddah Tower, Saudi Arabia

- Shanghai Tower, China
- Taipei 101, Taiwan
- Trump International Hotel and Tower, Chicago, USA
- Lotus Riverside Residential Apartment Complex Failure in 2009
- Petronas Tower, Kuala Lumpur, Malaysia
- Kinaxixi MXD Complex building, Launda, Angola

13.16 Review the following and discuss the failure of the 13-story Lotus Riverside Apartment Complex:

Tan, Y., Jiang, W-Z., Rui, H-S., Lu, Y., and Wang, D-L. (2020). "Forensic geotechnical analyses on the 2009 building-overturning accident in Shanghai, China: beyond common recognitions," *Journal of Geotechnical and Geoenvironmental Engineering*, ASCE, 146(7), 05020005.

References

Almeida, M. S. S., A. B. D. Fernando, and T. Lunne. "Use of the Piezocone Test to Predict the Axial Capacity of Driven and Jacked Piles in Clay," *Canadian Geotechnical Journal*, 33(1):23–41, 1996.

American Petroleum Institute (API). *Recommended Practice for Planning, Design, and Construction of Fixed Offshore Platforms, API RP2A*, 17th Edition, Washington, DC, 1987.

American Petroleum Institute (API). *Recommended Practice for Planning, Design, and Construction of Fixed Offshore Platforms, API RP2A*, 20th Edition, Washington, DC, 1993.

ARGEMA. P. L. Tirant (ed.), *Design Guides for Offshore Structures: Offshore Pile Design*, Éditions Technip, Paris, 1992.

ASTM D1143-20. Standard Test Method for Deep Foundation Elements under Static Axial Compressive Load.

ASTM D 4945-17. Standard Test Method for High-strain Dynamic Testing of Deep Foundations.

Berezantzev, V. G., V. S. Khristoforov, and V. N. Golubkov. Load Bearing Capacity and Deformation of Piled Foundations," *Proceedings of 5th International Conference on Soil Mechanics and Foundation Engineering*, Paris, 2, 1961, pp. 11–15.

Bhushan, K. Discussion of "New Design Correlations for Piles in Sand," *Journal of the Geotechnical Engineering Division*, ASCE, 108(GT11):1508–1510, 1982.

Bolin, H. W. "The Pile Efficiency Formula of the Uniform Building Code," *Building Standards Monthly*, 10(1):4–5, 1941.

Briaud, J.-L., L. Tucker, R. L. Lytton, and H. M. Coyle. *Behaviour of Piles and Pile Groups, Report No. FHWA/RD-83/038*, Federal Highway Administration, Washington, DC, US, 1985.

Burland, J. P. Closing Address, *Proceedings of Recent Large Scale Fully Instrumented Pile Tests in* Clay, Institution of Civil Engineers, London, 1993, pp. 590–595.

Carter, J. P., and F. H. Kulhawy. *Analysis and Design of Drilled Shaft Foundations Socketed into Rock*, Report EL-5918, Electric Power Research Institute, Palo Alto, 1988.

CGS. *Canadian Foundation Engineering Manual*, 3rd Edition, Canadian Geotechnical Society, 1992.

CGS. *Canadian Foundation Engineering Manual*, 4th Edition, Canadian Geotechnical Society, 2006.

Coates, D. F. *Rock Mechanics Principles*, Monograph 874, Dept of Energy, Mines, and Resources, Mines Branch, Queen's printer, Ottawa, Canada, 1967.

Coyle, H. M., and R. R. Castello. "New Design Correlations for Piles in Sand," *Journal of Geotechnical Engineering Division*, ASCE, 107(GT7):965–986, 1981.

Davisson, M. T. "High Capacity Piles," *Proceedings, Lecture Series, Innovations in Foundation Construction*, ASCE, Illinois, USA, 1973, pp. 81–112.

DeBeer, E. E. Proefondervindlijke Bijdrage tot de Studie van het Grensdraag Vermogen van Zand Onder Funderingen op Staal, *Tijdshift der Openbar Verken van Belgie*, No. 6, 1967 and No. 4, 5, and 6, 1968.

Decourt, L. "Prediction of Load Settlement Relationships for Foundations on the Basis of the SPT-T," *Ciclo de Conferencias Intern. "Leonardo Zeevaert,"* UNAM, Mexico, 1995, pp. 85–104.

Gates, M. "Empirical Formula for Predicting Pile Bearing Capacity," *Civil Engineering*, ASCE, 27(3):65–66, 1957.

Hiley, A. "A Rational Pile-driving Formula and Its Application in Piling Practice Explained," *Engineering (London)*, 119:657; 721, 1925.

Karlsrud, K., C. J. F. Clausen, and P. M. Aas. "Bearing Capacity of Driven Piles in Clay, the NGI Approach," *Proceedings of Frontiers in Offshore Geotechnics: ISFOG*, Perth, 2005, pp. 775–782.

Kolk, J., and E. van der Velde. "A Reliable Method to Determine Friction Capacity of Piles Driven into Clays," Paper OTC 7993, *Proceedings of Offshore Technology Conference*, Houston, TX, 1996.

Kulhawy, F. H. "Limiting Tip and Side Resistance, Fact or Fallacy," J. R. Meyer (ed.), *Symposium on Analysis and Design of Pile Foundations*, ASCE, San Francisco, 1984, pp. 80–98.

Kulhawy, F. H., and C. S. Jackson. "Some Observations on Undrained Resistance of Drilled Shafts," *Foundation Engineering: Current Principles and Practices Congress*, Evanston, IL, 2, 1989, pp. 1011–1025.

MELT. *Régles Techniques de Conception et de Calcul des Fondations des Ouvrages de génie Civil*, Fascicule No. 62, Titre V No 93-3 TO, Ministére de l'Equpment du Longment et des Transport, Paris, 1993.

Meyerhof, G. G. "Bearing Capacity and Settlement of Pile Foundations," 11th Terzaghi Lecture, *Journal of the Geotechnical Engineering Division*, ASCE, 102(GT3):197–228, 1976.

Olson, R. E., and K. S. Flaate. "Pile Driving Formulas for Friction Piles in Sand," *Journal of Soil Mechanics and Foundations Division*, ASCE, 93(6):279–296, 1967.

Poulos, H. G. *Tall Building Foundation Design*, CRC Press, Boca Raton, FL, 2017.

Poulos, H. G., and E. H. Davis. *Elastic Solutions for Soil and Rock Mechanics*, John Wiley & Sons, New York, 1974.

Randolph, M. F., and B. S. Murphy. "Shaft Capacity of Driven Piles in Clay," *Proceedings of 17th Annual Offshore Technology Conference*, Houston, TX, 1985, pp. 371–378.

Rausche, F., G. E. Likins, G. G. Goble, and M. Hussein. *The Performance of Pile Driving Systems Inspection Manual*, Federal Highway Administration, Report No. FHWA/RD-86/160, Washington, DC, 1986.

Reese, L. C., and M. W. O'Neill. "New Design Methods for Drilled Shafts from Common Soil and Rock Test," *Proceedings of Congress Foundation Engineering: Current Principles and Practices*, ASCE, 2:1026–1039, 1989.

Rowe, R. K., and H. H. Armitage. "A Design Method for Drilled Piers in Soft Rock," *Canadian Geotechnical Journal*, 24(1):126–142, 1987.

Skempton, A. W. "Bearing Capacity of Clays," *Proceedings of Building Research Congress*, 1:180–189, 1951.

Terazghi, K. *Theoretical Soil Mechanics*, John Wiley, New York, 1943.

Tomlinson, M. J. "The Adhesion of Piles Driven in Clay Soils," *Proceedings of 4th International Conference on Soil Mechanics and Foundation Engineering*, London, 2, 1957, pp. 66–71.

U.S. Army. *Design of Pile Foundations*, Technical Engineering and Design Guides, ASCE, 1993.

U.S. Navy. *Soil Mechanics, Foundations and Earth Structures*, NAVFAC Design Manual DM-7.02, Washington, DC, USA, 1986.

Vesic, A. S. *Design of Pile Foundations*, National Cooperative Highway Research Program Synthesis of Practice No. 42, Transportation Research Board, Washington, DC, USA, 1977.

Vijayvergia, V. N., and J. A. Focht, Jr. "A New Way to Predict Capacity of Piles in Clay," *Offshore Technology Conference, Paper 1718, 4th Offshore Technology Conference*, Houston, Texas, US, 1972.

Vipulanandan, C., A. Hussain, and O. Usluogulari. "Parametric Study of Open Core-hole on the Behaviour of Drilled Shafts Socketed in Soft Rock," W. Camp, R. Castelli, D. F. Laefer, and S. Paikowsky (eds.), *Proceedings of Geo-Denver 2007: Contemporary Issues in Deep Foundations*, Denver, Colorado, Geotechnical Special Publication 158 [CD Rom], ASCE, 2007, pp. 1–10.

Wellington, A. M. "Formulas for Safe Loads of Bearing Piles," *Engineering News*, December 29, 1888, Reprinted in Wellington (1893), 1888, pp. 22–23.

Zhang, L. "Predicting the End Bearing Capacity of Rock Socketed Shafts," *Proceedings of 33rd Annual and 11th International Conference on Piling and Deep Foundations*, New York, Deep Foundation Institute, Hawthorne NJ, 2008, pp. 301–316.

Zhang, L., and H. H. Einstein. "End Bearing Capacity of Drilled Shafts in Rock," *Journal of Geotechnical and Geoenvironmental Engineering*, ASCE, 124(7):574–584, 1998.

CHAPTER 14

Earth Retaining Structures

14.1 Introduction

So far we have been concerned with vertical loadings and designing structures such as foundations to carry them. While the loading at the ground level is often vertical, the stresses induced within the soil mass can be in all directions. These stresses are in addition to what is already acting within the soil mass due to the overburden. Figure 14.1 shows some examples where there are lateral earth pressures acting on the structure. Some earth retaining structures used to resist the lateral earth pressures are:

- Gravity retaining walls
- Cantilever retaining walls
- Cantilever sheet pile walls
- Anchored sheet pile walls
- Braced excavations
- Gabion walls
- Diaphragm walls
- Reinforced earth walls
- Crib walls
- Conjugate bored pile or secant pile walls

Figure 14.1a shows a deep excavation in Petaling Jaya, Malaysia, where the sides are supported by a wall consisting of overlapping bored piers, known as *secant pile wall*. Figure 14.1b shows an ancient *retaining wall* at Acropolis in Athens, Greece. Figure 14.1c shows a deep excavation at a busy road Bukit Bintang in Kuala Lumpur, Malaysia, using *braced excavations*, supported by sheet piles. Figure 14.1d shows a *gabion wall* in London, UK, at the bridge abutments. Gabion walls are made of wire baskets filled with stones.

Lateral earth pressures were discussed in Chap. 10, where we focused on three special cases: at-rest, active and passive states, and discussed the ways to compute the lateral loads. In this chapter, we will use these fundamentals in designing some earth retaining structures such as retaining walls, sheet pile walls, and braced excavations.

In an excavation within a soil mass, the vertical walls cannot stand unsupported. Especially when the exposed wall of the excavation is more than 1.5 m deep, it is a safety issue. There are several ways (e.g., using retaining walls, sheet piles, or braced

FIGURE 14.1 Earth retaining structures to resist lateral earth pressures: (a) deep excavation at Petaling Jaya, Malaysia, (b) an ancient retaining wall in Athens, Greece, (c) braced excavation at Kuala Lumpur, Malaysia, (d) gabion wall at Olympic Park, London. (Courtesy of Dr. Prapa Haran, Haran Engineering, New Jersey.)

excavations) to support this exposed vertical soil surface and resist the lateral earth pressures. A *gravity retaining wall* is made of plain concrete or masonry, with a large cross section. It is often unreinforced or nominally reinforced. The large cross section makes the wall heavy, and its stability relies on its self-weight. A *cantilever retaining wall* consists of a base and a stem, made of reinforced concrete, with a smaller cross section than a gravity wall. The bottom of the wall is partially fixed within the soil, with active and passive earth pressures from either side of the wall, making it act like a vertical cantilever partially fixed at the bottom. A sheet pile can be driven along the exposed wall of the excavation to retain the soil mass, with sufficient embedment within the underlying ground. This is called a *cantilever sheet pile*. For deeper excavations, the sheet pile is further supported by an anchor connected to a tie rod placed near the ground level. This is called *anchored sheet pile*. *Braced excavations* are used to enclose areas within which excavations are carried out followed by construction activity. *Diaphragm wall* is made by excavating a narrow trench and simultaneously filling it with bentonite slurry to maintain stability of the excavation, which is later replaced by reinforced concrete. *Reinforced earth walls* are used in situations where, layers of geosynthetic reinforcements such as geogrids or geotextiles, are used to support the exposed walls of the soil mass. *Crib walls* are made of interlocking concrete, steel, or timber elements that are filled with free draining soils. The width of a crib wall is typically 50–100% of the height of the wall. The self-weight of the system contributes to the stability of the wall. *Conjugate bored piles* or *secant piles* are closely spaced piles which act like a vertical wall and resist the lateral earth pressures (Fig. 14.1a).

14.2 Retaining Walls

The two major retaining wall types are gravity and cantilever retaining walls, shown in Figs. 14.2a and b, respectively. To avoid build-up of excess pore water pressure and additional loadings on the wall, the backfill placed behind the wall on the right side generally consists of free draining materials such as granular soils. Weep holes are provided on the wall to drain the water. The entire soil surrounding the retaining walls in Fig. 14.2 is a homogeneous granular soil with unit weight of γ and effective friction angle of ϕ'. The slope of the ground on the right side is at an angle β to the horizontal. In this figure, the soil height on the right side is significantly larger than that on the left side and the wall is pushed from the right, which is resisted by the soil on the left and the base friction. Let us assume that the soil on the right is at active state and the soil on the left is at passive state.

In the designs of retaining walls, Rankine's theory is widely used due to its simplicity. Rankine's theory is developed for smooth vertical wall. To apply Rankine's theory in the designs of retaining walls shown in Fig. 14.2, it requires some manipulation. Let us take the two thick dashed vertical lines passing through the toe and the heel of the wall. We will apply Rankine's theory to determine the active and passive thrusts on these vertical lines, and consider the equilibrium of the soil and concrete contained within the boundaries, which are assumed to act as a rigid body.

Figures 14.2a and b are in fact the free body diagrams for the two walls, showing the forces acting on the rigid body between the two thick dashed vertical lines, consisting of soil and concrete. The toe of the wall is O. The soil and concrete regions are divided into rectangular and triangular blocks for simplicity, and are numbered. W_i is the self-weight of the block i, where x_i is the horizontal distance from the centroid to the toe O. For clarity, only W_2 and x_2 are labeled. P_a and P_p are the active and passive thrusts, which act at $H/3$ and $h/3$, from the bottom. H and h, are the heights of the thick dashed

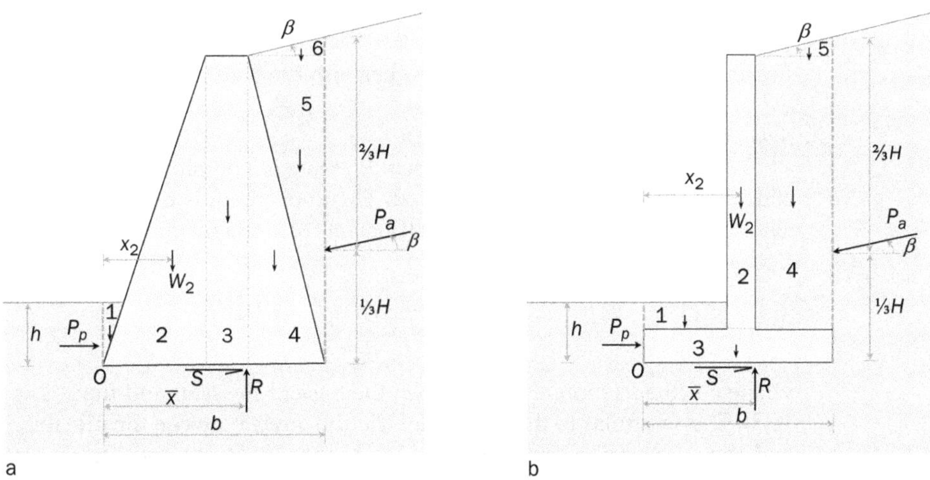

FIGURE 14.2 Free body diagrams: (a) gravity retaining wall, (b) cantilever retaining wall.

lines at the right and left edges of the rigid body under consideration. P_a and P_p can be determined as

$$P_a = 0.5K_a\gamma H^2 \tag{10.15}$$

$$P_p = 0.5K_p\gamma h^2 \tag{10.16}$$

where K_a is given by

$$K_a = \cos\beta\frac{\cos\beta - \sqrt{\cos^2\beta - \cos^2\phi'}}{\cos\beta + \sqrt{\cos^2\beta - \cos^2\phi'}} \tag{10.20}$$

and since the ground level on the left is horizontal, K_p is given by

$$K_p = \frac{1+\sin\phi'}{1-\sin\phi'} = \tan^2\left(45 + \frac{\phi'}{2}\right) \tag{10.13}$$

S is the shear resistance acting at the base of the wall and R is the vertical reaction, acting at a horizontal distance $\overline{x}$ from the toe O. The width of the base is b.

The rigid block is in equilibrium under the following forces:

- The self-weights of the blocks W_1, W_2, ...
- The active thrust P_a
- The passive resistance P_p
- Vertical reaction at the base R
- Shear resistance S

There are three possible modes the retaining wall can fail. They are:

a. Sliding along the base
b. Overturning about the toe, O
c. Bearing capacity failure into the soil beneath the wall

a. *Stability check for sliding along the base*

When P_a is significantly larger than P_p, there is a possibility that the wall may slide along the base from right to left. This movement is resisted by P_p and the sliding resistance at the base S, which can increase to a maximum of S_{max} given by

$$S_{max} = R\tan\delta = \left(P_a\sin\beta + \sum_{i=1}^{n} W_i\right)\tan\delta \tag{14.1}$$

where δ is the friction angle between the base of the wall and the soil (see Sec. 10.5). This is similar to the interfacial friction angle between the pile foundation and the surrounding soil, discussed in Chap. 13. Remember that for soil-soil interface we use ϕ'. When the soil is in contact with materials such as concrete, steel, or timber, the value of δ has is less. For concrete retaining wall,

δ varies in the range of 0.5–0.8 ϕ', and it is common to assume that $\delta = {}^2/_3\phi'$. For a perfectly smooth interface $\delta = 0$, and for a perfectly rough interface $\delta = \phi'$.

When $P_a \cos\beta$ exceeds $P_p + S_{max}$, the wall will slide. Here, $P_a \cos\beta$ is seen as the *driving force* that causes the slide. $P_p + S_{max}$ is seen as the *maximum possible resisting force*, which resists the slide. Therefore, the factor of safety with respect to sliding can be defined as,

$$F_{sliding} = \frac{\text{Maximum possible resisting force}}{\text{Driving force}} = \frac{P_p + S_{max}}{P_a \cos\beta} \tag{14.2}$$

It is necessary to ensure that $F_{sliding}$ is at least 1.5. When there is cohesion c', there is adhesion c'_a along the base of the wall which contributes toward the resistance against sliding. This c'_a is generally taken as 0.5–0.7 times the cohesion c'. In the numerator in the right-hand side of Eq. (14.2), the adhesion term bc'_a has to be included. Thus, when $c' > 0$, Eq. (14.2) becomes

$$F_{sliding} = \frac{P_p + \left(P_a \sin\beta + \sum_{i=1}^{n} W_i\right)\tan\delta + bc'_a}{P_a \cos\beta} \tag{14.3}$$

In the case of a reinforced concrete cantilever retaining wall, $F_{sliding}$ can be increased by providing a *key* at the base of the wall. Here, the stem of the wall is extended downward beneath the base of the wall into the ground. The soil on both sides of the key tends to act along with the wall as a rigid body. This increases the passive resistance significantly.

b. *Stability check for overturning about the toe O*

When H and/or P_a are high, there is a possibility that the retaining wall can overturn by rotating about the toe O. Since the wall is in equilibrium, the moment about the toe O has to be zero. Thus,

$$\sum_{i=1}^{n} W_i x_i + P_p \frac{h}{3} + P_a \sin\beta b - P_a \cos\beta \frac{H}{3} - R\overline{x} = 0$$

$$R\overline{x} = \sum_{i=1}^{n} W_i x_i + P_p \frac{h}{3} + P_a \sin\beta\, b - P_a \cos\beta \frac{H}{3} \tag{14.4}$$

For no overturning to occur, the base has to be in contact with the ground and hence $R\overline{x} > 0$. Therefore,

$$\sum_{i=1}^{n} W_i x_i + P_p \frac{h}{3} + P_a \sin\beta\, b > P_a \cos\beta \frac{H}{3}$$

The right-hand side of the above inequality is the *driving moment*, turning counterclockwise, which is trying to overturn about O and topple the wall. Each of the three terms on the left-hand side, turning clockwise, contributes toward the *resisting moment* in resisting the overturning. Therefore, the factor of safety with respect to overturning can be defined as

$$F_{\text{overturning}} = \frac{\text{Resisting moment}}{\text{Driving moment}} = \frac{\sum_{i=1}^{n} W_i x_i + P_p \frac{h}{3} + P_a \sin\beta b}{P_a \cos\beta \frac{H}{3}} \tag{14.5}$$

It is necessary to ensure $F_{\text{overturning}}$ is at least 2.0.

c. *Stability check for bearing capacity failure*

The base of the retaining wall can be treated as a strip footing of width b, carrying a line load which is inclined and eccentric. Here, it is necessary to determine the eccentricity e and the inclination α of the line load to vertical, based on equilibrium considerations.

For equilibrium,

$$R = \left(\sum_{i=1}^{n} W_i\right) + P_a \sin\beta \tag{14.6}$$

From Eqs. (14.4) and (14.6),

$$\bar{x} = \frac{\sum_{i=1}^{n} W_i x_i + P_p \frac{h}{3} + P_a \sin\beta b - P_a \cos\beta \frac{H}{3}}{\left(\sum_{i=1}^{n} W_i\right) + P_a \sin\beta} \tag{14.7}$$

The absolute value of the eccentricity of the applied line load is given by

$$e = |\bar{x} - 0.5b| \tag{14.8}$$

To ensure the entire width of the base is in compression, it is necessary that e is less than $b/6$ (see Sec. 12.6). The sliding resistance S is given by

$$S = P_a \cos\beta - P_p \tag{14.9}$$

The inclination α of the line load to vertical

$$\alpha = \tan^{-1}\frac{S}{R} \tag{14.10}$$

In Eq. (14.9), what happens if the value of P_p computed using Eq. (10.16) exceeds P_a cos β? Obviously S cannot become negative and hence reverse in direction. It just means that the passive resistance is not fully mobilized. In that case, P_a cos β can be distributed between P_p and S in a rational way (e.g., assuming the same degree of mobilization of passive and sliding resistance). In some situations, it is prudent to ignore the passive resistance and rely on the shear resistance along the base only.

The foundation depth in the bearing capacity equation must be taken as h. The maximum contact pressure beneath the base of the wall can be determined as

$$q_{\max} = \frac{Q}{b}\left(1 + 6\frac{e}{b}\right) \tag{12.17}$$

where the line load Q is the vertical reaction at the base R. The factor of safety with respect to bearing capacity is

$$F_{\text{bearing capacity}} = \frac{q_{\text{ult,net}}}{q_{\max}} \tag{14.11}$$

which has to be at least 3.0.

Example 14.1

Figure 14.3a shows a gravity retaining wall retaining a granular ($c' = 0$) backfill. The same soil is present at the bottom of the wall and on the left. The unit weight and the friction angle of the backfill are 18.5 kN/m³ and 35°, respectively. Unit weight of concrete is 24.0 kN/m³ and $\delta = {}^2/_3\phi'$. Determine the factors of safety with respect to sliding, overturning, and bearing capacity failures.

Solution Figure 14.3b shows the wall-soil block, with three soil zones and three concrete zones, all numbered 1–6.

The equilibrium of the rigid block is being analyzed here.

$$H = 1 + 5 + 2\tan 15 = 6.54 \text{ m}$$

For $\phi' = 35°$ and $\beta = 15°$, from Eq. (10.20), $K_a = 0.2968$

$$P_a = 0.5K_a\gamma H^2 = 0.5 \times 0.2968 \times 18.5 \times 6.54^2 = 117.4 \text{ kN/m}$$

For $\phi' = 35°$, with horizontal ground level, $K_p = 3.690$

$$P_p = 0.5K_p\gamma h^2 = 0.5 \times 3.690 \times 18.5 \times 1^2 = 34.1 \text{ kN/m}$$

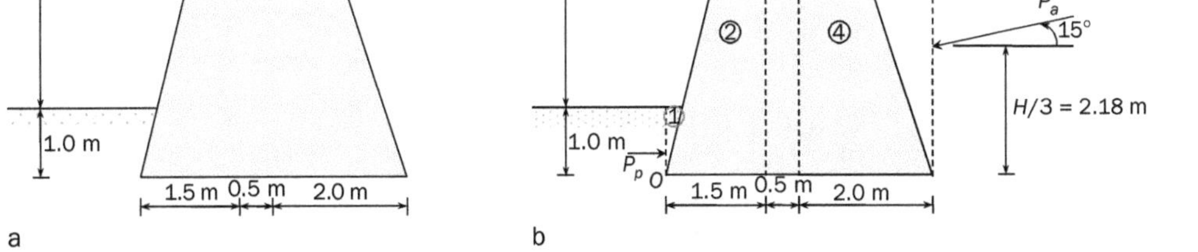

FIGURE 14.3 (a) Gravity retaining wall, (b) gravity retaining wall and soil divided into zones.

Section	Weight W_i (kN/m)	x_i (m)	Moment about *O* (kN·m/m)
1	(0.5)(1.0)(0.25)(18.5) = 2.3	0.08	0.18
2	(0.5)(1.5)(6.0)(24.0) = 108.0	1.0	108.0
3	(0.5)(6.0)(24.0) = 72.0	1.75	126.0
4	(0.5)(2.0)(6.0)(24.0) = 144.0	2.67	384.5
5	(0.5)(2.0)(6.0)(18.5) = 111.0	3.33	369.6
6	(0.5)(2.0)(0.54)(18.5) = 10.0	3.33	33.3
	$\sum W_i = 447.3$		$\sum W_i x_i = 1021.5$

Stability against sliding:

$$S_{\max} = \left(P_a \sin\beta + \sum_{i=1}^{n} W_i\right)\tan\delta$$

$$S_{\max} = ((117.4)\sin 15 + 447.3)\tan 23.3 = 205.7 \text{ kN/m}$$

$$F_{\text{sliding}} = \frac{P_p + S_{\max}}{P_a \cos\beta}$$

$$F_{\text{sliding}} = \frac{34.1 + 205.7}{117.4 \cos 15} = \mathbf{2.11}$$

Stability against overturning:

$$F_{\text{overturning}} = \frac{\sum_{i=1}^{n} W_i x_i + P_p \dfrac{h}{3} + P_a \sin\beta\, b}{P_a \cos\beta \dfrac{H}{3}}$$

$$F_{\text{overturning}} = \frac{1021.5 + (34.1)\left(\dfrac{1}{3}\right) + (117.4 \sin 15)(4)}{(117.4 \cos 15)\left(\dfrac{6.54}{3}\right)} = \mathbf{4.67}$$

Stability against bearing capacity failure:

$$R = \left(\sum_{i=1}^{n} W_i\right) + P_a \sin\beta = 447.3 + 117.4 \sin 15 = 477.7 \text{ kN/m}$$

$$S = P_a \cos\beta - P_p = 117.4 \cos 15 - 34.1 = 79.3 \text{ kN/m}$$

$$\alpha = \tan^{-1}\frac{S}{R} = \tan^{-1}\frac{79.3}{477.7} = 9.4°$$

$$\bar{x} = \frac{\sum_{i=1}^{n} W_i x_i + P_p \dfrac{h}{3} + P_a \sin\beta\, b - P_a \cos\beta \dfrac{H}{3}}{\left(\sum_{i=1}^{n} W_i\right) + P_a \sin\beta}$$

$$\bar{x} = \frac{1021.5 + (34.1)\left(\frac{1}{3}\right) + (117.4)(\sin 15)(4) - 117.4\cos 15\left(\frac{6.54}{3}\right)}{447.3 + 117.4\sin 15} = 1.900 \text{ m}$$

$$e = |\bar{x} - 0.5b| = |1.900 - (0.5)(4.0)| = 0.100 \text{ m}$$

The line load R lies within the mid-third of the strip (i.e., $e < b/6$) ensuring the entire base is in compression.

$$q_{max} = \frac{R}{b}\left(1 + 6\frac{e}{b}\right) = \frac{477.7}{4}\left(1 + 6\left(\frac{0.1}{4}\right)\right) = 137.3 \text{ kN/m}^2$$

Let us determine the ultimate bearing capacity q_{ult}, using general bearing capacity equation [Eq. (12.12)], with Meyerhof's factors.

$D_f = 1.0$ m (worst case)

$e = 0.1$ m and hence $B' = B - 2e = 4.00 - (2)(0.1) = 3.80$ m

$\phi' = 1.1 \times 35 = 38.5°$ [with plane strain correction—see Eq. (12.13)]

From Table 12.2, $N_q = 52.7$, $N_{\gamma(M)} = 71.1$

$s_q = s_\gamma = 1$ (for strip foundation)

$$d_q = d_\gamma = 1 + 0.1\frac{1.0}{4.0}\tan(45 + 38.5/2) = 1.05$$

$$i_q = \left(1 - \frac{\alpha°}{90}\right)^2 = \left(1 - \frac{9.4}{90}\right)^2 = 0.80$$

$$i_\gamma = \left(1 - \frac{\alpha°}{\phi}\right)^2 = \left(1 - \frac{9.4}{38.5}\right)^2 = 0.57$$

$$q_{ult,gross} = s_q d_q i_q \gamma D_f N_q + s_\gamma d_\gamma i_\gamma 0.5B\gamma N_\gamma$$

$$q_{ult,gross} = (1)(1.05)(0.8)(18.5)(1.0)(52.7) + (1)(1.05)(0.57)(0.5)(3.8)(18.5)(71.1)$$

$$q_{ult,gross} = 2314.7 \text{ kN/m}^2$$

$$q_{ult,net} = 2314.7 - (0.5)(18.5) = 2305 \text{ kN/m}^2$$

$$F_{\text{bearing capacity}} = \frac{q_{ult,net}}{q_{max}} = \frac{2305}{137.3} = \mathbf{16.8}$$

Generally, the factor of safety with respect to bearing capacity is very large. Sliding or overturning becomes the critical failure mode.

14.3 Cantilever Sheet Pile Walls

When a deep excavation is carried out, the exposed wall can collapse into the excavation if not properly supported. One of the ways to support the exposed wall is to use

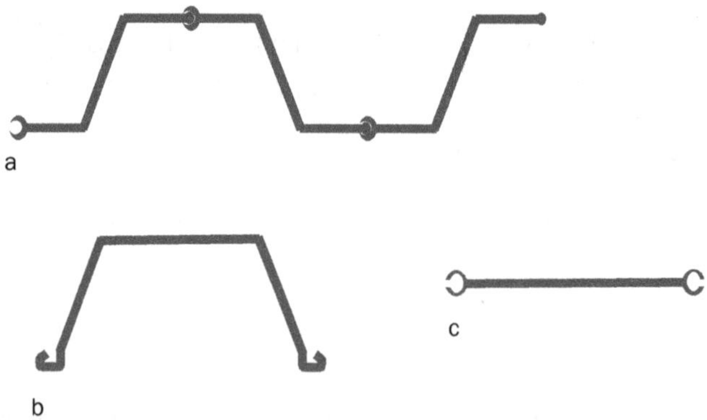

Figure 14.4 The three sheet pile sections: (a) Z-shape, (b) U-shape, (c) flat.

Figure 14.5 (a) Sheet pile wall separating the land from water, (b) vertical sections of sheet piles, driven into the ground, showing the joints.

sheet piles driven into the ground. Sheet piles are also used as *cofferdams* to isolate construction areas from surrounding water. Sheet piles are made of interlocking segments of sheets, making a continuous wall, which can retain water and/or soil (see Fig. 6.2). The joints are welded or sealants are used to make the wall water-tight. The sheet piles are commonly made of steel, and sometimes aluminum, vinyl, FRP, timber, or concrete. Sheet piles are generally driven into the ground using a vibratory hammer or hydraulic press. Sometimes, water jetting or preboring is used to assist penetration through hard layers. Steel sheet pile sections are generally made as flat (straight web), U-shape (arch web), and Z-shape. Z-shaped sections are the strongest and the most efficient. Figure 14.4a shows a sheet pile wall made of Z-shaped sections interconnected by ball and socket joints. Figures 14.4b and c show a U-shaped and flat sections, respectively. Figure 14.5a shows a sheet pile wall enclosure around a construction site which will be dewatered and excavated. Figure 14.5b shows the tight joint between two adjacent U-shaped sheet pile sections. The types of interlock joints for steel sheet piles include, double jaw, single jaw, double hook, thumb and finger, etc.

In the United States, sheet piling is specified using ASTM standards. The allowable stress of the sheet pile is selected to be about 65% (or 87% for earthquake loading) of the yield strength. The basic grade is ASTM A-328, with minimum yield strength of 265.5 MPa (38.5 ksi). ASTM A-572 is a higher strength section with minimum yield strength of 344.8 MPa (50 ksi). ASTM A-690 is developed as the corrosion resistance grade (e.g., marine applications) with minimum yield strength of 344.8 MPa (50 ksi). The allowable stresses for ASTM A-328, ASTM A-572, and ASTM A-690 are 172 (25 ksi), 224 (32.5 ksi), and 224 MPa (32.5 ksi), respectively. When they are driven deep into the ground and subjected to lateral loads, they develop high bending moments and hence high bending stresses. The appropriate sheet pile section is selected based on the minimum required section modulus value derived from the geometry of the section, bending stresses and the allowable stress for steel. Table 14.1 lists the properties of some popular steel sheet pile sections.

Timber sheet piles are used only for small excavations where water table is not present. Concrete sheet piles are used in sea walls and other waterfront structures, often as a part of the permanent structure. They are often precast sections with tongue and groove joints. The advantages of steel sheet piles include their resistance to high driving stresses, relatively light weight (compared to concrete), long service life, easy of increasing the length by welding or bolting, and the ability to be reused several times.

Type	Section	X-sectional area per m (m^2/m)	Weight per m^2 (kg/m^2)	Section modulus per m (m^3/m)	Moment of inertia per m (m^4/m)
Z shape	PZC 12	124.2×10^{-4}	97.6	12×10^{-4}	192.1×10^{-6}
	PZC 13	135.1×10^{-4}	106.0	13×10^{-4}	207.6×10^{-6}
	PZC 14	147.2×10^{-4}	115.5	14×10^{-4}	224.4×10^{-6}
	PZC 18	150.6×10^{-4}	118.2	18×10^{-4}	348.9×10^{-6}
	PZC 19	164.1×10^{-4}	128.8	19.5×10^{-4}	377.8×10^{-6}
	PZC 25	185.9×10^{-4}	145.9	24.5×10^{-4}	551.9×10^{-6}
	PZC 26	197.9×10^{-4}	155.4	26×10^{-4}	584.6×10^{-6}
	PZC 28	211.6×10^{-4}	166.1	27.5×10^{-4}	621.5×10^{-6}
Z shape	PZ 22	136.9×10^{-4}	107.4	9.7×10^{-4}	115×10^{-6}
	PZ 27	168.1×10^{-4}	131.8	16.2×10^{-4}	252×10^{-6}
	PZ 35	217.8×10^{-4}	170.9	26.1×10^{-4}	493×10^{-6}
	PZ 40	249.1×10^{-4}	195.3	32.6×10^{-4}	670×10^{-6}
Flat	PS 27.5	171.0×10^{-4}	134.2	1.0×10^{-4}	4.14×10^{-6}
	PS 31	192.9×10^{-4}	151.4	1.0×10^{-4}	4.14×10^{-6}

Note: PZC 14 has section modulus of 14×10^{-4} m^3/m; PZ 35 weighs 35 lbs per ft^2; PS 31 weighs 31 lbs per ft^2.

Table 14.1 Some Popular Steel Sheet Pile Sections

14.3.1 A Simplified Approach

Let us consider a sheet pile wall shown in Fig. 14.6a where an excavation is carried out in a granular soil to a depth of h. The effective friction angle and the unit weight of the granular soil are ϕ' and γ, respectively. To support the wall of the excavation, it is necessary to drive the sheet pile to a depth d below the excavation level. Let us determine d in terms of ϕ' and γ.

The original vertical position of the sheet pile is shown by the vertical dashed line AB. Due to the excavation, the point A moves to A' and B moves to B', with the sheet pile rotating about point O, which is at depth d_o below the excavation level. The horizontal line through O and the vertical dashed line through O divide the soil into four zones as shown in Fig. 14.6b, based on whether the wall is moving toward the soil or away from the soil. Figure 14.6c shows the values of the effective horizontal earth pressure σ'_h at critical locations and the variation of σ'_h with depth along the sheet pile on both sides.

Figure 14.7 shows a simplification made here to find the embedment depth d. Figure 14.7a shows the same excavation supported by the sheet pile which rotates about O, inducing active and passive pressures along the sheet pile as shown in Fig. 14.7b. The point of rotation O is often not very far from the tip. The lateral earth pressures beneath O on the two sides of the sheet pile are lumped into an equivalent single unknown horizontal force R in Fig. 14.7c.

The sheet pile in Fig. 14.7c is in equilibrium under the three forces: active thrust P_a, passive thrust P_p, and the horizontal force R at O. The active thrust P_a is given by

$$P_a = 0.5K_a\gamma(h + d_o)^2 \tag{14.12}$$

acting horizontally at a height of $(h + d_o)/3$ above O. The passive thrust P_p is given by

$$P_p = 0.5K_p\gamma d_o^{\,2} \tag{14.13}$$

acting horizontally at a height of $d_o/3$ above O. Taking moment about O,

$$P_a\left(\frac{h + d_o}{3}\right) - P_p\frac{d_o}{3} = 0$$

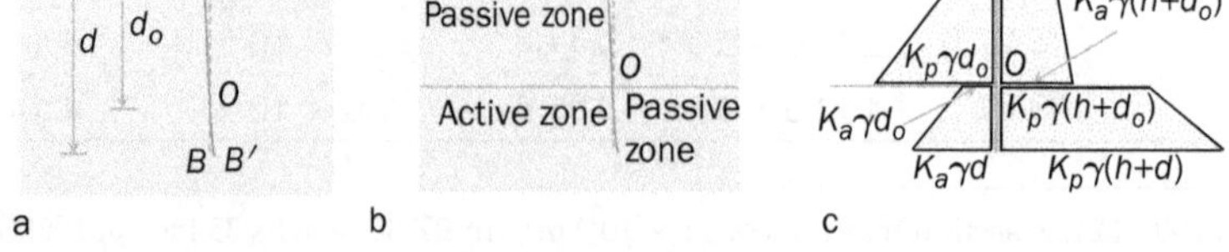

Figure 14.6 (a) Excavation supported by sheet pile, (b) active and passive zones, (c) variation of lateral earth pressure σ'_h with depth.

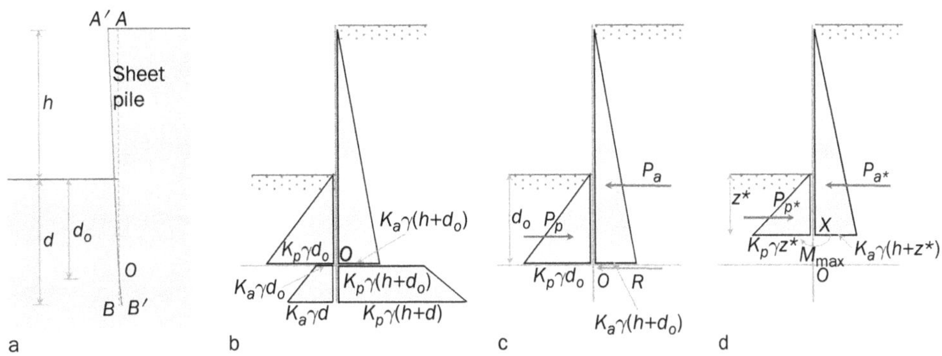

FIGURE 14.7 (a) Excavation supported by sheet pile, (b) variation of lateral earth pressure σ_h' with depth, (c) equilibrium under three forces, (d) free body diagram for the section above X.

$$0.5K_a\gamma(h+d_o)^2\left(\frac{h+d_o}{3}\right)-0.5K_p\gamma d_o^{\,2}\frac{d_o}{3}=0$$

$$K_a(h+d_o)^3 = K_p d_o^3$$

$$d_o = \frac{1}{\sqrt[3]{\frac{K_p}{K_a}}-1}h \tag{14.14}$$

Noting that (a) we have not yet incorporated a factor of safety and (b) the real d is larger than d_o, the sheet pile embedment depth d is determined by *one* of the following ways:

a. Increase the value of d_o by 20–40%; or
b. Rely on a reduced passive resistance by applying a factor of safety on K_p. Use K_p/F instead of K_p in the computations, where $F = 1.5$–2.0.

The second one gives a clear idea about the factor of safety provided and hence is recommended. Either of the two adjustments above would compensate for the simplifications and add an overall factor of safety on top of it.

The sheet pile is subjected to lateral loads from both sides, which induce bending moments and hence bending stresses on the sheet pile sections. The maximum bending moment occurs at depth z^* below the excavation line, where the shear force is zero (Fig. 14.7d). Therefore, for the lateral earth pressure diagram shown in Fig. 14.7d, $P_{a*} = P_{p*}$.

$$0.5K_a\gamma(h+z^*)^2 = 0.5K_p\gamma z^{*2}$$

$$z^* = \frac{1}{\sqrt{\frac{K_p}{K_a}}-1}h \tag{14.15}$$

Taking moment about point X at depth z^* in Fig. 14.7d, the maximum bending moment per unit length M_{max} can be determined as

$$M_{max} + P_{p^*}\left(\frac{z^*}{3}\right) - P_{a^*}\left(\frac{h+z^*}{3}\right) = 0 \tag{14.16}$$

P_{a^*} and P_{p^*} are given by

$$P_{a^*} = 0.5K_a\gamma(h+z^*)^2$$

$$P_{p^*} = 0.5K_p\gamma z^{*2}$$

Substituting these in Eq. (14.16), the maximum bending moment per unit length is

$$M_{max} = 0.5K_a\gamma(h+z^*)^2\left(\frac{h+z^*}{3}\right) - 0.5K_p\gamma z^{*2}\left(\frac{z^*}{3}\right)$$

$$M_{max} = \frac{1}{6}\gamma\left[K_a\left(h+z^*\right)^3 - K_p z^{*3}\right] \tag{14.17}$$

From the theory of bending discussed in Strength of Materials subject, the maximum bending stress caused in the sheet pile section is given by

$$\sigma_{max} = \frac{M_{max}}{I}y = \frac{M_{max}}{Z} \tag{14.18}$$

where I = moment of inertia of the sheet pile section about the neutral axis, y = distance to the furthest point on the section from the neutral axis where the maximum bending stress occurs, and Z = section modulus of the cross section of the sheet pile, generally provided by the sheet pile manufacturer. It is necessary to ensure that the maximum bending stress computed in Eq. (14.18) is less than the allowable stress. The sheet pile section is selected from the minimum required value of Z as determined from Eq. (14.18).

When the sand is not homogeneous or there is water table present, the variation of σ_h' with depth will not be a single straight line as shown in Fig. 14.7. Such a situation is discussed in Example 14.3.

Example 14.2

It is required to carry out a 5-m-deep excavation in sand where $\phi' = 33°$ and $\gamma = 17.5$ kN/m³, using cantilever sheet piles. Determine the required depth d below the excavation level using the two methods of incorporating factor of safety:

a. 20% increase of d_o, and

b. $F = 1.5$ on the passive resistance.

Solution For $\phi' = 33°$, $K_a = 0.295$, and $K_p = 3.392$.

Part a From Eq. (14.14),

$$d_o = \frac{1}{\sqrt[3]{\frac{K_p}{K_a}} - 1} h = \left(\frac{1}{\sqrt[3]{\frac{3.392}{0.295}} - 1} \right) (5.0) = 3.98 \text{ m}$$

Increasing d_o by 20%, d = **4.77 m**

Part b With factor of safety of 1.5, the reduced K_p is 3.392/1.5 = 2.261. From Eq. (14.14),

$$d = \frac{1}{\sqrt[3]{\frac{K_p}{K_a}} - 1} h = \left(\frac{1}{\sqrt[3]{\frac{2.261}{0.295}} - 1} \right) (5.0) = \mathbf{5.15\,m}$$

Example 14.3

In a site with sandy soil, the water table is at 2 m below the ground level. The unit weight of the sand above and below the water table are 16.0 and 19.3 kN/m³, respectively. The friction angle of the sand is 32°. It is required to carry out an excavation to 5 m depth, using cantilever sheet pile walls.

a. How deep would you drive the sheet pile, with a factor of safety of 1.75 on the passive resistance?
b. Determine the maximum bending moment in the sheet pile and the required section modulus of the steel section, assuming allowable stress of 225 MN/m².

Solution

Part a For $\phi' = 32°$, $K_a = 0.307$, $K_p = 3.257$, and $K_p/F = 1.86$.

The geometry of the excavation and the sheet pile are shown in Fig. 14.8a. The lateral earth pressures on the right and left sides of the sheet pile are computed as follows.

Right side of the sheet pile:

At the water table level, $\sigma'_v = 16 \times 2 = 32$ kN/m²

$$\sigma'_h = 0.307 \times 32 = 9.82 \text{ kN/m}^2$$

At the bottom of the sheet pile, $\sigma'_v = 16 \times 2 + (19.3 - 9.81)(d + 3) = 9.49d + 60.47$ kN/m²

$$\sigma'_h = 0.307(9.49d + 60.47) = 2.91d + 18.56 \text{ kN/m}^2$$

Left side of the sheet pile:

At the bottom of the sheet pile, $\sigma'_v = (19.3 - 9.81)\, d = 9.49d$ kN/m²

$$\sigma'_h = 1.86 \times 9.49d = 17.65d \text{ kN/m}^2$$

The lateral earth pressure diagrams on both sides of the sheet pile can be drawn as shown in Fig. 14.8a. They are divided into triangles and rectangles for the ease of computation, with the horizontal loads acting on the centroids of the zones.

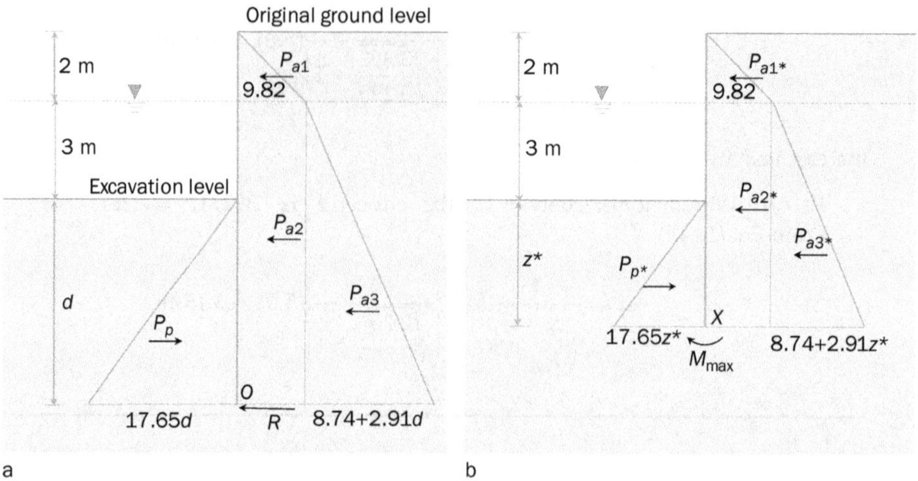

FIGURE 14.8 Lateral earth pressure variations: (a) up to O, (b) up to X.

Force	Magnitude (kN/m)	Height above O (m)	Moment (kN·m/m)
P_{a1}	$0.5 \times 9.82 \times 2 = 9.82$	$d + 3.67$	$9.82d + 36.04$
P_{a2}	$9.82(d + 3) = 9.82d + 29.5$	$0.5d + 1.5$	$4.91d^2 + 29.48d + 44.2$
P_{a3}	$0.5(2.91d + 8.74)(d + 3) = 1.46d^2 + 8.74d + 13.11$	$0.333d + 1$	$0.48d^3 + 4.37d^2 + 13.1d + 13.11$
P_p	$0.5 \times 17.65d \times d = 8.83d^2$	$0.333d$	$2.92d^3$

Taking moments about O,

$$(9.82d + 36.04) + (4.91d^2 + 29.48d + 44.2) + (0.48d^3 + 4.37d^2 + 13.1d + 13.11) - 2.92d^3 = 0$$

$$2.44d^3 - 9.28d^2 - 52.4d - 93.35 = 0$$

$$d = 7.4 \text{ m}$$

The total depth of the sheet pile = 7.4 + 5.0 = **12.4 m**

Part b Let us say the maximum bending moment M_{max} occurs at depth z^* below the excavation. The lateral earth pressure distribution above this depth (see Fig. 14.8b) would be very similar to the one shown in Fig. 14.8a. Therefore, the same expressions will hold, and we just have to replace d by z^*. Therefore,

$$P_{a1^*} = 9.82$$

$$P_{a2^*} = 9.82\, z^* + 29.5$$

$$P_{a3^*} = 1.46\, z^{*2} + 8.74z^* + 13.11$$

$$P_{p^*} = 8.83\, z^{*2}$$

For the shear force at X to be zero,

$$P_{a1*} + P_{a2*} + P_{a3*} - P_{p*} = 0$$

$$9.82 + 9.82z^* + 29.5 + 1.46z^{*2} + 8.74z^* + 13.11 - 8.83z^{*2} = 0$$

$$7.37z^{*2} - 18.56z^* - 52.43 = 0$$

$$z^* = 4.21 \text{ m}$$

Taking moment about X (Fig. 14.8b),

$$(9.82z^* + 36.04) + (4.91z^{*2} + 29.48z^* + 44.2) +$$
$$(0.48z^{*3} + 4.37z^{*2} + 13.1z^* + 13.11) - 2.92z^{*3} - M_{max} = 0$$

$$M_{max} = -2.44z^{*3} + 9.28z^{*2} + 52.4z^* + 93.35$$

Substituting $z^* = 4.21$ m,

$$M_{max} = -2.44\,(4.21)^3 + 9.28\,(4.21)^2 + 52.4\,(4.21) + 93.35 = \mathbf{296.4\ kN{\cdot}m/m}$$

Allowable stress of the sheet pile section is 225 MN/m^2.

Therefore, the required section modulus is

$$Z = \frac{M_{max}}{\sigma_{allowable}} = \frac{296.4 \times 10^3}{225 \times 10^6} = \mathbf{13.2 \times 10^{-4}\ m^3/m}$$

Section PZC 13 (see Table 14.1) would be suitable.

14.3.2 Using Net Lateral Pressure Diagram—in Dry Granular Soils

Let us consider an excavation to a depth h in a dry granular soil, supported by a cantilever sheet pile embedded to a depth d below the excavation level. In this method, we will draw the net effective lateral earth pressure diagram $AGDIFJ$ and use that to determine d.

At the bottom of the excavation, on the right side of C, $\sigma'_v = \gamma h$, where γ is the unit weight of the dry sand. Therefore, $\sigma'_h = K_a\gamma h = \sigma'_1$ (see Fig. 14.9a). Let us denote the depth below the excavation level as z (see Fig. 14.9a). At depth z below the excavation, and above the point of rotation, the active pressure σ'_{ha} on the right and the passive pressure σ'_{hp} on the left are given by

$$\sigma'_{ha} = K_a\sigma'_v = K_a\gamma(h + z)$$

$$\sigma'_{hp} = K_p\sigma'_v = K_p\gamma z$$

The net pressure σ'_{hn} below the excavation level, from right to left is

$$\sigma'_{hn} = \sigma'_{ha} - \sigma'_{hp} = K_a\gamma\,(h + z) - K_p\gamma z = K_a\gamma h - (K_p - K_a)\gamma z$$

$$\sigma'_{hn} = \sigma'_1 - (K_p - K_a)\gamma z \tag{14.19}$$

The depth z_0, at which the net pressure becomes zero, can be determined as

$$z_0 = \frac{\sigma'_1}{(K_p - K_a)\gamma} = \frac{K_a}{(K_p - K_a)}h \tag{14.20}$$

While z_0 and $\bar{z}$ are relatively easier to find, it requires some manipulation to determine z_1 and z_2, so that d can be computed as $d = z_0 + z_2$.

From Eq. (14.19), it can be seen that then net pressure σ'_{hn} below the excavation level decreases linearly with the depth z, along the line GH. The slope of the line GH is 1 vertical to $(K_p - K_a)\gamma$ horizontal. Therefore,

$$\sigma'_2 = \overline{HB} = z_2\left(K_p - K_a\right)\gamma = K_p\gamma d - K_a\gamma(h + d) \tag{14.21}$$

At the bottom of the sheet pile, there is active pressure on the left side and passive pressure on the right. They are given by

$$\sigma'_{\text{ha}} = K_a\sigma'_v = K_a\gamma d$$
$$\sigma'_{\text{hp}} = K_p\sigma'_v = K_p\gamma(h + d)$$
$$\sigma'_{\text{hn}} = \sigma'_3 = \sigma'_{\text{hp}} - \sigma'_{\text{ha}} = K_p\gamma(h + d) - K_a\gamma d$$
$$\sigma'_3 = K_p\gamma h + (K_p - K_a)\gamma d$$

Substituting $d = z_0 + z_2$,

$$\sigma'_3 = K_p\gamma h + (K_p - K_a)\gamma z_0 + (K_p - K_a)\gamma z_2$$

Let us define a known quantity σ'_4 as

$$\sigma'_4 = K_p\gamma h + \left(K_p - K_a\right)\gamma z_0 \tag{14.22}$$

Therefore,

$$\sigma'_3 = \sigma'_4 + \left(K_p - K_a\right)\gamma z_2 \tag{14.23}$$

A close look at Fig. 14.9a shows that there are two unknowns z_1 and z_2. Let us determine them by equilibrium considerations. The net pressures acting in the three regions as follows:

AGD: Right to left (resultant force *P*)

DFI: Left to right

FJB: Right to left

To simplify the computations, let us include the region *HIFB* to the net lateral pressures from both sides.

For equilibrium of forces in the horizontal direction,

$$P + \frac{1}{2}(\sigma'_2 + \sigma'_3)z_1 - \frac{1}{2}\sigma'_2 z_2 = 0 \tag{14.24}$$

where *P* is the area of the zone *AGD*. From Eq. (14.24),

$$z_1 = \frac{\sigma'_2 z_2 - 2P}{\sigma'_2 + \sigma'_3} \tag{14.25}$$

Taking moment about B,

$$P(z_2 + \bar{z}) + \frac{1}{2}(\sigma_2' + \sigma_3')z_1\frac{z_1}{3} - \frac{1}{2}\sigma_2' z_2 \frac{z_2}{3} = 0 \tag{14.26}$$

From Eqs. (14.21), (14.22), (14.23), (14.25), and (14.26), it can be shown that

$$z_2^4 + A_1 z_2^3 - A_2 z_2^2 - A_3 z_2 - A_4 = 0 \tag{14.27}$$

where

$$A_1 = \frac{\sigma_4'}{\gamma(K_p - K_a)} \tag{14.28}$$

$$A_2 = \frac{8P}{\gamma(K_p - K_a)} \tag{14.29}$$

$$A_3 = \frac{6P\left[2\bar{z}\gamma(K_p - K_a) + \sigma_4'\right]}{\gamma^2(K_p - K_a)^2} \tag{14.30}$$

$$A_4 = \frac{P(6\bar{z}\sigma_4' + 4P)}{\gamma^2(K_p - K_a)^2} \tag{14.31}$$

Equation (14.27) can be solved by trial and error process, or using a graphics calculator, and z_2 can be determined. Hence d becomes $z_0 + z_2$. A factor of safety can be incorporated by *one* of the following ways:

a. Increase the value of d by 20–40%; or

b. Rely on a reduced passive resistance by applying a factor of safety on K_p. Use K_p/F instead of K_p in the computations, where $F = 1.5$–2.0.

Let us determine the maximum bending moment in the sheet pile M_{max} (see Fig. 14.9b), which occurs at z^* below D where the net pressure from left to right is σ_5' given by

$$\sigma_5' = (K_p - K_a)\gamma z^*$$

M_{max} occurs when there is zero shear force. Therefore,

$$P = \frac{1}{2}\sigma_5' z^* = \frac{1}{2}(K_p - K_a)\gamma z^{*2}$$

$$z^* = \sqrt{\frac{2P}{(K_p - K_a)\gamma}} \tag{14.32}$$

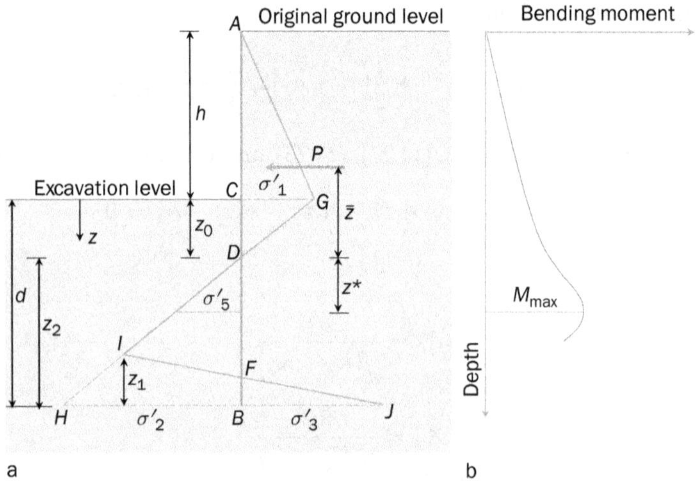

FIGURE 14.9 A cantilever sheet pile in dry granular soils: (a) net lateral earth pressure diagram, (b) bending moment diagram.

The maximum bending moment M_{max} can be determined as

$$M_{max} = P(\bar{z} + z^*) - \frac{1}{2}\sigma_5' z^* \left(\frac{z^*}{3}\right) \tag{14.33}$$

The maximum allowable stress and the maximum bending moment are related by

$$\sigma_{max} = \frac{M_{max}}{I} y = \frac{M_{max}}{Z} \tag{14.18}$$

where I = moment of inertia of the sheet pile section about the neutral axis, y = distance to the furthest point on the section from the neutral axis where the maximum bending stress occurs, and Z = section modulus of the cross section of the sheet pile, generally provided by the sheet pile manufacturer. The minimum value of Z required can be determined from Eq. (14.18) and the sheet pile can be selected.

Example 14.4

A 5-m-deep excavation is carried out in dry sand using cantilever sheet piles as in Example 14.2, where $\phi' = 33°$ and $\gamma = 17.5$ kN/m³. Using the net pressure diagram, determine

a. the required depth d of the sheet pile penetration below the excavation level, using a factor of safety of 1.5 on the passive resistance, and

b. the minimum section modulus required for the sheet pile, assuming the allowable stress is 225 MN/m².

Solution

Part a For $\phi' = 33°$, $K_a = 0.295$, and $K_p/F = 2.26$. We will use K_p/F instead of K_p, throughout.

$$\sigma_1' = K_a \gamma h = (0.295)(17.5)(5) = 25.8 \text{ kN/m}^2$$

$$z_0 = \frac{K_a}{(K_p - K_a)}h = \frac{0.295}{2.26 - 0.295}(5.0) = 0.75\,\text{m}$$

$$P = \frac{1}{2}(25.8)(5.0) + \frac{1}{2}(25.8)(0.75) = 74.2\text{ kN/m}$$

$$P\bar{z} = 0.5(25.8)(5)\left(0.75 + \frac{5.0}{3}\right) + 0.5(25.8)(0.75)(0.50)$$

$$\bar{z} = 2.17\text{ m}$$

$$\sigma_4' = K_p\gamma h + (K_p - K_a)\gamma z_0 = (2.26)(17.5)(5) + (2.26 - 0.295)(17.5)(0.75)$$

$$\sigma_4' = 223.5\text{ kN/m}^2$$

With these, we can determine the constants A_1, A_2, A_3, and A_4 as

$$A_1 = \frac{\sigma_4'}{\gamma(K_p - K_a)} = \frac{223.5}{17.5(2.26 - 0.295)} = 6.50\text{ m}$$

$$A_2 = \frac{8P}{\gamma(K_p - K_a)} = \frac{(8)(74.2)}{(17.5)(2.26 - 0.295)} = 17.26\text{ m}^2$$

$$A_3 = \frac{6P\left[2\bar{z}\gamma(K_p - K_a) + \sigma_4'\right]}{\gamma^2(K_p - K_a)^2} = \frac{6(74.2)\left[2(2.17)(17.5)(2.26 - 0.295) + 223.5\right]}{17.5^2(2.26 - 0.295)^2}$$

$$A_3 = 140.3\text{ m}^3$$

$$A_4 = \frac{P(6\bar{z}\sigma_4' + 4P)}{\gamma^2(K_p - K_a)^2} = \frac{(74.2)\left[(6)(2.17)(223.5) + (4)(74.5)\right]}{17.5^2(2.26 - 0.295)^2} = 201.3\text{ m}^4$$

$$z_2^4 + 6.50\,z_2^3 - 17.26\,z_2^2 - 140.3\,z_2 - 201.3 = 0$$

$$z_2 = 4.83\text{ m}$$

$$d = z_0 + z_2 = 0.75 + 4.83 = \mathbf{5.58\text{ m}}$$

This compares well with the d estimated as 5.15 m in Part b of Example 14.2 using $F = 1.5$ on the passive resistance.

Part b

$$z^* = \sqrt{\frac{2P}{(K_p - K_a)\gamma}} = \sqrt{\frac{2(74.2)}{(2.26 - 0.295)(17.5)}} = 2.077\text{ m}$$

$$\sigma_5' = (K_p - K_a)\gamma z^* = (2.26 - 0.295)(17.5)(2.077) = 71.4\text{ kN/m}^2$$

$$M_{max} = P(\bar{z} + z^*) - \frac{1}{2}\sigma_5' z^*\left(\frac{z^*}{3}\right) = (74.2)(2.17 + 2.08) - \frac{1}{2}(71.4)(2.08)\left(\frac{2.08}{3}\right)$$

$$M_{max} = 263.9\text{ kN·m}$$

The minimum required section modulus Z is given by

$$Z = \frac{M_{max}}{\sigma_{allowable}} = \frac{263.9 \times 10^3}{225 \times 10^6} = \mathbf{11.7 \times 10^{-4}\ m^3\ per\ m}$$

Section PZC 13 (see Table 14.1) would be suitable.

14.3.3 Using Net Lateral Pressure Diagram—in Submerged Granular Soils

When there is water table present within the soil, the net pressure diagram *AKGDIFJ* will be slightly different as shown in Fig. 14.10, but the method remains the same. Here, the water table is at depth h_1 below the original ground level.

At the water table level, on the right side of the wall, $\sigma'_v = \gamma_m h_1$ and hence $\sigma'_h = K_a \gamma_m h_1$.

At the excavation level, on the right side of the sheet pile at point C,

$$\sigma'_v = \gamma_m h_1 + \gamma' h_2$$

where γ' is the submerged unit weight.

$$\sigma'_h = K_a(\gamma_m h_1 + \gamma' h_2) = \sigma'_1$$

At depth z below the excavation level, with active pressure from the right and passive pressure from the left, the net pressure σ'_{hn} from right to left is given by

$$\sigma'_{hn} = K_a(\gamma_m h_1 + \gamma' h_2 + \gamma' z) - K_p \gamma' z = \sigma'_1 - (K_p - K_a)\gamma' z$$

which is the same as Eq. (14.19) where γ is replaced by γ'. The depth z_0 below the excavation level where the net pressure becomes zero is given by

$$z_0 = \frac{\sigma'_1}{(K_p - K_a)\gamma} \tag{14.34}$$

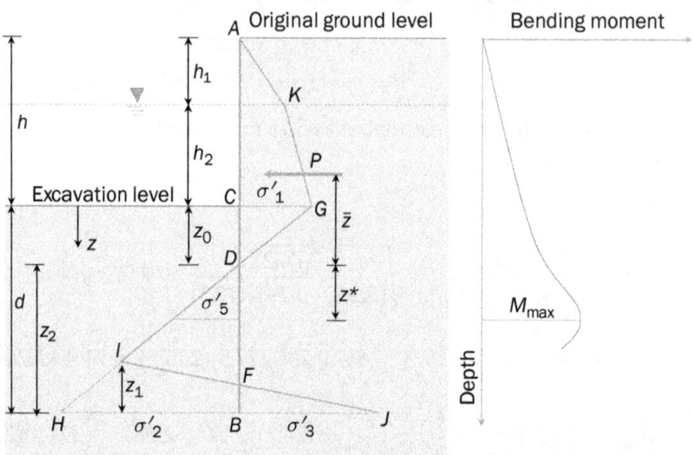

FIGURE 14.10 Net pressure diagram with water table above the excavation level.

As before, the net pressure decreases along the line *GDIH* at a slope of 1 vertical to $(K_p - K_a)\gamma'$ horizontal. Therefore,

$$\sigma_2' = \overline{HB} = z_2\left(K_p - K_a\right)\gamma'$$

At the bottom of the sheet pile, there is passive pressure on the right and active pressure on the left. In terms of effective stresses, they are

$$\sigma_{ha}' = K_a\gamma' d$$

$$\sigma_{hp}' = K_p\left(\gamma_m h_1 + \gamma' h_2 + \gamma' d\right)$$

The net horizontal effective stress σ_{hn}' acting from right to left is

$$\sigma_{hn}' = K_p\left(\gamma_m h_1 + \gamma' h_2 + \gamma' d\right) - K_a\gamma' d = \sigma_3' = \overline{BJ}$$

i.e.,

$$\sigma_3' = K_p\left(\gamma_m h_1 + \gamma' h_2\right) + \left(K_p - K_a\right)\gamma' d$$

Substituting $d = z_0 + z_2$,

$$\sigma_3' = K_p\left(\gamma_m h_1 + \gamma' h_2\right) + \left(K_p - K_a\right)\gamma'(z_0 + z_2)$$

Let us define a known quantity σ_4' as

$$\sigma_4' = K_p\left(\gamma_m h_1 + \gamma' h_2\right) + \left(K_p - K_a\right)\gamma' z_0 \tag{14.35}$$

It can be seen from equilibrium considerations, that Eqs. (14.24)–(14.27) still hold, but with a twist. In computing the four constants A_1 to A_4, and σ_4', appropriate unit weights have to be used. Note the difference between Eqs. (14.22) and (14.35), where the bulk unit weight γ_m above the water table and submerged unit weight γ' below the water table were used in computing σ_4'. In computing the four constants A_1 to A_4, using Eqs. (14.28)–(14.31), the bulk unit weight γ_m is replaced by the submerged unit weight γ'.

Example 14.5

Redo Example 14.3 using the net earth pressure diagram.

Solution

Part a For $\phi' = 32°$, $K_a = 0.307$, $K_p = 3.257$, and $K_p/F = 1.86$. Let us use the reduced value of 1.86 as K_p.

$$\gamma' = 19.3 - 9.81 = 9.49 \text{ kN/m}^3\text{, and } \gamma_m = 16.0 \text{ kN/m}^3 \text{ (above the water table)}$$

Let us refer to Fig. 14.10 while carrying out the calculations.

$$h_1 = 2 \text{ m},\ h_2 = 3 \text{ m, and } h = 5 \text{ m.}$$

At the water table level, on the right of the sheet pile,

$$\sigma'_h = (0.307)(2)(16.0) = 9.82 \text{ kN/m}^2$$

At the excavation level,

$$\sigma'_h = \sigma'_1 = K_a(\gamma_m h_1 + \gamma' h_2) = (0.307)(16.0 \times 2 + 9.49 \times 3) = 18.6 \text{ kN/m}^2$$

From Eq. (14.34),

$$z_0 = \frac{\sigma'_1}{(K_p - K_a)\gamma'} = \frac{18.6}{(1.86 - 0.307)(9.49)} = 1.26 \text{ m}$$

Let us compute P and $\bar{z}$ from the upper right portion of the net pressure diagram.

$$P = (0.5)(9.82)(2) + (9.82)(3) + (0.5)(18.6 - 9.82)(3) + (0.5)(18.6)(1.26)$$
$$= 9.82 + 29.46 + 13.17 + 11.72 = 64.17 \text{ kN/m}$$

$$P\bar{z} = (9.82)(1.26 + 3 + 0.667) + (29.46)(1.26 + 1.5) + (13.17)(1.26 + 1.0) + (11.72)(0.667 \times 1.26) = 169.31 \text{ kN}$$

Therefore, $\bar{z} = 2.64$ m.

From Eq. (14.35),

$$\sigma'_4 = K_p(\gamma_m h_1 + \gamma' h_2) + (K_p - K_a)\gamma' z_0$$

$$\sigma'_4 = (1.86)(16 \times 2 + 9.49 \times 3) + (1.86 - 0.307)(9.49)(1.26) = 131.0 \text{ kN/m}^2$$

$$A_1 = \frac{\sigma'_4}{\gamma'(K_p - K_a)} = \frac{131.0}{9.49(1.86 - 0.307)} = 8.89 \text{ m}$$

$$A_2 = \frac{8P}{\gamma'(K_p - K_a)} = \frac{8 \times 64.17}{9.49(1.86 - 0.307)} = 34.83 \text{ m}^2$$

$$A_3 = \frac{6P\left[2\bar{z}\gamma'(K_p - K_a) + \sigma'_4\right]}{\gamma'^2(K_p - K_a)^2} = \frac{(6)(64.17)\left[2(2.64)(9.49)(1.86 - 0.307) + 131.0\right]}{(9.49)^2(1.86 - 0.307)^2}$$

$$A_3 = 370.15 \text{ m}^3$$

$$A_4 = \frac{P(6\bar{z}\sigma'_4 + 4P)}{\gamma'^2(K_p - K_a)^2} = \frac{(64.17)(6 \times 2.64 \times 131 + 4 \times 64.17)}{(9.49)^2(1.86 - 0.307)^2} = 688.86 \text{ m}^4$$

From Eq. (14.27),

$$z_2^4 + A_1 z_2^3 - A_2 z_2^2 - A_3 z_2 - A_4 = 0$$

$$z_2^4 + 8.89 z_2^3 - 34.83 z_2^2 - 370.15 z_2 - 688.86 = 0$$

Solving this, $z_2 = 6.73$ m

$$d = z_0 + z_2 = 1.26 + 6.73 = 7.99 \text{ m}$$

Total depth of the sheet pile = 5.0 + 7.99 = **12.99 m.**

This compares well with the depth of 12.4 m determined in Example 14.3.

Part b From Eq. (14.32),

$$z^* = \sqrt{\frac{2P}{(K_p - K_a)\gamma'}} = \sqrt{\frac{2(64.17)}{(1.86 - 0.307)(9.49)}} = 2.95 \text{ m}$$

$$\sigma_5' = (K_p - K_a)\gamma' z^* = (1.86 - 0.307)(9.49)(2.95) = 43.48 \text{ kN/m}^2$$

$$M_{\max} = P(\bar{z} + z^*) - \frac{1}{2}\sigma_5' z^* \left(\frac{z^*}{3}\right) = 64.17(2.64 + 2.95) - \frac{1}{2}(43.48)(2.95)\left(\frac{2.95}{3}\right)$$

$$M_{\max} = \mathbf{295.6\ kN{\cdot}m}$$

The minimum section modulus required is

$$Z = \frac{M_{\max}}{\sigma_{\text{allowable}}} = \frac{295.6 \times 10^3}{225 \times 10^6} = \mathbf{13.1 \times 10^{-4}\ m^3/m}$$

This compares well with the value of 13.2×10^{-4} m^3/m determined in Example 15.3. Sheet pile section PZC 13 (see Table 14.1) would be suitable.

14.3.4 Using Net Lateral Pressure Diagram—Cantilever Sheet Piles Penetrating Clays

Let us consider a situation such as the one shown in Fig. 14.11, where the soil below the excavation is a clay. Immediately after the installation of the sheet pile, the clay is undrained and we will analyze it in terms of total stresses with c_u and $\phi_u = 0$ ($K_a = K_p = 1$). Above the excavation line, the water pressure is the same on both sides of the sheet pile and cancels out. The net pressure diagram *AKGCDIFJ* is shown in the figure.

In the granular soil above the excavation level, the effective active earth pressures can be computed and P and $\bar{z}$ can be determined as in Sec. 14.3.3. At the bottom of the excavation,

$$p_1 = K_{a,g}(\gamma_{m,g}h_1 + \gamma_g' h_2)$$

Here, the subscript g denotes granular soils. At depth z below the excavation, and above the point of rotation, the active earth pressure from right to left is

$$\sigma_{ha} = (\gamma_{m,g}h_1 + \gamma_g' h_2 + \gamma_{\text{sat},c} z) - 2c_u$$

The subscript c denotes clay. The passive earth pressure from left to right is

$$\sigma_{hp} = \gamma_{\text{sat},c} z + 2c_u$$

Therefore, the net pressure from left to right is

$$\sigma_{hn} = \sigma_{hp} - \sigma_{ha} = 4c_u - \left(\gamma_{m,g}h_1 + \gamma_g'h_2\right) = p_2 \tag{14.36}$$

At the bottom of the sheet pile, the passive earth pressure from right to left is

$$\sigma_{hp} = \left(\gamma_{m,g}h_1 + \gamma_g'h_2 + \gamma_{sat,c}d\right) + 2c_u$$

The active earth pressure from left to right is

$$\sigma_{ha} = \gamma_{\text{sat},c}d - 2c_u$$

Therefore, the net pressure from right to left is

$$\sigma_{hn} = \sigma_{hp} - \sigma_{ha} = 4c_u + \left(\gamma_{m,g}h_1 + \gamma_g'h_2\right) = p_3 \tag{14.37}$$

As before, we will include the region *IFBH* on both sides and consider equilibrium. In the horizontal direction,

$$P + \frac{1}{2}(p_2 + p_3)h_4 - p_2d = 0$$

$$P + \frac{1}{2}\left[4c_u - \left(\gamma_{m,g}h_1 + \gamma_g'h_2\right) + 4c_u + \left(\gamma_{m,g}h_1 + \gamma_g'h_2\right)\right]h_4 - \left[4c_u - \left(\gamma_{m,g}h_1 + \gamma_g'h_2\right)\right]d = 0$$

$$P + 4c_uh_4 = \left[4c_u - \left(\gamma_{m,g}h_1 + \gamma_g'h_2\right)\right]d$$

$$h_4 = \frac{\left[4c_u - \left(\gamma_{m,g}h_1 + \gamma_g'h_2\right)\right]d - P}{4c_u} \tag{14.38}$$

Equating the moment about the bottom of the sheet pile to zero,

$$P(d + \bar{z}) + \frac{1}{2}(p_2 + p_3)h_4\frac{h_4}{3} - p_2d\frac{d}{2} = 0$$

$$P(d + \bar{z}) + \frac{1}{2}(8c_u)\frac{h_4^{\ 2}}{3} - \left[4c_u - \left(\gamma_{m,g}h_1 + \gamma_g'h_2\right)\right]\frac{d^2}{2} = 0 \tag{14.39}$$

From Eqs. (14.38) and (14.39),

$$\left[4c_u - \left(\gamma_{m,g}h_1 + \gamma_g'h_2\right)\right]d^2 - 2Pd - \frac{P(P + 12c_u\bar{z})}{2c_u + \left(\gamma_{m,g}h_1 + \gamma_g'h_2\right)} = 0 \tag{14.40}$$

The unit weight of the clay does not appear in the equations. By solving Eq. (14.40), d can be determined. Due to higher uncertainty in clays, a larger factor of safety on passive resistance is used or d is increased by 40–60%.

The depth z^* at which the bending moment is the maximum (M_{max}) and the shear force is zero is determined as follows:

$$P - p_2z^* = 0$$

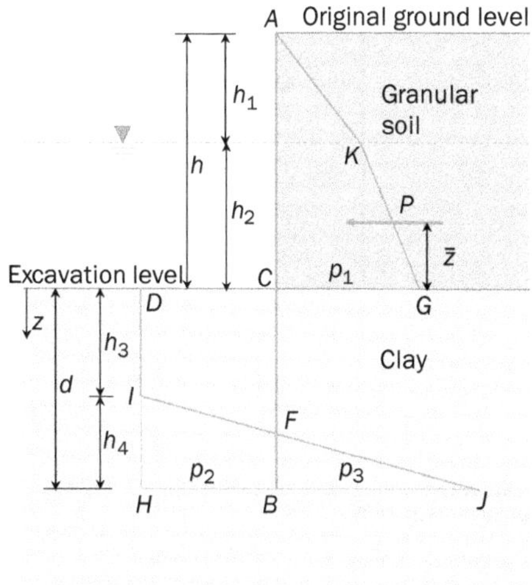

FIGURE 14.11 Net pressure diagram for a cantilever sheet pile penetrating clays.

$$z^* = \frac{P}{p_2} \tag{14.41}$$

$$M_{max} = P(z^* + \bar{z}) - p_2 z^* \frac{z^*}{2}$$

$$M_{max} = P(z^* + \bar{z}) - p_2 \frac{z^{*2}}{2} \tag{14.42}$$

From the maximum bending moment, the required section modulus can be determined.

14.4 Anchored Sheet Piles

With cantilever sheet piles, when the excavation gets deeper, the embedment depth d becomes larger, requiring longer sections of sheet pile. Generally, h and d are in the same order of magnitude. In addition, with larger excavation depth and embedment, the earth pressures are larger and the moments get larger. This requires sheet piles with higher section modulus, which increases the overall cost. For excavation depths more than about 6.0 m, anchored sheet piles are more economical. Here, as shown in Fig. 14.12, close to the ground level, a tie rod is attached to the sheet pile at O and fixed to a rigid anchor away from the sheet pile. A popular form of anchor is a concrete block known as *deadman*, which can also be in the form of continuous beam. The lateral support for the sheet pile is derived from the passive resistance and the pull by the tie rod.

There are two ways to analyze the anchored sheet pile: (a) the simpler *free-earth support method* and (b) the more complex *fixed earth support method*. In the free-earth support method, it is assumed that the sheet pile embedment is not large enough to

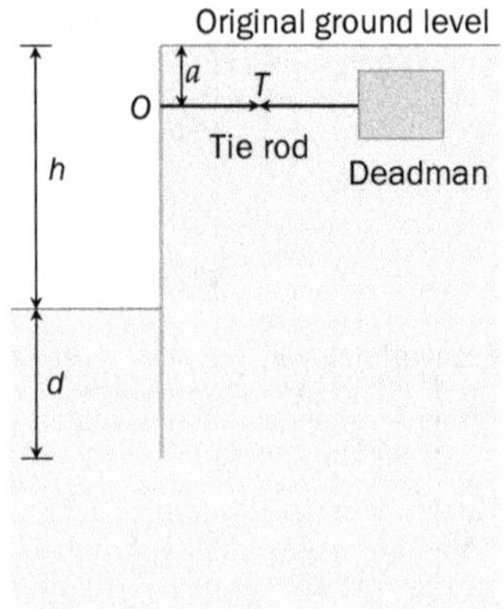

Figure 14.12 Anchored sheet pile.

provide fixity at the bottom and allows the sheet pile to rotate about the bottom tip. In the fixed earth support method, it is assumed that the embedment of the sheet pile is large enough to provide fixity at the bottom, making the analysis a little more complex. Only the free-earth support method is discussed here.

14.4.1 Free-Earth Support Method—Simplified Approach

Let us consider an excavation in a granular soil, supported by an anchored sheet pile as shown in Fig. 14.13a. The depth of excavation is h, and the embedment depth of the sheet pile is d. The tie rod, tied to the sheet pile at O, is located at depth a below the original ground level and fixed to a rigid deadman anchor. Figure 14.13b shows the free-body diagram of the sheet pile subjected to the lateral earth pressures on both sides and the tensile force T from the anchor. Since the sheet pile is rotating about the bottom tip, the entire sand on the right side is assumed to be in active state, and the entire sand on the left is assumed to be in passive state. The resultant active and passive thrusts are given by

$$P_a = \frac{1}{2}K_a\gamma(h + d)^2 \tag{14.43}$$

acting at a height of $(h + d)/3$ from the bottom of the sheet pile, and

$$P_p = \frac{1}{2}K_p\gamma d^2 \tag{14.44}$$

acting at a height of $d/3$ from the bottom of the sheet pile. To provide a factor of safety, K_p in Eq. (14.44) can be replaced by K_p/F, where F is in the range of 1.5–2.0.

The sheet pile is in equilibrium under three forces T, P_a, and P_p. There is no more R we had at the bottom of the cantilever sheet pile to maintain equilibrium. Taking moment about O at the tie rod,

$$P_a\left[\frac{2}{3}(h + d) - a\right] - P_p\left[h - a + \frac{2}{3}d\right] = 0$$

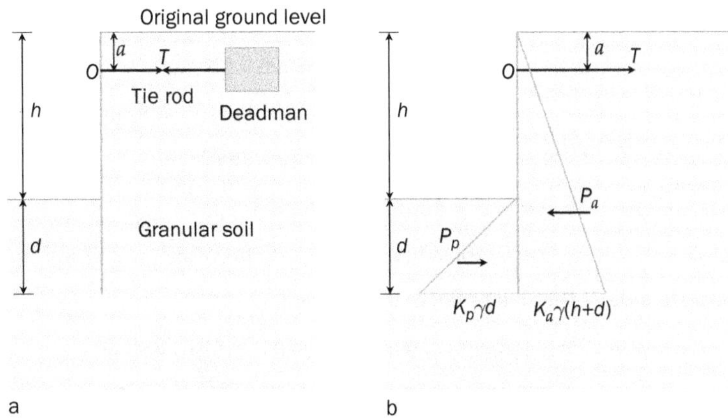

FIGURE 14.13 Analysis of an anchored sheet pile: (a) geometry, (b) free body diagram.

Substituting for P_a and P_p from Eqs. (14.43) and (14.44),

$$\left[\frac{1}{2}K_a\gamma(h+d)^2\right]\left[\frac{2}{3}(h+d)-a\right]-\left[\frac{1}{2}K_p\gamma d^2\right]\left[h-a+\frac{2}{3}d\right]=0 \tag{14.45}$$

Solving Eq. (14.45), d can be determined.
For equilibrium,

$$T = P_a - P_p \tag{14.46}$$

From the known values of P_a and P_p, the tension in the tie rod T can be determined from Eq. (14.46). The maximum bending moment M_{max} occurs above the excavation level. As before, the location can be determined by locating where the shear force is zero. From this M_{max}, and the required section modulus of the sheet pile can be determined. See Example 14.7.

Example 14.6

The top 7 m of a site consists of sands ($\gamma = 18.0$ kN/m^3, $\phi' = 34°$) underlain by clayey sand ($\gamma_{sat} = 20.0$ kN/m^3, $\phi' = 36°$, $c' = 5.0$ kN/m^2). The water table is at a depth of 7 m. It is required to carry out excavation to the depth of the water table, using anchored sheet pile, where the tie rods are placed at 1 m depth below the original ground level. How deep would you drive the sheet pile, assuming a safety factor of 1.5 on the passive resistance? What would be the load on the tie rods if they are placed at 3 m horizontal intervals?

Solution The lateral earth pressure distributions on both sides of the sheet pile are shown in Fig. 14.14. The values shown in the diagram are computed as follows.

For the sand, $\phi' = 34°$ and hence $K_a = \tan^2(45 - 17) = 0.283$

For the clayey sand, $\phi' = 36°$ and hence $K_a = \tan^2(45 - 18) = 0.260$ and $K_p/F = 2.57$

On the right side of the sheet pile, just above the water level,

$$\sigma'_v = 7 \times 18 = 126.0 \text{ kN/m}^2$$

$$\sigma'_h = (0.283)(126.0) = 35.7 \text{ kN/m}^2$$

On the right side of the sheet pile, just below the water level,

$$\sigma'_v = 7 \times 18 = 126.0 \text{ kN/m}^2$$

$$\sigma'_h = (0.260)(126.0) - (2)(5.0)(\sqrt{0.260}) = 27.7 \text{ kN/m}^2$$

On the right side of the sheet pile, at the bottom,

$$\sigma'_v = 7 \times 18 + (20 - 9.81)d = 126.0 + 10.19\, d \text{ kN/m}^2$$

$$\sigma'_h = 0.260(126.0 + 10.19\, d) - (2)(5.0)(\sqrt{0.260}) = 27.7 + 2.65\, d \text{ kN/m}^2$$

On the left side of the sheet pile, at the excavation level

$$\sigma'_v = 0$$

$$\sigma'_h = (2)(5.0)(\sqrt{2.57}) = 16.0 \text{ kN/m}^2$$

On the left side of the sheet pile, at the bottom,

$$\sigma'_v = (20 - 9.81)d = 10.19\, d \text{ kN/m}^2$$

$$\sigma'_h = (2.57)(10.19\, d) + (2)(5.0)(\sqrt{2.57}) = 26.2\, d + 16.0 \text{ kN/m}^2$$

Force	Magnitude (kN/m)	Height to O (m)	Moment (kN·m/m)
P_{a1}	$0.5 \times 35.7 \times 7 = 124.95$	3.67	458.6
P_{a2}	$27.7(d) = 27.7d$	$6.0 + 0.5d$	$13.85\, d^2 + 166.2\, d$
P_{a3}	$0.5(2.65d)(d) = 1.33d^2$	$6.0 + 0.67d$	$0.89\, d^3 + 8.0\, d^2$
P_{p1}	$16.0 \times d = 16.0d$	$6.0 + 0.5d$	$8.0\, d^2 + 96.0\, d$
P_{p2}	$0.5(26.2d)(d) = 13.1d^2$	$6.0 + 0.67d$	$8.78\, d^3 + 78.6\, d^2$

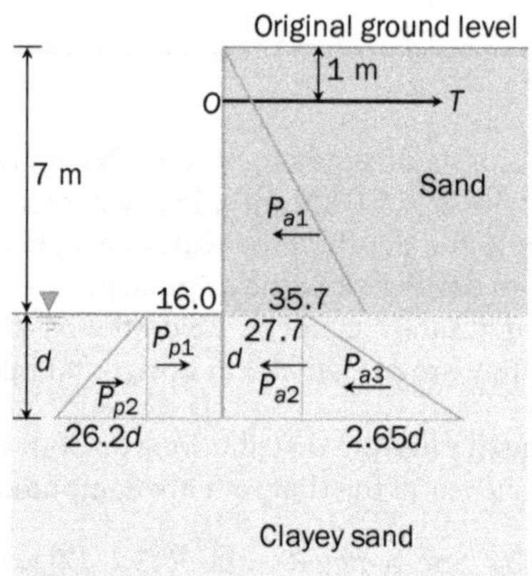

FIGURE 14.14 Lateral earth pressures.

Taking moment about O,

$$458.6 + (13.85\,d^2 + 166.2\,d) + (0.89\,d^3 + 8.0\,d^2) - (8.0\,d^2 + 96.0\,d) - (8.78\,d^3 + 78.6\,d^2) = 0$$

$$7.89\,d^3 + 64.75\,d^2 - 70.2\,d - 458.6 = 0$$

$$d = 2.75\text{ m}$$

and hence the **sheet pile has to be 9.75 m deep.**

Substituting $d = 2.75$ m,

$$P_a = P_{a1} + P_{a2} + P_{a3} = (124.95) + (27.7 \times 2.75) + (1.33 \times 2.75^2) = 211.2\text{ kN/m}$$

$$P_p = P_{p1} + P_{p2} = 16.0 \times 2.75 + 13.1 \times 2.75^2 = 143.1\text{ kN/m}$$

$$T = P_a - P_p = 211.2 - 143.1 = 68.1\text{ kN/m}$$

If the tie rods are spaced at 3.0 m intervals horizontally, load per tie rod is 3 × 68.1 = **204.3 kN.**

It is likely that the maximum bending moment occurs above the excavation level. As we did for the cantilever sheet piles, the location where the shear force is zero can be determined and from this the maximum bending moment can be determined.

Example 14.7

In Example 14.6, determine the maximum bending moment in the sheet pile.

Solution Let us say the maximum bending moment M_{max} occurs at depth z^* below the *original ground level*, where the shear force is zero.

For equilibrium in the horizontal direction for the sheet pile above this location,

$$T - \frac{1}{2}K_a\gamma z^{*2} = 0$$

$$68.1 - \frac{1}{2}(0.283)(18)z^{*2} = 0$$

$$z^* = 5.17\text{ m}$$

Taking moment about this location,

$$M_{max} - T\left(z^* - 1.0\right) + \frac{1}{2}K_a\gamma z^{*2}\left(\frac{z^*}{3}\right) = 0$$

$$M_{max} = 68.1(5.17 - 1.0) - \frac{1}{2}(0.283)(18)(5.17)^2\left(\frac{5.17}{3}\right) = \mathbf{166.7\ kNm\ per\ m}$$

14.4.2 Free-Earth Support Method Using Net Pressure Diagram—Granular Soils

The idea of using net pressure diagrams is similar to what we did with cantilever sheet piles. Let us consider an anchored sheet pile in a granular soil as shown in Fig. 14.15. The water level is at a depth of h_1 below the original ground level. Let us measure the depth z from here.

At $z = h_1$,

$$\sigma'_v = \gamma_m h_1$$
$$\sigma'_h = K_a \gamma_m h_1 = \sigma'_0$$

At $z = h_1 + h_2$,

$$\sigma'_v = \gamma_m h_1 + \gamma' h_2$$
$$\sigma'_h = K_a(\gamma_m h_1 + \gamma' h_2) = \sigma'_1$$

Below the excavation level, where, $z > h_1 + h_2$, the net pressure is shown by the line *GDH* which is at a slope of 1 vertical to $(K_p - K_a)\gamma'$ horizontal. Therefore, h_3 is given by

$$h_3 = \frac{\sigma'_1}{(K_p - K_a)\gamma'} \tag{14.47}$$

From these, P and $\bar{z}$ can be determined as in the case of cantilever sheet piles.

At the bottom of the sheet pile, the net pressure σ'_2 is given by

$$\sigma'_2 = (K_p - K_a)\gamma' h_4 \tag{14.48}$$

For equilibrium,

$$T + \text{Area } HDB \text{ of the pressure diagram} - P = 0$$

$$T + \frac{1}{2}\sigma'_2 h_4 - P = 0$$

$$T = P - \frac{1}{2}(K_p - K_a)\gamma' h_4^2 \tag{14.49}$$

To determine T from Eq. (14.49), it is necessary to find h_4. Let us take moment about O.

$$P[h_1 + h_2 + h_3 - a - \bar{z}] - \left(\frac{1}{2}(K_p - K_a)\gamma' h_4^2\right)\left(h_1 + h_2 + h_3 - a + \frac{2}{3}h_4\right) = 0 \tag{14.50}$$

With all other parameters in Eq. (14.50) known, h_4 can be determined. Therefore, d becomes: $d = h_3 + h_4$. As in the case of cantilever sheet piles in granular soils, d has to be increased by 20–40% or a factor of safety of 1.5–2.0 is used on the passive resistance.

The maximum bending moment occurs where the shear force is zero. This is generally between the water level and the excavation levels (i.e., $h_1 < z < h_1 + h_2$). Other possibilities can be checked for different situations. Let us say the maximum bending moment and zero shear force in the sheet pile occur at $z = z^*$.

For equilibrium in the horizontal direction for the section above z^*,

$$\frac{1}{2}\sigma'_0 h_1 + \sigma'_0(z^* - h_1) + \frac{1}{2}K_a\gamma'(z^* - h_1)^2 - T = 0 \tag{14.51}$$

From Eq. (14.51), z^* can be determined. From this, the maximum bending moment and the required section modulus can be determined.

14.4.3 Free-Earth Support Method Using Net Pressure Diagram—Anchored Sheet Piles Penetrating Clays

Let us consider the situation shown in Fig. 14.16 where the sheet pile is penetrating into the clay. Immediately after the sheet pile is installed, the clay is under undrained conditions and hence we will analyze the clay in terms of total stresses with c_u and $\phi_u = 0$ ($K_a = K_p = 1$). The granular soil is always drained and hence we will use the effective stress parameters for the granular soils. The lateral earth pressure diagram in the granular soil is the same as in Fig. 14.15.

From the values of σ_{ha} from right to left and σ_{hp} from left to right within the clay, the net pressure p [similar to the way we determined p_2 in Eq. (14.36)], in terms of total stresses, is given by

$$p = 4c_u - \left(\gamma_{m,g}h_1 + \gamma'_g h_2\right) \tag{14.52}$$

which remains the same within the clay as shown in the figure. The unit weight of the clay does not appear in the equation. The subscript g in the unit weights denotes the granular soils.

For the equilibrium of the forces in the horizontal direction,

$$T - P + pd = 0 \tag{14.53}$$

Taking moment about O,

$$P[h_1 + h_2 - a - \bar{z}] - pd\left[h_1 + h_2 + \frac{1}{2}d - a\right] = 0$$

$$pd^2 + 2p[h_1 + h_2 - a]d - 2P[h_1 + h_2 - a - \bar{z}] = 0 \tag{14.54}$$

Solving Eq. (14.54), d can be determined.

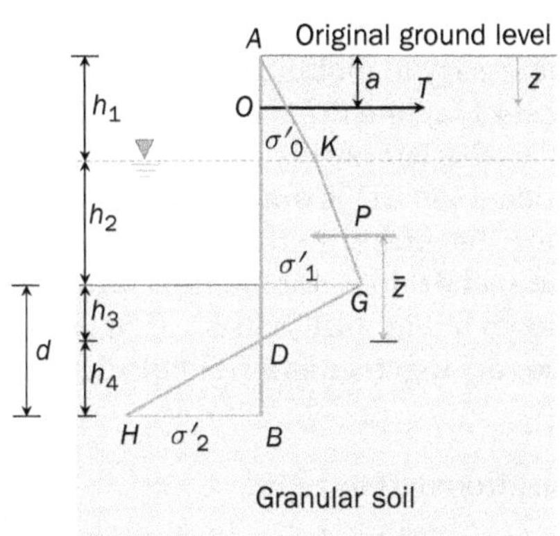

FIGURE 14.15 Net pressure diagram for anchored sheet pile in granular soils.

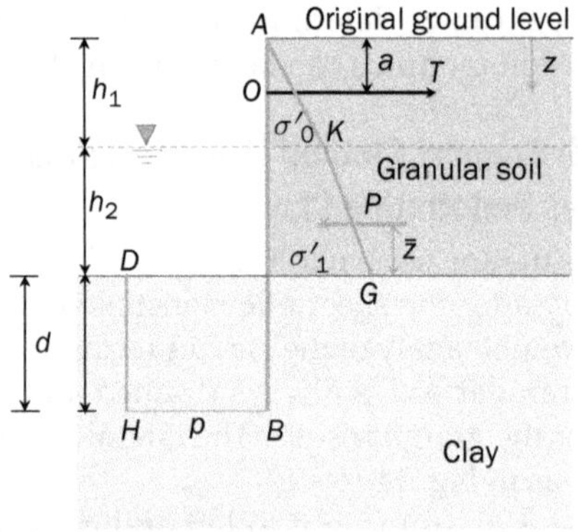

FIGURE 14.16 Anchored sheet pile penetrating clays.

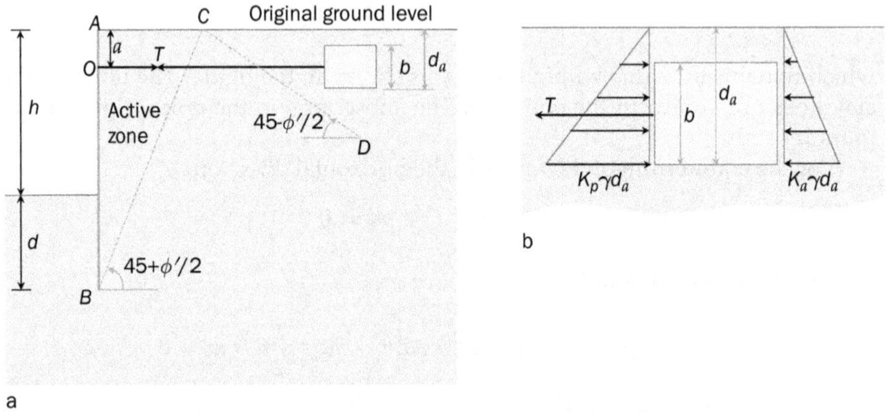

FIGURE 14.17 Deadman anchor: (a) suggested location, (b) force equilibrium.

14.4.4 Deadman Anchor

A simple approach to the design of deadman anchor is discussed here. For the anchor to be effective, it is necessary to mobilize the full passive resistance from the soil in front of the anchor. In Fig. 14.17a, line *BC* is inclined at $45 + \phi'/2$ to the horizontal, defining the active zone *ABC* to the left of this line. Obviously, the anchor must not be located within this failure zone. Line *CD* is drawn such that it is at an angle of $45 - \phi'/2$ to the horizontal. By locating the anchor to the right of *CD*, it can be ensured that the passive zone to the left of the anchor does not interfere with the active zone.

The anchor is shown in Fig. 14.17b. If $b > 0.5d_a$, it can be assumed the soil above the anchor and the anchor act together as a rigid block. The anchor is in horizontal equilibrium under

(a) passive thrust from left to right,

(b) active thrust from right to left, and

(c) the force on the tie rod.

Allowing for a factor of safety $F = 1.5$–2.0 on the tie rod force T, where the tie rods spaced horizontally at s,

$$F \times T + \frac{1}{2}K_a\gamma d_a^2 s - \frac{1}{2}K_p\gamma d_a^2 s = 0$$

$$F\,T = \frac{1}{2}(K_p - K_a)\gamma d_a^2 s \tag{14.55}$$

With known value of T and assumed factor of safety F, d_a can be determined from Eq. (14.55). The anchor can be in the form of a continuous beam to which the tie rods are connected at spacing s, or separate anchors attached to tie rods at spacing s.

14.5 Braced Excavations

To install pipe lines or other services below the ground, it is required to dig trenches to bury them. These trenches can be deep and it becomes necessary to protect the sides of the excavations from failure. Figure 14.18a shows a trench of depth H, excavated into the ground, supported by the two sheet piles on the sides. As the excavation progresses, to keep the sheet piles in place *wales* and *struts* are placed as necessary. Wales are like beams placed longitudinally along the length of the excavation at different levels. The horizontal struts, held between two wales, are in compression. The bracing system, consisting of sheets, wales, and struts are designed based on the lateral earth pressures on the two sides of the excavation.

Based on some *in situ* measurements of the strut loads in several excavations in Berlin, Munich, Toronto, New York, Chicago, and other areas, Peck (1969) suggested the apparent lateral earth pressure envelopes shown in Figs. 14.18b, c, and d for sands, stiff clays, and soft to medium clays, respectively. He defined a dimensionless *stability number* $\gamma H/c_u$ to separate stiff clays from soft to medium clays. A clay is considered stiff if $\gamma H/c_u < 4$. The clay is soft to medium when $\gamma H/c_u > 4$.

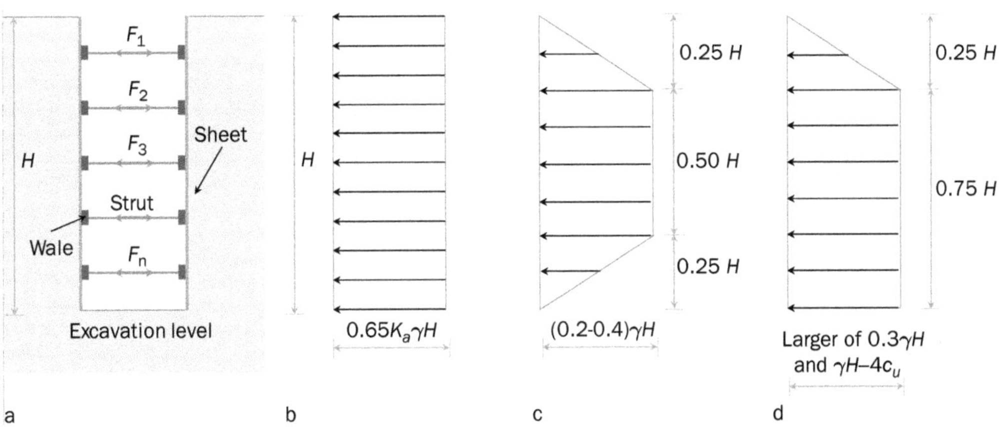

Figure 14.18 Braced excavations: (a) geometry, and apparent earth pressure envelopes in (b) sands, (c) stiff clays, (d) soft to medium clays.

14.5.1 Determination of Strut Loads

The strut loads F_1, F_2, to F_n are determined based on the pressure envelope for the specific soil. For the right wall of the excavation in Fig. 14.18a, the lateral earth pressures from right to left are defined by the assumed pressure envelope. These loadings are resisted by the strut loads. To simplify the analysis, it is assumed that except for the top and bottom wall-strut joints, all other wall-strut joints are treated as hinges, which cannot carry moments.

Figure 14.19a shows a braced excavation in soft clay where the undrained shear strength is c_u and the unit weight is γ. The middle three wall-strut joints are shown as hinges. The pressure envelope shown in Fig. 14.19b is separated along the hinges into four blocks A, B, C, and D (see Fig. 14.19c where the blocks are staggered for clarity). The strut load F_2 acting on the topmost hinge is separated into two components F_2' and F_2'', acting on blocks A and B, respectively. Similarly, F_3 and F_4 are also separated into two components.

By considering equilibrium of each of the four blocks, with two equilibrium equations for each block, the eight unknowns F_1, F_2', F_2'', F_3', F_3'', F_4', F_4'', and F_5 can be determined. $F_2 = F_2' + F_2''$, $F_3 = F_3' + F_3''$, and $F_4 = F_4' + F_4''$.

14.5.2 Design of Sheet Piles and Wales

As in the case of cantilever or anchored sheet piles, the minimum required section modulus of the sheet pile is determined from the maximum bending moment M_{max} acting on the sheet pile. Each block in Fig. 14.19c can be seen as a simply supported beam. For each block, the maximum bending moment has to be determined, and the largest of these maximums is selected as the M_{max}. The minimum required section modulus for the sheet pile is determined as

$$Z = \frac{M_{max}}{\sigma_{all}} \tag{14.56}$$

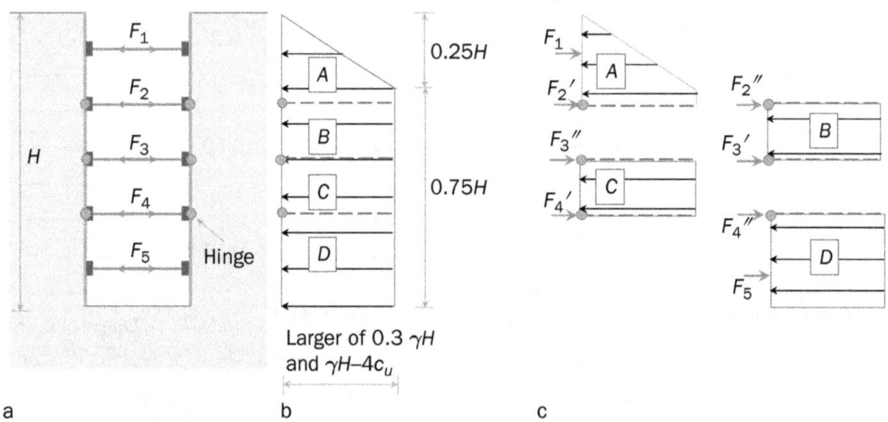

FIGURE 14.19 Determination of strut loads: (a) geometry showing the hinges, (b) pressure envelope, (c) pressure envelope separated along the hinges into blocks.

The longitudinal wales are designed as uniformly loaded beams pinned at the strut-wale joints, with the maximum moment M_{max} occurring at the mid-span. M_{max} is given by

$$M_{max} = \frac{ws^2}{8}$$

where w = uniformly distributed load along the wale in kN/m, and s = strut (support) spacing. The load at the ith support (i.e., the strut) is given by

$$F_i = ws$$

Therefore,

$$M_{max} = \frac{F_i s}{8} \tag{14.57}$$

The section modulus of the wale can be determined from Eq. (14.56).

14.5.3 Bottom Heave in Clays

When braced excavations are made in soft clays, the bottom of the excavation between the walls can fail, resulting in soft clay getting pushed into the excavation through the bottom, endangering safety. This is known as *bottom heave*. Such failures occur during the early stages of the excavation, and take place under undrained conditions. Therefore, bottom heave can be analyzed using the undrained shear strength parameters c_u and $\phi_u = 0$, and by considering bearing capacity failure similar to that of a shallow foundation in soft clays.

In Figs. 14.20a and b, the failure surface shown on the right by a dashed line *abcd* consists of a vertical surface *ab*, a cylindrical surface *bc* with the axis through *e*, and an

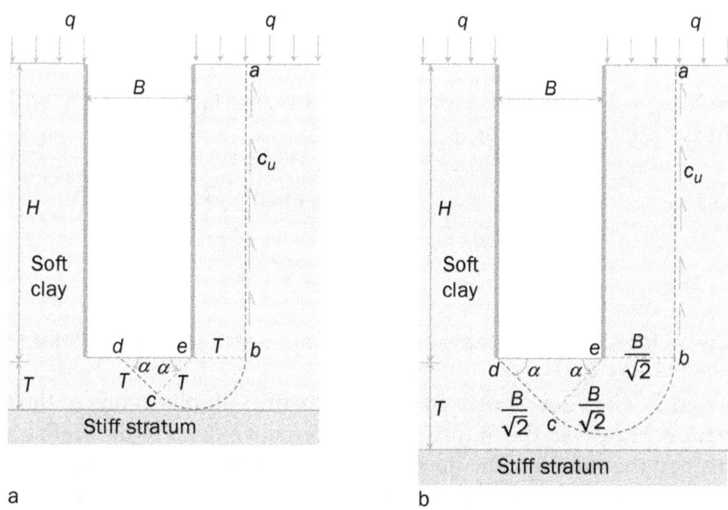

FIGURE 14.20 Bottom heave in soft clays: (a) $T < B/\sqrt{2}$, (b) $T > B/\sqrt{2}$.

inclined plane *cd*. The failure mechanism assumed here is a form of bearing capacity failure of a shallow foundation in reverse. Here, a simple failure mechanism suggested by Terzaghi (1943) is assumed. In deriving the bearing capacity equation for the shallow foundation, we noted that the angle α is $45 + \phi/2$ (see Fig. 12.5). In undrained conditions in clays, $\alpha = 45°$. Let us consider two situations as follows:

a. the stiff stratum close to the excavation level ($T < B/\sqrt{2}$), and

b. the stiff stratum is not present within the vicinity of the excavation level ($T > B/\sqrt{2}$).

Here, T is the depth below the excavation to the stiff stratum.

Case I: $T < B/\sqrt{2}$

In Fig. 14.20a, the stiff stratum is at depth T below the excavation level and $T < B/\sqrt{2}$. Since *bc* is a circular arc, $eb = ec = T$. Also, $cd = ce = T$ and $de = \sqrt{2}T$, which is less than B. Therefore, the failure does not extend to the full width of the excavation.

The failure can occur when the soil column between *ab* and the sheet pile can get pushed down and then upward through the bottom of the excavation. The slide is resisted by the shear strength c_u along the failure surface. Let us check the possibility for bearing capacity failure along *eb*.

The net applied pressure at *eb* is

$$q_{\text{app,net}} = \frac{\text{Load}}{\text{Area}} = \frac{(q \times T \times L) + (\gamma \times T \times H \times L) - (c_u \times H \times L)}{L \times T}$$

$$q_{\text{app,net}} = q + \gamma H - c_u \frac{H}{T} \tag{14.58}$$

where q = surcharge pressure at the ground level, and L = length of the excavation. The net ultimate bearing capacity at *eb* is given by

$$q_{\text{ult,net}} = c_u N_{c,\text{strip}} \left(1 + 0.2\sqrt{2}\frac{T}{L}\right) \tag{14.59}$$

where $N_{c,\text{strip}} = 5.14$ (from Chap. 12). Therefore, the factor of safety with respect to bottom heave can be defined as

$$F = \frac{c_u N_{c,\text{strip}} \left(1 + 0.2\sqrt{2}\frac{T}{L}\right)}{q + \gamma H - c_u \frac{H}{T}} \tag{14.60}$$

The factor of safety with respect to bottom heave must be greater than 1.5.

Case II: $T > B/\sqrt{2}$

When $T > B/\sqrt{2}$ as shown in Fig. 14.20b, the failure extends to the full width of the excavation. However, the failure surface does not extend to the stiff stratum, which has no effect on the stability. The derivations are similar to the previous scenario and T has to be replaced by $B/\sqrt{2}$ in Eqs. (14.58), (14.59), and (14.60). In other words, the smaller value of T and $B/\sqrt{2}$ must be used instead of T in Eq. (14.60).

Example 14.8

Figure 14.21a shows a 3.2-m-wide braced excavation in a clay where $\gamma = 18.0$ kN/m^3, and $c_u = 20$ kN/m^2. The struts are spaced horizontally at 2.5 m intervals.

a. Find the strut loads.

b. Find the factor of safety with respect to bottom heave, assuming the depth to the stiff stratum is very large. Assume a very long excavation, and neglect the surcharge at the ground level.

c. Determine the minimum section modulus required for the third wale, assuming allowable stress of 225 MN/m^2.

Solution

Part a: Strut loads

$$\frac{\gamma H}{c_u} = \frac{18 \times 10}{20} = 9 > 4$$

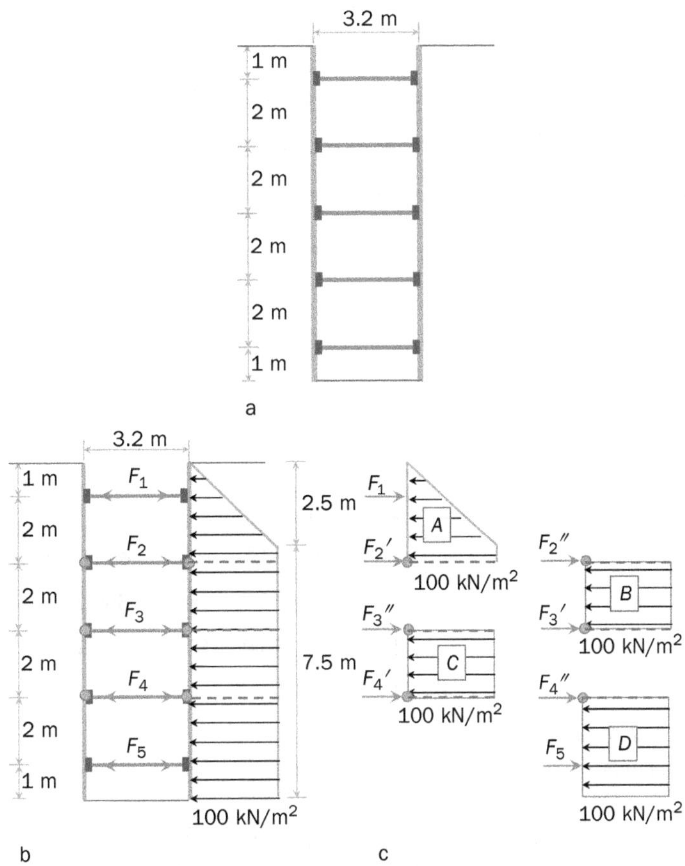

FIGURE 14.21 (a) Geometry, (b) Pressure envelope, (c) pressure envelope separated into blocks along the hinges.

Therefore, the clay must be treated as soft to medium, with pressure envelope as shown in Fig. 14.18d.

$$0.3\gamma H = (0.3)(18.0)(10) = 54.0\ \text{kN/m}^2$$

$$\gamma H - 4c_u = (18.0)(10) - (4)(20) = 100.0\ \text{kN/m}^2 \leftarrow \text{Use this larger value}$$

The strut loads and the pressure envelopes are shown in Fig. 14.21b. The pressure envelope is separated into four blocks (*A*, *B*, *C*, and *D*) along the three hinges as shown in Fig. 14.21c.

For equilibrium of block *A* in the horizontal direction,

$$F_1 + F_2' = (100)(0.5)(2.5) + (0.5)(100)(2.5)(2.5) = 125.0 + 312.5 = 437.5\ \text{kN}$$

Taking moment for the forces in block *A* about the hinge,

$$F_1 \times 2.0 - (125.0)(0.25) - (312.5)\left(0.5 + \frac{2.5}{3}\right) = 0$$

$$F_1 = 224.0\ \text{kN and } F_2' = 213.5\ \text{kN}$$

For equilibrium of the horizontal forces in block *B* and by symmetry,

$$F_2'' = F_3' = 0.5(100)(2)(2.5) = 250\ \text{kN}$$

For equilibrium of the horizontal forces in block *C* and by symmetry,

$$F_3'' = F_4' = 0.5(100)(2)(2.5) = 250\ \text{kN}$$

For equilibrium of block *D* in the horizontal direction,

$$F_4'' + F_5 = (100)(3.0)(2.5) = 750.0\ \text{kN}$$

Taking moment for the forces in block *D* about the hinge,

$$F_5 \times 2 - 750(1.50) = 0$$

$$F_5 = 562.5\ \text{kN and } F_4'', = 187.5\ \text{kN}$$

Therefore, the strut loads are:

$$F_1 = \mathbf{224.0\ kN}$$

$$F_2 = F_2' + F_2'' = 213.5 + 250.0 = \mathbf{463.5\ kN}$$

$$F_3 = F_3' + F_3'' = 250.0 + 250.0 = \mathbf{500.0\ kN}$$

$$F_4 = F_4' + F_4'' = 250.0 + 187.5 = \mathbf{437.5\ kN}$$

$$F_5 = \mathbf{562.5\ kN}$$

Part b: Bottom heave

From Eq. (14.60),

$$F = \frac{c_u N_{c,\text{strip}}\left(1 + 0.2\sqrt{2}\frac{T}{L}\right)}{q + \gamma H - c_u \frac{H}{T}}$$

With depth to the stiff stratum being large, T has to be replaced by $B/\sqrt{2} = 3.2/\sqrt{2} = 2.26$ m.

$$F = \frac{c_u N_{c,\text{strip}}\left(1 + 0.2\sqrt{2}\dfrac{T}{L}\right)}{q + \gamma H - c_u \dfrac{H}{T}} = \frac{20 \times 5.14}{(18)(10) - (20)\left(\dfrac{10}{2.26}\right)} = \mathbf{1.12}$$

Part c: Third wale from the top

From Eq. (14.57),

$$M_{\text{max}} = \frac{F_3 s}{8} = \frac{(500 \times 10^3)(2.5)}{8} = 156.25 \times 10^3 \text{N·m}$$

$$Z = \frac{M_{\text{max}}}{\sigma_{\text{all}}} = \frac{156.25 \times 10^3}{225 \times 10^6} = 0.69 \times 10^{-3} \text{m}^3$$

14.6 Retaining Walls Made of Piles

14.6.1 Bored Pile Walls

Bored piles can be used for constructing retaining walls. They are effective for construction of basements and in limited spaces where ground movement has to be minimized. With low construction noise and vibration, they can be installed close to existing structures in any ground conditions. Two popular bored pile retaining wall systems are *secant pile walls* and *contiguous bored pile* (or *tangent pile*) *walls*. These piles can be as large as 1500 mm in diameter and can be made using continuous flight auger (CFA) or by rotary drilling.

A contiguous bored pile (CBP) wall consists of a line of single piles almost touching the adjacent ones. When constructed, there is a small gap of about 100–150 mm between the adjacent piles, thus exposing the soil during excavation, as shown in Fig. 14.22a. The soil will generally stay in place due to arching mechanism. The gap between the piles can be grouted to minimize seepage. CBP walls are suitable when the ground water table is below the excavation level and hence seepage is not a problem.

Secant pile wall system is preferable when the wall is required to be water-tight. Here, primary piles (also known as female piles) are constructed first, followed by the secondary piles (called male piles) placed between the primary ones, with an overlap of about 150 mm as shown in Fig. 14.22b. The primary piles can be reinforced or unreinforced. When the secondary piles are bored, they cut into the adjacent primary piles, to form a continuous wall. For the ease of construction, generally the primary piles are unreinforced or made of lower grade concrete or cement-bentonite mix. Cutting into a high-grade concrete pile can be relatively difficult. The secondary pile is generally reinforced and is of high-grade concrete (e.g., 40 MPa). They are installed while the primary pile is still soft and not hardened. A secant pile can be a cost-effective alternative to a diaphragm wall, for depths less than about 12 m. For larger depths, diaphragm walls are economical.

CBP walls are more economical than secant pile walls. The exposed soils in CBP walls can self-support temporarily, due to arching mechanism. However, it is often

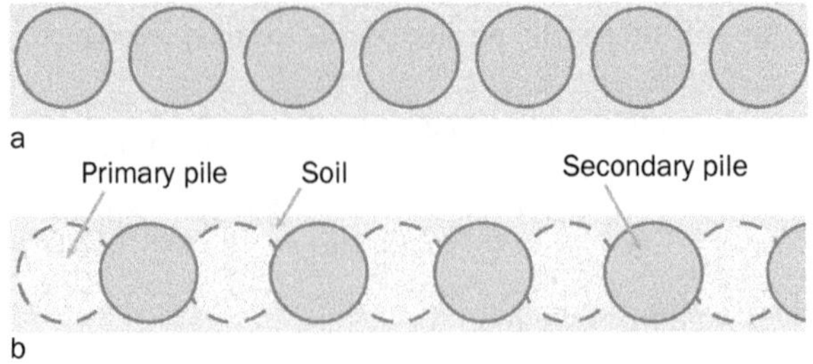

FIGURE 14.22 (a) Contiguous bored pile wall, (b) secant pile wall.

FIGURE 14.23 Contiguous bored pile wall.

necessary to have a concrete lining to hold the soil in place and to prevent erosion and any ground water seepage. Any possible erosion can lead to settlement of the soil that is retained by the wall. The exposed pile walls can be relatively rough and therefore the concrete panel lining will make it look better. Figure 14.23 shows the photograph of a contiguous bored pile wall where the gaps between the piles and the rough surfaces can be seen.

Secant piles and CGPs are generally limited to 15 m in depths due to the accuracy of installation and alignment problems. For larger depths, diaphragm walls are better alternatives. All three can be considered as rigid retaining walls, which would limit the wall movements. When the distance between the piles exceeds the pile diameter, the arrangement is known as *intermittent bored pile wall* which is possible only in overconsolidated or cemented soils. Bored piles walls are generally made a permanent structure. The top of the bored pile wall is generally capped by a reinforced concrete beam to see that the wall distributes the lateral thrust from the soil.

14.6.2 Pipe Pile Walls

Figure 14.24a shows 1200-mm-diameter pipe piles, with wall thickness of 14–19 mm, being driven into the ground. Here, large pipes are installed in the ground, very close

a b

FIGURE 14.24 (a) Pipe pile wall, (b) contiguous pipe pile wall.

to each other, with steel connectors between the adjacent piles. Figure 14.24b shows the finished wall, with the steel connectors between the piles. The steel pipe piles can be removed when required for later reuse, and hence form temporary retaining walls. There are also specially made interlocking pipe piles which act like sheet piles.

14.7 Main Points

1. It is easier to design retaining walls, sheet piles, and braced excavations using Rankine's earth pressure theory than Coulomb's.
2. Retaining walls must be checked for stability against sliding and overturning. It is also a good practice to check for bearing capacity.
3. Cantilever sheet piles are adequate for excavations up to about 6.0 m. For deeper excavations, anchored sheet piles may be economical.
4. The sheet pile sections are selected based on the minimum section modulus required, determined using the maximum bending moment and the allowable stress.
5. In braced excavations, the strut loads are computed from pressure envelopes.
6. Bottom heave occurs mainly in braced excavations in soft clays. Use smaller of T and $B/\sqrt{2}$ instead of T in computing the factor of safety.

Review Exercises

14.1 Surf the Web and prepare a two-page essay on the different types of earth retaining structures. Include pictures.

14.2 Surf the Web and identify some commonly used sheet pile sections and list the properties including the moment of inertia and the section modulus. Distinguish straight web, U-shape (arch web), and Z-shape sheet piles.

14.3 Determine the factor of safety of the retaining wall shown in Fig. RE14.3 with respect to sliding, overturning, and bearing capacity failure. Unit weight of concrete is 24.0 kN/m^3, and $\delta = {}^2/_3\phi'$. Being unsure of the soil conditions in the front of the wall in the long term, it is suggested by the client to neglect the passive resistance in the computations.

Answers: 1.70, 4.82, 7.7

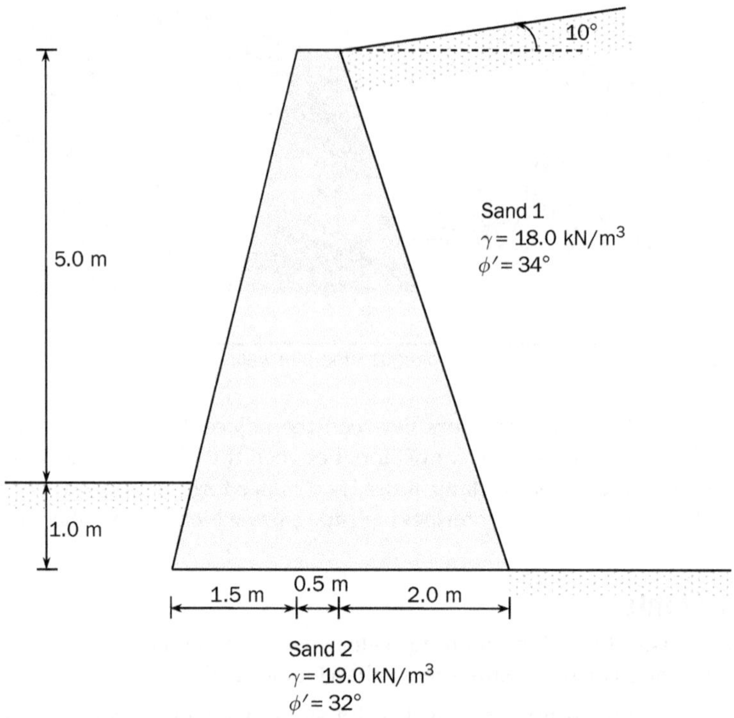

FIGURE RE14.3 Gravity retaining wall.

14.4 Figure RE14.4 shows a retaining wall constructed on an existing ground where $\gamma = 18.0$ kN/m^3 and $\phi' = 36°$. The wall retains a 4.0-m-high backfill consisting of loose sand with $\gamma = 18.0$ kN/m^3 and $\phi' = 30°$. Unit weight of concrete is 24.0 kN/m^3, and $\delta = \frac{2}{3}\phi'$.

a. Assuming the soil on the right side of the wall is in active state, determine the magnitude and the location of the active thrust on the wall.

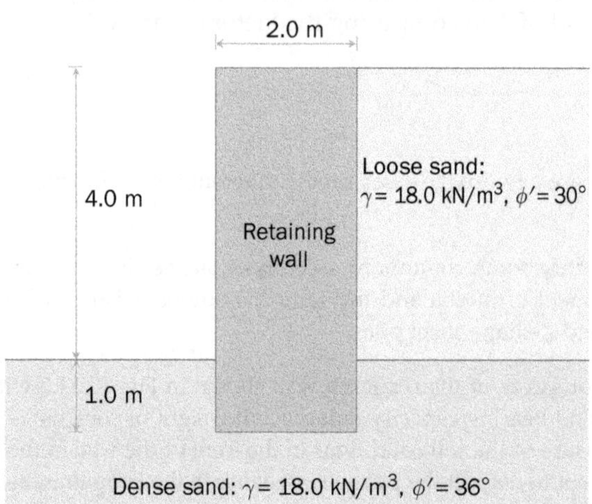

FIGURE RE14.4 Retaining wall.

b. Evaluate the stability of the wall with respect to sliding and overturning.

c. Check whether the reaction at the base lies within the mid-third.

Answer: 69.1 kN/m and 1.77 m above the base; 2.05, 2.06; outside the mid-third

14.5 A gravity retaining wall of 6 m high is shown in Fig. RE14.5, retaining a sandy backfill with $\phi' = 34^\circ$ and $\gamma = 17.5$ kN/m^3. The base is resting on a gravelly sand where $\phi' = 38^\circ$ and $\gamma = 18.0$ kN/m^3. Find the minimum possible value for the dimension x such that the factor of safety with respect to sliding and overturning are at least 1.5 and 2.0, respectively. Unit weight of concrete = 24 kN/m^3 and $\gamma = {}^2\!/_3\phi'$.

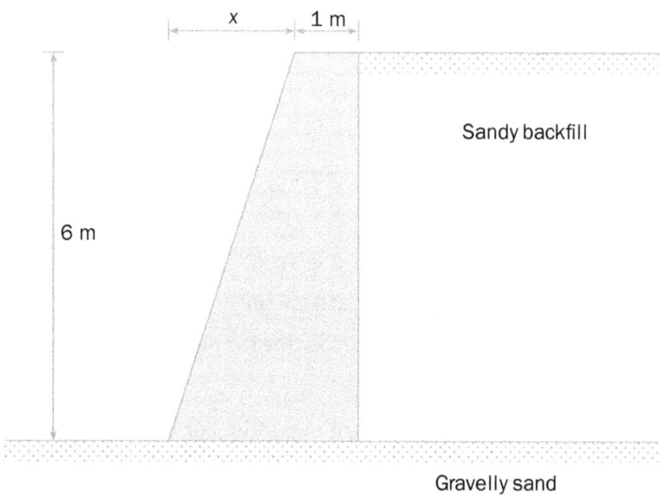

FIGURE RE14.5

Answer: 1.92 m

14.6 A cantilever retaining wall shown in Fig. RE14.6 retains a sandy backfill with $\phi' = 32^\circ$ and $\gamma = 17.5$ kN/m^3. The same soil is present beneath the wall and in front of the wall. Unit weight of concrete = 24 kN/m^3 and $\delta = {}^2\!/_3\phi'$. Fid the factors of safety with respect to sliding and overturning.

Answer: 1.27, 3.47

14.7 The water table at a site lies at depth of 5 m below the ground level, and it is required to excavate to this level. The soil profile consists of a thick bed of sand where the unit weight is 17.0 kN/m^3 above the water table and 20.0 kN/m^3 below. The friction angle of the sand is 37°. How deep would you drive the sheet pile for carrying out this excavation, such that the safety factor on passive resistance is 1.5? Determine the maximum bending moment in the sheet pile and the required section modulus for a steel section, assuming allowable stress of 190 MN/m^2. Use the simplified method.

Answer: 10.6 m, 230.6 kN·m/m, 12.1 $\times$ 10^{-4} m^3/m

14.8 The soil profile at a site consists of a 5-m-thick sand deposit underlain by clayey sand. The water table lies at the top of the clayey sand layer. It is proposed to carry out an excavation up to the water table depth, supported by cantilever sheet pile, with a safety factor of 1.5 on the passive resistance. The design parameters for the two soils are as follows.

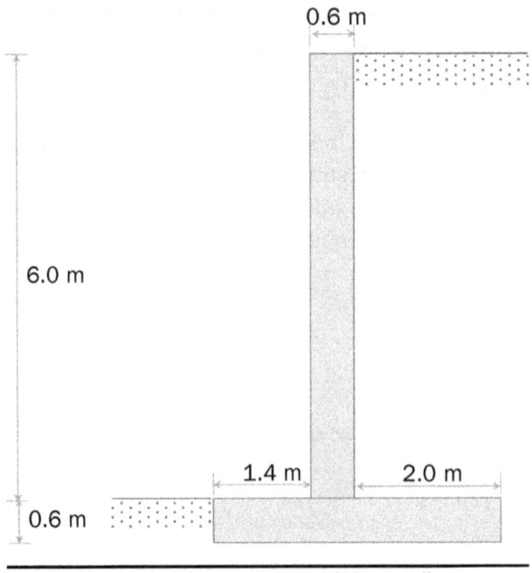

FIGURE RE14.6 Cantilever retaining wall.

Sand: Bulk unit weight, $\gamma_m = 17.5$ kN/m^3; $\phi' = 35°$

Clayey sand: Saturated unit weight $\gamma_{sat} = 20.0$ kN/m^3; $c' = 5$ kPa; $\phi' = 38°$

How deep should the sheet pile be driven? Use the simplified method.

Answer: 9.3 m

14.9 Refer to the net pressure diagram shown in Fig. 14.10. For an excavation where $h_1 = 2$ m, $h_2 = 4$ m, $h = 6$ m, $\gamma_m = 17.5$ kN/m^3, $\gamma_{sat} = 19.0$ kN/m^3, and $\phi' = 35°$, how deep would you drive the sheet pile into the ground? Determine the depth d and increase by 20%.

Answer: 12.83 m

14.10 There is 5.0 m of sand below the ground level, underlain by a thick clay deposit. The water table is at 1.5 m below the ground level. It is required to excavate 5.0 m, to the top of the clay deposit. How deep would you drive the sheet pile? Allow for 50% increase on the embedment depth d. The soil properties are as follows. Use net pressure diagram.

Sand: Bulk unit weight = 16.5 kN/m^3, saturated unit weight = 20.0 kN/m^3

Friction angle = 32°

Clay: Saturated unit weight = 19.5 kN/m^3, undrained cohesion = 30.0 kN/m^2

Answer: 9.9 m

14.11 The water table lies at a depth of 3 m below the ground level at a site that consists entirely of sands. A 7-m-deep excavation is to be made with the use of anchored sheet piles to support the excavation. The tie rod of the anchor will be placed 1 m below the ground, with horizontal spacing of the tie rods being 3 m. The unit weight of the sand above and below the water table are 17.0 and 20.0 kN/m^3, respectively. The friction angle of the sand is 35°. Using the simplified method, answer the following.

a. How deep would you drive the sheet pile, allowing a safety factor of 1.5 on the passive resistance?

b. What is the load on the tie rods?

Answer: 10.25 m, 183 kN

14.12 Redo Problem RE14.11 using the net pressure diagram and referring to Fig. 14.15. Compare the results between RE14.11 and RE14.12.

Answer: 10.15 m, 198 kN

14.13 In Problem RE14.12, determine the location of the maximum bending moment, the magnitude of the maximum moment, and the minimum required section modulus if the allowable stress of the sheet pile is 190 MN/m^2.

Answer: 5.60 m below the original ground level, 174.4 kN·m/m, 0.92 $\times$ 10^{-3} m^3/m

14.14 A 9.0-m-deep and 3.0-m-wide braced excavation is made in a clay where $c_u = 45$ kN/m^2 and $\gamma = 19.0$ kN/m^3. The struts are placed at four levels as shown in Fig. RE14.14, at 2.5 m apart horizontally along the trench.

a. Determine the strut loads.
b. Determine the minimum required section modulus for the second wale from the top. Assume allowable stress of 225 MN/m^2.

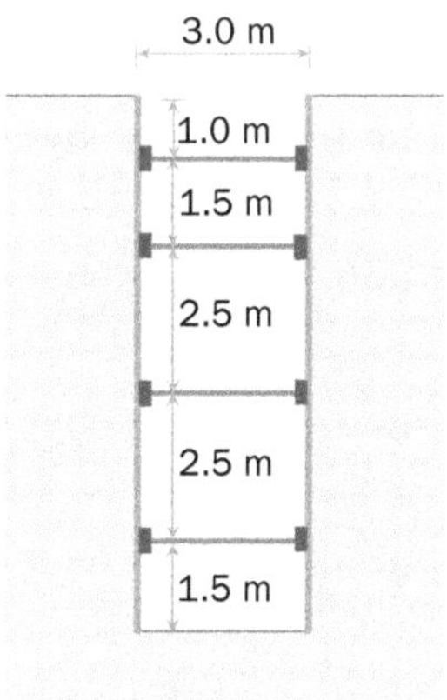

FIGURE RE14.14 Braced excavation.

Answer: 98.9 kN, 141.6 kN, 198.8 kN, 314.2 kN; 0.20 $\times$ 10^{-3} m^3

14.15 In a thick clay deposit where $\gamma = 17.0$ kN/m^3, and $c_u = 16$ kN/m^2, a 3.0-m-wide and 20-m-long braced excavation will be carried out. How deep can the excavation go before bottom heave occurs? Assume that the stiff stratum is located very deep.

Answer: 9.0 m

14.16 Figure RE14.16 shows a braced excavation in sand where $\gamma = 18.0$ kN/m^3 and $\phi' = 32°$. The struts are spaced at 3.0 m intervals along the trench.

a. Determine the strut loads.
b. Determine the maximum bending moment in the sheet pile and find the minimum required section modulus of the sheet pile. Assume allowable stress of 170 MN/m^2.

Answer: 275.5 kN, 137.0 kN, 275.5 kN; 32.3 kN·m/m, 0.19 $\times$ 10^{-3} m^3/m

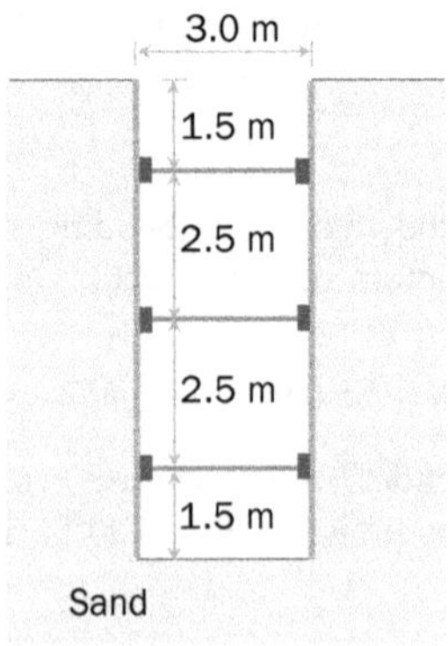

FIGURE RE14.16 Braced excavation.

References

Peck, R. B. "Deep Excavations and Tunnelling in Soft Ground," *Proceedings of the Seventh International Conference on Soil Mechanics and Foundation Engineering*, Mexico City, State-of-the-art volume, 1969, pp. 225–290.

Terzaghi, K. *Theoretical Soil Mechanics*, John Wiley & Sons, New York, 1943.

CHAPTER 15
Slope Stability

15.1 Introduction

Slopes can be natural or artificial. Natural slopes are found mostly in hilly terrains or formed by the nature through geological processes such as landslides, erosion, ground subsidence, or settlements. Artificial slopes are made by humans while carrying out excavations or building embankments or similar structures. When the ground is not horizontal, it is possible that part of the soil mass from the higher ground may slide downward. The extent of damage depends on the volume of the soil that has displaced and the consequences. Some of the factors triggering slope failures include rainfall, weathering, dynamic loads (e.g., earthquake, pile driving), erosion, scouring, and fluctuations of water table.

Figure 15.1 shows the rockslide in the small mining town Frank in Alberta, Canada, in 1903, displacing 110 million tonnes of limestone rock boulders, cobbles, and aggregates which buried a part of the town, with an estimated death toll of about 80. The failure was so sudden, the debris avalanche took just a few minutes and the rock materials were reported to have traveled at a speed of 110 km/h.

Figure 15.2 shows an embankment and an excavation on a stable horizontal ground. In both cases, the stability is governed by the height H, slope angle with horizontal β, and the shear strength parameters cohesion c and friction angle ϕ of the soils. The larger the height H or the slope β, the greater are the chances for instability. Stability of the slope is also influenced by the shear strength parameters of the existing ground and the embankment. The larger the shear strength parameter, the greater is the stability.

In this chapter, we will treat slope stability as a two-dimensional problem, where the third dimension perpendicular to the plane is large. Therefore, only a transverse section of the slope is considered in the analysis. The observations of failed slopes suggest that the failure surface can be approximated as a circular arc such as the one shown in Fig. 15.3a. In reality, the slope failure can be a three-dimensional problem but the analysis becomes too complex.

Figure 15.3 shows three possible failure modes. The first one shown in Fig. 15.3a is known as *rotational failure*, which is common in homogeneous soils. The failure surface is cylindrical, where the soil contained within it rotates about the horizontal axis through the center O. When a relatively thin layer of soil is underlain by a stiff stratum such as bed rock, as shown in Fig. 15.3b, the soil mass slides along the slope, parallel to the stiff stratum. This is known as *translational failure*. Here, the failure surface is the interface between the soil and the stiff stratum. In situations such as the one shown in Fig. 15.3c, the failure arc developed from the ground level on the right side cannot pass through the underlying stiff stratum. The soil mass in the middle slides along the

FIGURE 15.1 Frank slide in Canada.

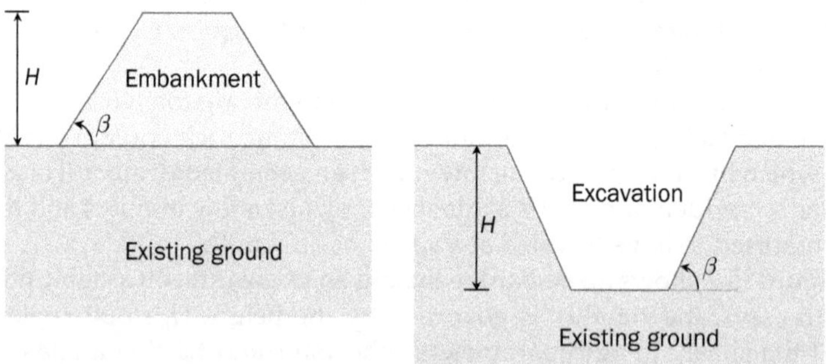

FIGURE 15.2 Factor affecting the stability of slopes.

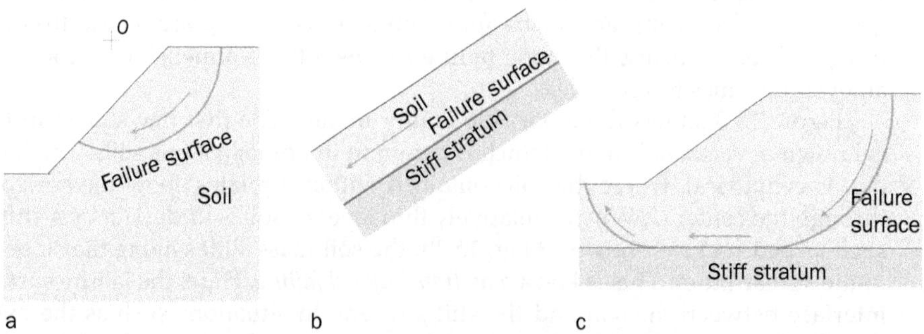

FIGURE 15.3 Failure modes: (a) rotational, (b) translational, (c) compound.

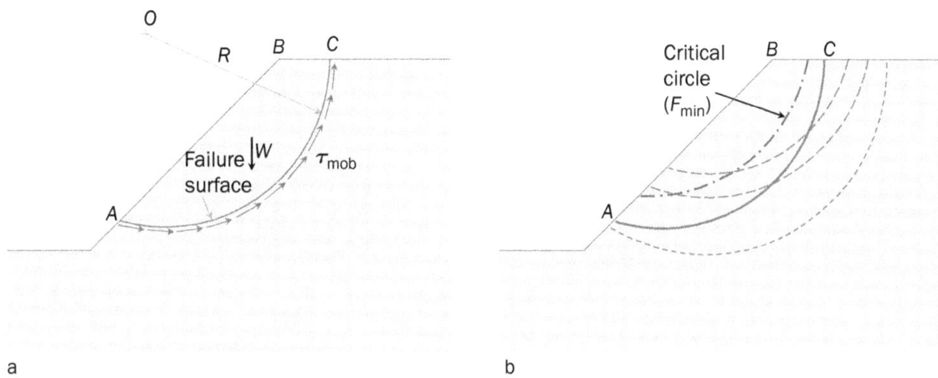

FIGURE 15.4 (a) Equilibrium of sliding mass, (b) potential slip surfaces.

base and then develops another failure arc on the left. Here, rotational and translational failures are combined.

15.2 Factor of Safety

Figure 15.4a shows a sliding mass of soil *ABC* with self-weight of *W*, enclosed within the circular arc *AC*. Due to its self-weight, the soil mass tends to slide down along the circular arc *AC*, which is the potential failure surface. To resist the slide and to maintain equilibrium, a fraction of the shear strength along the failure surface is *mobilized*, which is denoted by τ_{mob}. If this shear stress required to maintain equilibrium (i.e., *mobilized shear strength* τ_{mob}) is less than the shear strength τ_f, the slope does not fail along this arc. If the shear stress required to maintain equilibrium is greater than the shear strength (i.e., $\tau_{mob} \geq \tau_f$), the slope fails along this arc. The factor of safety for slope failure occurring along this particular arc *AC* is given by

$$F = \frac{\tau_f}{\tau_{mob}} \tag{15.1}$$

For simplicity, we have assumed that the shear strength τ_f and the mobilized shear strength τ_{mob} are constants along the arc *AC*. In reality, they vary along the failure surface.

For a slope such as the one shown in Fig. 15.4a, there are hundreds of potential failure surfaces, all in the form of circular arcs as shown in Fig. 15.4b. However, failure may occur only along one *critical circle*, which has to be identified. For each randomly selected trial failure circle, it is possible to define the factor of safety as shown in Eq. (15.1). The one with the minimum factor of safety F_{min}, is identified as the critical circle. If $F_{min} > 1$, the slope is safe. Depending on the situation, different minimum values for the factor of safety in the range of 1.0–1.5 can be specified to ensure the slope is safe.

15.3 Stability of Homogeneous Undrained Clay Slopes

Figure 15.5 shows a homogeneous undrained clay slope where the undrained shear strength is c_u ($\phi_u = 0$) and the unit weight is γ. In undrained clays, we always analyze in

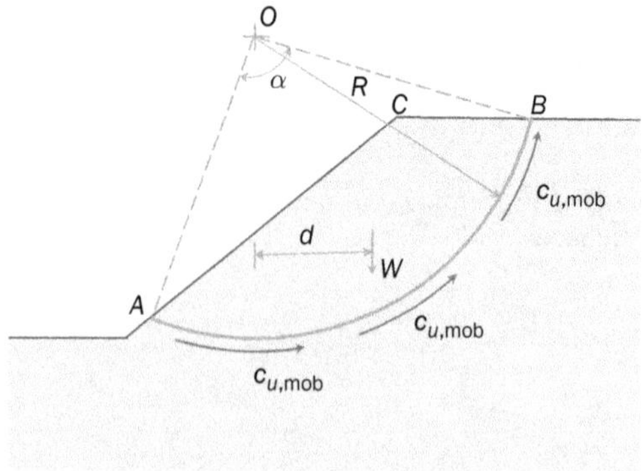

FIGURE 15.5 A potential failure circle in a homogeneous undrained clay.

terms of total stress and the failure envelope is horizontal. Therefore, the shear strength τ_f is the same as c_u at all normal stresses.

Let us consider the potential failure arc *AB* shown in the figure, subtending an angle α at the center *O*. The soil wedge *ACB* tends to slide down due to the self-weight *W*. The slide is resisted by the mobilized shear strength $c_{u,\text{mob}}$ along the failure arc *AB*. The radius of the failure arc is *R*, and hence the arc length *AB* is given by $l_{AB} = R\alpha$, where α is in radians. The self-weight of the soil *W* acts at the centroid of the wedge, at horizontal distance *d* from the center *O*. With $\phi_u = 0$, the stresses normal to the failure arc *AB* pass through the center *O*. The soil wedge is in equilibrium under the following three forces:

a. Self-weight *W*
b. Shear resistance along the arc *AB*, due to $c_{u,\text{mob}}$
c. Normal loads due to the stresses normal to the arc *AB* and hence passing through *O* (not shown in the figure)

Considering a unit width, and taking moment about *O*,

$$Wd - c_{u,\text{mob}} l_{AB} R = 0$$

$$c_{u,\text{mob}} = \frac{Wd}{l_{AB} R} \tag{15.2}$$

Once the mobilized shear strength $c_{u,\text{mob}}$ is determined, the factor of safety for this potential failure circle can be determined from Eq. (15.1), where $\tau_f = c_u$.

Example 15.1

Figure 15.6 shows a potential failure circle for an undrained clay slope. The weight of the soil enclosed within the arc is estimated as 376 kN per unit width. The undrained shear strength of the clay is 25 kN/m². What is the factor of safety for failure along this arc?

Solution The length of the arc is given by

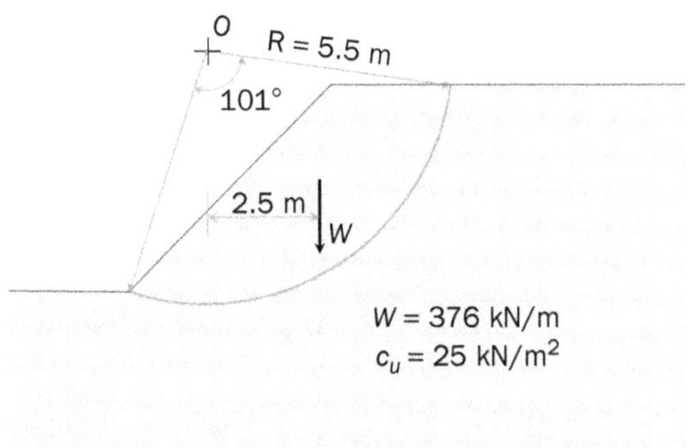

FIGURE 15.6 Potential failure circle in Example 15.1.

$$5.5 \times \left(\frac{\pi}{180} \times 101\right) = 9.695 \text{ m}$$

Taking moment about the center O,

$$376 \times 2.5 - c_{u,\text{mob}} \times 9.695 \times 5.5 = 0$$

$$c_{u,\text{mob}} = 17.63 \text{ kN/m}^2$$

Therefore, the factor of safety for failure occurring along this specific arc is

$$F = \frac{c_u}{c_{u,\text{mob}}} = \frac{25}{17.63} = \mathbf{1.42}$$

Example 15.2

A clay with $c_u = 25$ kN/m^2 is underlain by another clay with $c_u = 35$ kN/m^2. Both clays have unit weight of 17.5 kN/m^3. An excavation is carried out to the top of the lower layer as shown in Fig. 15.7. Assess the possibility of slope failure along the arc *ABCD*. The area of the wedge *ABCDE* is 50.8 m^2, and the centroid is at a horizontal distance of 2.7 m from *O* (see Fig. 15.7).

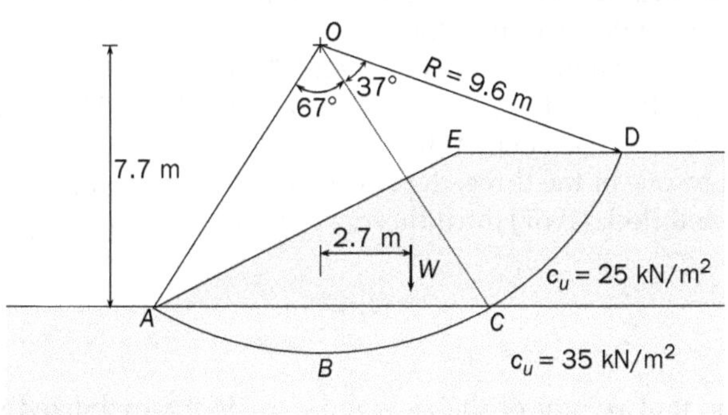

FIGURE 15.7 Potential failure circle for Example 15.2.

Solution The arc length *ABC* is

$$l_{ABC} = 9.6 \times \left(\frac{\pi}{180} \times 67\right) = 11.23 \text{ m}$$

The arc length *CD* is

$$l_{CD} = 9.6 \times \left(\frac{\pi}{180} \times 37\right) = 6.20 \text{ m}$$

Let us assume a factor of safety *F* and that it is the same along the entire arc *ABCD*. Thus, the mobilized shear resistance in the upper and lower clays are $25/F$ and $35/F$, respectively. The self-weight of the soil wedge *ABCDE* is

$$W = (50.8 \times 1)(17.5) = 889.0 \text{ kN/m}$$

Considering a unit width and taking moment about *O*,

$$889.0 \times 2.7 - \frac{35}{F} \times 11.23 \times 9.6 - \frac{25}{F} \times 6.20 \times 9.6 = 0$$

$$F = \mathbf{2.19}$$

15.4 Taylor's Stability Chart for Undrained Clays

As we realized from Fig. 15.4b, one can draw hundreds of potential failure circles for a slope stability problem. Through so many trials, it may require significant effort to locate the critical circle which gives the minimum factor of safety F_{min}. Taylor (1937) proposed a simple technique with design charts to determine the minimum factor of safety F_{min} for clay slopes under undrained conditions, without going through any trials.

Figure 15.8 shows a slope in undrained clay underlain by a stiff stratum. Taylor identified three groups of failure circles shown in the figure. *Toe circles* are the ones that pass through the toe of the slope. *Slope circles* are the ones that cut through the slope. The third type are the *base circles*, which pass through the base of the clay layer, tangent to the stiff stratum. They are also known as *mid-point circles* since their centers are directly above the middle of the slope.

When the slope angle β is greater than 53°, the critical circle is always a toe circle passing through the toe of the slope. When β is less than 53°, and n_d is greater than 4.0, the critical circle is a base circle. When β is less than 53°, and n_d is less than 4.0, the critical circle will be one of the three, depending on the slope angle β.

Terzaghi and Peck (1967) introduced a dimensionless *stability number* N_s defined as

$$N_s = \frac{\gamma H}{c_{u,\text{mob}}} \tag{15.3}$$

where γ is the unit weight of the clay and $c_{u,\text{mob}}$ is the mobilized shear strength along the critical circle. This stability number is similar to the $\gamma H/c_u$ we used in Chap. 14 with

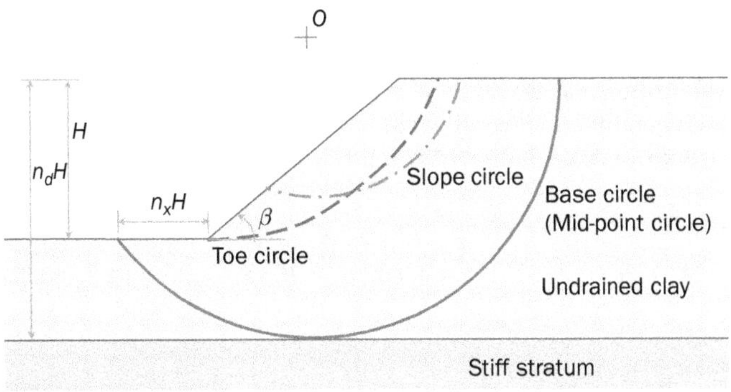

FIGURE 15.8 Critical failure circles.

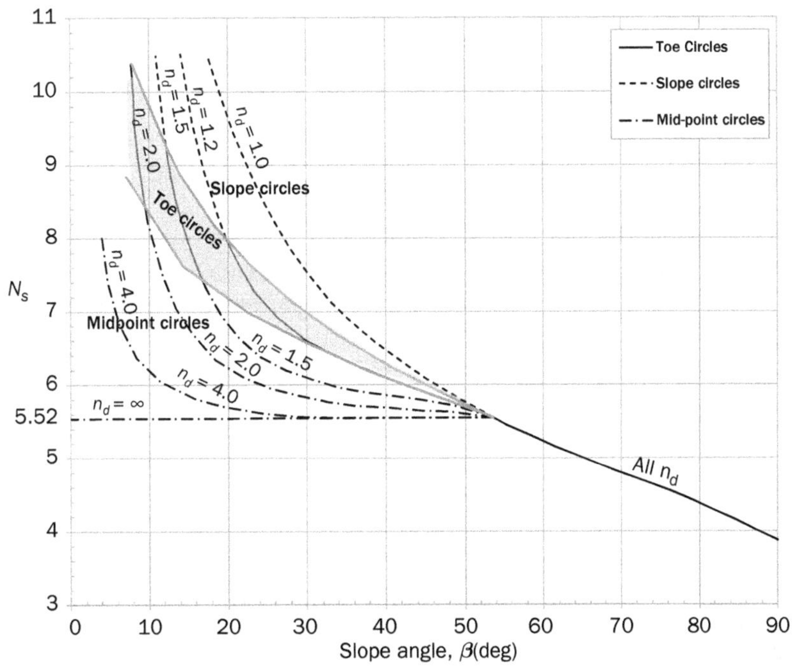

FIGURE 15.9 Taylor's stability chart for undrained clay slopes.

braced excavations to separate soft to medium clays from stiff clays. The interrelationship among β, n_d, and N_s is presented as Taylor's stability chart for undrained clay slopes in Fig. 15.9. From N_s, the mobilized undrained shear strength $c_{u,\mathrm{mob}}$ can be determined, and the minimum factor of safety corresponding to the critical circle can be determined as

$$F_{\min} = \frac{c_u}{c_{u,\mathrm{mob}}} \tag{15.4}$$

Example 15.3

A very stiff stratum underlies a 10-m-deep clay deposit. The undrained shear strength of the clay is 25 kN/m² and its unit weight is 18.5 kN/m³. If an excavation is carried out to 8 m depth at a slope of 30°, what would be the factor of safety of this slope?

Solution

$$H = 8.0 \text{ m}, n_d = 10/8 = 1.25, \text{ and } \beta = 30°$$

From Fig. 15.9, for $\beta = 30°$ and $n_d = 1.25$

$$N_s = \frac{\gamma H}{c_{u,\text{mob}}} = 6.5$$

$$c_{u,\text{mob}} = \frac{18.5 \times 8}{6.5} = 22.8 \text{ kN/m}^2$$

$$F_{\text{min}} = \frac{c_u}{c_{u,\text{mob}}} = \frac{25}{22.8} = \mathbf{1.10}$$

It will be a base circle failure.

15.4.1 Locating the Critical Circles

For $\beta > 53°$, the critical toe circle can be located as shown in Fig. 15.10. For the specific slope angle β, the values of α and θ (in degrees) can be determined from Fig. 15.10a. As shown in Fig. 15.10b, the line AB can be drawn with an angle α to the horizontal, passing through the toe. The line AB becomes a chord in the critical circle, with A and

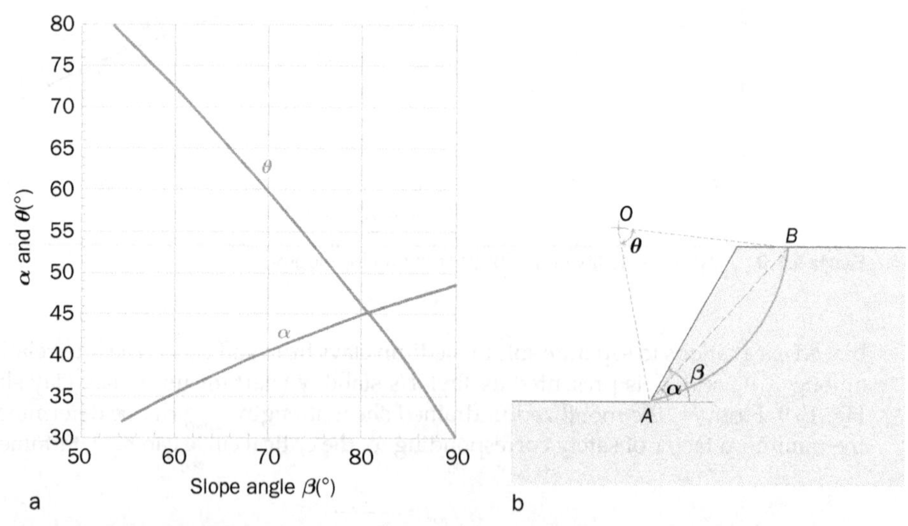

FIGURE 15.10 Locating the critical toe circle when $\beta > 53°$ (after Fellenius, 1927).

B being the two ends of the critical arc. Since the angle θ is known, the center O can be located.

For locating the critical toe circles when $\beta < 53°$, the following procedure suggested by Fellenius (1927) can be used. The procedure is illustrated in Fig. 15.11. For a given slope, with known value of m, the angles α_1 and α_2 can be read off from the table. The two lines AO and BO can be drawn to intersect at O, defining the center. This procedure is a little rough and may not always locate the true critical circle.

The critical base circle, tangent to the stiff stratum and with the center above the middle of the slope has the parameters as shown in Fig. 15.8. The interrelationship among the slope angle β, n_d, and n_x is shown in Fig. 15.12. For a given set of values β and n_d, n_x can be determined from this figure and the critical circle can be located.

m	α_1	α_2
1.0	28	37
2.0	26	35
3.0	25	35
4.0	25	35
5.0	25	37

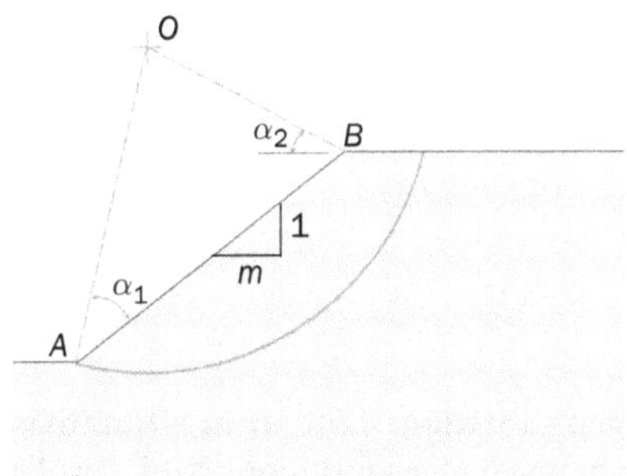

FIGURE 15.11 Locating the critical toe circle when $\beta < 53°$ (after Fellenius, 1927).

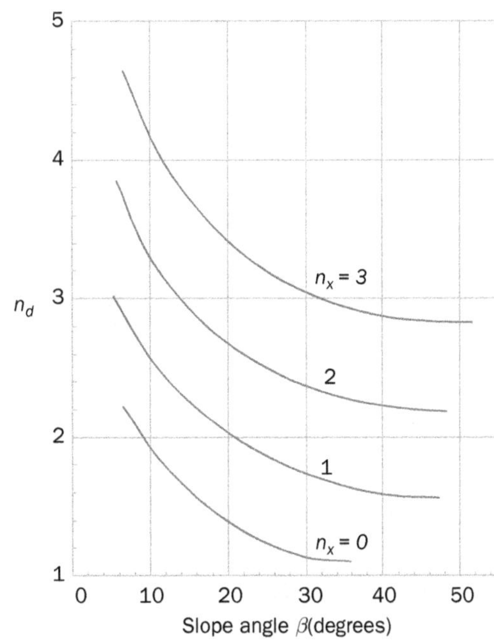

FIGURE 15.12 Locating the critical base circle (after Fellenius, 1927).

15.5 Taylor's Stability Chart for $c' - \phi'$ Soils

In Sec. 15.4, we dealt with clays in undrained condition where the shear strength has only the cohesion component but no friction. This has made the analysis simpler. For clayey sands or clays in drained situations, it is possible to have both cohesion and friction contributing to the shear strength. Let us discuss these in terms of effective stress parameters c' and ϕ' in this section, with

$$\tau_f = \sigma' \tan \phi' + c' \tag{15.5}$$

where τ_f is the shear strength and σ' is the effective normal stress on the failure plane. τ_f has two separate components coming from cohesive and frictional strengths.

Taylor (1937) proposed the chart shown in Fig. 15.13, which can be used to determine the minimum factor of safety associated with the critical circle for a given slope. In a stable slope ($F_{min} > 1$), even along the critical circle, only a fraction of the available shear strength is mobilized. The mobilized shear strength is denoted by τ_{mob}. Therefore, the factor of safety is

$$F = \frac{\tau_f}{\tau_{mob}} \tag{15.1}$$

When a fraction of the shear strength is mobilized, only fractions of the cohesive and frictional resistances are mobilized too. The mobilized shear strength can be written as

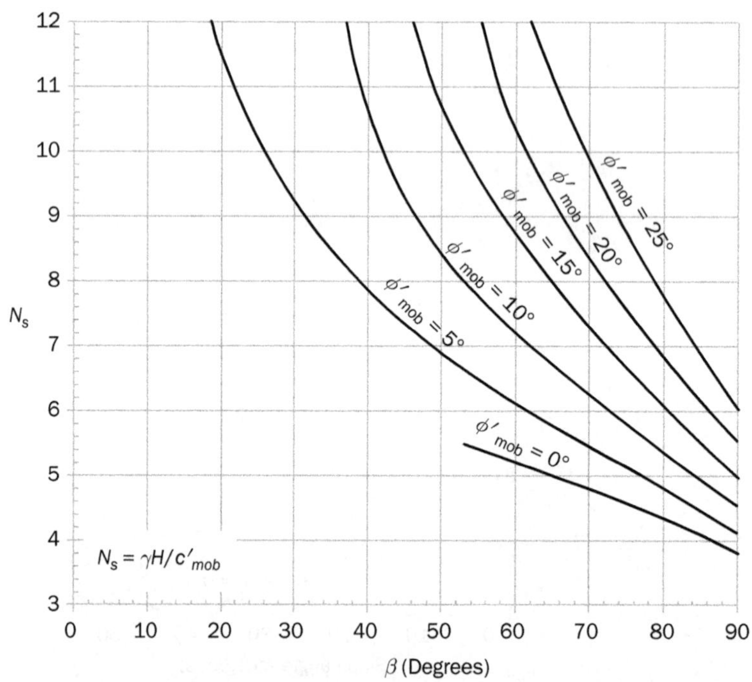

Figure 15.13 Taylor's stability chart for $c' - \phi'$ soils.

$$\tau_{\text{mob}} = \sigma' \tan \phi'_{\text{mob}} + c'_{\text{mob}} \tag{15.6}$$

where ϕ'_{mob} is the mobilized friction angle and c'_{mob} is the mobilized cohesion. In terms of cohesion, the factor of safety can be defined as

$$F_c = \frac{c'}{c'_{\text{mob}}} \tag{15.7}$$

In terms of friction angle, the factor of safety can be defined as

$$F_c = \frac{\tan \phi'}{\tan \phi'_{\text{mob}}} \tag{15.8}$$

It can be shown that when c'_{mob} and ϕ'_{mob} are determined using Fig. 15.13 such that $F_c = F_\phi$, then it turns out to be $F_c = F_\phi = F$. It requires an iterative procedure to identify the values of c'_{mob} and ϕ'_{mob} such that $F_c = F_\phi$. This is described through Example 15.4.

Example 15.4

A 5.5-m-high embankment with a slope of 60° is made of a soil with $c' = 18$ kN/m², $\phi' = 23°$, and $\gamma = 18.0$ kN/m³. What is the factor of safety?

Solution

Trial 1: Let's try $F_c = 2.0$

$$F_c = \frac{c'}{c'_{\text{mob}}} = \frac{18}{c'_{\text{mob}}} = 2.0$$

Therefore, $c'_{\text{mob}} = 9.0$ kN/m²

$$N_s = \frac{\gamma H}{c'_{\text{mob}}} = \frac{18.0 \times 5.5}{9} = 11.0$$

From Fig. 15.13, for $\beta = 60°$ and $N_s = 11.0$, $\phi'_{\text{mob}} = 21.5°$

$$F_\phi = \frac{\tan\phi'}{\tan\phi'_{\text{mob}}} = \frac{\tan 23}{\tan 21.5} = 1.08$$

The assumed F_c of 2.0 does not match the calculated F_ϕ of 1.08.

Trial 2: Let's try $F_c = 1.7$

$$F_c = \frac{c'}{c'_{\text{mob}}} = \frac{18}{c'_{\text{mob}}} = 1.7$$

Therefore, $c'_{\text{mob}} = 10.6$ kN/m²

$$N_s = \frac{\gamma H}{c'_{\text{mob}}} = \frac{18.0 \times 5.5}{10.6} = 9.3$$

From Fig. 15.13, for $\beta = 60°$ and $N_s = 9.3$, $\phi'_{mob} = 17°$

$$F_\phi = \frac{\tan\phi'}{\tan\phi'_{mob}} = \frac{\tan 23}{\tan 17} = 13.9$$

The assumed F_c of 1.7 still does not match the calculated F_ϕ of 1.39.

Trial 3: *Let us try $F_c = 1.6$*

$$F_c = \frac{c'}{c'_{mob}} = \frac{18}{c'_{mob}} = 1.6$$

Therefore, $c'_{mob} = 11.25 \text{ kN/m}^2$

$$N_s = \frac{\gamma H}{c'_{mob}} = \frac{18.0 \times 5.5}{11.25} = 8.8$$

From Fig. 15.13, for $\beta = 60°$ and $N_s = 8.8$, $\phi'_{mob} = 15°$

$$F_\phi = \frac{\tan\phi'}{\tan\phi'_{mob}} = \frac{\tan 23}{\tan 15} = 1.58$$

The assumed F_c of 1.6 is very close to the calculated of 1.58. Therefore, the factor of safety for the slope is about 1.6. An alternate way is to plot F_c against F_ϕ and locate the point where the curve intersects the $F_c = F_\phi$ line.

15.6 Cousins' Stability Chart

Cousins (1978) developed a design chart for homogeneous slopes of height H, angle β to horizontal and with the soil parameters c, ϕ, and γ. For a specific slope parameters, a stability number N_F can be computed from Fig. 15.14. The factor of safety F can be determined from this stability number as

$$F = N_F \frac{c}{\gamma H} \tag{15.9}$$

The chart and the procedure can be used to carry out the analysis in terms of total stress or effective stress parameters.

Example 15.5

An 8.0-m-high and 1(V):1.5(H) slope is made of soil with following properties: $c' = 10 \text{ kN/m}^2$, $\phi' = 23°$, and $\gamma = 18 \text{ kN/m}^3$. What is its factor of safety?

Solution

$$\beta = \tan^{-1}(1/1.5) = 33.7°$$

$$\frac{\gamma H \tan\phi'}{c'} = \frac{18 \times 8 \times \tan 23}{10} = 6.11$$

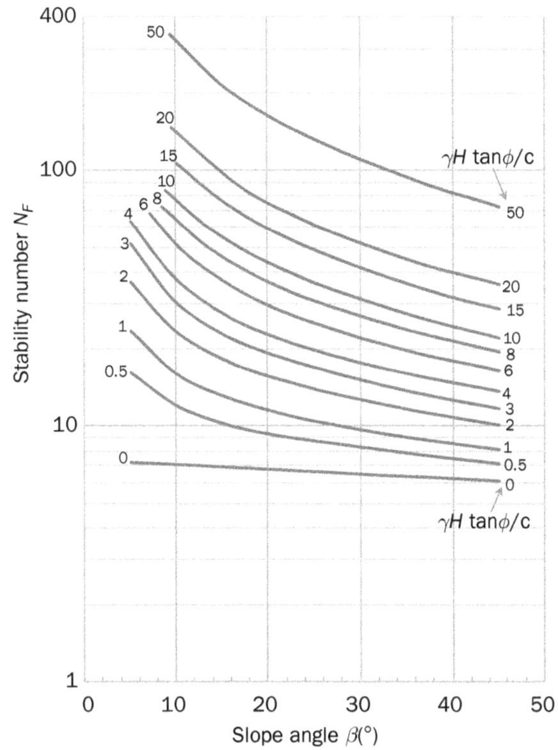

Figure 15.14 Cousins stability chart.

From Fig. 15.14, stability number $N_F = 21$

$$F = N_F \frac{c'}{\gamma H} = 21 \times \frac{10}{18 \times 8} = \mathbf{1.46}$$

15.7 Michalowski's (2002) Stability Charts for Slopes Subjected to Pore Water Pressures

Michalowski (2002) presented design charts shown in Fig. 15.15, for slopes where pore water pressure is present, including steady seepage conditions. Here, β is the inclination of the slope to the horizontal and the *pore water pressure ratio* r_u is defined as

$$r_u = \frac{u}{\gamma h} \tag{15.10}$$

where γ = unit weight of the soil, u = pore water pressure, and h = depth of the failure surface below the top of the ground.

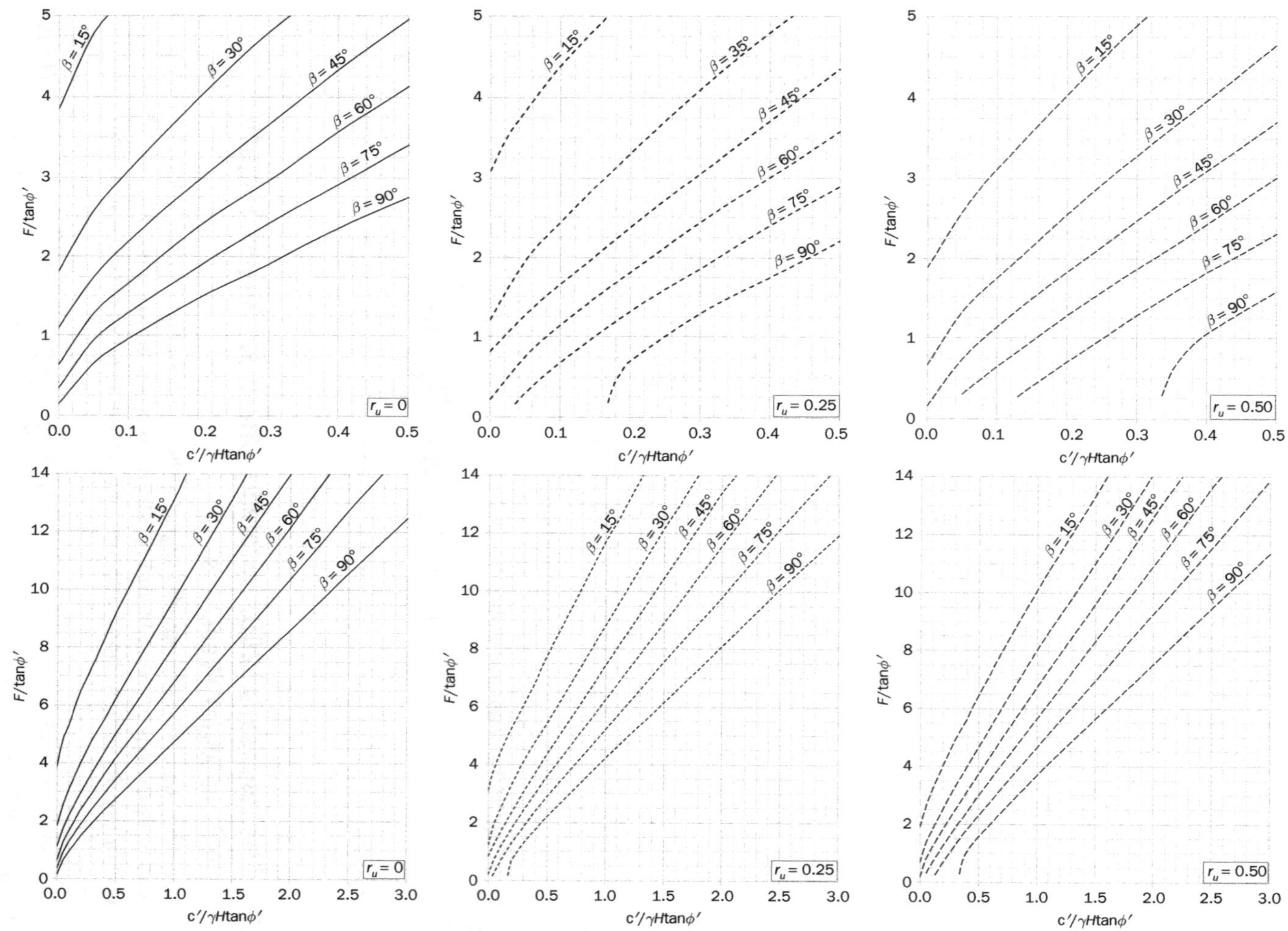

FIGURE 15.15 Michalowski's design charts.

Example 15.6

A 10-m-high and 2H:1V slope has the following soil parameters: $\phi' = 24°$, $c' = 10\ \text{kN/m}^2$, and $\phi = 19.0\ \text{kN/m}^3$. The pore water pressure can be specified as $r_u = 0.25$.

Solution

$$\beta = \tan^{-1} 0.5 = 26.6^{\circ}$$

$$\frac{c'}{\gamma H \tan\phi'} = \frac{10}{19 \times 10 \times \tan 24} = 0.118$$

From Fig. 15.15, for $r_u = 0.25$ and $\beta = 26.6°$,

$$\frac{F}{\tan 24} = 3.1$$

Therefore, $F = \mathbf{1.38}$.

Example 15.7

Redo Example 15.4 using Michalowski's chart.

Solution

$$\frac{c'}{\gamma H \tan\phi'} = \frac{18}{18 \times 5.5 \times \tan 23} = 0.428$$

For $r_u = 0$ and $\beta = 60°$,

$$F/\tan\phi' = 3.7$$

$$F = \mathbf{1.6}$$

15.8 Method of Slices

The methods discussed in the previous sections are for homogeneous slopes. In reality, the soil mass enclosed within the failure arc may have different soils. In addition, there may be water table that submerges some of the soil mass. Such slopes can be analyzed by a more rigorous analytical method, known as *method of slices*, where the slope is divided into several vertical slices.

Figure 15.16a shows the soil within the failure arc divided into several (n) vertical slices of width b. The radius of the circular arc is R and the center is O. The phreatic surface passes through the slices and the soil below this line is submerged. The ith slice is in equilibrium under the following forces (see Fig. 15.16b):

- Self-weight of the slice, W_i
- Tangential force resisting the slide, T_i

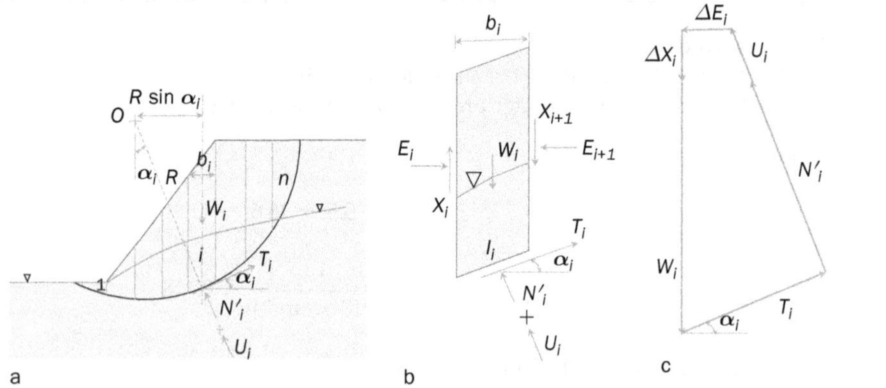

FIGURE 15.16 Method of slices: (a) Failure zone divided into slices, (b) free body diagram for *i*th slice, (c) force polygon for ith slice.

- Normal load due to the effective normal stress, N'_i
- Normal load due to pore water pressure, U_i
- The horizontal end forces on the vertical sides, E_i and E_{i+1}, the difference being ΔE_i
- Vertical shear forces between the slices, X_i and X_{i+1}, the difference being ΔX_i

The circular arc at the bottom of the *i*th slice is inclined at an angle of α_i to horizontal. Figure 15.16c shows the force polygon for these forces in equilibrium.

Let us consider a unit width (perpendicular to the plane), and take moment about *O* for all the slices and add them up.

$$\sum_{i=1}^{n} T_i R = \sum_{i=1}^{n} W_i R \sin \alpha_i$$

$$\sum_{i=1}^{n} T_i = \sum_{i=1}^{n} W_i \sin \alpha_i \tag{15.11}$$

In Fig. 15.16b, l_i is the arc length measured at the bottom of the slice, and hence

$$l_i = \frac{b_i}{\cos \alpha_i} \tag{15.12}$$

The mobilized shear strength along the length l_i is given by

$$\tau_{\text{mob},i} = \frac{\tau_{f,i}}{F} \tag{15.13}$$

where $\tau_{f,i}$ is the shear strength at the bottom of the *i*th slice and *F* is the factor of safety.

Note that the shear strength varies along the failure arc. We are only referring to the strength along the arc at the *i*th slice.

From Eqs. (15.11) and (15.13),

$$\sum_{i=1}^{n}(\tau_{\text{mob},i} l_i) = \sum_{i=1}^{n} W_i \sin\alpha_i$$

$$\sum_{i=1}^{n}\frac{\tau_{f,i}}{F} l_i = \sum_{i=1}^{n} W_i \sin\alpha_i$$

$$F = \frac{\sum_{i-1}^{n}(\tau_{f,i} l_i)}{\sum_{i-1}^{n} W_i \sin\alpha_i} \tag{15.14}$$

The shear strength at the bottom of the slice $\tau_{f,i}$ can be expressed as

$$\tau_{f,i} = c_i' + \sigma_i' \tan\phi_i' \tag{15.15}$$

From Eqs. (15.14) and (15.15),

$$F = \frac{\sum_{i-1}^{n}(c_i' l_i + \sigma_i' \tan\phi_i' l_i)}{\sum_{i-1}^{n} W_i \sin\alpha_i}$$

$$F = \frac{\sum_{i=1}^{n} c_i' l_i + \sum_{i=1}^{n} N_i' \tan\phi'_i}{\sum_{i=1}^{n} W_i \sin\alpha_i} \tag{15.16}$$

The factor of safety F can be determined from Eq. (16.16) if the variables with subscript i are determined for each slice. The variables c'_i, l_i, ϕ'_i, W_i, and α_i can be easily determined for every slice, but not N'_i. This makes the solution quite complex, with N'_i depending on the interslice forces E_i and X_i. Figure 15.16c shows the force polygon for the ith slice, where $\Delta E_i = E_{i+1} - E_i$, and $\Delta X_i = X_{i+1} - X_i$.

There are few different methods to determine the factor of safety from Eq. (15.16) by making some assumptions about the interslice forces X_i and E_i. They are discussed below.

15.8.1 Ordinary Method of Slices

Out of the several different methods of slices discussed in the literature, the simplest is the ordinary method of slices, also known as Swedish or Fellenius method of slices. Here, it is assumed that the interslice forces are equal and opposite and hence $\Delta E_i = 0$ and $\Delta X_i = 0$ (Fellenius, 1936). Therefore, the force polygon shown in Fig. 15.16c becomes a simpler force triangle shown in Fig. 15.17. From the figure, it can be seen that

$$W_i \cos\alpha_i = N_i' + U_i$$

Therefore,

$$N_i' = W_i \cos\alpha_i - U_i = W_i \cos\alpha_i - u_i l_i \tag{15.17}$$

Now that N'_i is determined, by substituting Eq. (15.17) in the expression for the factor of safety [Eq. (15.16)],

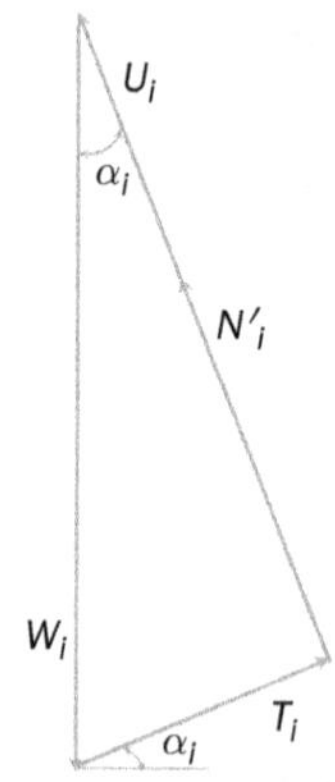

FIGURE 15.17 Force triangle for the ordinary method of slices.

$$F = \frac{\sum_{i=1}^{n} c'_i l_i + \sum_{i=1}^{n} (W_i \cos \alpha_i - u_i l_i) \tan \phi'_i}{\sum_{i=1}^{n} W_i \sin \alpha_i} \tag{15.18}$$

When there is no pore water present within the slope, Eq. (15.18) simplifies to

$$F = \frac{\sum_{i=1}^{n} c'_i l_i + \sum_{i=1}^{n} (W_i \cos \alpha_i) \tan \phi'_i}{\sum_{i=1}^{N} W_i \sin \alpha_i} \tag{15.19}$$

Since all the parameters on the right-hand side of Eq. (15.18) can be determined for every slice and summed, the factor of safety F can be determined. Since the interslice forces are assumed equal and opposite, the method does not satisfy force equilibrium but satisfies moment equilibrium. Depending on whether the failure surface curves to the right or left, the angle α_i can be positive or negative.

15.8.2 Bishop's Simplified Method of Slices

Bishop's (1955) simplified method of slices is more rationale than the ordinary method. He only assumed that $\Delta X_i = 0$ for each slice. The force polygon is shown in Fig. 15.18. It can be seen from the force polygon that

$$W_i = T_i \sin \alpha_i + N'_i \cos \alpha_i + U_i \cos \alpha_i \tag{15.20}$$

$$T_i = \frac{\tau_{f,i} l_i}{F} = \frac{c'_i l_i + N'_i \tan \phi'_i}{F}$$

where $\tau_{f,i}$ is the shear strength along the failure surface at the bottom of the ith slice. Substituting for T_i in Eq. (15.20), and noting that $U_i = u_i l_i$,

$$W_i = \left(\frac{c'_i l_i + N'_i \tan \phi'_i}{F} \right) \sin \alpha_i + N'_i \cos \alpha_i + u_i l_i \cos \alpha_i$$

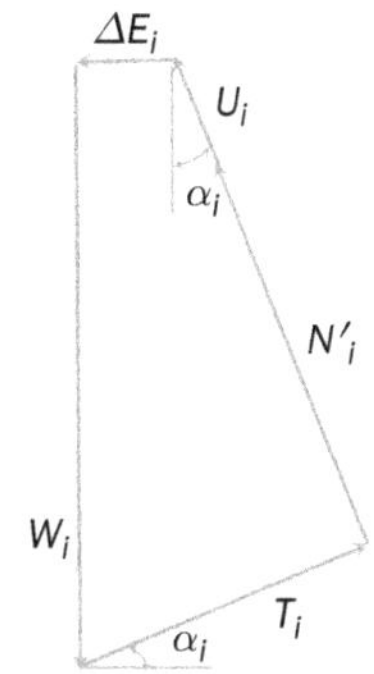

FIGURE 15.18 Bishop's simplified method of slices.

$$W_i - \frac{c_i' l_i \sin\alpha_i}{F} - u_i l_i \cos\alpha_i = N_i'\left(\cos\alpha_i + \frac{\sin\alpha_i \tan\phi_i'}{F}\right)$$

$$N_i' = \frac{W_i - \dfrac{c_i' l_i \sin\alpha_i}{F} - u_i l_i \cos\alpha_i}{\cos\alpha_i + \dfrac{\sin\alpha_i \tan\phi_i'}{F}}$$

Substituting for l_i from Eq. (15.12),

$$N_i' = \frac{W_i - \dfrac{c_i' b_i \tan\alpha_i}{F} - u_i b_i}{\cos\alpha_i + \dfrac{\sin\alpha_i \tan\phi_i'}{F}} \tag{15.21}$$

The factor of safety F is given by Eq. (15.16). Substituting for l_i from Eq. (15.12),

$$F = \frac{\sum_{i=1}^{n} c_i' l_i + \sum_{i=1}^{n} N_i' \tan\phi_i'}{\sum_{i=1}^{n} W_i \sin\alpha_i} = \frac{\sum_{i=1}^{n} (c_i' b_i / \cos\alpha_i) + \sum_{i=1}^{n} N_i' \tan\phi_i'}{\sum_{i=1}^{n} W_i \sin\alpha_i}$$

$$F\sum_{i=1}^{n} W_i \sin\alpha_i = \sum_{i=1}^{n}\left\{\frac{c_i' b_i}{\cos\alpha_i} + \left(\frac{W_i - \dfrac{c_i' b_i \tan\alpha_i}{F} - u_i b_i}{\cos\alpha_i + \dfrac{\sin\alpha_i \tan\phi_i'}{F}}\right)\tan\phi_i'\right\}$$

$$F\sum_{i=1}^{n} W_i \sin\alpha_i = \sum_{i=1}^{n}\{c_i' b_i + (W_i - u_i b_i)\tan\phi'\}\left\{\frac{\sec\alpha_i}{1 + \dfrac{\tan\alpha_i \tan\phi_i'}{F}}\right\}$$

$$F = \frac{\sum_{i=1}^{n}\{c_i' b_i + (W_i - u_i b_i)\tan\phi'\}\left\{\dfrac{\sec\alpha_i}{1 + \dfrac{\tan\alpha_i \tan\phi_i'}{F}}\right\}}{\sum_{i=1}^{n} W_i \sin\alpha_i} \tag{15.22}$$

The problem with Eq. (15.22) for determining the factor of safety F is that it appears on both sides of the equation. Therefore, it requires a trial and error process to find the factor of safety by iteration.

The pore water pressures within the slices are generally expressed through a pore water pressure ratio r_u, defined in Eq. (15.10) as

$$r_u = \frac{u}{\gamma h}$$

Therefore,

$$u_i b_i = r_u \gamma h_i b_i = r_u$$

Substituting $W_i(1 - r_u)$ for $W_i - u_i b_i$ in Eq. (15.22),

$$F = \frac{\sum_{i=1}^{n} \{c_i' b_i + W_i(1 - r_u)\tan\phi'\}\left\{\dfrac{\sec\alpha_i}{1 + \dfrac{\tan\alpha_i \tan\phi_i'}{F}}\right\}}{\sum_{i=1}^{n} W_i \sin\alpha_i} \tag{15.23}$$

The ordinary method of slices is generally very conservative, underestimating the factor of safety. Bishop's simplified method gives a slightly higher value for the factor of safety. There are more sophisticated methods such as Spencer's method, Janbu's method, Morgenstern-Price method and more, which are not discussed here.

Few decades ago, it was a common practice to compute the factor of safety from the method of slices, manually in the class rooms and consulting practices. Each trial failure circle analyzed requires significant computational time. This makes it difficult to consider too many trial circles in locating the critical failure circle with the minimum factor of safety. These are better analyzed by a computer program.

Nowadays, there are quite a few software packages available commercially for slope stability analysis with many advanced features. They can carry out hundreds of trials in a matter of minutes and locate the critical circle more precisely. Some of these programs come with the free student versions that can be used in the university teaching. Example 15.8 is solved using the free student version SLOPE/W, part of GeoStudio software suite.

Example 15.8

Figure 15.19 shows a soil profile consisting of three different soils, where a slope is made. The soil properties are as follows:

	γ (kN/m³)	c' (kN/m²)	ϕ' (°)
Top layer	18.0	20	24
Middle layer	19.0	15	26
Bottom layer	19.5	10	22

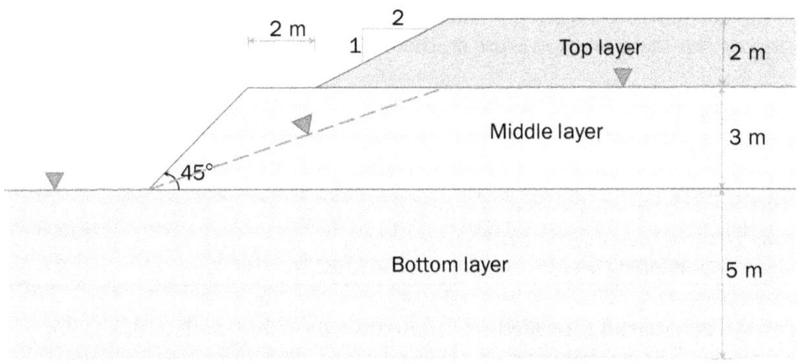

FIGURE 15.19 Slope profile.

Evaluate the minimum factor of safety for the slope using Bishop's method and locate the critical circle. Also, compare these with the values determined from other methods.

Solution The slope was analyzed using the student version of SLOPE/W. The trial circles were divided into 30 slices (default). Hundreds of trial circles were considered by varying the radius and the locations of the center. The critical circle, with factor of safety of 1.88, is shown in Fig. 15.20. The coarse gird at the top of the figure shows the locations of the centers for the trial circles, and the factor of safety contours.

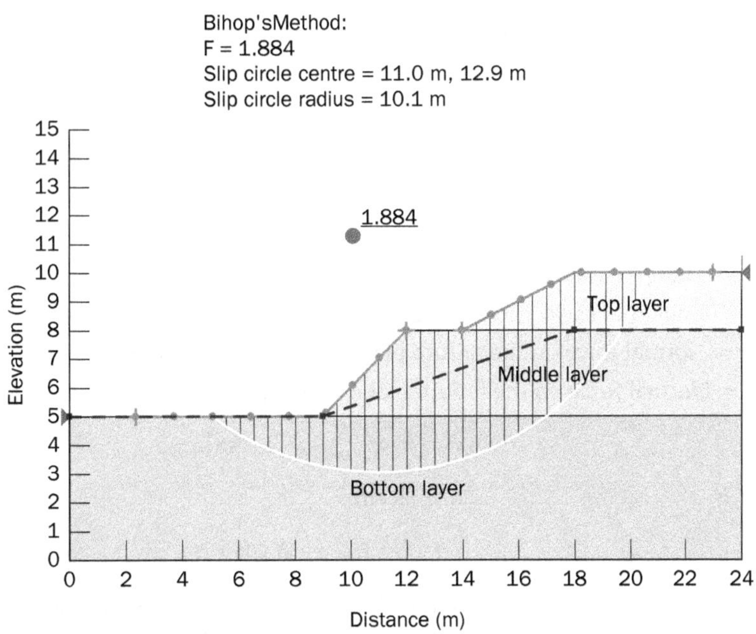

FIGURE 15.20 Critical failure circle from Bishop's method.

In addition to Bishop's method, the Ordinary method, Janbu's method, and Morgenstern-Price method were studied. The values are summarized below.

Method	F_{min}	Center coordinates (x,y) (m)	Radius (m)
Bishop	1.88	11.0, 13.0	10.1
Ordinary	1.78	11.7, 11.9	9.6
Janbu	1.76	11.1, 12.0	9.8
Morgenstern-Price	1.88	12.1, 12.5	9.4

It can be seen that all four methods compare reasonably well, with a factor of safety of about 1.9 and the failure circle radius of about 10 m. The center coordinates are approximately $x = 11$ m and $y = 12$–13 m.

In a typical slope stability analysis, several different scenarios are analyzed. *Rapid drawdown* is often the worst case scenario when analyzing earth dams. Here, the water level in a reservoir or river suddenly lowered, which makes it necessary to use the saturated density instead of submerged density in the analysis. The gravitational load gets almost doubled during this process while there is no change in the shear strength. This reduces the factor of safety by 50%.

15.9 Infinite Slopes

Infinite slopes are slopes where the overburden soil thickness is relatively low compared to the length of the slope, and hence the end effects can be neglected. To simplify the problem, it is assumed that the slope, water table, and the failure surface are parallel, inclined at an angle β to horizontal, as shown in Fig. 15.21. The slip is purely translational, where the entire soil mass above the failure plane slides down the slope.

In Fig. 15.21, the height of the soil is z, and the water table is at a height of mz above the failure plane. Let us consider the equilibrium of a vertical slice of unit thickness shown in the figure. The slice is in equilibrium under the following forces:

W = Self-weight of the slice
R = End forces acting on both vertical sides of the slice
T = Tangential shear force along the failure plane, resisting the slide
N' = Normal force on the failure plane due to the effective stresses
U = Normal force on the failure plane due to pore water pressure

$$W = bmz\gamma_{sat} + b(1-m)z\gamma_m \tag{15.24}$$

$$T = W\sin\beta \tag{15.25}$$

$$N = N' + U = W\cos\beta \tag{15.26}$$

N is the total normal load that includes the contributions from the effective stresses and pore water pressures. From Eqs. (15.24) and (15.26),

$$N = N' + U = \left(bmz\gamma_{sat} + b(1-m)z\gamma_m\right)\cos\beta$$

$$N' + U = bmz\cos\beta\left(\gamma' + \gamma_w\right) + b(1-m)z\gamma_m\cos\beta$$

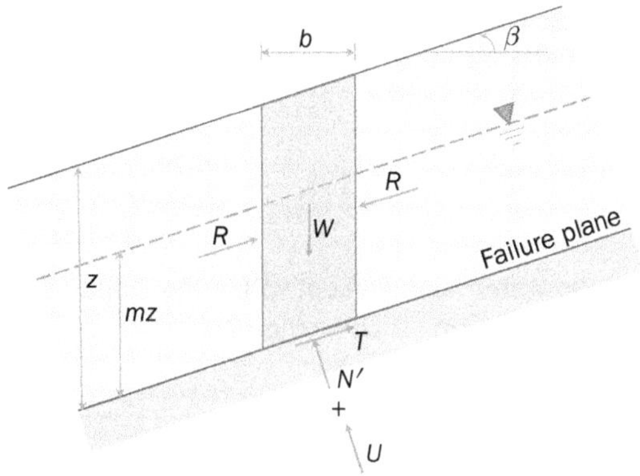

FIGURE 15.21 Infinite slope.

Therefore,

$$N' = bmz\gamma'\cos\beta + b(1-m)z\gamma_m\cos\beta$$

$$U = bmz\gamma_w\cos\beta$$

The effective normal stress and the pore water pressure on the failure plane can be written as

$$\sigma' = \frac{N'}{b\ /\ \cos\beta} = mz\gamma'\cos^2\beta + (1-m)z\gamma_m\cos^2\beta \tag{15.27}$$

$$u = \frac{U}{b\ /\ \cos\beta} = mz\gamma_w\cos^2\beta \tag{15.28}$$

The shear strength along the failure plane is given by

$$\tau_f = c' + \sigma'\tan\phi' = c' + [m\gamma' + (1-m)\gamma_m]z\cos^2\beta\tan\phi'$$

The shear stress mobilized on the failure plane is given by

$$\tau_{\text{mob}} = \frac{T}{b\ /\ \cos\beta} = \frac{W\sin\beta}{b/\cos\beta} = [m\gamma_{\text{sat}} + (1-m)\gamma_m]z\sin\beta\cos\beta$$

Therefore, the general expression for the factor of safety against sliding becomes

$$F = \frac{\tau_f}{\tau_{\text{mob}}} = \frac{c' + [m\gamma' + (1-m)\gamma_m]z\cos^2\beta\tan\phi'}{[m\gamma_{\text{sat}} + (1-m)\gamma_m]z\sin\beta\cos\beta} \tag{15.29}$$

Let us consider some special cases.

Case I: Dry Granular Soils For dry granular soils, $c' = 0$ and $m = 0$. Equation (15.29) becomes

$$F = \frac{\tan\phi'}{\tan\beta} \tag{15.30}$$

Case II: Fully Submerged Granular Soil with Steady Seepage Down the Slope When the slope is fully submerged with seepage occurring down the slope, $c' = 0$ and $m = 1$. Equation (15.29) becomes

$$F = \frac{\gamma'}{\gamma_{sat}} \frac{\tan\phi'}{\tan\beta} \tag{15.31}$$

Case III: Clays with No Water Table Present within the Slope In clay slopes with no water table present, $m = 0$. Equation (15.29) becomes

$$F = \frac{c'}{\gamma_m z \sin\beta\cos\beta} + \frac{\tan\phi'}{\tan\beta} \tag{15.32}$$

Case IV: Fully Submerged Clay Slopes with Seepage Down the Slope In fully submerged clay slopes with seepage down the slopes, $m = 1$. Equation (15.29) becomes

$$F = \frac{c'}{\gamma_{sat} z \sin\beta\cos\beta} + \frac{\gamma'}{\gamma_{sat}} \frac{\tan\phi'}{\tan\beta} \tag{15.33}$$

Example 15.9

An infinite slope is made of 4.0-m-thick sandy soil with friction angle of 33°, underlain by bedrock. The bedrock and the ground level are inclined at 15° to horizontal.

(a) What is the factor of safety against sliding if the sand is dry?

(b) What is the factor of safety if there is steady seepage occurring down the slope with water table coinciding with the ground level? Assume saturated unit weight of 21.0 kN/m³.

Solution

Part a

$$F = \frac{\tan\phi'}{\tan\beta} = \frac{\tan 33}{\tan 15} = \mathbf{2.42}$$

Part b

$$F = \frac{\gamma'}{\gamma_{sat}} \frac{\tan\phi'}{\tan\beta} = \frac{(21.0 - 9.81)}{21.0} \times \frac{\tan 33}{\tan 25} = \mathbf{1.29}$$

15.10 Main Points

1. The failure of a homogeneous slope occurs along a circular arc. Infinite slopes fail along planes parallel to the slope.

2. While one can try hundreds of possible failure arcs, the one that gives the minimum factor of safety is the critical failure arc.
3. Separate analysis is required for the same slope under drained and undrained conditions. The drained analysis uses effective stress parameters and the undrained analysis uses total stress parameters.
4. While there are a number of charts to analyze stability of homogeneous slopes, they give comparable values for the factor of safety.

Review Exercises

15.1 An undrained clay slope shown in Fig. RE15.1 has $c_u = 25$ kN/m^2 and $\gamma = 18.0$ kN/m^3. The area of the soil enclosed within the potential failure arc AB shown in the figure is 19.3 m^2, and the centroid is at a horizontal distance of 3.30 m from the center *O*. Determine the factor or safety against a possible slide along this arc.

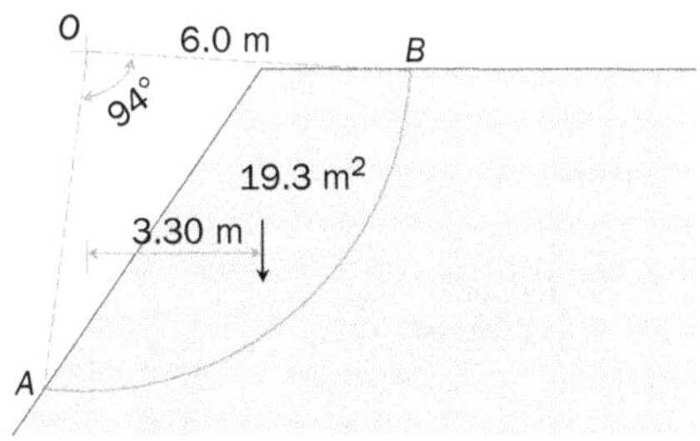

FIGURE RE15.1 Potential failure circle.

Answer: 1.29

15.2 A 3.5-m-high embankment is proposed for construction on existing ground ($c_u = 20$ kN/m^2) using a local clay soil. The expected undrained shear strength of the embankment will be 40 kN/m^2. What would be the factor of safety against failure along the arc AFED shown in Fig. RE15.2? The weight of the soil wedge enclosed within the arc is 430 kN/m.

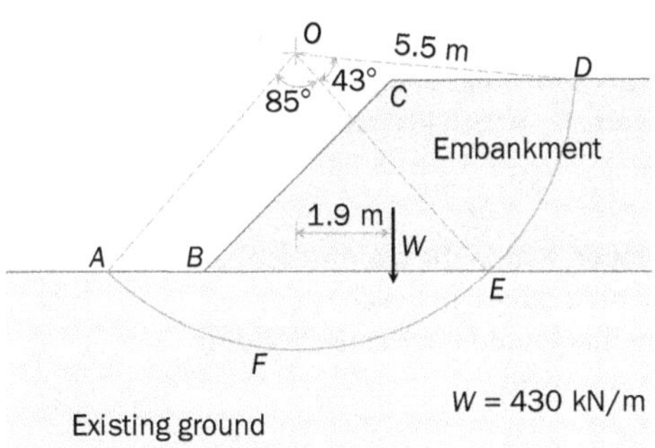

FIGURE RE15.2 Potential failure circle.

Answer: 2.21

15.3 A 10-m-thick clay deposit ($c_u = 30$ kN/m^2, $\gamma = 18.5$ kN/m^3) is underlain by a stiff stratum. It is required to carry out an excavation at a slope of 1.0(V):2.0(H) with a factor of safety of at least 1.4. What is the maximum possible depth for the excavation?

Answer: 7.5 m

15.4 A 6.0-m-high homogeneous slope with constructed at 50° to the horizontal has the following soil properties: $c' = 15$ kN/m^2, $\phi' = 24°$, and $\gamma = 18$ kN/m^3. Determine the factor of safety using Taylor's stability chart.

Answer: 1.55

15.5 Redo Problem RE15.4 using Michalowski's charts.

Answer: 1.56

15.6 Figure RE15.6 shows an undrained clay slope.

a. Determine the factor of safety for the slope using Taylor's chart.
b. Determine the factor of safety using a software.

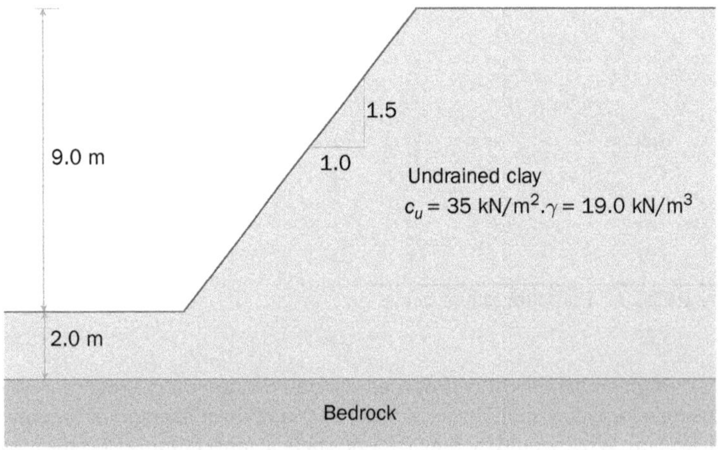

Figure RE15.6 Undrained clay slope.

Answer: 1.10, 1.11

15.7 A cohesive infinite soil slope has 2.5 m soil overburden above the underlying stiff stratum. The slope is inclined at 20° to the horizontal and there is no water table within the overburden soil. The soil properties are: $\gamma_m = 19.0$ kN/m^3, $\phi' = 15°$, $c' = 35$ kPa. Determine the following:

a. Safety factor against sliding
b. Maximum shear stress developed within the overburden soil
c. Shear strength along the potential failure plane above
d. Critical height of overburden that would have caused sliding

Answer: 3.03, 15.3 kN/m^2, 46.2 kN/m^2, 21.7 m

References

Bishop, A. W. "The Use of Slip Circle in the Stability Analysis of Earth Slopes," *Geotechnique*, 5(1):7–17, 1955.

Cousins, B. F. "Stability Charts for Simple Earth Slopes," *Journal of the Geotechnical Engineering Division*, ASCE, 104(GT2):267–282, 1978.

Fellenius, W. *Erdstatische Berechnungen*, Revised Edition 1939, W. Ernst u. Sons, Berlin, 1927.

Fellenius, W. "Calculation of the Stability of Earth Dams," *Proceedings, Second Congress of Large Dams*, 4:445–463, 1936.

Michalowski, R. L. "Stability Charts for Uniform Slopes," *Journal of Geotechnical and Geoenvironmental Engineering*, ASCE, 128(4):351–355, 2002.

Taylor, D. W. "Stability of Earth Slopes," *Journal of Boston Society of Civil Engineers*, 24:197–246, 1937.

Terzaghi, K., and R. B. Peck. *Soil Mechanics in Engineering Practice*, 2nd Edition, Wiley, New York, 1967.

CHAPTER 16

Reliability-Based Design

16.1 Introduction

The variability in geomaterials is significantly larger than that in the other civil engineering materials. With the uncertainty coming from the loadings and the soil, there is good value in going for probabilistic design methods than the traditional deterministic approaches.

In the past few decades, there has been significant developments to introduce the *load and resistance factor design* (LRFD) concepts in the designs of foundations, retaining walls, bridges, etc. There are advantages compared to the traditional *allowable stress design* (ASD) approaches. LRFD comes from the *limit state design*, widely used in structural engineering for much longer.

Limit state design involves reliability concepts and partial factors of safety. The two main requirements in the limit state designs are:

a. The structure or facility must be safe against *failure* during the design life, and
b. The structure or facility must serve the purpose without undergoing excessive deformations during the design life.

The first is known as the *ultimate limit state* which includes bearing capacity failure of a foundation, sliding or overturning of a retaining wall, and piping beneath a dam. The second is the *serviceability limit state*, which includes settlements, deflections, tilt, cracking, vibrations, and noise.

Figure 16.1 shows the variation of cost versus the factor of safety in the case of a foundation, retaining wall, or similar. When a higher level of safety is required, the cost increases significantly. The required level of safety depends on the uncertainty in the design parameters, design life, importance of the structure, probability of the failure, and the consequences associated with possible failure. In civil engineering, we use factor of safety slightly more than 1.0 for temporary structures, two to three for foundations, and significantly higher values for more important facilities such as nuclear power plants where failure can be catastrophic.

There are two different approaches to the design. They are:

a. ASD or *working stress design* (WSD), and
b. LRFD.

ASD is the older of the two and is incorporated in most textbooks, including this book. It is gradually superseded by the LRFD, which is widely used in Europe, North

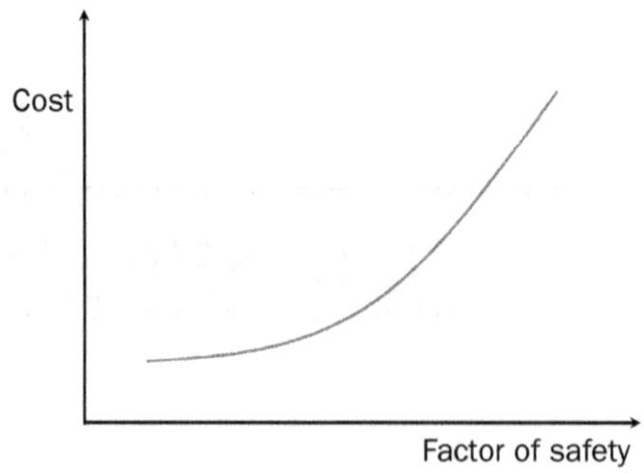

FIGURE 16.1 Cost versus factor of safety.

America, and Australia especially in the designs of foundations, retaining walls, slopes, bridges, etc. The first objective of this chapter is to introduce the LRFD, considering the bearing capacity of a shallow foundation as an example. The method is the same when it is applied to a pile foundation or other geotechnical applications. The second objective is to introduce the concept of reliability-based design as a probabilistic approach for handling geotechnical problems.

16.2 Capacity-Demand Model

A civil engineering design can be seen as a struggle between capacity (*C*) and demand (*D*). The designer strives to ensure that the capacity exceeds the demand at all times during the design life. Table 16.1 identifies the capacity and demand in three different geotechnical engineering applications. Capacity refers to strength or resistance, and depends on the material properties. Demand refers to the loadings, which includes dead loads and live loads considering different combinations.

Due to the variabilities associated with the material properties and loadings, both capacity and demand have a probabilistic distribution as shown in Fig. 16.2, with mean values of $\bar{C}$ and $\bar{D}$, respectively.

16.3 Allowable Stress Design

ASD or WSD has been practiced in civil engineering for centuries. Here, the mean values of $\bar{C}$ and $\bar{D}$ are estimated from the material properties and the loadings. Considering

Problem	Capacity	Demand
Sliding of retaining wall	Resistance to sliding due to passive load and base friction	Horizontal component of active load
Overturning of retaining wall	Moment resisting overturning	Driving moment in opposite direction, causing overturning
Piping due to seepage	Critical hydraulic gradient	Exit hydraulic gradient
Bearing capacity failure of a foundation	Ultimate bearing capacity	Pressure applied to the underlying soil

TABLE 16.1 Some Simple Capacity-Demand Analogies

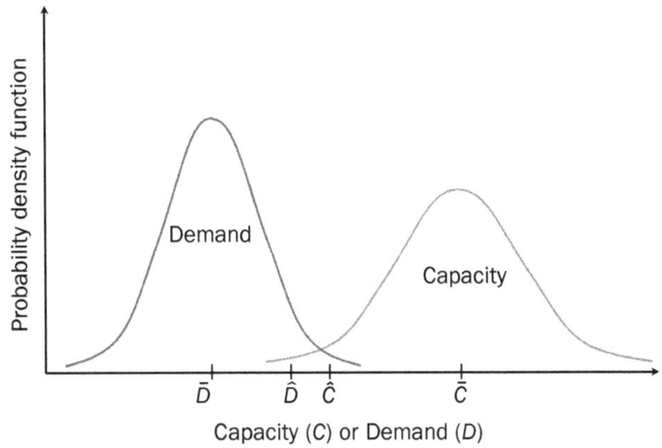

Figure 16.2 Capacity-demand model.

the spatial variabilities of the two, the approximations or simplifications in the design methodology, and other factors, a conservative factor of safety is used as

$$F = \frac{\bar{C}}{\bar{D}} \tag{16.1}$$

This single lumped safety factor takes into account all the uncertainties. It is impossible to see what the relative contributions are from the different sources of uncertainties. Table 16.2 summarizes the typical values used for the factor of safety used in practice.

In the ASD, the loading (S) is similar to demand, and the resistance (R) is similar to capacity. Using the *mean values* of S and R, the factor of safety is defined as

$$F = \frac{\bar{R}}{\bar{S}} \tag{16.2}$$

The factor of safety is a reflection of the probability of failure during the design period. For example, it is expected that the probability of failure during the service life is in the order of 1×10^{-2} for earthworks, 1×10^{-3} for earth retaining structures, and 1×10^{-4} for foundations (Meyerhof, 1984).

16.4 Load and Resistance Factor Design

A slightly different way of looking at the capacity-demand model forms the basis for the LRFD. The capacity $\hat{C}$ (see Fig. 16.2) under the worst-case scenario is estimated such that there is very little likelihood (e.g., 2% probability) that the actual capacity will be below this value. Similarly, the demand $\hat{D}$ is estimated such that there is very little likelihood that actual demand will exceed this value. The design is deemed satisfactory if $\hat{C}$ is still greater than $\hat{D}$. Since the underestimated capacity $\hat{C}$ and the overestimated demand $\hat{D}$ already have built-in factors of safety, there is no further factor of safety required here.

LRFD is a limit state design approach. The loading (S) is similar to demand and resistance (R) is similar to capacity. They are factored conservatively. The loads are

Structure	F	Reference
Shallow foundations		
Spread footings in compression	2.0–3.0	Bowles, 1988
Mat foundations	1.7–2.5	Bowles, 1988
Uplift	1.7–2.5	Bowles, 1988
Pile foundations in compression		
Verified by static load tests	2.0	USACE, 1991
Verified by pile-driving analyzer (PDA)	2.5	USACE, 1991
Not verified by static load tests or PDA	3.0	USACE, 1991
From pile-driving equations	3.0–6.0	
Retaining walls		
Sliding—if passive resistance neglected	1.5	Goodman and Karol, 1968
Sliding—if passive resistance included	2.0	Goodman and Karol, 1968
Overturning—granular backfill	1.5	Teng, 1962
Overturning—cohesive backfill	2.0	Teng, 1962
Seepage		
Piping	3.0–5.0	Bowles, 1988
Heave or uplift	1.5–2.5	Bowles, 1988
Earthworks	1.2–1.6	Bowles, 1988
Temporary braced excavations	1.2–1.5	Bowles, 1988
Sheet piling coffer dams	1.2–1.6	Bowles, 1988

Table 16.2 Typical Values of the Factor of Safety

Category	Item	Factor
Loads:	Dead loads	$\Psi = 1.25$
	Live, wind, or earthquake load	$\Psi = 1.50$
	Water pressure	$\Psi = 1.25$
Shear strength:	Cohesion (stability, earth pressure)	$\Phi = 0.65$
	Cohesion (foundations)	$\Phi = 0.50$
	Friction (tan ϕ')	$\Phi = 0.80$

Table 16.3 Typical Values for Load Factor (Ψ) and Resistance Factor (Φ) (After Canadian Foundation Engineering Manual, 1992)

multiplied by factors greater than unity, known as *load factors* (Ψ). The resistances are multiplied by factors less than unity, known as *resistance factors* (Φ). The load and resistance factors vary with the application, and are specified by the codes and standards. Some load and resistance factors suggested by the Canadian Foundation Engineering Manual (1992) are given in Table 16.3.

	United States				Canada		Europe	
Load type	**AASHTO**	**ACI**	**AISC**	**API**	**MOT**	**NRC**	**DGI**	**ECS**
Dead	1.25–1.95	1.4	1.2–1.4	1.1–1.3	1.1–1.5	1.25	1.0	1.0–1.35
	(0.65–0.9)	(0.9)	(0.9)	(0.9)	(0.65–0.95)	(0.85)	(0.85)	(0.95)
Live	1.35–1.75	1.7	1.6	1.1–1.5	1.15–1.4	1.5	1.3	1.3–1.5
				(0.8)				
Wind	1.4	1.3	1.3	1.2–1.35	1.3	1.5	1.3	1.3–1.5
Seismic	1.0	1.4	1.0	0.9	1.3	1.0	1.0	1.0

Note: Values in parentheses apply when the load is favorable and resists failure.
AASHTO = American Association of State Highway Transportation Officials
ACI = American Concrete Institute
AISC = American Institute of Steel Construction
API = American Petroleum Institute
MOT = Ministry of Transportation
NRC = National Research Council of Canada
DGI = Danish Geotechnical Institute
ECS = Eurocode

TABLE 16.4 Load Factors for Foundation Designs

For any geotechnical or foundation engineering problem, the factored loading S^* can be a function of more than one type of load (e.g., dead load, live load). It is computed as

$$S^* = \sum_{i=1}^{n} \psi_i S_i \tag{16.3}$$

where S_i are the different load types and Ψ_i are the corresponding load factors. The load types include permanent (dead load, water thrust), variable (live load, wind load, snow load), and accidental loads (seismic load, explosion, vehicle impact).

The load effects are computed by considering different loading scenarios. Some of the common load combinations that are considered in the designs are:

$$Q = 1.25\,Q_D + 1.75\,Q_L \tag{16.4}$$

$$Q = 1.25\,Q_D + 1.35\,Q_L + 0.4\,Q_W \tag{16.5}$$

where Q = design load, Q_D = dead load, Q_L = live load, and Q_W = wind load. The load factors used in the United States, Canada, and Europe, by the different agencies are summarized in Table 16.4.

The factored resistance R^* is computed as

$$R^* = \sum_{i=1}^{n} \phi_i R_i \tag{16.6}$$

The design is deemed satisfactory if $S^* \leq R^*$. There is no further factor of safety required. We have already used partial factors of safety in the form of load factors and resistance factors.

Structure	Resistance factor Φ
Bearing capacity of shallow foundations	0.45–0.55
Bearing capacity of driven piles	
From load tests	0.75–0.80
From pile-driving formulae	0.10–0.40
From static analysis: Clay and mixed soils	
α-method	0.35
β-method	0.25
λ-method	0.40
From static analysis in sands	0.45

TABLE 16.5 Resistance Factors Suggested by AASHTO (2012)

The factored resistance R^* is computed in two slightly different ways. They are demonstrated through Examples 16.1 and 16.2.

Method 1: Partial Resistance Factors

In this method, separate resistance factors are applied to each of the strength (or resistance) parameter such as cohesion, friction angle, etc. The factored strength parameters are used in computing the factored resistance R^*.

Method 2: Total Resistance Factor

In this method, the material parameters are not reduced in computing the resistance, which is multiplied by a total resistance factor Φ, to give the factored resistance R^*. Such total resistance factors recommended by AASHTO bridge design specifications are given in Table 16.5.

The range of possible load combinations to be considered in the design, the magnitudes of the load and resistance factors, and how they are applied vary between countries and the different agencies.

Example 16.1

A square pad foundation of 2.0 m width is placed at a depth of 1.0 m in a granular soil where $\phi' = 36°$ and $\gamma = 19.0$ kN/m³. It is expected that the live load is 75% of the dead load. Determine the maximum dead load that can be applied to the footing based on

a. ASD with $F = 3$, and
b. LRFD using load and resistance factors in Table 16.3.

Use the bearing capacity equation

$$q_{\text{ult}} = 1.3cN_c + \gamma D_f N_q + 0.4B\gamma N_\gamma$$

with Meyerhof's bearing capacity factor N_γ.

Solution

Part a: Allowable stress design

For $\phi' = 36°$, from Table 12.2, $N_q = 37.9$ and $N_\gamma = 44.7$

$$q_{ult} = (19.0 \times 1)(37.9) + 0.4(19.0)(2.0)(44.7) = 1399.5 \text{ kN/m}^2$$

$$q_{all} = \frac{1399.5}{3} = 466.5 \text{ kN/m}^2$$

Let the dead load be P kN, and hence the live load is 0.75 P kN. The total load on the foundation is 1.75 P kN. The pressure applied to the ground is

$$\frac{1.75\,P}{2 \times 2} = 0.4375\,P \text{ kN/m}^2$$

$$0.4375\,P \leq 466.5$$

Therefore, $\mathbf{P \leq 1066.3}$ **kN**

Part b: Limit state design

Applying the resistance factors from Table 16.3,

$$\tan\phi' = 0.8 \times \tan 36 = 0.5812$$

$$\phi' = 30.2°$$

From Table 12.2, $N_q = 18.9$ and $N_\gamma = 16.3$

$$q_{ult} = (19.0 \times 1)(18.9) + 0.4(19.0)(2.0)(16.3) = 606.9 \text{ kN/m}^2$$

Applying the load factors from Table 16.3,

$$\text{Total load} = 1.25\,(P) + 1.5\,(0.75\,P) = 2.375\,P$$

The pressure applied to the soil is

$$\frac{2.375\,P}{2 \times 2} = 0.5938\,P \text{ kN/m}^2$$

$$0.5938\,P \leq 606.9$$

Therefore, $\mathbf{P \leq 1022.1}$ **kN**

Both methods give similar results.

Example 16.2

Based on LRFD, determine the maximum dead load that can be applied on the 2.0-m-wide footing from Example 16.1. Assume the following load/resistance factors.

Load factor for dead load = 1.25

Load factor for live load = 1.75

Strength reduction factor on the ultimate bearing capacity = 0.50

Solution

$$\text{Total load} = 1.25\,(P) + 1.75\,(0.75\,P) = 2.563\,P$$

Pressure applied on the soil is

$$S^* = \frac{2.563P}{2 \times 2} = 0.641\,P\ \text{kN / m}^2$$

For $\phi' = 36°$, from Table 12.2, $N_q = 37.9$ and $N_\gamma = 44.7$

$$q_{ult} = (19.0 \times 1)(37.9) + 0.4(19.0)(2.0)(44.7) = 1399.5\ \text{kN/m}^2$$

$$\text{The factored resistance } R^* = 0.5 \times 1399.5 = 699.8\ \text{kN/m}^2$$

For $S^* \leq R^*$

$$0.641\,P \leq 699.8$$

$$\mathbf{P \leq 1091.7\ kN}$$

16.5 A Probabilistic Approach

It can be seen from Fig. 16.2 that the system fails when the demand (D) exceeds the capacity (C). The probability of failure (p_f) can thus be defined as the probability that demand exceeds capacity. Denoting probability by P,

$$p_f = P[C < D] = P[C - D < 0] \tag{16.7}$$

Let us introduce a new parameter *safety margin* (SM) defined as $SM = C - D$. Equation (16.7) becomes

$$p_f = P[SM < 0] \tag{16.8}$$

The probability density function of the safety margin is shown in Fig. 16.3. Here, $\overline{SM}$ and s_{SM} are the mean and standard deviation of the safety margin. The hatched area on the figure is the probability of failure, implying the demand exceeds capacity. The distance of from the origin is a measure of the factor of safety, defined by Eq. (16.1). It is quantified by the *reliability index* β defined as

$$\beta = \frac{\overline{SM}}{s_{SM}} \tag{16.9}$$

The larger the reliability index β, the larger the factor of safety F. Reliability of the system is defined by the probability the system does not fail and hence

$$\text{Reliability} = 1 - p_f \tag{16.10}$$

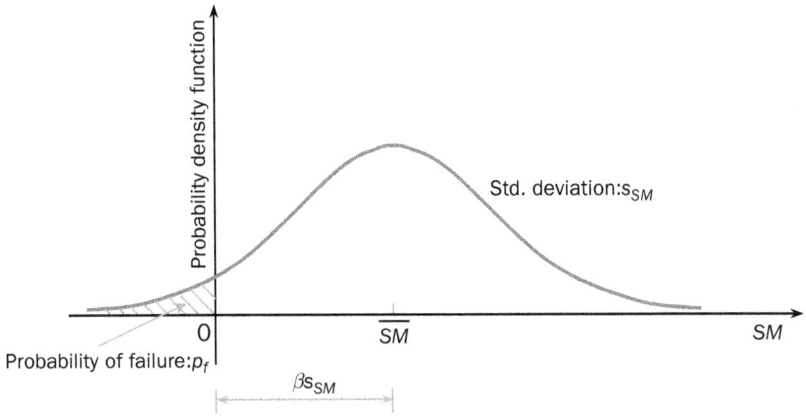

FIGURE 16.3 Probability density function of the safety margin.

The reliability index β and the probability of failure p_f are closely related. In other words, the reliability index is a good measure of the probability of failure. The values of β and corresponding p_f are given in Table 16.6, assuming the safety margin follows a Gaussian normal distribution. Table 16.6 is a standard normal distribution table found in any statistics textbook.

In the capacity-demand model,

$$SM = C - D \tag{16.11}$$

Therefore, the mean of the safety margin is given by

$$\overline{SM} = \bar{C} - \bar{D} \tag{16.12}$$

The standard deviation of the safety margin s_{SM} is given by

$$s_{SM} = \sqrt{s_C^2 + s_D^2 - 2\,\mathrm{Cov}(C,D)} \tag{16.13}$$

where Cov (C, D) is the *covariance*, a measure of the interdependence between the random variables C and D. Assuming C and D are independent and hence statistically uncorrelated, Cov $(C, D) = 0$. Thus, Eq. (16.13) becomes,

$$s_{SM} = \sqrt{s_C^2 + s_D^2} \tag{16.14}$$

Example 16.3

The capacity of a system is defined by a normal distribution, with mean and standard deviation of 4000 and 500. The demand, defined by a normal distribution, has mean and standard

β	0.00	0.01	0.02	0.03	0.04	0.05	0.06	0.07	0.08	0.09
0.0	0.5000	0.4960	0.4920	0.4880	0.4840	0.4801	0.4761	0.4721	0.4681	0.4641
0.1	0.4602	0.4562	0.4522	0.4483	0.4443	0.4404	0.4364	0.4325	0.4286	0.4247
0.2	0.4207	0.4168	0.4129	0.4090	0.4052	0.4013	0.3974	0.3936	0.3897	0.3859
0.3	0.3821	0.3783	0.3745	0.3707	0.3669	0.3632	0.3594	0.3557	0.3520	0.3483
0.4	0.3446	0.3409	0.3372	0.3336	0.3300	0.3264	0.3228	0.3192	0.3156	0.3121
0.5	0.3085	0.3050	0.3015	0.2981	0.2946	0.2912	0.2877	0.2843	0.2810	0.2776
0.6	0.2743	0.2709	0.2676	0.2643	0.2611	0.2578	0.2546	0.2514	0.2483	0.2451
0.7	0.2420	0.2389	0.2358	0.2327	0.2296	0.2266	0.2236	0.2206	0.2177	0.2148
0.8	0.2119	0.2090	0.2061	0.2063	0.2005	0.1977	0.1949	0.1922	0.1894	0.1867
0.9	0.1841	0.1814	0.1788	0.1762	0.1736	0.1711	0.1685	0.1660	0.1635	0.1611
1.0	0.1587	0.1562	0.1539	0.1515	0.1492	0.1469	0.1446	0.1423	0.1401	0.1379
1.1	0.1357	0.1335	0.1314	0.1292	0.1271	0.1251	0.1230	0.1210	0.1190	0.1170
1.2	0.1151	0.1131	0.1112	0.1093	0.1075	0.1056	0.1038	0.1020	0.1003	0.0985
1.3	0.0968	0.0951	0.0934	0.0918	0.0901	0.0885	0.0869	0.0853	0.0838	0.0823
1.4	0.0808	0.0793	0.0778	0.0764	0.0749	0.0735	0.0731	0.0708	0.0694	0.0681
1.5	0.0668	0.0655	0.0643	0.0630	0.0618	0.0606	0.0594	0.0582	0.0571	0.0559
1.6	0.0548	0.0537	0.0526	0.0516	0.0505	0.0495	0.0485	0.0475	0.0465	0.0455
1.7	0.0446	0.0436	0.0427	0.0418	0.0409	0.0401	0.0392	0.0384	0.0375	0.0367
1.8	0.0359	0.0341	0.0344	0.0336	0.0329	0.0322	0.0314	0.0307	0.0301	0.0294
1.9	0.0287	0.0281	0.0274	0.0268	0.0262	0.0256	0.0250	0.0244	0.0239	0.0233
2.0	0.0228	0.0222	0.0217	0.0212	0.0207	0.0202	0.0197	0.0192	0.0188	0.0183
2.1	0.0179	0.0174	0.0170	0.0166	0.0162	0.0158	0.0152	0.0150	0.0146	0.0143
2.2	0.0139	0.0136	0.0132	0.0129	0.0125	0.0122	0.0119	0.0116	0.0113	0.0110
2.3	0.0107	0.0104	0.0102	0.0099	0.0096	0.0092	0.0091	0.0089	0.0087	0.0084
2.4	0.0082	0.0080	0.0078	0.0075	0.0073	0.0071	0.0069	0.0068	0.0066	0.0064
2.5	0.0062	0.0060	0.0059	0.0057	0.0055	0.0054	0.0052	0.0051	0.0049	0.0048
2.6	0.0047	0.0045	0.0044	0.0043	0.0041	0.0040	0.0039	0.0038	0.0037	0.0036
2.7	0.0035	0.0034	0.0033	0.0032	0.0031	0.0030	0.0029	0.0028	0.0027	0.0026
2.8	0.0026	0.0025	0.0024	0.0023	0.0023	0.0022	0.0021	0.0021	0.0020	0.0019
2.9	0.0019	0.0018	0.0018	0.0017	0.0016	0.0016	0.0015	0.0015	0.0014	0.0014
3.0	0.0013	0.0013	0.0013	0.0012	0.0012	0.0011	0.0011	0.0011	0.0010	0.0010
3.1	0.0010	0.0009	0.0009	0.0009	0.0008	0.0008	0.0008	0.0008	0.0007	0.0007
3.2	0.0007	0.0007	0.0006	0.0006	0.0006	0.0006	0.0006	0.0005	0.0005	0.0005
3.3	0.0005	0.0005	0.0005	0.0004	0.0004	0.0004	0.0004	0.0004	0.0004	0.0003
3.4	0.0003	0.0003	0.0003	0.0003	0.0003	0.0003	0.0003	0.0003	0.0003	0.0002
3.5	0.00023									
4.0	0.000032									

Table 16.6 Values of p_f Assuming Normal Distribution

deviation of 2000 and 600. Find the factor of safety, reliability index, probability of failure, and the reliability of the system.

Solution

$$\bar{C} = 4000,\ s_C = 500,\ \bar{D} = 2000, \text{ and } s_D = 600.$$

From Eq. (16.1),

$$F = \frac{\bar{C}}{\bar{D}} = \frac{4000}{2000} = 2.0$$

$$\overline{SM} = \bar{C} - \bar{D} = 4000 - 2000 = 2000$$

$$s_{SM} = \sqrt{s_C^2 + s_D^2} = \sqrt{500^2 + 600^2} = 781.0$$

Therefore, the safety margin follows a normal distribution similar to the one shown in Fig. 16.3, with $\overline{SM} = 2000$ and $s_{SM} = 781.0$.

$$\text{Reliability index } \beta = \frac{\overline{SM}}{s_{SM}} = \frac{2000}{781} = \mathbf{2.56}$$

From Table 16.6, for $\beta = 2.56$

The probability of failure p_f = **0.0052**

The reliability of the system is 1 − 0.0052 = **0.9948**

Example 16.4

A 2-m-wide strip foundation carries a wall load with mean value of 360 kN/m and standard deviation of 50 kN/m. The ultimate bearing capacity of the foundation has a mean value of 800 kN/m² and standard deviation of 200 kN/m². Determine the factor of safety, reliability index, and the probability of failure. Assume normal distributions.

Solution The applied pressure q follows a normal distribution with mean $\bar{q} = \frac{360}{2} = 180.0\ \text{kN/m}^2$ and standard deviation $s_q = \frac{50}{2} = 25\ \text{kN/m}^2$.

$$F = \frac{\text{Mean of the ultimate bearing capacity}}{\text{Mean of the applied foundation pressure}} = \frac{800}{180} = \mathbf{4.4}$$

$$\overline{SM} = \bar{C} - \bar{D} = 800 - 180 = 620\ \text{kN/m}^2$$

$$\overline{SM} = \sqrt{s_C^2 + s_D^2} = \sqrt{200^2 + 25^2} = 201.6\ \text{kN/m}^2$$

$$\text{Reliability index } \beta = \frac{\overline{SM}}{s_{SM}} = \frac{620}{201.6} = \mathbf{3.075}$$

From Table 16.5, for $\beta = 3.075$, $p_f = 0.0011$

$$\text{Reliability} = 1 - 0.0011 = \mathbf{0.9989}$$

16.6 Determination of the Mean and Standard Deviation of Capacity and Demand

To determine the mean and standard deviation of the safety margin *SM*, it is necessary to determine the mean and standard deviation of the capacity *C* and the demand *D*. Capacity (or resistance) and demand (or loading) are functions of the other random variables (e.g., soil parameters, loadings) following specific probability distributions having their own mean and standard deviation values. For example, the ultimate bearing capacity of a foundation is a function of three random variables: cohesion, friction angle, and the unit weight of the soil, with other parameters such as foundation dimensions being constants. It is necessary to determine the mean and standard deviation of the ultimate bearing capacity from the values of mean and standard deviation of cohesion, friction angle, and unit weight.

The procedure to determine the mean and standard deviation is the same for both capacity and demand. Let us denote capacity or demand as the random variable Y which depends on $X_1, X_2, \ldots, X_n$, all of which are random variables. For simplicity, let us assume all the variables follow a normal distribution. Three methods to determine the mean and standard deviation of Y from the known values of mean and standard deviation for $X_1, X_2, \ldots, X_n$ are discussed in this section. These methods are:

a. First-order second moment (FoSM) method

b. Point estimate method

c. Monte Carlo simulation

Application of the methods is illustrated through a practical example related to compaction of clays. The same problem is solved by the three methods, which give similar results. It is useful to refresh here that, the standard deviation s_Y of a random variable Y can be determined from the m trials as

$$s_Y^2 = \overline{Y^2} - \left(\bar{Y}\right)^2 \tag{16.15}$$

where $\bar{Y} = \dfrac{\sum_{i=1}^{m} Y_i}{m}$ and $\overline{Y^2} = \dfrac{\sum_{i=1}^{m} Y_i^2}{m}$. Here, $\bar{Y}$ is the mean of Y and $\overline{Y^2}$ is the mean of Y^2.

16.6.1 First-Order Second Moment (FoSM) Method

For simplicity, it is assumed that the variables $X_1, X_2, \ldots, X_n$ are independent and hence uncorrelated.

$$Y = Y(X_1, X_2, \ldots X_n)$$

Here, $X_1, X_2, \ldots, X_n$ and Y are random variables. Y is a function of $X_1, X_2, \ldots,$ and X_n. The mean value of Y can be determined using the mean values of $X_1, X_2, \ldots, X_n$ as

$$\bar{Y} = Y\left(\overline{X_1}, \overline{X_2}, \ldots, \overline{X_n}\right)$$

The standard deviation of Y can be determined from the Taylor series expansion where the higher-order terms are neglected and suggested by Harr (1977) as

$$s_Y^2 = \left(\frac{\partial Y}{\partial X_1}\right)^2 s_{X_1}^2 + \left(\frac{\partial Y}{\partial X_2}\right)^2 s_{X_2}^2 + \cdots + \left(\frac{\partial Y}{\partial X_n}\right)^2 s_{X_n}^2 \tag{16.16}$$

Here, the partial derivatives $\frac{\partial Y}{\partial X_1}, \frac{\partial Y}{\partial X_2}, \ldots, \frac{\partial Y}{\partial X_n}$ are evaluated at their mean values.

Example 16.5

The degree of saturation S of a compacted clay is given in terms of three random variables unit weight γ, moisture content w, and the specific gravity of the soil grains G_s by

$$S = \frac{\gamma w G_s}{(1+w)G_s\gamma_w - \gamma}$$

The mean and standard deviation of the three variables are given as (Schmertmann, 1989).

Variable	Mean	Std. deviation
γ (kN/m³)	20.90	0.3135
w (%)	10.95	0.8213
G_s	2.75	0.0275

Determine the mean and standard deviation of S, using FoSM.

Solution Substituting the mean values of γ, w, and G_s in the equation for S,

$$\bar{S} = \frac{20.90 \times 0.1095 \times 2.75}{(1+0.1095)(2.75)(9.81) - 20.90} = \mathbf{0.6968} \text{ or } \mathbf{69.68\%}$$

$$\frac{\partial S}{\partial \gamma} = \frac{(1+w)G_s^2 w\gamma_w}{[(1+w)G_s\gamma_w - \gamma]^2}$$

$$\frac{\partial S}{\partial w} = \frac{G_s\gamma(G_s\gamma_w - \gamma)}{[(1+w)G_s\gamma_w - \gamma]^2}$$

$$\frac{\partial S}{\partial G_s} = -\frac{\gamma^2 w}{[(1+w)G_s\gamma_w - \gamma]^2}$$

Therefore, from Eq. (16.16),

$$s_S^2 \approx \left(\frac{(1+w)G_s^2 w\gamma_w}{[(1+w)G_s\gamma_w - \gamma]^2}\right)^2 s_\gamma^2 + \left(\frac{G_s\gamma(G_s\gamma_w - \gamma)}{[(1+w)G_s\gamma_w - \gamma]^2}\right)^2 s_w^2 + \left(\frac{\gamma^2 w}{[(1+w)G_s\gamma_w - \gamma]^2}\right)^2 s_{G_s}^2$$

Substituting the mean values of γ, w, and G_s in the above expression,

$$s_S^2 \approx 0.002697$$

$$s_S = \mathbf{0.0519} \text{ or } \mathbf{5.19\%}.$$

16.6.2 Point Estimate Method

Point estimate method was developed by Rosenblueth (1975, 1981) and introduced in geotechnical applications by Harr (1977). Let us consider a variable Y which is a function of random variables X_1, X_2, …, and X_n that follow a normal distribution with specific mean and standard deviation values for each of them. The first step is to determine the values of the independent variables X_1, X_2, …, and X_n at one standard deviation away from the mean on both sides and denote them as X_{i+} and X_{i-} defined as

$$X_{i+} = \overline{X_i} + s_{X_i} \tag{16.17a}$$

$$X_{i-} = \overline{X_i} - s_{X_i} \tag{16.17b}$$

The next step is to evaluate the different values of Y for all possible combinations of $X_{1+}, X_{1-}, X_{2+}, X_{2-}, \ldots, X_{n+}, X_{n-}$. The mean and standard deviation of these 2^n different values of Y thus computed are approximately equal to the mean and standard deviation of Y. Here, it is assumed that X_1, X_2, …, and X_n are independent and hence uncorrelated. If they are correlated, the coefficient of correlation comes in the picture (see Harr, 1977).

Example 16.6

Redo Example 16.5 using the point estimate method.

Solution The mean and standard deviations of the three independent variables are:

$$\bar{\gamma} = 20.90 \text{ kN/m}^3, s_\gamma = 0.3135 \text{ kN/m}^3$$

$$\bar{w} = 10.95\%, s_w = 0.8213\%$$

$$\bar{G}_s = 2.75, s_{G_s} = 0.0275$$

Let us evaluate the three variables at one standard deviation away from mean on both sides.

$$\gamma_+ = 20.90 + 0.3135 = 21.2135 \text{ kN/m}^3, \gamma_- = 20.90 - 0.3135 = 20.5865 \text{ kN/m}^3$$

$$w_+ = 10.95 + 0.8213 = 11.7713\%, w_- = 10.95 - 0.8213 = 10.1287\%$$

$$G_{s+} = 2.75 + 0.0275 = 2.7775, G_{s-} = 2.75 - 0.0275 = 2.7225$$

Substituting these values in the equation for γ, eight (2^3) different values are obtained as follows.

$$S = S(\gamma, w, G_s)$$

$$S_{+++} = S(21.2135 \text{ kN/m}^3, 11.7713\%, 2.7775) = 0.7505$$

$$S_{++-} = S(21.2135 \text{ kN/m}^3, 11.7713\%, 2.7225) = 0.7870$$

$$S_{+-+} = S(21.2135\ \text{kN/m}^3, 10.1287\%, 2.7775) = 0.6787$$

$$S_{+--} = S(21.2135\ \text{kN/m}^3, 10.1287\%, 2.7225) = 0.7134$$

$$S_{-++} = S(20.5865\ \text{kN/m}^3, 11.7713\%, 2.7775) = 0.6821$$

$$S_{-+-} = S(20.5865\ \text{kN/m}^3, 11.7713\%, 2.7225) = 0.7121$$

$$S_{--+} = S(20.5865\ \text{kN/m}^3, 10.1287\%, 2.7775) = 0.6148$$

$$S_{---} = S(20.5865\ \text{kN/m}^3, 10.1287\%, 2.7225) = 0.6432$$

From the eight values above, $\overline{S} = \mathbf{0.6977}$ or **69.77%** and $s_S = \mathbf{0.0521}$ or **5.21%**. The two values of mean and standard deviation determined here match well with those determined from FoSM in Example 16.5.

16.6.3 Monte Carlo Simulation

Monte Carlo simulation is a brute force technique where m sets of random numbers are generated for each of the variable $X_1, X_2, \ldots, X_n$ following their specific probability distribution with the given mean and standard deviation values. As before, we will assume that Y is a function of $X_1, X_2, \ldots, X_n$. For each set of random values generated for $X_1, X_2, \ldots, X_n$, the value of Y can be computed. Using the m different values of Y thus computed, the mean $\overline{Y}$ and standard deviation s_Y can be determined. Random number generators are available in Excel™ where this exercise can be carried out on a spreadsheet. For the data given in Examples 16.5 and 16.6, Schmertmann (1989) carried out Monte Carlo simulation with 5000 trials and obtained, $\overline{S} = 70.24\%$ and $s_S = 5.20\%$. The following random number generator is adequate for all practical purposes.

Step 1: In Microsoft Excel™, develop random numbers using the function RAND. This gives a random number in the range 0–1 based on a *uniform* distribution. We have to convert this to a Gaussian normal distribution.

Step 2: Subtract 0.5 from the value in step 1 which gives a random number z between −0.5 and 0.5.

Step 3: The required random number is computed as $x = \sqrt{12}zs_x + \overline{x}$, where $\overline{x}$ and s_x are the mean and standard deviation of the variable X_i assumed in generating the random numbers, and z is the random number from step 2.

In Excel™, the function RAND() returns a random number following a uniform distribution in the range 0-1. The statistical function NORM.INV(RAND(), a, b) gives a random number from a Gaussian normal distribution, with specified mean of a and standard deviation b.

Example 16.7

Redo Example 16.5 using 2000 trials in Excel™ for the Monte Carlo simulation. Use the random number generator suggested above. List the mean and standard deviation values of the random numbers generated.

Solution

$$\bar{\gamma} = 20.90\ \text{kN/m}^3,\ s_\gamma = 0.3135\ \text{kN/m}^3$$

$$\bar{w} = 10.95\%,\ s_w = 0.8213\%$$

$$\bar{G}_s = 2.75,\ s_{G_s} = 0.0275$$

From 2000 trials carried out in Excel™, $\bar{S} = 0.6988$ or 69.88%, and $s_S = 0.0520$ or 5.20%.

Parameters	Mean	Std. deviation
γ (kN/m³)	20.906 (20.9)	0.3180 (0.3135)
w (%)	10.970 (10.95)	0.8149 (0.8213)
G_s	2.751 (2.75)	0.0273 (0.0275)
S (%)	69.881	5.201

While the mean values of the 2000 random numbers generated for γ, w, and G_s matched the mean values of the random numbers used in the simulation, the standard deviation values differ very slightly. The original values of the mean and standard deviation used in the simulation are shown in parentheses for comparison.

The mean and standard deviations of the degree of saturation S, computed by the three methods, are summarized in Table 16.7, showing an excellent agreement between the different methods.

The histogram developed from the 2000 values of the degree of saturation is shown in Fig. 16.4.

	FoSM	Point estimate	Monte Carlo
$\bar{S}$	69.68	69.77	69.88
s_S (%)	5.19	5.21	5.20

Table 16.7 The Mean and Standard Deviation of S from the Three Methods

---

Example 16.8

A 2.0-m-wide square foundation is placed in a sand at 1.0 m depth to carry a column load Q. The mean and standard deviation of the friction angle and the unit weight are as follows.

$$\bar{\phi} = 32°, s_\phi = 3°; \bar{\gamma} = 18.0\ \text{kN / m}^3, s_\gamma = 1.0\ \text{kN / m}^3$$

Using the point estimate method, and the simplified bearing capacity equation given by

$$q_{\text{ult}} = \gamma D_f N_q + 0.4 B \gamma N_\gamma$$

determine the mean and standard deviation of the ultimate bearing capacity of the foundation. The bearing capacity factors are given by

$$N_q = e^{\pi \tan\phi} \tan^2\left(45 + \frac{\phi}{2}\right)$$

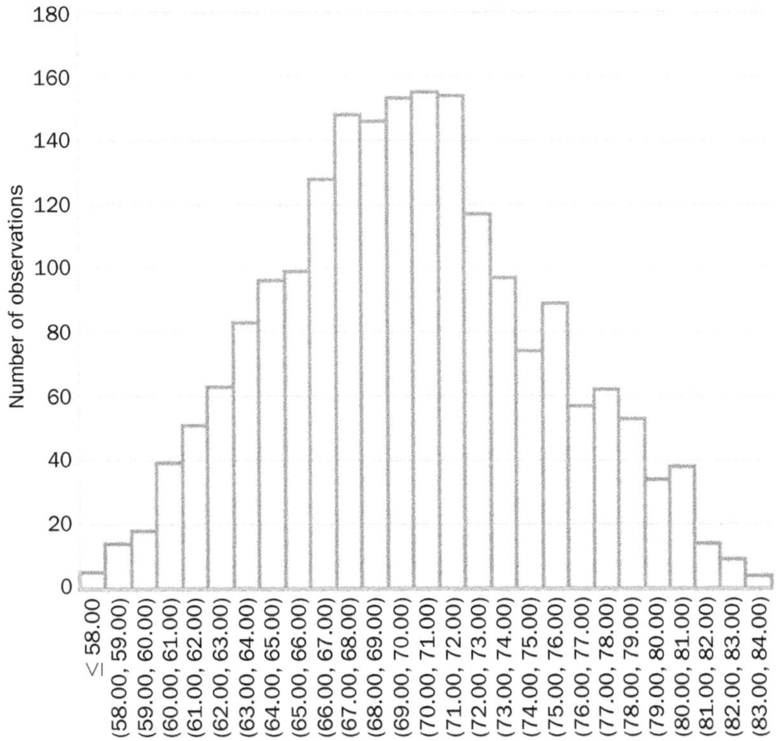

FIGURE 16.4 Histogram for the 2000 values of S (%).

$$N_\gamma = \left(N_q - 1\right)\tan(1.4)\phi$$

Assuming that the ultimate bearing capacity follows a Gaussian normal distribution, find the probability that it will be less than 500 kN/m².

Solution

$$B = 2.0 \text{ m}, D_f = 1.0 \text{ m}$$

Let us compute the four possible values of $q_{\text{ult}}(\phi, \gamma)$.

$$\phi_+ = 32 + 3 = 35°, \phi_- = 32 - 3 = 29°$$

$$\gamma_+ = 18.0 + 1.0 = 19.0 \text{ kN/m}^3, \gamma_- = 18.0 - 1.0 = 17.0 \text{ kN/m}^3$$

ϕ (°)	N_q	N_γ
35	33.4	37.3
29	16.5	13.3

$$q_{\text{ult},++} = 19.0 \times 1.0 \times 33.4 + 0.4 \times 2.0 \times 19.0 \times 37.3 = 1201.6 \text{ kN/m}^2$$

$$q_{\text{ult},+-} = 17.0 \times 1.0 \times 33.4 + 0.4 \times 2.0 \times 17.0 \times 37.3 = 1075.1 \text{ kN/m}^2$$

$$q_{\text{ult},-+} = 19.0 \times 1.0 \times 16.5 + 0.4 \times 2.0 \times 19.0 \times 13.3 = 515.7\ \text{kN/m}^2$$

$$q_{\text{ult},--} = 17.0 \times 1.0 \times 16.5 + 0.4 \times 2.0 \times 17.0 \times 13.3 = 461.4\ \text{kN/m}^2$$

From these four values,

$$\overline{q_{\text{ult}}} = 813.5\ \text{kN/m}^2$$

$$\overline{q^2_{\text{ult}}} = 769{,}629.8\ \text{kN}^2/\text{m}^4$$

From Eq. (16.15),

$$s^2_{q_{\text{ult}}} = 769{,}629.8 - 813.5^2 = 107{,}847.5$$

$$s_{q_{\text{ult}}} = 328.4\ \text{kN/m}^2$$

The mean and standard deviation of the ultimate bearing capacity are 813.5 and 328.4 kN/m², respectively.

To determine the probability that the ultimate bearing capacity is less than 500 kN/m², it is possible to use the standard normal distribution table, which is the same as Table 16.5. Alternatively, the function NORM.DIST in Excel™ can be used.

$$\frac{813.5 - 500}{328.4} = 0.955$$

$$P[q_{\text{ult}} < 500\ \text{kN/m}^2] = \mathbf{0.1698}\ \text{(see Fig. 16.5)}$$

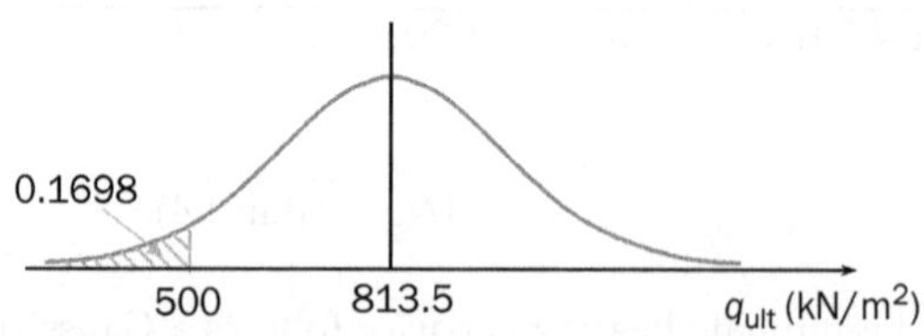

FIGURE 16.5 Probability density function.

It will be shown in Example 16.9 that the ultimate bearing capacity does not follow normal distribution.

Example 16.9

For the foundation in Example 16.8, using Monte Carlo simulation, and assuming that the friction angle and unit weight of the soil follow normal distribution, determine the mean and standard deviation of the ultimate bearing capacity. Show the histogram and discuss.

Find the probability that the ultimate bearing capacity will be less than 500 kN/m².

Solution In Excel™, 3000 random values were generated for ϕ and γ using the procedure described in Sec. 16.6.3. The mean and standard deviation of these random values of ϕ and γ are

	Mean	Std. deviation
ϕ (degrees)	31.94	3.03
γ (kN/m³)	17.97	1.00

which match the given values that were used in generating the random numbers.

For each set of ϕ and γ, N_q and N_γ were determined, and finally q_{ult} was calculated. From the 3000 values of q_{ult}, the mean and standard deviation were determined as 805.7 and 348.9 kN/m², respectively. These values compare well with those determined by the point estimate method.

The histogram is shown in Fig. 16.6. It appears that it does not follow the Gaussian normal distribution. It may better fit a log normal or beta distribution.

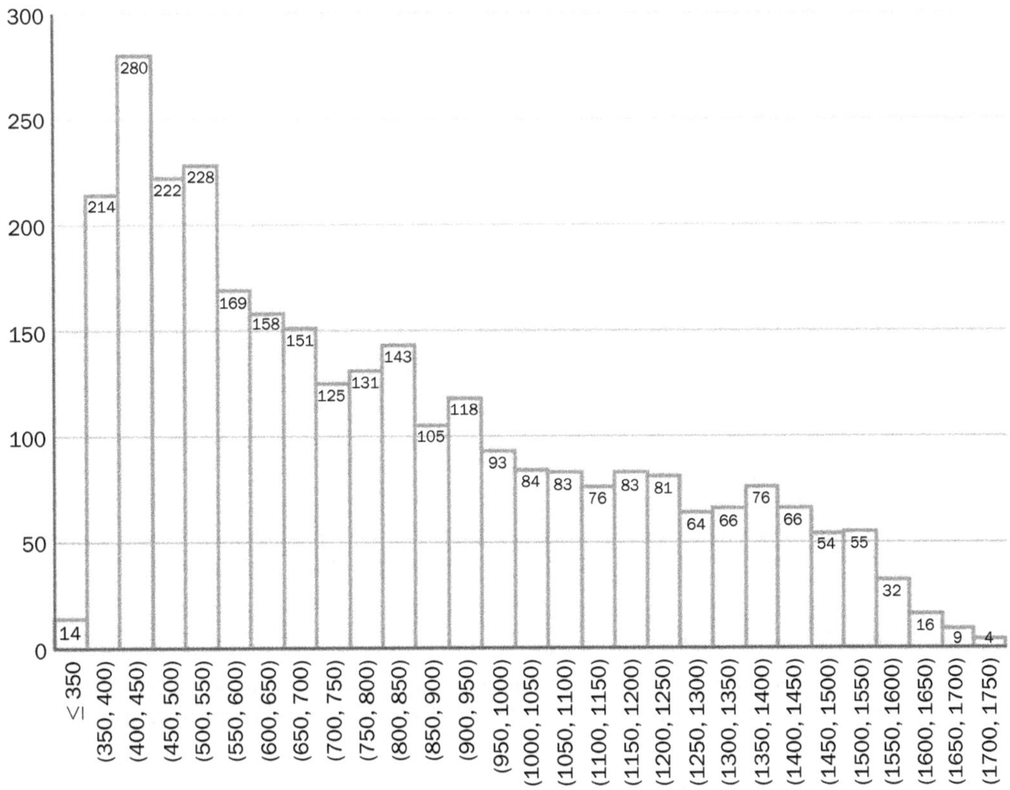

FIGURE 16.6 Histogram.

From the number of observations shown in the histogram, the probability that the ultimate bearing capacity will be less than 500 kN/m² is given by

$$\frac{14 + 214 + 280 + 222}{3000} = \mathbf{0.24}$$

This value is different to 0.1698 determined in Example 16.8 assuming that q_{ult} follows a normal distribution.

16.7 Main Points

1. The ASD and the LRFD are two different ways of designing a structure.
2. ASD has been used for centuries and the current trend is to go for the LRFD.
3. LRFD is a limit state design, which incorporates separate load factors (Ψ) and resistance factors (Φ), with no further factor of safety. In the ASD, all the uncertainties were lumped into a single global factor of safety.
4. There is notable difference in the way the *resistance factors* can be applied in the LRFD: (a) multiple resistance factors applied to the strength parameters such as cohesion and friction angle; and (b) one single resistance factor applied to the final soil resistance computed using unfactored strength parameters.
5. In LRFD, the factored load has to be less than the factored resistance.
6. Reliability-based design is a probabilistic approach that considers the statistical distribution of the variables involved.
7. First-order second moment, point estimate, and Monte Carlo simulation are three alternate techniques to determine the mean and standard deviation of the capacity or demand.

Review Exercises

16.1 A 2 m × 2 m square footing is placed at a depth of 0.5 m in a clayey sand where the cohesion and friction angle in terms of effective stresses are 5 kN/m² and 34°. Average unit weight of the soil is 19 kN/m³. Find the maximum load Q that can be safely applied on the footing, by:

a. Allowable stress method, and

b. LRFD method (assuming $\psi = 1.25$ for load; $\Phi = 0.50$ for cohesion and $\Phi = 0.80$ for tan of friction angle).

For simplicity, neglect the difference between net and gross bearing capacities and applied pressures, and use the following bearing capacity equation with Meyerhof's N_γ.

$$q_{\text{ult}} = 1.3cN_c + \gamma D_f N_q + 0.4B\gamma N_\gamma$$

Answer: 1376 kN, 1324 kN

16.2 A square footing at 0.5 m depth is required to carry a column load of 1000 kN in a clayey sand where $c' = 8$ kN/m², $\phi' = 35°$, and $\gamma = 19.5$ kN/m³. Determine the footing width by:

a. ASD, and

b. LRFD method, assuming $\Psi = 1.3$ for the column load, $\Phi = 0.6$ for cohesion, and $\Phi = 0.75$ for tan of the friction angle.

Use the following bearing capacity equation with Meyerhof's N_γ.

$$q_{\text{ult}} = 1.3cN_c + \gamma D_f N_q + 0.4B\gamma N_\gamma$$

Answer: 1.55 m, 1.77 m

16.3 A 1.75-m-wide square footing is placed at 0.5 m depth in the ground where $c' = 10$ kN/m², $\phi' = 33°$, and $\gamma = 19.0$ kN/m³. The column load consists of 60% dead load and 40% live load. Determine the maximum allowable column load by LRFD, assuming the following:

$\Psi = 1.25$ for dead load

$\Psi = 1.50$ for live load

$\Phi = 0.45$ for the ultimate bearing capacity

Use the following bearing capacity equation with Meyerhof's N_γ.

$$q_{ult} = 1.3cN_c + \gamma D_f N_q + 0.4B\gamma N_\gamma$$

Answer: 1126 kN

16.4 The degree of saturation of a compacted clay is given in terms of three random variables γ, w, and G_s by

$$S = \frac{\gamma w G_s}{(1+w)G_s\gamma_w - \gamma}$$

The mean and standard deviation of the three variables are as follows (Schmertmann, 1989).

Variable	Mean	Std. deviation
γ (kN/m^3)	21.64	1.2984
w (%)	14.86	4.458
G_s	2.75	0.1100

Determine the mean and standard deviation of S, using FoSM.

Answer: 94.6%, 26.3%

16.5 Redo Problem 16.4 using the point estimate method.

Answer: 96.5%, 27.3%

16.6 Redo Problem 16.4 using Excel to carry out Monte Carlo simulation with 2000 trials and using the procedure suggested in Sec. 16.6.3. List the mean and standard deviation of the random numbers generated.

Answer: 96.9%, 28.0%

Parameters	Mean	Std. deviation
γ (kN/m^3)	21.65 (21.64)	1.3054 (1.2984)
w (%)	14.84 (14.86)	4.4788 (4.458)
G_s	2.75 (2.75)	0.1086 (0.1100)
S (%)	96.93	27.95

The assumed values of the mean and standard deviation for γ, w, and G_s, in generating the random numbers are given in parentheses.

References

Harr, M. E. *Mechanics of Particulate Media*, McGraw Hill, New York, 1977.

Rosenblueth, E. "Point Estimates for Probability Moments," *Proceedings of the National Academy of Science, USA*, 72(10):3812–3814, 1975.

Rosenblueth, E. "Two-point Estimates in Probabilities," *Applied Mathematical Modelling*, 5(5):329–335, 1981.

Schmertmann, J. H. "Density Tests Above Zero Air Void Line," *Journal of Geotechnical Engineering, ASCE*, 115(7):1003–1018, 1989.

APPENDIX A
Unsaturated Soil Mechanics

Most of the chapters in this book dealt with soils that are either dry or saturated. While this is true for soils that are really dry or soils below the water table, in many situations soils are unsaturated. For example, in compacted soils and soils above the water table, there is substantial moisture held within the voids by capillary effects or *suction*, and they are *unsaturated soils*. In saturated soils, the pore water pressure $u_w \geq 0$, where the water within the voids is in compression; in unsaturated soils $u_w < 0$ where the water within the voids is in tension. Saturated soils contain only two phases. Unsaturated soils contain three phases, making them more complex to study. In fact, the 5-Å-thick "contractile skin" formed at the air-water interface is sometimes treated as the fourth phase.

In the 1930s, Terzaghi introduced effective stress $(\sigma - u_w)$ for defining the state of stress of a saturated soil. In the 1950s and the early 1960s, Bishop introduced an effective stress parameter χ for unsaturated soils as

$$\sigma' = (\sigma - u_a) + \chi(u_a - u_w) \tag{A.1}$$

where χ is a measure of saturation and takes the value of 1.0 for saturated soils and 0 for dry soils. There are two components to σ' in Eq. (A.1). The first term on the right-hand side of Eq. (A.1) is the net normal stress and the second term is the contribution from suction.

The suction within an unsaturated soil has two independent components, *matric suction* and *osmotic suction*. Matric suction is the difference between the pore air (u_a) and pore water (u_w) pressures. It is due to the capillary effects and surface tension, and depends on the pore-size distribution within the soil and the moisture content. Drier the soil, higher is the matric suction. Osmotic suction, also called solute suction, is caused by the dissolved ions and salts present in the pore water. Salt in the pore fluid attracts water and hence induces osmotic suction. Higher the salt content, higher is the osmotic suction. It can be a significant fraction of the total suction, and is denoted by π. Thus, total suction can be written as

$$\text{Total suction } (\psi) = \text{Matric suction } (u_a - u_w) + \text{Osmotic suction } (\pi) \tag{A.2}$$

At high values of total suction (e.g., $> 1500\ \text{kN/m}^2$), the contributions from osmotic suction are negligible and total suction and matric suction are approximately equal. Osmotic suction is present in both unsaturated and saturated soils. However, the strength of an unsaturated soil is controlled by the matric suction.

Suction is sometimes expressed using a pF (called picofarad) scale (Schofield, 1935), defined as

$$pF = \log_{10} h \tag{A.3}$$

where h is the height (in cm) of a water column that gives the same suction pressure. It was one of the early units used for suction, and was used widely in agronomy and soil physics. Lately, there is a tendency use more metric or SI units. At a point, 1 m above the water table in a saturated clay has suction of 9.81 kN/m^2 or 2.0 pF.

Example A.1

What is 150 kN/m^2 suction in pF scale?

Solution

$$\begin{aligned} 150 \text{ kN/m}^2 \text{ suction} &= 150/9.81 = 15.29 \text{ m of water} \\ &= 1529 \text{ cm of water} \\ &= \mathbf{3.18\ pF} \end{aligned}$$

Soil-Water Characteristic Curve

Soil-water characteristic curve (SWCC) is the graphical relationship between the moisture held within the voids and the soil suction. The moisture held within the voids can be represented by the gravimetric moisture content (w), volumetric moisture content (θ), or the degree of saturation (S).

Gravimetric moisture content (w) is defined as

$$w = \frac{\text{Mass of water}}{\text{Mass of soil grains}} = \frac{M_w}{M_s} \tag{A.4}$$

It is the traditional moisture content used in geotechnical engineering.

Volumetric moisture content (θ) is defined as

$$\theta = \frac{\text{Volume of water}}{\text{Total volume}} = \frac{V_w}{V_s + V_w + V_a} \tag{A.5}$$

Both moisture contents are generally multiplied by 100 and given as percentages. The two are related by

$$\theta = w\left(\frac{\rho_d}{\rho_w}\right) \tag{A.6}$$

where ρ_d = dry density of the soil and ρ_w = density of water.

Soil suction is represented by the total suction or the matric suction. The soil suction is plotted in logarithmic scale since it varies over several orders of magnitude. Nevertheless, in geotechnical engineering we are interested in suction pressures only up to about 1000 kN/m^2. All soils give a similar reverse S-type curve (see Fig. A.1) on a semilog scale, irrespective of what is on either of the two axes. Most commonly volumetric water content (θ) is plotted against the matric suction.

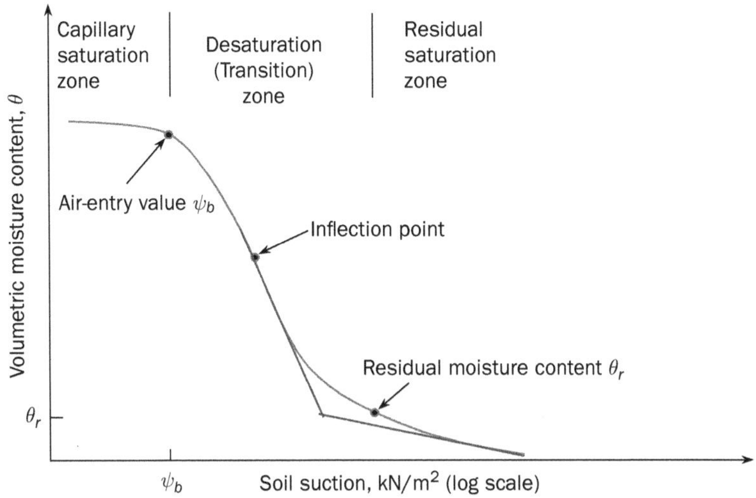

FIGURE A.1 A typical soil-water characteristic curve.

SWCC, sometimes called *soil-water retention curve* (SWRC), was originally used in soil science and agriculture to quantify the water storage capacity of soils. SWCC is influenced by the grain size distribution, clay mineralogy, soil density, and the pore fluid characteristics. It can be used to determine many geotechnical parameters associated with strength, volume change, and permeability of unsaturated soils.

One of the key parameters of SWCC is the *air-entry value* (AEV). As the soil dries, it shrinks initially while remaining saturated. Further drying will increase the suction to an extent that it allows air to enter the soil matrix. Air-entry value is the matric suction value (ψ_b) that has to be exceeded before air can get into the soil voids. Firstly, the air gets into the largest pores. Air-entry value is called the *displacement pressure* in petroleum engineering and *bubbling pressure* in ceramics engineering. AEV is inversely proportional to the pore diameter. Finer the grains higher is the AEV. Therefore, clays have larger air-entry values than sands. The logarithmic scale for the soil suction helps in determining the air-entry value ψ_b more precisely.

Residual moisture content (θ_r) is the moisture content below which it requires large change in suction to remove any extra water from the pores. Figure A.1 shows a typical soil-water characteristic curve, with the air-entry value and the residual moisture content. The residual moisture content is defined in the SWCC by the intersection of the tangent drawn through the inflection point and the straight-line portion of the tail end.

The SWCC in Fig. A.1 is divided into three zones. In the *capillary saturation zone*, the soil is saturated but the pore water is in tension. In a saturated soil, matric suction is almost zero, which is the lowest limit of soil suction. The water in this zone is free draining.

In *desaturation zone*, water is displaced by air. At residual moisture content, pore water becomes discontinuous and the permeability is greatly reduced. The water in this zone is available to plants, which can apply high suction. In *residual saturation zone*, the pores contain tightly held adsorbed water. When dried at 105°C in the oven, to zero moisture content and relative humidity, the soil suction is in the order of 1,000,000 kN/m^2,

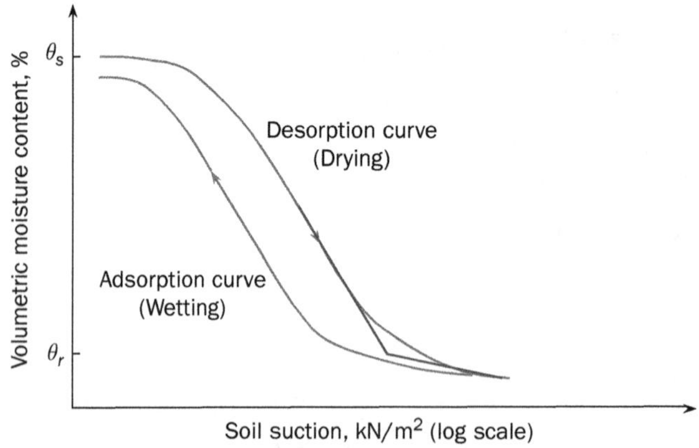

FIGURE A.2 Adsorption and desorption curves of SWCC.

which is the upper limit of soil suction. In some very fine clay fabrics (e.g., montmorillonite) the suction pressures can reach such high values. Most expansive soils have field suctions in the range of 100–10,000 kN/m^2 (Nelson and Miller, 1992). The lower limit of soil suction is zero.

At the same matric suction, soils can have different moisture contents depending on whether they are going through wetting or drying cycles. Due to hysteretic nature, the wetting (adsorption) and drying (desorption) curves are different, as seen in Fig. A.2. In other words, at the same suction, the solid-water interfacial area is larger, resulting in larger average intergranular stress while the soil specimen is on the drying path. The air-entry value and the residual water content are determined from the desorption curve. Air-entry value ψ_b is the suction pressure at the right end of the horizontal segment of the SWCC where the volumetric moisture content θ starts dropping (see Fig. A.1). The moisture content corresponding to zero soil suction is the saturation water content θ_s. The residual moisture content θ_r is defined by the intersection of the two straight-line segments shown in Fig. A.2.

Equations to Describe SWCC

Many simple mathematical expressions have been suggested in the literature to describe the soil-water characteristic curve. These empirical equations often ignore the hysteresis shown in Fig. A.2. One of the earliest empirical relationship to describe a SWCC was proposed by Brooks and Corey (1964) as

$$\Theta = \left(\frac{\psi_b}{\psi}\right)^{\lambda} \qquad \text{for } \psi > \psi_b \tag{A.7a}$$

$$\Theta = 1 \qquad \text{for } \psi \leq \psi_b \tag{A.7b}$$

where ψ = suction and ψ_b = air-entry value of suction, Θ = dimensionless volumetric moisture content defined as

$$\Theta = \frac{\theta - \theta_r}{\theta_s - \theta_r} \qquad \text{for } \theta \geq \theta_r \tag{A.8}$$

and λ is the *pore-size distribution index* which can theoretically take any positive value. Fredlund et al. (2012) defined λ as the negative slope of the *effective* degree of saturation (S_e) versus matric suction. Here, at degree of saturation of S, the effective degree of saturation is defined as

$$S_e = \frac{S - S_r}{1 - S_r}$$

where S_r is the residual degree of saturation.

Narrower pore size gives larger λ and vice versa, with soils having values in the range of 0.2–0.5 and values larger than 1 for rocks. When θ_r is very small, the parameter Θ is similar to the degree of saturation S.

Williams et al. (1983) suggested that

$$\ln\Theta = a_1 + b_1 \ln\psi \tag{A.9}$$

where a_1 and b_1 are curve-fitting parameters. McKee and Bumb (1987) and Bumb (1987) suggested

$$\Theta = \frac{1}{1 + \exp\left[\frac{\psi - a_2}{b_2}\right]} \tag{A.10}$$

where a_2 and b_2 are curve-fitting constants. This equation gives good estimates at the low suction range but not at high suction values.

Van Genuchten (1980) suggested that

$$\Theta = \left[\frac{1}{1 + (\alpha\psi)^n}\right]^m \tag{A.11}$$

where α, m, and n are positive curve-fitting constants. A common simplification is to assume that $m = 1 - 1/n$. Here, α can be crudely approximated as the reciprocal of the air-entry value ψ_b, and n is a parameter reflecting the pore-size distribution.

Example A.2

A silty clay loam has the following parameters for van Genuchten model: $\theta_r = 0.098$, $\theta_s = 0.550$, $\alpha = 0.27\ \text{kPa}^{-1}$, and $n = 1.41$.

a. Develop the soil-water characteristic curve.
b. Determine the air-entry value.

Solution

$$m = 1 - \frac{1}{n} = 1 - \frac{1}{1.41} = 0.291$$

Using Eq. (A.11), the following table is prepared, and the SWCC is developed (Fig. A.3).

Ψ (kN/m²)	Θ	θ	θ (%)
0.01	0.9999	0.5500	54.9969
0.03	0.9997	0.5499	54.9852
0.1	0.9982	0.5492	54.9195
0.3	0.9917	0.5463	54.6268
1	0.9582	0.5311	53.1128
3	0.8507	0.4825	48.2524
10	0.6264	0.3800	38.0034
30	0.4176	0.2868	28.6768
100	0.2579	0.2146	21.4581
300	0.1647	0.1724	17.2441
1000	0.1005	0.1434	14.3444
10000	0.0391	0.1157	11.5669
100000	0.0152	0.1049	10.4869
1000000	0.0059	0.1007	10.0671

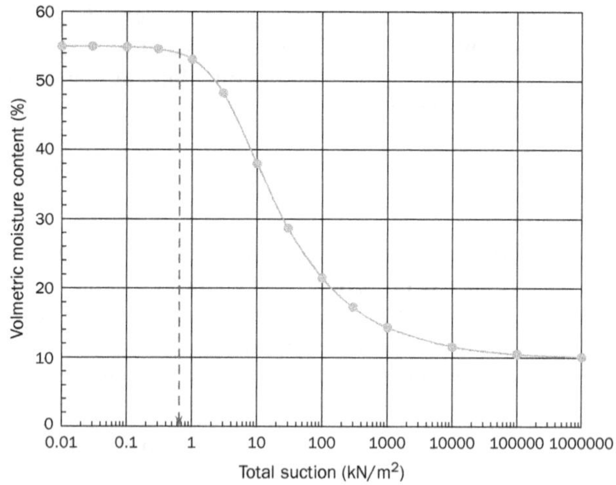

FIGURE A.3 SWCC.

From the SWCC, the air-entry value is 0.6 kN/m².

Figure A.4 shows some typical SWCCs for different soils. It can be seen that clays have higher air-entry values than sands.

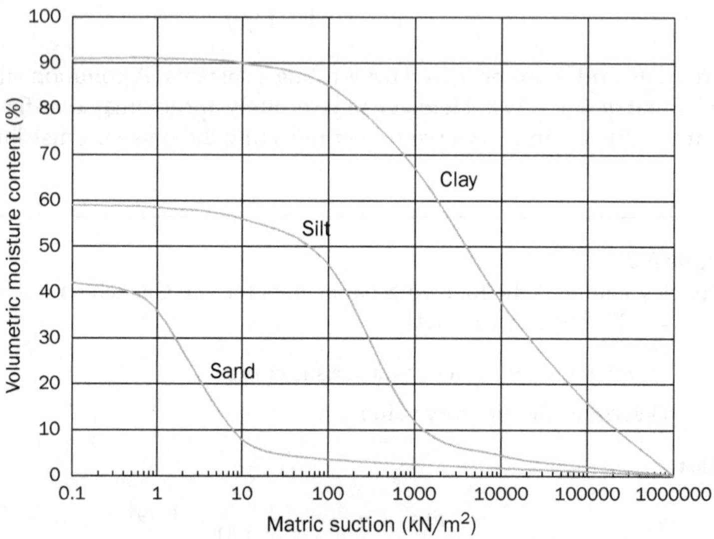

FIGURE A.4 Typical SWCC for clayey, silty, and sandy soils.

Example A.3

A simplified soil-water characteristic curve is shown in Figure A.5, where the saturation volumetric moisture content is 48% and the air-entry value of the soil is 8.5 kN/m^2. The straight-line segment of the SWCC suggests that there is 25% drop in the volumetric moisture content for every log cycle of soil suction. What would be the total suction at volumetric moisture content of 20%?

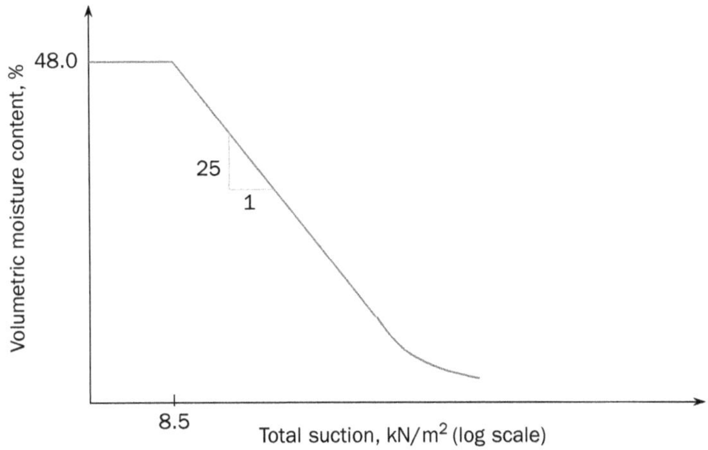

FIGURE A.5 SWCC.

Solution

$$\text{Reduction in water content} = 48.0 - 20.0 = 28.0\%$$

$$\text{No of log cycles of suction pressure} = 28/25 = 1.12$$

Let the suction pressure at 20% moisture content be ψ.

$$\log_{10}\left(\frac{\psi}{8.5}\right) = 1.12$$

$$\psi = 13.1826 \times 8.5 = \mathbf{112.1\ kN/m^2}$$

Thermodynamic Relationship for Soil Suction

Free-energy state of soil water can be related to the partial vapor pressure of the soil water through thermodynamic relationship. Substances, including water have a tendency to change from higher energy to lower energy states. In other words, water molecules move from wet areas to dryer areas (i.e., higher energy to lower energy). The difference between free energy of soil water and pure water in a standard reference state is termed soil water potential or total suction. Dry air has about 78% nitrogen by volume and 21% oxygen. Water molecules can enter the air voids and be present in the air and reach equilibrium, exerting a vapor pressure that depends on the temperature. Each gas in the air (i.e., nitrogen, oxygen, water) exerts a partial vapor pressure, which adds up to the total pressure. Relative humidity (R_h) is defined as the ratio of the partial vapor pressure in the air to the vapor pressure in the air when it is fully saturated. In a fully saturated environment, the relative humidity is 100% and the total suction is zero.

The total suction is given by (Fredlund and Rahardjo, 1993)

$$\psi = \frac{\rho_w RT}{M} \ln R_h \tag{A.12}$$

where ρ_w = density of water (1000 kg/m³), M = molecular weight of water (0.01802 kg/mol), T is the absolute temperature in Kelvin (note: Kelvin = Celsius + 273), R is the universal gas constant (8.31432 N·m/mole·K), R_h is the relative humidity expressed as a decimal number rather than a percentage, and ψ = total suction in N/m². Substituting these values, at 20°C (68°F),

$$\psi(\text{kN/m}^2) = \frac{(1000)(8.31432)(293)}{0.01802} \ln R_h = 135{,}188 \ln R_h$$

Example A.4

What is the total suction when the temperature is 25°C and the relative humidity is 90%?

Solution

$$\psi = \frac{(1000)(8.31432)(298)}{0.01802} \ln 0.90 = -14{,}486 \times 10^3\,\text{N/m}^2 = -14{,}487\ \text{kN/m}^2$$

Measurement of Soil Suction

There are several ways of measuring the soil suction. The simplest of these is the filter paper method. The other methods include using psychrometer, tensiometer, or pressure plates. A psychrometer uses a thermocouple to measure relative humidity, which is related to total suction. It can be used to measure 100–8000 kN/m² of total suction. Tensiometer measures up to 100 kN/m² of matric suction.

Filter Paper Method

Filter paper method (ASTM D 5298-16) can be used to measure suction pressures covering a wide range of 0.01–100,000 kN/m². A filter paper (e.g., Whatman No. 42 or Schleicher & Schuell No. 589-WH), weighed dry, is placed in a covered container along with the moist soil for 1–2 weeks, until it reaches moisture equilibrium with the soil specimen. When at equilibrium, the moisture content of the filter paper is calculated, which is an indirect measure of the suction pressure. The filter papers must be weighed to 0.0001 g. From the moisture content of the filter paper, the suction can be determined using a calibration curve obtained beforehand. Such calibration curves, separately for total and matric suction, for different filter papers are also available in the literature including ASTM D 5298-16. Typical calibration curve is bilinear, following equations like

$$\log(s) = A - Bw \tag{A.13}$$

where s = suction (kN/m²), w = moisture content, and A, B = calibration constants. ASTM D 5298-16 suggests using the gravimetric moisture content in the calibration of the filter paper. In developing the calibration curves, the suction values are calculated from the relative humidity using Eq. (A.12).

There are two ways of placing the filter paper: (a) In contact with the soil to measure matric suction, and (b) Not in contact with the soil to measure the total suction. When in contact with the soil, there is only matric suction and no osmotic suction, since the water in the filter paper and the soil is the same. Thus, the suction measured is matric suction only. When the filter paper is not in contact with the soil sample, the water in the filter paper is pure and does not contain any ions. The suction process would have allowed both matric and osmotic suction to take place and hence this method measures total suction. When the filter paper comes into equilibrium with the soil either through vapor or liquid flow, the suction value of the filter paper and the soil are the same.

Hydraulic Conductivity

In unsaturated soils, the hydraulic conductivity can be significantly lower than when they are saturated. When water flows through unsaturated soils, the large voids are filled with air, leaving only the smaller voids to hold and transmit water. Similar to the soil grains, the entrapped air also provides resistance to water flow. The degree of saturation will have significant influence on the hydraulic conductivity, especially in fine-grained soils. With decreasing degree of saturation, the hydraulic conductivity can reduce by few orders of magnitude.

While hydraulic conductivity of saturated soils is related to the void ratio, in case of unsaturated soils, it is related to any two of the three variables, namely, void ratio, degree of saturation, and moisture content.

Relative permeability (or hydraulic conductivity) is the ratio of the unsaturated permeability to the saturated permeability (k_s), denoted by k_r. It can be approximated as

$$k_r = \Theta^n \tag{A.14}$$

where n is a constant and Θ is the dimensionless volumetric moisture content defined in Eq. (A.7). Brooks and Corey (1964) suggested that $n = (2 + 3\lambda)/\lambda$, where λ is the pore-size distribution index introduced in Eq. (A.7a). They also suggested that

$$k_r = \left(\frac{\psi_b}{\psi}\right)^{2+3\lambda} \quad \text{for } \psi > \psi_b \tag{A.15a}$$

$$k_r = 1 \quad \text{for } \psi \le \psi_b \tag{A.15b}$$

Van Genuchten (1980) suggested that

$$k_r = \frac{\left\{1 - (\alpha\psi)^{nm}\left[1 + (\alpha\psi)^n\right]^{-m}\right\}^2}{\left[1 + (\alpha\psi)^n\right]^{m/2}} \tag{A.16}$$

where α, m, and n are the same van Genuchten constants as in Eq. (A.11), and ψ is the soil suction in kN/m². Van Genuchten (1980) also showed that

$$k_r = \sqrt{\Theta}\left[1 - \left(1 - \Theta^{1/m}\right)^m\right]^2 \tag{A.17}$$

Fredlund et al. (1994) and Leong and Rahardjo (1997) summarized the empirical relationships between the relative permeability k_r and soil suction ψ or volumetric moisture content θ.

Vanapalli and Lobbezoo (2002) suggest that

$$k_r = S^{7.9\gamma} \tag{A.18}$$

where $\gamma = 0.75$ for nonplastic soils and increases with plasticity to values larger than 2.

Example A.5

The silty clay loam in Example A.2 has saturated permeability of 2.5×10^{-5} cm/s. Using the same van Genuchten parameters given in the example, find the permeability using Eq. (A.16) when the suction is 2 kN/m².

Also, find the permeability at the same suction of 2 kN/m², using Eq. (A.17).

Solution From Eq. (A.16),

$$k_r = \frac{\left\{1 - (\alpha\psi)^{nm}\left[1 + (\alpha\psi)^n\right]^{-m}\right\}^2}{\left[1 + (\alpha\psi)^n\right]^{m/2}}$$

$$= \frac{\left\{1 - (0.27 \times 2)^{1.41\times0.29}\left[1 + (0.27 \times 2)^{1.41}\right]^{-0.29}\right\}^2}{\left[1 + (0.27 \times 2)^{1.41}\right]^{0.29/2}} = 0.0843$$

Therefore, the permeability is $0.0843 \times (2.5 \times 10^{-5}) = \mathbf{0.21 \times 10^{-5}}$ **cm/s**

From Eq. (A.17),

$$k_r = \sqrt{\Theta}\left[1 - \left(1 - \Theta^{1/m}\right)^m\right]^2$$

At $\psi = 2$ kN/m², $\theta = 50.5\%$ (from Fig. A.3).

This gives

$$\Theta = \frac{\theta - \theta_r}{\theta_s - \theta_r} = \frac{0.505 - 0.098}{0.550 - 0.098} = 0.900$$

Therefore,

$$k_r = \sqrt{\Theta}\left[1 - \left(1 - \Theta^{1/m}\right)^m\right]^2 = \sqrt{0.900}\left[1 - \left(1 - 0.900^{1/0.29}\right)^{0.29}\right]^2 = 0.0809$$

Therefore, the permeability is $0.0809 \times (2.5 \times 10^{-5}) = \mathbf{0.20 \times 10^{-5}}$ **cm/s.**

Summary

An unsaturated soil is a three-phase medium, with soil grains, water, and air. When the soil is unsaturated, the water in the voids is held by negative pressures, known as suction. The soil suction has two components, namely, matric suction and osmotic suction.

Matric suction is induced by the nature of the soil skeleton and osmotic suction is due to the salts present in the pore fluid.

The hydraulic conductivity of an unsaturated soil can be significantly less than that of the same soil when saturated. Lower the water content, lower is the hydraulic conductivity. The unsaturated hydraulic conductivity can be determined indirectly using mathematical models, which relate hydraulic conductivity to the soil suction or moisture content.

SWCC or soil water retention curve is the graphical representation of how the soil suction varies with the moisture content. It is a very valuable tool in the indirect determination of the geotechnical properties of the unsaturated soils. This can be determined using a simple filter paper method in the laboratory, and can be mathematically described using simple equations such as the ones suggested by Brooks and Corey (1964) or van Genuchten (1980). Such equations are also available for determining the relative permeability of unsaturated soils.

References

ASTM 5298-16. Standard Test Method for Measurement of Soil Potential (Suction) using Filter Paper.

Bishop, A. W. "The Use of Pore Water Coefficients in Practice." *Géotechnique*, 4(4):148–152, 1954.

Bishop, A. W. "The Principle of Effective Stress." *Teknisk Ukeblad I Samarbeide Med Teknikk*, Oslo, Norway, 106(39):859–863, 1959.

Bishop, A. W., and G. E. Blight. "Some Aspects of Effective Stress in Saturated and Unsaturated Soils." *Géotechnique*, 13(3):177–197, 1963.

Brooks, R. H., and A. T. Corey. *Hydraulic Properties of Porous Medium*, Hydrology Paper 3, Colorado State University, Fort Collins, Colorado, 1964.

Bumb, A. C. *Unsteady-state Flow of Methane and Water in Coalbeds*, Ph.D. thesis, Department of Chemical Engineering, University of Wyoming, Laramie, 1987.

Fredlund, D. G., and H. Rahardjo. *Soil Mechanics for Unsaturated Soils*, John Wiley & Sons, New York, 1993.

Fredlund, D. G., H. Rahardjo, and M. D. Fredlund. *Unsaturated Soil Mechanics in Engineering Practice*, Wiley, New York, 2012.

Fredlund, D. G., A. Xing, and S. Huang. "Predicting the Permeability Function for Unsaturated Soils Using Soil-water Characteristic Curve," *Canadian Geotechnical Journal*, 31(4):533–546, 1994.

Leong, E. C., and H. Rahardjo. "Permeability Functions for Unsaturated Soils," *Journal of Geotechnical and Geoenvironmental Engineering*, 123(12):1118–1126, 1997.

McKee, C. R., and A. C. Bumb. "Flow-testing Coalbed Methane Production Wells in the Presence of Water and Gas," *SPE Formation Evaluation*, December, 2(4):599–608, 1987.

Nelson, J. D., and D. J. Miller. *Expansive Soils: Problems and Practice in Foundation and Pavement Engineering*, John Wiley & Sons, Inc., New York, 1992, p. 259.

Schofield, R. K. "The pF of the Water in Soil." *Transaction of the 3rd International Congress on Soil Science*, Oxford, 2, 1935, pp. 37–48.

van Genuchten, M. Th. "A Closed-form Equation for Predicting the Hydraulic Conductivity of Unsaturated Soils", *Soil Science Society of America Journal*, 44:892–898, 1980.

Vanapalli, S. K., and J. P. Lobbezoo. A Normalized Function for Predicting the Coefficient of Permeability of Unsaturated Soils. Juca, J. F. T., T. M. P. de Campos, and F. A. M. Marinho (eds.), *Proceedings of the Third International Conference on Unsaturated Soils*, Recife, Brazil, 2, AA Balkema, 2002, pp. 839–844.

Williams, J., R. E. Prebble, W. T. Williams, and C. T. Hignett. The Influence of Texture, Structure and Clay Mineralogy on the Soil Moisture Characteristic. *Australian Journal of Soil Research*, 21:15–32, 1983.

APPENDIX **B**

Vesic's (1973) Factors for Eq. (12.11)

The shape factors and the depth factors are the same for Vesic's (1973) and Brinch Hansen's (1970) methods.

Shape Factors

The shape factors are given by

$$s_c = 1 + \left(\frac{B}{L}\right)\frac{N_q}{N_c} \tag{B.1}$$

$$s_q = 1 + \left(\frac{B}{L}\right)\tan\phi \tag{B.2}$$

$$s_\gamma = 1 - 0.4\left(\frac{B}{L}\right) \tag{B.3}$$

Depth Factors

The depth factors are given by

$$d_c = d_q - \frac{1 - d_q}{N_c \tan\phi} \qquad \text{for } \phi > 0 \tag{B.4a}$$

$$d_q = 1 + 2\tan\phi\,(1 - \sin\phi)^2\left(\frac{D_f}{B}\right) \tag{B.5a}$$

For the special case of $\phi = 0$, Eq. (B.4a) becomes (see Vesic, 1973)

$$d_c = 1 + 0.4\left(\frac{D_f}{B}\right) \tag{B.4b}$$

and Eq. (B.5a) becomes

$$d_q = 1 \tag{B.5b}$$

For all values of ϕ,

$$d_\gamma = 1 \tag{B.6}$$

Inclination Factors

The inclination factors are given by

$$i_c = 1 - \frac{mH}{BLcN_c} \qquad \text{for } \phi = 0 \tag{B.7a}$$

$$i_c = i_q - \left(\frac{1 - i_q}{N_q - 1}\right) \qquad \text{for } \phi > 0 \tag{B.7b}$$

$$i_q = \left(1 - \frac{H}{BLc \cot \phi}\right)^m \tag{B.8}$$

$$i_\gamma = \left(1 - \frac{H}{V + BLc \cot \phi}\right)^{m+1} \tag{B.9}$$

In Eqs. (B.7)–(B.9), H and V are the horizontal and vertical components of the foundation load. The parameter c is the adhesion at the foundation-soil interface and hence a reduced value of the cohesion must be used. Depending on whether the horizontal component of the foundation load is parallel to the length or the breadth, m can be taken as m_L or m_B defined as

$$m = m_L = \frac{2 + L/B}{1 + L/B} \tag{B.10a}$$

$$m = m_B = \frac{2 + B/L}{1 + B/L} \tag{B.10b}$$

Soil Compressibility Factors

Terzaghi's original bearing capacity equation was developed assuming the soil is incompressible. To allow the compressibility the shear strength parameters must be reduced. Vesic (1973) proposed the factors C_c, C_q, and C_γ to use in Eq. (12.11) to take soil

ϕ'	B/L										
(°)	0	0.1	0.2	0.3	0.4	0.5	0.6	0.7	0.8	0.9	1
0	13.6	13.0	12.4	11.8	11.3	10.8	10.3	9.9	9.5	9.0	8.6
5	18.3	17.4	16.6	15.8	15.1	14.3	13.6	13.0	12.4	11.8	11.2
10	25.5	24.2	22.9	21.7	20.6	19.5	18.5	17.5	16.6	15.8	14.9
15	36.9	34.8	32.8	30.9	29.2	27.5	25.9	24.5	23.1	21.8	20.5
20	55.7	52.2	49.0	45.9	43.1	40.4	37.9	35.5	33.3	31.2	29.3
25	88.8	82.8	77.1	71.9	67.0	62.4	58.1	54.2	50.5	47.0	43.8
30	151.8	140.4	129.9	120.2	111.1	102.8	95.1	88.0	81.4	75.3	69.6
35	283.2	259.7	238.2	218.5	200.4	183.8	168.6	154.6	141.8	130.1	119.3
40	592.1	537.7	488.2	443.3	402.5	365.5	331.9	301.3	273.6	248.4	225.6
45	1442.0	1293.5	1160.4	1040.9	933.7	837.6	751.4	674.0	604.6	542.4	486.6
50	4330.9	3827.2	3382.1	2988.8	2641.2	2334.0	2062.6	1822.7	1610.7	1423.4	1257.9

TABLE B.1 Values of the Critical Rigidity Index $I_{r(cr)}$

compressibility into account. He defined the rigidity index I_r at a depth of $B/2$ below the base of the foundation as

$$I_r = \frac{G}{c' + q' \tan \phi'} \tag{B.11}$$

where $G = \dfrac{E}{2(1+v)}$ = shear modulus of the soil, q' = effective overburden stress at $B/2$ below the foundation.

Vesic (1973) defined the critical rigidity index $I_{r(cr)}$ as

$$I_{r(cr)} = \frac{1}{2}\exp\left\{\left(3.3 - 0.45\frac{B}{L}\right)\cot\left(45 - \frac{\phi'}{2}\right)\right\} \tag{B.12}$$

Values of $I_{r(cr)}$ computed for different values of B/L and ϕ' are summarized in Table B.1. The variation of $I_{r(cr)}$ as a function of friction angle and the foundation shape is shown in Fig. B.1. If $I_r > I_{r(cr)}$, the soil can be assumed as incompressible and $C_c = C_q = C_\gamma = 1$, requiring no correction for soil compressibility.

If $I_r < I_{r(cr)}$, the soil compressibility has to be taken into account. Vesic (1973) suggested that for all practical purposes $C_q = C_\gamma$, and they are given by

$$C_q = C_\gamma = \exp\left\{\left(-4.4 + 0.6\frac{B}{L}\right)\tan\phi' + \left[\frac{3.07 \sin\phi' \log 2I_r}{1 + \sin\phi'}\right]\right\} \tag{B.13}$$

Variation of C_q and C_γ with the rigidity index and friction angle is shown in Fig. B.2.

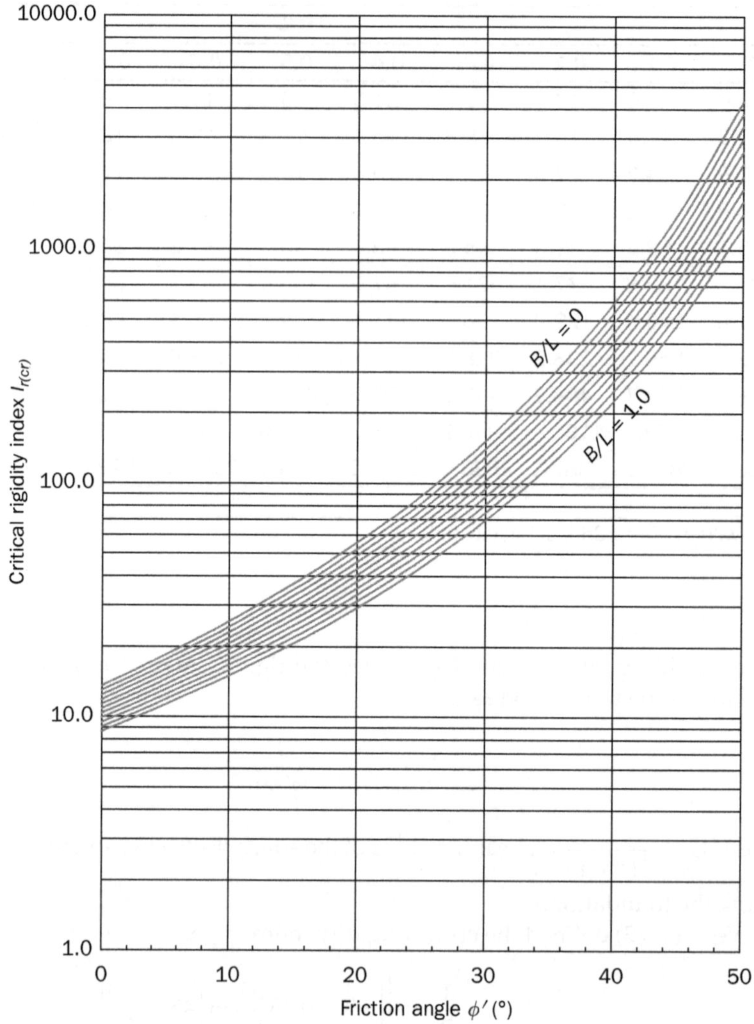

FIGURE B.1 Critical rigidity index values.

For $\phi' > 0$,

$$C_c = C_q - \frac{1 - C_q}{N_c \tan \phi'} \tag{B.14}$$

For $\phi = 0$,

$$C_c = 0.32 + 0.12\frac{B}{L} + 0.6 \log I_r \tag{B.15}$$

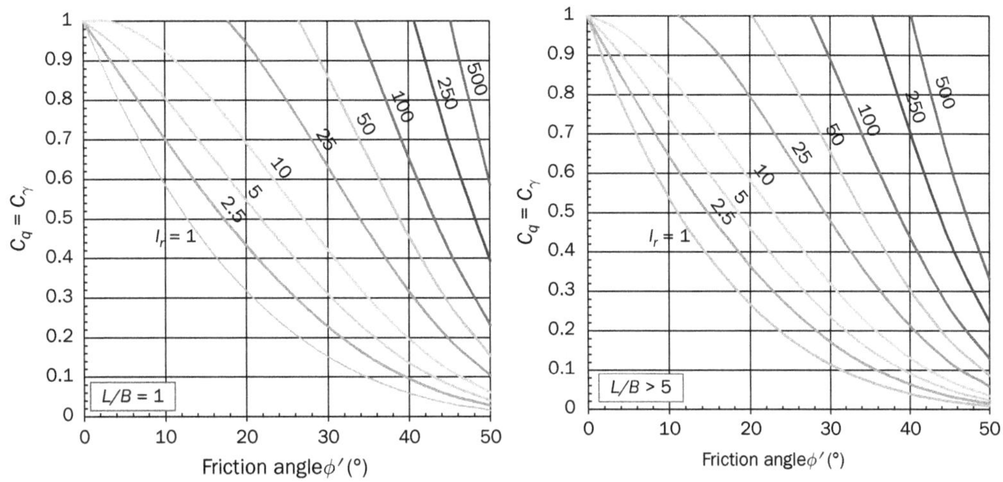

FIGURE B.2 Variations of C_q and C_γ with friction angle ϕ'.

Example B.1

A 2.0-m-wide strip foundation is placed within a sandy clay at 1.0 m depth. The properties of the clayey sand are: $c' = 5\ \text{kN/m}^2$, $\phi' = 27°$, $\gamma = 19.0\ \text{kN/m}^3$, $E = 3.2\ \text{MN/m}^2$, and $\nu = 0.2$. Determine the ultimate bearing capacity.

Solution For $\phi' = 27°$, $N_q = 13.2$, $N_c = 24.0$, and $N_{\gamma(\text{Vesic})} = 14.5$.

For strip foundation, shape factors are unity.

Since the wall load is vertical, inclination factors are unity.

Depth factors:

$$d_q = 1 + 2\tan\phi\,(1 - \sin\phi)^2\left(\frac{D_f}{B}\right)$$

$$d_q = 1 + 2\tan 27\,(1 - \sin 27)^2\left(\frac{1}{2}\right) = 1.15$$

$$d_c = d_q - \frac{1 - d_q}{N_c \tan\phi} = 1.15 - \frac{1 - 1.15}{24\tan 27} = 1.16$$

$$d_\gamma = 1.0$$

Compressibility factors:

$$G = \frac{E}{2(1+\nu)} = \frac{3.2}{2(1+0.2)} = 1.33\ \text{MN/m}^2$$

$$I_r = \frac{G}{c + q'\tan\phi'} = \frac{1333.3}{5 + 38\tan 27} = 54.7$$

From Table B.1, for $B/L = 0$ and $\phi' = 27°$, the critical rigidity index $I_{r(cr)} = 114$. Since $I_r < I_{r(cr)}$, we have to compute the compressibility factors.

$$C_q = C_\gamma = \exp\left\{\left(-4.4 + 0.6\frac{B}{L}\right)\tan\phi' + \left[\frac{3.07\sin\phi' \log 2I_r}{1 + \sin\phi'}\right]\right\}$$

$$C_q = C_\gamma = \exp\left\{(-4.4 + 0.6 \times 0)\tan 27 + \left[\frac{3.07\sin 27 \log(2 \times 54.7)}{(1 + \sin 27)}\right]\right\} = 0.75$$

$$C_c = C_q - \frac{1 - C_q}{N_c \tan\phi'} = 0.75 - \frac{1 - 0.75}{24\tan 27} = 0.73$$

From the general bearing capacity equation [Eq. (12.11)], assuming the ground inclination and base inclination factors as unity,

$$q_{\text{ult}} = s_c d_c i_c g_c b_c C_c c N_c + s_q d_q i_q g_q b_q C_q \gamma D_f N_q + s_\gamma d_\gamma i_\gamma g_\gamma b_\gamma C_\gamma 0.5 B \gamma N_\gamma$$

$q_{\text{ult}} = (1.16 \times 0.73 \times 5 \times 24) + (1.15 \times 0.75 \times 19 \times 1 \times 13.2) + (1 \times 0.75 \times 0.5 \times 2 \times 19 \times 14.5) =$ **590.4 kN/m²**

APPENDIX C

Units and Conversions

1 Angstrom (Å) = 10^{-10} m	
1 foot (ft) = 0.3048 m	1 m = 3.281 ft
1 inch (in.) = 25.4 mm = 2.54×10^{-2} m	1 m = 39.37 in. = 1.094 yd
1 mile = 1.6093 km	1 km = 0.6214 mile
1 yard (yd) = 3.00 ft = 0.9144 m	
1 mil = 1/1000 in. = 2.54×10^{-5} m	

Length

1 ft^2 = 929.03×10^{-4} m^2	1 m^2 = 10.764 ft^2 = 1.196 yd^2
1 acre = 4840 sq. yd = 4046.9 m^2	1 sq. mile = 2.59 sq. km
1 hectare (ha) = 10,000 m^2	1 ha = 2.471 acres
1 perch = 25.293 m^2	
1 square = 9.290 m^2	

Area

1 gallon (U.S.) = 3.785×10^{-3} m^3 = 3.785 liters	
1 gallon (U.K.) = 4.546×10^{-3} m^3 = 4.546 liters	
1 liter = 1.0×10^{-3} m^3	
1 ounce (U.S. fluid) = 2.957×10^{-5} m^3	
1 pint (U.S. fluid) = 4.732×10^{-4} m^3	
1 ft^3 = 28.317×10^{-3} m^3	1 m^3 = 35.32 ft^3
1 in.^{3} = 16.387 cm^3	

Volume

1 British ton (short) = 2000 lbm = 907.2 kg	
1 British ton (long) = 2240 lbm = 20 hundredweights	
1 hundredweight (U.K.) = 112 lbm	
1 hundredweight (U.S.) = 100 lbm	
1 metric ton (or tonne) = 1000 kg = 1.0 Mg	
1 U.S. ton = 2000 lbm = 907.2 kg	
1 ounce = 1/16 lbm = 28.35 g	
1 pound mass (lbm) = 0.4536 kg	1 kg = 2.2046 lbm

Mass

1 dyne (dyn) = 1.0×10^{-5} N	
1 pound-force (lbf) = 4.448 N	1 N = 0.2248 lbf
1 kilogram-force (kgf) = 9.81 N	1 kgf = 2.2046 lbf
1 kip-force (1000 lbf) = 4448.22 N	1 kN = 0.2248 kip = 0.1124 U.S. ton
1 U.S. ton-force = 8.896 kN	
1 lbf/ft = 14.593 N/m	1 N/m = 0.06853 lbf/ft

Force

1 atmosphere (760 mm Hg) = 101.3 kPa	
1 bar = 100 kPa	
1 kgf/cm^2 = 98.07 kPa	
1 lbf/in.^{2} (psi) = 6.895 kPa	1 kPa = 0.1450 psi
1 kip/in.^{2} = 6.895 MPa	
1 tonf (short)/sq. ft = 95.76 kPa	1 kPa = 0.01044 tonf (short)/sq. ft
1 tonf (long)/sq. ft = 107.25 kPa	
1 lbf/ft^2 = 47.88 Pa	1 kPa = 20.886 lbf/ft^2

Pressure/Stress

1 lbf/ft^3 = 0.1572 kN/m^3	1 kN/m^3 = 6.361 lbf/ft^3

Unit Weight

1 pcf = 16.02 kg/m^3	1 kg/m^3 = 0.0624 pcf

Density

1 lbf·ft = 1.3558 N·m	1 N·m = 0.7375 lbf·ft
1 lbf·in. = 0.11298 N·m	1 N·m = 8.851 lbf·in.

Moment

1 in.^{3} = 1.6387 × 10^4 mm^3	1 mm^3 = 6.102 × 10^{-5} in.^{3}

Section Modulus

1 in.^{4} = 4.162 × 10^5 mm^4	1 mm^4 = 2.402 × 10^{-6} in.^{4}

Moment of Inertia

1 ft/min = 30.48 cm/min	1 cm/min = 0.0328 ft/min
1 ft/s = 30.48 cm/s	1 cm/s = 0.0328 ft/s
1 in./s = 2.54 cm/s	1 cm/s = 0.3937 in./s

Hydraulic Conductivity

1 in.^{2}/s = 6.452 cm^2/s = 20.35 m^2/year	1 cm^2/s = 0.155 in.^{2}/s
1 ft^2/s = 929.03 cm^2/s	1 cm^2/s = 1.0764 × 10^{-3} ft^2/s
1 in.^{2}/s = 20.346 × 10^3 m^2/year	1 m^2/year = 4.915 × 10^{-5} in.^{2}/s

Coefficient of Consolidation

1 lbf·ft = 1.3558 N·m	1 N·m = 0.7375 lbf·ft

Work or Energy

1 poise (P) = 0.1 Pa·s	
1 centipoise (cP) = 0.001 Pa·s	

Dynamic Viscosity

1 stoke (St) = 1.0 cm^2/s	

Kinematic Viscosity (Dynamic Viscosity Divided by Density)

Gravity = 9.81 m/s^2 = 32.2 ft/s^2	
Density of water = 1.0 g/cm^3 = 1000 kg/m^3 = 1 Mg/m^3 = 62.4 lb/ft^3	
Unit weight of water = 9.81 kN/m^3	
Dynamic viscosity of water = 0.001 Pa·s	

Universal Constants

Index

C

D

E

F

L

M

Q

R

S

T

Milton Keynes UK
Ingram Content Group UK Ltd.
UKHW010213020924
447683UK00001B/1

9 781260 468489